ADVANCES IN **CLIMBING** AND **WALKING ROBOTS**

ADVANCES IN **CLIMBING** AND **WALKING ROBOTS**

Proceedings of 10th International Conference (CLAWAR 2007)

Singapore 16 – 18 July 2007

editors

Ming Xie

Nanyang Technological University, Singapore

Steven Dubowsky

Massachusetts Institute of Technology, USA

Jean-Guy Fontaine

Massachusetts Institute of Technology, USA

M Osman Tokhi

University of Sheffield, UK

Gurvinder S Virk

University of Sheffield, UK

World Scientific

NEW JERSEY • LONDON • SINGAPORE • BEIJING • SHANGHAI • HONG KONG • TAIPEI • CHENNAI

Published by

World Scientific Publishing Co. Pte. Ltd.

5 Toh Tuck Link, Singapore 596224

USA office: 27 Warren Street, Suite 401-402, Hackensack, NJ 07601

UK office: 57 Shelton Street, Covent Garden, London WC2H 9HE

British Library Cataloguing-in-Publication Data
A catalogue record for this book is available from the British Library.

ADVANCES IN CLIMBING AND WALKING ROBOTS
Proceedings of 10th International Conference (CLAWAR 2007)

ISBN-13 978-981-270-815-1
ISBN-10 981-270-815-4

Printed in Singapore by World Scientific Printers (S) Pte Ltd

PREFACE

Robotics is an exciting field in engineering and natural sciences. Robotics has already made important widespread contributions and impact in industrial robots for tasks such as assembly, welding, painting, and material handling. In parallel, we have also witnessed the emergence of special robots, which perform valuable tasks, in non-industrial environments such as in search and rescue, de-mining, surveillance, exploration, and security missions. Furthermore, research and development works are currently in progress in the robotics technology for use in the domestic and professional service sector. The emergence of mobile machines, such as climbing and walking robots, for these missions in un-structured environments, has significantly broadened challenges that must be considered by robotics research. This includes not only the technological and engineering aspects including standardisation, but also socio-economic and ethical aspects.

CLAWAR 2007 is the tenth in a series of international conferences organised annually since 1998 with the aim to report on latest research and development findings and to provide a forum for scientific discussion and debate within the mobile service robotics community. The series has grown in its popularity significantly over the years and has attracted researchers and developers from across the globe. The CLAWAR 2007 proceeding reports on the latest scientific and developmental achievements, future challenges and exciting applications of mobile machines in general, in particular climbing and walking robots. Eighty-five technical presentations by authors from 22 countries covering five continents were presented at the CLAWAR 2007 conference, held in Singapore between 16-18 July 2007. The text of the proceedings is organized into five parts: Plenary Introduction, Advances in Climbing Robots, Advances in Walking Robots, Advances in Humanoid Soccer Playing Robots, and Supporting Technologies.

The editors would like to thank members of the International Programme Committee, International Technical Advisory/Organising Committee and National Organising Committee for their hard work in creating a well-run and productive meeting and for their efforts in reviewing the submissions. We would like to thank the authors for their quick response to comments and suggestions of the reviewers. It is hoped that this edition of the CLAWAR conference proceedings will form a valuable addition to the scientific and developmental knowledge in mobile robotics.

M. Xie
Nanyang Technological University, Singapore
General Chair

S. Dubowsky
Massachusetts Institute of Technology, USA
General Co-Chair

J. G. Fontaine
Italian Institute of Technology, Italy
General Co-Chair

M. O. Tokhi
University of Sheffield, U.K.
Programme Chair

G. S. Virk
University of Leeds, U.K.
Programme Co-Chair

15 May 2007

CONFERENCE COMMITTEES

Honorary Advisor, General Chairs and Programme Chairs

for the Tenth International Conference on Climbing and Walking Robots

K. Y. Lam (Honorary Advisor)	– NTU, Singapore
M. Xie (General Chair)	– NTU, Singapore
S. Dubowsky (General Co-Chair)	– MIT, USA
J. G. Fontaine (General Co-Chair)	– IIT, Italy
M. O. Tokhi (Program Chair)	– University of Sheffield, U.K.
G. S. Virk (Program Co-Chair)	– University of Leeds, U.K.

INTERNATIONAL ADVISORY COMMITTEE

for the Tenth International Conference on Climbing and Walking Robots

Manuel Armada	– Spain
Yvan Baudoin	– Belgium
Kasten Berns	– Germany
Philippe Bidaud	– France
Bryan Bridge	– U.K.
Chan-Hin Kam	– Singapore
Krzysztof Kozlowski	– Poland
Giovanni Muscato	– Italy
Lakmal Seneviratne	– U.K.

INTERNATIONAL PROGRAMME COMMITTEE

for the Tenth International Conference on Climbing and Walking Robots

A K M Azad	– USA	A T de Almeida	– Portugal
K Althoefer	– UK	R Arkin	– USA
C Balaguer	– Spain	D P Barnes	– UK
J Billingsley	– Australia	C Bostater	– USA
K Bouazza-Marouf	– UK	M Buehler	– USA
G Bugmann	– UK	S Cameron	– UK
C Chevallerau	– France	H Cruse	– Germany
S Cubero	– Australia	J S Dai	– UK
B Davies	– UK	R Dillmann	– Germany
L Fortuna	– Italy	T Fukuda	– Japan
P Gonzlez de Santos	– Spain	S S Ge	– Singapore
V Gradetsky	– Russia	A Halme	– Finland
R Hasselvander	– France	E Herrera	– Colombia
N Heyes	– UK	S Hirose	– Japan
M A Hossain	– UK	D Howard	– UK
O Kaynak	– Turkey	P Kiriazov	– Bulgaria
P Kopacek	– Austria	K Kwok	– Singapore
D Lefeber	– Belgium	J Lopez-Coronado	– Spain
C Melhuish	– UK	R Molfino	– Italy
N K MSirdi	– France	D E Okhotsimsky	– Russia
E Dupuis	– Canada	H R Pota	– Australia
R Quinn	– USA	M Ribeiro	– Portugal
J Sa da Costa	– Portugal	M A Salichs	– Spain
T P Sattar	– UK	M H Shaheed	– UK
G Seet	– Singapore	L Steinicke	– Belgium
A Vitko	– Slovakia	K J Waldron	– USA
R Walker	– UK	H Worn	– Germany
A Yigit	– Kuwait	A Zomaya	– Australia

NATIONAL ORGANISING COMMITTEE

for the Tenth International Conference on Climbing and Walking Robots

Z. W. Zhong (Publicity)	– NTU, Singapore
K. H. Low (Workshops)	– NTU, Singapore
C. J. Zhou (Exhibitions)	– SP, Singapore
C. M. Chew (Technical Visits)	– NUS, Singapore
D. Hsu (Technical Visits)	– NUS, Singapore
A. P. New (Awards)	– DSO, Singapore
H. Y. Yu (Awards)	– DSO, Singapore

SECRETARIAT

for the Tenth International Conference on Climbing and Walking Robots

Shirley Soh – CCE, NTU, Singapore

CONTENTS

Advances in Walking Robots

Advances in Humanoid Soccer Robots

Supporting Technologies

Plenary Introduction

BIPEDAL HUMANOID ROBOTICS AND ITS APPLICATIONS

ATSUO TAKANISHI†

Department of Modern Mechanical Engineering
Waseda University, Japan

1. Introduction

Even though the market size is still small at this moment, applied fields of robots are gradually spreading from the manufacturing industry to the others in recent years. One can now easily expect that applications of robots will expand into the first and the third industrial fields as one of the important components to support our society in the 21st century. There also raises strong anticipations in Japan that robots for the personal use will coexist with humans and provide supports such as the assistance for the housework, care of the aged and the physically handicapped, since Japan is one the fastest aging societies in the world. Consequently, humanoid robots and/or animaloid robots have been treated as subjects of robotics researches in Japan such as a research tool for human/animal science, an entertainment/mental-commit robot or an assistant/agent for humans in the human living environment.

Over the last couple of years, some manufactures including famous global companies started to develop prototypes or even to sell mass production robots for the purposes mentioned above, such as SONY, TMSUK, ZMP, TOYOTA, HONDA, etc. Most of those robots have two legs for its mobility . On the other hand, Waseda University, where we belong to, has been one of the leading research sites on bipedal walking robot and humanoid robot research since the late Prof. Ichiro Kato and his colleagues started the WABOT (WAseda roBOT) Projects and developed the historical humanoid robots that are WABOT-1 and WABOT-2 done in the early 70s and 80s respectively.

One of the most important aspects of our research philosophy is as follows: By constructing anthropomorphic/humanoid robots that function and behave like a

† www.takanishi.mech.waseda.ac.jp

human, we are attempting to develop a design method of a humanoid robot having human two legs to coexist with humans naturally and symbiotically, as well as to scientifically build not only the physical model of a human but also the mental model of it from the engineering view point. Based upon the philosophy, I and my colleagues have been doing researches on bipedal humanoid robots. In my plenary speech I will introduce the research philosophy of bipedal humanoid robotics introducing current bipedal walking robots WABIAN-2R and WL-16RIII as shown in Fig. 1 and 2 respectively, the design and the control of the robots and its applications collaborating with robotics companies

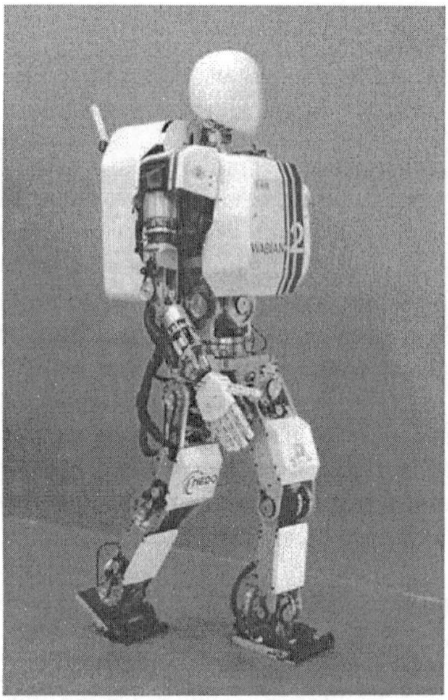

Fig. 1 Bipedal Humanoid Robot WABIAN-2R that walks with 0.96 s step time and 0.5 m step length.

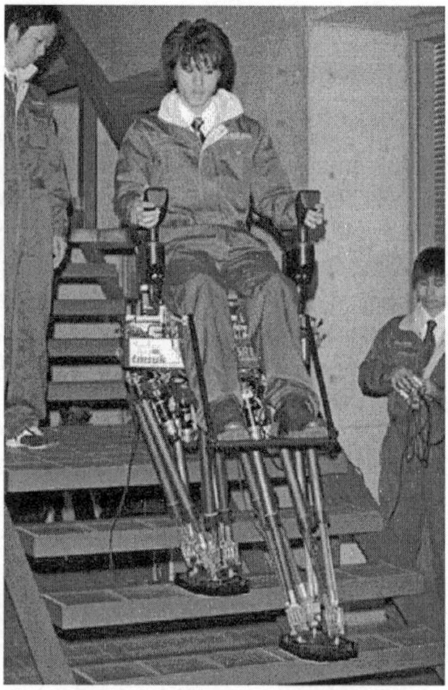

Fig. 2 Bipedal Walking Vehicle WL-16RIII that carries 75 kg human and walks up/down on 0.25 height stairs.

CLIMBING UP THE WALL

JOHN BILLINGSLEY

University of Southern Queensland
Toowoomba, Australia

Reminiscences cover over two decades of legged robot research. From Robug I, a six-legged walker that was overweight, the story leads through a succession of wall-climbers both simple and elaborate, some of which have found commercial applications in the nuclear industry.

1. Introduction

The robotics research lab of Portsmouth Polytechnic, as it was then, was transformed in June 1985. Of course, work on the Craftsman Robot and manufacturing cells continued, but in that month Arthur Collie joined the Group with a Royal Society grant to indulge his passion for legged robotics. The first project resulted in Robug I, an overweight six-legged machine with a body in the shape of a coffin. Later robots such as Robug II were designed with weight very much in mind, although they still favoured pneumatic actuation.

While some projects pursued intelligence (or rather cunning), others such as Zig-Zag sought to improve mechanical simplicity. However the nature of the research was transformed when the nuclear industry took an interest.

Early collaboration with Arthur had been through his role as Development Manager of Turnright Controls Ltd, part of the Tube Investments Group. The interest then had been in controls for domestic cookers, a far cry from walking robots. A realignment of the Tube Investments structure led to a management buy-out and Turnright's transformation into a new company, Portsmouth Technology Consultants, or PorTech for short.

The new company needed products to sell. When the nuclear industry expressed a readiness to place an order for a robot to enter a reactor, the opportunity was seized and a succession of Neros resulted.

By 1992 I had had my fill of winter arthritis and all-too-brief English summers. I deserted Portsmouth for the warmth of Queensland. At the University of Southern Queensland on top of the Great Dividing Range the work on legged robots continued.

Within a year, Michael Bellmann had completed Hydra, a 'somersaulting' wall-climber. Michael Rook developed the Toad for his master's degree and by 1997 Sam Cubero had completed STIC, a lightweight four-legged ceiling-walker.

Some projects are waiting in limbo, half completed. Nantha Kumar's roller-skating robot still occupies considerable space in the lab, but is a striking icon that gets photographed on Open Days.

Graham McLatchey has pulled most of the legs off Robug 4, which we scavenged from the demise of Portech a few years ago. He is seeking to improve the efficiency of the actuation, but will present his own account of it.

2. The first Robug

We did not enter the task blindly.

Arthur's funding was by the Royal Society under their industrial fellowship scheme, an innovative initiative to increase the exposure of universities to industry. The fellowship gave the full facilities of the Royal Society for its two-year duration, paying his salary at a reduced amount. Funding for the project itself came from SERC. Arthur relates that the SERC representative said, very grudgingly, "I suppose under the circumstances we shall have to give you a grant."

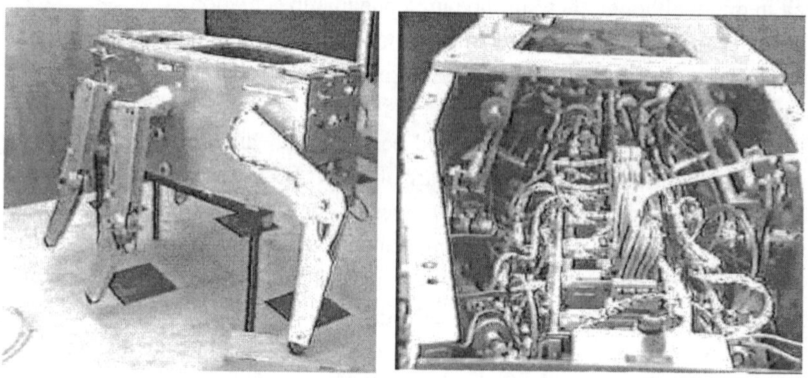

Figure 1. Robug 1 was shaped like a coffin, filled with circuit boards

There was also a small grant from the Ministry of Defence on the grounds that the machine could be used either as a sort of mine clearing machine or as a single mission device. As a mobile antitank mine for use in rugged terrain it would lie in wait in a ditch, clear of the mine clearing flails, then jump up, run towards the sound of vehicles and go out in a suicidal blaze of glory.

Accordingly we set off to inspect the DARPA legged robot in Columbus, Ohio. Some massive engineering brute force made the task look easy. It was not.

Robug I's body was shaped rather like a coffin. Inside it, six single-board BBC Micros (the embedded processor of choice at that time) provided the computing power, one for each leg. It is interesting to compare the sizes of trial legs of Robug and DARPA. Their budgets were equally disproportionate.

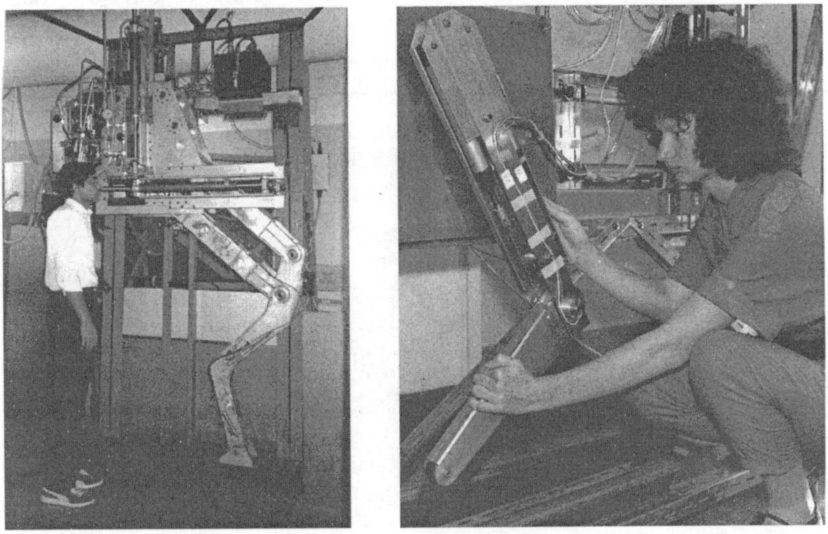

Figure 2. Legs of different sizes, DARPA and Robug I

Since it must be self-powered, it was intended that pneumatic power should eventually be provided by an explosive gas generator and an accumulator. Pneumatic actuation would also give the compliance that would allow the load to be shared evenly between the legs.

Much was learned from Robug I. Compliance and stability are uneasy bedfellows. Without an inertial sensor, Robug's early steps were somewhat faltering. We also learned how easy it is to underestimate the load each leg would have to bear, not only from movement of the centre of gravity but also from inertial forces.

In those days, before mechatronics became recognised as an integrated discipline, we discovered a fault of human nature. During some holidays, the project's programmer had been left in charge of the machine. We returned, to be told that the time had been spent writing a wonderful new algorithm that identified dynamic errors in one of the angle sensors and adaptively compensating for them. Arthur showed great restraint in the way he told the young lady in question that using an Allen key to tighten the sensor's grubscrew would have been an altogether better solution.

3. Robug II

In 1988, another grant allowed the work to continue. Armed with plenty of hindsight, we aimed for low weight. To improve stability, the robot followed an insect-based scheme with its knees higher than its centre of gravity. Wall-climbing looks more dramatic than walking, but it allows stability to be simplified even further. A "Shufflebottom" version with two arms/legs with pneumatic grippers, mounted on a base with two more suckers, showed that the principle worked. Another two legs were added to make the full Robug II.

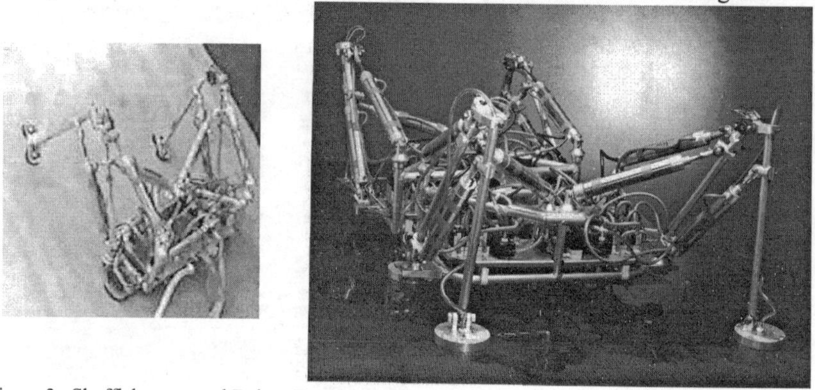

Figure 3. Shufflebottom and Robug II

With 'belly suckers' augmenting the legs, a strange but safe gait was possible. Each foot was released and advanced in turn, then the body was released and heaved forwards. Strategies of exploring for a safe foothold impressed the viewers of BBC Television's 'Tomorrow's World'.

Figure 4: Robug II on "Tomorrow's World" and climbing on a building

We then wanted to make a transition from floor to wall, or from wall to ceiling, so the body gained a hinge so that the front part could rear up. Bing Lam Luk joined the project and took the software in a new direction.

4. Zig-Zag

A single pneumatic cylinder flexes a square jointed frame to become a rhombus. Four suckers at the corners hold it to a wall. The grippers on one side are released, the shape is flexed and Zig-zag has taken a step. Arthur boasted this as the simplest form of walking robot. However it has the disadvantage that the unattached corners merely 'flop' against the wall and it cannot operate on an overhang. Nevertheless in 1990 Arthur and Bing were proud to take it to the first Robot Olympics, where it won its event on a high vertical wall.

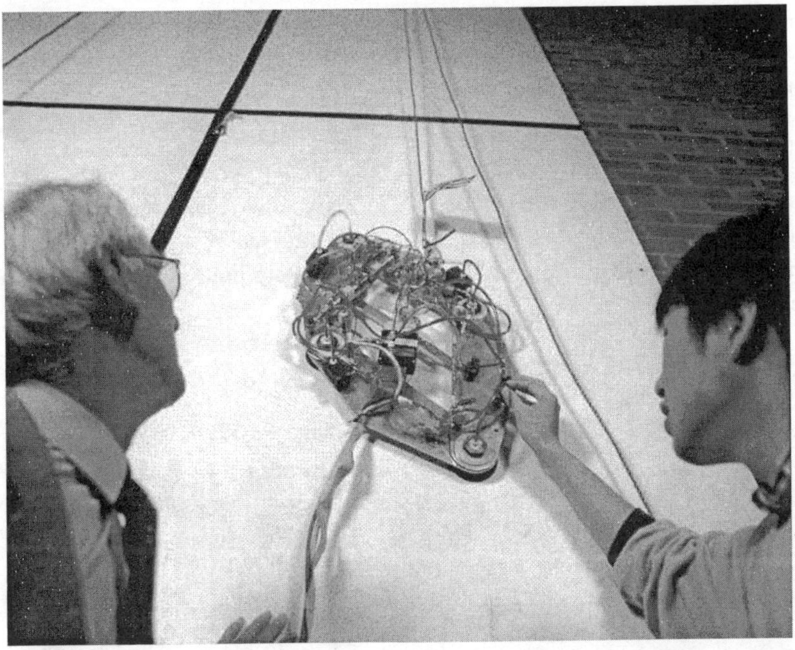

Figure 5. Zig-zag with Arthur and Bing at the 1990 Robot Olympics

5. Hydra

By now I had made my escape to Australia. I had an idea for a simple walker that could make the transition from floor to wall. It had just two large suction feet and one joint. It would stand doubled over, like an athlete gripping his ankles, then straighten into a series of 'flicflac' somersaults that would take it across the floor. The addition of swivelling joints in the 'ankles' would give navigation and with careful positioning the Hydra could transfer to a wall.

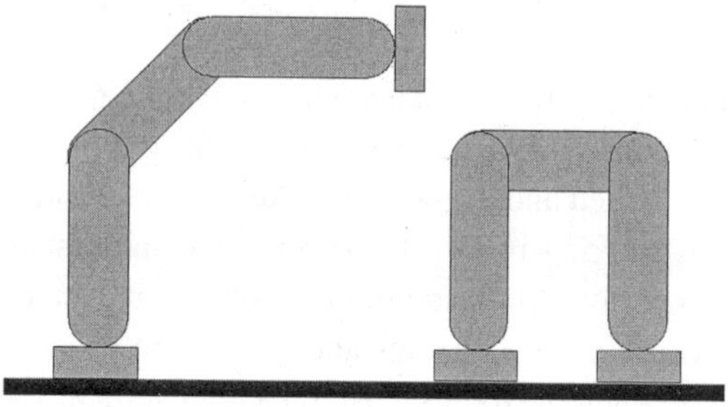

Figure 6. The principle of Hyda

Michael Bellmann joined me in Toowoomba and within months Hydra was completed as part of his project work for TU Muenchen.

6. Toad

I wanted a more versatile competitor for Zig-Zag and the result was Toad. Two front suckers are mounted at the ends of a horizontal bar. Two rear suckers are mounted at the ends of a second bar. The two bars are joined by a link that can be extended or shortened by a pneumatic cylinder.

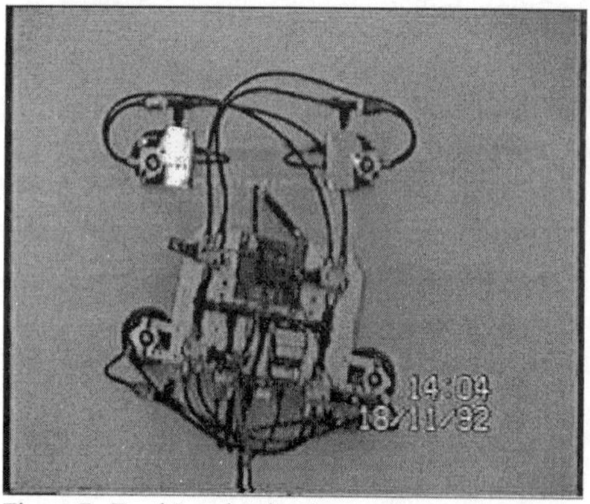

Figure 7. Toad I, and a simpler version running on the ceiling on Tomorrow's World

If the front left foot is released and the link is extended, Toad takes one step forwards. The foot grips again, the rear left foot is released and the link retracts to take the next step. This is repeated on the right-hand side and Toad has advanced by a full pace.

Some simulation showed that the rear pivot of the cylinder must be behind the feet to ensure that the rear legs are stable in lining up with the front ones. Extra actuation was added to twist the link, so that when a foot is released it can be actively lifted or pressed against the surface to take a new grip.

Michael Rook developed the machine for his Master's project, and Tomorrow's World viewers saw a Toad running across the studio ceiling.

7. Nero

Meanwhile in England, Portech had taken up the challenge of making a robot for the nuclear industry. The first robot task was to climb on a reactor pressure vessel in order to put a line of temperature sensors in place and Nero I had this single purpose.

Whereas we had struggled to incorporate intelligence into Robug, the last thing the clients wanted was a robot that moved on its own initiative. Each pace must be verified by an operator. If a mishap occurred and the robot had to be retrieved by a human, reactor shut-down costs would amount to six figures.

The design was in the form of two rectangular frames, each with a gripper at each corner. One was long and thin, inside which a carriage could move up and down. The second was square, also carried four grippers, was mounted on the carriage and could be made to swivel. Frames were released and lifted in turn, while the carriage operated either to carry the smaller frame forward, or the larger frame if that was lifted.

Figure 8. Nero and its working environment

The success of Nero 1 led to several further Neros. Soon they were working upside-down in ducts of flowing hot gas, with grinding wheel attachments for tidying up the metalwork.

8. STIC

Sam Cubero was determined to make a rival for Robug, to be lighter and more agile. Once again the actuation was to be pneumatic.

Figure 9. Sam Cubero proudly demonstrated STIC

A vital part of the control was in the valves used to control the cylinders. Poppet valves released huge amounts of air as they controlled pressure with mark-space operation. Could a proportional gas valve be devised that would permit more sophisticated control? A combination of the Coanda effect and planar solenoids seemed to be the answer. Indeed STIC walked on a ceiling, rather more steadily than on a floor, and is now in retirement in Sydney's Powerhouse Museum.

9. Robugs III and IV

The collaboration between Portech and Portsmouth University now gained strength with two major projects from the European Community. To some extent, the Robugs were seen as a demonstrator for a new communication protocol. Robug IV has two dozen processors, three for each of its eight legs.

Figure 10. Robug 4

A few years ago, the supply of orders for walking robots dried up. Although Portech had diversified into the supply of general mechatronic test equipment, its finances were found to be no longer viable. The consolation was that we were able to buy Robug IV for the University of Southern Queensland.

With eight legs, concerns for the robot's 'standing strength' should be minimal. Indeed there is video of someone sitting on Robug III. However weight was not so strictly constrained and when operated at a laboratory pressure of 4 bar, as opposed to the design pressure of 10 bar, actuator force again becomes an issue. Graham McLatchey has been working on solving it.

10. Conclusion

As robots become more commonplace, public interest is waning. Autonomous vacuum cleaners and lawnmowers have not taken the market by storm. When a robot has legs, however, it still seems to keep its cachet. Walking and climbing robots will continue to be great fun for many years.

Acknowledgment

Arthur Collie has given me substantial help in refreshing my memory of those early days. Bing Luk has also retained an archive.

References

1. The development of a pneumatically powered walking robot base, A A Collie, J Billingsley, L Hatley, Proc IMechE C377/86, Conf. UK Research in Advanced Manufacture, London, Dec. 1986, pp137-143.
2. Design and performance of the Portsmouth climbing robot, A A Collie, J Billingsley, S F Mo, E von Puttkamer, Proc 7th International Symposium on automation and robotics in construction, June 1990, Bristol.
3. Design and performance of the Portsmouth climbing robot, A A Collie, J Billingsley, S F Mo, E von Puttkamer, Mechatronic Systems Engineering, 1, pp 125-130
4. A climbing robot with minimal structure, J Billingsley, A A Collie, B L Luk, Proc IEE conference Control 91, March 1991, Edinburgh, 2, pp 813-815.
5. Robug II: an intelligent wall climbing robot, B L Luk, A A Collie, J Billingsley, IEEE conference on Robotics and Automation, April 1991, Sacramento, USA, pp 2342-2349.
6. Real time software control system for the Nero wall-climbing robot, B L Luk, A A Collie, J Billingsley, T White, N Bevan, Proc 4th Euromicro Workshop on Real Time Systems, Athens, June 1992, pp 74-78
7. An improved Wall Climbing Robot with Minimal Actuation, J. Billingsley, A A Collie, M J Rook, Proc 1st IFAC International Workshop on Intelligent Autonomous Vehicles, Southampton UK, 18-21 April 1993, pp 14-19.

8. An Articulated Limb Climbing Vehicle with Autonomous Floor to Wall Transfer Capability, B L Luk, A A Collie, N Bevan, J Billingsley, Proc 1st IFAC International Workshop on Intelligent Autonomous Vehicles, Southampton UK, 18-21 April 1993, pp 20-24.
9. Toad - a wall-climbing robot, J Billingsley, M J Rook, A A Collie, Australian Robot Association Conference, Robots for Competitive Industries, Brisbane, July 14-16 1993.
10. Design Aspects of a Leg for a Walking Robot, J Billingsley, A Collie, Proc Mechatronics and Machine Vision in Practice, Toowoomba Australia, September 13-15 1994, pp 162-165.
11. Automatic surface transition adaptation for a quadrupedal space frame robot, S N Cubero, J Billingsley, Proc Second International Conference on Mechatronics and Machine Vision in Practice, Hong Kong, September 12-14 1995, pp 113-118.
12. A novel proportional gas valve for mechatronics applications, S N Cubero, J Billingsley, Proc Second International Conference on Mechatronics and Machine Vision in Practice, Hong Kong, September 12-14 1995, pp 179-184.
13. Command and compliance of a leg for a walking robot, J Billingsley, A A Collie, Proc Second International Conference on Mechatronics and Machine Vision in Practice, Hong Kong, September 12-14 1995, pp 256-259.
14. Automatic control of a surface adapting, four legged wall climbing robot, S J Cubero, J Billingsley, Mechatronics 96 With Mechatronics & Machine Vision in Practice 96, University of Minho, Geuimaraes Portugal, September 1996 pp 1.135-1.142.
15. S Cubero, J Billingsley, Force, Compliance, and Position Control for a Space Frame Manipulator, 4th Mechatronics and machine vision in practice, Toowoomba; Australia, Sep 1997. pp. 124-130.
16. N K Kanesen, J Billingsley, A Roller Skating Robot, Mechatronics and Machine Vision, Research Studies Press, England, September 2000, ISBN 0-86380-261-3, pp 283-292.
17. McLatchey, G.J. and Billingsley J, Force and Position Control Using Pneumatic Cylinders, 9th International Conference on Climbing and Walking Robots, 2006, Belgium.

From Micro to Nano and Swarm Robots

Heinz Wörn, Ramon Estaña, Heiko Hamann, and Marc Szymanski

Institute for Process Control and Robotics, Universität Karlsruhe,
Engler-Bunte-Ring 8, 76131 Karlsruhe, Germany
E-mail: {woern, estana, hamann, szymanski}@ira.uka.de
wwwipr.ira.uka.de

Current research in Micro, Nano and Swarm Robots as results of the European projects **Miniman**, **MiCRoN** and **I-SWARM** will be presented. First, the design and the control of 5 to $10cm^3$ sized mobile micro robots with five degrees of freedom will be shown.They can handle miniaturized parts as for example an optical component or a biological cell with a size in the micrometre-area with an accuracy of $100nm$ under a microscope or a raster-electron microscope. Second, the design and the control of a $1cm^3$-sized mobile untethered micro robot will be demonstrated. Here, the robot consists of five parts: the Piezzo locomotion module, the micro control unit, the communication unit, the navigation system and the micro gripper. The mobile robot can be guided and positioned in an arena with an accuracy of 5 micrometre and can be programmed and controlled over the wireless communication unit. Third, the design and the control of $3 \times 3 \times 3\ mm^3$ sized micro-/nanorobots with 2 degrees of freedom will be presented. The transmission of energy and the communication between the robots is realized via infrared. The robot controller is fully integrated and has limited functionalities. Via basic sensors communication functions and elementary rules and behaviours the micro robot can act in a swarm consisting of hundreds and thousands of robots. Future applications could be monitoring-, inspection-, exploring-tasks etc. of big areas or objects.

Keywords: Micro Robots, High precise Navigation, Swarm Intelligence, Micro Handling

1. The Miniman Project

This section gives a brief overview over the EU-sponsored **Miniman** project. The main result of the research activity consists of the design and fabrication of a set of components which enhance and make feasible the micro manipulation and end-effectors, force and tactile sensors, gripper systems and their integration in a flexible micro robot system, thus resulting in an essential tool for very different micro manipulation tasks. The micro manipulation workstation comprises a few versions of robots with different overall size; the robot prototypes consist of a mobile positioning unit with two translatory and one rotational degree of freedom (DOF), which carries a manipulation unit with a micro tool. The robots can operate at high speed (up to 30 mm/s) and have a motion resolution of about 10 nm. In Fig. 1 two types of **Miniman** micro robots are depicted.

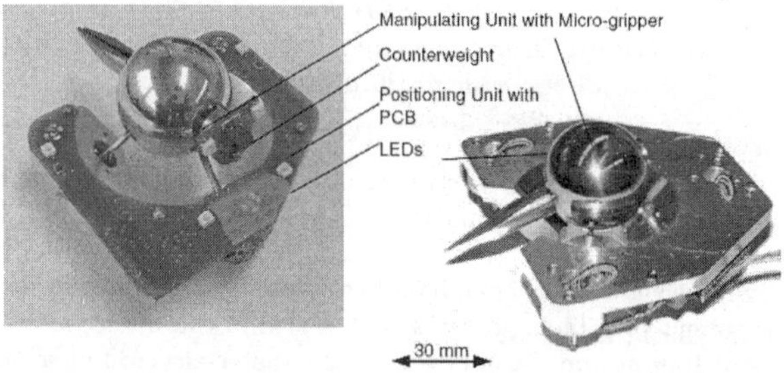

Fig. 1. Two types of the **Miniman** micro robot as a result of the project.

The positioning units of the robots consist of three tube-shaped piezo legs, which can be actuated independently by a control software, which includes all necessarry algorithms for moving the robot over the smooth glass plate of the working plate (300×200 mm in size). The movement is based onto the well established slip-stick-principle used already in other micro robot projects like Nanowalker.[1] The **Miniman** is controlled via an 6D space mouse and can be used for example as an intelligent tweezer in a SEM (see Fig. 2) as well as a tool for micro lens assembly and cell inspection.
As already mentioned, specially designed micro robots can also act as a flexible assembly facility for prototype micro systems in a Scanning Elecctron Microscope (SEM), as probing devices for in-situ tests in various applications or just as a helpful teleoperated tool for the SEM operator when examining samples. Several flexible micro robots of this kind have been developed and tested. Driven by piezoactuators, these few cubic centimeters small mobile robots perform manipulations with a precision of up to 20nm and transport the gripped objects at speeds of up to 3cm/s.
As described in this section, several new tools have been developed to master the requirements of the micro world, i.e. micro lens assembling. The system is now in a state for continuous usage, and it is a helpful tool for various micro manipulation tasks. The results obtained give clear indication about parameters and steps which should be important for further miniaturisation, for example the use of miniature actuators. At the same

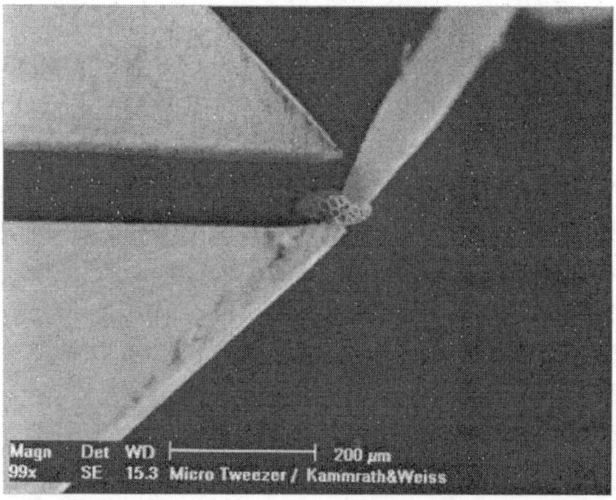

Fig. 2. A **Miniman** Gripper (left) and a **Miniman** needle as a helping hand in a SEM.

time, the system proves that the development of mobile micro robots is a promising approach to realise very small and flexible tools useful for different applications. By means of its intuitive teleoperation mode, the system enables the user to work in the micro world.

2. The MiCRoN Project

2.1. *Introduction*

The EU-sponsored **MiCRoN** project aimed at the development of a completely new micro robot system based on flexible mobile, 1 cm^3 sized robots acting autonomouslyand and fully untethered. Eight European project partners started in March 2002 with the development of a system based on the results from the **Miniman** Project. The **MiCRoN** project's outcome is a major contribution in the field of micro mechatronic components. The new micro robot is using low voltage piezo actuators and hybrid on-board electronics. Several micro tools have been developed:

- Millimetre-sized grippers for the 3D assembly of meso-scale objects
- A robot-mounted micro syringe chip for the injection of substances into living cells
- A robot-mounted micro syringe chip for the injection of substances into living cells
- AFM tools for standard AFM imaging and using functionalised nano tips for biological experiments.

The robot actuators have been developed using a novel rapid prototyping process for multilayer piezoceramics; the robot hardware consists of an on-board VLSI electronics module. Another major contribution of the project are methods for wireless micro robot operation. A power floor has been built as a working prototype, which transmits electrical energy to mobile units operating on a $250 \times 250\ mm^2$ area. The developed infrared communication schemes are used for controlling a small group of robots. An advanced control system has been realized including an implementation of the progressive Kalman Filtering Theorem. This software package includes all software needed for robot control, navigation, planning, simulation and user interfacing. For the integrated on-board vision system, a camera, which can be mounted on mobile micro robots, has been developed into a working prototype. This system can generate 3D data of micro manipulation scenarios. The computer vision system developed offers a broad range of stable recognition algorithms for micro handling applications. In the field of object localization, the global localisation system represents a major achievement. This system has reached final prototype state with a position resolution of about 5 microns over the complete size of the workspace. The system components key figures can be summarized as follows:

- Mobile micro robots: platform with holonomic three degrees of freedom (DOF) x, y and ϕ, equipped with wireless powering,micro grippers, SyringeChip for cell manipulations, on-board Atomic Force Microscope (AFM)-tools, piezoactuated miniature motor, Onboard electronics module and wireless IR-based communication module
- Handling experiments: multi-robot scenarios for cell-handling experiments, cell-injection experiments and involving the 3D assembly of meso-scale products and objects using a micro soldering process
- Vision and Position Sensors: These are sensor systems which provide the global localisation, object recognition, scene interpretation and local images to the system
- System control: this is the brain of the complete system

Together with the high resolution navigation system,[2] a highly reliable and precise handling tool is available (see Fig. 3).
More informations about the **MiCRoN** and the **Miniman** robot can be found at `http://microrobotics.ira.uka.de/`

Fig. 3. The visionary project objectives for the **MiCRoN** project.

3. The I-SWARM Project

The EU-sponsored **I-SWARM** project aims to develop and build a robot swarm of up to 1,000 robots.[3,4] Each one equipped with limited, pre-rational on-board intelligence. These swarm should be capable of performing tasks that are not possible with either a single micro robot, or with a small group of micro robots. The expected outcome shall be an observable self-organisation effect in the robot swarm similar to that seen within ecological systems like ant states, bees colonies and other insect aggregations. Further benefits are a greater flexibility and adaptability of the swarm to the environment, robustness to failures, etc. utilizing emergent effects. Within the project the suitability of the strategies of animals and especially insect colonies and swarm intelligence for managing micro robot swarms are investigated. One of the major goals of the project is to transform knowledge of eusocial insect behaviour, into the swarm intelligence of robots and to apply this to the completely new swarm of micro robots.

During the **I-SWARM** project, the *Jasmine* robot was designed and fabricated for the software test purposes (see Fig. 4, `www.swarmrobot.org`). It has a size of $3 \times 3 \times 3\ cm^3$ and supports almost all concepts the **I-SWARM** robot uses. Equipped with six infra-red emitters and receivers the robot is a perfect platform due to its excellent communication abilities, which are essential for swarm robot research. Different kinds of bio-inspired swarm algorithms have been implemented and investigated with *Jasmine* robots.

The final **I-SWARM** micro robots with a size of $3 \times 3 \times 3\ mm^3$ have a design depicted in Fig 5. The locomotion platform has three legs based on a piezoelectric polymer actuator material. A vibrating needle utilizing

Fig. 4. Swarm of Jasmine robots.

the same actuator material acts as touch sensor. An ASIC comprises a 8051 based micro controller kernel, the power managment including amplifiers, the communication interfaces for infra-red and the solar cells. The ASIC is also the robot's backbone and is connected via an FPC with the four channel infra-red communication device and the solar cells. The solar cells deliver the energy for the whole system and are also used as an ego-positioning sensor.

This robot design is far beyond the current state-of-the-art micro robot research. Indeed those micro robots may have a worse resolution regarding the manoeuvrability and controllability than the current micro robots. But there is no state-of-the-art micro robot swarm reaching the **I-SWARM** constraints. Furthermore, the computational, communicational and sensorial constraints the on-board micro controller and sensors dictate lead to a more basic robot swarm research. Swarm robotics will benefit from this fundament research, as fancy algorithms using a whole personal computer to control a single swarm robot could not be implemented on the **I-SWARM** robots. The self-organisation abilities have to clearly rise from the pure number of robots in the swarm and simple but intelligent algorithms.

Several bio-inspired algorithms like tropholaxis strategies have been transferred from nature to artificial swarms and tested with the *Jasmine* robot or in simulation.[5] A motion description language (MDL2ϵ) for rule based swarm behaviour has been designed that is similar to a regular language.[6] The description language is used to describe the robots behaviour in a sim-

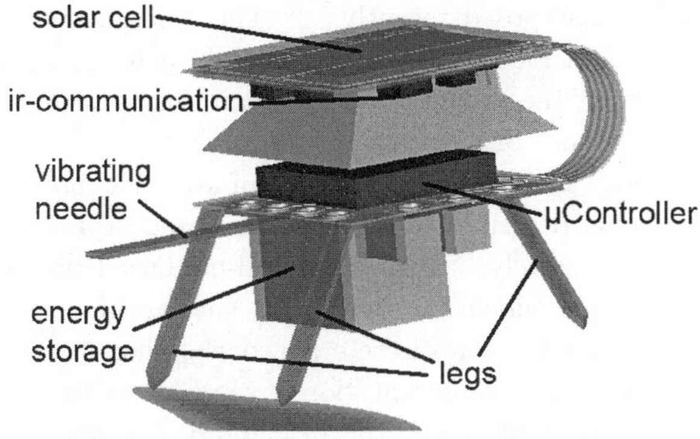

Fig. 5. The design of the final $3 \times 3 \times 3\ mm^3$ sized **I-SWARM** robot.

ulation and could also be directly transferred to the robot. The language describes the robot controller based on primitive action so called atoms. The description language is also used to generate controllers for emergent behaviour using the paradigm of *Genetic Programming.*

3.1. *Modeling Swarm Robotic Systems to Support the Controller Design*

Designing and implementing artificial self-organizing systems is in general a challenging task since they typically behave non-intuitive and only little theoretical foundations exist. Predicting a system of many components with a huge amount of interactions is beyond human skills. A connected complex of problems is the micro-macro link – determining the global behavior resulting from local behavior patterns and vice versa.[7] This applies also to swarm robotic systems in particular. The currently common use of simulations for design support is not satisfying, as it is time-consuming and the results are most likely suboptimal.
The problem is to find an effective and efficient control algorithm for the robot swarm to accomplish a predefined task. To minimize the complexity of the entire system and due to the limited capabilities of an individual swarm robot, the development targets simple rules and, in an allusion to nature, one hopes for emergent behavior of the robot group that leads to the solution of the given task. However, the design of self-organizing and emergent behavior is very difficult. There are no general methods available that support the engineer in designing and optimizing the control algorithm. First steps to establish such a theory of self-organizing behavior in swarm

robotics have been proposed recently.[8–10] The fundamental substance of these approaches is a model that approximately predicts the behavior of the swarm. However, the prediction of emergence is considered to be hard or even impossible as a contradiction in terms.[11,12]

In ongoing work the authors expand the set of available models by an analytical approach that describes space continously and explicitly rather than approximating it discretely.[13,14] In many self-organized swarms the effectivity of the behavior depends primarily on spatial inhomogeneities that are hard to approximate by a discrete approach. Thus, it can be inferred that an explicit and exact representation of space is useful in these cases.

Our model describes, first of all, the probability density function of the robot's positions. This macroscopic description is derived from a microscopic description of an individual robot's movement. This is done by regarding the swarm robotics system as a system showing Brownian motion with drift and using statistical physics to derive a model. Hence, the micro-macro link is solved almost for free by established physical methods. However, currently our model depends on heuristic reasonings as well to model the communication by means of stigmergy or direct communication because physics does not provide the theory for autonomous and explicitly communicating particles.

At first, we model microscopically the movements of an individual robot using a Langevin equation of the form

$$\dot{\mathbf{r}} = A(\mathbf{r}) + C(\mathbf{r})\mathbf{F}(t), \tag{1}$$

with a noise term $\mathbf{F}(t)$, $A(\mathbf{r})$ describing the deterministic proportion of the robot's motion, and $C(\mathbf{r})$ the nondeterministic proportion. From this equation, a partial differential equation (PDE), the Fokker-Planck equation, can be derived modeling the swarm robotic system macroscopically.

Omitting the derivation[15,16] here, the Fokker-Planck equation under certain assumptions is

$$\frac{\partial f}{\partial t} = \frac{\partial}{\partial \mathbf{r}}(A(\mathbf{r})f) + \frac{1}{2}\frac{\partial^2}{\partial \mathbf{r}^2}(C(\mathbf{r}))^2 f, \tag{2}$$

with the product $f(\mathbf{r}, t)dr_x dr_y$ determining the probability of encountering a robot at position $\mathbf{r}$ within the rectangle defined by dr_x and dr_y at time t. See Fig. 6 as an example of f in an aggregation scenario.

We model direct or indirect communication between the robots, e.g. virtual pheromones, by a potential field on which $A(\mathbf{r})$ and $C(\mathbf{r})$ depend. The developement over time of this potential field can be a PDE as well extending the model to a system of PDEs.

Once the model is derived it is possible to predict the swarm behavior for a given set of parameters. If the algorithm is properly parameterized it is feasible to scan whole parameter intervals to find regions of optimal behavior. For a detailed description and a positive validation of two special scenarios against a simulated robot swarm see[13,14]

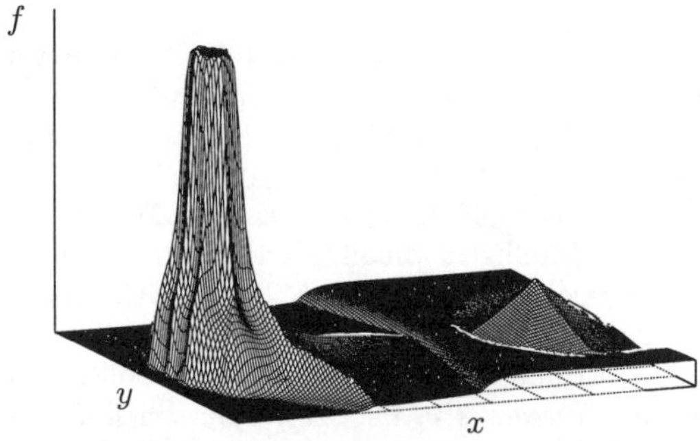

Fig. 6. The robot density over space for an aggregation scenario as an example.

4. Summary and Outlook

In this paper we described the development of three different kinds of micro robots. With the $5-10cm^3$ **Miniman** micro robot, it was proved that the development of small micro manipulation units is a positive approach to realise tools that can handle parts of the micro world. The $1cm^3$ sized untethered **MiCRoN** robot shows the higest miniaturisation of a micro robot equipped with tools at the moment It enables autonomous cell handling as well as a new kind of micro part assembly. The results of the **I-SWARM** project are expected to have a huge impact in the longer term: realisations of robotic swarms are conceivable in diagnostic tasks in a medical or technical context. Such a swarm could be inserted into an engine, pipeline system or patient, gather distributed data, diagnose faults or illnesses, transmit information to the outside and act accordingly. However, methods investigated in the project can also be applied in the near future: self-organisation and configuration becomes increasingly interesting for technical devices or systems.

References

1. S. Martel, M. Sherwood, C. Helm, W. G. de Quevedo, T. Fofonoff, R. Dyer, J. Bevilacqua, J. Kaufman, O. Roushdy, and I. Hunter, Three-legged wireless miniature robots for mass-scale operations at the sub-atomic scale, in *roc. of the 2001 IEEE Int. Conf. on Robotics and Automation*, ed. E. SahinLNCS 3342 (Springer Verlag, Berlin, Germany, 2001).
2. R. Estaña and H. Wörn, Moire-based positioning system for micro robots, in *SPIE's Int. Conference on Optical measurement Systems for Industrial Inspection III*, (SPIE, 2003).
3. R. Estaña, M. Szymanski, N. Bender and J. Seyfried, Towards a Real Micro Robotic Swarm, in *Ant Colony Optimization and Swarm Intelligence. 4th International Workshop, ANTS 2004*, ed. M. D. et al.LNCS 3172 (Springer Verlag, Berlin, Germany, 2004).
4. J. Seyfried, M. Szymanski, N. Bender, R. Estana, M. Thiel and H. Wörn, The I-SWARM project: Intelligent Small World Autonomous Robots for Micro-Manipulation, in *Swarm Robotics*, ed. E. SahinLNCS 3342 (Springer Verlag, Berlin, Germany, 2005).
5. T. Schmickl, C. Möslinger and K. Crailsheim, Collective perception in a robot swarm, in *Proceedings of the Second Swarm Robotics Workshop*, eds. E. Sahin, W. M. Spears and A. F. T. Winfield, LNCS, Vol. 4433, 2007.
6. M. Szymanski and H. Wörn, JaMOS - A MDL2ϵ based Operating System for Swarm Micro Robotics, in *IEEE Swarm Intelligence Symposium*, eds. Y. Shi and M. D. (Ed.), 2007.
7. D. Yamins, Towards a theory of "local to global" in distributed multi-agent systems, in *AAMS'05, Utrecht, Netherlands*, 2005.
8. A. Martinoli, K. Easton and W. Agassounon, *Int. Journal of Robotics Research* **23**, 415 (2004).
9. S. Kornienko, O. Kornienko and P. Levi, Swarm embodiment – a new way for deriving emergent behavior in artificial swarms, in *Autonome Mobile Systeme*, eds. P. Levi, M. Schanz, R. Lafrenz and V. Avrutin, 2005.
10. K. Lerman, C. Jones, A. Galstyan and M. Mataric, *Int. J. of Robotics Research* **25**, 225 (2006).
11. V. Darley, Emergent phenomena and complexity, in *Artificial Life 4*, eds. R. Brooks and P. Maes, 1994.
12. C. Müller-Schloer, Organic computing - on the feasibility of controlled emergence, in *CODES+ISSS'04, Stockholm, Sweden*, 2004.
13. H. Hamann and H. Wörn, A space- and time-continuous model of self-organizing robot swarms for design support, in *First IEEE International Conference on Self-Adaptive and Self-Organizing Systems (SASO'07), Boston, USA*, July 2007.
14. H. Hamann and H. Wörn, An analytical and spatial model of foraging in a swarm of robots, in *2nd SAB 2006 International Workshop*, eds. E. Sahin, W. Spears and A. Winfield, LNCS, Vol. 4433 (Springer, April 2007).
15. H. Haken, *Synergetics – an introduction* (Springer, 1977).
16. N. G. van Kampen, *Stochastic processes in physics and chemistry* (North-Holland, Amsterdam, 1981).

CLIMBING ROBOTS FOR NONDESTRUCTIVE TESTING: HISTORICAL PERSPECTIVE AND FUTURE TRENDS

BRYAN BRIDGE*

Co-Director, Research Centre for Automated and Robotic NDT
Faculty of Engineering, Science and the Built Environment
London South Bank University, United Kingdom

1. Introduction

It is perhaps not well known that mobile robots were already deployed in Nondestructive Testing (NDT) in the early 1970's, well before fixed robots came into use on manufacturing plant. The need for NDT robots first arose with the growth in exploitation of the world's oil and gas supplies. Vast length of pipeline are involved in transporting these supplies from source to multiple destinations. Sediment and chemicals in the unrefined products cause rapid cause inner wall thinning from both erosion and chemical attack. Pipe rupture at just one point, anywhere along a pipe can cause a major environmental disaster so there was always a need to inspect complete lengths of long runs of pipe. The handling of inspection sensors by human operators would thus always involve vast and costly numbers of personnel. This situation begged for sensor handling by mobile robot. A second reason for robotic deployment arose from the fact that many pipelines were buried underground, under concrete or ran along the sea bed before rising to the surface (risers). Human placement of sensors is impossible in these cases.

2. Pipelines

Five kinds of mobile and climbing robots were developed to deal with the above pipeline inspection problems. By definition a generic requirement will be that the robot will be adaptable in some way to curved surfaces.

* bridgeb@lsbu.ac.uk

(i) Pipeline pigs were cylindrical vehicles tailored to fit snugly inside pipes and move along a pipe under pressure hand having a collar of electromagnetic or ultrasonic sensors which automatically scanned the internal pipe walls for corrosion and erosion thinning. British Gas invested several Million pounds (at todays's prices) on the R &D of such a vehicle. It won a major prize for industrial innovation and was exhibited at the Science museum, which was testament to the perceived importance of and need for this technology.

(ii) Internal pipe crawlers were wheeled vehicles designed for pipe weld inspection. They carried panoramic X ray tubes inside pipes with co-moving detectors (film, phosphor screens image intensifiers etc) on the outside. This arrangement, only possible when external access to the pipe was possible allowed the preferred single wall through transmission inspection method to be deployed. The first crawler to be made small enough to fit inside 8 inch pipes, produced by Oilfield Inspection Services at Great Yarmouth, was reported on the front page of one edition of the Financial Times in the late 1970's.

(iii) External pipe crawlers. These robots, usable when there is external access to the pipe surface, were developed in the early eighties. Usually they had limited movement in which ultrasonic sensors performed orbits round circumferential weld. At is simplest the motion was simply achieved by a chain drive and the robot moved manually from one weld to the next. The pioneer of this technology in Europe was the Danish Welding Institute (Densk Norsk Veritas).

(iv) Small wheeled vehicles which could move both around and along pipes with magnetic adhesion emerged soon after (iii) as an alternative way of deploying ultrasonic sensors on accessible pipe surfaces.

(v) Submarine (swimming) inspection robots

The exploitation of offshore subsea oil and gas wells arose in the early 1970's from the rapid inflation caused by the step increase in the price of oil from land imposed by OPEC. So the need to inspect offshore oil risers arose at this time. Manned submarines for sea bed and riser inspection were rapidly followed by unmanned i.e. robot submarines. In 1976, the SONAMARINE robot underwent initial trials in a deep water lake at Stoney Cove, Leicestershire. The robot had a multi-axis manipulator to handle an ultrasound probe and the umbilical down which the ultrasound signals were sent and returned exceeded 600m, permitting operation in the deepest parts of the North Sea gas fields. The author was in charge of the ultrasonic data collection and to his knowledge, was the first to report robotic ultrasonic NDT in deep water.

3. Further Developments

Wheeled climbing vehicle technology was further adapted to the inspection of storage tanks and ship hulls involving greater climbing heights and thus greater payload requirements because of the longer umbilical service cables required. DNV, RTD and Silver Wing are examples of EU companies that pioneered this technology. Traditionally magnetic wheels were used and the vehicles were prone to slip on dirty or greasy surfaces. Thus use of electromagnets separate from the wheels reduced the slipping problem but led to heavier and bulky robots. The latest state of the art is represented by the OCTOPUS robot of Cybernetix which uses powerful magnetic plates to generate more than 600Kg of contact pressure on the climbing surface, benefiting from the latest improvements in rare earth magnet technology. Originally designed for ship hull cleaning it has recently been fitted with ultrasound probes for hull inspection.

In later years, probably the first use of a climbing vehicle for inspection of ship hulls dates back to the early 1970's when a robot was custom designed to ultrasonically inspect the Polaris Nuclear Submarines in dry Dock. An intrinsically safe system was specified and this gave rise to a climbing vehicle using pneumatics both for propulsion and foot suction. In the late 1980's Portech developed this technology further for cleaning/inspection work in nuclear power plant.

4. State of the Art

The rapid increase in computer processing power in the last 15 years has allowed a commensurate increase in sophistication in automated and robotic NDT. However there was and still is a tendency in the NDT industry for automated systems to be custom designed too carry out just one task. One example is complex manipulators for inspecting a specific nodal joint in a specific design of coolant circuit in a nuclear power plant. After a single joint inspection at one outage the expensive system would lie in a cupboard somewhere until the next outage at the power plant. Therefore in the early 1990's much research has been carried out, especially under European funded programmes, in attempting to find more generic solutions to robotic NDT. The Research Centre in Automated and Robotic NDT (RCART) at London South Bank University (LSBU) were the first to report a climbing robot which could climb to 30 metres carrying a 6 axis PUMA arm, able to scan an ultrasound probe through an continuously variable set of scanning routines, thus allowing multiple task NDT with one system. Later this work was extended to include a 7 - axis arm designed by Ansaldo in the late 1990's which at 22kg, at the time was the world's lightest arm available for NDT.

At RCART since 1992 Climbing and swimming robot NDT systems have been prototyped to cover all the key demand areas for remote automated NDT, namely (i) aircraft fuselages and wings, (ii) nodal joint inspection in nuclear power plant in a radioactive environment, (ii) container ship hulls, (iii) storage tank floors whilst still full of oil (iv) inside nuclear pressure vessels full of radioactive coolant . Key members of these partnerships have been TWI, and Zenon in the projects RRNDT,

ROBTANKINSPEC, ROBAIR FPSOINSPECT and RIMIMI. The ROBTANK vehicle for tank floor inspection and wall inspection under oil won the CLAWAR 2004 prize for industrial innovation. In PIPESCAN, ZENON designed a novel inspection robot that could walk and climb round pipes with thick protection thermal lagging, carrying out not contact inspection underneath

In these projects interesting solutions were implemented in making robots that could change surfaces under liquid or could climb round curved surfaces of very small radius whilst deploying a heavy payload of scanned sensors in a stable manner. A number of mechanisms involving flexible feet, ankle and thigh joints were successfully developed to transport these heavy loads.

5. The Future

Looking to the future RCART, with ZENON , CYBERNETIX, BISSIACH and CARRU, NAFTOSOL and TERNA are protoyping a climbing robot cell of three cooperating robots for simultaneous welding and weld NDT in ship storage tan and wind turbine manufacture (Figure 1). This project takes the world of robotic manufacturing out of the factory floor and into the field. With TWI , RCART are also involved in the prototyping of a robot for in situ inspection of turbine blades in offshore wind farms using the novel technique of nanofocal computed axial Xray tomography (Figure 2).

6. Conclusions

Nondestructive testing is an industrial area in which climbing and walking robots can truly be said to have made a practical impact. This impact continues on a rapidly climbing curve. A particularly likely development is to extend the concept of networked robots carrying NDT tools and other working tools to smaller and smaller robots so that operation in highly confined spaces such as arise in nuclear power stations can be carried out.

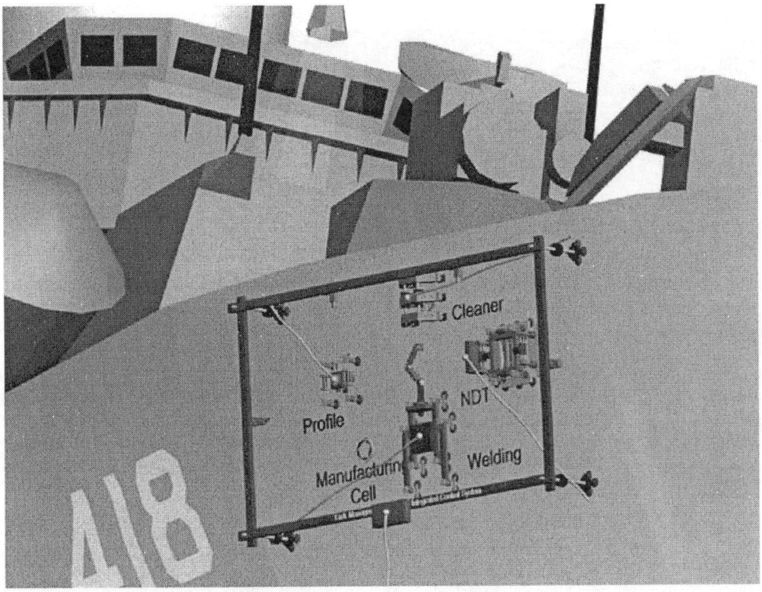

Figure 1. The CROCELLS project (Climbing robot cells) . By distributing a payload of working tools over several climbing and cooperating robots it is possible to carry out complex manufacturing tasks in situ on large structures such as ship hulls, storage tanks, wind turbines and over civil engineering structures. The distributed payload means that the robots can be kept small so that they can climb on highly curved surfaces. In the CROCELLS project real time NDT of long weld runs will be carried out on long weld runs simultaneously with the welding. So the NDT data can be used to control the weld process parameters in real time and instant weld repairs carried out if the NDT data indicate substandard welding

Figure 2 The CONCEPT project: *nanofocal compute X ray tomography for three dimensional imaging of defects in fibre reinforced wind turbine blades.* **A ring shaped climbing robot with magnet adhesion over a large area provides the necessary support for a heavy payload with large overturning moments. The payload consists of an x ray generator and detector array scanned in x and y and with angular scans provided by blade rotation to achieve the 3D inage reconstruction**

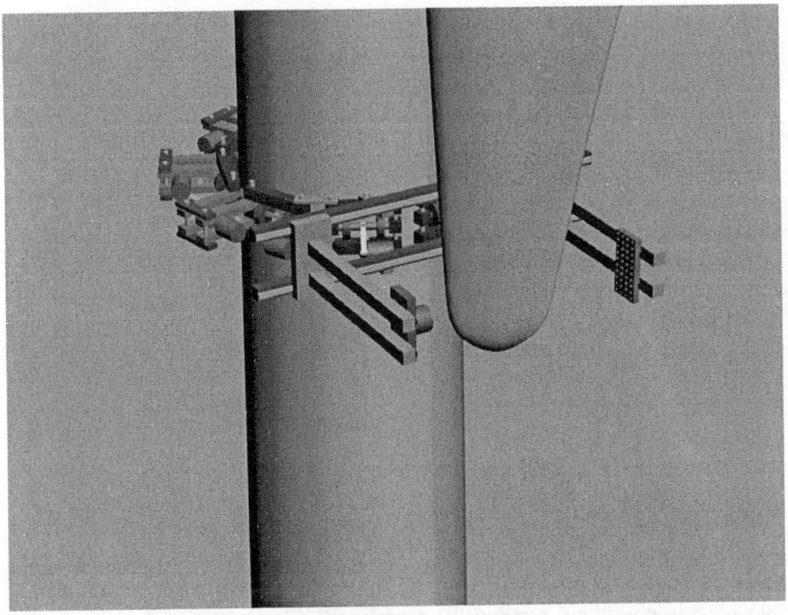

Figure 3. As Figure 2 with an enlarged view of the climbing ring robot and scanner

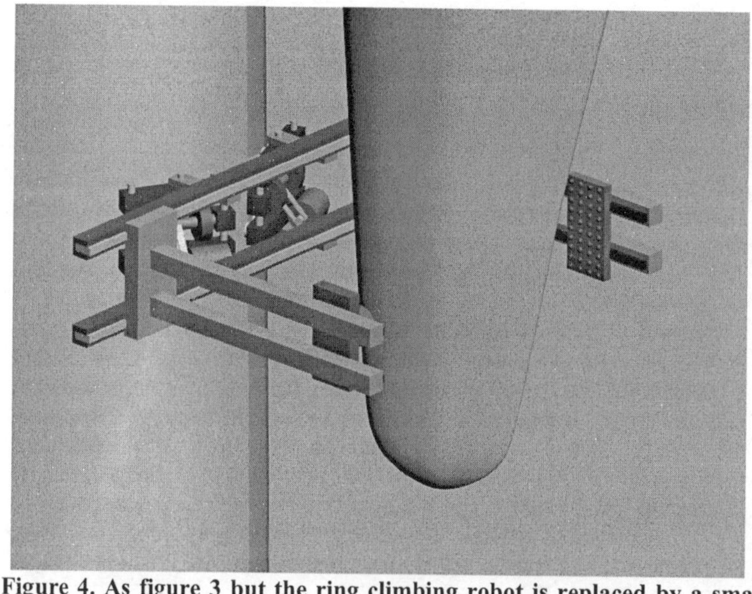

Figure 4. As figure 3 but the ring climbing robot is replaced by a smaller robot with less payload capbability which still adapts to the circumference of the turbine tower. The final choice between a ring design and the smaller robot will depend on the final design payload presented by the scanner

BIOMECHANICS AND ROBOTICS

NEVILLE HOGAN[†]
Department of Mechanical Engineering,
Department of Brain and Cognitive Sciences
Newman Laboratory for Biomechanics and Human Rehabilitation
Massachusetts Institute of Technology, USA

1. Introduction

Robots are well on their way to becoming commonplace domestic appliances but to realize their true potential requires the perfection of *contact robotics*, machines that physically cooperate with humans. One pioneering application requiring close physical cooperation between robots and humans is the delivery of physiotherapy to facilitate recovery after orthopedic and especially neurological injury. However, controlling robots to interact effectively with humans presents unique challenges. A quantitative knowledge of human motor and sensory performance is important to optimize machines for human contact. I will briefly review our success with robotic treatment of upper-limb motor disorders and how it has profited from the availability of relatively detailed quantitative knowledge of unimpaired biomechanics and neural control of arm movement.

Robotic treatment of lower-limb motor disorders presents unique challenges as the control of locomotion is substantially more difficult and less well understood. On the engineering side, locomotion presents far greater challenges than arm movement. The dynamic process is fundamentally *nonlinear*, including prominent discontinuities. It is a *hybrid* dynamic system, switching between regimes governed by different dynamic equations. Add to these the paramount challenge of safely managing physical contact and interaction with frail human subjects. On the biological side, the relative importance of centrally-specified limb trajectories and semi-autonomous peripheral neural oscillators such as the putative locomotor pattern generator remains unclear. Furthermore, while dynamics is largely subordinated to kinematic goals in upper-limb motor

[†] neville@mit.edu

behavior, musculo-skeletal mechanics appears to play a central role in controlling the lower limbs.

However, the relative importance of musculo-skeletal mechanics suggests that minimal actuation may provide substantial control authority; motors at every joint are almost certainly unnecessary. While this has obvious advantages for minimally-encumbering machine design, it makes the control problem even more difficult as the system is *underactuated.* I will discuss our initial investigations of the feasibility of controlling locomotion and assisting its recovery using distal actuation, in particular a two degree-of-freedom robot interacting with the ankle.

Advances in Climbing Robots

A CPG WITH FORCE FEEDBACK FOR A STATICALLY STABLE QUADRUPED GAIT*

JOSÉ CAPPELLETTO, PABLO ESTÉVEZ, GERARDO FERNANDEZ-LOPEZ AND JUAN C. GRIECO

Mechatronics Group. Electronics Department, Simón Bolívar University, Sartenejas, Caracas 1020, Venezuela.

Abstract. In this paper it is described the design and test of a system for statically stable gait generation for legged robots using a Central Pattern Generator (CPG) and normal force feedback. The CPG, based on Amplitude Controlled Phase Oscillator (ACPO), coordinates the generation of temporal references for each robot leg. The ACPO phases are passed through a parametric function which transforms them into the desired rectangular leg trajectories. Another parametric function generates a normalized force reference for each leg, which is employed to compute three different desired Center of Gravity (COG). Those references were used to enhance platform static stability margin (SSM), by the addition of a simple PI controller and ground contact forces measurements. Simulations were performed using a dynamic robot simulator (Webots) for a quadruped and hexapod platform, and a real quadruped platform with 3-DOF per leg was also used. Simulated and experimental results obtained for different controlled environments are shown, and analyzed in order to identify the stability improvement of the CPG gait generator by the addition of normalized contact force feedback to the spatial reference generator.

Keywords: Central Pattern Generator, Amplitude Controlled Phase Oscillators, Force Feedback, Static Stability, Legged Robots.

1. Introduction

Recent works in legged robots includes biologically inspired concepts as the Central Pattern Generator (CPG). In animals, this system is able to generate coordinate articulations movements based on local feedback and high level instructions. Several authors have used recurrent neural networks to act as CPG, behaving as dynamic oscillators to control leg movements (CTRNN) [1]. Despite the ability of CTRNN to model dynamic systems, the proposed training methods based in Genetic Algorithms (GAs) and modified real time back propagation does not provide an analytical method to synthesize desired gaits for robots with different leg configurations. J. Buchli proposed to use Amplitude Controlled Phase Oscillators (ACPO) to describe the dynamics in robot

* This work was supported in part by Simon Bolivar University Research & Development Deanship.

locomotion [2]. This approach provides a direct way to control temporal coordination between robot legs. However, it remains unsolved the relation between those phase references and spatial leg references. It has been proposed the utilization of simple feed forward neural networks to perform nonlinear spatial transformation between different temporal references into motor position references, as described by Cappelletto *et al*[3]. By using feed forward neural networks it is also possible to perform soft transitions between different leg trajectories. It has been reported the appearance of irregularities on leg trajectory due to over fitting during neural network training; those undesired behaviors can reduce stability during robot walking.

The work described in this paper proposes a solution for the problem previously described replacing the spatial transformation subsystem by a parametric transformation into a controlled rectangular trajectory. A direct way to control the leg support factor (β) through a companding curve is also added. Using this approach it is possible to control the desired gait mode and the desired leg trajectory in direct and separated way. It is included a static stability enhancement subsystem based on geometrically generated references for the center of gravity, thus improving the robot walking for some marginally stable gaits.

This paper is organized as follows: Section 2 describes the general architecture for the gait generator and force based stability controller; and includes a detailed description of the spatial and temporal generator subsystems. The three COP generators, the stability enhancement and controller are presented in the Section 3. Experimental setups for the simulation and real platform are shown in Section 4; Section 5 and 6 shows the obtained results and analysis and the main paper conclusions.

2. General Architecture

The CPG is implemented through two separated subsystems, making possible to cope with possible changes in the mechanical platform with minor modifications in the system. The walking adaptation process can be performed by acting in the desired subsystem, thus allowing synthesizing different gait modes with the same CPG model. Figure 1 shows a general description of the main subsystems present in the pattern generator model here described.

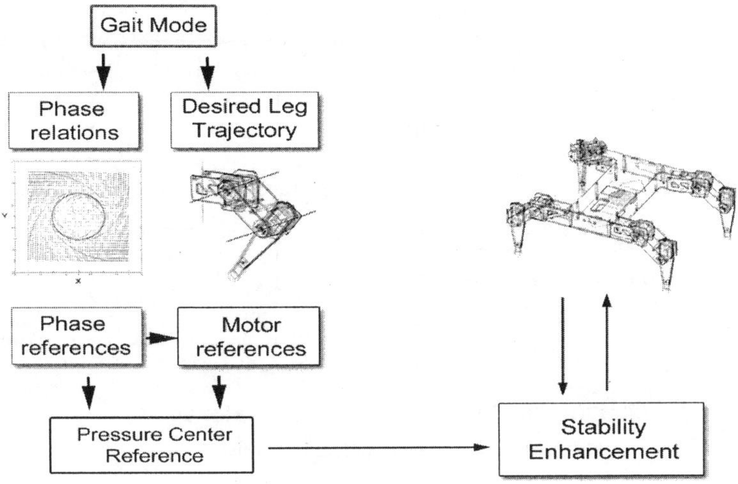

Fig. 1. General System Architecture

Each subsystem was implemented as follows:

2.1. *Temporal reference subsystem:*

As stated before, this subsystem generates the temporal references according to the desired gait mode. Each walking mode can be described as a set of phase relations between each leg, which can be translated in a direct way as the rotation angles for the coupling matrix between all the oscillators' nodes [2]. In the solution here described, it is used only the phase of the state vector for each node. By this way is possible to neglect amplitude perturbations due to abrupt switching between phase references corresponding to different gait modes. For the quadruped model it was employed a fully interconnected architecture, and for the hexapod model it were used connections between neighbors legs. It was discarded the use of full interconnections between all legs for the hexapod model because the excess of redundant connections. The interconnections for each robot model are illustrated in Figure 2.

A companding curve is applied to each oscillator state vector, acting over the phase according to the desired support factor (β). This technique is described with more detail in [4].

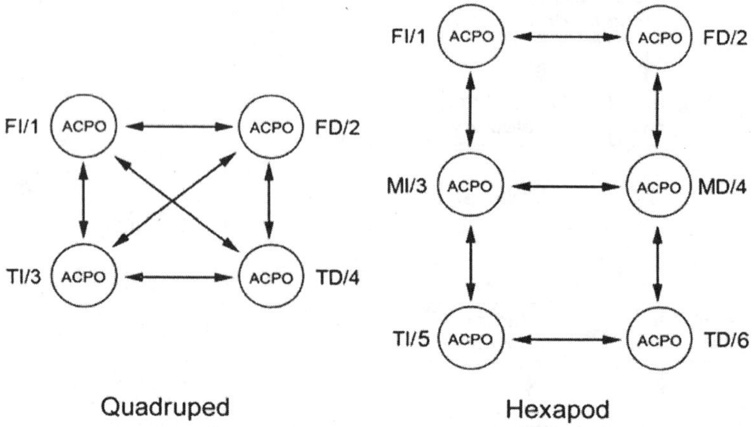

Fig. 2. ACPO connections for Quadruped and Hexapod

2.2. *Spatial reference subsystem:*

To transform the temporal references given by the phases of the output vectors for the previous subsystems, it was employed a parametric description of the desired spatial leg trajectory. The trajectory here employed correspond to a rectangle described by its height (**H**) that controls the elevation of the leg between support and transfer phase, the length (**L**) that is equivalent to the stride length thus giving control over the platform speed. It was also included a displacement factor to control the separation between the leg and the main body. This trajectory is parameterized for the variable **t** (0 to 1), and the resulting space reference point (X, Y, Z) for the leg is converted into actuators space through inverse kinematics.

3. Stability enhancement

The approach previously described generates coordinated trajectories for all the legs of a walking platform, given the geometrical parameters for the leg trajectories, and inter legs phase relationships. Nevertheless, these trajectories do not ensure the static stability of the platform since no attention is paid to its weight distribution. This results on the platform falling periodically as a consequence of its center of gravity (COG) being projected out of the support polygon generated by its legs. Also, irregularities in the terrain or changes in the weight distribution of the platform may yield to static instability.

In order to increase the platform Static Stability Margin (SSM) a control loop is included at the end of the spatial references generation system. The control action is to displace the robot main body from its original position, keeping its COG inside of the support polygon and far from its borders, and thus increasing the SSM according McGhee and Frank's criterion [5][6]. Such displacement is done in a plane parallel to the support one.

3.1. *CG reference generator*

Based on the phase reference generator, a force reference generator system produces a normalized variable indicating how much force is the leg ready to bear on each part of its gait cycle. This variable has a positive value (100) when the leg is in support phase and a 0 value for the transfer phase. A short slope is generated between these values to allow soft transition when possible. Using this force reference, three algorithms are utilized to generate an appropriate COG reference for the stability enhancement system.

The first one, named Balanced Forces Point (BFP), calculates an average of all supporting leg tips positions using their referential forces as weights (Eq. 1). The legs on transfer phase are naturally ruled out because of their null force reference, and the slopes in the force references allows for soft transitions between changes of the BFP. The BFP is always located inside the convex hull of the support polygon, and gives a balanced distribution of effort between the legs.

P_i : Referential Pressure in leg i.

(Xi, Yi) : Position of the tip of leg i.

$$X_{BFP} = \frac{\sum_{i=1}^{N} X_i \cdot P_i}{\sum_{i=1}^{N} P_i} ; Y_{BFP} = \frac{\sum_{i=1}^{N} Y_i \cdot P_i}{\sum_{i=1}^{N} P_i} \quad (1)$$

It is easy to obtain support legs distributions yielding to a location of the BFP with sub-optimal SSM. Nevertheless, experiments show that for the kind of support distributions usually found in legged platforms and for small number of legs, the BFP shows and acceptable performance.

The second algorithm identifies the desired convex support polygon based on the referential leg forces, and calculates its Area Centroid (AC) (Eq. 2). Given the convexity of the support polygon, this point is also always inside of it. By ensuring a balanced distribution of the support polygon area around itself, the AC locates the reference at a balanced distance of the polygon borders.

$$\vec{r}_{CA} = \frac{\int\limits_{polygon} \vec{r} \cdot da}{\int\limits_{polygon} da} \qquad (2)$$

The third algorithm tries to reduce the computing requirements of the previous one, while taking into account the most important condition: distance to the borders. The convex support polygon Contour Centroid (CC) is calculated by averaging all points in the polygon contour (Eq. 3). It tends to balance the distance to all its borders.

$$\vec{r}_{CC} = \frac{\oint\limits_{contour} \vec{r} \cdot d\ell}{\oint\limits_{contour} d\ell} \qquad (3)$$

3.2. *Control loop*

The control loop consist of a COG projection reference generated from the trajectories and force references of each leg, a scale factor (**sf**) required to avoid references outside the robots work area and a PI controller generating the required horizontal displacements for the platform.

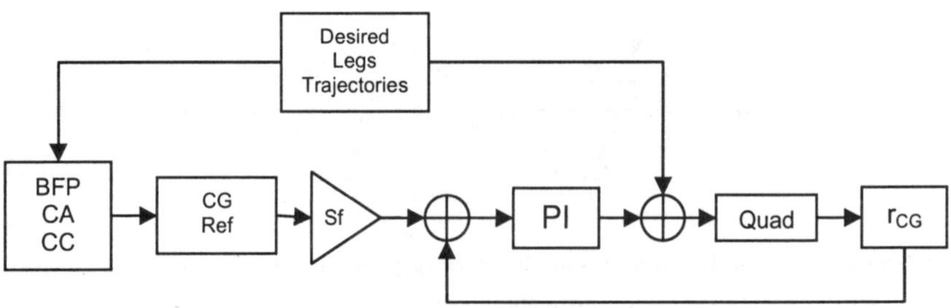

Fig. 3. Stability Enhancement Control Loop

The feedback is obtained from normal contact force sensors located at the tip of each leg. Those forces are combined with leg positions to compute the vertical projection of the COG into the support polygon given static equilibrium conditions. Because the SSM enhancement can be described as a trajectory tracking problem a PI controller was employed.

4. Experimental setup

This section describes the three main tests applied to the system architecture previously described. In the first test were compared the 3 COG references shown in Section 3.1, using spatial and temporal references that correspond to valid walking movements in the quadruped and the hexapod. The second test consisted in dynamic simulations of both robot models using Webots™ under different environments, including uneven terrain and uneven weight distribution, with and without the control loop. Finally, several tests were performed on the real platform in order to illustrate the impact of the force based feedback on the SSM for this CPG-based model.

4.1. *Reference tests*

Once the base gaits were synthesized, simulations were made for the COG reference generation algorithms in the quadruped and hexapod platforms with different values of β, and the standard McGhee criterion was used to evaluate them. This way, standard walking support polygons were utilized for the tested.

4.2. *Simulations*

To study with greater detail the system here described, a hexapod and a quadruped platform, both with 3 DOF legs, were simulated using Webots™. The dynamic model is similar to the real quadruped robot, built on acrylic and polycarbonate, and metal-plastic actuators (servomotors). The simulated quadruped dimensions were 240x190 mm, and 335x150mm for the hexapod.

The CPG phase relations were configured to generate a crawling gait in the hexapod robot, and the scaling factor for the COG position was set to 0.4 for both coordinates (X-Y) to ensure leg positions within their working space. With this setup, different β values were used to test the performance of the platform, both with and without stability enhancement, and using the three proposed methods. The vertical projection of the center of gravity of the robot was measured using the force sensors information for all the tests where the platform showed a statically stable behavior, and this data was combined with the measurements of the real legs position to estimate the SSM.

An equivalent test was conducted on the simulated quadruped robot, resembling the Lynxmotion quadruped platform.

4.3. *Real quadruped platform*

In this test it was employed the Lynxmotion Quadruped-3 with 3 degrees of freedom per leg, with servo motors as actuators. The robot dimensions are

240x190 mm, and the maximum range for each leg is 133mm. The platform is equipped with Flexiforce® sensors in a rotationally compliant base tied to the tip of each leg, measuring the normal force between the support surface and the robot. The sensor data was processed under a Win32 C/C++ based GUI that performed real time filtering and platform control with a base operating rate of 50 Hz.

Several tests were done with the real platform in order to replicate some cases evaluated with the simulated quadruped, including uneven weight distribution.

5. Results and Analysis

Proceeding with the experiments previously listed, it was obtained the following simulated and real data.

Computing the SSM of the three COG references for different β values on the hexapod platform, is produced the data shown in Table 1.

Reference	McGhee (mm)	β = 0.5	β = 0.6	β = 0.7	β = 0.8	β = 0.83
BFP	Average	92.2	79.2	90.1	93.3	92.5
	Minimum	10.5	39.5	43.8	29.1	0.3
AC	Average	97.4	102.2	108.5	126.6	131.2
	Minimum	49.1	46.3	44.3	61.2	0.2
CC	Average	96.4	100.3	106.5	124.1	128.7
	Minimum	39.5	37.3	35.8	63.8	0.2

Table 1 McGhee Statistics for CG references on the hexapod platform.

A similar test was ran for the quadruped and the obtained data is shown in Table 2.

Reference	SSM (mm)	β = 0.75	B = 0.8	β = 0.85	β = 0.9
BFP	Average	109.1	108.0	107.8	109.7
	Minimum	69.4	82.5	94.5	94.5
AC	Average	119.3	119.5	119.9	120.1
	Minimum	76.7	77.0	106.9	107.5
CC	Average	119.3	119.5	119.9	120.2
	Minimum	69.7	69.8	107.2	107.7

Table 2 McGhee Statistics for CG references on the quadruped platform.

The data of Table 1 and Table 2 shows how the average SSM for all COG references tends to increase with the value of β, as could be expected for larger support polygons. The BFP method provide a fast way to compute the desired COG, without involving convex hull computation, and gives SSM pretty close to those obtained with AC and CC for quadruped robot. However, for the hexapod platform it doesn't show any important improvement and reaches moderate SSM compared to those obtained with AC and CC: 21.2% and 19.5% lower average SSM for all β.

Testing the whole system with the PI controller for several setups, the resulting SSM are shown in Table 3. The controller constants were P=0.6 and I=0.05, and sf=0.3. The step length was L=95mm, and β=0.85 for crawling mode. For the real quadruped the controller constants were P = 0.3 and I = 0.08.

Average SSM (mm)	BFP	AC	CC	No control
Quadruped β=0.75	33.54	31.66	32.55	28.76
Quadruped β=0.85	39.48	43.91	42.54	21.73
Hexapod β=0.5	67.69	68.71	68.25	60.31
Hexapod β =0.8	86.44	81.59	84.02	77.39
Hexapod β =0.8 (w/uneven weight)	93.13	n/a	n/a	78.33
Quadruped β =0.8 (w/0.02 rad slope)	52.15	n/a	n/a	20.58
Real Quadruped β =0.85 (w/uneven weight)	52.5	n/a	n/a	47.28

Table 3. Average measured SSM for different experimental and simulated setups

It can be observed the similar response for the three COG for all the cases. The addition of the control loop increased the average SSM for the performed tests, compared to the performance of the platform without stability enhancement. For the last three tests, that included perturbations like uneven weight distribution or ground slope, the simulated and the real platforms weren't able to walk in a stable way with the stability control disabled.

6. Conclusions

Based on the experimental results previously described, it was shown that the proposed CPG architecture is able to produce statically stable gaits for quadruped and hexapod robots. The decomposition of the walking problem into separated subsystems simplifies the adaptation of the CPG for different gait modes, and kinematical robot architecture by modifying only a specific component of the model while keeping the others unmodified.

All the three COG generation algorithms provided references that increase the platform SSM under different operative conditions. It was also observed that for higher support factors the SSM also increased.

The BFP-COP generation method provides a soft transition between COG references positions, however it exhibits a SSM lower than those obtained for the other two geometrically based methods: Area Centroid and Contour Centroid.

References and Bibliography

1. J. Hillel, Randall D. Beer and J. C. Gallagher. "Evolution and Analysis of Model CPGs for Walking: I. Dynamical Modules". *Journal of Computational Neuroscience*. pp 99-118, 1999.
2. J. Buchli and A. J. Ijspeert, "Distributed Central Pattern Generator Model for Robotics Application Based on Phase Sensitivity Analysis", *In Proc. Of 1st Intl. Workshop Bio-ADIT*. 2004
3. J. Cappelletto, P. Estévez, W. Medina, L. Fermín, J. M. Bogado, J. C. Grieco and G. Fernández-López. "Gait Synthesis and Modulation for Quadruped Robot Locomotion using a Simple Feed-Forward Neural Network". *Lecture Notes in Computer Science*, Vol 4029, pp 731-739, 2006.
4. J. Cappelletto, P. Estévez, J. C. Grieco, G. Fernández-López and M. Armada. "A CPG-Based Model for Gait Synthesis in Legged Robot Locomotion". *In Proceedings of CLAWAR06*. pp 59-64, 2006.
5. McGhee, R. B. and Frank, A. A. "On the stability properties of quadruped creeping gaits". *Mathematical Bioscience*, Vol. 3, pp. 331-351, 1968.
6. E. García, J. Estremera and P. González Dos-Santos. "A Classification of Stability Margins for Walking Robots". *5th Int. Conf. Climbing and Walking Robots*, Paris, France, 25-27 Sept 2002.
7. Y. Fukuoka , H. Kimura and A. H. Cohen. "Adaptive Dynamic Walking of a Quadruped Robot on Irregular Terrain Based on Biological Concepts". *Intl. Journal of Robots*. Vol 22, No 3-4, pp 187-202, 2003.
8. H. Kimura, Y. Fukuoka, Y. Hada and K. Takase. "Three-Dimensional Adaptive Dynamic Walking of a Quadruped – Rolling Motion Feedback to CPGs Controlling Pitching Motion". *In Proc. of ICRA02*, pp 2228-2233. 2002.
9. S. Hirose, H. Tsukagoshi and K. Yoneda. "Normalized Energy Stability Margin and its Contour of Walking Vehicles on Rough Terrain". *In Proc. ICRA 2001*. Vol 1, pp 181-186, 2001.
10. P. González de Santos, J. Estremera, E. García and M. Armada. "Including Joint Torques and Power Consumption in the Stability Margin of Walking Robots". *Autonomous Robots*, Vol. 18, pp 43-57, 2005.

A Sliding Sock Locomotion Module for a Rescue Robot

Luca Rimassa, Matteo Zoppi and Rezia Molfino

University of Genoa, PMAR Robotics research group,
Genoa, 16145, Italy, www.dimec.unige.it/PMAR
E-mail: {rimassa;zoppi;molfino}@dimec.unige.it

Rescue robotics for disaster response deserves practical importance. Reference tasks are inspection and monitoring of collapsed structures and disaster sites with searching/localization of victims. Mobile robots using unconventional locomotion strategies are developed and applied (crawling, peristaltic, serpentine). The identification of effective locomotion mechanisms and especially their practical realization is one major research issue; often reproducing effectively bioinspired mechanisms is challenging with today's available technologies.

This paper presents a development step toward highly efficient crawling locomotion: A continuous membrane sliding along the entire body of the module provides distributed trust with minimum risk to get stuck. At one extremity of the module, the membrane rolls in, slides back and rolls out again at the other extremity.

The design is discussed and a physical prototype is presented.

Keywords: Rescue Robot; Sliding Sock Locomotion; Serpentine Robot

1. Introduction

Mini and micro rescue robots are used in calamity response operations for search and localization of victims in collapsed buildings and disaster sites.

Three classes of rubble environments have been identified based on structure and size of voids available:[1] • *semi-structured* (partially fallen buildings whose original layout is globally preserved); • *confined* (fully collapsed constructions with voids comparable to human size); • *sub-human confined* (fully collapsed structures bring into very compact debris with very tight voids).

Confined and sub-human confined spaces can be explored only using worm rescue robots. Robot reconfigurability to adapt to mission/environment requirements can be obtained by modular architectures: a suitable sequence – number and assortment – of modules with same or different sets of functionalities (locomotion, steering, carrying sensors).

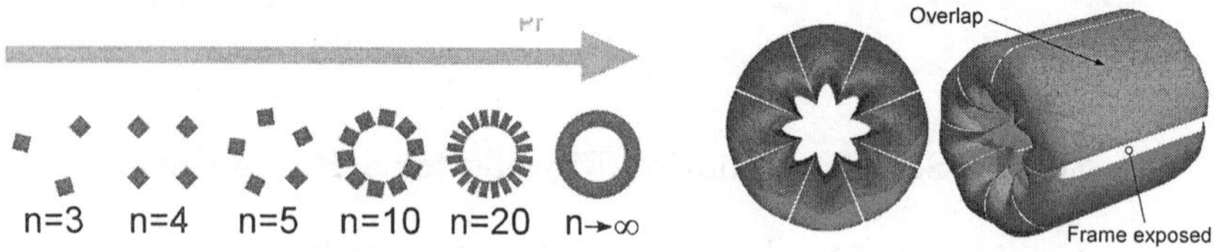

(a) (b)

Figure 1. Increase of PR with the number of tracks (a) and PR= 1 with flexible belt track (b)

One main research issue is locomotion. Several methods have been proposed to move effectively in rubble debris. The more the trusting action is distributed around the body of the robot, the higher is the overall advancing capability of the robot and the lower the risk that it gets stuck because some passive part of its body (not generating trust) touches the environment.

The efficiency of a locomotion mechanism can be ranked based on a *propulsion ratio* (PR)[2] defined as the ratio between surface providing trust and total boundary surface.

No rescue robots developed have propulsion ratio really close to one, which is the ideal PR realized when all body boundary surface contributes to locomotion. Some systems with the highest propulsion ratio are: *Gembu*[3] using wheeled modules; *Kohga*[4] composed of tracked locomotion modules connected by passive and active joints; *ACM*[5] merging snake-like locomotion to traditional locomotion in wheeled joint modules; *MOIRA*[6] and *Omnitread*[2] with tracks distributed around the body, making the robot indifferent to roll over, common to serpentine robots over rugged terrain.

The research has moved in the direction of PR= 1 by progressively increasing the number of tracks around the body of locomotion modules, Fig. 1(a). Such increase cannot truly lead to unit PR and the design complexity becomes soon very high. 5-6 tracks appear a limit difficult to overcome in small robots.

Design can be simpler using tracks with flexible belts, Fig. 1(b). PR= 1 is achieved in principle but, especially with long modules, strips can overlap and part of the frame can contact the environment, with practical reduction of the PR.

2. Continuous membrane

The real step forward is the realization of a fully moving body envelope.

This is challenging and only two development experiences can be found in the very recent state of the art. The former,[7,8] more at a conceptual stage, addresses the actuation of an elastic toroid membrane with oblong section by expanding and contracting transversal rings integrated in the membrane. It is proven that locomotion can be generated but no materials are available today to realize the rings. The second development,[9] with a physical prototype realized, uses a rolling donut with circular toroid section and wire actuation. An oblong section with external surface cylindrical, required to move effectively in debris, cannot be obtained.

With this research background, we have started a two fold activity aimed at the development of two locomotion modules with PR= 1 and oblong section for rescue robots: the former is with rigid frame and uses traditional solutions for the actuation of the sliding membrane; the second, still under conceptual development, will be compliant to adapt to the shape of the channel where it will advance. The following presents the main design steps and the final set up of the locomotion module with rigid frame, for which a prototype has been realized.

3. Design issues with continuous membrane

The design of a continuous membrane track system with internal rigid frame is challenging. Assessed that it is not reasonable to imagine that the membrane material can split and recombine during functioning, then the membrane divides two fully separate domains, inside and outside. Two main problems are faced: • the connection of the group composed of membrane and its internal frame to the notch frame of the module; and • the actuation of the membrane. The two problems are strictly interrelated and they have been faced together.

Actuation splits in two further subproblems: how to make the membrane moving and were placing the actuators: provided that a moving mechanism is identified, if it operates on the membrane from inside and actuators are inside as well, then the problem is how to provide actuation power. Locomotion by anything embedded in the membrane body was not taken into account at this stage because of complexity.

The connection of the membrane-frame group to the notch frame is necessary to transmit the traction force from the membrane to the rest of the robot. Independently from the membrane moving mechanism adopted, no material bonding is possible to transmit this traction force. Two ways can be identified instead: a coupling that can generate bonding forces through the membrane and geometric bonding with distributed contact forces. Both

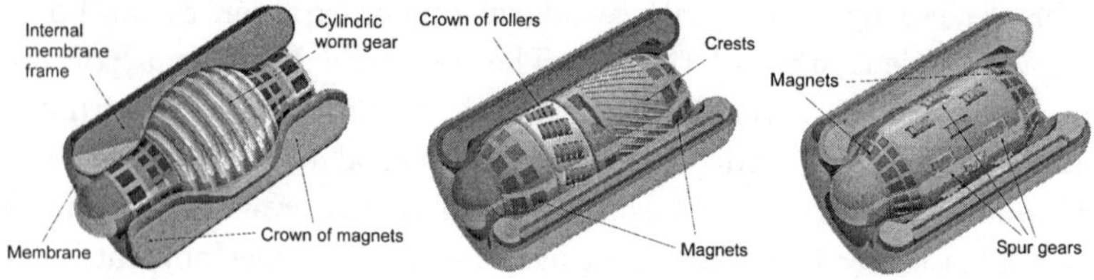

(a) (b) (c)

Figure 2. Actuation with worm gear (a); rollers (b); spur gears (c)

directions have been investigated.

For the former, we have studied combinations of magnetic fields generating repulsion forces and a stable equilibrium configuration for the notch frame inside the membrane-frame assembly. Strong Neodymium magnets are cheap and can be used to generate the fields; the considered magnets layouts are shown in Fig. 2. The major drawback of this solution is the quantity of ferromagnetic dust, always present in soil and construction material, that would enter and lock the locomotion module very soon in real application.

Then we decided to realize a geometric coupling by suitably shaping the extremities of the frame inside the membrane. The design, shown in the detailed view of Fig. 3, is further explained in Section 5.

4. Alternative membrane actuation solutions

Once defined the way to join the notch frame to the group membrane–frame, we move to the discussion of the alternative actuations considered. As mentioned in the Introduction, for this locomotion module conventional mechanical solutions are adopted to have low cost and development time.

4.1. *Actuation with worm gears*

A suitably shaped worm gear, Fig. 2(a), rotating about the longitudinal axis of the module, on the surface of the notch frame, is in contact with the membrane. Some interference between the worm of the gear and the membrane (which is supposed compliant) allows the gear worm to pull the membrane. Rounds at the extremities of the gear make gearing progressive. The large contact area results in safe and distributed pulling pressure. The direction of motion can be easily inverted by inverting the direction of rotation of the worm gear.

Main drawbacks are the stretching in the membrane caused by the friction forces tangent to the worm gear and the high counter torque between internal and notch frames, which is not easy to transmit if we use a geometric coupling with contact forces to join notch and internal frames and this coupling has to satisfy the axisymmetric geometry of the module.

A counter rotating worm gear can be added to balance most of the coupling torque but this does not solve or makes even worse the problem of membrane stretching caused by worm gear friction forces.

To reduce both friction stretching and counter torque, the worm gear is replaced by a crown of rolling elements, Fig. 2(b), with longitudinal crests on the notch to balance the residual counter torque. Also this alternative design has drawbacks: it is complex and the rolling elements have small contact area to pull the membrane with risk of damage.

4.2. *Actuation with spur gears*

To overcome the drawbacks of worm gears, design with spur gears with axes orthogonal to the longitudinal axis of the module have been developed, Fig. 2(c). The teeth of each gear etch the membrane and pull it tangentially with no friction torque generated. The continuous membrane in between the gears distributes the localized pulling forces so that finally an equivalent continuous traction of the membrane is realized. A long worm gear coaxial to the notch frame can transmit motion to all spur gears together.

The main design issue is finding an effective disposition of the gears on the notch frame.

The limitations on minimum radius and thickness of the gears not to damage the membrane result in a limit on the maximum number of gears that the notch frame can host (minimum angular spacing of gears at each transversal section of the notch). A distributed pulling action is obtained by more sets of gears at different transversal sections, so that an angular offset can be introduced for a more homogeneous distribution of the pulling forces along the membrane.

5. Final module design

The final design, Fig. 3, meets the advantages of spur gears and geometric joining of membrane and notch frames.

The membrane frame is cylindrical with two inward lips at the sides. The membrane is first winded and bonded to realize a tube; then the tube is winded around the membrane frame and bonded again to get the closed

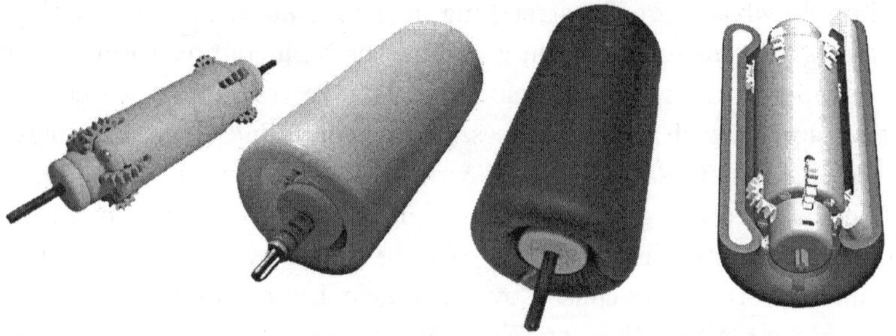

Figure 3. Final design: virtual model and pictures of the realized prototype

final topology.

After attentive evaluation of alternative elastomeric materials with suitable softness, compliance and mechanical and fretting resistance characteristics, a polychlorophene micro foam is selected. This material is commonly used for diving wetsuit and is very low cost.

The membrane slides along the frame. A low friction factor is obtained by suitable selection of materials and injecting a film of lubricant inside the closed membrane. For the membrane frame Teflon is used; a nylon fabric sheet is glued to the internal side of the membrane with weaving directions oriented along the sliding direction to further reduce friction. As lubricant a silicon based fluid is adopted.

The notch frame, realized in PVC, embeds two sets of 5 gears at each side. It is realized in 4 parts (two shells and two covers) plus gears, transmissions and motor. Two worm gears provide motion to the spur gears and actuation is provided by an electric motor in the center of the notch frame. When the spur gear rotates the notch moves in the desired direction of motion up to come in contact with the lip, then it starts pulling the membrane and the higher is the resistance of the membrane to slip (related to the resistance to advance of the locomotion module in the environment) the more the teeth of the spur gears penetrate in the membrane pressed against the lip, so getting higher trust.

The cabling used to interface the different modules of the complete robot (electric power and signal bus) move on one side of the motor and in the space available between the spur gears.

Compliance and stiffness of the membrane allow large geometric tolerances on the notch module, whose manufacturing can be easy and cheap (e.g. standard moulding). All is designed for manufacturing with rapid prototyping equipment. For minimum cost, gears, some axes and other small components of the module are commercial *LEGO* parts.

The final size of the locomotion module is 110 mm length and 50 mm diameter plus membrane. Mass is about 200 g.

At the sides of the locomotion module mechanical/signal interfaces can be arranged for fast and easy combination of modules.

6. Combined robot architectures

The locomotion module presented can be used to assemble worm robots with various architectures. Distributed steering for small bending radius of the worm and high steering capability can be realized with short locomotion modules and steering modules between every two of them.

A compact design variant also suitable for inspection of pipes and structured channels is with one crown of gears instead of two on the notch frame and with the membrane frame internally rounded as if the two lips described in Section 5 had moved and become tangent. This rounded shape matches the profile of the gears with minimum longitudinal backlash when the direction of motion is inverted.

Externally the membrane is straight and the module has oblong section although shorter than in the version with more crowns of gears, as required to generate effective locomotion.

Membrane pulling is not as well distributed as in the case of multiple crowns of spur gears, but sliding friction is much lower due to the shorter membrane frame.

Due to the compact shape, the motor cannot be integrated in the module anymore. One motor for all modules is placed at one side of the robot and a flexible shaft is used to transmit motion to each and every module.

For pipe inspection the steering modules can be passive and the robot is guided along the pipe by the pipe walls. Smaller bending radii are then achieved.

The same robot architecture with some redesign of the modules can be adapted to intestinal inspection. The main change should be in the membrane, to be replaced or covered so that the external surface is rough and with good grip on the intestinal walls.

7. Conclusions

A locomotion mechanism to realize worm robot locomotion modules with PR= 1 has been developed and is presented in detail. The concept of continuous sliding membrane is developed and it is proven that conventional actuation solutions can be used to move the membrane. Preliminary tests of a physical prototype are confirming the effectiveness of the design proposed.

This work belongs to a larger research frame whose final goal is a sliding membrane rescue robot with compliant body.

Design of steering modules will be also involved: even if the PR of each locomotion module is 1, conventional steering modules open lateral windows when the robot steers, reducing the overall PR. The step beyond is masking these windows by adopting one continuous membrane through all locomotion units. Modularity is still there because still a chain of identical locomotion and steering modules is used.

The same rethinking of the robot will be possible when the compliant locomotion module will have been developed.

Bibliography

1. R. Murphy, *IEEE Trans. On Systems, Man, And Cybernetics Part C: Applications And Reviews* **34** (2004).
2. J. Borenstein, G. Granosik and M. Hansen, The omnitread serpentine robot - design and field performance, in *Proc. SPIE Defense and Security Conf., Unmanned Ground Vehicle Technology VII*, (Orlando, Florida, 2005).
3. H. Kimura and S. Hirose, *IEEE/RSJ Int. Conf. on Intelligent Robots and System* **1** (2002).
4. T. Kamegawa, T. Yamas, H. Igarashit and F. Matsuno, Development of the snake-like rescue robot kohga, in *Proc. 2004 IEEE Int. Conf. on Robotlcs & Automation*, (New Orleans, LA, 2004).
5. M. Mori and S. Hirose, Development of active cord mechanism acm-r3 with agile 3d mobility, in *Proc. IEEE/RSJ Int. Conf. on Intelligent Robots and Systems IROS*, (Maui, Hawaii, 2001).
6. K. Osuka and H. Kitajima, Development of mobile inspection robot for rescue activities: Moira, in *Proc. 2003 IEEWRSI Int. Conf. on Intelligent Robots and Systems*, (Las Vegas, Nevada, 2003).
7. M. Ingram and D. Hong, Whole skin locomotion inspired by amoeboid motility mechanisms, in *Proc. 29th ASME Mechanisms and Robotics Conf.*, (Long Beach, California, USA, 2005).
8. M. Ingram and D. Hong, Mechanics of the whole skin locomotion mechanism concentric solid tube model: The effects of geometry and friction on the efficiency and force transmission characteristics, in *Proc. IDETC/CIE ASME Int. Design Eng. Techn. Conf. & Computers and Information in Eng. Conf.*, (Philadelphia, Pennsylvania, USA, 2006).
9. P. Breedveld, Development of a rolling stent endoscope, in *2006 IEEE / RAS-EMBS Int. Conf. on Biomedical Robotics and Biomechatronics*, (Pisa, Italy, 2006).

A WHEELED WALL-CLIMBING ROBOT WITH A CLIMBING LEG*

YILI.FU, ZHIHAI.LI, HEJIN.YANG, SHUGUO.WANG
Robotics Institute, Harbin Institute of Technology
Harbin, Heilongjiang Province, China

In resent years, wall-climbing robots used for special tasks such as rescue, inspection, surveillance and reconnaissance have been studied hotly. But the technologies used in the previous wall-climbing robots cannot meet the needs of these challengeable missions mainly in mobile capability. A new prototype of self-contained wall-climbing robot is developed, which integrate the advantages of wheeled robots and that of legged robots. It is composed of a base body and a mechanical leg with 3 DOF. The base body is a big flat suction cup with three-wheeled locomotion mechanism in it, and there is a small flat sucker in the end of the mechanical leg. It can achieve quick motion on wall surface, as well as obstacle- spanning on wall surfaces and smooth wall-to-wall transitions. The new designed chamber seal has simple structure and has steady and reliable performance. Kinematics model of the robot is established to analyze the robot's motion on the wall. Furthermore, the locomotion gait of the robot is discussed.

1. Introduction

Wall-climbing robots integrate the mobile technology and adhesion technology, it can adhere to wall surface and move around carrying kinds of tools. Therefore it is expected to replace human to execute some dangerous missions in hazardous environments. Since now, wall-climbing robots have been developed and used in many industries, for example, maintaining nuclear facilities, inspection of stage tanks, cleaning high-rise building, painting, and so on [1][2].

In resent years, wall-climbing robots used for special tasks such as rescue, inspection, surveillance, reconnaissance, counter terrorism and so on, has become a new branch of wall-climbing robots. These new tasks demand the robot have small volume, low weight, excellent mobility and maneuverability. But the technologies used in the previous wall-climb robots cannot meet the need of these new special tasks. They always have big body, high weight, and low agility. So, new prototypes of self-contained wall-climbing robot need to be developed to execute these special tasks [3][4].

---

* This work is supported by National Natural Science Foundation of China (60675051)

Since now, wall-climbing robots developed for surveillance and reconnaissance tasks mostly use vacuum adhesion mechanism to make the robots sticking on wall surfaces, because this method can suit walls such as ceramic tile, glass, cement and brick walls. Although there have been some little wall-climbing robot prototypes developed, there is also some problems required to be settled. For example, the wheeled wall-climbing robot has high moving speed, but its obstacle-spanning capability is poor. In other side, the legged wall-climbing robot can span obstacles on the walls, such as bars, barriers and ledges, but its moving speed is low.

Here, a new type of wall-climbing robot is proposed, which integrates the advantages of wheeled robots and legged robots. It can achieve quick motion as well as obstacle- spanning on wall surfaces, and even smooth wall-to-wall transition by applying integrated locomotion mechanism. Vacuum based adhesion mechanism is used to adapt to versatile surfaces.

2. Mechanical Design

This novel wall-climbing robot is composed of two parts: a base body, which is a big flat suction cup; and a 3-DOF mechanical climbing leg with a little suction cup in the end, as shown in Fig.1. Design of the robot's mechanical system mainly includes adhesion mechanism, sealing mechanism, wheeled locomotion mechanism, and mechanical leg.

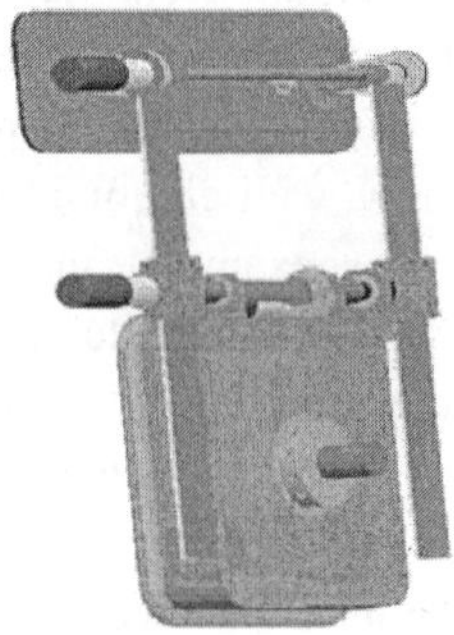

Fig. 1 Outlook of the novel wall-climbing robot

The base body takes on a cuboid frame. It contains adhesion mechanism, sealing mechanism and wheeled locomotion mechanism. Thus, the base body can be an integrated robot alone. Besides, it needs to carry batteries, controller, sensors and other tools. And when the climbing leg is idle, it will be a load to the base body. Therefore, the base body needs to have high payload capability.

2.1. *Adhesion mechanism*

An excellent adhesion mechanism is the guarantee for the robot's reliable working on wall surfaces. Main limitations of selecting adhesion method are the materials and state of wall surface. In order to enable the robot to adapt to versatile surfaces, vacuum adhesion technique is chosen. Considering working surfaces may be porous and bumpy, vacuum rotor package is selected since it has more advantageous than micro-pump in this condition. The vacuum rotor package consists of DC motor, impeller, and exhaust cowling.

When the robot sticks on the wall, a vacuum chamber is formed between the base body and surface. The chamber and vacuum rotor package constitute the vacuum adhesion mechanism. Under the high speed rotation of the impeller, air in the chamber is pumped out via the exhaust cowling. A low air pressure under atmospheric pressure is generated in the vacuum chamber, so that a force due to the pressure difference presses the robot attached to wall surface reliably.

2.2. *Sealing mechanism*

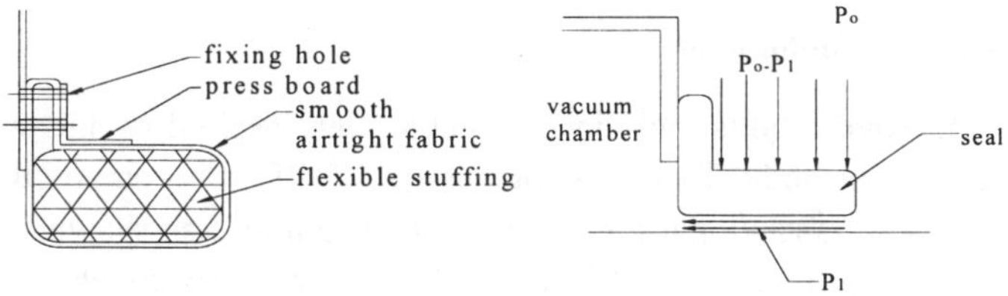

Fig.2. Sketch and Force analysis of the new vacuum chamber seal

The negative pressure generated in the vacuum chamber need to be maintained by sealing the gap between suction cup and wall surfaces. Two sealing methods are always used: inflated rubber tube skirt seal and flexible bristle seal. The former which is always composed of some springs and an inflated rubber tube filled with air has perfect performance, but the effect is correlative to the air pressure in the inflated tube. The performance is not steady considering air leak.

In order to improve the stability of seal, a new sealing skirt is designed as shown in Fig.2, which is composed of inner flexible stuffing, outer smooth airtight fabric. It can realize perfect effect as well as inflated tube skirt seal, furthermore it has other advantages-self-press capability. When the vacuum motor works, air flows from the outside into the chamber via the gap between seal skirt and surface. A low pressure area is generated in the gap. Therefore, the flexible skirt can be pressed against wall surface under air pressure automatically. Additionally, this novel seal has a simple and firm structure; its stable parameters make it have stable performance.

2.3. *Wheeled locomotion mechanism*

In the past, wall-climbing robots with a single suction cup always use 2 or 4 wheeled drives, but 3-wheeled drive is rarely referred. Comparing with the former two modes, 3-wheel drive is both simple and steady, thus 3-wheel drive is adopted. As shown in Fig.3, the locomotion mechanism consists of two drive wheels and a castor wheel. The two driving wheels are driven by DC motors independently, which are controlled by differential principle. Tires mounting on the driving wheels heighten friction coefficient, and enhance damp between robot and surface. Wheeled drive guarantees the robot have high moving speed.

Fig.3. 3-wheeled locomotion mechanism

2.4. *Mechanical climbing leg*

In order to improve robot's mobility, a climbing leg inspired by inchworm is equipped to the base body. This climbing leg has 3 DOF. It is composed of two rotational joints, a translational joint and a flat suction disc as shown in Fig.1. The two rotational joints are driven by DC motor respectively, and the translational joint is realized by rack and pinion mechanism driven by a DC motor. The climbing leg can help the robot climb over obstacles since there is always protrusion or gap on the building wall [5]. The robot can also imitate the inchworm's movement on the tree.

Weight of the robot is 9 *kg*, the base body is 300(W)×300(L)×150(H) in *mm*. Fully extended it is 900 *mm* long. The wheeled mechanism's speed is 0~10*m/s*; step size of the legged mechanism is 0~450 *mm*.

3. Kinematics analysis

In order to facilitate the analysis and control of the robot's movement, kinematics model of the robot is established. Considering the wheeled locomotion mode and legged locomotion mode works independently in different conditions, kinematics models of the two modes are settled respectively. They are illustrated as follow.

For wheeled locomotion mode, the base body coordinate is established, as shown in Fig.6. In the world coordinates, the position/orientation of the base body can be expressed by $(x, y, \theta)^T$, where ϕ_1 and ϕ_2 denote the two driving motors' rotational angle respectively, and i denotes transmission ratio. The wheeled locomotion mechanism's kinematics equation is expressed as Eq. (1).

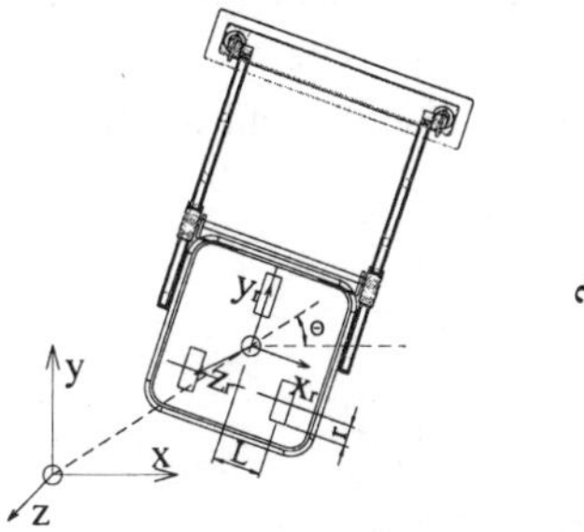

Fig. 4 Coordinates frames of the base body

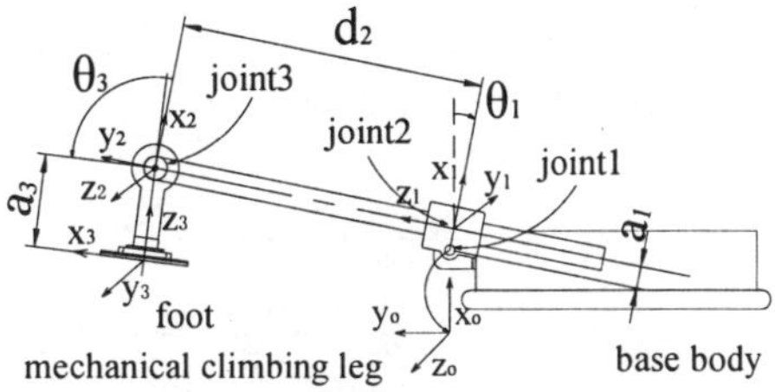

Fig. 5 Coordinates frames of the mechanical leg

$$\begin{bmatrix} \dot{x} \\ \dot{y} \\ \dot{\theta} \end{bmatrix} = \frac{ir}{2} J_1(\theta, L) \begin{bmatrix} \dot{\phi}_1 \\ \dot{\phi}_2 \end{bmatrix}, \qquad J_1(\theta, l) = \begin{bmatrix} \cos\theta & \cos\theta \\ \sin\theta & \sin\theta \\ -1/L & 1/L \end{bmatrix} \tag{1}$$

For wheeled locomotion mode, the assignment of coordinate frames based on the Denavit-Hartenberg (D-H) method is illustrated in Fig.5. The link parameters are shown in Table 1, θ_1, d_2 and ϕ_3 are joint variables. Transformation matrix ${}^0T_3$ (Eq.2) expresses the small sucker's position/orientation in coordinate 0. On the other hand, matrix ${}^3T_0$ (${}^3T_0 = {}^0T_3^{-1}$) expresses the base body's position/orientation in coordinate 3.

Table 1. Link coordinate parameters of the mechanical leg.

Joint	θ	d	a	α
Joint1	θ_1	0	a_1	–90°
Joint2	0	d_2	0	+90°
Joint3	θ_3	0	a_3	+90°

$${}^0T_3 = \begin{bmatrix} -C(\theta_1+\theta_3) & 0 & S(\theta_1+\theta_3) & a_1\cos\theta_1 - d_2\sin\theta_1 - a_3\sin(\theta_1+\theta_3) \\ S(\theta_1+\theta_3) & 0 & C(\theta_1+\theta_3) & a_1\sin\theta_1 + d_2\cos\theta_1 + a_3\cos(\theta_1+\theta_3) \\ 0 & 1 & 0 & 0 \\ 0 & 0 & 0 & 1 \end{bmatrix} \tag{2}$$

4. Force analysis on the wall

If a wall climbing robot can reliably move on vertical surfaces, it needs to fulfill some force constraints. Static force analysis is used to analyze these constraints, which is executed in two cases: wheeled motion mode and legged motion mode.

4.1. *Force analysis of the wheeled locomotion mechanism*

When the robot works in wheeled motion mode, sticking and driving of the robot rely on the friction force between the drive wheels and wall surface. The robot can move in any orientation under differential control. They can be classified in two main categories-driving wheels underside and driving wheels upside, as shown in Fig.6 a) and b).

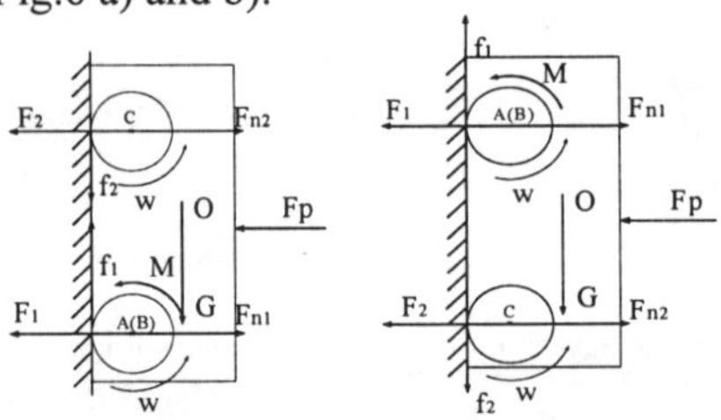

a) Driving wheels underside b) Driving wheels upside

Fig. 6 Forces and torques of the robot at two conditions

As shown in Fig.6 a), forces acting on the robot include gravity, negative pressure adhesion force, supporting force by surface, and friction forces between wheels and surface, they are denoted by G, F_p, F_{NA} and f_{NA} respectively. Here friction between the seal and surface is ignored. When the driving wheels is underside, and wheels doesn't slip. We have,

$$\begin{cases} \sum X = 0 \ , & (F_{NA} + F_{NB}) + F_{NC} - F_p = 0 \\ \sum Y = 0 \ , & (f_A + f_B) - f_C - G = 0 \\ \sum M(B) = 0, & F_p \cdot l - F_{NC} \cdot 2l - G \cdot h = 0 \end{cases} \tag{3}$$

Since there is no slipping, friction between wheels and surface is static friction. $\mu_{1\max} \approx 0.8$, $\mu_{2\max} \approx 0.05$ .So, the friction forces can be calculated by,

$$f_A = f_B = \mu_1 \cdot F_{NA} \le \mu_{1\max} \cdot F_{NA} , \quad f_C = \mu_2 \cdot F_{NC} \le \mu_{2\max} \cdot F_{NC} \tag{4}$$

By Eq. (3) and Eq. (4), we have,

$$F_p = \frac{2l - (\mu_1 + \mu_2) \cdot h}{(\mu_1 - \mu_2) \cdot l} G \tag{5}$$

$$F_{NA} = F_{NB} = (F_p \cdot l + G \cdot h) / 4l \ , \quad F_{NC} = (F_p \cdot l - G \cdot h) / 2l$$

Eq. (5) indicates the relationship between negative pressure force and general gravity. It can be rewritten as,

$$F_p = H_{underside} \cdot G \ , \quad H_{underside} = \frac{2l - (\mu_1 + \mu_2) \cdot h}{(\mu_1 - \mu_2) \cdot l}$$

H is a function of μ_1 and μ_2 . As shown in Fig.7, H increases along with μ_1 increasing and μ_2 decreasing. And $\min(H_{downside}) = 1.751$. Along with H decreasing, F_p decreases. The needed negative pressure $F_p^{'}$ in practice can be calculated by,

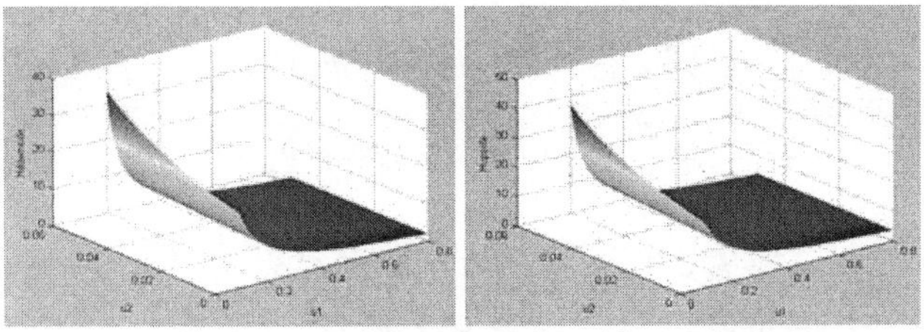

Fig.7 H_{upside} and $H_{downside}$ varying with μ_1 and μ_2

$$F_p^{'} = H \bullet G \bullet [S] \tag{6}$$

Where, $[S]$ is secure coefficient.

If the robot's upper wheel is attached to the surface, we have,

$$F_{NC} = (F_p \bullet l - G \bullet h) / 2l \geq 0$$

It can be rewritten as, $F_p \geq G \bullet h / l$

So, the needed negative pressure $F_p^{'}$ can be calculated by,

$$F_p^{'} \geq \max(H_{downside} \bullet G \bullet [S], G \bullet h / l) \tag{7}$$

When the driving wheels are upside as shown in Fig.6 b), force analysis can be executed by the same procedure. We can get that,

$$H_{upside} = \frac{2l + (\mu_1 + \mu_2) \bullet h}{(\mu_1 - \mu_2) \bullet l}, \; F_p^{'} \geq \max(H_{upside} \bullet G \bullet [S], G \bullet h / l) \tag{8}$$

From Fig.7, it can be found that load capacity is higher when driving wheels are downside than when they are upside. Therefore, load capacity should be calculated by the analysis when the wheels are upside.

4.2. *Force analysis of the climbing leg mechanism*

The obstacle-spanning of the robot relies on the lifting capability of the small sucker in the end of the mechanical leg. Therefore, force analysis should be used to analyze the lifting capability.

As shown in Fig.8, when the small sucker keeps staying on the surface, it bears such forces as negative pressure F_p, normal force N, friction force f, and load force F_y,, which includes gravity of base body and climbing leg. We have,

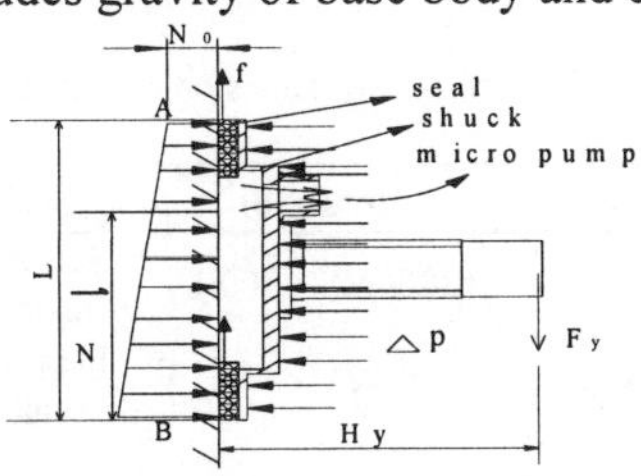

Fig. 8 Forces analysis of the small sucker

$$
\begin{cases}
\sum X = 0 \ , & \int_0^L \left[N_0 + n \cdot (L - l) \right] \cdot dl - F_p) = 0 \\
\sum Y = 0 \ , & \mu \cdot \int_0^L \left[N_0 + n \cdot (L - l) \right] \cdot dl - F_y = 0 \\
\sum M(B) = 0, - \int_0^L \left[N_0 + n \cdot (L - l) \right] \cdot l \cdot dl - F_p \cdot \frac{L}{2} - F_y \cdot H_y = 0 &
\end{cases} \tag{9}
$$

By Eq. (9), we have, $$F_p = F_y / \mu \tag{10}$$

$$N_0 = \frac{F_p}{L} - \frac{6H_y}{L^2} F_y \tag{11}$$

Considering $\mu \le \mu_{\max}$, by Eq. (10) we have, $F_y \le \mu_{\max} \cdot F_p$ (12)
If the small sucker is attached to surface, $N_0 \ge 0$, Eq. (11) can be rewritten as,

$$F_p \ge F_y \cdot 6H_y / L \tag{13}$$

By Eq. (12) and Eq. (13), the load is given by,

$$F_y \le \min(\mu_{\max} \cdot F_p, \ \frac{F_p \cdot L}{6H_y}) = \min(\mu_{\max} \cdot \Delta P \cdot L \cdot M, \ \frac{\Delta P \cdot L^2 \cdot M}{6H_y}) \tag{14}$$

Relationship among F_y, H_y, and ΔP is indicated by Fig.9. P_{diff} represents the vacuum degree ΔP. F_y increases along with ΔP increasing with the constant H_y. Load capacity of the small sucker becomes higher with ΔP increasing.

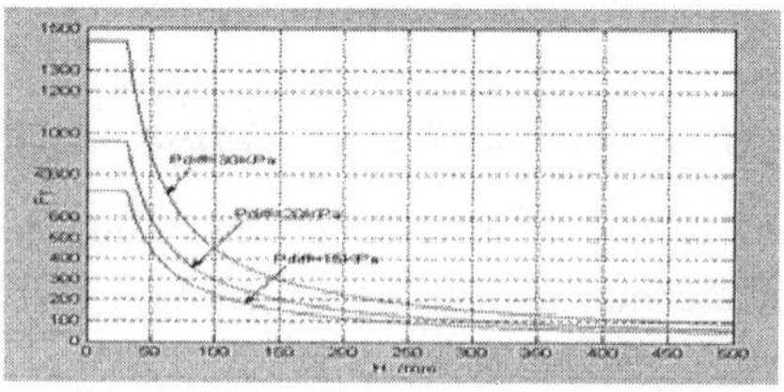

Fig. 9 Relationship between F_y, H and P_{diff}

The specifications of the two adhesion mechanisms are shown in table 2.

Table 2. Specifications of the two adhesion mechanisms.

	Vacuum degree	Size (mm)	Seal method	Load capacity
Big sucker of base body	9000~18000Pa	300(W) × 300(L)	Self-designed flexible seal	207~422N
Small sucker of climbing leg	15K~30Ka	150(W) × 540(L)	Flexible Rubber seal	150~300N (H=150mm)

5. Analysis and Simulation of robot's movement on wall surfaces

The integrated locomotion mechanism improves the robot's mobility evidently, since it enable the robot to have two motion modes. The wheeled locomotion mechanism is used to realize high speed movement. And legged mechanism is used to span obstacles. Selection between the two modes is determined by the condition of wall surface. The two motion modes are analyzed and simulated by establishing 3D model and executing kinematics and dynamics analysis.

If the wall surface is smooth, the wheeled mechanism works. The adhesion force press the robot against the surface steadily, actuators are controlled in a differential way to drive the 3-wheeled locomotion mechanism to realize robot's omni-directional movement. As shown in Fig.10, the robot can move upwards, downwards and along arbitrary directions in normal or inverse orientations.

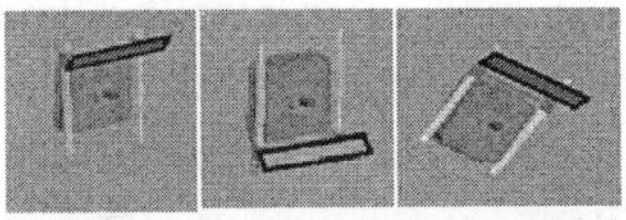

Fig.10 Robot in wheeled motion mode

When the robot is faced with obstacles in its routine, the mechanical leg is controlled to realize obstacle-spanning movement. In this mode, the Fig.10 shows an example of the motion sequence for spanning ledge.

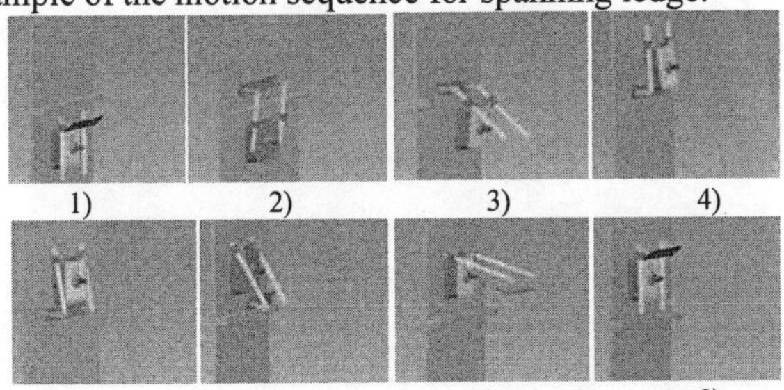

Fig. 11 Locomotion sequence to span a ledge on wall surface

1. The robot gets close to the ledge. Then base body sticks on the surface firmly and keep still, as shown in Fig.11.1).
2. Extent the mechanical leg and drive it over the ledge. Then align the small sucker in the end of the mechanical leg with wall surface and move it until a secure grip is verified with micro air pumps enabled, as shown in Fig.11.2).
3. Base body released from the wall, as shown in Fig.11.3).
4. The mechanical leg lifts the base body over the ledge, as shown in Fig.11.4).
5. When base body lands on surface reliably, air pump works. After verifying the base body attached to the surface securely, the small sucker released, as shown in Fig.11.5) and 11.6).

6. The mechanical leg is retracted and placed on the back of the base body, as shown in Fig.11.7) and 11.8). Obstacles-spanning movement is finished.

The above sequences are assumed to be realized by the on-board controller autonomously. When the predefined strategy does not work, each joint and sucker may also be controlled manually to assist the robot to achieve secure movement. Besides, the mechanical climbing leg enables the robot has the ability to transit between two walls, wall and ground or between wall and ceiling. The alternate use of the two locomotion modes enables the robot to be adapted to more surfaces.

6. Conclusion

This paper presents a novel wall-climbing robot, which integrate two locomotion modes. This principle overcomes the limitations of previous wall-climbing robots, and this robot prototype is better in term of capability, mobility, and adaptability than previous ones. Integration of the wheeled locomotion mode and the legged-climbing locomotion mode evidently improves the mobility of the robot, so that the robot has both high moving speed and excellent obstacle-spanning capability. The performance makes the robot could be competent for versatile missions. The next step is to design embedded controller board and build sensory system to realize a self-contained robot system. Based on these, autonomous capability can be implemented to improve the performance of the robot.

References

1. A. Nishi, Development of wall-climbing robots, Computer and Electrical Engineering, 22(2): 123 (1996).
2. Y.Wang, S.L.Liu, D.G.Xu, Y.Z.Zhao, H.Shao and X.S.Gao, Development & Application of Wall-Climbing Robots, Proceedings of the 1999 IEEE International Conference on Robotics & Automation, 1207 (1999).
3. J. Xiao, J.Z. Xiao, N. Xi, R.Lal Tummala, and Ranjan Mukherjee, Fuzzy Controller for Wall-Climbing Microrobots, IEEE TRANSACTIONS ON FUZZY SYSTEMS, VOL. 12, NO. 4, 466 (2004).
4. J. Z. Xiao, A. Sadegh, M. Elliott, A. Calle, A. Persad and H. M. Chiu, Design of Mobile Robots with Wall Climbing Capability, Proceedings of the 2005 IEEE/ASME International Conference on Advanced Intelligent Mechatronics, 438 (2005).
5. B.L luk, A.A Collie and J.Billingsley, ROBUG II: AN INTELLIGENT WALL CLIMBING ROBOT, Proceedings of the 1991 IEEE International Conference on Robotics and Automation, 2342 (1991).

AN EVOLVED NEURAL NETWORK FOR FAST QUADRUPEDAL LOCOMOTION

IRENE MARKELIC

Bernstein Center for Computational Neuroscience, Bunsenstrasse 10
Goettingen,,37073, Germany

KEYAN ZAHEDI

Fraunhofer Institute for Autonomous Intelligent Systems (AIS), Schloss Sankt Augustin
Sankt Augustin, 53757,Germany

This paper presents a modular neural network controller for fast locomotion of a quadruped robot. It was generated by artificial evolution techniques using a physical simulation of the Sony Aibo ERS-7. Co-evolution was used to develop neuromodules controlling the single legs as well as the coordination between the four legs. The final neurocontroller utilizes a central pattern generator and does not make use of available sensory inputs. In experiments with the physical robot a top walking speed of 47.34 cm/s was measured, where lateral leg movement contributed considerably to the achieved high velocity.

1. Introduction

The control of legged locomotion in robots is still a challenging problem, especially for fast locomotion. Apart from some exceptions like Raibert's robots [1] most legged machines move rather slowly [2]. It has been found that locomotion in many organisms is mostly driven by central pattern generators (CPGs) [3]. These are neural networks producing a rhythmic pattern without the need of sensory feedback [4]. For walking machines a CPG can be realized by an oscillatory artificial neural network. Such oscillators were used for instance by Billard and Ijispeert to realize a continuous passage between a walking, scratching and lying down behavior of an Aibo robot [5]. Kimura used neural oscillators to realize dynamic walking and running in a quadruped [6]. To find appropriate controllers for a hexapod, Beer used genetic algorithms and a simulation of the robot [7]. Finally he transferred the generated controllers to a physical machine [8].

Cruse et al. developed neural networks by means of a simulation and implemented them in a hexapod [9]. In this paper a recurrent neural network for fast quadrupedal walking of a Sony Aibo ERS-7 robot is presented. It was derived by artificial evolution techniques and is for use in the international competition of robot soccer. Since velocity is a key feature, many teams are concerned with fast locomotion. Published velocities of Aibo walking using the ERS-7 model are between 39.7 - 43.0 cm/s [10,11,12,13]. The paper is organized as follows: In section 2 (The Experimental Setup) the neuron model is introduced, and the tools employed for evolution and composition of the network are explained. In the third section (Results) the resulting controller is presented and analyzed. In the final section (Discussion) the results are discussed and related to other works.

2. The Experimental Setup

All experiments were conducted using the time discrete dynamics of networks with standard additive neurons, using the hyperbolic tangent as transfer function ; i.e.,

$$a_i(t+1) = \Theta_i + \sum_{j=1}^{N} \omega_{ij}\sigma(a_i(t)), \quad i=1,..N \tag{1}$$

where a_i denotes the activation, Θi the bias term of the ith neuron, t denotes a discrete time step, ω_{ij} refers to the weight of synapse rom neuron j to i, and N is the total number of neurons in a given network. Input neurons were treated as buffers for sensor signals. For evolution the program package ISEE (Integrated Structure Evolution Environment) was used, which was designed to realize Artificial Evolution experiments. It implements the ENS^3 algorithm [14] and offers the physical simulator YARS based on ODE (Open Dynamics Engine). ISEE permits to influence the evolutionary process and the network by various parameters that can be set during runtime, for example, to foster small networks by assigning cost terms to neurons and synapses. It also supports co-evolution, i.e. the simultaneous evolution of several populations. Furthermore, the generation of network structure as well as parameter optimization is possible.

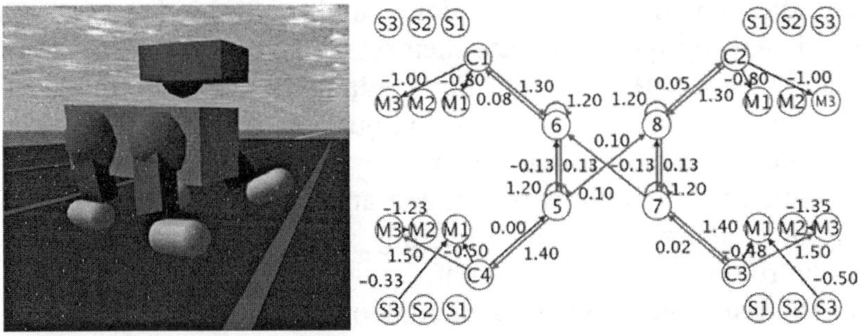

Figure 1.Left: The simulated robot. Right: The neurocontroller generating the fastest walking behaviour. Circles denote neurons and are connected via synapses. Neurons C1 to C4 are connector neurons which form the junction between leg modules and coordination module.

A dog-like robot, the Aibo ERS-7 was used as physical platform for the evolved controller. Each of its four legs has three degrees of freedom realized by the three motors, M1, M2 and M3. The first motor, M1, moves the leg back and forth, whereas M2 moves it sideways. M3 moves only the lower limb of the leg back and forth. For each of these motors a sensor (S1, S2, and S3), is detecting its deflection angle.

The physical simulation of the robot, as shown in Fig. 1 left, was realized with YARS. As relevant features to be modelled for a walking behavior were considered: body parts (head, trunk and limbs) defined by their dimensions and weights as well as the 12 leg motors which were each defined by a deflection angle, strength, and velocity, and finally the according 12 deflection angle sensors. Since hinge torque and velocity of the motors used for the ERS-7 are not published they had to be measured experimentally. For maximal force 0.43 Ncm were used, and 286 °/s for maximal angular velocity. To instantiate the controller in the robot it was rewritten as C++ program and incorporated into the freely available German Team Software which provides access to motor control and sensory information.

Based on the hypothesis that control of insect walking can be considered hierarchical and modular [16], and assuming that both fore legs and both hind legs each carry out the same step cycle, a modular approach, and thus co-evolution, was chosen as an appropriate method. It was decided to use one network per leg, where the networks for both fore- and both hind legs were identical. All leg modules were then connected to a coordination module. A minimal leg controller contained seven neurons, one input neuron per sensor and one output neuron per motor. The seventh neuron was a hidden neuron by which the connection to the coordination module was realized and which is referred to as connector neuron. The resulting network formed the neurocontroller, comp. Fig. 1, right.

When a reasonable walking behavior was observed during simulation the corresponding controller was transferred to the hardware, and the speed of the resulting gait measured. Therefore a test course was set up with two labels, indicating one meter and the robot had to cover this distance from a flying start and with fully charged batteries while being filmed. Each neurocontroller was tested 10 times, and its speed was set to be the average value of these runs.

Various experiments were conducted, and the best results were achieved when using a CPG as coordination module. The employed CPG, depicted in Fig. 2A, was adopted from [17]. Its output are four sinusoidal signals with equal phase shift (see Fig. 2, bottom left), of which each leg module received one in the order: right hind leg, right foreleg, left hind leg and left foreleg. This, together with the corresponding phase shift, generated a reasonable walking gait. An evolution run was started by providing initial network structures for the leg modules in order to avoid the bootstrap-problem. They caused a pendulum like movement of the legs which was triggered by sensory input. Parameter settings of the evolutionary ENS^3 program enabled an alteration of structure, bias and synaptic weights of these networks. Structure and synaptic weights of the CPG were chosen to be fixed, because already small changes often had undesirable effects for the output signals, e.g. the loss of periodicity. The fitness function F was given by

$$F = \sum_{t=0}^{T} x(t)^{2} - c(t) * \alpha \qquad (2)$$

where t denotes discrete time, and x the position along the x-axis of the simulated robot in world coordinates, which was equal to the covered forward distance. It was squared to increase selection pressure. To punish individuals with high frequency oscillating legs, a problem that frequently occurred, the change of direction in leg movement was counted, denoted by *c*. It was then multiplied by a factor, α, which was adjustable during runtime. In addition, low obstacles were built into the simulated environment which could not be overcome by individuals with fast oscillating leg movements, so that they could not gain high fitness values. After 4320 generations a neurocontroller leading to a speed of 39.56 cm/s was generated.

3. Results

The resulting network structure is shown in Fig.2, right. It is observable that no further hidden neurons, apart from the connector neuron, were part of the leg networks and that the only driving force of the controller was the CPG. Although sensory input from 12 sensors was available during evolution, only one sensor in the hind leg modules was effciently used. It was found that this sensor input was important for the initial behavior of the simulated robot. It modulated the trajectory of the signals that controlled M1 of both hind legs in such a way that the simulated robot could get in a position from which it could start walking. Because the physical robot was not started from such a position (it was put on the floor when moving) it was not affected by this mechanism. Thus, a controller for walking without any incorporation of sensory feedback had been evolved.

Signals were fed back into the CPG coming from the connector neurons. Although the according synapse weights were small, between [-0.08, 0.08] they had impact on the CPG's output signals, as can be observed when comparing Fig. 2, bottom middle and bottom right. Changes in amplitude (C1, C4), period (C1, C2, C3, C4), phase shift (C4, C3) and signal shape (C3) are observable.

With this controller the robot did not walk in a straight line, which was due to the different signals the motors received. Therefore synaptic weights were carefully adjusted so that all legs received a similar signal (see Fig. 2 top, right).

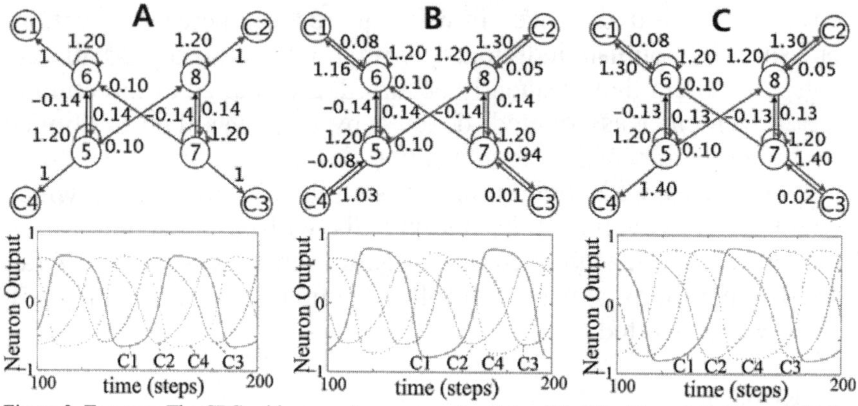

Figure 2. Top row: The CPG with connector neuron of each leg (C1-C4). The originally used CPG, left, the CPG with evolutionary altered parameters, middle, and the manually adjusted CPG, right. The corresponding bias values are listed in table 1. Bottom row: The plot below each CPG shows the output of the four connector neurons. Each curve has is labelled with the corresponding neuron.

Furthermore parameters were altered to find out which stride frequency and length were most appropriate. This was done manually, but could have been done using evolution. As was already indicated by the longer period time of CPG signals after evolution, the best results were attained with a lower CPG signal frequency of 1.8Hz. The initial frequency in comparison was 2.2Hz. In Fig. 3 the signal of M1 of the right foreleg is plotted along with the according sensor signal. It is observable that the motor lags behind the given signal and never reaches the target amplitude. Thus, with a larger period the motor can follow the given signal better, which finally results in a larger step length. In contrast, a higher frequency automatically led to smaller steps and due to the inertia of the motors to a lower speed of the walking behavior.

Leg posture was determined by the bias values of the according output neurons. Changing these values enabled a straighter posture of the legs which caused larger steps. The final neurocontroller (see Fig 2, right) generated a walking behavior with a speed of 45.95 cm/s, with a top speed of 47.34 cm/s. Videos of the behaviour are available under www.fraunhofer.de/~zahedi/aibo.html. The obtained gait is a walk with duty factor $\beta = 0.58 \pm 0.2$ for all legs. Where duty factor is the ratio between the duration that a leg has ground contact and the duration when it is lifted off the ground during one stride. One major difference to other Aibo walking behaviors was the incorporation of M2 in the hind leg modules, which resulted in a sideways movement of these legs. It caused the robot to rock from one side to the other while walking (compare video). Lesion experiments showed that this resulted in a gain of speed of 11.8%.

Table 1. A comparison of bias values of the different CPGs. The enumeration of neurons is according Fig. 2, nn = neuron number, bo = bias of original CPG, ba = bias of evolutionary altered CPG, and bop = bias of optimized CPG.

nn	bo	ba	bop	nn	bo	ba	bop
C1	0	-0.003	0	5	0.01	0.017	0.0175
C2	0	0.0016	0	6	0	0.0053	0.0053
C3	0	-0.002	0	7	0	0.0037	0.0037
C4	0	0.0065	0	8	0	-0.0044	-0.0044

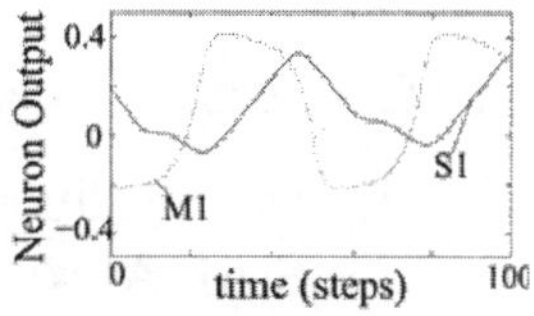

Fig. 3. Output of M1 and S1. S1 shows the actual deflection of the motor. Whereas M1 can be interpreted as a setpoint signal.

4. Discussion

Modular neurocontrollers were evolved to control a physical Aibo-ERS-7 robot. Using a physical simulator and additional adjustments, the final controller was able to generate a fast straight walking behavior for this robot of about 47,3 cm/s. It was found that a sideway movement of hind legs resulted in a higher walking velocity. This can be explained by the shift of the robot's center of gravity from side to side, caused by the lateral hind leg movement. When one bodyhalf was tilted towards the ground, carrying most of the weight, the other side was moved upwards, leaving more time and force for legs on this side to be moved.Thus, it is concluded that a pace gait in addition to the horizontal body tilt can lead to a higher walking speed for this particular robot, than a trot or a walk Experiments by Beer suggest that sensory input is always incorporated when available [8]. That this does not happen in this case may result from the fact that sensory input was simply not needed for the task at hand, because of the driving CPG. Evolution took place in an environment that indeed contained obstacles, but obstacle height was relatively low. Once it was provided that legs were lifted up high enough there was no reason for a more sophisticated behavior. That no sensory input was fed into the network supports an hypothesis by Full and Koditschek [19] that rapid locomotion relies rather on feedforward control than on continuous sensorimotor feedback. This is thought to be due to the little time available to process proprioceptive input [19]. For future work, it is desirable to steer the walking direction of the robot. For this, visual sensory input could be used as a tropism (e.g. using the robot's camera to follow a colored ball).

References

1. M.H. Raibert & I.E. Sutherland, Sci. Am., **248,** 44 (1983).
2. R.M. Alexander, *Phys. Rev.*, **69,** 1199 (1989).
3. S. Grillner, *Curr. Opin. Neurobiol*, **1(4)**, 231, (1991).
4. F. Delcomyn, *Science*, **210(4469)**, 492 (1980).
5. A. Billard & A.J. Ijspeert, *Proc. IJCNN*, **VI**, 637 (2000)
6. H. Kimura et al, *Aut. Rob.*, **7(3)**, 247 (1999).
7. R.D. Beer & J.C. Gallagher, *Adapt. Behav.,* **1**, 91 (1992)
8. J.C. Gallagher et al., *Robot. Auton. Syst.*, **19**, 95 (1996).
9. H. Cruse et al., *Adap. Behav.*, **3(4)**, 385 (1995).
10. A. Shinohara & A. Ishino, www.jollypochie.org/papers/JollyPochie2004TR.pdf, (accessed 2006).
11. M. Quinlan *et al.*, www.robots.newcastle.edu.au/publicationns/NUbotFinalReport2005.pdf,(accessed 2006).
12. D. Cohen et al., *www.cis.upenn.edu/robocup/UPenn04.pdf*, (accessed 2006).
13. T. Röfer, 8th International Workshop on RoboCup, (2004).
14. F. Pasemann *et al.*, *Theory Biosc.,* **120**, 311 (2001).
15. M. Fujita & H. Kitano, *Aut. Rob.,* **5**, 7 (1998).
16. F. Delcomyn, *Aut. Rob.*, **7**, 259 (1999).
17. B. Klaassen *et al.*, *VDI-Berichte*, **1841**, 633 (2004).
18. F. Pasemann et al., Proc. IWANN, **2686**,144 (2003).
19. R. Koditscheck, *J. Exp. Biol*, **22**, 3325 (1999).

Autonomous Climbing Motions for Connected Crawler Robots

Sho YOKOTA*, Yasuhiro OHYAMA,
Hiroshi HASHIMOTO and Jin-Hua SHE

School of Bionics, Tokyo University of Technology,
Tokyo, Japan
** E-mail: yokota@bs.teu.ac.jp*

Kuniaki KAWABATA
RIKEN,
Saitama, Japan
E-mail: kuniakik@riken.jp

Hisato KOBAYASHI
Faculty of Eng.,
Hosei University,
Tokyo, Japan
E-mail: h@k.hosei.ac.jp

Pierre BLAZEVIC
Institut des Sciences et
Techniques des Yvelines,
Mantes, France
blazevic@lisv.uvsq.fr

The purpose of our research is to develop the mobile system for rough terrain. Our mobile system adopts the connected crawler mechanism. In this paper, we proposed the operation strategy for autonomous motions. We also made verification experiment of proposed operation strategy. For this verification, we prepared 2 kinds of experiment. These experimental results showed that the robot could pass over the different heights and different structures of obstacles by using only (same) strategy. Moreover the sensors which realize proposed strategy were very simple, and the number of sensor was very small. Therefore it could be concluded that proposed strategies had an extremely high usefulness.

Keywords: Crawler, Rough terrain, Step climbing, Autonomous motion, Connected crawler

1. Introduction

There is a great meaning to use crawler mechanisms as a mobile function on rough terrain. Because, in general, the crawler mechanism can obtain big impulsion due to its wider grounding bearing areas than the leg mechanism and the wheel mechanisms. On the contrary, the crawler also has weak points as a poor stability in complex geographical features. And the mobility on the area like the stairs is inferior to that of the leg.[1]

Therefore, a lot of researches have been done to supplement these weak points. The main theme common to those researches is to improve the mobility. Generally the method which changes the form of crawler is adopted as an approach for this main theme. In order to realize this transformation, many researches proposed the connected crawler mechanism which crawler stages were connected by active joints. Lee et al[2] designed two stages one active joint type resucue robot. "Souryu-III"[3] is 3 stages 2 joints type for resuce operations. "MOIRA"[4] is also rescue robot which is 4 stages 3 joints type connected crawler.

Although we can see such researches, there are no robots which show autonomous operations. The one of the most important purpose to introducing rescue robots to disaster places is to automate sufferer searching in place of the manpower searching. If many rescue robots can search the sufferers automatically, it is enable to search wider and faster than conventional manpower searching. However current rescue robots don't realize the autonomous operations, therefore that is not achieved the above mentioned important purpose which is to introduce robots to disaster places.

Thus this research proposes the rough terrain mobile robot which can realize autonomous motion in disaster places. Especially, this paper proposes operation strategies for passing over obstacles autonomously.

2. The prototype

First, let see the outline of our prototype. The mobile function of our prototype adopts crawler mechanism. Because The crawler mechanism shows the high mobile ability on various terrains; moreover it is simple mechanism and easy to control. But conventional single track mechanism has mobility limitations; the limitation is determined by attacking angle, radius of sprockets, and length of crawler. In order to improve its mobility, we add some active joints to conventional crawler tracks, namely that is connected crawler mechanism. The specifications of the prototype are as follows: Length(maximum) is 354.0 mm, Length(minimum) is 118.0 mm, Width is 125 mm, Mass is 0.608 kg, Radius of the sprockets is 20.0 mm.

2.1. *Mechanical structure*

Our mobile mechanism has 3 connected stages with the motor-driven crawler tracks on each side(Fig. 1). RC-servo motors are used for driving joints between the stages. The left and right crawlers are driven by 2 DC motors independently. Three crawlers on each side are driven by same motor simultaneously.

Fig. 1. The overview of Connected crawler robot

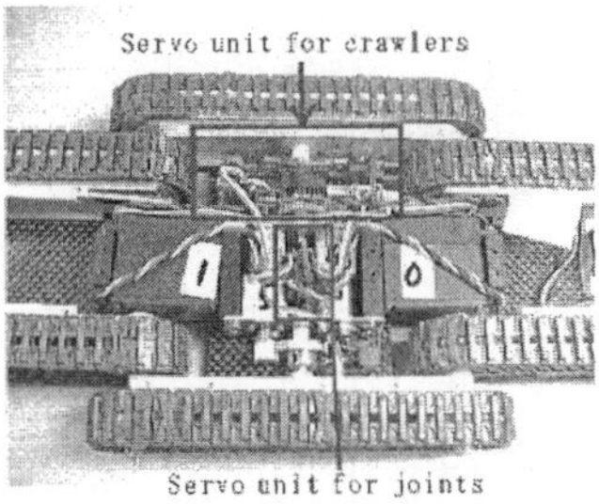

Fig. 2. The overview of the servo unit

2.2. *Control structure*

We adopt a hierarchical control structure by installing an intelligent servo driver to each actuator. We connect each of them to the master contoll unit by UART serial line. The parts marked by red line in Fig. 2 are servo drivers.

Fig. 3 shows the control structure of this system. The master unit is equipped with several sensors which are increnometers, PSD distance sensor and photo reflector. The usage of these sensors will be shown in Chapter 3. Master unit culclates high level task (setting trajectory, sensing environment, etc), and servo driver works for low level tasks. The master unit processes the high level task, and derive the data to low level task (crawler velocity, joint angel and so on), and send them to the servo drivers. After receiving these data, the servo drivers control their motor by conventional feedback control low.

3. Operation strategies

The shpae of rough terrain such as disaster places is various. It is difficult to derive the each autonomous motions relative to each terrain. Therefore, in this chapter, we assume that the environment is one bump, and consider about the operation strategies for autonomous motion to pass over the bump. Because the climbing bump ability is important factor as one of the most fundamental mobility index,[5] climbing bump experiment is taken by many researches as an evaluation experiment of mobility on rough terrain. The proposed operation sterategy has 7 steps (Fig. 4). Following sections show the details of each steps. Fig. 5 is the difinition of the parameters. Here θ_c is the orientation of the center stage. θ_1 and θ_2 are the 1st and 2nd joint angle related to the θ_c.

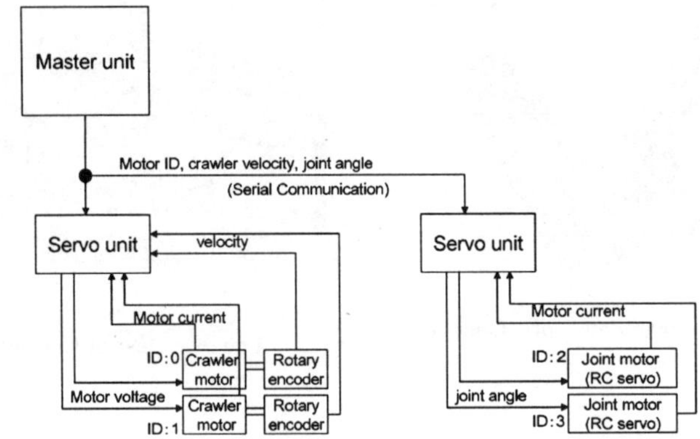

Fig. 3. The control system

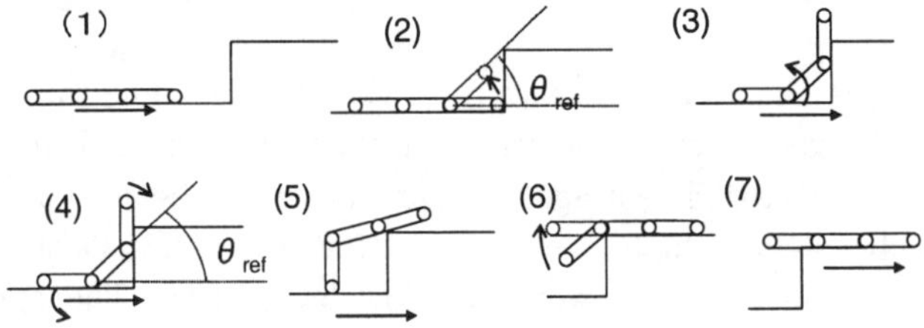

Fig. 4. Proposed operation strategies

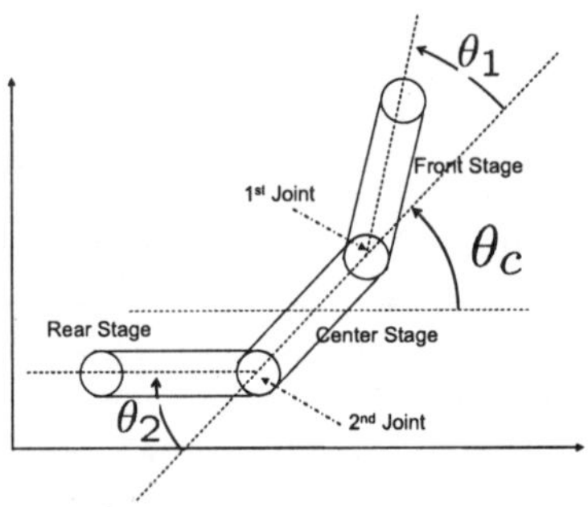

Fig. 5. The definition of the parameters

3.1. *First step*

First, the robot goes forward until detecting the wall. If the robot faces the wall, then robot stops moving. PSD distance sensor which is attached to the 1st stage for detecting the wall (Fig. 6). The output of the PSD sensor is managed by the main controller (Fig 3).

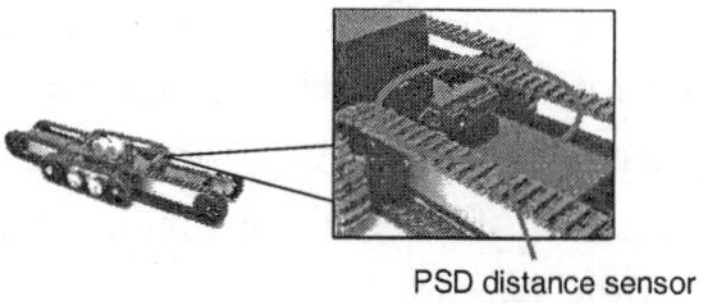

Fig. 6. The PSD distance sensor

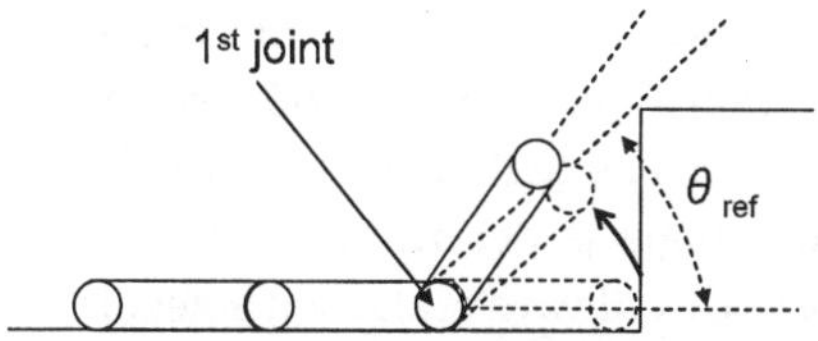

Fig. 7. The definition of θ_{ref}

3.2. *Second step*

In this step, 1st joint are driven to detect θ_{ref}. θ_{ref} is the 1st joint angle when the tangent of front stage meets the edge of the bump (Fig. 7).

3.3. *Third step*

In third step, 2nd joint is driven while the robot goes forward. The purpose of this step is to get the traction forces by keeping a grounding of rear stage. If the robot goes forward without driving 2nd joint, then robot could not get enough traction forces due to the lift of rear stage. In order to keep the grounding of rear stage, 2nd joint angle should be set to angle of center stage, namely the 2nd joint is driven in the following condition.

$$\theta_2 = \theta_c$$

Here, the inclinometers are attached to the center stage for detecting the angle of the center stage (Fig. 8).

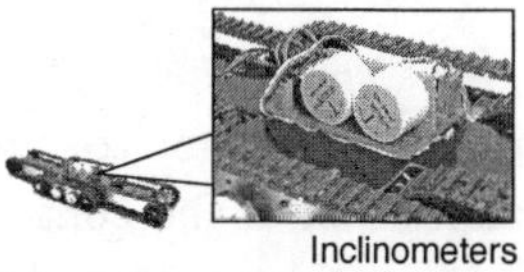

Fig. 8. Inclinometers for detecting θ_c

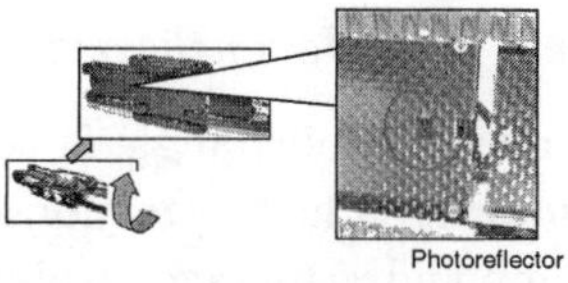

Fig. 9. Contact detection device

3.4. *Fourth step*

In this step, 1st joint angle is set to 0 rad, 2nd joint is driven to let the angle between rear stage and ground be right angle. At this moment, the robot continues moving. There are two purpose in this step. One is to obtain the traction forces, that is the role of 1st joint motion. The other is to lift up the robot as high as possible, that is the purpose of 2nd joint motion. 2nd joint angle is determined by following condition. By this condition, rear stage can always stand with keeping right angle to the ground.

$$\theta_2 = \frac{\pi}{2} - \theta_c$$

The trigger to shift third step to fourth step is the θ_{ref}. In the third step, when the orientation of center stage θ_c is equal to θ_{ref}, operation step is shifted.

3.5. *Fifth step*

The robot goes forward with keeping above mentioned conditions. When the center of gravity of the robot is in the right side of the bump edge, then the clock wise moment is generated around the edge, the robot can climb a bump. Fig. 10 shows the situation of this case.

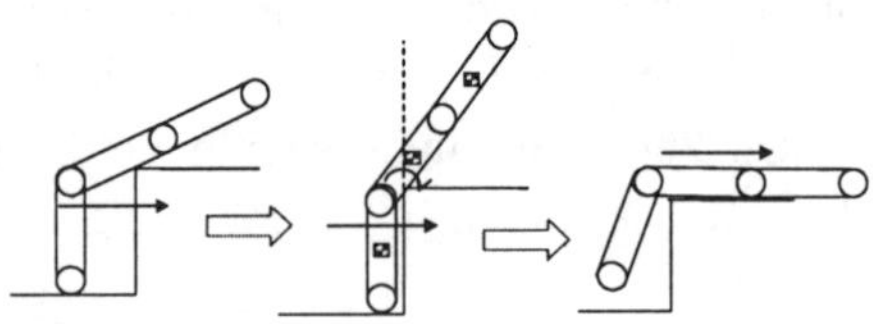

Fig. 10. Inclinometers for detecting θ_c

3.6. *Sixth and Seventh step*

At the end, 2nd joint angle is set to the initial position, not to interfere robot's moving. The trigger for shifting to this motion is the contact between bump and rear stage. A photoreflector is attached to root and bottom of the rear stage for detecting the contact(Fig. 9).

By above steps, climibing a bump is completed.

4. Experiments

In order to confirm proposed operation strategy,We prepare two kinds of experiment. One is that the robot passes over two bumps with different heights. The other is stairs ascending. There are remarkable points in these experiments. These experiments verify whether the robot can pass over the different heights and different structures of obstacles by using only proposed strategy. Moreover the sensors which realize proposed strategy is very simple, and the number of sensor is very small. Therefore if these experiments success, it can be concluded that proposed strategy has extremely high usefulness.

4.1. *Passing over the different hight bumps*

In this section, the robot passes over the different heights of bumps. The heights of bumps are 150 mm and 40 mm. We made the experiment by implementing proposed strategy to main controller. The result is shown in Fig. 11. This Figure shows that the robot can pass over the different hights of bumps autonomously.

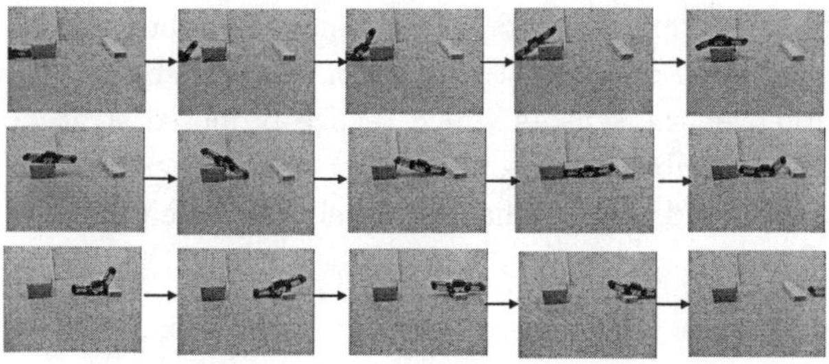

Fig. 11. The experimental results of passing over bumps

4.2. *Stairs ascending*

Next experiment is stairs ascending. The height between stairs is 150 mm, that is conventional stairs. The implemented software to main controller is the same as experiment in 4.1, namely we do not add any modification, that is completely same. The result is Fig. 12. From this Figure, it is turned out that the robot could ascend stairs autonomously with driving joints.

5. Conclusions

The purpose of our research is to develop the rough terrain mobile robot which can realize autonomous motion in disaster places. Especially, this

Fig. 12. The experimental results of stairs ascending

paper have proposed operation strategy for passing over obstacles autonomously. This operation strategy is consisted of 7 steps, and it needs only 3 simple sensors which were PSD distance sensor, inclinometer and photoreflector. We have also made verification experiment of proposed operation strategy. For this verification, we prepared 2 kinds of experiment. One was that the robot passes over the different heights of bumps. The other was stairs ascending.

Both experiments had a success. There were remarkable points in these experiments. These experiments showed that the robot can pass over the different heights and different structures of obstacles by using only (same) strategy. Moreover the sensors which realize proposed strategy were very simple, and the number of sensor was very small. Therefore it can be concluded that proposed strategy has extremely high usefulness.

References

1. S. Hirose, *Mechanical Designe of Mobile Robot for External Environments (in Japanese)*, vol.18, no.7, pp904-908 edn. (Journal of Robotics Society of Japan, Tokyo, 2000).
2. C. Lee, *Double -track mobile robot for hazardous environment applications*, vol. 17, no. 5, pp 447-495, 2003 edn. (Advanced Robotics, Tokyo, 2003).
3. T. Takayama, Development of connected crawler vehicle souryu-iii for rescue application, in *Proc. of 22nd conference of Robotics Society of Japan CD-ROM*, (Robotics Society of Japan, 2004).
4. K. Osuka, Development of mobile inspection robot for rescue activities:moira, in *Proceedings of the 2003 IEEE/RSJ Intl. Conference on Intelligent Robots and Systems*, (IEEE, 2003).
5. T. Inoh, Mobility of the irregular terrain for resucue robots, in *10th Robotics symposia*, (pp 39-44, 2005).

CONTROLLING AN ACTIVELY ARTICULATED SUSPENSION VEHICLE FOR MOBILITY IN ROUGH TERRAIN

SIDDHARTH SANAN

Mechanical Engineering, Indian Institute of Science-Bangalore
Bangalore, Karnataka 560012, India

SARTAJ SINGH

Centre for Artificial Intelligence and Robotics, DRDO-Bangalore
Bangalore, Karnataka 560093, India

K MADHAVA KRISHNA

Robotics Research Laboratory, IIIT-Hyderabad
Hyderabad, Andhra Pradesh 500032, India

This paper studies the problem of traversing a rough terrain by wheeled vehicles. The criterion for mobility of a wheeled vehicle in any terrain is formally developed, providing insights into the mechanical structure requirements of the vehicle. A vehicle structure with an actively articulated suspension is found as a solution to improved rough terrain mobility. The contact forces of the vehicle with the surface being traversed are identified as the critical factor in determining the traversability of the surface. Hence a control strategy involving the control of the contact forces (normal and traction) is proposed. The key feature of the locomotion strategy, thus, developed is that it provides a solution for mobility in terrain which cannot be traversed using a solution involving assumptions that ignore the dynamics of the main body of the vehicle.

1. Introduction

The problem of finding a suitable mechanical structure for a mobile platform that can move efficiently on rough terrain not known *a priori* has been long standing. The main disadvantage of the passively suspended and fixed suspension vehicles is that the robustness of these systems under varied terrain conditions is not certain. To enhance the performance of such systems, a class of robots with actively articulated suspensions called the Wheeled and Actively Articulated Vehicles (WAAVs) has been developed, the terminology first used by Srinivasan *et al.* [1]. The Hybrid Wheel-Legged Vehicles (HWLV) [2] is a class of vehicles that consists of any combination of wheeled and legged mechanisms. The Hylos [2] and WorkPartner [3] are examples of such vehicles.

A static analysis of the contact forces on an HWLV such as Hylos would reveal that the contact forces are functions of body-weight, payload, posture and contact angles [4], [5]. Hence previous work [6], [2] has aimed at controlling the posture, defined by the internal configuration parameters, with the purpose of optimizing the contact forces for maximizing traction and stability. Algorithms for traction optimization and power efficient mobility in rough terrain are presented in [4]. However [4] does not speak of conditions where forward motion is not possible for a slow-moving fixed suspension vehicle.

The approach in this paper which we term as the force-control methodology can be considered inverse of the approach as stated in previous work [2], [7] with respect to traction for mobility in rough terrain. In that, we are able to control the posture of the main body by controlling the contact forces rather than vice versa. We aim at directly controlling contact forces that result in the desired traction.

We develop an analysis of the surface conditions leading to the complete loss of ability to move forward in a given terrain for any rigidly suspended vehicle which provides the motivation for an actively articulated suspension vehicle structure and also to develop our control methodology. The possible advantages that such a system might offer as compared to the posture control methodology based on the vehicle kinematics is also studied.

2. Related Works

Ch. Grand *et al.* at the Laboratoire de Robotique de Paris have developed the Hylos robot [2]. The central theme for the Hylos locomotion is to achieve a posture of the main body which maximizes stability and traction using the posture control algorithm that uses the velocity model to set velocities at the various joints based on the posture error.

The critical assumptions made in the analysis of the Hylos [2] are: (i) sufficient contact forces exist to allow for pure rolling of the wheels, also implying that wheel-ground contact exists at all times, and (ii) the dynamic effects are neglected. The authors believe that the method of altering the contact forces to control the posture and improve traction performance in an actively articulated suspension would be a first attempt and hope to solve the problems prevalent due to assumptions stated in [2].

3. Problem Definition

We need to find a method that controls the contact forces F^c at the various contact points between the wheels and the surface such that the robot successfully traverses a terrain. Hence the primary motive of the vehicle is to be able to traverse a given terrain. Since our aim is to study the traction or the forward motion problem a planar analysis will suffice as also in [4], however extensions to the 3D problem can be done. To start the development, we first understand the need for WAAV as compared to a rigid suspension vehicle and why it offers a better possibility to traverse a terrain. From this analysis, we understand the role the contact forces play in helping the vehicle traverse a given terrain. We then use a generic platform consisting of a main body and two actuated wheels in the sagittal plane, each wheel connected to the main body through a Linear Force Actuator for development of our control methodology. We call this vehicle LFA-V (Figure 1).

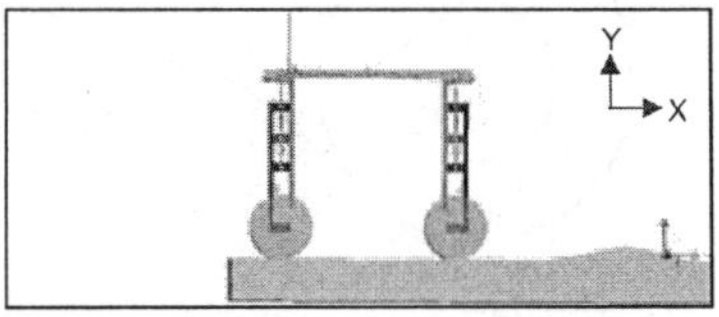

Figure 1. The LFA-V mechanical structure

The following conditions are unsuitable for control methodology developed in prior work [2]: (i) the vehicle is moving fast, (ii) large surface discontinuities exist (surface does not belong to C^1), and (iii) surface geometry that does not permit mobility under static conditions of the main body of the vehicle. The basic reason for the failure under these conditions is that any posture control methodology based on the kinematic analysis of the system fails to sense its interaction with the environment since all sensory data is local to the vehicle itself. Interaction with the environment is necessary for maintaining near ideal conditions such as pure rolling and no lateral slippage used for developing the kinematic model to control the posture. This is one way in which the force control methodology differs from the others.

In general, the normal force and traction are related by:

$$\mu N > T > R + I \tag{1}$$

where T is the tractive force, N is the normal contact force, R is the resistance in the direction of motion and I is the inertia force. In most cases, I is neglected. We therefore have to decide the required normal contact forces N_i at each wheel surface contact that limit the traction T_i. It should be made clear that resistance includes potential fields like gravity as well as non potential fields like friction. We also need to device a control to achieve the desired value of contact force $F^C = [T \quad N]^T$ using an articulated suspension. This issue has been addressed previously for industrial robots in a comprehensive manner [8], [9] and the work described in this paper does not focus on this issue. The control strategy developed in the paper is shown to be useful for application to terrain conditions that are infeasible for traverse by fixed suspension vehicles. For all further analysis, we assume the wheels have only single point contact in the plane of analysis and that the wheels are cylindrical.

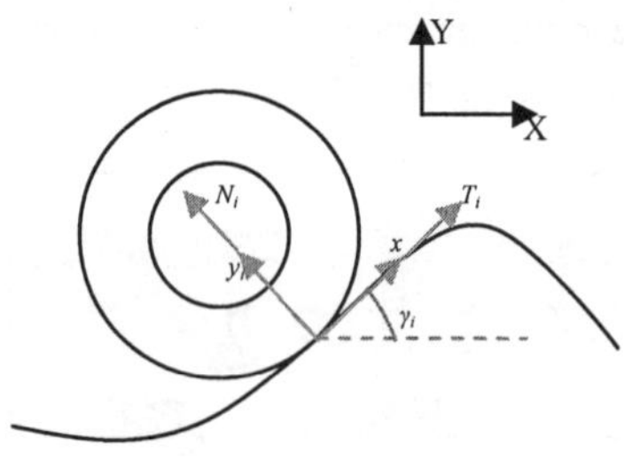

Figure 2. Wheel terrain contact forces and parameters

4. Fixed Suspension versus WAAV

Traction control systems are generally used in fixed suspension vehicles to maximize the traction it gets from a surface, to prevent being rendered motionless. The quasi-static equations that relate the normal contact and traction forces to the forces on platform for a planar two-wheeled vehicle are given by (2). As shown in Figure 2., the global fixed frame is $\{XYZ\}$, where Y is parallel to the gravity force; $\{xyz\}_i$ is the contact frame at the ith contact point where T_i is along x and N_i along y; V_i^j is the component of the vector from the ith contact point to the CG point in the j direction; and γ_i is the contact angle at the ith contact point. The index i increments as we move from the rear to the front wheels.

$$A \cdot C = D \tag{2}$$

For a vehicle with two wheels in the plane, $A = A_G$.

$$A_G = \begin{bmatrix} \cos\gamma_1 & -\sin\gamma_1 & \cos\gamma_2 & -\sin\gamma_2 \\ \sin\gamma_1 & \cos\gamma_1 & \sin\gamma_2 & \cos\gamma_2 \\ V_1^y & -V_1^x & V_2^y & -V_2^x \end{bmatrix}$$

$$C = [T_1 \quad N_1 \quad T_2 \quad N_2]^T \; ; \; D = [F_X \quad F_Y \quad M_Z]^T$$

These set of equations are redundant and hence infinite solutions exist under the following constraints:

$$N_i > 0 \; \forall \, i, i = \{1,2\} \tag{3}$$

$$\Gamma_i^{min} \leq (T_i \cdot r) \leq \Gamma_i^{max} \quad \forall \, i, i = \{1,2\} \tag{4}$$

$$T_i \leq \mu N_i \; \forall \, i, i = \{1,2\} \tag{5}$$

$\Gamma_i^{max}, \Gamma_i^{min}$ are the maximum and minimum torques that the motor at the ith wheel can generate. For the vehicle to be able to move forward, it must at least remain in equilibrium. Therefore, we set $F_X \geq 0$, $F_Y = Wg$ and $M_Z = 0$, neglecting effects like rolling friction, inertial forces. W is the mass of the vehicle in kilograms and M_Z is the moment in the plane of analysis. The above problem can be viewed as a linear programming problem, where a solution to all variables exists if the solution space is not a null set. The objective function is set as:

$$\min(S), S = T_1 \tag{6}$$

to evaluate the regions for γ_1 and γ_2, where a solution to the above set of equality and inequality constraints (2)-(5) does not exist. To demonstrate the existence of regions with no solutions, we set $\gamma_1 = 0$ and vary γ_2 from 0 to $\pi/2$. Also the pitch angle ψ is varied from 0 to $\pi/3$.

Figure 3 shows the plot of $\min(T_1)$ as a function of the contact angle γ_2 and pitch angle ψ. The values of γ_2 and ψ corresponding to the brown (darkened and flat) region indicates the region of infeasibility where no solution for the given values of γ_2 and ψ exists. In this region, no values of traction forces at the wheels can maintain the system in equilibrium with the vehicle having a rigid or fixed suspension. Hence if the vehicle is not moving fast, it cannot overcome this state and cannot traverse the terrain. It is in these regions of infeasibility that actively articulated vehicles such as the HWLV and the LFA-V demonstrate their advantage.

A number of strategies have been identified with the actively articulated class of vehicles to traverse regions that offer no solution with the rigid or fixed suspension vehicles. In further sections we introduce the control methodology devised and its use to implement these strategies.

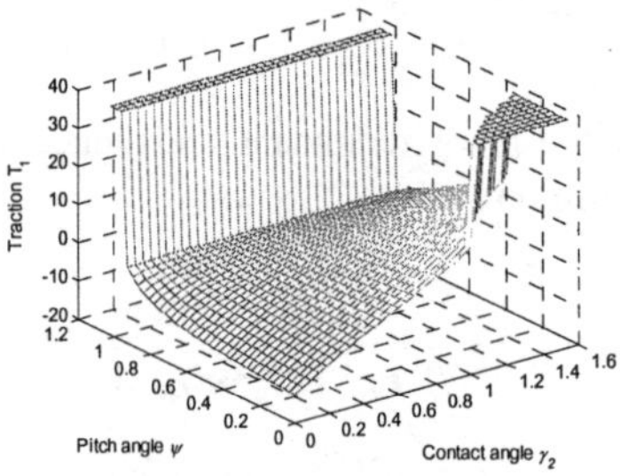

Figure 3. Plot of min(T_1) showing regions of infeasibility (dark flat surface)

5. Force Control Methodology

We use the planar model of the LFA-V in Fig.1 described earlier to develop the Force Control Methodology. The mechanical structure for analysis consists of 2 wheels, each pinned to an outer slide link which is connected to an inner slide link through a prismatic joint. The inner slide is fixed to the main body. The prismatic joint is actuated through a linear actuator, mounted on the main body, to which a desired force F^A can be commanded. This force F^A acts between the main body and the output slide. Although the input and output slides have a finite mass, we consider this to be negligible compared to the mass of the body and therefore neglect them in our analysis.

The 3 DOF's of the system can be identified as the height h of the main body and its pitch angle ψ and the position along the horizontal direction X_v. The wheel ground contacts are assumed to be no-slip contacts for the kinematic analysis.

The initial aim of our control shall be to ensure that the vehicle is successful in moving over a terrain for a particular posture (h, ψ) of the main body. The position X_v is defined by the motion of the vehicle. To ensure that the vehicle moves in the forward direction, the force on the main body should be greater than or equal to zero $(F_X \geq 0)$. This is the main criterion for the system to traverse a given terrain.

We can set arbitrary requirements for the two parameters (h, ψ) assuming they lead to solutions for T_i and N_i under constraints (3)-(5). In effect we try to control the posture of the main body and its forward acceleration by controlling the contact forces T_i and N_i using the motor torques and the linear force actuators.

We shall write down the quasi-static force balance equations for the LFA-V. $\overline{V}_i^J$ represents the effective moment arm of the force at the ith contact point acting perpendicular to the j direction. The fourth row in (7) has been included to set equal traction forces at both the wheels which is a strategy used by most automobiles. The equations below are written by considering the force actuator force $F_i^A = N_i/\cos(\gamma_i - \psi)$. These equations are different from that for a rigidly suspended vehicle (2) shown earlier. C, D are defined previously in (2)

$$\begin{bmatrix} A \\ B \end{bmatrix} \cdot C = D \tag{7}$$

For the LFA-V specific mechanical structure, $A = A_L$.

In the following, c is a cosine and s is a sine function such that $c_3 \equiv \cos\psi$; $c_i^3 \equiv \cos(\gamma_i - \psi)$; and other notations follow. Also

$$\overline{V}_1^y = k_1 \sin(\theta_1)\cos(\gamma_1 - \psi) + r\cos(\gamma_1 - \psi);\ \overline{V}_1^x = k_1 \cos(\theta_1)/\cos(\gamma_1 - \psi)$$

$$\overline{V}_2^y = k_2 \sin(\theta_2)\cos(\gamma_2 - \psi) + r\cos(\gamma_1 - \psi);\ \overline{V}_2^x = -k_2 \cos(\theta_2)/\cos(\gamma_2 - \psi)$$

$$\theta_i = \operatorname{atan}(2l_i/w);\ k_i = \sqrt{\frac{w^2}{4} + l_i^2}$$

$$A_L = \begin{bmatrix} c_1^3 \cdot c_3 & -\frac{s_3}{c_1^3} - s_1^3 \cdot c_3 & c_2^3 \cdot c_3 & -\frac{s_3}{c_1^3} - s_1^3 \cdot c_3 \\ c_1^3 \cdot s_3 & \frac{c_3}{c_1^3} - s_1^3 \cdot s_3 & c_2^3 \cdot s_3 & \frac{c_3}{c_2^3} - s_2^3 \cdot s_3 \\ \overline{V}_1^y & -\overline{V}_1^x & \overline{V}_2^y & -\overline{V}_2^x \end{bmatrix};\ B = \begin{bmatrix} 1 & 0 & -1 & 0 \end{bmatrix}$$

Here r is the radius of the wheel, w is the wheel base of the vehicle and l_1 and l_2 are the wheel-to-frame distance along the direction of the actuator of the first and second wheel. We can control the posture and motion of the vehicle by defining F_Y, M_Z and F_X.

$$F_Y = k_v \dot{e}_h + k_p e_h + Wg \tag{8}$$

where $e_h = h_d - h$, and

$$M_Z = \overline{k}_v \dot{e}_\psi + \overline{k}_p e_\psi \tag{9}$$

where $e_\psi = \psi_d - \psi$

F_X can be set to any arbitrary positive value for a desired acceleration.

As the next level of development to include constraints (3)-(5) an alternate solution for $\{T_1, N_1, T_2, N_2\}$, which is the solution of the minimization problem

$\min(S)$, $S = T_1$ with the constraint equations $A \cdot C = D$ and inequalities (3)-(5) is chosen. If a feasible solution space does not exist to the minimization problem, then row 2 of (2) containing the equality constraint with F_Y can be dropped and the minimization problem can be set with the constraints:

$$\hat{A} \cdot C = \hat{D} \tag{10}$$

and constraints (3)-(5). Use of constraint (10) is possible for the LFA-V since it allows for dynamics at the main body in the vertical direction.

Figure 4 shows a plot of T_1 as the solution of (6) with constraints (10) and (3)-(5) using A_G. It reveals that there is no infeasible set of conditions in the scanned region for $\gamma_1 = 0$. Hence theoretically it is possible to traverse all terrain by using the modified constraint equation (10). The solution would result in dynamics of the main body in the vertical direction while ensuring the vehicle moves forward.

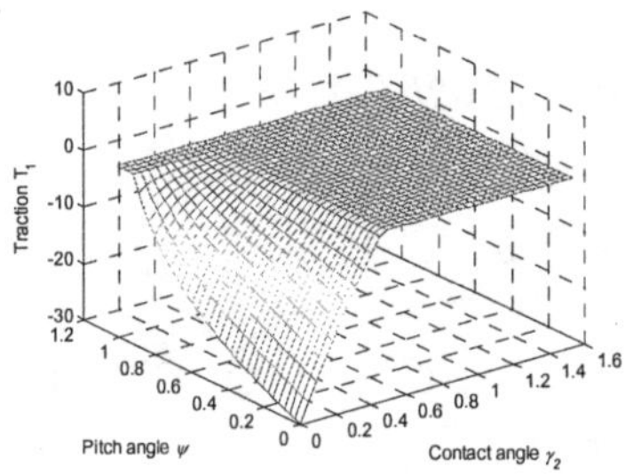

Figure 4. Plot of min(T_1) showing no regions of infeasibility

6. Simulation Results

The simulation was done using Matlab Simulink and MSC Visual Nastran. The surfaces used as terrain were higher-order cubic continuous surfaces such as B-splines.

The main body maintained its height h_d and pitch ψ_d and the surface variations showed little effect on the dynamics of the platform. Hence by the use of force control, the dynamics/posture of the main body and therefore the entire vehicle could be controlled. A minimum acceleration of 0.16 m/s^2 was needed to initially start the system which was commanded using F_X in (5). This is due to the faceted geometry of the rigid wheel used by the simulation package which contributes in an analogous way to rolling friction in the physical world. Maximum pitch angle deviation from zero was found to be –1.57° (Figure 5a).

The rms of the pitch angle, ψ through the motion on the surface was found to be 0.6727° as can be seen in Figure 5a. The performance was compared with a rigid suspension. The maximum pitch angle deviation from zero was found to be –5.73° in this case (Figure 5b) with rms value of 2.135°. Hence a large reduction in pitch angle variation is possible by using force control in the LFA-V. The spiked nature of the graph in Figure 5a is partly due to the faceted geometry of the wheel mentioned earlier which causes oscillation of the wheel centre. This is not visible in Figure 5b due to the scale of observation.

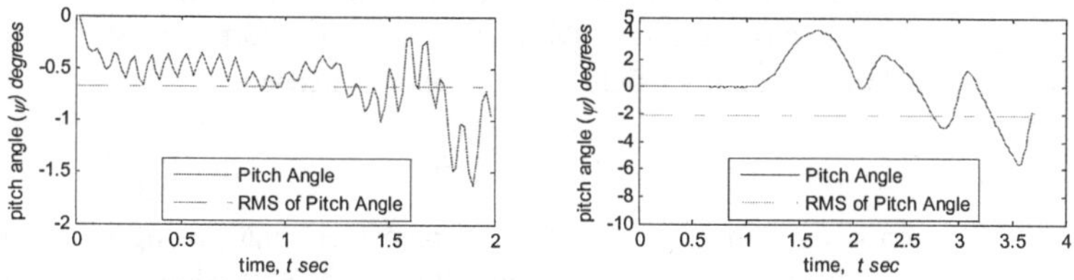

a. LFA-V with Force Control b. Rigid Suspension

Figure 5. Variation of pitch angle in the simulation

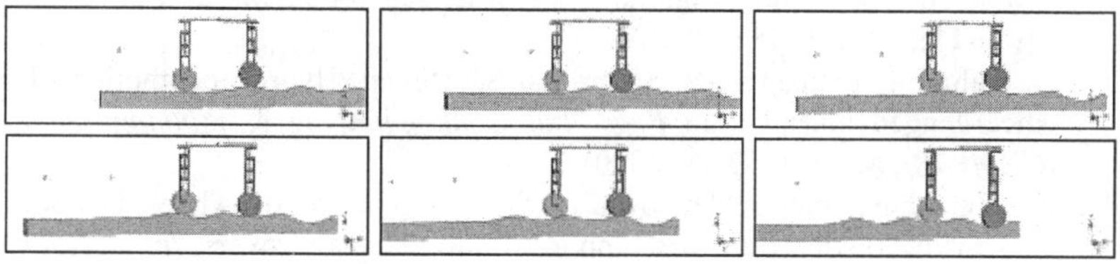

Figure 6. Time frames of the simulation with the LFA-V traversing an uneven surface*

The simulation time frames can be seen in Figure 6. Hence the force control methodology using the normal contact force and traction forces work satisfactorily to maintain posture and ensure the vehicle traverses the terrain. Thus an alternate solution to control the posture of a WAAV such as the LFA-V is provided, based on the contact forces at the contact points, which can also be utilized in terrain conditions difficult to traverse.

7. Conclusions and Future Work

It is formally understood that rigid or fixed suspension vehicles are unsuited for successful operation under certain terrain conditions. This brings out the primary motivation for using WAAVs to maximize traversability of the terrain.

---

* Simulation videos available at http://research.iiit.ac.in/~viswanath/robotics/Force_control.html.

Therefore a method based on controlling the contact forces responsible for causing the motion of a WAAV like the LFA-V is developed. This provides a strategy for successfully traversing a terrain and also controlling the posture of the vehicle.

As future work, we plan to build a physical implementation of a system utilizing the Force Control methodology. The force actuator element to generate the required contact forces for the physical implementation has been identified as a 2-dof leg with a wheel (leg-wheel) as the end effecter. Control for this leg-wheel has been formulated to generate required normal contact forces, which has been studied in past research. Also extension to deformable surfaces of contact, particularly for outdoor field environments is needed.

References

1. S. V. Sreenivasan and K. J. Waldron, Displacement analysis of an actively articulated wheeled vehicle configuration with extensions to motion planning on uneven terrain, *J. Mech. Des.* **118(2)**, 312–317 (1996).
2. Ch. Grand, F. BenAmar, F. Plumet and Ph. Bidaud, Stability and traction optimization of a reconfigurable wheel-legged robot, *Int. J. Robot. Res.* **23(10-11)**, 1041-1058 (2004).
3. A. Halme, I. Leppänen, S. Salmi and S. Ylönen, Hybrid locomotion of a wheel-legged machine, in *Proc. Int. Conf. Climbing & Walking Robots (CLAWAR)*, Madrid, Spain (2000).
4. K. Iagnemma and S. Dubowsky, Traction control of wheeled robotic vehicles in rough terrain with application to planetary rovers, *Int. J. Robot. Res.* **23(10-11)**, 1029-1040 (2004).
5. M. H. Hung, D. E. Grin and K. J. Waldron, Force distribution equations for general tree-structured robotic mechanisms with a mobile base, in *Proc. IEEE Int. Conf. Robot. Automat. (ICRA)*, Detroit, MI, USA, **4**, 2711-2716 (1999).
6. K. Iagnemma, A. Rzepniewski, S. Dubowsky and P. Schenker, Control of robotic vehicles with actively articulated suspensions in rough terrain, *Auton. Robot.* **14(1)**, 5-16 (2003).
7. Ch. Grand, F. BenAmar, F. Plumet and Ph. Bidaud, Decoupled control of posture and trajectory of the hybrid wheel-legged robot Hylos, in *Proc. IEEE Int. Conf. Robot. Automat. (ICRA)*, New Orleans, LA, **5**, 5111-5116 (2004).
8. J. T. Wen and S. Murphy, Stability analysis of position and force control for robot arms, *IEEE Trans. Automat. Contr.* **36(3)**, 365-371 (1991).
9. L. L. Whitcomb, S. Arimoto, T. Naniwa and F. Ozaki, Adaptive model-based hybrid control of geometrically constrained robot arms, *IEEE Trans. Robot. Automat.* **13(1)**, 105-116 (1997).

DESIGN AND CONSTRUCTION OF A ROPE CLIMBING ROBOT

Juan Pablo Martínez Esponda*

Robotics, Automation and Educational Technology Research Group
Department of Mechatronics
Tecnológico de Monterrey Campus Santa Fe, Carlos Lazo 100, Col. Santa Fe, Delegación Álvaro Obregón C.P.01389, Mexico City, Mexico

We present details of a rope climbing robot developed as an undergraduate project. The design was initially tested in a national level robotic competition but later modified and improved significantly. The current working model is almost completely autonomous. It can detect the rope automatically, closes its hands to grab the rope and can move up and down using ultrasonic sensors to detect the arrival of ceiling and floor. All the power sources are on-board and the system can sustain this load for a sufficient amount of time to make it a valuable prototype for further work and improvements.

1. Introduction

Intelligent robots for the household or industrial repetitive tasks have been under development for quite a long time with many commercial models now available off the shelf. For example there are robots to clean the external windows of sky rise buildings, robots to clean carpets in houses and offices, and robots to carry load from one place to the other in manufacturing plants, etc. A robot can replace many of the boring and tedious tasks disliked by human operators. The cleaning of the external windows in large buildings is one such area where considerable effort has been made to replace humans with robots. The risk to the lives of human workers can also be eliminated provided we can find robust automatic systems to replace humans in this type of work. In this paper we present a robot developed specifically to climb up a rope and descend automatically. The work was initiated in response to the Xth national mini-robot

* juanpas2k@yahoo.com.mx

competition held at Tecnologico de Monterrey – Campus Queretaro[†], in May 2005.

There were two categories in this competition in which one could participate:

1. The participant could manually position the rope inside the robot system and the robot was only supposed to move up and down the rope. The rope was about 11 mm in diameter (meant for professional hiking and climbing). In this case the maximum points were only 200, or

2. The participant could leave the robot at the extreme of a circle with rope at the center. The robot in this case had to be autonomous to move towards the rope, detect it, grab it, and finally ascend and descend it. The targeted height was 2 meters. Such robots had 200 extra points from the beginning and hence could go up to 400 points in total.

There were five participating groups in the year 2005 and the robot designed by the author was the only one with the capability to move towards, and detect the rope autonomously. However, the robot had some other problems at the time of the competition (which have now been resolved) and hence could not ascend the rope. The problems were identified as follows:

1. Improper distribution of weight of the robot.
2. Badly designed relay based control system.
3. Heavy relays, contributing to the total weight.
4. Lack of experience, both in electronics and programming.
5. Inability to cope with the rope thickness. It was not of a fixed diameter and had variations between 11mm to 16 mm.

The rethinking and redesigning needed to remove the above mentioned problems and critical areas, took almost a year. The big advantage though was a significantly increased knowledge of mechanics, analog electronics, digital electronics, microcontrollers and their programming, control, etc[‡].

[†] www.qro.itesm.mx

[‡] It is important to mention that the author was starting his second semester in engineering when he developed the first model and was still in the fourth semester when this paper was initially submitted. Hence no formal training in the area of engineering design and robotics was available. This is evident from the paper which is, for obvious reasons, scant in this area. The author will start taking these courses starting from next (spring 2007) semester.

The plan of the paper is as follows. In section 2 we review the current literature. Section 3 discusses our proposed design. We give our conclusions and recommendations in section 4.

2. Review of current literature

This mini review contains two types of systems. First are the robots designed to clean windows, walls, etc. These systems perform the same task as we ultimately intend to do with our robot, though use many different strategies rather than focusing on rope climbing. The second set of references discus automatic climbers of different types. Since the goals are partly common in both of these categories, we think such a review is justified, though not all the techniques of either group serve our purpose.

The window cleaning robot WICTOR [1] uses a rail to move the system across the window whereas the second subsystem cleans the window. The robot remains hanging during this process. Hence part of the problem of sticking to the window always without falling, is solved by hanging the robot thorough a rope. The Fraunhofer Institute for Manufacturing Engineering and Automation IPA in Stutgart, has designed various models for this purpose including Raccoon. It can start from one corner of the window and can autonomously move up and down and sideways to clean the whole window. It too uses suction pads for this purpose. When idle in its base camp, it can also recharge its batteries. Quirl [2] is a similar robot that uses suction pads to stick to the glass window. It is a lighter version (only 600 grams compared to 6.5 kilograms of original Raccoon) and has a footprint no larger than a postcard.

A lot of similar work in this area has been done by many other groups around the world. For example Ref [3] discusses the ROBOCLIMBER spider which can climb with weight on it. The applications of climbing robots in maritime industries have been explored in Ref [4]. References [5-6] discuss the effects of gravity and stability of climbing robot, respectively. Similar applications, for example in the area of window cleaning and wall climbing/inspection have been discussed in Ref [7] and Ref [8-9] respectively.

References [10-12] discuss the possibility of using a parallel robotic manipulator as a climbing robot. The interest in this case is not in cleaning windows but to develop a robot that can ascend complicated structures describing unknown spatial trajectories, such as palm trunks, tubes, bridges, etc. These papers suggest the use of a Stewart-Gough platform as a climbing robot with the capability of avoiding structural nodes.

3. The New Modified Design

Based on the problems encountered we have made the following changes and additions to the system:

1) Weight of the robot: The initial design was too heavy (with an improper distribution of weight) and hence had difficulty in moving up the rope. To rectify this, the chassis was perforated to the maximum extent to reduce about 400 gms of weight. The circuits were redesigned and built to reduce further 300 gms of weight. To distribute the weight correctly, some of the circuitry was relocated to make the system more homogeneous.

2) Ultrasonic sensors: To make the robot more robust and autonomous, ultrasonic sensors were added. The robot can now "see" the floor and the roof and can also detect transparent surfaces.

3) Control system: Initially, the control system used very heavy relays and needed a very high voltage to function properly. Hence this subsystem was not optimized. The original controller was replaced by a Polulu[§] system which is much smaller and controls the new motors. The voltage requirement goes down to 12 volts. The new relays are much smaller in size and need only 6 volts to operate and can transmit up to 30 volts at 10 amperes. The relays have been connected in series with diodes so that they switch ON only when the voltage crosses a certain level. The energy requirements go down to a level that can be managed by normal battery cells.

4) The original motor has also been replaced by a better 18000 rev/min motor which can generate sufficient torque with the use of a velocity reducer. The final velocity goes down to 300 rev/min. This motor helps the triangular part (Part 3 of Figure 1) to move up-

[§] http://www.pololu.com/

5) The new circuitry uses a Basic Stamp[**] 2p40 and hence need a regulator (Lm2940 in this case) so that it can work at the needed 5 volts. Problems with the serial communication between the micro and the control circuitry were removed to make it work correctly.

After corresponding adjustments, the weight of the current model of the robot is about 2.9 Kg including batteries. The angular velocity of the rotors is about 0.73 rev/sec without load. The load, however, decreases the velocity significantly to about 0.35 rev/sec.

Figures 1 and 2 show the back and front view of the robot and have been explained in detail in the figures section of the article. One of the major parts of the system, the pulley system which allows the system to grab the rope firmly in place and helps it in moving up and down the rope is visible in figure 2. Considerable amount design and construction time was needed to make the system perfect for this type of application. The completed system has been tested on many occasions including an on-campus demonstration.

4. Conclusions

In this paper we have presented and discussed the results of a project initiated by a local robotic competition. It demonstrates a concept that can be extended further and can be converted into a commercial product. The robot can only ascend and descend at the moment and needs additional sub-systems for cleaning purposes. With some modifications it can be used in various other applications. The list includes fire rescue work in high rise buildings, vertical mining, support in hiking/rope climbing with extra weight, maintenance of electric power lines, window cleaning and repair work, etc. Though the current system is limited in its capabilities we hope to do a lot of extension work in the current model in the future.

5. Acknowledgements

The author wishes to thank Tec de Monterrey – Santa Fe Campus for providing the resources needed to complete this work.

References

[**] http://www.parallax.com/

1. See, <http://www.iti.mu-luebeck.de/Research/NC/IES/CleaningRobots/>
2. See <www.fraunhofer.de/fhg/Images/magazine1-2005-34f_tcm6-14061.pdf>
3. R. Molfino, M. Armada, F. Cepolina, M. Zoppi, "ROBOCLIMBER the 3 ton spider", Industrial Robot. IF: Vol. 32 (2), 2005.
4. M. Armada, M. Prieto, T. Akinfiev, R. Fernández, P. González de Santos, E. García, H. Montes, S. Nabulsi, R. Ponticelli, J. Sarria, J. Estremera, S. Ros, J. Grieco, G. Fernández, "On the Design and Development of Climbing and Walking Robots for the Maritime Industries", Journal of Maritime Research Vol. 1 (2) April 2005
5. T. Akinfiev (;) M. Armada, "The influence of Gravity on Trajectory Planning for Climbing Robots with Non-rigid Legs", Journal of Intelligent and Robotic Systems, 35 (3), pp. 309-326. Kluwer, 2002. ISSN 0921-0296.
6. T. Akinfiev, M. Armada, M. Prieto, M. Uquillas, "Concerning a Technique for Increasing Stability of Climbing Robots", Journal of Intelligent and Robotic Systems, vol 27, 2000, pp. 195-209.
7. Houxiang Zhang; Jianwei Zhang; Guanghua Zong; Wei Wang; Rong Liu, "Sky Cleaner 3: a real pneumatic climbing robot for glass-wall cleaning," Robotics & Automation Magazine, IEEE, Volume 13, Issue 1, March 2006 Page(s):32 – 41.
8. Longo, D.; Muscato, G., "The Alicia/sup 3/ climbing robot: a three-module robot for automatic wall inspection," Robotics & Automation Magazine, IEEE, Volume 13, Issue 1, March 2006 Page(s):42 – 50.
9. Lal Tummala, R.; Mukherjee, R.; Ning Xi; Aslam, D.; Dulimarta, H.; Jizhong Xiao; Minor, M.; Dang, G., "Climbing the walls [robots]," Robotics & Automation Magazine, IEEE, Volume 9, Issue 4, Dec. 2002 Page(s):10 – 19.
10. Roque Saltaren, Rafael Aracil, Oscar Reinoso and Maria Antonieta Scarano, "Climbing Parallel Robot: A Computational and Experimental Study of its Performance around Structural Nodes," IEEE Transactions on Robotics, VOL. 21, NO. 6, DECEMBER 2005.
11. M. Almonacid, R. J. Saltarén, R. Aracil, and O. Reinoso, "Motion Planning of a Climbing Parallel Robot," IEEE Transactions on Robotics and Automation, Vol. 19, No. 3, June 2003 485.
12. Aracil, R.; Saltaren, R.J.; Reinoso, O., "Climbing parallel robot: a robot to climb along tubular and metallic structures," Robotics & Automation Magazine, IEEE, Volume 13, Issue 1, March 2006, Page(s):16 – 22.

Figures

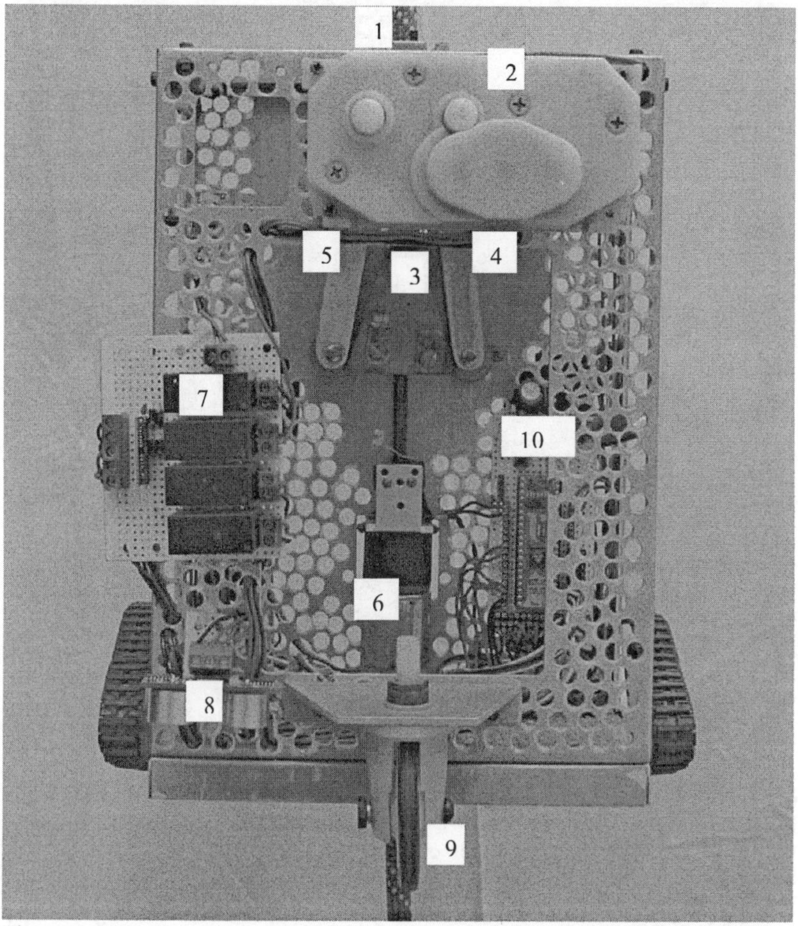

Figure 1: The robot moves up and down the rope (1) using the motor (2). The triangular part (3) moves up and down via the motor (6) and helps opening and closing the hands (4) and (5) which have a strong spring attached on the upper side to return to the original (closed) position. (7) is the part of the control circuitry whereas the sensors to detect the ground are shown in (8). The back wheel (9) is fixed in one direction though in principle it can be adjusted in any desired direction. The BasicStamp is visible in (10).

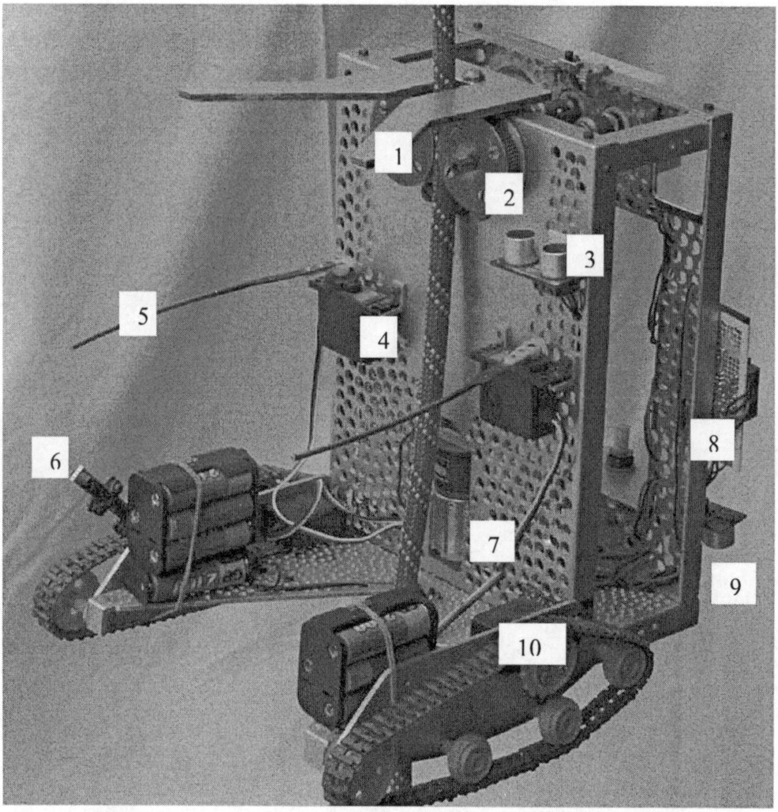

Figure 2: This is the final form of the robot. The rope gets directed towards the pulleys (2) using the system shown on the top (1). The sensors (3) detect the arrival of the floor or some other obstacle while the robot is ascending. The servos (4) control the folding hands (5) (shown opened here). These help the rope stay in its position and start closing a few seconds after detecting the rope so that the robot has moved further in the direction of the rope and the rope is now in its position (1). This time has been measured and programmed according to the servo speed. The sensor (6) shown as a small extended hand detects the rope. The motor to open and close the rollers (2) is shown in (7), though it is hardly visible. The electronic circuitry for the same is behind (8). The sensors (9) detect the arrival of the floor or any other obstacle. The orange roller (10) helps the system move forward and backward. The other four rollers are there to facilitate the movement and maintain tension in the belt. This belt helps the system avoid slipping.

DESIGN OF NEW LEG MECHANISM FOR A WHEELED WALL CLIMBING ROBOT *

YI-LI FU, HE-JIN YANG

Mechanical Engineering School, Robotics Institute, Harbin Institute of Technology Harbin, 150001, China

High mobility and strong surface adaptability are two basic requirements for a counter-terrorism wall climbing robot. However, traditional wall climbing robots fails to meet both of the two requirements. This paper designed a new leg mechanism and by fixing it onto the front of a wheeled wall climbing robot, obtained a novel wall climbing robot which can move quickly on both smooth and rough planar surfaces, make transitions between them and stride obstacles when necessary. Various aspects about the leg mechanism's design are introduced, including the design requirements, mechanical design and gait design .Simulation and experiment results demonstrate the validity of the design concept.

1. Introduction

In recent years, there have been strong demands for counter-terrorism wall climbing robots to perform reconnaissance and surveillance missions as terroristic activities increase. Because we never know when and where the crime is going to happen but have to respond to it quickly, we need the robot should be of high mobility and strong surface adaptability, which offers big challenges to the locomotion mechanism design. Traditional wall climbing robots, no matter what applications they were designed for, can be divided into two major categories according to their locomotion mechanism: continuous motion robots [1--3] and alternate motion robots [1,4--7]. The continuous motion robots have high mobility but poor surfaces adaptability because they usually move on wheels. On the contrary, the alternate ones can be adapted to irregular surfaces but very slow, for the simple reason that they use their legs or other adhesion mechanisms in turn. Therefore, traditional wall climbing robots are not completely satisfactory for counter-terrorism application.

This paper proposes a new concept wall climbing robot different from traditional ones: It combined a wheeled single suction cup wall climbing robot with an innovatively designed leg mechanism, created a robot that can move

* This work is supported by the National Natural Science Foundation of China (60675051).

both continuously and alternately, i.e. the robot uses its wheels on planar wall surfaces with a fast speed, when comes across on obstacle, the leg mechanism will be used to stride it. We envision that this robot be used in urban environments for surveillance purpose. This proposal is reasonable because most of the building surfaces are flat except for some obstacles such as gaps, window frames and/or steps. The main job of this paper is to introduce the design of the leg mechanism.

This paper is organized as follows: Section 2 gives general requirements to the leg mechanism design. In section 3, detailed descriptions about the mechanical design are introduced. Section 4 illustrates how the robot strides obstacles and makes surface transitions. In section 5, we present the experiment and simulation results. Conclusions with future works are drawn in section 6.

2. Design Requirements

As mentioned in section 1, mobility and surface adaptability is two essential requirements to meet. Our strategy is to employ an existed wheeled robot and design a leg mechanism for it, so the mobility will not be a problem. Considering the wheeled robot's payload and that the new robot will be self-contained, controlled wirelessly, with half-autonomous intelligence, the requirement for the leg mechanism design are:

- Be as an independent module
- Be able to work on both smooth and rough surfaces
- Strong striding obstacle ability
- Can make transitions between walls
- Light
- Simple structure
- Energy saving
- Easy to control

3. Mechanical Design

Great efforts have been made to meet all the design requirements. Figure 1 shows the final model of the new robot. It consists of two parts: the body and the leg mechanism. The body is actually a typical wheeled single suction cup robot as we mentioned earlier. The locomotion and adhesion mechanisms are similar to the robot in Ref.[2].

The leg mechanism is designed depending on the size and weight of the body listed in Table 1. As an independent module, it did little changes to the body. The leg mechanism is made up of 3 links and 3 joints. In robotics theory, it is a 3-DOF planar manipulators thus is very easy to control. The specification of the leg mechanism is also listed in Table 1.

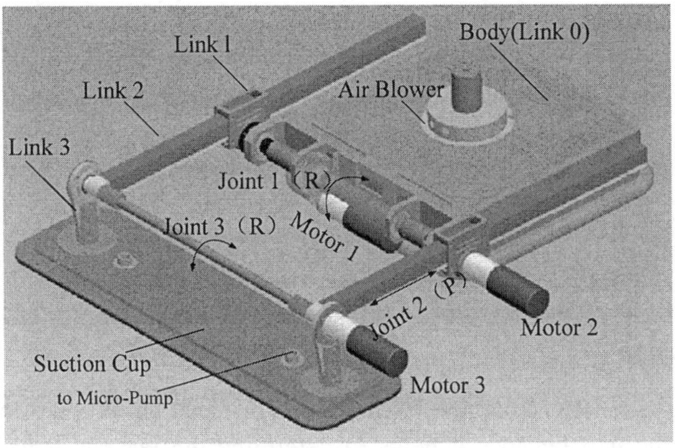

Figure1. A CAD model of the new robot with the leg mechanism

Table 1.specifications of the body and the leg mechanism

Body	Size	300mm*300mm*85mm
	Weight	5kg
	Payload	5kg
	Speed	0—10m/s
Leg	Link 1	64mm (Length)
	Link 2	500mm (Length)
	Link 3	80mm (Length)
	Suction Cup	500mm*150mm*10mm
	Motor 1	10rpm, 6.8Nm , 317g
	Motor 2	50rpm, 1.7Nm, 263g
	Motor 3	7.5rpm, 7.6Nm, 317g
	Micro-Pump	2W,5L/min,130g
	Weight	3.8kg

The suction cup located at the end of the manipulator, which is part of link 3, is special designed for rough surfaces. As shown in Fig.2, when the suction cup is covered on a rough wall, two types of apertures can be shaped: open aperture and closed aperture. Because of the suction cup's large size and wide edge, most apertures belong to the closed ones. However, some open apertures still exist. It is the flexible rubber strip that crams them so that a vacuum chamber can be formed. If the apertures are too large or too deep to be crammed, they could be treated as gaps and will be bypassed or stridden.

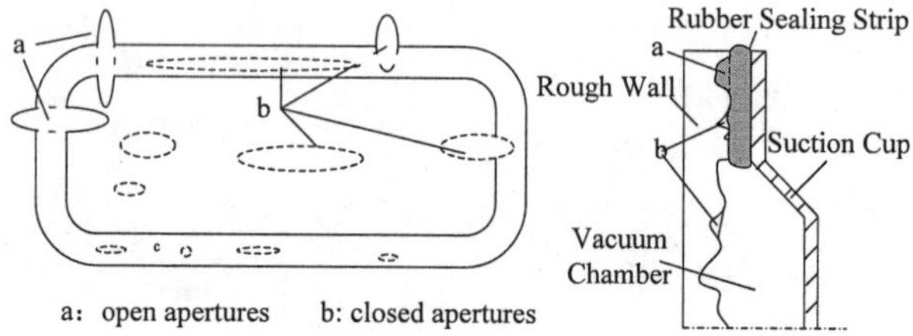

Figure2. Illustration of the suction cup's sealing mechanism

4. Gait Design

"Gait" refers to how a legged robot moves its legs. It is very important to investigate the gait because it helps to determine the leg's structure and size, select joints' driving motors and to know the obstacle dimensions that this robot can deal with. The gait was designed following two principles:

1. Safe: the mass center of the robot should be close to the wall so that the robot would not fall down under the gravity;
2. Less power consumption: less power consumption means longer working time and lighter batteries, which is very important to a self-contained wall climbing robot.

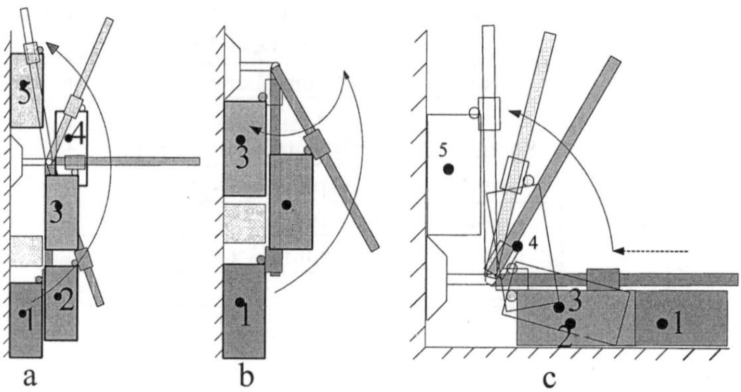

Figure3. The process of striding a obstacle and making transitions

Figure 3(a) illustrates a typical process of striding a step or a gap. If the distance between the small suction cup and obstacle is longer than the length of body, the process could be simplified as shown in Fig.3 (b).

Figure 3(a) illustrates a way of making wall to wall transitions. It is similar to the process of striding obstacles but the joints' driving torque is much smaller. This process could also be regarded as ground-wall, wall-roof, wall-ceiling transitions. But the wall-ceiling transition is not realized because the body is too heavy.

The prismatic joint in the leg mechanism offers the robot powerful abilities to stride obstacles. Figure 4 illustrates several different situations and the admissible conditions. The numbers are based on the robot size in Table 1.

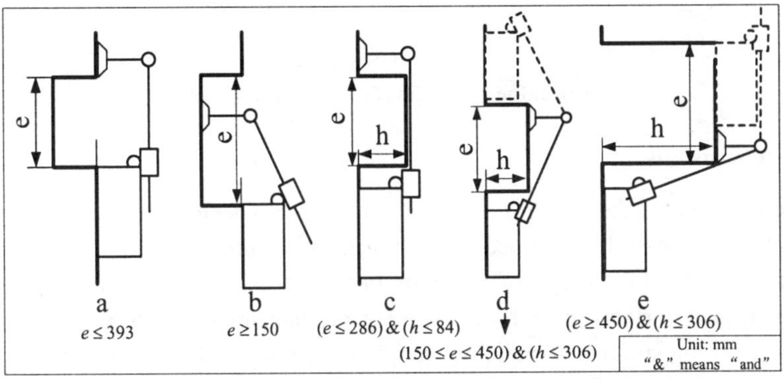

Figure4. Illustration of striding obstacles and admissible conditions

5. Simulations and Experiments

Both kinematic and dynamic simulations were made to assist the leg mechanism's design and to verify the design concept. The simulation model we built is much identical with the real robot. Here we present one of those results as shown in Fig.5, which records the distance between each link's mass center and the wall surface during the process illustrated in Fig.3 (a).The result indicates the succeed of the planned path and the satisfaction to the safety rule in Section 3.

Some simple experiments were made to test the adhesion ability of the suction cup designed in Section 3.We used the micro-pump in Table 1 to produce vacuum and got a maximum negative pressure of -38kpa on glass wall, and -32kpa on brick wall. According to the size of the suction cup, a minimum of 2595N force and a 194 Nm moment could be produced in theory. In fact, a 15kg object was hung at joint 3 and the suction cup never slipped or fell down. The results indicate that the suction cup can adhere to both smooth and rough walls reliably.

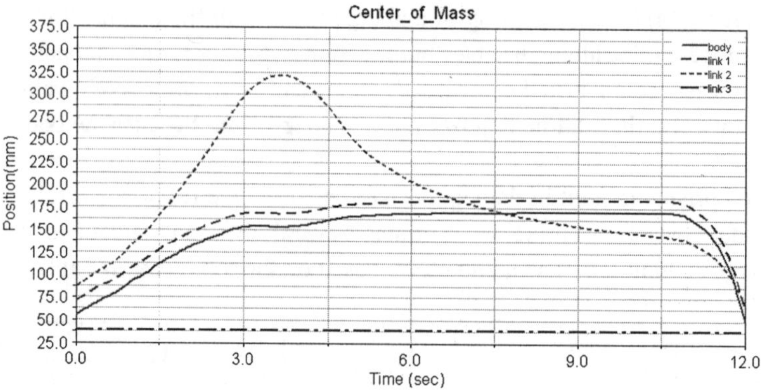

Figure5. The trajectories of each link's mass center

6. Conclusions

A new concept robot which is a combination of the wheeled wall climbing robot and the successfully designed leg mechanism is proposed in this paper. It overcomes the limitations of previous ones and surpasses them in terms of mobility and surface adaptability. The leg mechanism succeeds in many ways such as strong surface adaptability, simple structure, modularity and so on. The future work is to design an embedded controller board and integrate the electrical system with mechanical design to realize the surveillance purpose for counter-terrorism missions.

References

1. A. Nishi, *Development of wall-climbing robots. Computers Elect. Eng. 22(2): 123-149* (1996).
2. Y. Wang, S.L. Liu, *et al, Development & Application of Wall-Climbing Robots. Proc.IEEE Int. Conf. Robotics & Automation.* (1999).
3. J.Z.Xiao, A. Sadegh, *et al. Design of Mobile Robots with Wall Climbing Capability. Proc.IEEE/ASME Int.Conf. Adv. Intel. Mechatronics* (2005).
4. B. Luk, A. Collie and J. Billingsley, ROBUG II: An intelligent wall climbing robot, *Proc. Int. Conf. Robotics and Automation*, (1991)
5. A. Nagakubo and S. Hirose, *Walking and running of the quadruped wall-climbing robot. Proc. IEEE Int. Conf. robotics and Automation.* (1994).
6. H. R. Choi and S. M. Ryew, *et al. A Wall Climbing Robot with Closed Link Mechanism.Proc. IEEE/RSJ,Int.Conf.Intelligent Robots and Systems.*(2001)
7. R. L. Tummala, *et al. Climbing the Walls. IEEE Robotics and Automation Magazine. pp 10-19, Vol. 9,No. 4* (2002).

DEVELOPMENT OF A CLIMBING ROBOT FOR INSPECTION OF LONG WELD LINES

JIANZHONG SHANG, BRYAN BRIDGE, TARIQ P. SATTAR, SHYAMAL MONDAL, ALINA BRENNER

Centre for Automated and Robotic NDT, London South Bank University, 103 Borough Road, London, SE1 0AA, UK

This paper presents a wheeled robot designed for climbing on ferrous surfaces for the inspection of real time inspection of long weld lines simultaneously with the welding process. Neodymium permanent magnets are used for adhesion, which are capable of producing a maximum adhesion pressure of $4x10^4Nm^{-2}$ at a 20mm air gap. The strong neodymium magnets make the robot have a high payload carrying capability. The arrangement of the magnet array increases its performance at large air gaps so that the robot has excellent capability to overcome obstacles, such as weld caps. The design of the wheeled robot with two sections jointed by a hinge joint takes the advantages of high speed and good manoeuvrability, as well as working on curved surfaces and transferring between angled surfaces.

1. Introduction

1.1. *Objectives of the project*

The project Climbing Robot Cell for Fast and Flexible Manufacture of Large Scale Structures (CROCELLS) seeks to modernise and take into the future the technology of the manufacture of large fixed welded structures such as box girder bridges, storage tanks, ships and other steel fabrications which arise on construction sites, in large chemical and foodstuff plants, on offshore oil platforms etc. It creates a transportable manufacturing cell consisting of a team of cooperating climbing robot work tools whose activities are coordinated and integrated through a central intelligence. Each robot will be dedicated to a different task to optimise overall system performance[1].

The system in this project consists of three cooperated robots, a welding robot holding a welding torch and its control unit, a utility robot that carries the weld wire spool and navigational sensors and an NDT robot for carrying weld inspection tools. As part of the project, the NDT robot for weld inspection is developed in the Centre for Automated and Robotic NDT, London South Bank University.

The main purpose of the NDT robot is the real time inspection of long weld lines with 100% volume coverage, simultaneously with the welding process. Such NDT data is then be used to control the welding process parameters including authorisation of weld repairs in real time. Some examples of the real work environment, assembly of Blocks and/or Mega Blocks to form the ship, ship hull of container ships, are shown in Figure 1.

(a)

(b)

Figure 1 Examples of work environments

1.2. *General Requirements of the NDT robot*

The requirements of the robot mainly include:

- The robot should be able to climb on steel walls up to 30m height.
- Climbing performance on curved surfaces where the minimum diameter is down to 2.5m.
- System weight for an individual robot platform, including the robot weight and payload, is required to be less than 70kg.

- To surmount surface obstacles, such as ridges on ship hulls, of up to 20mm height. Higher obstacles, for example stiffeners and nozzles will be circumnavigated.
- To transfer between angled surfaces (135 degree) such as those which arise on container ships.
- To move between stiffeners of 700mm separation such as those on container ships, therefore the width limitation of the robot with NDT probes mounted is 600mm.
- The robot is required to move at constant speed to follow the welding process as well as move at high speed when locating itself to different places to save time. The welding speed for ship hulls and storage tanks are about 10 to 15m per pass per hour.

1.3. *Current Technology and the Challenge*

There are a number of ways for the adhesion of a wall-climbing robot, such as vacuum suction cups, magnets, sticky materials, etc. For a steel-wall-climbing robot, the obvious adhesion technology is magnetic force. A great number of robots with magnet adhesion have been developed by researchers in different organisations. They are commonly featured as robots with magnet tracks[2,3,4], robots with magnetic wheels/disks[5] or legged robots with magnet feet[6,7], etc. However, none of them can meet the requirements of the particular application in this project. A new robot has been designed and the real challenge of the robot development is the capability that the magnets work over large air gaps for the purpose of overcoming obstacles and working on curved surfaces.

2. Design of the NDT Robot

In order to achieve the smooth and continuous movement, as well as excellent manoeuvrability of the robot, a wheeled robot with magnet adhesion is developed.

2.1. *Magnets investigation*

The payload of the NDT robot is approximately 10kg. To surmount the obstacle of 20mm height, the minimum air gap between the magnets and the surface is 20mm, which requires bigger magnets to generate the sufficient holding force over the air gap. However, big magnets will increase the robot weight. In consequence, the payload carrying capability is decreased. The trade-off of the size, weight and magnetic force are considered carefully.

A comparison of a 1 by 4 magnet array with different gap between the magnets is illustrated in Figure 2. The attraction force is between the magnet array and a big steel plate with 10mm thickness. The result shows that when the magnets are closer to each other, more attraction force is created at further distance, and the force variation from tested distance (5mm to 35mm) is less, which is our desired result.

From Figure 2, we also see that the use of a back plate significantly increases the attraction force, especially at further distance. The use of mild steel back plate has the following advantages:

- Flux concentration. With the steel back plate, more magnetic lines are concentrated between the back plate and the steel wall, so that the attraction force is increased.
- Reduce the magnetic field strength in the back, so that there is less interference to the electronic equipments carried on the robot, which makes the shielding of the magnetic field easy.
- Ease the management of the magnets. The adhesion force can fix the magnets on the steel plate so that there is no additional fixing mechanism required.

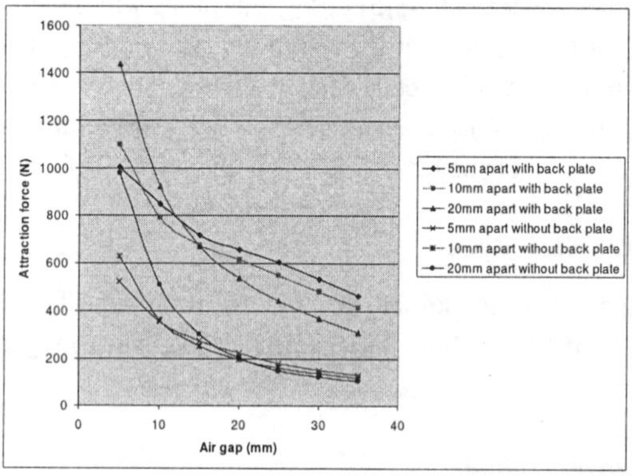

Figure 2 Comparison of attraction force at different magnets arrangement

Based on the analysis above, two blocks of magnet array are built as shown in Figure 3. Instead of using one single big block magnet, which is extremely difficult to handle, a number of smaller block magnets are placed on a mild steel plate to form a magnet array that covers big area. Each of the magnets is made of neodymium 42 and has the dimension of 50mm x 50mm with 25mm thickness.

In each magnet array, the magnets are placed against each other as close as possible, so that more magnetic lines are forced to the working surface.

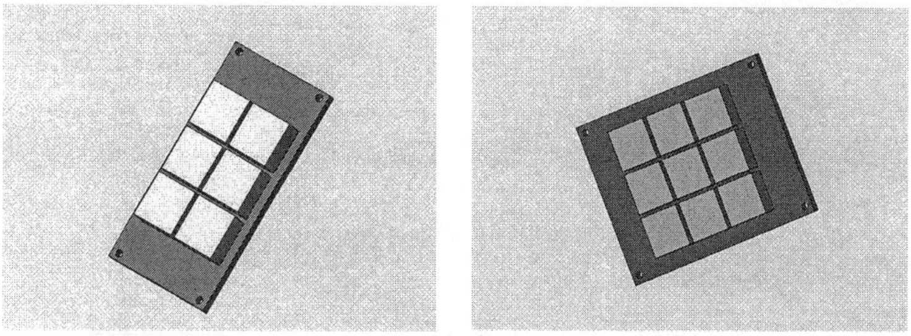

Figure 3 Magnet arrays with neodymium block magnets on mild steel plates

The magnetic force of the designed magnet array against the increasing of the air gap is shown in Figure 4.

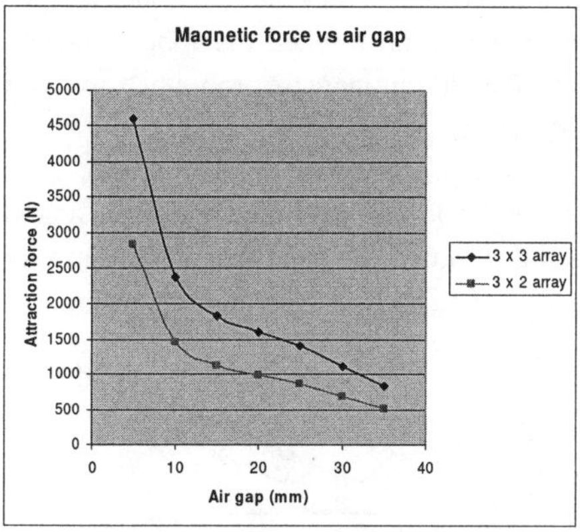

Figure 4 Attraction force changing against air gap

2.2. *Robot Structure*

The NDT robot is designed to have two sections connected by a hinge joint, with two wheels to drive the robot, and two omniwheels, one in the front and one in the back, to support the robot. The driving wheels are driven by two servo motors differentially, so that the robot can be turned easily and driven along any direction. The structure of the robot is shown in Figure 5.

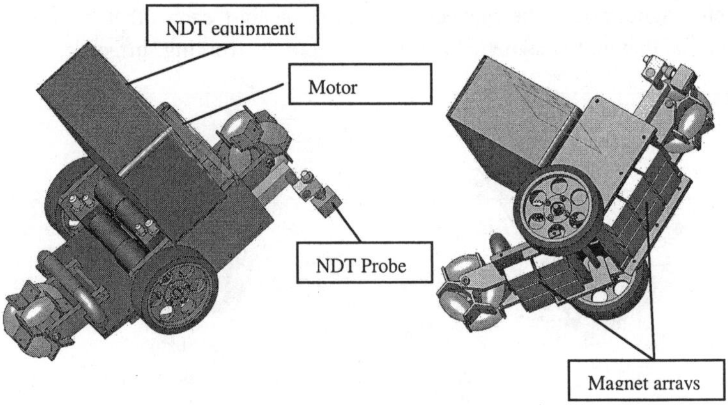

Figure 5 Robot structure

The two-section design is concerned for the robot travelling through the sharp angled corners presented in ship hulls. In the design without the hinged structure as shown in Figure 6(a), when the robot transfer from one plane to another at the corner, the air gap increases too much so that the magnet force is too low to support the robot.

With the hinged design as shown in Figure 6(b), when the front half of magnet is lifted up, the back half still remains strong holding force. When the driving wheels reach the corner, the front magnets resume strong holding force, then the back magnet is lifted up to complete the transfer.

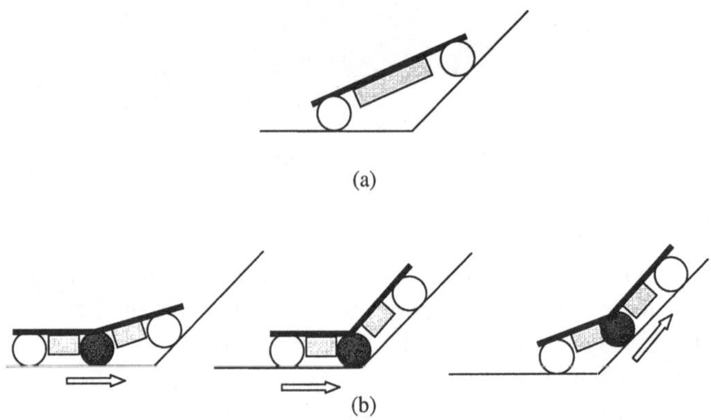

Figure 6 Illustration of the hinged structure that can transfer between angled surfaces

The driving wheels are made of aluminium hubs bonded with solid 65 IRHD polyurethane tyres. The polyurethane material has the coefficient of friction of 0.9 on steel walls. The key benefits of the material are

- Abrasion Resistance (Non-Marking)
- Impact Resistance
- Cut Resistance
- Heat and cold resistance
- Load Bearing
- Long Working Life
- High Grip-Good Drive Material
- Oxygen & Ozone Resistant

The robot is featured as follows:

- Size: 600mm x 375mm x 340mm
- Weight: 30kg (Including payload)
- Speed: 0-10m/min
- Minimum gap between magnets and work surface (flat): 20mm

2.3. *Robot Control*

The NDT robot control unit is developed as illustrated in Figure 7.

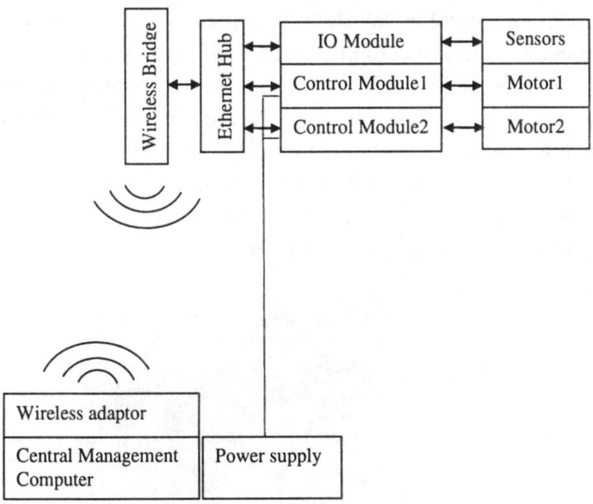

Figure 7 Robot control unit

The on-board robot controllers include two motor control modules and one interface conversion module. All the modules have Ethernet interface, so that they can be controlled via Ethernet network. The interface conversion module converts Ethernet to serial, IO, AD and I2C interfaces, allows the connection of different sensors and equipments, enables monitor or control them via standard TCP-IP protocol, makes the combination with other equipments easy.

All the on-board modules with the standard Ethernet protocols are plugged to an Ethernet hub which is carried on the robot. The uplink of the hub is connected to the wireless bridge for the wireless communication with the central task manage computer. The wireless communication is based on the Wi-Fi technology. The application of Wi-Fi wireless communication eliminates the communication cable, so that the robot umbilical is reduced. The only umbilical for the robot is the power supply cable, which cannot be eliminated at this stage due to the robot is required to work for long time.

2.4. *Robot Guidance and cooperation with the welding robot*

For the robot travelling along the stiffeners such as those which arise on container ships, there are two infrared distance sensors facing side way, one in the front and one in the back of the robot. The sensors measure the parallelness and distance between the robot and the stiffener. According to the sensor feedback, the controller calculates the speed of the two driving wheels to keep the robot moving parallel to the stiffener and maintain a constant distance.

In the case of hot weld inspection, the robot needs to follow the welding robot. A thermal array sensor, as in Figure 8 is used together with the infrared distance sensors to follow the hot welding point. The thermal array sensor has an array of eight thermopiles arranged in a row. It can measure the temperature of 8 adjacent points simultaneously. The sensor reads infra-red in the 2um to 22um range, which is the radiant heat wavelength.

Figure 8 Thermal array sensor for guiding the robot

To follow the hot welding point, the NDT robot is placed next to the welding robot as shown in Figure 9, with a reflective object raised on the welding robot so that the NDT robot can 'see' the welding robot. The infrared distance sensors read the distance from the welding robot and the central controller keeps the NDT robot parallel to the welding robot and remains constant distance. The thermal array sensor is placed with its row of the thermopiles parallel to the weld. By controlling the robot so that the hottest point always stays at one particular sensor element, the robot will always follow the hot point where the welding torch points to.

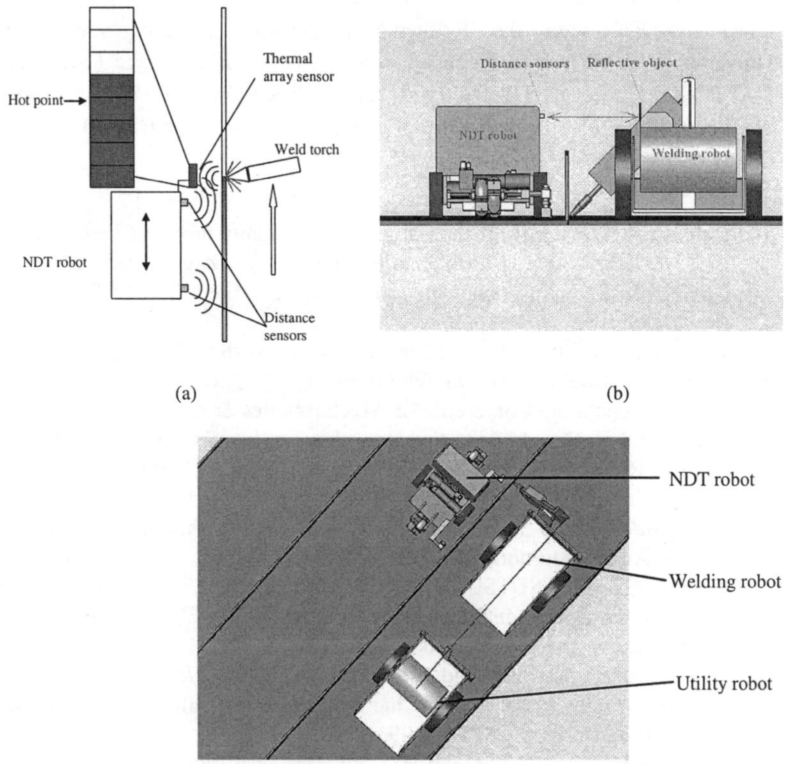

Figure 9 The NDT robot follows the welding robot for hot weld inspection

3. Conclusion

The NDT robot has been developed with the capability of climbing on steel walls carrying the specified payload and the capability of overcome the obstacles. The robot is also able to climb on curved surfaces with excellent manoeuvrability, and transfer between angled surfaces.

Acknowledgments

This work is sponsored by the European Community through the project Climbing Robot Cell for Fast and Flexible Manufacture of Large Scale Structures (CROCELLS), 6th Framework Programme, project No. STRP017509[1] with the following partnership: London South Bank University(UK), Cybernetix(France), Zenon(Greece), Bisiach and Carrù(Italy), NAFTOSOL(Greece), TERNA and GEK Group(Greece). The project is coordinated by Professor Bryan Bridge of London South Bank University.

References

1. Deliverable D2 report for the European Community project *Climbing Robot Cell for Fast and Flexible Manufacture of Large Scale Structures* (CROCELLS), 6th Framework Programme, project No. STRP017509. (Restricted access to the partnership and the EC only).
2. W. Shen, J. Gu and Y. Shen, *Proposed Wall Climbing Robot with Permanent Magnetic Tracks for Inspecting Oil Tanks*, Proceedings of the IEEE, International Conference on **Mechatronics & Automation** (2005).
3. Z. Xu, and P. Ma, *A wall-climbing robot for labelling scale of oil tank's volume*, **Robotica**, ISSN:0263-5747, Pages: 209 – 212 (2002).
4. L. P. Kalra and J. Gu, *An autonomous self contained wall climbing robot for non-destructive inspection of above-ground storage tanks*, **Industrial Robot: An International Journal**, Vol. 34, No. 2, pp. 122-127 (2007).
5. S. HIROSE, H. TSUTSUMITAKE, *Disk Rover: A Wall-Climbing Robot Using Permanent Magnet Disks,* Proceedings of the IEEE/RSJ International Conference on **Intelligent Robots and Systems** (1992).
6. B. L. Luk, A. A. Collie and J. Billingsley, *ROBUGII: An intelligent wall climbing robot*, Proceedings of the International Conferance on **Robotics and Automation**, pp. 2342-7 (1991).
7. T. Kang, H. Kim, T. Son and H. Choi, *Design of quadruped walking and climbing robot*, IEEE/RSJ International Conference on **Intelligent Robots and Systems**, Vol. 1, pp. 619-24 (2003).

DEVELOPMENT OF A SEALING SYSTEM FOR A CLIMBING ROBOT WITH NEGATIVE PRESSURE ADHESION

C. HILLENBRAND*, D. SCHMIDT and K. BERNS

Robotics Research Lab,
Department of Computer Science,
University of Kaiserslautern,
67653 Kaiserslautern, Germany
**E-mail: cahillen@informatik.uni-kl.de*
http://agrosy.informatik.uni-kl.de

T. LEICHNER, T. GASTAUER and B. SAUER

Institute of Mechanical Engineering and Gear Technology,
Department of Mechanical and Process Engineering,
University of Kaiserslautern,
67653 Kaiserslautern, Germany
http://megt.mv.uni-kl.de

The non-destructive inspection of large concrete walls is still an unsolved problem. One possible technique is to use driven wheels for the propulsion and a vacuum system for the adhesion. The seals for the vacuum chambers are slipping over the rough surface, therefore it is not guaranteed that the chambers are always airproof. Especially over concrete walls a special seal construction must be found to make the adhesion more safe. On the other side the propulsion system must be able to produce enough force for carrying and accelerating the robot to a suitable velocity.

This paper will present the climbing robot CROMSCI which uses the described techniques. The propulsion system consists of three omni directional driven wheels which are airproof and completely rotatable and has been presented in earlier papers before. For adhesion a vacuum system of seven controllable vacuum chambers and one reservoir chamber is used. This system including chambers and seals will be discussed in more detail. The rough and sharp-edged surface of concrete walls cause strong requirements to the sealing concerning leak tightness and attrition. Therefore, each sealing must be flexible to allow a good adaption to the ground but also let the robot slip when the wheels are turning.

Keywords: Climbing Robot, Adhesion System, Negative Pressure, Sealings.

1. Introduction

Regular inspections and repairings of concrete buildings like motorway bridges and dams are very extensive. Among high expenses in money and time the technical stuff has to work in hazardous environment by using complex access devices. For a better objectivity and reproducibility of inspections as well as for safe working conditions a climbing robot has been build up which can check the building remote controlled. During different experiments with the prototypes the importance of sealings became obvious. Although we got promising results onto smooth concrete or wooden surfaces the sealing failed on rough surfaces which can be found on every concrete building. For this the existing adhesion concept was overhauled and new sealing materials and constructions were developed.

In literature other climbing robots using negative pressure adhesion can be found, although not all of them can be used for this application [1] [2] [3] [4]. Some of them are suitable only for flat surfaces like glass or use legs for locomotion, which will result in slow movement. For climbing on concrete surfaces an active vacuum system driven by wheels seems to be the best solution because of fast continuous motion and a simple mechanical structure.

2. Adhesion Concept

The adhesion concept of our climbing robot CROMSCI[a] consists of seven single vacuum chambers which are supported by one large reservoir chamber at the top of the robot. Each chamber receives its negative pressure from the reservoir chamber which is evacuated by three suction engines and can be controlled by valves. If one or more working chambers are loosing negative pressure they can be isolated from the vacuum system to avoid the propagation of normal pressure to the other chambers which will result in loss of adhesion. The air-pressure in each chamber and the reservoir is measured by pressure sensors. These informations are given to a pressure controller which opens and closes the valves to evacuate the chambers separately depending on the actual leak tightness. The components of the adhesion system are:

- one large reservoir chamber (including valve to outside)
- three suction engines evacuating the reservoir chamber

[a]Climbing RObot with Multiple Sucking Chambers for Inspection tasks

- seven working chambers (each including valve, pressure sensor and sealing)
- electronics for low-pressure control

Beside the real adhesion mechanism a thermodynamical model has been created to simulate the airflow and pressure variation with modelled leakage areas and cracks. It shows that the usage of multiple sucking chambers are as important as a good driving strategy if the cracks are too large for the sealing. More details on the vacuum control system can be found in [5].

2.1. *New Sealing Design*

An important aspect is the concept of the sealings between the negative pressure inside the vacuum chambers and the normal pressure outside. To guarantee that the climbing-robot keeps the contact to the wall while moving on the surface, it is neccessary to develop a seal which needs to be wear-resistant, leak-proof and easy sliding. To reach this intention a prototype demonstrator has been developed which enables to apply several different sealing materials and shapes to find the optimum in friction and wear.

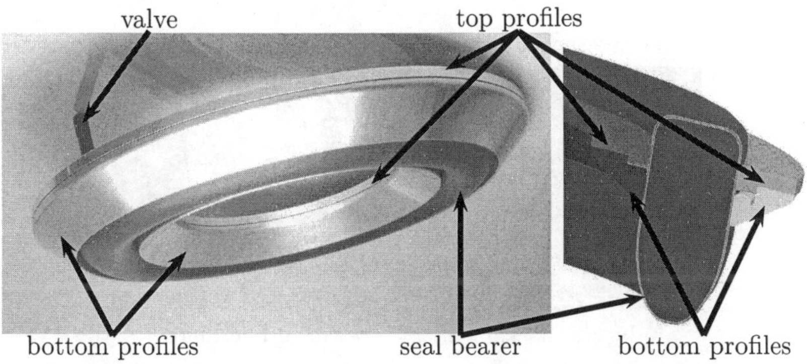

Fig. 1. Prototype and drawing of the basic construction

It is an assembly of one seal-bearer and two aluminium profiles, in which the bearer is clipped (fig. 1). It is possible to vary the height of the seal-bearer by changing the air pressure in the tube made of Butyl. Onto the surface of the tube the different samples of seal material and friction can be applied.

Another important point with regard to practical application is an easy changing of the seals if they become porous due to abrasion. With the shown sealing this replacement can be done fast and simple.

2.2. *Test Stand*

To detect the lift/drag ratio and the value of friction between the sample and the surface vis-a-vis, a Pin-on-Disc test stand is used which fulfills DIN 53516 [6]. With this test bench it is possible to adjust different velocities of the turning disk to simulate the relative movement between seal and wall. The normal force, which is affected because of the vacuum to pull the robot against the wall, can be simulated with the help of some weights, which are placed on top of the pin. The sliding friction is detected by using a beam in bending, which is coevally the reference surface for the inductive sensor to detect the amplitude of the beam. To determine the wear across the time of testing, a second inductive sensor is applied on the opposite site of the beam in bending. They are both placed on one rocker, shown in Figure 2. The sliding friction and the wear will be detected while using a dry, wet and steamed contact between pin and disk. In this way, all conditions, which appear on the seal of the climbing robot, can be simulated.

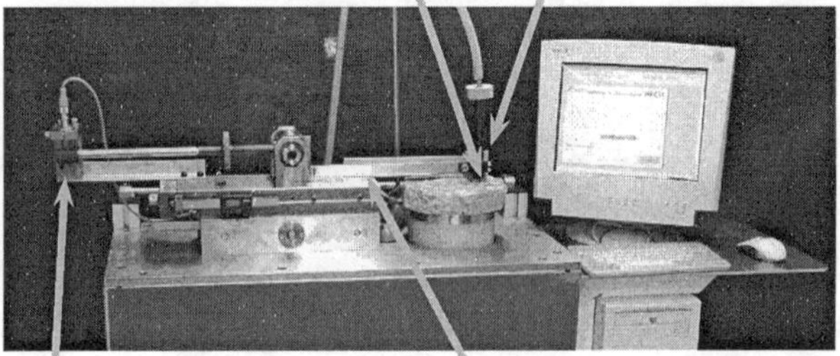

Fig. 2. Pin on Disc test stand

3. Locomotion

As already mentioned before our robot CROMSCI is driven by three omnidirectional wheels which let it slip over the surface. The interaction between adhesion system and drive is important due to the fact that the robot should neither stuck to the wall nor fall down. If the sealings are too soft the robot would suck itself to the wall, if they are too hard they lose leak tightness. So we have to make a crompromise and allow controlled leakage.

To measure all occuring forces a load cell is integrated into each wheel so that the robot can recognize if the negative pressure is too high for movement or dangerously low. The drive and the adhesion system will soon be integrated into one control mechanism which allows a safe driving onto rough surfaces. More informations about the propulsion system can be found in [7].

4. Experiments and Test Results

In the first studies with the Pin on Disc stand the friction between several samples of seals are tested. In the following chart, the different parameters can be seen.

Test No.	rel. Velocity $\left[\frac{m}{s}\right]$	Temp-erature $[^{\circ}C]$	Lubri-cation	Contact Pressure $\left[\frac{10^{-3}N}{mm^2}\right]$	Material Pin	Material Disc
1	0,1-0,03	20	-	9,4	Rubber	Glass
2	0,1-0,03	20	-	9,4	Rubber	Marble
3	0,1-0,03	20	Dust	9,4	Rubber	Concrete
4	0,1-0,03	20	-	9,4	Carpet	Glass
5	0,1-0,03	20	-	9,4	Carpet	Marble
6	0,1-0,03	20	Dust	9,4	Carpet	Concrete

The tests no. 1 to 3 are made with rubber on glass, plane marble and plane concrete. They provide a basis to compare well known materials with the samples of carpet. In all experiments the temperature and the contact pressure were kept constant. It could be possible, that the temperature departs in a field of ± 1 kelvin but this is irrelevant for the result of the friction coefficient. It was attempted to run the test without any kind of lubrication, but this was not possible with the samples of concrete. The concrete disc wears very fast, because the pin in contact grinds the fine sand on the surface. This fine sand works like a kind of lubrication and tempers the dry friction. However all pins, which stay in contact with the concrete disc, excite this special wear, so they are comparable among each other. Three different materials were used in the Pins: rubber, carpet (short fibres) and vertofloor (long fibres) as shown in figure 3.

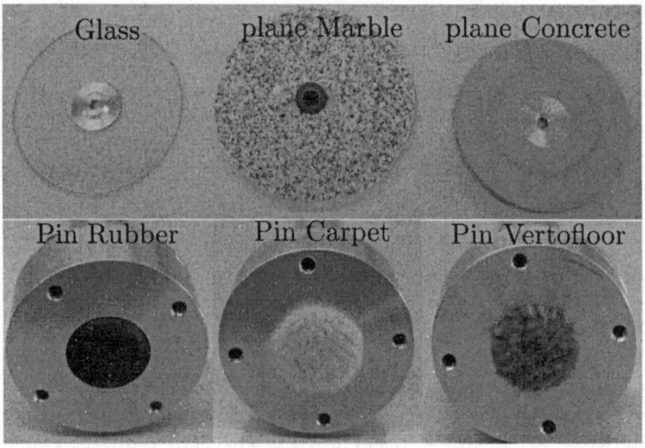

Fig. 3. Disc materials (top) and pin materials (bottom) used for the tests

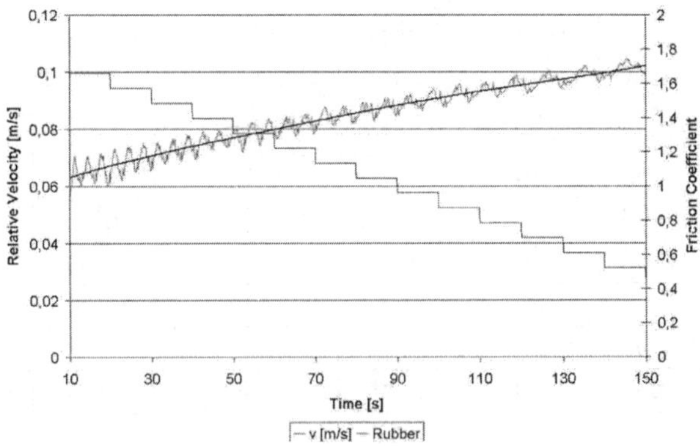

Fig. 4. Dynamic friction: rubber to glass

Figure 4 shows the graphs of the velocity steps and the value of the friction coefficient between rubber and glass. The dynamic friction coefficient starts with a value of approx. 1 at the highest relative velocity of 0,1 meters per second and rises slowly to 1,7 while the velocity decreases to 0,03 meters per second. The irregularities in the graph are generated by disturbances in the contact, resulting by the symmetry of the disc.

Figure 5 shows similar to figure 4 a friction coefficient. In opposition the current graph does not follow a clear line, so the dynamic friction cannot be clearly seen. Because of the distinctive swinging of the results, it has to be assumed, that there is more static than dynamic friction. That's why the trend line is not in the middle of the amplitudes, but on the maximums.

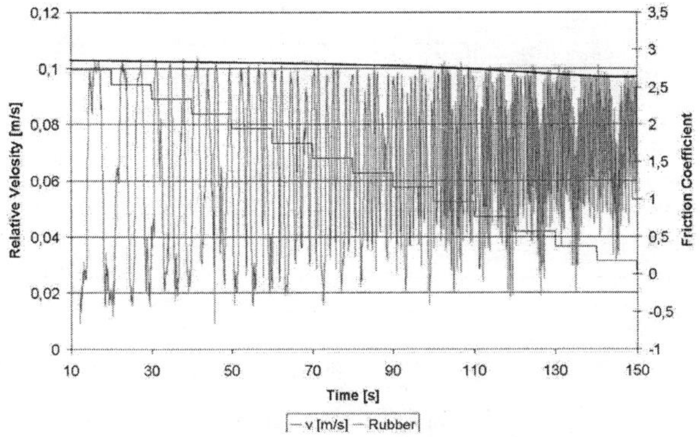

Fig. 5. Static friction: rubber to plane marble

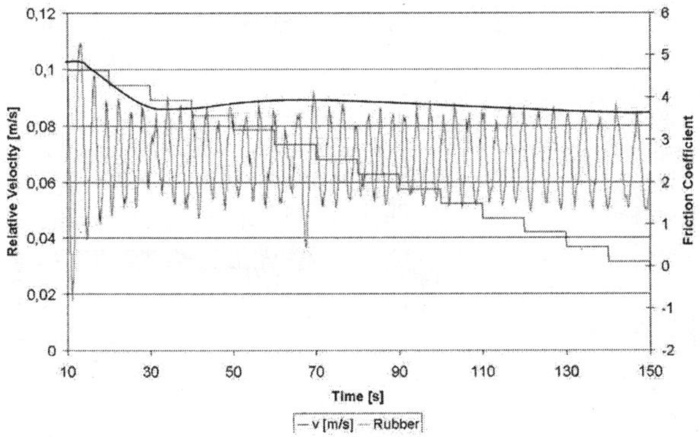

Fig. 6. Static friction: rubber to concrete

The contact between rubber and concrete (fig. 6) seems like a mixture of figure 4 and 5. The results of the experiments are also swinging, but the dynamic and frequency is much lower than in figure 5. Similar to the friction coefficient in the graph before, the dynamic friction coefficient cannot clearly be detected. Again the static friction has to be used, to get a comparable result.

The graphs in figure 7 are the first results, which can be compared to the graph in figure 4. Two distinctive differences can be seen: all values are much lower. They vary from 10% to nearly 30%. The second conspicuousness in the graph is the inverse pitch: the rubber friction rises with falling velocity, the carpet frictions are nearly constant and fall at the end of the experiences. The carpet slides much easier on glass than rubber, whereas

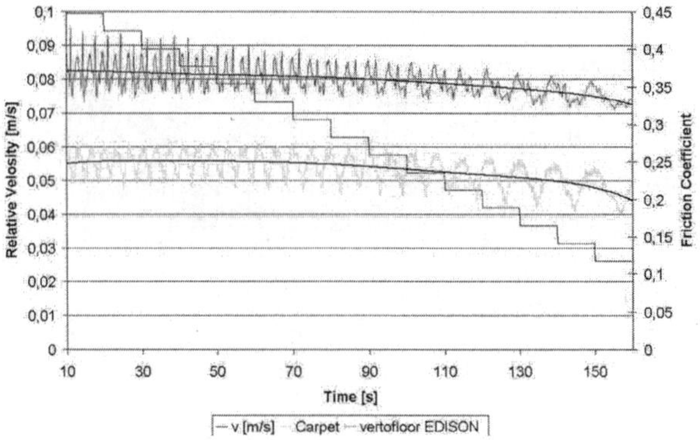

Fig. 7. Dynamic friction: carpet to glass

the carpet with the short fibers has a lower friction coefficient.

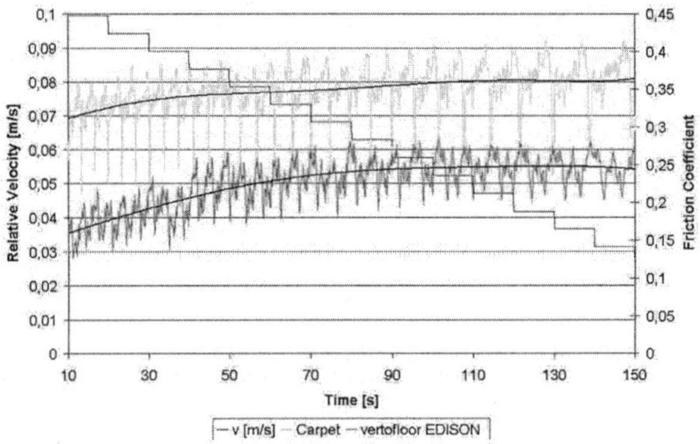

Fig. 8. Dynamic friction: carpet to plane marble

By comparison figure 5 with figure 8 the friction and the swing frequency between adhesion and sliding is much lower. With the graph in the current figure it is possible to find a dynamic friction, which is significantly lower than the static friction of the rubber in contact with the plane marble. The difference is again nearly up to 10

The course of the graphs in figure 9 shows a distinctly static friction. It is necessary to use the maximums of the amplitudes to compare it with figure 6. The course of the friction for both kinds of carpets is approx. from a third up to a quarter. Summary of results:

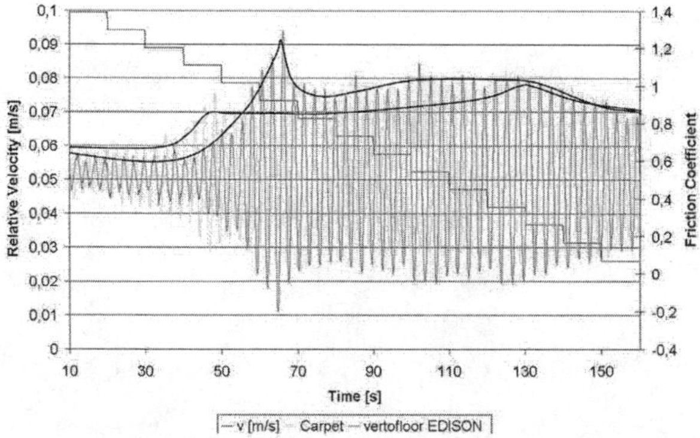

Fig. 9. Static friction: carpet to concrete

Test No.	Static Friction	Dynamic Friction
1 rubber	-	1,1 - 1,7
2 rubber	2,3 - 2,7	-
3 rubber	3,7 - 4,9	-
4 carpet	-	0,33 - 0,37
4 vertofloor	-	0,20 - 0,25
5 carpet	-	0,17 - 0,25
5 vertofloor	-	0,32 - 0,37
6 carpet	0,70 - 1,00	-
6 vertofloor	0,60 - 1,22	-

5. Conclusion

In this paper we introduced a novel sealing system which meets high demands. A sealing prototype is presented and different applicable facings have been tested concerning friction and abrasion onto several surfaces like concrete or glass. These experiments pointed out, that the sliding characteristic strongly depends on the facing material and will have a high influence on robot movement and adhesion.

Future Work mainly consists of experiments under real conditions. For this the adhesion system of our robot CROMSCI (see figure 10) has to be completed so that it will cling to the wall. Furthermore more experiments concerning the leak tightness of the sealing facings have to be carried out. These results will be compared with simulated experiments and evaluated to achieve best adhesion attributes.

Fig. 10. CROMSCI platform as presented at the Hannover Messe 2007 with seven vacuum chambers, mounted sealings (left) and reservoir chamber (right).

References

1. Y. D. Zhang, K. L. Fan, B. L. Luk, Y. H. Fung and S. K. Tso, Mechanical design of a climbing cleaning robot, in *8th IEEE Conference on Mechatrinics and Machine Vision in Practice*, (Hong Kong, 2001).
2. D. Longo and G. Muscato, Design of a single sliding suction cup robot for inspection of non porous vertical wall, in *ISR 2004: 35th International Symposium on Robotics. Proceedings.*, (Paris, France, 2004).
3. M. Hägele, Assistive robots in everyday's environments Europe's Information Society - ICT Riga 2006(June, 2006).
4. B. L. Luk, D. Cooke, S. Galt, A. A. Collie and S. Chen, *Robotics and Autonomous Systems 53* , pp. 142(October 2005).
5. J. Wettach, C. Hillenbrand and K. Berns, Thermodynamical modelling and control of an adhesion system for a climbing robot, in *20th IEEE International Conference on Robotics and Automation (ICRA)*, (Barcelona, Spain, 2005).
6. Oenorm din 53516: Testing of rubber and elastomers - determination of abrasion (1987).
7. C. Hillenbrand and K. Berns, The force controlled propulsion and adhesion system for a climbing robot, in *9th International Conference on Climbing and Walking Robots (CLAWAR)*, September 12-14 2006.

Development of a Suction Type Miniature Climbing Robot with Minimal Actuators

M.V.Vignesh
Research Scholar, Department Of Mechanical Engineering, Anna University, Chennai-25, India.

L.Karunamoorthy
Professor, Department Of Mechanical Engineering, Anna University, Chennai-25, India.

ABSTRACT

A wall-climbing robot intended for inspection of tall structures alike has been developed. The robot has characteristic features of kinematical design and is capable of moving a tool at a specific speed to follow complex paths on surfaces. In real field conditions, the labor intensive inspection demands great attention to details by inspection personnel and is subjected to human errors and limited in reliability. The primary objective in the design of this vertical crawling robot is focused on its ability to travel vertically on a wall. The robot totally uses two actuators, one for actuating the trunk and the legs to get the forward motion, backward motions and lifting up and down the suction cups. The other one is to give the rotary motions for changing the directions. Thus the robot uses totally two actuators for navigating on the curved surface as well as along the surface of the vertical wall. In this work, it has been successfully tried to actively employ the body as an extra leg.

1. Introduction

Wall- climbing robots are modeled to do their jobs in place of physical labor. Miniature climbing robot developed for inspection in harsh conditions can reduce the labor-intensive work and man-hours. Such robots should be small enough for easy handling and capable of traveling on vertical, inclined and curved surfaces of any structure. Hence, an attempt has been made to develop a novel multifunctional automated crawling system, which can carry the miniature instrumentation to perform a wide range of inspection tasks while attached to the surface of the structure of interest.

A large size robot of heavy weight with permanent magnetic disk wheels, which is capable of steady and smooth omni- directional locomotion on the surface of a flat or curved iron wall, has been developed by Shigeo et al [1]. A robot, whose motions are accomplished by means of flexible pneumatic cylinders made of rubber, has been developed by Leoncio et al [2]. A robot of larger size with a heavy weight capable of moving a tool at a specific speed to follow a complex path is developed by Lin Guo et al [3]. A multifunctional automated crawling system to carry miniature instrumentation to perform a wide variety of tasks while being attached to an aircraft's surface is developed by Paul et al [4]. A self contained robot developed by Tomoaki et al [5] uses suction for gripping. A robot with a magnetic suction arrangement, which can work in extremely hazardous environments, but with the limits of many factors has been developed by Wang Yan et al [6]. A robot which uses a caterpillar concept to perform 3D complex movements to inspect a high number of dangerous manual operations especially in large industrial environments has been developed by Abderrahim et al [7]. Ming Chen and Song Huat Yeo [8] have developed a robot planar walker, based on a novel planar 8-bar mechanism using planar rigid motion group. Two climbing robots considering under-actuated kinematic structures have been developed by Mark A. Minor and Ranjan Mukherjee [9]. A quadruped robot designed by Taehun Kang et al [10] has used an integrated approach to carry an ultrasonic tool for inspection of large surfaces in industrial utilities. Jizhong Xiao et al [11] have developed a gait controlled miniature bipedal climbing robot with under actuated mechanism. A quadruped robot that uses wheel legs for locomotion on surfaces regardless of the presence of gravity or direction was developed by Daltorio et al [12]. But most of the previous wall-climbing robots have utilized a number of actuators which are generally composed of complicated bulky mechanisms for driving and adhesion which lead to the increase in volume and weight accordingly.

Therefore, an attempt has been successfully made to design a miniature-climbing robot with minimum number of actuators. The robot thus developed uses only two actuators. The work presented in this paper describes only a part of the ongoing program on developing miniature robot, with critical mechanism to make the robot climb vertically with minimum number of actuators deployed in it.

2. Mechanical Structure

The mechanical arrangement of the robot shown as an exploded view in Fig.1a consists of three main assemblies, namely a frame, a drive mechanism and two legs.

The photographic view of the robot is shown in Fig.1b. The robot is 165x188x90 mm in size made of aluminum weighing 1500 g capable of carrying a pay load of 800 g. The crawler employs stepper motor for mobility and suction cups for surface adhesion. The crawler has legs for linear motion and a rotary element for curvilinear motion.

The main frame, namely the trunk is fitted with two actuators one for actuating the trunk and the legs to get the forward and backward motions of the robot and the other one to give the rotary motions for changing directions. Thus the robot uses totally two actuators for navigating on vertical wall. The trunk and the legs have been fitted with separate suction pads on them. Suction pads help in surface adhesion for robot. Robot uses totally seven suction cups, three for the trunk and the remaining four for the legs, having two suction cups on each leg. The location and the placement of the suction cups are shown in fig.2. Vacuum generator activates the suction cups, which uses compressed air for functioning. In this work, it has been tried to actively employ the body as an extra leg. This idea becomes more effective because it can produce a simple gait pattern. The suction pads of the trunk are activated and fixed to the wall. By rotating the actuator, the legs move in a semicircular trajectory and close to the wall as described in fig.3 and at the same time the vacuum sensors detect whether the vacuum at the legs is established or not.

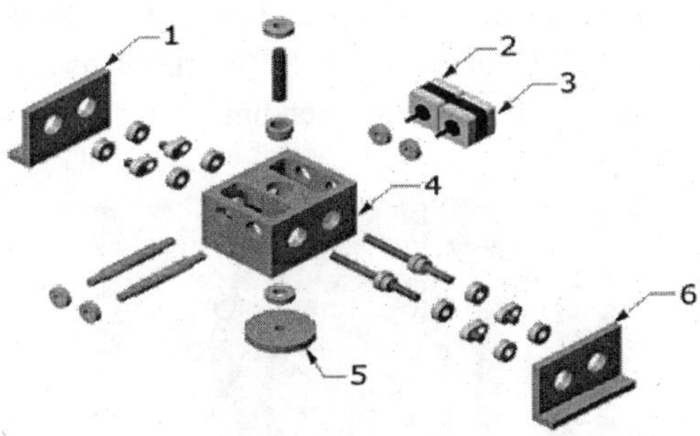

1.Leg one, 2.Motor one, 3.Motor two, 4.trunk,5. Rotary element, 6. Leg two

Figure 1a. Exploded view of the robot

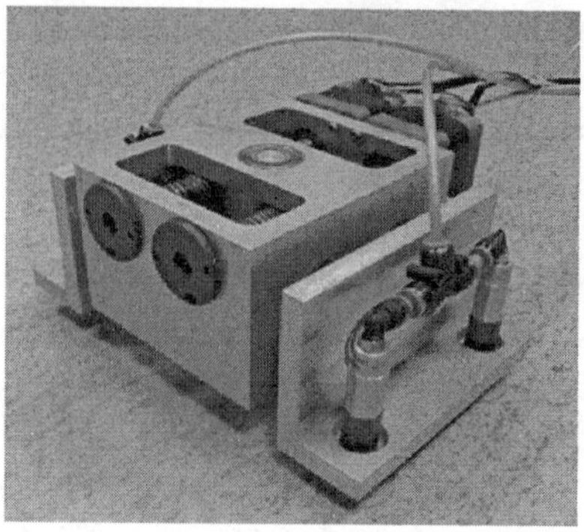

Figure 1b. Photographic view of the actual robot

Once the legs are firmly fixed to the wall due to the vacuum created in them, the suction pads of the trunk are released. The ongoing rotation of the motor makes the trunk to step forward to have a forward motion, as shown in Fig.3 and now again the suction pads in the trunk will have a contact with the wall. Forward motion of the robot is produced by successively repeating this sequence of operations. The same actuator which is used for forward and backward motion of the robot is used for lifting up and down of the suction cups. This lifting up and down of the suction cups is very much necessary to avoid dragging of them

As the robot is developed to perform specific tasks like inspection, the mechanical structure includes even special fittings on it and suitable NDT sensors intended to inspect the structural flaws can be fitted on to the robot body. These fittings are custom built for different types of sensors depending on their size and application.

Figure 2. Bottom view of the robot showing the location of the suction cups

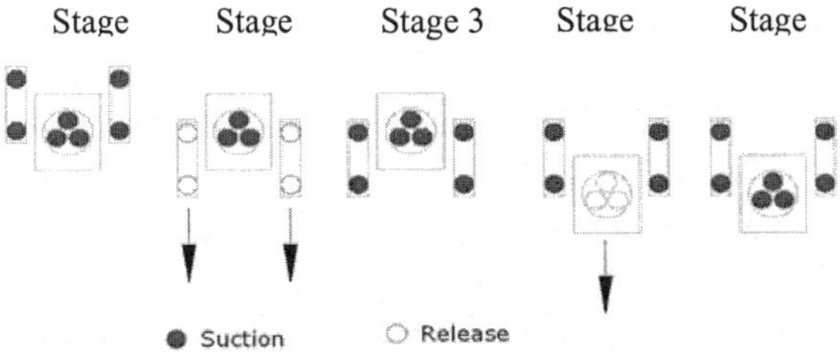

Figure 3. Gait control for linear motion

3. Gait Control

The sequences of translation and rotary motions are shown in Figs.3 and 4. To obtain a forward motion, the trunk is initially made to stick to the surface (Fig.3). Gait control for rotary motion then the two legs are simultaneously stepped forward and are made to fix to the surface. Once the legs are fixed to the surface the trunk releases from the surface and it moves forward. The forward motion of the robot is obtained by successively repeating this sequence of operations. To make the robot steer on the wall, the trunk is made to attach to the surface. While the two legs are detached from the ground, they are in a raised position. Now an actuator is used to rotate the trunk about a point, which enables the robot to turn around and make its move. The Fig.4 shows the stages and the positions of the suction pads at the time the rotary motion is taking place. Thus the robot body will negotiate a directional change using the steering actuator, when the suction pads of the trunk are attached to and the legs detached from the wall.

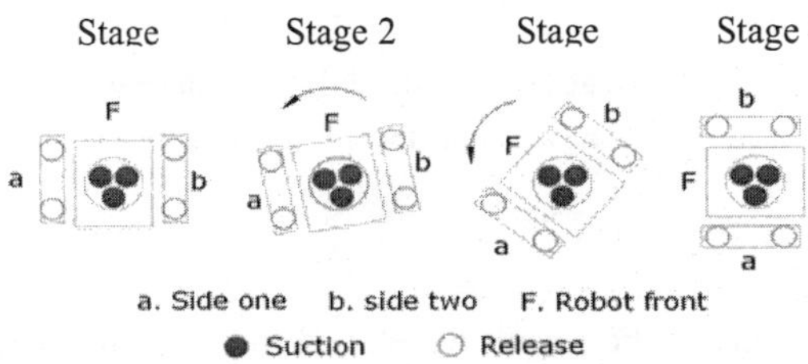

Figure 4. Gait control for rotary motion

4. Walking Algorithm to Scan an Area

A climbing robot can be used for inspection of surfaces. The quality of the results obtained from the scanning of an area, depends upon the resolution of motion of the robot. A 1 mm x 1 mm scanning resolution is considered to be optimal .To obtain this, the robot motion should be made more precise and accurate. By moving the trunk and the legs, the robot is capable of obtaining a forward motion as shown in fig.3. A spring loaded probe holder is used to obtain constant contact with the test surface. This makes it possible to get 1.0 mm resolution along one direction. To get the same 1.0 mm resolution along the other orthogonal direction, the robot is made to follow a set of movements as shown in Fig.5. After keeping the trunk of the robot firmly made to stick to the wall, firstly the robot legs are lifted from the ground. Secondly the robot is made to rotate through an angle of 2.388° followed by two steps. The forward motion of the trunk covers a linear distance of 24mm. The resultant of these two motions gives the required 1.0 mm linear distance in the orthogonal direction. By reversing the same procedure, which is by rotating the robot through 2.388° in the reverse direction, the robot is brought to the original straight position. Thus similar operations are repeated to get the linear movements in the co-ordinate directions.

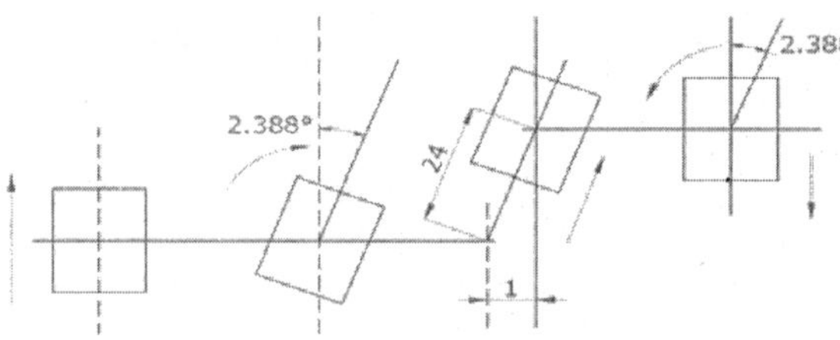

Stage 1 Stage 2 Stage Stage 4

Figure 5. Scanning algorithm for orthogonal movement

5. Pneumatic Control

The Fig.6 shows the pneumatic circuit developed for wall-climbing of the robot. This pneumatic circuit is used only for gripping the surface. The pneumatic circuit is separated into two parts; the suction cups and the vacuum efficiency valve are placed on the robot while the circuit components are placed in the ground station. The separation of circuit components is shown in fig.7. These separated circuit parts are connected by flexible tubes.

Compressed air is let into the system by opening a shut off valve. The air from the compressor is filtered before it is made to use, thus let air pressure is regulated and maintained using regulator. Here the air channel is split into two routs one will take care of the suction cups in the legs and other will take care of suction cups in the trunk. In each channel, electrically operated solenoid valves will regulate the air flow. These solenoid valves are controlled by pc based software. Vacuum in the suction cups is achieved, depending upon the opening or closing of the solenoid valves. Once the solenoid valve is opened the vacuum generator is made to work. Suction is achieved through the vacuum generator based on the Venturi effect. Compressed air is forced into the vacuum generator and it is made to pass through and out of the exhaust. On its way through, it draws air in from a perpendicular pipe causing suction at this pipe. This results in gripping of the suction pad. Vacuum efficiency valve are put to use, when any of the suction cups fail to have grip on the surface. That is, in some situations a suction cup may not be able to stick to the surface. Under these situations, the robot motion will be stopped. To avoid such situation, vacuum efficiency valve will shut the mouth of the suction cup to prevent any more air leakage. Even if any one of the suction cup fail to stick to the surface, with the help of other suction cups the robot can be put in motion. The airflow is regulated through solenoid-actuated valves. Depending upon the position of legs and trunk, the valves act accordingly. The controller guides the stepper motor and the solenoids in a sequential manner to achieve the vertical motion of the robot.

6. System Control

The overall system layout sketch is shown in fig.7. Labview based software is used for controlling the robot. The software controls both the stepper motor and the pneumatic circuit. The robot can be semi automatically made to reach the scanning location from where the robot can be fully automated by software to scan the desired surface. In case of any failure in achieving the surface grip, the software can be intimated through vacuum sensors so that the robot does not move any further. The solenoid valves and the vacuum sensor are controlled through homemade hardware which is too software controlled.

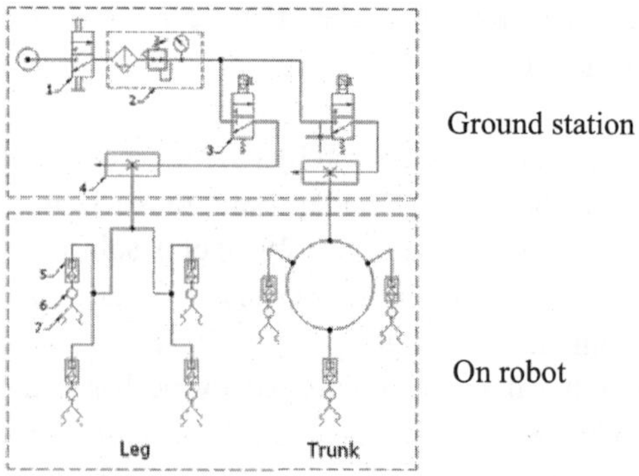

1. Shut off valve 2.Filter regulator 3.Solenoid valve 4. Vacuum generator

5. Vacuum efficiency valve 6. Vacuum Sensor 7.Suction cups

Figure 6. Pneumatic circuit

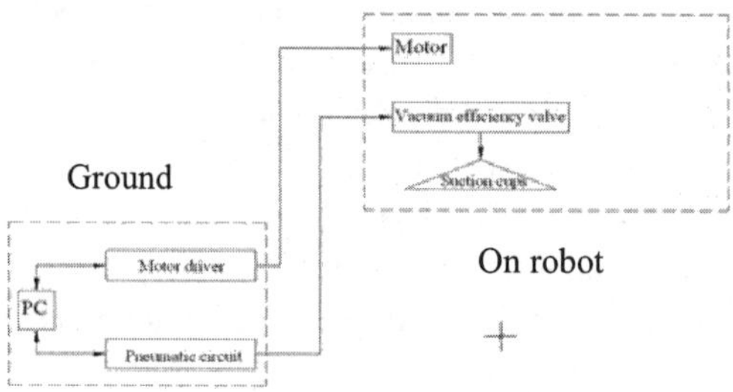

Figure7. Overall system layout

7. Conclusion

In this work, a wall-climbing robot is designed and made with compact and robust mechanism. Compared with the previous wall climbing robots that are available in literature, the present robot has the unique advantage of having reduced number of actuators and other parts as well as with an easy controlling module. It is developed in such a way that the robot can produce all the elementary motions like the forward, backward and the rotational with use of only two actuators. The same actuator which is used for forward and backward motions of the robot is used for lifting up and down of the suction cups. The special mechanical structure allows this robot to explore environment using climbing gaits. The pneumatics is used only for gripping the surface, while the stepper motors are used for mobility in this robot. The dynamics of the robot is electrically controlled through a computer using a code developed in Labview for this purpose. Such a crawler can travel on an aircraft structure and automatically perform inspection tasks that can be controlled remotely. Effectively, this robot can be used to integrate a suite of sensors and to perform intelligent scanning and testing for corrosion and other defects. The crawler is made at significantly more affordable cost by employing commercially available components while allowing the NDE industry to concentrate on making dedicated sensor modules as plug-and-play boards. The crawler is developed to operate autonomously and search for flaws on structures without human interference.

The components of the robot namely, main body, worm gear drives, crank drives, legs, hub drive and relevant smaller parts like bushes have been fabricated. The stepper motors and all other components like compressor, suction cups, solenoid valves, vacuum generator required for pneumatic system have been purchased from the market.. The software for the drive has been developed using Labview to suit the requirement. The assembly of the robot has been carried out using all these components and successfully demonstrated for its smooth mobility in crawling with a sufficient pay load.

8. Reference

1. Shigeo Hirose, Hiroshi Tsutsumitake and Disc rover: “A wall-climbing robot using permanent magnet disks”, Proceedings of IEEE, International Conference on Robotics and Automation, Raleigh, NC, July 1992. pp. 2074 – 2080.

2. Leoncio Briones, Paul Bustamante and Miguel A. Serna, “Wall-climbing robot for inspection in nuclear power plant”, Proceedings of IEEE, International Conference on Robotics and Automation, May 1994, Vol. 2, pp. 1409-1414.

3. Lin Guo, Kevin Rorers and Robin Kirkham, “A climbing robot with continuous motion”, Proceedings of IEEE, International Conference on Robotics and Automation, Santiago, May 1994, Vol. 3, PP. 2495-2500.

4. Paul G. Backes, Yoseph Bar-Cohen and Benjamin Joffe, “The multifunction automated crawling system (MACS)”, Proceedings of IEEE, International

Conference on Robotics and Automation, New Mexico, April1997, Vol. 1, pp. 335-340.

5. Tomoaki Yano, Shinji numao and Yukio Kitamura, "Development of self-contained wall climbing robot with scanning type suction cup", IEEE, International Conference on Robotics and Automation, Canada, Oct 1998, pp. 244 – 259.

6. Wang Yan, Shao Hoa, Lliu Shuliang, Xu Dianguo and Gao Xueshan, "Development and application of wall-climbing robots", IEEE, International Conference on Robotics and Automation, Michigan, May 1999, Vol.2, pp. 1207 – 1212.

7. M.Abderrahim, C.Balaguer, A Gimenez, J.M.Pastol and V.M.Padron, "ROMA: A climbing robot for inspection operations", Proceedings of IEEE. International Conference on Robotics and Automation, May 1999, Vol. 3, pp. 2303 – 2308.

8. I.Ming Chen and Song Huat Yeo, "Locomotion and navigation of a planar walker based on binary actuation", Proceedings of IEEE. International Conference on Robotics and Automation, Washington, DC, May 2002, pp. 329-334.

9. Mark A. Minor and Ranjan Mukherjee, "Under-actuated kinematic structures for miniature climbing robot", ASME Journal of Mechanical Design, June 2003, Vol. 125, pp. 281 – 291.

10. Taehun Kang, Hyungseok Kim, Taeyoung Son and Hyoukryeol Choi, "Design of quadruped walking and climbing robot", Proceedings of IEEE. International Conference on Robotics and Automation, Las Vegas, Oct 2003. pp. 619 - 625.

11. Jizhong Xiao, Ning Xi, Jun Xiao and Jindong Tan, "Multi-sensor referenced gait control of a miniature climbing robot", Proceeding of IEEE, International Conference on Robotics and Automation, Las Vegas, Oct 2003, PP. 3656-3661

12. Kathryn A.Daltorio, Andrew D.Horchler and Stanislav Gorb, Roy E.Ritzmann and Roger D.Quinn, "A small wall-walking robot with compliant, adhesive feet", IEEE, International Conference on Robotics and Automation, July 2005, pp. 34 – 39.

DEVELOPMENT OF AN AMPHIBIOUS HEXAPOD ROBOT BASED ON A WATER STRIDER

SOH FUJII

Major of Precision Engineering, Graduate School of Science and Engineering, Chuo University, 1-13-27 Kasuga, Bunkyo-ku, Tokyo 112-8551, Japan

TARO NAKAMURA

Major of Precision Engineering, Graduate School of Science and Engineering, Chuo University, 1-13-27 Kasuga, Bunkyo-ku, Tokyo 112-8551, Japan

There is an increasing demand for mobile robots that can operate in a variety of environments, including across the ground, on water, and over rough terrain. In this study, we focused on a water strider, which is an amphibious creature. The water strider moves on water, supporting its body on four legs and propelling its body with two other legs. On land, water striders walk using all six legs. We developed an amphibious hexapod robot based on the water strider, and confirmed that this robot can move on the ground and on water. This robot can move smoothly over shoals because it does not use a screw propeller.

The locomotion mechanism of the water strider robot has 12 servomotors. It is equipped with a radio camera and controller, which allows operation from a remote location. Furthermore, this robot can change its direction while remaining in place.

Experimental evaluation of the performance of the water strider robot, moving on both ground and water, showed it to be similar to that of an actual water strider.

1. Introduction

Today's mobile robots are expected to function in a variety of environments, including land, water, and irregular terrain. Mobile robots, both bipedal walking robots and those with wheel drive, are coming into practical use; however, their movement mechanisms limit their use in more difficult environments. There is still a demand for robots that can move freely to disaster locations and places that are difficult or dangerous for humans to enter. For instance, if a robot can freely move on both water and land, it would be useful for rescue operations and research in desert locations.

In this study, we examined a water strider, which is an aquatic insect that is able to move easily on water and on land. In addition, it can move smoothly in shallow and shoal areas, and make sharp turns. We developed a water strider robot, which is a mobile robot with the locomotion characteristics of the water strider, with the goal of realizing an amphibious movement mechanism that affords both high mobility and stability [1][2].

2. Locomotion of a living water strider

A water strider has fuzzy legs covered with wax that makes them water resistant. The high surface tension of these legs makes the water strider float on the surface of the water. The water strider is small, and weighs from 0.01to 0.1g. The slim leg creates a wide depression on the surface of the water, isolating the water strider's body from the surface of the water. Figure 2.1 shows the depressions the legs create on the surface of the water.

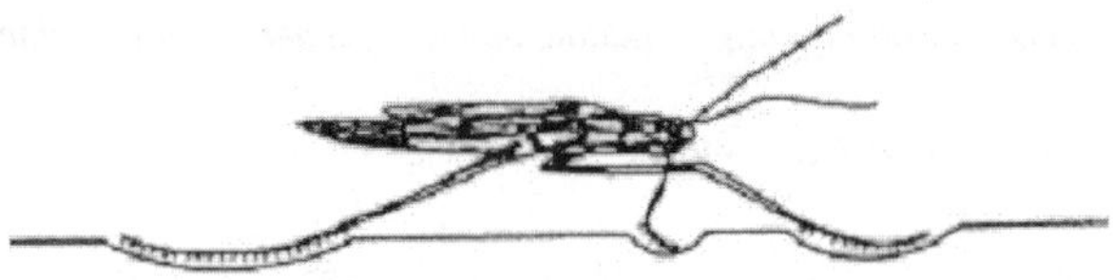

Fig. 2.1. A water strider on water.

A water strider puts its fore and hind legs on the water to support itself and uses the two mid legs for propulsion. These mid legs scratch the water like the oars of a boat. Changing the angle of the legs controls the direction of movement and speed. In addition to supporting the body, the hind legs are used as a rudder, providing minute steering action. The front legs, in addition to providing support, can grasp an object for predation.

3. Experimental plan

We evaluated the locomotion mechanism of the water strider and developed an amphibious robot. The robot has six legs, and each leg has the same specifications; however, the tips of the front and hind legs are equipped with floats for floatation and the mid legs have oars. We simplified the water strider's movement mechanism. During water movement, the front and hind legs are locked in place, and the mid legs are used as the drive legs—moving the mid legs drives the robot. On land, however, the robot raises the mid legs, fixes them in place, and uses the front and hind legs as its drive legs. Moving the front and hind legs provides the mobility. Independent servomotors drive each leg. A PIC controls the servomotor's angle of rotation. In addition, the robot has a power supply, built-in control system, and radio control. Figure 3.1 shows the robot, and its specifications are given in Table 3.1.

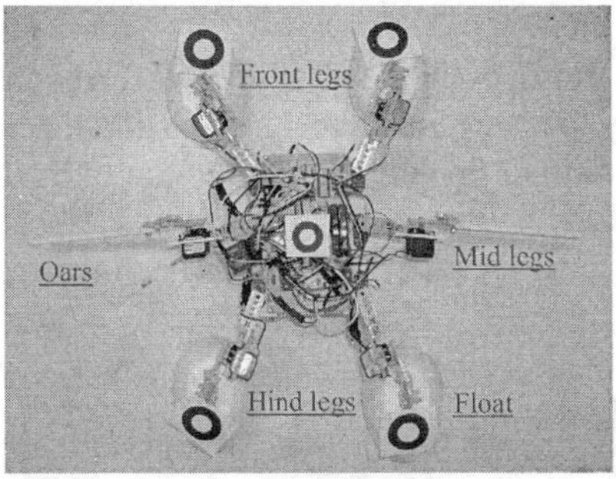

Fig. 3.1. An amphibious robot.

Table 3.1. Amphibious robot specifications.

Weight[g]	600
Length[mm]	410
Width[mm]	460
Height[mm]	130

We examined mechanisms for lifting the legs and for moving the legs back and forth. The mechanism that moves a leg from front to back consists of the turning of a servomotor. Figure 3.2 shows a schematic of the leg mechanism. The mechanism for lifting the leg was realized with a parallel linkwork, such as ② in Fig. 3.2.

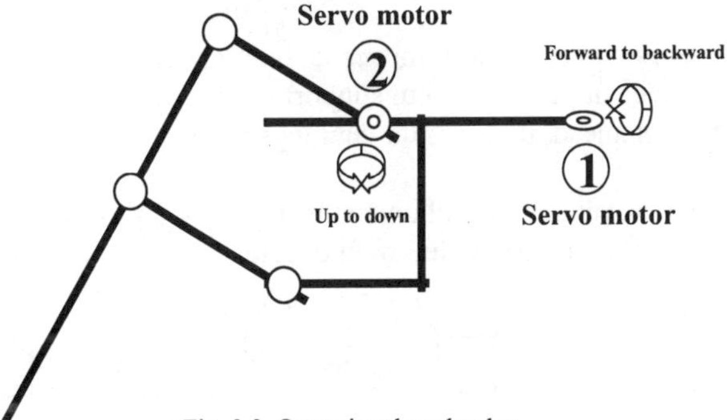

Fig. 3.2. Operating the robot leg.

Servomotor ② lifts (or lowers) the leg, and servomotor ① moves the leg from front to back. In other words, servomotor ① allows the leg to rake the water.

4. Locomotion in water

4.1. *Gait model*

In water, the robot fixes its two front and two hind legs, and moves only on the mid legs. The same method is used by the water strider. Here, we implemented four movements: forward, backward, turn to the left, and turn to the right. These correspond to the front, rear, left, and right commands on radio controller sticks.

Figure 4.1 shows the movement of each leg as the robot moves on water.

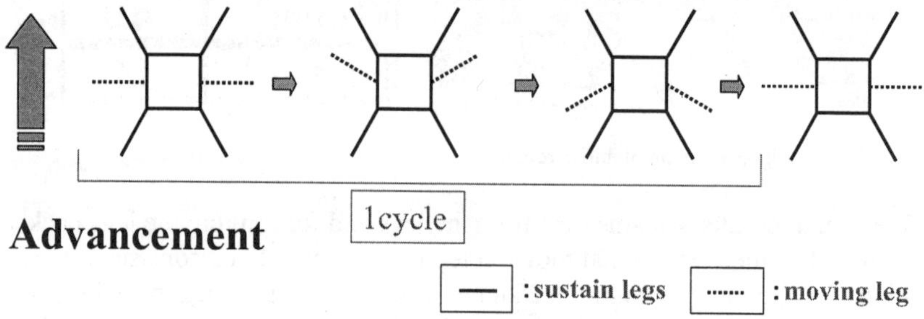

Fig. 4.1. Advancement on water.

As the robot moves on the water, it moves its mid legs from front to back. The front and hind legs do not move, but support the body.

Once the robot moves ahead, it moves the mid legs in reverse motion to that shown in Fig. 4.1.

Next, I examined movement in shallow areas. Figure 4.2 shows the movement of each leg when turning left on the water.

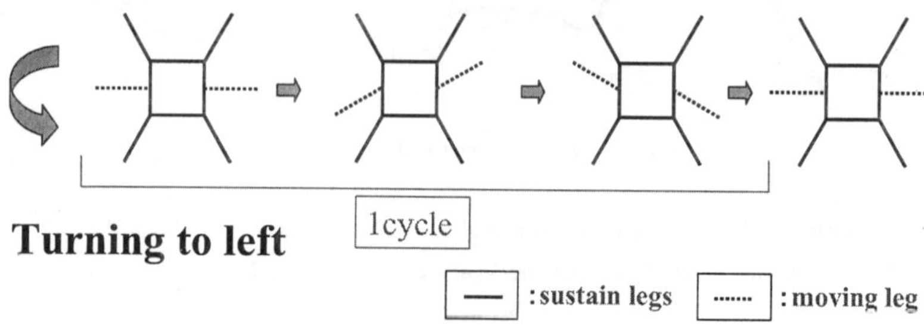

Fig. 4.2 Turning left on water.

To turn left on the water, the left mid leg moves from back to front and the right mid leg moves from front to back, as shown in Fig. 4.2. The front and hind legs do not move, and only support the body.

To turn to the right, the robot moves its mid legs in the opposite direction.

4.2. *Experimental apparatus*

We performed experiments to evaluate the robot's motion on the water. Figure 4.3 shows the experimental setup. I fixed a camera with a tripod and photographed the robot's water movements from directly above. We used MOVIAS pro3D for analysis, just as we did for the water strider.

We analyzed the movement of my new amphibious robot, and compared its performance with that of an actual water strider.

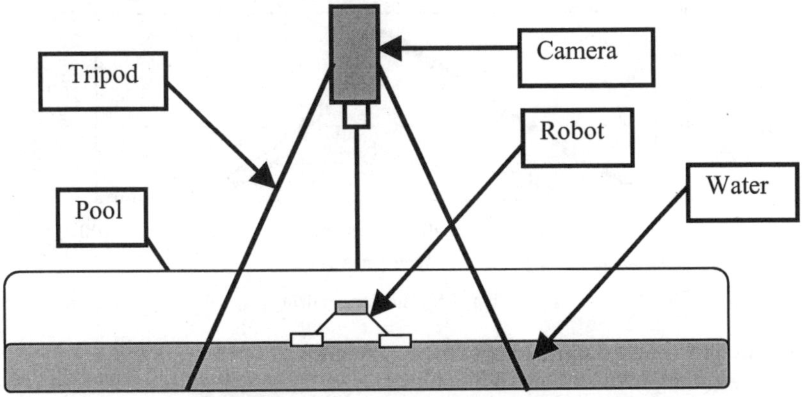

Fig. 4.3. Experimental apparatus for evaluating locomotion on water.

4.3. *Experimental results*

Figure 4.4 shows the movement distance and angle of the amphibious robot's mid legs.

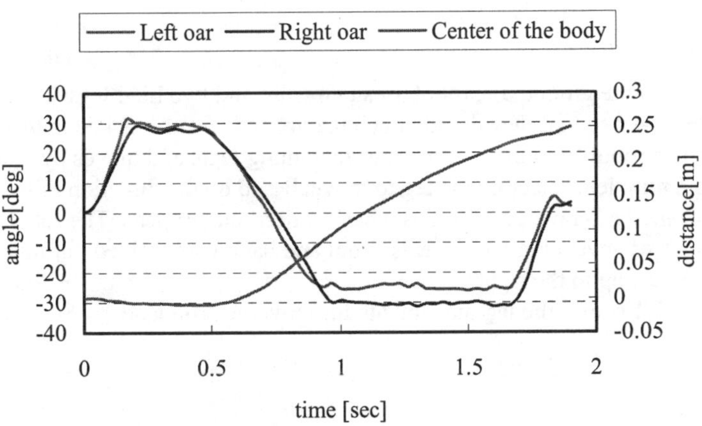

Fig. 4.4 Amphibious robot performance data.

This Figure shows that mid legs move forward and backward by 30 degrees. For this progress, the right and left mid legs move in a motion that is linearly symmetric to the body axis. The body retreats slightly in reaction to the forward motion of the mid legs, but once the mid legs row, the robot moves forward.

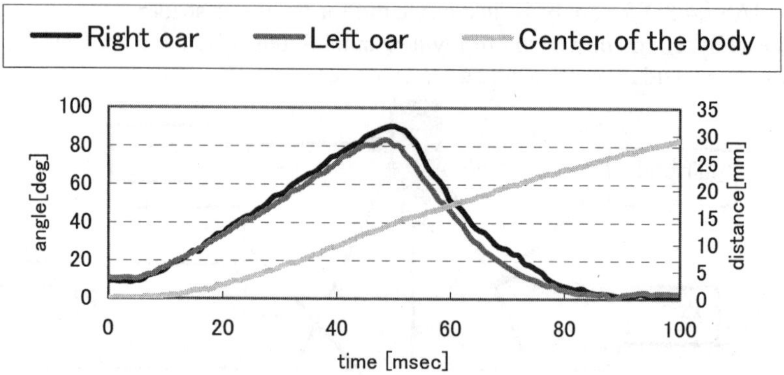

Fig. 4.5 Water strider data.

Figure 4.5 shows the movement of an actual water strider.

The robot moves 30 mm with one stroke of its mid-leg oars. The robot body is 410-mm long and the body of the living water strider is 14-mm long. The increased power of the robot is no larger than the increase in resistance from upsizing the volume, and the differences in its floating method produce the surface tension ("the ascending force").

5. Locomotion on the ground

5.1. *Gait model*

To move on the ground, the robot uses two front and two hind legs. The mid legs are raised so that they do not touch the ground and are locked in place [3][4][5]. The robot fixes three of the four remaining legs, and moves by operating the remaining leg, selecting the leg to move, based on the direction of travel. This is called a “crawl gait,” and is a basic gait of quadrupeds. The robot has four kinds of advancing movements: front and back legs is fixed, turning to the left, and turning to the right.

Figure 5.1 shows the leg movements for movement on land.

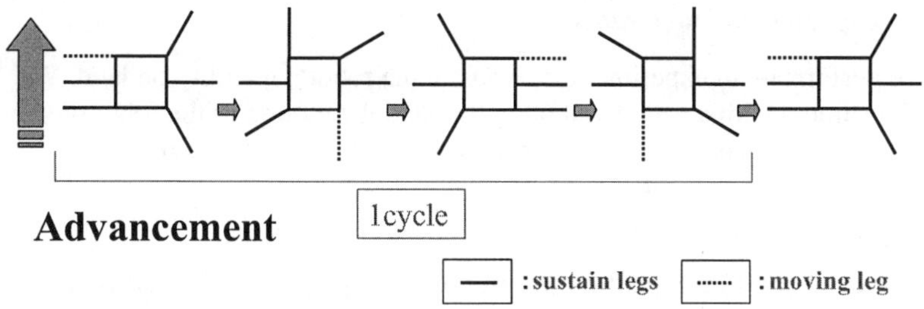

Fig. 5.1. Leg movement on ground when advancement

When walking on four legs and turning to the left, the mid legs are fixed (above the ground) and three of the front and hind legs provide balance while the last leg provides propulsion. The moving leg changes in rotation.

Figure 5.1 shows that the moving leg changes from left front to right hind, right front, and finally left hind. These represent one cycle.

Figure 5.2 shows leg movement for turning left on the ground.

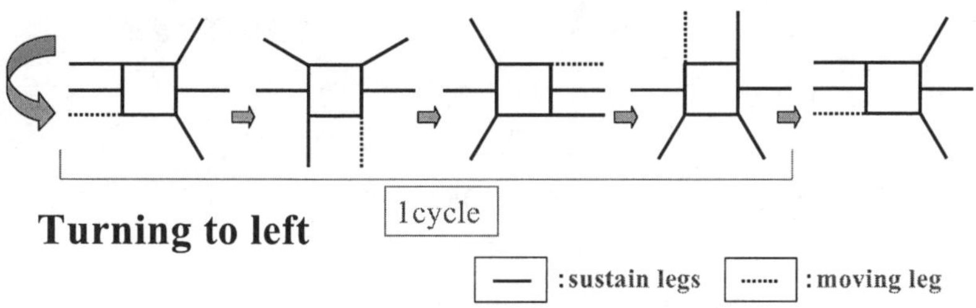

Fig. 5.2. Leg movement on ground when turning left.

When walking on four legs and turning to the left, the mid legs are fixed (above the ground) and three of the front and hind legs provide balance while the last leg provides propulsion. The moving leg changes in rotation.

Figure 5.2 shows that the moving leg changes from left hind to right hind, right front, and finally left front to complete one cycle.

Turning to the right involves the reverse rotation of these motions.

5.2. *Experimental apparatus*

We performed an experiment to evaluate the robot's mobility on land. We fixed a camera with a tripod and photographed the motion of the robot on land from directly above. The analysis method is similar to that in water.

5.3. *Experimental results*

Figure 5.3 shows the position transition of each leg and the body's center of gravity as the robot moves.

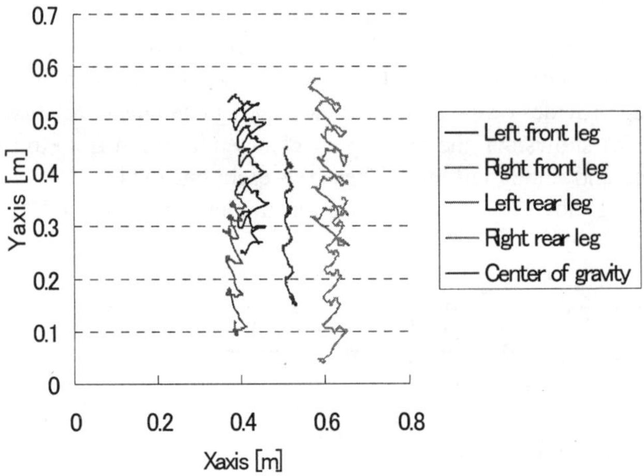

Fig.5.3. Analysis of movement on land.

Each leg lifts, moves forward, and then drags backward slowly. Figure 5.3 shows that the whole robot goes straight without deviation, but the motion of each leg blurs the center of the body. This blurring was about 1 cm to the right and left for a movement of about 30 cm by actual survey [6][7].

Figure 5.4 shows the position transitions of each leg, the center of gravity of the body, and positions of the head and tail, as the amphibious robot turns left.

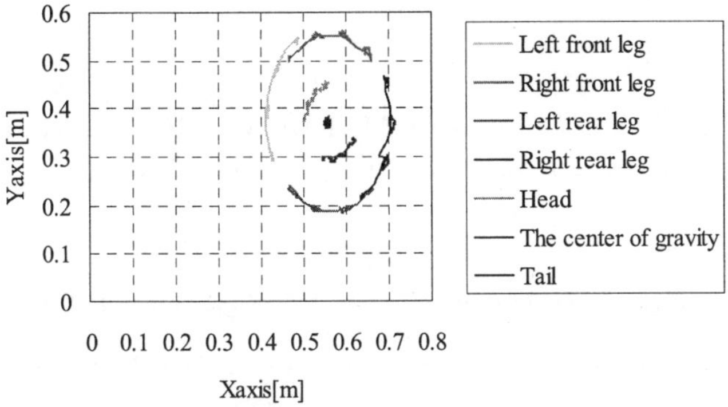

Fig. 5.4. Experimental analysis of turning to left.

Figure 5.4 shows three inputs of an order to turn left. One turn command provides a turn of about 30 degrees; therefore, the robot turns 90 degrees in the figure. A moving leg draws a circle around the center of gravity and moves, while the center of gravity does not change. Thus, the robot can change direction while remaining in place. This allows it to work effectively within a small domain, where other robots do not have the space necessary to change direction.

References

1. Taro NAKAMURA, Takahiro OMATA, Keisuke IINUMA, Soh FUJII, "Development of a water strider robot which moves surface on the water," JSME Conference on Robotics and Mechatronics, 2P1-A18, 2006
2. Keisuke IINUMA, Soh FUJII, Taro NAKAMURA, "Development of an Amphibious Robot based on a Water Strider," JSME Conference on Robotics and Mechatronics, 1P1-C05, 2007
3. Shigeo HIROSE, Hiroyuki IWASAKI, Yuji UMETANI, "The Basic Study of the Intelligent Gait Control for Quadrupedal Walking Vehicle," The Society of Instrument and Control Engineers (SICE), Japan, 1980.
4. Hideki HAYASHI, Toshiyuki KONDO, Koji ITO, "Biomimetic Hexapod Robot Using Mechanical Compliance," The Society of Instrument and Control Engineers (SICE), Japan, 2002.
5. Noriho KOYACHI, Hironori ADACHI, Tatsuo ARAI, Makoto IZUMI, Takeshi HIROSE, Naofumi SENJO, Ryoji MURATA, "Walk and Manipulation by a Hexapod with Integrated Limb Mechanism of Leg & Arm –Configuration Change of Body and Limbs for Dual Arm Manipulation," The Robotics Society of Japan, Vol.22, No.3, pp.411-421, 2004.
6. Shigeo HIROSE, Yuji UMETANI, "The Synthesis of Basic Motion Regulator System for Quadrupedal Walking Vehicle and Its Experiments," The Society of Instrument and Control Engineers (SICE), Japan, 1979.
7. Seiji MASAKADO, Takayuki ISHII, Kazuo ISHII, "A Gait-Transition with Stability Secured for a Quadruped Walking Robot," The Robotics Society of Japan, Japan, 2004.

DEVELOPMENT OF AN OMNI-DIRECTIONAL MOBILE ROBOT BASED ON SNAIL LOCOMOTION

KUNIAKI SATOH

Major of Precision Engineering, Graduate School of Science and Engineering, Chuo University,
1-13-27 Kasuga, Bunkyo-ku, Tokyo 112-8551, Japan

TARO NAKAMURA

Major of Precision Engineering, Graduate School of Science and Engineering, Chuo University,
1-13-27 Kasuga, Bunkyo-ku, Tokyo 112-8551, Japan

In recent years in the field of bio-mechatronics, a robot that imitates a creature's motion has been developed. In this study, we present a locomotion robot based on locomotion of snail. The snail moves by propagating traveling waves from tail to head. If it is possible to propagate a traveling wave in many directions, an omni-directional mobile robot could be realized. Since the locomotion mechanism of the snail involves moving a larger area than in the case with other creatures, it is able to move not only on irregular ground such as swamps, but also on walls and ceilings. We have developed an omni-directional mobile robot that makes use of a traveling wave and have confirmed that the robot can move in many directions using a traveling wave.

1. Introduction

At present, the wheel mechanism is the main mechanism for moving over ground. However, this has a disadvantage that it is difficult to move freely over bad ground such as swamps and wasteland. In addition, it is difficult for the wheel to move in a diagonal direction. Therefore, to realize omni-directional locomotion, a special tire is necessary such as omni-wheel. However, this has the disadvantage that the wheel slides axially on a steep gradient.

On the other hand, in recent years in the field of bio-mechatronics, a robot that imitates a creature's motion has been developed[1][2]. Here, we report a study on the locomotion mechanism of snail. The snail has an elastic motion in the direction of movement, and this is called a pedal locomotion mechanism. In this mechanism, a traveling wave propagates from tail to head. Since the locomotion mechanism of the snail involves moving a larger area than in the case with other creatures, it is able to move not only on irregular surfaces such as swamps, but also on walls and ceilings. Also, because of the mechanism to propagate the wave in the direction of movement, omni-directional motion may be expected.

In this research, we developed an omni-directional mobile robot using the locomotion mechanism of the snail and studied the basic properties of the robot.

2. The locomotion mechanism of the snail

The common belief is that the snail moves by a traveling wave locomotion mechanism. This produces a traveling wave in the gastropod's foot that propagates in the direction of movement[3]. The state of the traveling wave seen from a section of the gastropod's foot is shown in Fig.2.1, and the state of the traveling wave when moving forward is shown in Fig.2.2. The snail has two kinds of muscles: the anterior oblique and the posterior oblique muscles (Fig.2.1). These produce a traveling wave that propagates from tail to head (Fig.2.2). When the traveling wave occurs, the part that is not forming the traveling wave in the gastropod's foot remains in contact with the ground. The ground generates a frictional force with the grounded foot of the gastropod.

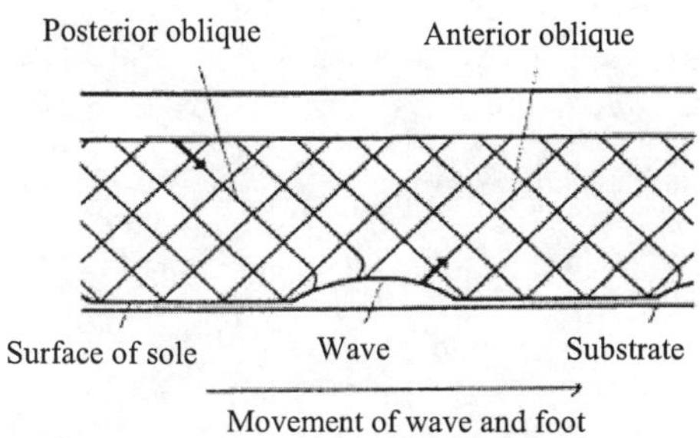

Fig.2.1 Diagrammatic section of snail's foot[4].

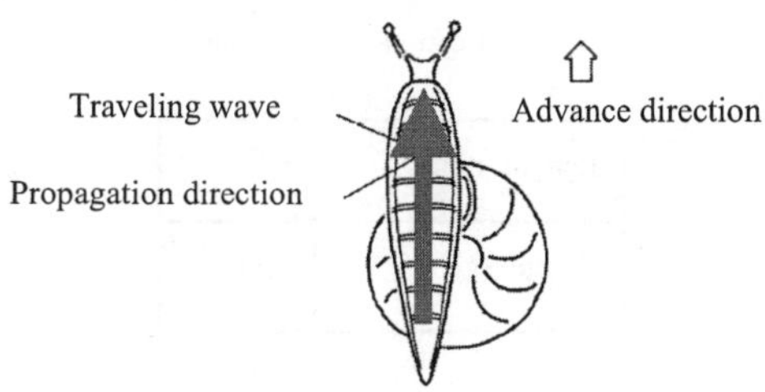

Fig.2.2 Propagation of traveling wave of snail[5]

3. The omni-directional mobile robot

3.1. *Overview of the robot*

We have developed an omni-directional mobile robot that uses the locomotion mechanism of the snail (Fig.3.1). The specifications of the robot are shown in Table 3.1. The omni-directional mobile robot is composed of eight units arranged in a circular shape and each abutting unit is connected by a spring.

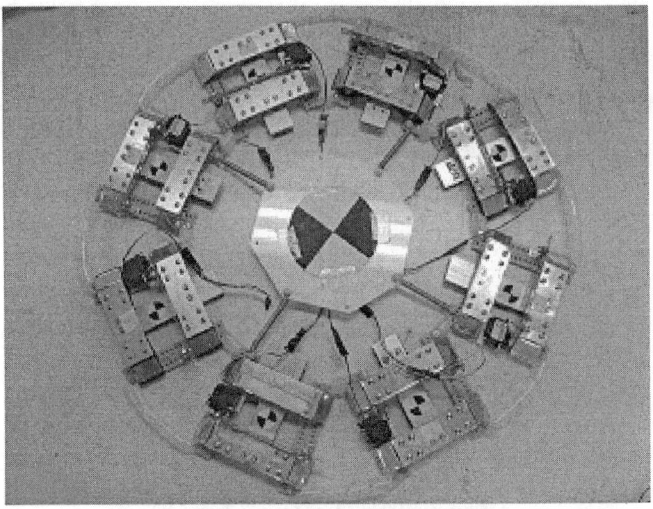

Fig.3.1 Omni-directional mobile robot[6]

Tab.3.1 Specifications of the robot

Length (mm)	500
Weight (g)	1600

3.2. *Unit*

A diagram of the unit used in the omni-directional mobile robot is shown in Fig.3.2. It is composed of two substrate parts (side: 20 mm, depth: 105 mm): the friction part (side: 25 mm, depth: 105 mm) and a servomotor.

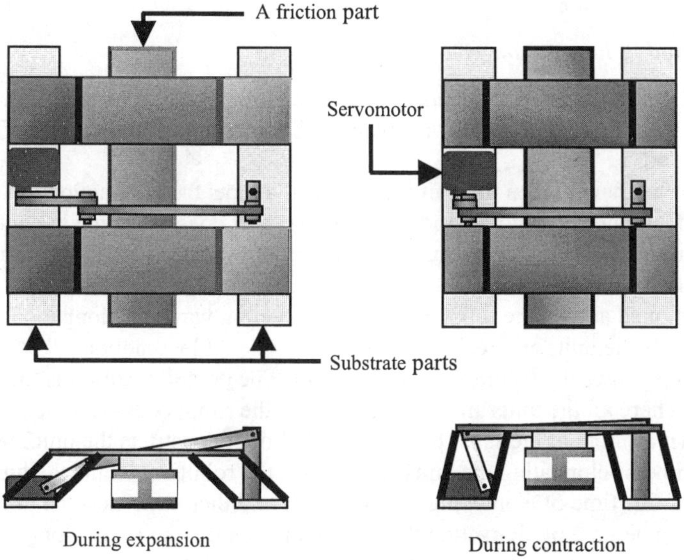

Fig.3.2 Mechanism of the unit[7].

Tab.3.2 Specification of the unit

Length (mm)	Min	88
	Max	112
Width (mm)		105
Weight (g)		200

The substrate part is always in contact with the ground. In the condition where the substrate part is grounded, the unit has elasticity in the direction of movement.

The frictional part is the part that holds the unit stationary when the unit is elongating. A frictional sheet is attached to the contact area of the frictional part. Then, this frictional part is connected to the substrate part using parallel linkages.

By using a parallel linkage, during elongation, the frictional part connects to the ground horizontally, and during contraction, the frictional part leaves the ground horizontally.

We fixed a servomotor to the substrate part, which is connected to the other substrate part using a crank mechanism.

3.3. *Method of locomotion*

When the unit is stationery, it adopts the elongation condition from the servomotor, the frictional part is in contact with the ground and the unit is fixed on the ground.

On the other hand, when the unit is moving, it adopts the contraction condition from the servomotor, and the frictional part does not come in contact with the ground. As a result, because the friction of the unit is small, it is able to move freely.

As mentioned above, the robot generates a traveling wave by elongation and contraction of the units and realizes an omni-directional locomotion.

Figure 3.3 shows the locomotion of the robot. The default position is the condition where all the units are elongated. First, the robot constricts the pair of units shown as unit1 at the same time. Then, the robot constricts the unit2 pair at the same time of elongating the unit1 pair. Later, the robot constricts the unit3 pair at the same time of elongating the unit2 pair and then constricts the unit4 pair at the same time of elongating the unit3 pair. Finally, the robot elongates the unit4 pair and we call this complete cycle 1 period. Because of this movement, the robot generates the traveling waves shown in Fig.3.3 and moves forward in the advance direction. To change the advance direction, the robot changes to a different unit1 pair appropriate for the new direction of travel. These units constrict first and this changes the direction of propagation of the traveling wave.

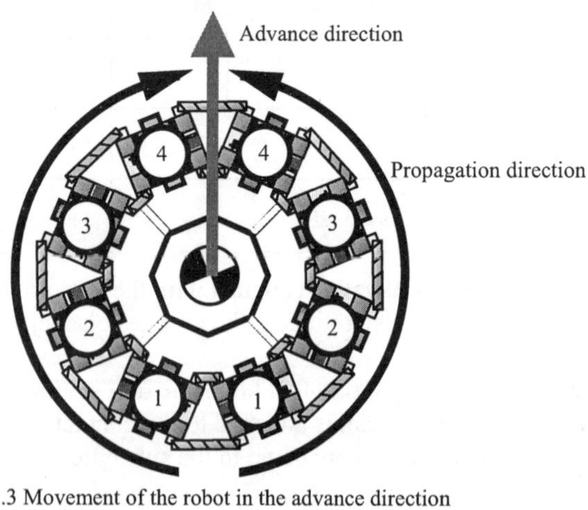

Fig.3.3 Movement of the robot in the advance direction

4. The locomotion experiment

4.1. *Confirmation experiment for traveling wave locomotion*

First, we experimented to confirm whether the robot moved using the traveling wave that is the locomotion mechanism of the snail. The gage points in the center of the units are shown in Fig.4.1. Then, we made the robot move in the y axial direction and confirmed the traveling wave locomotion. The experimental result is shown in Fig.4.2. This graph shows the relationship between the time and the axial displacement of y. The most backward units from the direction of movement constrict first and we found that the constriction of the unit propagates in the direction of movement. From this result, we confirmed that the robot moved by the traveling wave.

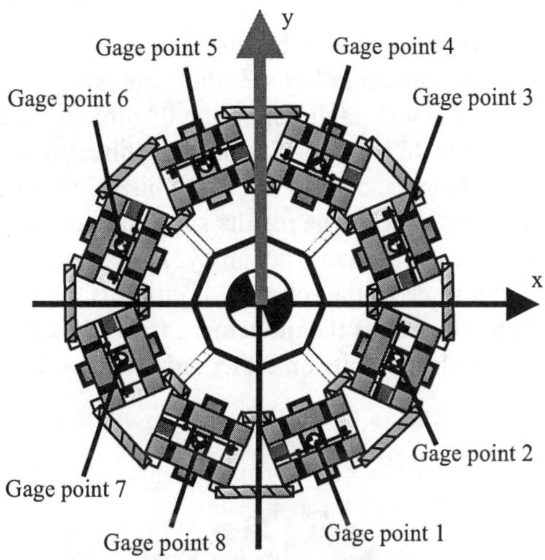

Fig.4.1 Points attached at each unit of the robot

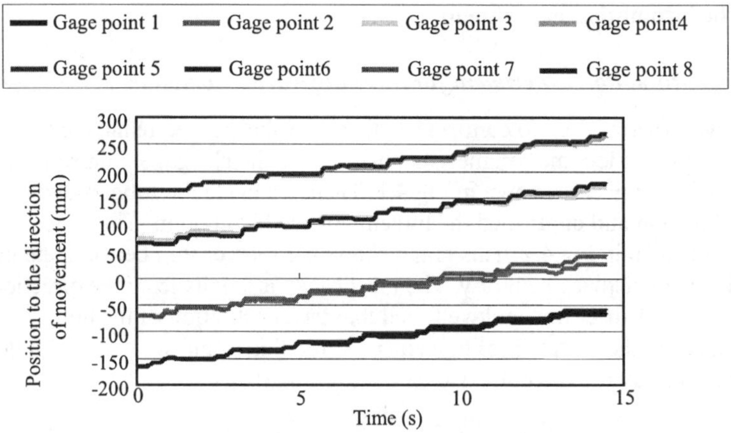

Fig.4.2 Confirmation of the traveling wave

4.2. *Direction control experiment*

We experimented to discuss whether the robot moved in the expected direction. In experiment, as shown in Fig.4.3, the gage point is at the center of the robot. Next, we made the y axial direction the 0° direction and made the robot move clockwise in the 45°, 90°, 135° and 180° directions, respectively.

The experimental results are shown in Fig.4.4. This shows the track that the robot moved to and the ideal track. The results suggest that the robot moved in the direction of travel with a reasonable accuracy. However, in the 45° and 135° directions, the movement track of the robot is slightly off the ideal track. It is thought that the robot has come off the movement track because the frictional forces are different in the elastic units on either side of the robot.

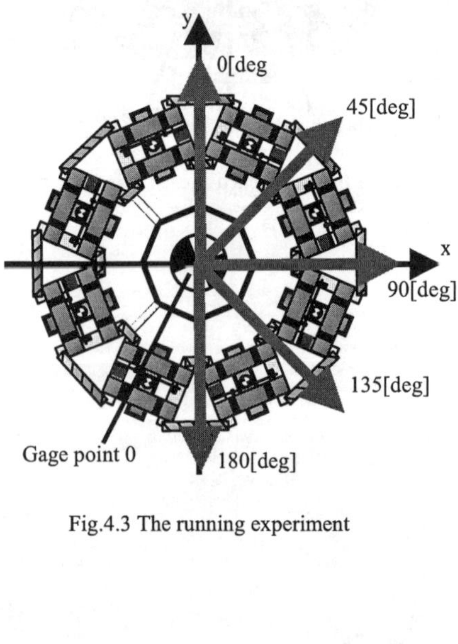

Fig.4.3 The running experiment

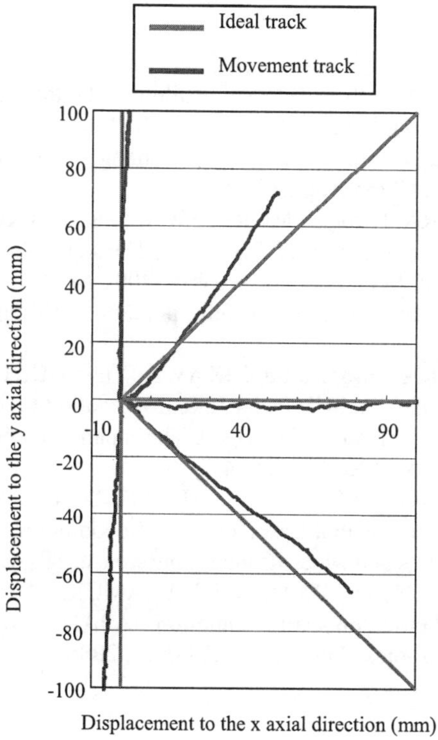

Fig.4.4 Track of the robot

5. Conclusions

In this research, we have developed an omni-directional mobile robot using the locomotion mechanism of the snail and have reached the following conclusions.

- We constructed an omni-directional mobile robot using the locomotion mechanism of the snail and moved the robot on a flat surface.
- We moved the robot and confirmed that it moves using the traveling wave locomotion.
- With the locomotion experiment, we confirmed that the robot moved in the expected direction with reasonable accuracy.

References

1. Brian Chan, N.J.Balmforth, A.E.Hosoi, Building a better snail: Lubrication and adhesive locomotion, PHYSICS OF FLUIDS, 2005
2. Bai Chen, Yinsheng Zhou, Daqiang Gu, Jie Zhong, A Novel Snail-Like Micro Robot, ICIMA, China August 2004
3. Ryousen FUJIHARA, Hirohisa MORIKAWA, Shun-ichi KOBAYASHI, The Mechanism of Pedal Locomotion of Gastropod, The Japan Society of Mechanical Engineers, JAPAN, vol67 No.658 (2001-6)
4. H.D.Jones, Circulatory pressures in Helix pomatia, L.Comp,Biochem.Physiol, 39A, 289,(1971)
5. Ryousen FUJIHARA, Hirohisa MORIKAWA, Yuya HUKUYA, Hiroshi SAKAI, Shunichi KOBAYASHI, Pedal-Like Locomotion Mechanism Modeled on Pedal Crawling of Terrestrial Gastropod, The Japan Society of Mechanical Engineers, JAPAN, Vol70 No.695 (2004-7)
6. Shohei IMAI, Kuniaki SATOU, Taro NAKAMURA, Development of an Omni-Directional Mobile Robot Based on Snail Locomotion, JAPAN, JSME Conference on Robotics and Mechatronics Conference, 1P1-C06, 2007
7. Yuki MITAMURA, Syohei IMAI, Taro NAKAMURA, Development of a traveling wave robot based on snail locomotion, JAPAN, JSME Conference on Robotics and Mechatronics Conference, 2P1-A19, 2006

GAIT PARAMETER ADAPTATION TO ENVIRONMENTAL PERTURBATIONS IN QUADRUPEDAL ROBOTS

E. GARCIA*, J. ESTREMERA, P. GONZALEZ DE SANTOS and M. ARMADA

Industrial Automation Institute - CSIC,
28500 La Poveda, Madrid, Spain
**E-mail: egarcia@iai.csic.es*
www.iai.csic.es/users/egarcia

Quadrupedal robots working outdoors are very slow robots prone to tumble down in the presence of perturbations. This paper presents a novel gait-adaptation method that enables walking-machine gaits to autonomously adapt to environmental perturbations, including the slope of the terrain, by finding the gait parameters that maximize robot's dynamic stability. Experiments with the SILO4 quadruped robot are presented and show how robot stability is more robust when the proposed approach is used for different external forces and sloping terrains.

Keywords: Quadruped robot; Gait adaptation; Dynamic stability margin.

1. Introduction

Walking robots designed for field and service applications usually perform statically stable gaits, which have been designed to optimize a static stability margin. Even when a dynamic stability margin is used to control the robot's motion, the gait pattern is a statically-stable one. As a result of this contradiction, the robot will still tumble down when confronted with any perturbing effect. Although an accurate dynamic stability margin measures robot stability, unstable situations are merely observed, not avoided, unless the gait pattern is modified. Existing dynamic gaits like the trot and the gallop require light, simplified robot designs with elastic actuators, which differ greatly from the heavy-limbed machines[1,2] used for field and service applications. Therefore, statically stable robot gaits should be modified based on a more suitable dynamic stability margin to improve gait control.

Only a few researchers have attended to partially solve this problem, achieving adaptation to uneven terrain, yet they did not consider robot

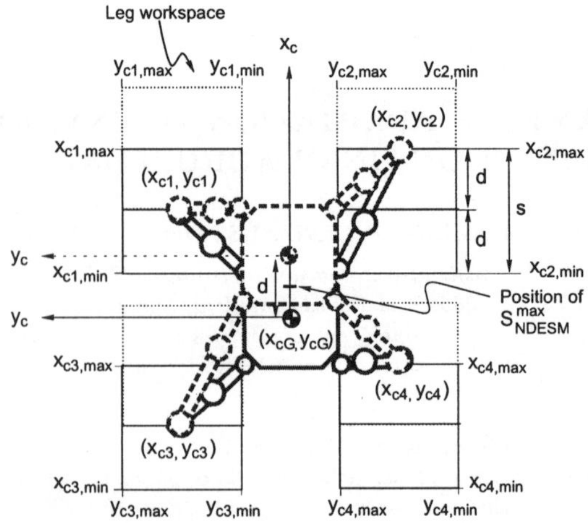

Fig. 1. Outlining of gait-parameter adaptation for a quadruped in the body-motion phase of a two-phase discontinuous gait

dynamics nor environmental perturbations.[3,4] This paper presents a new method for gait adaptation to the environment based on the Normalized Dynamic Energy Stability Margin, S_{NDESM}.[5] The method maximizes the dynamic stability margin during the transfer and support phases and thus we get a more robust gait against external perturbations. Section 2 describes the gait-parameter adaptation approach. Experiments using the SILO4 walking robot (see Figure 5) are described in Section 3 to show the improvement of the gait-parameter adaptation method when external forces are applied.

2. Gait Adaptation

Let us assume a quadruped robot walking along the direction of its longitudinal body axis. Let us define an external, fixed reference frame $\{x, y, z\}$ and a body-fixed reference frame, $\{x_c, y_c, z_c\}$ centered at the robot's center of gravity (CG) so that robot motion is along the x_c axis and the z_c axis is orthogonal to the body plane. The terrain can be inclined, forming an angle Ψ with the x-axis (pitch angle) and an angle θ with the y-axis (roll angle). The gait optimization procedure is divided into the following two phases:

2.1. *First phase: Leg-stroke and foothold calculation*

In this phase, foot and body CG positions in the body plane and the leg stroke that optimize the gait are determined. Figure 1 outlines a top view of leg workspaces, footholds and leg stroke for a quadruped during the body-motion phase in a two-phase discontinuos gait.[5] The four legs are in support, thus propelling the body forward. The initial position of the robot has been plotted in thick, solid line, while the final position of the robot has been plotted in thick, dashed line. Let us name x_{ci} and y_{ci} the x_c-coordinate and y_c-coordinate respectively of foothold i referring to the body frame $\{x_c, y_c, z_c\}$ just before the body-motion phase starts. The CG trajectory has been also plotted (x_{cG}, y_{cG}), finding the maximum-stability position at the middle of the trajectory, as explained before. The workspace of leg i is delimited by $x_{ci_{max}}$, $x_{ci_{min}}$, $y_{ci_{max}}$, $y_{ci_{min}}$. Notice that the length of the body motion is half the leg stroke, that is, $s = 2d$. Therefore, maximizing s is achieved by maximizing d, which can be expressed as a function of front legs' footholds:

$$d(x_{c1}, x_{c2}) = x_{c2} - x_{c1}. \tag{1}$$

The two rear legs' footholds can be expressed in terms of the two front legs' footholds:

$$x_{c3} = x_{c3_{min}} + x_{c2} - x_{c1} \tag{2}$$

$$x_{c4} = x_{c3_{min}} + 2x_{c2} - 2x_{c1}. \tag{3}$$

The stroke pitch in the y_c-axis direction, P_y, provides the following relation to foothold y_c components:

$$y_{c2} = y_{c1} - P_y \tag{4}$$

$$y_{c4} = y_{c3} - P_y. \tag{5}$$

To obtain the optimized gait, x_{c1}, y_{c1}, x_{c2}, y_{c3}, x_{cG} and y_{cG} have to be obtained so that S_{NDESM} is maximum at the middle of the body-motion phase. This condition will equal the probabilities of losing robot stability when a rear leg is lifted or when a front leg is lifted, thus decreasing the overall probability of tumbling. Therefore, S_{NDESM} should be expressed as a function of footholds and CG position in the body reference frame, that is (see Ref. 6 for a more detailed explanation):

$$S_{NDESM} = \mathcal{F}(x_{c1}, y_{c1}, x_{c2}, y_{c3}, x_{cG}, y_{cG}, z_{cG}). \tag{6}$$

Taking into consideration that in this first phase of the optimization process the CG height is considered constant, $z_{cG} = h_0$. S_{NDESM} has

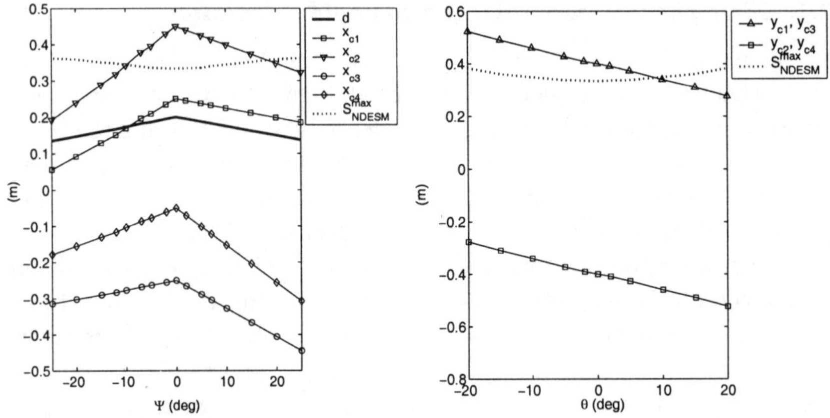

(a) (b)

Fig. 2. Results of first-phase gait optimization for sloping terrain (a) Effect of pitch angle, (b) Effect of roll angle.

to be maximum when the body CG is located at the middle of the CG trajectory; therefore the objective function is:

$$J_1 = \mathcal{F}\left(x_{c1}, y_{c1}, x_{c2}, y_{c3}, x_{cG} + \frac{d}{2}, y_{cG}\right). \tag{7}$$

However, another objective function exists, because the leg stroke must be maximum too:

$$J_2 = d(x_{c1}, x_{c2}) \tag{8}$$

with the following constraints:

$$x_{c1_{min}} \leq x_{c1} \leq x_{c1_{max}} \tag{9}$$

$$x_{c2_{min}} \leq x_{c2} \leq x_{c2_{max}} \tag{10}$$

$$y_{c1_{min}} \leq y_{c1} \leq y_{c1_{max}} \tag{11}$$

$$y_{c3_{min}} \leq y_{c3} \leq y_{c3_{max}} \tag{12}$$

$$4x_{c1} - 3x_{c2} \geq \xi_1 - x_{c1_{max}} \tag{13}$$

Constraints (9) to (12) are given by the leg-workspace limits, while constraint (13) avoids leg-workspace overlapping, where $\xi_1 \geq 0$ is a constant.

To solve the multi-objective problem, the ε-Constraint Method is used.[7] The problem has been solved numerically for the SILO4 robot in an iterative manner starting from an initial estimate of foot positions x_1 and x_2. Figures 2 and 3 show results of the first phase optimization for the medium-sized

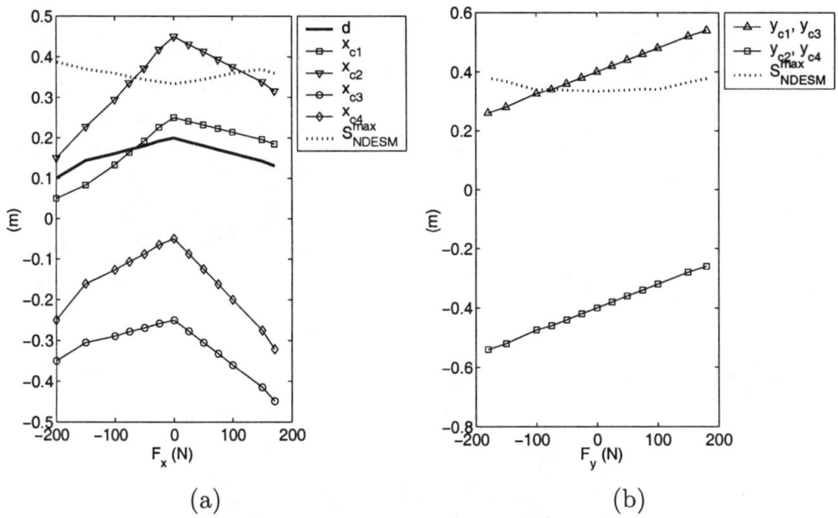

Fig. 3. Results of first-phase gait optimization for external forces (a) x component of force, (b) y component of force.

quadruped robot SILO4. Figure 2 shows footholds, leg stroke and maximum S_{NDESM} obtained for a robot walking on sloping terrain. As the terrain angle increases the robot has to modify foothold x_c and y_c components linearly to decrease the leg stroke (see d) and increase S_{NDESM}^{max}. Similar results have been found at the first phase optimization when external forces are applied (see Figures 3(a) and (b)). Note that the direct consequence of the linear reduction of leg stroke for increasing perturbations is a restriction in leg workspace. Thus, the bigger the disturbance, the bigger the reduction of leg workspace.

2.2. *Second phase: CG-height calculation*

When, as a result of the first phase of the optimization approach the leg stroke is reduced to adapt the walk to a stepped ground or an external force and enhance stability during the body-motion phase, stability is reduced during the transfer phase. This problem can be solved by reducing the CG height. Thus, the following condition is stated:

$$\mathcal{F}(h) - S_{NDESM}^{0} = 0 \tag{14}$$

where S_{NDESM}^{0} stands for a constant value of the gait stability margin in nominal conditions, that is, over flat terrain when there are no external disturbances. The function $\mathcal{F}(h)$ is the result from the first phase of the

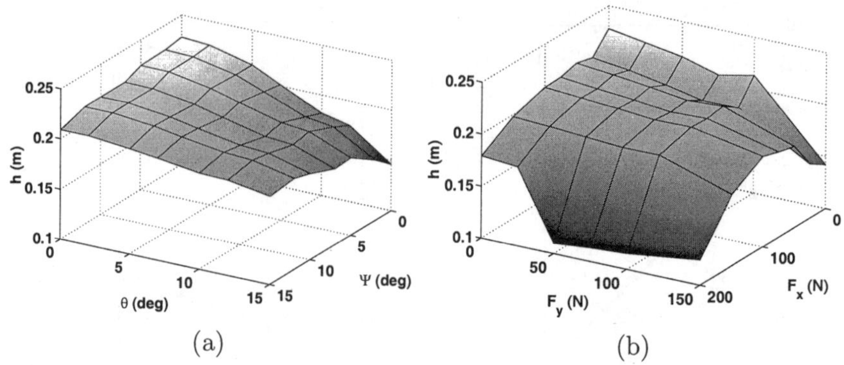

Fig. 4. Results of second-phase gait optimization for (a) sloping terrain, (b) external force.

Fig. 5. Quadruped robot SILO4 pulling a 50-N load

optimization procedure, where every gait parameter has been obtained. Therefore, the only variable at this phase is CG height (h). This problem is a nonlinear equation of a single variable that has been solved by least-squares optimization.

Figure 4 shows results of CG height for the SILO4 walking (a) in sloping terrain and (b) when external forces are applied.

3. Experiments with SILO4

The improvement of gait stability obtained by the use of the gait-optimization method herein proposed has been proved experimentally using the SILO4 robot[5,8] shown in Figure 5. A real service application has been simulated. The SILO4 robot walks while it pulls a load, thus being affected by an external force $F_x = -50\,\mathrm{N}$. As a consequence of the perturbation the robot finally tumbles down. Then, the experiment is repeated using the gait-optimization approach, which modifies the gait pattern for the -50-N

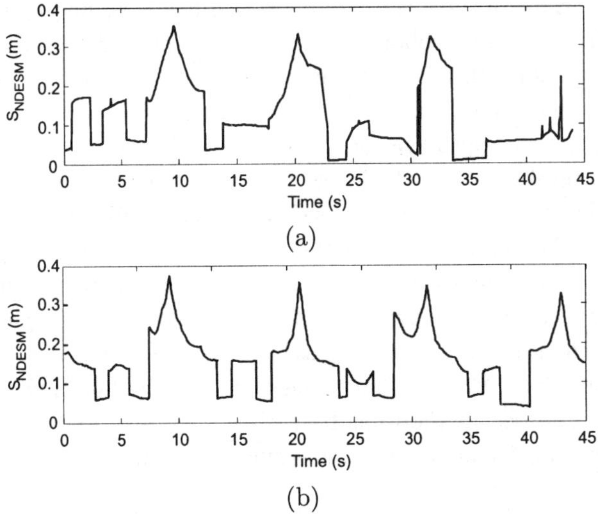

Fig. 6. Time evolution of S_{NDESM} for the SILO4 robot walking against a 50-N disturbance. (a) Standard gait, (b) Optimized gait using the proposed approach.

external force. In this experiment the robot pulls successfully the same load that made the robot tumble down in the previous experiment.

Figure 6(a) shows the evolution of the stability margin during the first experiment, when the SILO4 robot walks against the disturbance using a standard two-phase discontinuous gait. The stability margin drops during the transfer phase of the rear legs (because the destabilizing effects of leg transfer and force disturbance combine). Let us compare Figure 6(a) with Figure 6(b) which shows the evolution of the stability margin during the second experiment, when the SILO4 robot walks using the results of the gait optimization. As a result of adapting the gait to a -50-N force, the stability margin is symmetrical for the transfer phases of both the rear and the front legs. As a result of the adaptation to a -50-N force, the gait is more stable than in Figure 6(a), increasing the gait stability margin by 70% in this experiment. As shown in the gait optimization results, the use of the proposed gait-parameter adaptation approach increases the robot's robustness to external disturbances from a 50-N load up to a 200-N load, that is, four times the perturbation in this particular robot.

4. Conclusions

A new gait-adaptation method has been presented in this paper to cope with external perturbations. The method enables quadrupedal gaits to adapt to the slope of the terrain and external forces by finding the gait parameters that maximize robot stability. The resulting footholds of the gait optimization approach are linear functions of external forces and terrain-inclination angles. This reduces the computation time and enables the application of the gait-optimization method in real time.

Experimental results with the SILO4 robot have shown how robot stability is enhanced when the proposed approach is used to counteract external forces. The improvement in gait stability margin when applying the proposed gait-adaptation method to the SILO4 robot has been proved to be 70 percent for a -50-N external force. Also, the resulting gait increases the SILO4 robot's robustness to external disturbances from a 50-N load up to a 200-N load, that is, four times the perturbation in this particular robot. The proposed method has been shown to be of major relevance for the use of walking robots in field and service applications, where robots and environment interact.

Acknowledgements

This work has been funded by the Spanish Ministry of Education and Science through Grant DPI2004-05824.

References

1. P. Gonzalez de Santos, E. Garcia, J. Estremera and M. Armada, *International Journal of Systems Science* **36**, 545 (2005).
2. T. Doi, R. Hodoshima, Y. Fukuda, S. Hirose, T. Okamoto and J. Mori, *Journal of Robotics and Mechatronics* **18**, 318 (2006).
3. D. Wettergreen and C. Thorpe, Developing planning and reactive control for a hexapod robot, in *Proc. IEEE Int. Conf. Robotics and Automation*, (Atlanta, Georgia, 1996).
4. H. Tsukagoshi and S. Hirose, Intermittent crawl gait for quadruped walking vehicles on rough terrain, in *Int. Conf. Climbing and Walking Robots*, (Brussels, Belgium, 1998).
5. P. Gonzalez de Santos, E. Garcia and J. Estremera, *Quadrupedal Locomotion: An Introduction to the Control of Four-Legged Robots* (Springer-Verlag, London, 2006).
6. E. Garcia and P. Gonzalez de Santos, *IEEE Transactions on Robotics* **22**, 1240 (2006).
7. V. Chankong and Y. V. Haimes, *Multiobjective Decision MakingTheory and Methodology* (Elsevier, New York, 1983).
8. Industrial Automation Institute, C.S.I.C., Madrid, Spain, *The SILO4 Walking Robot*, (2002). Available: `http://www.iai.csic.es/users/silo4/`.

INTELLIGENT SPIDER* WALKING ROBOT FOR ROUGH TERRAIN

MICHAEL MCCREADY, LIQIONG TANG AND GURVINDER VIRK

Institute of Technology and Engineering
Massey University
New Zealand

This paper presents the design of a legged walking robot capable of traversing rough terrain that would be difficult to negotiate using wheeled vehicles. The body of the robot looks like a spider but there are only six legs in order that various insect gaits can be studied. Each leg is driven by a worm gear which allows each leg to make a vertical movement individually thereby realizing easy control of the leg height. The design incorporates two cam systems to synchronize the horizontal movement of the legs on both sides of the robot's body. Each cam is driven by a stepper motor and only controls the three legs on one side. Such a design allows steering of the robot body to be controlled by the speed difference of the two motors. The mechanical system of the robot and the motor driving systems have been built. An intelligent control system is still under development. On completion, it is expected that the robot will be smart enough to walk around and collect useful information for its applications. The paper presents the details of the mechanical system, the walk patterns, and the driving system. The intelligent features to be added to the robot are also discussed.

* SPIDER – Special Purpose Insectoid Drive Experimental Rover

1. Introduction

The aim of this project is to design and build a legged walking robot, much like a spider, that would be capable of traversing terrain that would otherwise be difficult or unstable for a wheeled or tracked vehicle. Unlike other types of systems, walking robots shown in Figure 1 are more complicated to design due to their relative complexity, as they normally require complex mechanical systems to support the walking algorithm [1],[2]. Most existing walking robot requires many motors to realize the movements and are often suited for a sole task. Without major modification to the design, it is difficult to use for another job. Therefore a general purpose walking robot, which could be easily adapted depending on the job required, is of interest. This project emphasises simplicity in the design with mechanical components that are easy to manufacture. It incorporates modularity in the design concept so that the later development of add-on units such as vision, sound, and other mechanical and electronic control units which would use their own processor can be added to enhance the control and movement commands to the controller for the legs.

NASA Spider-bot – Space Exploration

Plustech Walking Forest Machine - Forestry

Figure 1. Walking robots

The robot presented is a six-leg walking robot as shown in Figure 2. Such a design gives better proportions of width to length and also reduces the number of components and weight. The body of the robot are built by layers, Each layer has a base plate and all the base plates have the same profile and dimensions. Such a body structure provides an easy add-on platform for future developments to be added. A cam system is employed to control the robot's forward and backward leg movements. Leg lift is individually controlled through six small motors, which makes the robot able to walk on uneven and sloped surfaces. The control system is under development. The system will provide an easy platform for future development to integrate with new software and hardware and add-on new functionalities to the robot.

Figure 2. Massey's SPIDER Robot

2. Mechanical Driving System

The SPIDER robot has six legs. It is possible to let three legs on the ground to support the body while moving the other legs forward or backward. This project prefers using four legs to support the robot. There are different methods and algorithms to control the robot legs to achieve different walking gaits [3], [4], [5]. After a good study and research on leg movement, the decision was made to incorporate a cam system to realise the horizontal leg movement. The cam system is required to make four of the six legs, two on each side, stay on the ground at a time giving maximum support to the robot body and also obtain a good stability while two of the legs move forward or backward. A leg lifting system is designed to work closely with the cam system. Only until the two legs are lifted, can the two legs move in the desired direction. The cam system developed consists of six cams, three on each side of the robot. All six cams have identical profile that has been specifically designed, which moves the leg forward within the first third of a revolution. During the other two third revolution, the leg stays on the ground as shown in Figure 3. With a 120° offset between the three cams on one side of the robot, it is possible to have two legs on the ground at a time. Combining the legs on both sides, the cam system is able to move two legs forward while the other four legs stay on the ground to support the robot body. With different offset between the two sides, it is possible to make any two legs as a pair. This provides more choices for studying the moving algorithm.

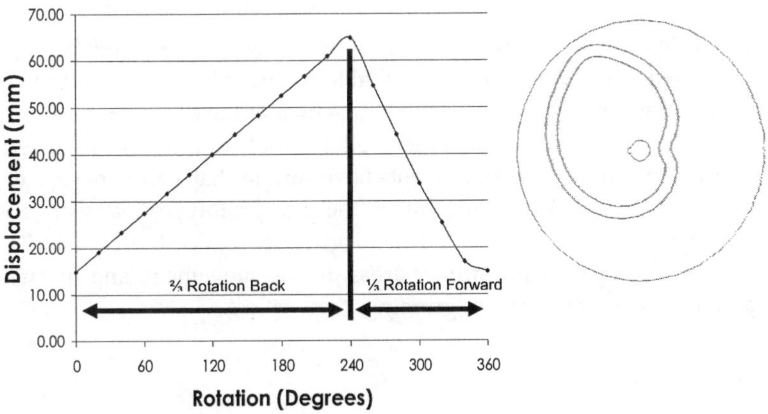

- Takes ⅓ rotation to lift leg forward
- Then ⅔ rotation to pull robot forward
- Cams 120° out of phase so four legs always on ground

Figure 3. Cam profile and offset

The cam driving and linkage system employs two stepper motors where each motor controls the three cams on one side of the robot. The mechanical arrangement is presented in Figure 4. Through the bevel gears, each stepper motors is able to drive the three cams underneath the second layer board which linked to the three robot legs. The cam profile is designed based-on the cam driving system and the dimension of the linkage. The distance required to move the leg forward and backward is also one of the main inputs for the cam profile design.

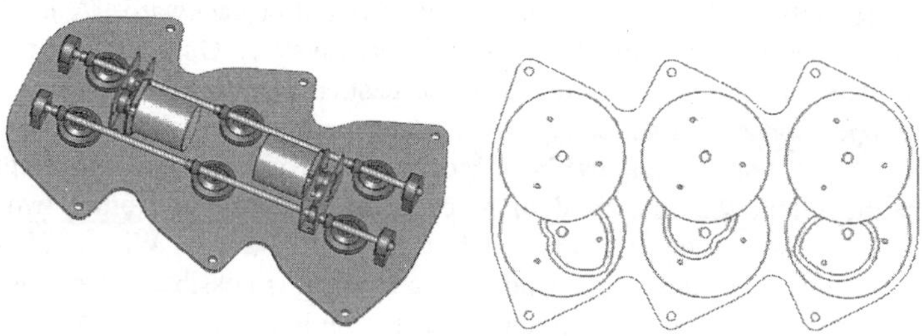

Figure 4. Cam driving and linkage system

3. Leg Structure and Control

The robot leg structure, apart from realising forward and backward movement, is also required to be able to make the leg move up and down. The leg design is made on the basis that the leg components must be easily manufactured with the available tools in a typical engineering workshop. All the leg components have simple shape and are easy to be made and assembled. Since this is the first prototype it was preferred not to use bearings in the joints in order to reduce the cost, but add washers to separate the surface areas during movement, and use nylon at the waist joint to prevent grinding.

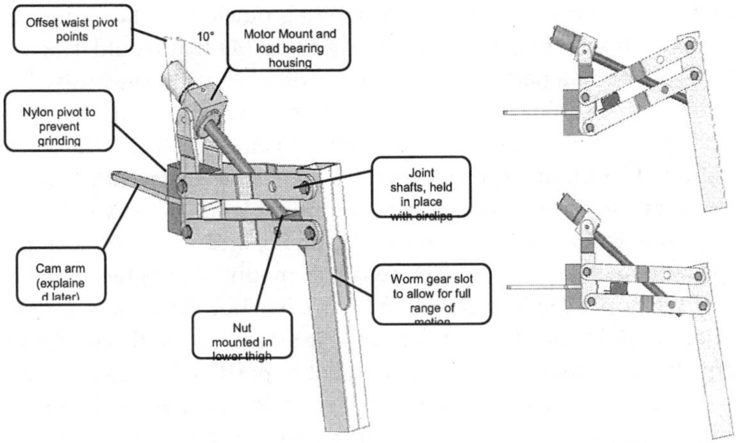

Figure 5. Robot leg assembly

In the leg assembly shown in Figure 5 the pivot point from the waist is designed to have an angle of 10° from the vertical. This makes the leg always be splayed at an angle of 10° to make it more stable. To complement the lifting movement, the motor is placed above the waist and the nut mounted in the lower section of the thigh, close to the knee. This means the motor's output torque would be utilised to a much greater potential.

A test model was made and produced to check the design. The model was capable of making the required movements. The prototype also revealed there was too great a stress on the output shaft of the motor. The modified version incorporates a bearing to the mount to take the load off the motor. Such a modification worked well and a further five legs were built to complete the full prototype model. Figure 5 shows the complete leg design with the key parts highlighted.

The robot leg can move up and down and forward and backward. To synchronise these movements such as the leg lift must be activated before any forward movement, a sensing system must be implemented to give the appropriate feedback to the robot controller. Such a sensing system can also prevent parts from moving outside the limit positions. Two limit sensors are installed in the leg assembly that would 'catch' the leg if it moved to a particular limit and send a signal to the controller to stop driving. This also prevents the worm gear winding out of the nut if the leg was lowering, and the motor mount being crushed if the leg was lifting up. The mechanical structure of the robot leg provides an ideal location for the sensors, which makes it possible to use only one sensor per leg to detect both upper and lower limits, and to keep tracking the leg position using software. The sensor is mounted in between the two rails of the upper thigh as shown in Figure 5. Floor detection sensors are planned to be installed in the robot's feet to get the feedback when the leg touches the ground, allowing the robot to position its legs at different levels on uneven terrain. The robot feet design is another design task. The floor detection sensors will be installed at the bottom of the robot feet.

4. Robot Walking Patterns

4.1 *Forward and Backward Movement*

The cam drive systems synchronise the robot to move forward and backward. Table 1 is an example to move the robot to make forward movement. As stated above, the cam system can only make two legs move forward at a time and the other four will stay on the ground to pull the robot body. Through offsetting the cams, it is possible to make any two legs as a pair to move forward. Table 1 illustrates the movement of the legs when the cams rotate one revolution. The symbol "▲" represents the legs moving forward and "■" means that the legs stay on the ground to pull the body forward.

Table 1. Robot Forward and Backward Movement

	1° -120°		
Left Leg 1	■	▲	Right Leg 1
Left Leg 2	▲	■	Right Leg 2
Left Leg 3	■	■	Right Leg 3
	121° -240°		
Left Leg 1	■	■	Right Leg 1
Left Leg 2	■	▲	Right Leg 2
Left Leg 3	▲	■	Right Leg 3
	241° -360°		
Left Leg 1	▲	■	Right Leg 1
Left Leg 2	■	■	Right Leg 2
Left Leg 3	■	▲	Right Leg 3

4.2 *Left and Right Movement*

The left and right movement is controlled by DC motors. The six DC motors mounted on each of the robot legs can move the robot leg individually to different height. When the robot is under the left and right movement mode, the cam system is locked. Each move on the left or right will only lift three legs. The other three are kept on the ground to support the robot's body. Table 2 illustrates one of the possible ways to make the robot walk to the right.

Table 2. Robot Left and Right movement

Move to right			
	Step 1		
Left Leg 1	Down	Up	Right Leg 1
Left Leg 2	Up	Down	Right Leg 2
Left Leg 3	Down	Up	Right Leg 3
	Step 2		
Left Leg 1	Down	Right stretch	Right Leg 1
Left Leg 2	Right stretch	Down	Right Leg 2
Left Leg 3	Down	Right stretch	Right Leg 3
	Step 3		
Left Leg 1	Up	Down	Right Leg 1
Left Leg 2	Down	Up	Right Leg 2
Left Leg 3	Up	Down	Right Leg 3
	Step 4		
Left Leg 1	Right stretch	Down	Right Leg 1
Left Leg 2	Down	Right stretch	Right Leg 2
Left Leg 3	Right stretch	Down	Right Leg 3
	Step 5		
Left Leg 1	Down	Down	Right Leg 1
Left Leg 2	Down	Down	Right Leg 2
Left Leg 3	Down	Down	Right Leg 3

5. Robot Control System

Two stepper motors are employed to drive the robot's legs move forward and backward and six DC motors for individually lifting the legs up and down. The motor driver circuits for the two stepper motors are built using two ICs, L297 and L298. Additionally there is an on-board chopper circuit, designed to take current sensing signals from the driver, and a reference voltage input, and generate a PWM output to the driver. This allows the use of a higher voltage supply than the motor is rated at, and keeps a higher current through its windings to maximise the output power. The six DC motors are controlled through a board made using L293 driver ICs. The prototype control system uses a microcontroller from Silicon Laboratories. The 8051 microprocessor has the features the project required. After integrating the motor drive circuit boards and the microcontroller, the robot is able to move each of the legs up and down and drive forwards and backwards by the cams. Further developments are still under way. Once the project is completed, the SPIDER should be able to move around with a wireless user interface.

6. Conclusion

The spider robot presented integrates mechanical design, electronic circuit board design, control algorithms development, and programming. It produces a good platform for study and testing different walking patterns and control algorithms. The layered mechanical structure also makes it easy for future add-ons and to integrate new electronic, mechanical and software modules.

References

1. P. G. de Santos, E Garcia and J Estremera, "Quadrupedal locomotion : an introduction to the control of four-legged robots", *Springer* (2006).
2. S. Shoval, E. Rimon and A. Shapiro, "Design of a Spider-Like Robot for Motion with Quasistatic Force Constraints", *Proceedings of the 1999 JEEE International Conference on Robotics & Automation*, Michigan (1999).
3. John J. Craig, "Introduction to robotics: mechanics & control", *Addison-Wesley Pub. Co.* (1986).
4. R. Quinn, G. Nelson, R. Bachmann, D. Kingsley, J. and R. Ritzmann, "Insect Designs for Improved Robot Mobility", *Proc. 4th Int. Conf. On Climbing and Walking Robots, Berns and Dillmann eds., Prof. Eng. Pub.*, 69-76 (2001).
5. R. Bachmann, G. Nelson, R. Quinn, etc., "Construction of a Cockroach-like Hexapod Robot", *In Proc. Sixth IASTED Int. Conf. on Robotics and Automation*, Banff, Canada, pp. 22-27 (1998).

KINEMATICS, SENSORS AND CONTROL OF THE FULLY AUTOMATED FACADE CLEANING ROBOT SIRIUSC FOR THE FRAUNHOFER HEADQUARTERS BUILDUNG, MUNICH

NORBERT ELKMANN, MARIO LUCKE, TINO KRÜGER, DIETMAR KUNST, THOMAS STÜRZE

Robotic Systems, Fraunhofer Institute for Factory Operation and Automation, Sandtorstrasse 22, 39106 Magdeburg, Germany

The Fraunhofer Institute for Factory Operation and Automation IFF has developed the automatic facade cleaning robot SIRIUSc for use on the Fraunhofer-Gesellschafts headquarters, a high-rise building in Munich, Germany. The building has a height of 80m, its facade an area of 4000 m². Apart from the robot that moves along and cleans the facade, the complete, fully automated system consists of a fully automated gantry that secures, supplies energy to and above all positions the robot. Part of the project involved completely automating a standard gantry, which is an integral part of the complete facade cleaning robot system. This article presents an overview of the significant basic functions of the robot and the gantry, emphasizing the kinematics, the control and sensor systems for navigation and the cleaning sequence that employs the extensive fully automatic functions of the robot and gantry.

1. Introduction

The complete façade cleaning robot system consists of more than the robot alone (Figure 1). The rooftop gantry positions the robot at the top of the facade on every pane path to be cleaned.

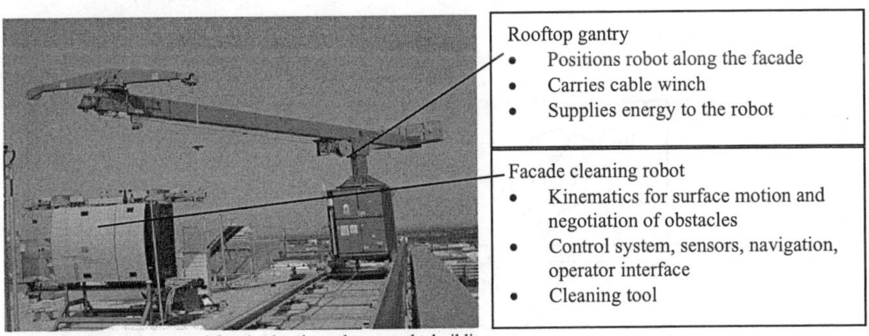

Figure 1. Rooftop gantry and facade cleaning robot atop the building

The robot then descends the facade vertically and cleans when it ascends. Four cables connected to the gantry atop the building safeguard the robot against falling. Since the cables must be taut to ensure the robot is secure, the cables also used to position the robot as the winding and unwinding of the winch on the rooftop gantry moves it vertically along the building as well as to bear the load of the robot. Cables transmit data and supply power too. The robot is one of the first fully automatic systems of this kind used in a public setting. The robot weighs 450 kg, the gantry 5000 kg.

2. Rooftop Gantry Crane

The gantry (Figure 2) possesses three degrees of freedom relevant for positioning the robot on the facade: Movement along the rails and two rotary cantilever arms.

Measuring systems help position the gantry on the roof for its start on the pane path to be cleaned. The gantry has two asynchronous drives to move along the rails. Two encoders determine the gantry's position. The gantry is positioned by a continuous controller that reads the pane path positions taught out of a data module of the control system. Another redundant measuring system has been installed because of the slip.

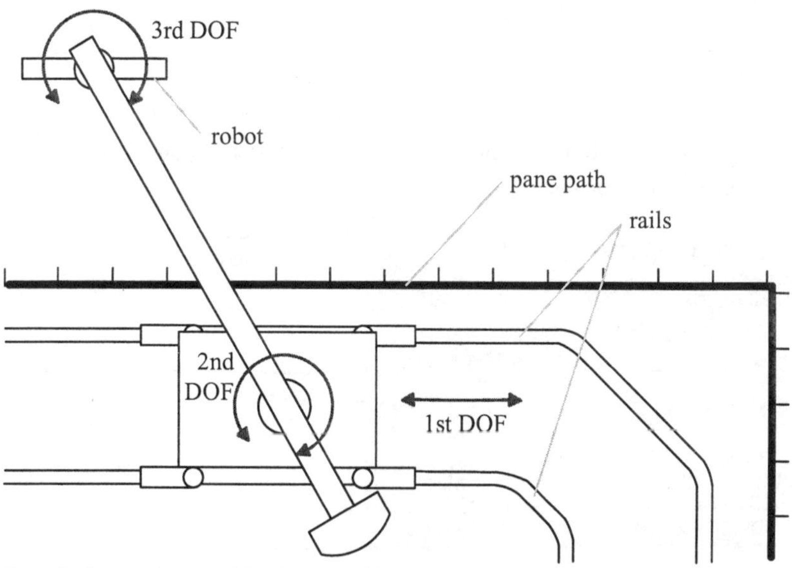

Figure 2. Gantrys degrees of freedom to position and deposit the robot on the facade

When the encoder target position is reached, the transponders mounted on each of the pane path target positions must be matched with the transponder system. If the transponder assigned to each pane path can be read out successfully, the target position is reached and the positioning of the cantilever arm can begin. Otherwise an error code is generated and the start on the pane path must be repeated.

A synchronous movement of the two rotary cantilever arm drives sets the robot on the facade. To achieve a high level of security when positioning the robot, its orientation parallel to the facade during motion must be guaranteed. A cascaded controller in the gantry control system ensures this.

A particular challenge is positioning the robot and thus the gantry on the corners of the facade since the rails executes a curve and do not run parallel to the facade. Here the gantry's traveling mechanism (translation) and the cantilever arms (rotation) must run sequentially synchronized to prevent any collision of the robot with the facade. In order to reduce the robot's distance to the facade and be able to deposit the robot on it, the gantry must proceed along the rails. This simultaneously influences the robot's orientation plane-parallel to the facade negatively. To simplify the positioning process, the traveling mechanism is guided in a second step to the window pane target position and then the cantilever arms are moved again so that the robot reaches its ideal position (distance of approximately 100mm and parallel to the facade). All operations are fully automated. The gantry is a standard piece of equipment that was completely automated in the project.

3. SIRIUSc Kinematics

Above all, the modular kinematics (Figure 4) ensures the robot remains in constant contact with the facade and is able to navigate a multitude of typical obstacles and move quickly along a facade. The kinematics is based on a structure of two pairs of linear modules, the so-called "advanced sliding module mechanism" [8,9]. Two linear modules constitute one pair that performs the same linear movement, thus ensuring secure and stable contact with the facade.

Servo drives move each of the outer or inner sucker units to the next position on a pane within a frame. The linear modules with their suckers are positioned in such a way that the suckers are located above or below the horizontal pane frames when docking onto the facade. Once a sucker unit has suctioned on, the drive's brake is deactivated and the other unit's suckers are released and retracted. The winding and unwinding of the securing cables on the gantry lift winch produces the robot's upward and downward motion on the facade. The alternation of the outer and inner sucker units each with 2x3 suckers produces the walking motion of the robot. While the robot is lifted or lowered to the desired position, the other activated linear module pair is moved to the next free position.

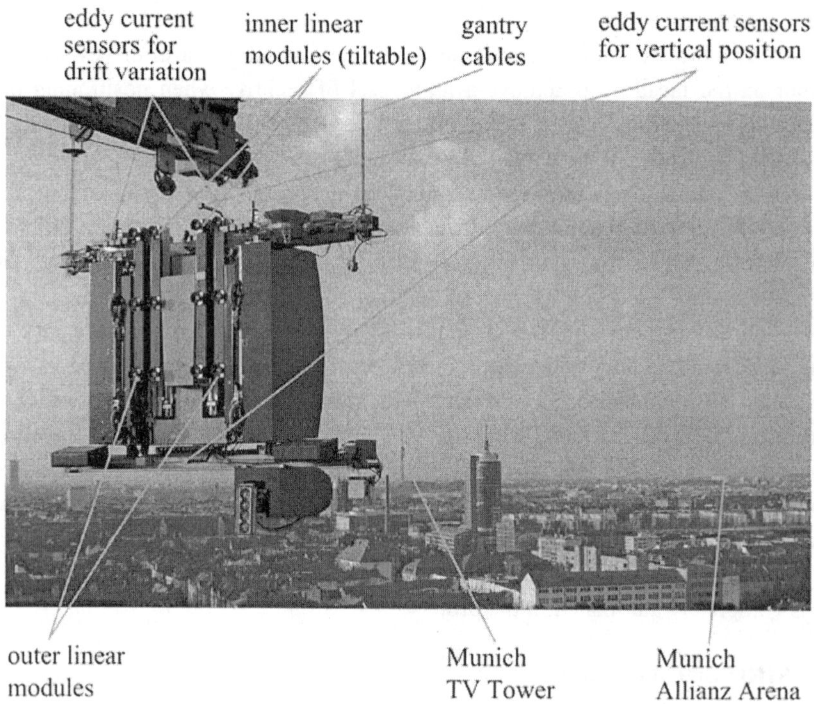

Figure 3. Kinematics and navigation sensors of SIRIUSc

Eddy current sensors mounted on the top and bottom of the robot's body detect and store the positions of horizontal pane frames.

One of the requirements for outdoor operation in high winds is that the robot is able to correct its direction of motion, should it drift a little off course. To this end, the inner pair of linear modules is tilt adjustable to enable small steering movements to keep the robot on a straight path. Two drift scanners have been mounted on the robot to control drift compensation. The scanner's job is to detect the robot's position in relation to pane frames (vertical pane jambs) in order to systematically control the inner linear modules. A drift scanner consist of 2x4 eddy current sensors (spaced 50mm), which a controller assembly switches off and on in succession. This is necessary since the eddy current sensors would otherwise interfere with one another. Depending on the robot's direction of movement and which eddy current sensors detect the metallic pane frames, the pneumatic drift compensation drive turns the inner linear guide right, to the middle or left.

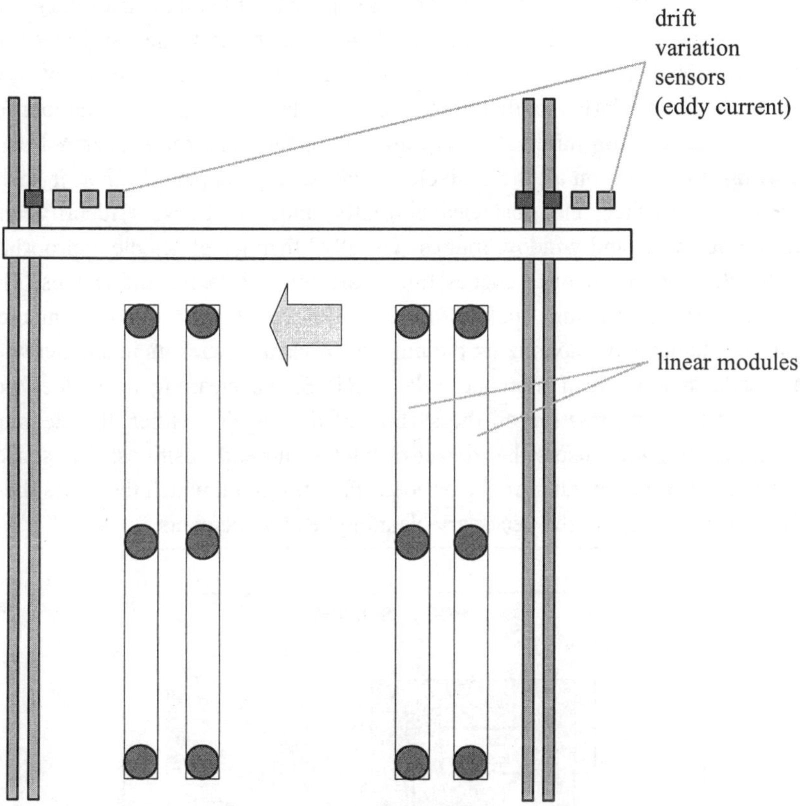

Figure 4. Schematic diagram of the detection of drift variation

The pane frames are made of various materials and profiles. This represents a special challenge. In extensive testing with various types of sensors, only novel eddy current sensors with a detection distance of 0-50mm were sufficiently reliable. None of the optical or other sensors were sufficiently reliable under the given conditions (rain, reflected sunlight on the panes).

4. Control System and Navigation

The heart of the system is its control system, which receives and combines sensor data and operator instructions to generate robot actions. Selected for it is stability and modularity, a programmable logic controller (PLC) is used for the robot and the gantry control system. The PLC is on board and controls the entire system. It synchronizes the walking mechanism with the trolley and cleaning head. All robot motions and actions are fully automated. The robot was programmed modularly so that it can be transferred to a large number of different facades with minimal reprogramming effort. The robot does not start out with information on all the obstacles it will face in its path. Rather, it keeps track of the surface and obstacles currently under it. Sensors identify and measure obstacles and window frames. The PLC then generates the appropriate step lengths for the robot to successfully walk over obstacles and frames. The PLC ensures that vacuum suckers directly over an obstacle are not engaged while simultaneously maximizing the number of vacuum suckers in contact with the facade at any given moment. Like SIRIUS, the cleaning head does not require detailed information on the surface of the facade. Rather, the cleaning system has its own sensors that detect obstacles and end positions. The sensor signals are also incorporated in the onboard PLC program, which then uses the information to generate the necessary cleaning head movements.

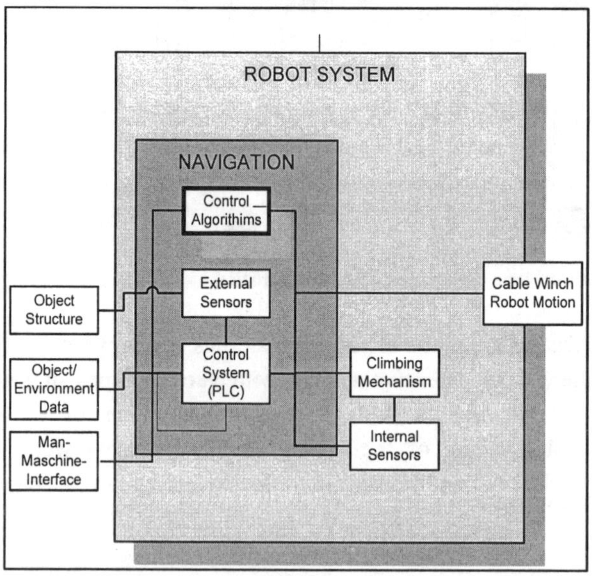

Figure 5. Control system concept

The only manual input information the robot requires before starting off on a given surface is end point data such as the height of the structure or building, off limit zones and the maximum length of obstacles. The robot automatically detects any other necessary surface information during operation. An operator master display is also located in the building so that an operator can monitor the robot's progress. A remote maintenance module allows downloading the robot's status and sensor data, uploading new program modules and executing simple operator commands such as the motion commands for a pair of linear modules, all over the Internet.

Little knowledge about the general structure of a building's surface is needed before robot movement can be generated. The input data includes end positions, moving distances and path characteristics. This a priori data is supplemented by online sensors that detect the facade surface and search for possible obstacles. In addition to identifying obstacles, the external sensor technology also corrects the direction of motion. Sensors detect where the robot must deviate from a path, e.g. girders or window and panel seals. The robot control system communicates with the building control system, making sure all windows are closed in areas being cleaned.

Figure 6. SIRIUSc on the high-rise building of the Fraunhofer-Gesellschafts headquarters in Munich

5. Conclusion

The Fraunhofer IFF has developed a fully automated facade cleaning robot for the headquarters of the Fraunhofer-Gesellschaft in Munich. SIRIUSc has been developed for vertical facades and consists of the main components robots mechanics and kinematics, rooftop gantry, sensor systems to detect facade shape, frames and obstacles, control technology and navigation, power supply and integrated cleaning unit. SIRIUSc was delivered to the facility management of the Fraunhofer-Gesellschafts headquarters building in 2006. The facility management staff is now able to clean the facade without any technical support from the researchers at the Fraunhofer IFF.

References

1. S. Hirose, K. Kawabe: Ceiling Walk Climbing Robot Ninja-II. Proceedings of CLAWAR 1998, First International Symposium on Mobile, Climbing and Walking Robots, Brussels 26.-28. October, 1998, pp. 143-147
2. M. M. Cusack, J. G. Thomas: Robotics for the Inspection of Vertical Surfaces of Buildings and Structures. In 25th ISIR, pp. 287-295.
3. N. Elkmann, T. Felsch, M. Sack, T. Boehme: Modular Climbing Robot for Outdoor Operations. Proceedings of CLAWAR 1999, Second International Conference on Climbing and Walking Robots, Portsmouth 13.-15. September, 1999, pp. 413-419
4. N. Elkmann, T. Felsch, M. Sack, T. Boehme, J. Saenz: SIRIUS: Modular Climbing Robot for Facade Cleaning and Other Service Jobs. International Conference on Field and Service Robotics FSR 2001, Helsinki
5. N. Elkmann, T. Felsch, M. Sack, J.Saenz, J. Hortig: Innovative Service Robot Systems for Facade Cleaning of Difficult-to-Access Areas. International Conference on Intelligent Robots and Systems IROS 2002, Zurich
6. N. Elkmann, Lucke M., Krüger T., Kunst D, Stürze T.: SIRIUSc: Fully Automatic Facade Cleaning Robot for a High-rise Building in Munich, Germany. ISR/ROBOTIK 2006, Munich

ON FOUR LEGS TOWARDS FLEXIBLE AND FAST LOCOMOTION

C. KARA, C. HECKHOFF, T. BRANDT, D. SCHRAMM

Chair of Mechatronics, University of Duisburg-Essen,
47057 Duisburg, Germany
E-mail: {kara, heckhoff, brandt, schramm}@imech.de
www.imech.de

The breathtaking development of electronic components is one key factor in the fast evolution of mobile robots. Depending on the mission, robots might operate autonomously or in cooperation with an human operator. In either case the locomotion of the robot is a key element. In general, legged systems are more flexible compared to wheel based systems and can be deployed in impractical terrain. Quadruped locomotion has the particular advantage to feature static as well as dynamic walking. Compared to $n > 4$ legs quadruped motion can be very agile, which is impressively demonstrated by cheetahs being the fastest mammal runners. On the other hand, quadrupeds feature static walking, which is beneficial for large scale walking machines such as legged excavators. In this paper a static free-gait pattern for quadrupeds allowing interaction with an human operator is discussed. Since experiments with large scale walking machines are very expansive, a scaled prototype is presented. However, the scaled prototype differs from the original system in many details. To guarantee the portability of the free-gait algorithm, the scaled workspace of the feet of the walking machine is proposed as criterion.

Keywords: motion generation, gait pattern, free-gait, quadrupedal walking, scaled prototypes, walking machines ALDURO and ADONIS

1. Introduction

Mobile robots evolve very fast. Sensors, actuators and electronic control units are among the key elements promoting this development. However, mobile robots are mainly characterized by their way of locomotion. The most general classification are wheel based and leg based systems. Wheel based systems allow a very energy efficient and fast motion, but are not applicable in impractical terrain. Here, legged systems are beneficial.

Systems with $n > 3$ legs are particular flexible. They can walk dynamically, but besides, walk statically and, if indicated, stand still. However, focusing on dynamic motion the number of legs should be limited. For $n > 4$ legs the additional mass of the system generally reduces the dynamic performance. This observation is affirmed by the cheetah being the fastest of all land animals. It can reach speeds up to 110 km/h in short sprints up to 460m and it is able to accelerate from 0 to 100 km/h in 3.5s.

For these reasons quadruped systems seem to be an interesting approach to flexible as well as to agile locomotion. As a first step towards quadruped locomotion including static as well as dynamic walking, this paper addresses a static free-gait pattern. Here, emphasis is put on flexible motion. Hence, this pattern does not rely on precomputed motion sequences. The gait can easily adapt to environmental changes or commands of an human operator. Besides that, the control software is developed in a way, that it is generic for any quadruped walking machine. Due to the characteristic of the free-gait pattern, the criterion of portability is the workspace of the feet. This is shown by means of two quadruped walking robots ALDURO[1] (Amphiboreptilomorphically Legged and Wheeled Duisburg Robot, see Fig. 2) and ADONIS (ALDURO Demonstrator with Onboard Navigation and Intelligent Step Generation, see Fig. 3).

2. Motion Control Software for Generic Quadrupeds

Thinking of a generic system includes the conception of a flexible software. This issue was one of the main interests during the development of ALDURO. The result is a division of the control software into three main layers as shown in Fig. 1.[2]

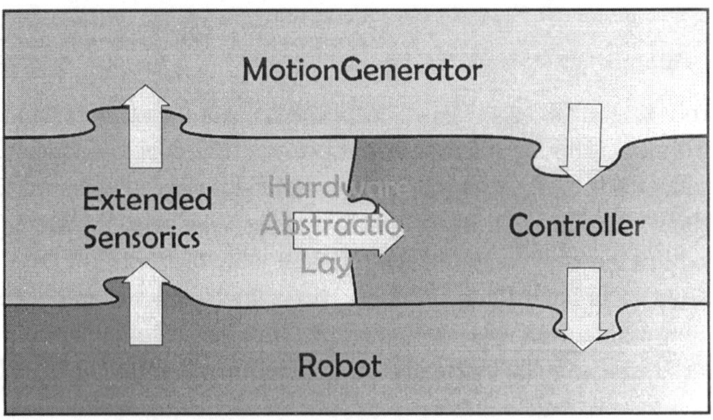

Fig. 1. Structure of proposed control software

Independent of the particular hardware in use, the *MotionGenerator* determines the motion of the central body, decides which foot performs the next step and derives the velocities for this foot according to velocity and direction demanded by the operator. This procedure is independent of the hardware due to the abstraction of sensor data and control inputs. The latter is done by the *Controller* module that transforms the given feet and central body velocities into actuator velocities using the inverse kinematics of the robot and, by comparing the actual and the desired actuator velocities, determines the actuator signals. In case of ALDURO the actuator settings are servo valve voltages. In contrast, the *ExtendedSensorics* module transforms the sensor data into dimensionless and robot structure independent quantities. These two modules are combined into the second layer, the *HardwareAbstraction*. The remaining third layer represents the interface to the particular robot hardware. The software can easily be adapted to other quadruped walking machines. An exchange of the *Robot* module also affects the *Hardware Abstraction* layer, which has to be modified according to the specifications of the robot. However, the *MotionGenerator*, and thus, the part of main interest here, can be left unchanged.

3. Flexible statically stable motion

Apart from the flexibility of the software, the goal of the development process was the construction of a highly flexible walking machine, that can be utilized independent of the location. Thus, the use of a flexible gait pattern is necessary. Characteristic for flexible gait pattern is, that the motion of the platform is governing the motion of the legs. While the platform is moving according to the instantaneous operator's commands, the instantaneous state of the robot prescribes the choice of the foot that should perform the next step.[3] This can be done following different basic concepts. One possibility consists in regarding the proximity of the feet to the limits of the corresponding workspace.[4] If $\delta_{eot,f}$ represents an abstracted measurement of the proximity and $v_{eot,f}$ the velocity towards the boundary of the workspace of the corresponding foot $f \in \mathbb{F}, \mathbb{F} \hat{=}$set of the four feet, the foot can be determined by

$$f_{step} = \{f \in \mathbb{F}_{\text{FreeToLift}} \mid \max(\delta_{eot,f} \cdot v_{eot,f})\}\,, \tag{1}$$

where $\mathbb{F}_{\text{FreeToLift}}$ is the set of feet, that can be lifted without disturbing the stability of the robot. For this foot, step length and direction are determined. The step direction is directly prescribed by the operator. In a first implementation, the step length is a combination of the instantaneous proximity of the foot to the boundary of workspace and the desired velocity:

$$l_{step} = k_{prox}\delta_{eot} + k_{vel} f(\dot{\hat{x}}, \dot{\hat{y}}, \dot{\hat{\psi}}_z), \tag{2}$$

where k_{prox} and k_{vel} are scaling factors and $\dot{\hat{x}}$, $\dot{\hat{y}}$ and $\dot{\hat{\psi}}_z$ the desired central body velocities. To ensure a continuous motion of the central body without deadlocks and to guarantee the possibility to lift the feet, a corrective velocity in opposite direction of the desired velocity and proportional to the maximal proximity measurement is applied:

$$\dot{x}_{corr} = -(\max(\delta_{eot,f} \cdot v_{eot,f}))^{\kappa_{legmove_x}} \dot{\hat{x}}, \tag{3}$$

$$\dot{y}_{corr} = -(\max(\delta_{eot,f} \cdot v_{eot,f}))^{\kappa_{legmove_y}} \dot{\hat{y}}, \tag{4}$$

$$\dot{\psi}_{z_{corr}} = -(\max(\delta_{eot,f} \cdot v_{eot,f}))^{\kappa_{legmove_z}} \dot{\hat{\psi}}_z. \tag{5}$$

Step length and direction must be transformed in a foot velocity. The step can be divided into three phases. In the first phase the foot is lifted, the second phase lasts until the necessary step length is reached, during the third phase the foot is lowered. Each phase generates velocities to fulfill the specific function. To receive a smooth foot motion, the three phases overlap.

4. Quadruped prototypes: ALDURO and ADONIS

The software was initially developed for the quadruped walking machine ALDURO (see Fig. 2). The robot mainly consists of a 3.5x2.0x2.0m^3 (LxWxH) steel structure. With the four legs straightened it reaches a height of 3.5m. Each leg consists of an upper and a lower part, which are connected by a rotational joint in the knee. In the hip, each leg has three degrees of freedom (d.o.f.). The legs are actuated by hydraulic differential cylinders. The hydraulic system operates at a pressure of 160bar. However, as the system weights about 1.8t, tests are expansive due to safety measures. To simplify experiments the small-scale quadruped mechanism ADONIS was desired. It consists of a 30x40cm^2 aluminum structure and four equal legs as depicted in Fig. 3 and Fig. 4. The main body is holding the electronic control units, sensors and batteries. It weights about 13kg and is able to carry an additional payload of 5kg. The main characteristics of ADONIS are listed in Tab. 1.

In contrast to ALDURO, the d.o.f. of each leg of ADONIS are reduced to two rotational joints in the hip and one in the knee. The first joint rotates around a vertical axis that is mounted on the body frame, the other two

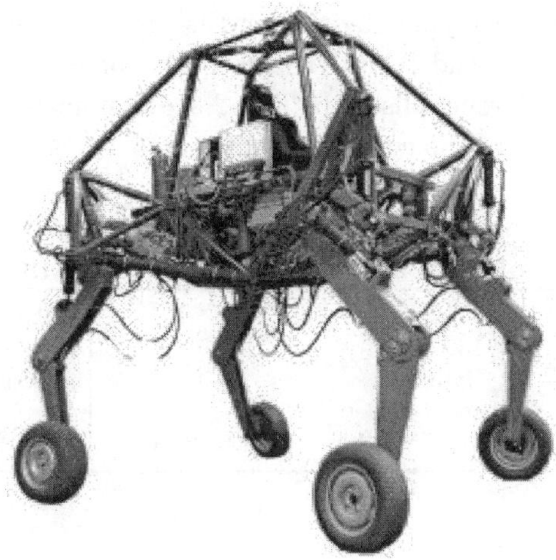

Fig. 2. ALDURO

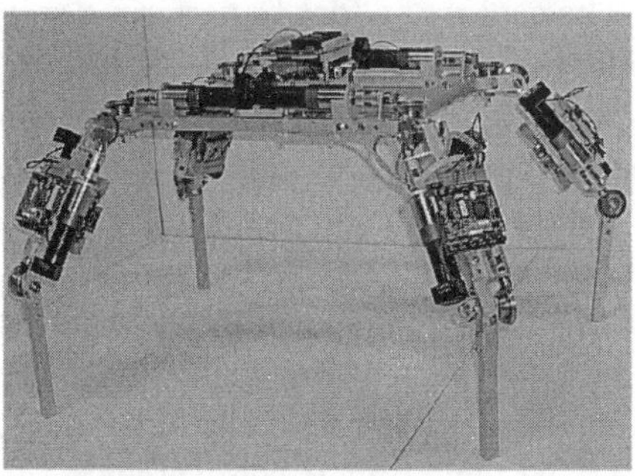

Fig. 3. ADONIS

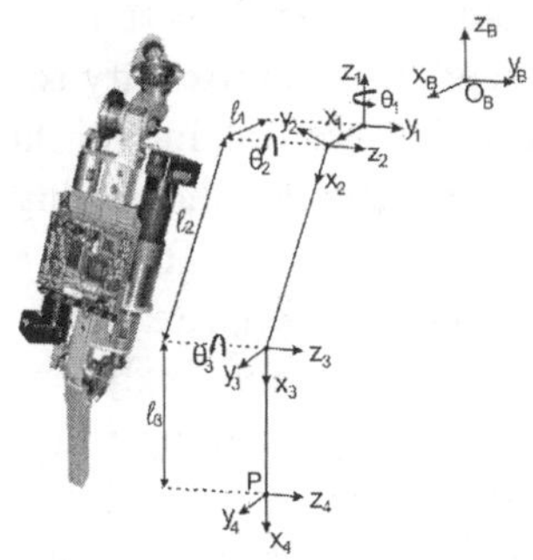

Fig. 4. The leg mechanism of ADONIS

Table 1. Main features of ADONIS

LxWxH	1.0x0.5x0.5m^3	Total Weight	13kg
Material	Aluminium	Power Supply	5V DC/24V DC
DC Motor	28Nmm	Hip Length	55mm
Planetary Gearbox	4.5Nm	Upper Leg Length	220mm
Bevel Gear ratio	1:2.5	Lower Leg Length	170mm

joints rotate around two parallel horizontal axes. The joints are actuated by means of DC motors.

Both robots carry a computer similar to a normal PC, a linux system with embedded real time environment is installed supporting the control software. ALDURO and ADONIS are both quadruped walking machines. However, the differences listed in Tab. 2 are significant, which emphasis the generic character of the control software used for both machines.

Table 2. Main differences of ALDUROand ADONIS

	ALDURO	ADONIS
Dimension LxWxH	3.5x2x3.5m^3	1.0x0.5x0.5m^3
Total weight	1600kg	13kg
Engine	Hydraulic	Electric
Degrees of Freedom	16	12
Leg Structure	3 DOF at hip 1 DOF at knee	2 DOF at hip 1 DOF at knee

5. Portability Criterion for Static Gaits

Although both systems differ, the relevant characteristics to use the introduced *free-gait* algorithm are similar. The choice of a foot to perform a step depends on its proximity to the boundary of the corresponding workspace. Thus, for two walking machines using this algorithm, the workspace of the feet and the distance of hip to the ground should only differ by a scalar factor. For ALDURO we assume a walking height of about $1.3 - 1.4$m, for ADONIS approximately $0.25 - 0.3$m. That leads to a scaling factor of $\lambda \approx 5$. Correspondingly, the dimensions of the workspaces in the walking planes must differ by the same factor λ. As shown in Fig. 5, the shape of the planes is not surprisingly different due to the different kinematics, but the scaled workspace plane of ALDURO lays almost completely in the plane supported by an ADONIS leg. To be able to transfer the experimental resuts gained with ADONIS to ALDURO, the workspace of the feet of ADONIS is artificially limited by the control software during the abstraction of the sensor information.

6. Conclusion and Future Work

Quadruped walking machines offer a compromise between fast and flexible, robust locomotion.

In this paper a generic *free-gait* algorithm for flexible, statically stable walking was proposed. This particular algorithm can be used to control different quadruped machines. It is independent of the actuation type used for a particular machine and independent of the kinematic structure of the legs. As a criterion for the portability of the gait pattern, the scaled workspace of the feet in the walking plane is proposed. The application of this criterion is shown for the experimental prototypes ALDURO and ADONIS.

In future research also dynamic gaits for quadrupeds should be addressed, Here, the question whether portability criterion also for dynamic gaits exist is of special interest.

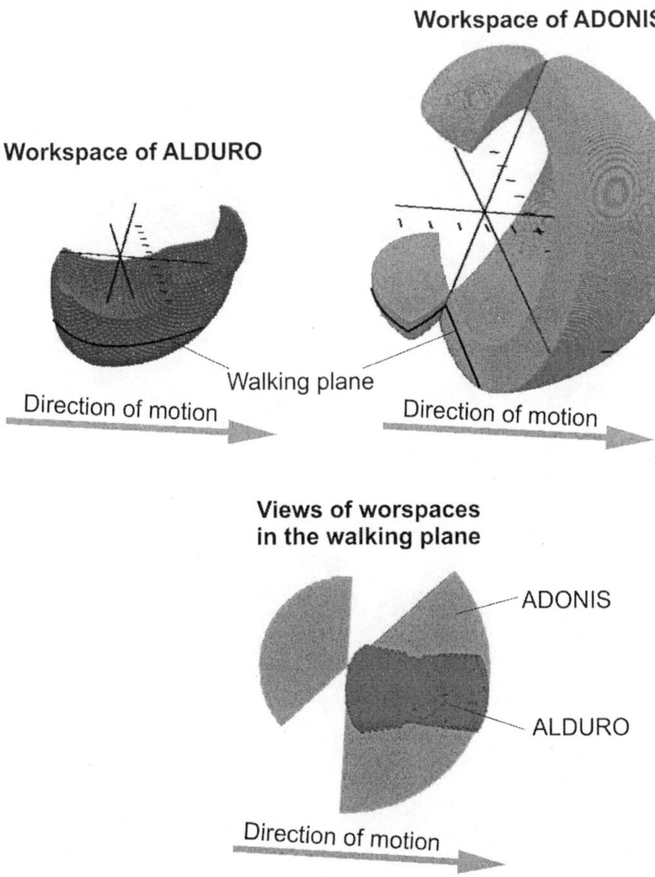

Fig. 5. Comparison of workspaces for ALDURO and ADONIS

References

1. J. Müller, *Entwicklung virtueller Prototypen des hydraulisch angetriebenen Schreitfahrwerks ALDURO*, Dissertation, University of Duisburg, April 2001.
2. D. Germann, M. Hiller, D. Schramm, *Design and Control of the Quadruped Walking Robot ALDURO*, ISARC 2005, FERRARA, ITALY, Sept. 11-14, 2005.
3. D. Germann, *A Modular Controller Structure for the Quadruped Robot ALDURO*, Dissertation, University Duisburg-Essen, to be published 2007.
4. S. Song, Y. Chen, *A free gait algorithm for quadrupedal walking machines*, Journal of Terramechanics, 28(1), 1991.

ON THE DESIGN OF A FOUR-BAR MECHANISM FOR OBSTACLES CLIMBING WHEELS

ANTONIO GONZALEZ

School of Industrial Engineering, University of Castilla-La Mancha, Av. Camilo José Cela s/n, Ciudad Real, 13071, Spain

ERIKA OTTAVIANO, MARCO CECCARELLI

LARM: Laboratory of Robotics and Mechatronics. DiMSAT, University of Cassino, Via di Biasio 43, Cassino (Fr), 03043, Italy

This paper describes the design of a mechanism that allows a wheel to climb obstacles and stairs. The proposed four-bar linkage can be installed on each wheel of a vehicle, which can be capable to climb stairs with suitable smooth motion. An algorithm is formulated to design an optimum solution of the mechanism fulfilling task requirements. A suitable trajectory for the centre of the wheel can be ensured through an easily controlled motion, and the compactness of the mechanism design makes it suitable for staircase-climbing wheelchairs for people with disability.

1. Introduction

The number of applications for mobile robots increases continuously. One of the most interesting and useful application is related to powered wheelchairs. These vehicles greatly improve the mobility of the people with disability but they may become useless when face with obstacles, which still exist in many cities and buildings. In general these obstacles can be modeled as steps, slopes, and stairs, with different characteristics.

Several systems have been developed to make a mobile robot capable of climbing and descending steps, slopes and stairs. Conventional wheeled vehicles have a limited capability for climbing steps and therefore, several solutions can be used for having efficient and comfortable vehicles. The most used solutions make use of tracks [1] or cluster of wheels [2] or legged systems [3].

Hybrid systems have been also developed as combination of wheeled and legged solutions to exploit the advantages of both mobile systems. In [4] a leg-wheel hybrid stair climbing wheelchair is proposed, which consists of eight independent prismatic-joint legs, four active wheels, and four passive casters. A prototype reported in [5] consists of a four wheels wheelchair with two legs for

obstacle climbing.
The CALMOS-Wheelchair prototype (**CA**stilla-**La** **M**ancha **O**bstacle **S**urpassing **W**heelchair) has been built at School of Engineering of Castilla-La Mancha University and it belongs to the category of hybrid locomotion vehicles. This prototype has been developed to solve drawbacks of existing step-climbing systems though a new design presented in [7]. The main features of CALMOS-Wheelchair are high load capability and a considerable adaptability to a wide variety of obstacle geometries.
In this paper we attach the problem of designing a suitable mechanism, which can drive wheels with a smooth motion in climbing obstacles and steps. This paper is focused in the synthesis of a climbing mechanism, whose preliminary solution has been conceived for CALMOS-Wheelchair prototype. The aim of the new mechanism proposed in this paper is to increase the adaptability of the CALMOS-Wheelchair prototype to different obstacle geometries. The system is also designed in order to be compact and easily controlled.

2. The step-climbing problem for a wheel

The problem of step-climbing can be defined as driving a wheeled vehicle to climb stairs with suitable smooth motion and efficient power transmission-mechanisms. The aim can be identified in enhancement of wheelchair functionality for people with disability. Main characteristics must be preserved in terms of lightweight design and easy controllability.
It is well known that wheels have limited capacity to climb obstacles. The maximum obstacle height that a wheel can climb depends on the geometry of the obstacle and the friction coefficient μ. Figure 1a shows the trajectory of the centre of a wheel while climbing a step of height H < r. Figure 1b shows the forces and torques acting on a climbing wheel at the initial configuration of the climbing process.

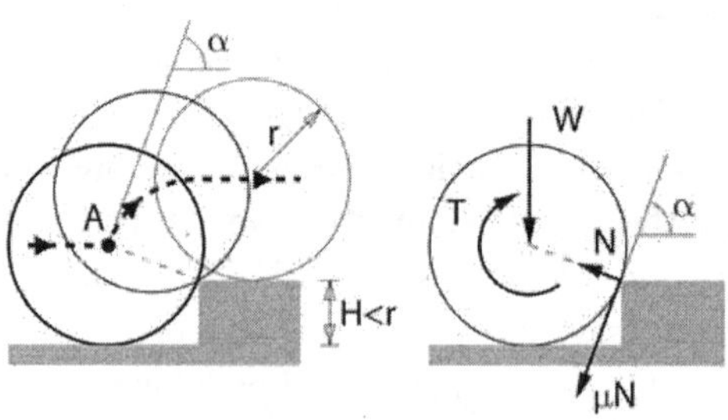

Figure 1: Wheel climbing a step with height H < r: a) Trajectory of the Wheel centre b) Forces acting on the wheel.

N is the normal force at the contact wedge and μ**N** is the tangential force. The weight supported by the wheel is **W** and the traction torque is **T**. The needed friction coefficient μ to equilibrate **W** during the climbing motion can be obtained from the vertical equilibrium force equations in the wheel. This value is $\mu = \tan(\alpha_{max})$, where α_{max} can be computed as

$$\alpha_{max} = \cos^{-1}\left(1\text{-}H/R\right) \tag{1}$$

and it is the slope in the initial point of the trajectory point A in Fig. 1a. Therefore, the maximum step height H that a wheel of radius r can climb with a friction coefficient μ, can be evaluated by considering Fig. 1a as

$$H = r - r\left(1+\mu^2\right)^{-1/2} \tag{2}$$

Similar expressions can be obtained for slopes and H > r steps. The illustrated situations show that it is necessary to use extra mechanisms when using a wheel for step climbing. This mechanism will be called Climbing Mechanism.

3. A solution for powered wheels

A Climbing Mechanism must provide a way to maintain the contact of the wheel with the obstacle and to ensure the needed traction at each configuration of the climbing process. Therefore, the Climbing Mechanism must fulfill the conditions: 1) to guarantee traction force for the trajectory without depending on the friction; 2) to adapt the trajectory of the wheel centre to the obstacle geometry.

The proposed climbing mechanism in Fig. 2 is composed of a frame attached to the wheelchair chassis, a four bar linkage, and a sliding support. Friction coefficients at the wheel contact for usual surfaces in urban environment do not allow high heights and large slope angles in obstacle-climbing. In order to overcome this problem, a sliding support can be used, as shown in Fig. 2b). The so-called sliding support ensures the traction during the obstacle climbing and it is connected to the chassis by a motorized prismatic joint. The wheel will remain in contact with the obstacle surface while an additional sliding support is deployed. This is necessary in order to retransfer the wheelchair weight to the wheel when the sliding support has terminated its work. Therefore, the wheel axle cannot be fixed to the chassis. For this reason, an additional DOF must be considered for the wheel axle to allow its backward movement for avoiding the obstacle interference and ensuring surface adaptability. Dashed line in Fig. 2b) shows the relative path of point M as referred to the Climbing Mechanism frame, and δ is the slope angle for a desirable trajectory for M.

This trajectory of the wheel centre should be as close as possible to a straight line in order to obtain a smooth and easily controlled movement of the chassis in the first phase of the climb. The slope of the trajectory must be also set with descendent, horizontal or ascend direction to ensure the desired behavior of the mechanism. An ascending trajectory of the wheel has been used for the CALMOS-Wheelchair in Fig.2a) to show the operation of the proposed climbing mechanism. The slope angle of this trajectory is δ. The main advantage of this slope is that the climbing or descending process can begin without help when the added DOF is unlocked. A drawback is that the free barrier wheel position of the Climbing Mechanism can be movable. Thus, in this configuration the component of the reaction force in the direction of the trajectory, may push the wheel to move forward and, therefore the wheel axle must be locked when the system is in free barriers environments. The four-bar mechanism is also equipped with a spring to ensure the backward movement of the wheel to the initial position when the obstacle has been surpassed.

Climbing Mechanism operation in step ascending can be described through the following phases. The wheelchair stops when the wheel reaches an obstacle and the sliding support is deployed once the head of the sliding support has reached the top of the obstacle, the wheel axle DOF is unlocked and the weight is transferred from the wheel to the sliding support. The sliding support is being deployed while the added DOF allows the wheel to avoid the interference with the step. The axle DOF is blocked when wheel goes back to its original position. Then the sliding support can be folded back. This operation transfers the weight from the sliding support to the wheel again.

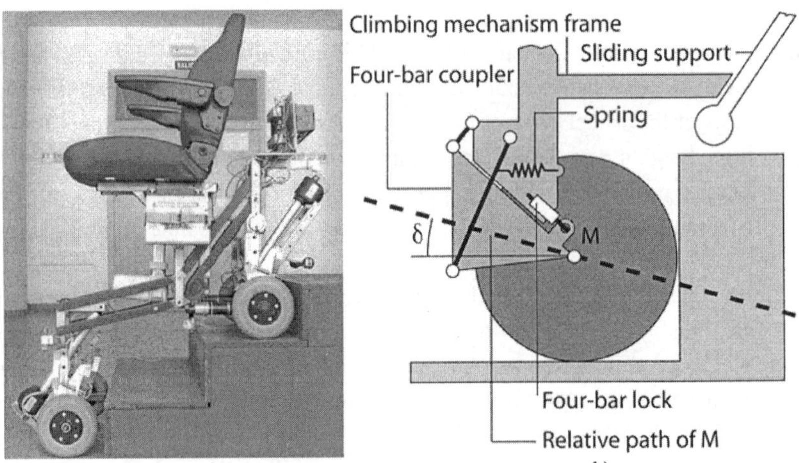

a) b)

Figure 2: CALMOS-Wheelchair with a climbing system: a) a prototype; b) a scheme of the proposed climbing mechanism.

4. A four-bar mechanism for the Climbing Mechanism

A four-bar mechanism has been poposed for the Climbing Mechanism because of its robustness and trajectory generation ability, besides the characteristics for lightweight compact design. The proposed four-bar mechanism can be synthesized to give a desired trajectory of a reference point M of the coupler. The wheel centre has been chosen as coincident to point M, and its trajectory must be as close as possible to a straight line with a slight ascending slope δ, as shown in Fig.2b).

A symmetrical four-bar mechanism has been selected with an approximate straight-line trajectory. A design scheme is shown in Fig. 3 with b = c = q as to give a Chebyshev mechanism [8]. Referring to the sizes of the CALMOS-Wheelchair, Fig. 2, a suitable straight-line segment for M path will be of L = MM' = 100 mm. This value allows surpassing the most usual obstacles in urban environments. The path deviation from a straight-line can be evaluated by the area denoted by S that can be computed as

$$S = \int_{\alpha_0}^{\alpha_f} \left| y(\alpha) - y_0 \right| \frac{\dot{x}}{\dot{\alpha}} \, d\alpha \tag{3}$$

where α_0 and α_f are the values of input α angle at the initial and final configurations; x and y are the coordinates of point M with respect to the OXY frame, being y(α) a general value depending of α and y_0 is a value at the initial configuration of the mechanism. When of the origin of the reference frame is chosen to be coincident with C then y_0 is coincident with V parameter, and the X coordinates of center of the segment MM' is 0, Fig.3.

The compactness of the mechanism is expressed through a function u, which is the maximum of the distances among points A, B, C, D and M, in the form

$$u = \max(b, CM, CD, MB, MD) \tag{4}$$

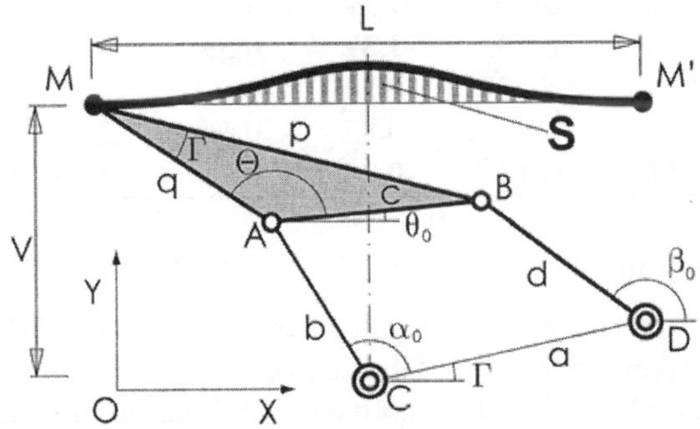

Figure 3: A design scheme for a four-bar linkage whose coupler point M traces an approximated straight-line trajectory between M and M'.

A maximum range $\Delta\theta = \theta_{fin} - \theta_0$ for the swinging motion of the coupler can be also considerer as a constraint to avoid interference among links of the mechanism.

Two objective functions can be considered for an optimum design of a four-bar linkage as a climbing mechanism for Climbing Mechanism in CALMOS-Wheelchair by using the above-mentioned criteria. The main design variables, are V, b and Γ. In order to have dimensionless objective functions, S and u can be suitably divided by S_{Cheb} and u_{Cheb}, which are the values of functions S and u for a Chebyshev mechanism having b/a = 2 and d/a = 2.6. Thus, the optimization design problem for a synthesis of a four bar linkage can be formulated as

$$\min F = \left[K_1 \frac{S}{S_{Cheb}} + K_2 \frac{u}{u_{Cheb}} \right] \tag{5}$$

subjected to

$$G = \Delta\theta_{max} - 2\tan^{-1}\frac{L}{2V} < 0 \tag{6}$$

where K_1 and K_2 are weighting factors that can be conveniently chosen according to the relationship $K_1 + K_2 = 1$; $\Delta\theta_{max}$ is the prescribed maximum swinging rotation of the coupler during its operation, and condition G is computed by taking into account of the coupler rotation angle as L/2V.

Table 1. Optimal design results of the optimization design process of Eqs. (5) and (6) for different values of the coefficients K_1

K_1	Δθ (rad)	b (mm)	Γ (rad)	F	S (mm²)	u (mm)
0.9	0.41	247.8	-1.39	0.24	5.4	247.8
0.7	0.58	175.8	-1.39	0.53	16.0	175.8
0.5	1.25	86.2	-0.84	0.51	12.5	117.4
0.3	1.45	75.6	-0.85	0.64	24.8	104.5
0.1	1.57	70.8	-0.86	0.71	35.9	98.5

5. Numerical example

Table 1 shows results of the optimization design process for different values of K_1, when the maximum rotation of the coupler is limited to π/2. Because of the low number of design variables a thorough search has been performed in order to avoid local minima of F. The optimal design solution has been selected with $K_1 = K_2 = 0.5$ to have equal consideration of the optimality criteria.

Figure 5 shows the evolution of the objective function F and its components S and u, together with the design constraint in (6). In the reported example, the optimization process takes 140 iterations to converge to the optimal solution. The computation time was 11.23s with a 1.7 GHz Pentium IV using the code *fmincon* of Matlab Optimization Toolbox [9]. Figure 6 shows a scheme for the optimal design solution that can be applied to CALMOS-Wheelchair. The approximate straight line path MM' is reported with a mechanism simulation that has been carried out for the layout in Fig. 2b.

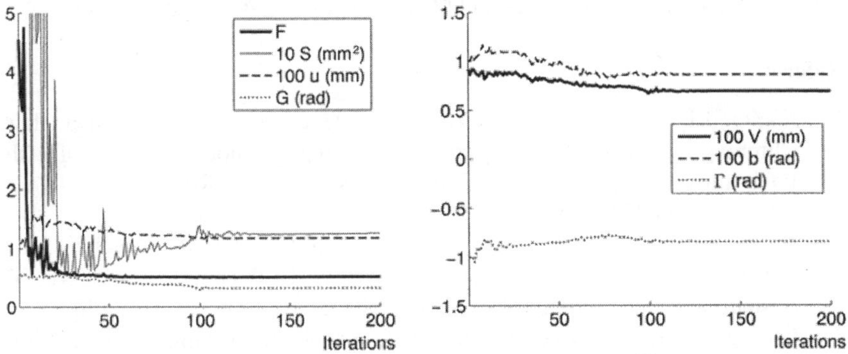

a) b)

Figure 5: Evolution of the solution for $K_1 = K_2 = 0.5$: a) the objective function F with its components S and u and constraint G; b) the design variables H, b, Γ.

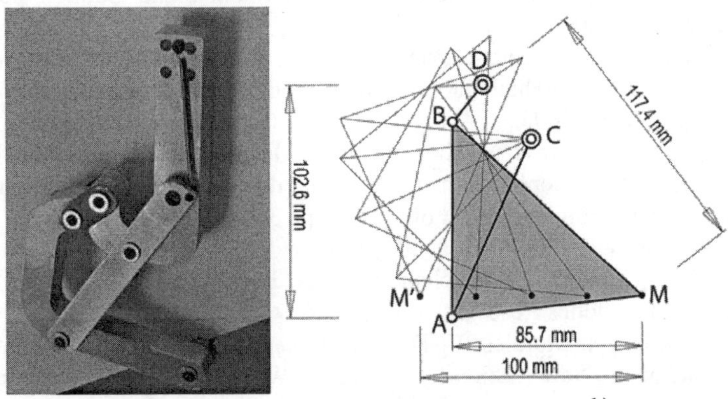

a) b)

Figure 6: Optimal mechanism for the relative motion in figure 2b) for the optimization design procedure in (5) and (6) for the case with $K_1 = K_2 = 0.5$: a) picture; b) scheme

6. Conclusions

In this paper a Climbing Mechanism for wheelchairs has been proposed with features of compact design and easy operation. A four bar linkage for the climbing Mechanism has been designed as attached to each wheel to give suitable smooth motion during the step climbing. An optimization problem has been formulated to synthesize a proper linkage for the CALMOS-Wheelchair with satisfactory optimal features in terms of path generation and design sizes

Acknowledgments

The first author wishes to thank the Regional Government of Castilla-La Mancha (Spain) for the grant that has permitted him to spend a post-doctoral period at Laboratory of Robotics and Mechatronics in Cassino during 2006.

References

1. K YONEDA, Y OTA, and S HIROSE. Development of a Hi-Grip Stair Climbing Crawler with Hysteresis Compliant Blocks. Proc. of 4th International Conference on Climbing and Walking Robots (CLAWAR 2001), pp.569-576 Karlsruhe, 2001
2. M. J. Lawn, T. Ishimatzu. Modelling of a Stair-Climbing Wheelchair Mechanism with High single-Step Capability. IEEE Transaction on Neural Systems and Rehabilitation Research. Vol 11. No. 3. pp 323-332. 2003.
3. Giuseppe Carbone, Marco Ceccarelli, Yusuke Sugahara, Hun-ok Lim, Atsuo Takanishi. Stiffness Experimental Monitoring for WL-16RII Biped Locomotor During Walking. ROMANSY 16. Proc. of the 16th CISM-IFToMM Symposium on Robot Design. Dynamics and Control. pp.105-112. Warsaw, 2006.
4. J. Yuan, S. Hirose. Zero Carrier: A Novel Eight Leg Wheels Hybrid stair climbing Mobile Vehicle. Journal of Robotics and Mechatronics. Vol.17. No.1. pp. 44-51. 2005.
5. P. Wellman, V. Krovi, V. Kumar and W. Harwin. Design of a wheelchair with legs for people with motor disabilities. IEEE Transactions on Rehabilitation Engineering, Vol.3. No.4. pp 343-353. 1995.
6. R. Morales, V. Feliu, A. González, P. Pintado. Kinematic Model of a New Staircase Climbing Wheelchair and its Experimental Validation. Int. Journal of Robotics Research. Vol. 25. No. 9. pp. 825-841. 2006
7. A. González, R. Morales, V. Feliu, P. Pintado. Improving the Mechanical Design of a New Staircase Climbing Wheelchair. Proc. of the 9th International Conference on Climbing and Walking Robots (CLAWAR 2006). pp. 14-18. Brussels. 2006
8. M. Ceccarelli A. Vinciguerra. Approximate four-bar circle-tracing mechanisms: classical and new synthesis. Mechanism and Machine Theory. Volume 35. Issue 11, pp. 1579-1599. 2000
9. Grace A., "Optimization Toolbox User's Guide", The Matlab Works Inc., 1992.

PATH PLANNING FOR THE "3DCLIMBER"

MAHMOUD TAVAKOLI
LINO MARQUES
ANÍBAL T. DE ALMEIDA
Institute for Systems and Robotics, University of Coimbra, Portugal
Email: mahmood.tavakoli@gmail.com, [lino, anibal]@isr.uc.pt

3DCLIMBER is a running project in the University of Coimbra to develop a climbing robot for climbing and manipulating over 3D human-made structures. This paper mainly discusses the path planning algorithm of the robot.

1. Introduction

Development of climbing robots was a challenging area during last decade. Different types of climbing robots were developed for climbing over flat or curved surfaces. For holding the robot attached to the surface they used suction cups [1,2,3,4,5] or magnets [6,7]. Also robots whose end-effectors match engineered features of the environment like fences or porous materials or bars [8,9,10,11,12] were developed. Different kind of robots also developed for climbing inside pipes or ducts [13,14] or climbing over poles [15,16,17,18].

3DCLIMBER is a running project in the University of Coimbra to develop a pole climbing robot which is not only able to climb over 3D structures with bends and branches, but is also able to perform practical and complex tasks like painting and performing non destructive tests on 3D structures. The robot takes advantage of a dedicated 4-DOF serial mechanism which enables it to climb over poles, overcome the bent section and alter between different branches of the pole. This paper mainly focuses on the path planning and simulation of the 3DCLIMBER, also a summary of the design and kinematic analysis will be presented.

2. The Robot Design

As M.Tavakoli et. Al. demonstrated before, locomotion and manipulation along circular 3D structures with bents and branches requires a minimum of 4 DOFs include 2 translational and 2 rotational DOFs [19].

The proposed pole climbing robot consists of two main parts (Fig. 1) which are: 4-DOF climbing module and gripping module. The robot was designed for

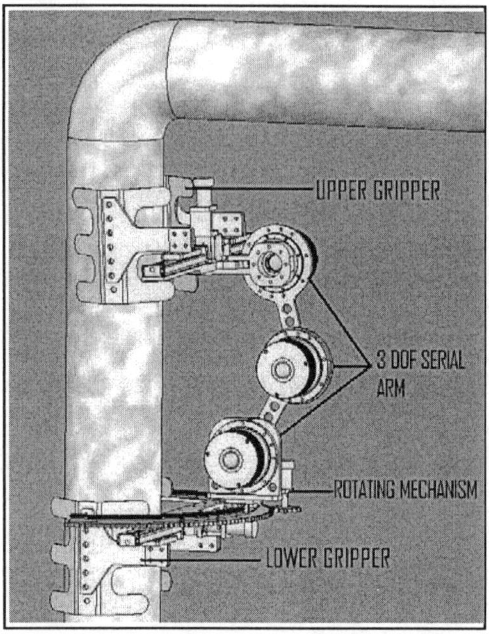

Figure 1. Detailed design of the 3DCLIMBER

a step by step movement. One of the grippers is attached to a manipulator, and the other one is attached to the base of a rotating platform. This configuration provides four DOF between grippers, allowing the movement along poles with different cross sections and geometric configurations. The proposed design takes advantage of novelties in the design of all parts: the climbing module, the gripping module and the rotating platform. The climbing module consists of a 3-DOF planar Serial arm and a Z axis rotating mechanism. Combining the 3-DOF arm with the rotating mechanism provides two rotations and two translations, which are necessary to achieve the design objectives as explained previously.

There are 2 gripping modules. One of them attached to the base of the 4-DOF climbing module which is indeed the Z axis rotation mechanism and a second gripper is attached to the manipulator of 3-DOF arm. Each gripper consists of two unique multi-fingered V-shaped bodies, a motor, one right hand and one left hand ball screws and 2 linear guides. Table 1 shows the estimated characteristics of the robot.

3. Kinematics of the Manipulator

Kinematics analysis is necessary in order to perform the workspace analysis, path planning, optimization and control of the robot.The study of manipulator kinematics is divided into two parts, inverse (or reverse) kinematics and forward (or direct) kinematics. The inverse kinematics problem involves mapping a known pose (position and orientation) of the moving platform of the manipulator to a set of input joint variables that will achieve that pose. The forward kinematics problem involves mapping from a known set of input joint variables to a pose of the moving platform which result from those given inputs.

For performing kinematics analysis, a simplified model of the robot was considered and Denavit-Hartenberg table was established (fig. 2).

For the direct kinematics analysis, transform matrices were obtained. The result of the direct kinematics is the transform matrix between the 4th and the 0th coordinate systems. Here detailed calculations are not presented but only the final transformation matrix is presented.

$$ {}^{0}_{4}T={}^{0}_{1}T\times{}^{1}_{2}T\times{}^{2}_{3}T\times{}^{3}_{4}T $$

$$ {}^{4}_{0}T=\begin{bmatrix} 0.5\cos(V)+0.5\,C_{0123} & 0.5\sin(V)-0.5*S_{0123} & S_0 & P_X \\ 0.5\sin(V)-0.5\,S_{0123} & 0.5\,C_{0123}-0.5\cos(V) & C_0 & P_Y \\ S_{123} & C_{123} & 0 & P_Z \\ 0 & 0 & 0 & 1 \end{bmatrix} $$

Letting:

$V= -\theta_3-\theta_2+\theta_0-\theta_1$

$P_X=0.5L_2\times\cos(-\theta_2+\theta_0-\theta_1)+ 0.5L_2\times C_{012} +0.5L_1\times\cos(\theta_0-\theta_1)+ 0.5L_1\times C_{01}+ C_0\times L_0$

$P_Y= 0.5L_2\times\sin(-\theta_2+\theta_0-\theta_1)+ 0.5L_2\times S_{012} +0.5L_1\times\sin(\theta_0-\theta_1)+ 0.5L_1\times S_{01}+ S_0\times L_0$

$P_Z = S_{12}\times L_2+ S_1\times L_1$

For inverse kinematics problem the position matrix $[P_X\ P_Y\ P_Z]^T$ of the 4th coordinate system and also θ which is the angle of X_4 and X_1, around Z_4, were considered to be known. The values of θ_0 to θ_3 should be obtained. For a known pose 2 different set of answers obtained, while for both of them θ_0 is same.

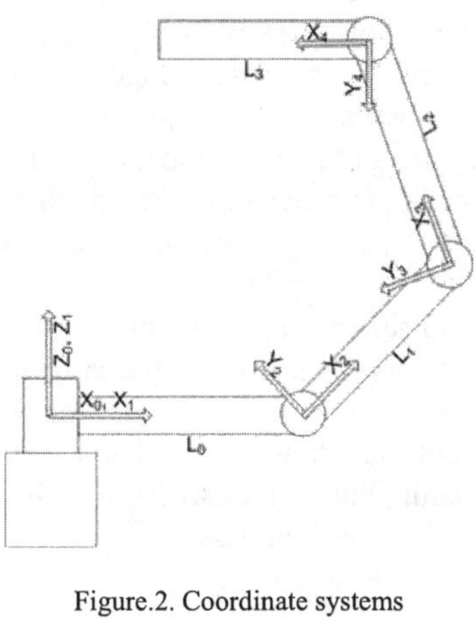

Figure.2. Coordinate systems

TABLE II
MAIN CHARACTERISTICS OF THE DETAILED DESIGNED OF THE ROBOT

Degrees of Freedom	4
Quantity of Motors	6
Climbing Procedure	Step by Step
Approximate Weight (kg)	40
Robot Size (cm)	50*60*50
Extended Robot Size (cm)	50*60*100
Theoretical Climbing Speed (m/min)	12

$$\theta_0 = A\tan 2\frac{P_Y}{P_X}, \theta_{21} = A\tan 2(+\sqrt{1-Q^2}, Q), \theta_{22} = A\tan 2(-\sqrt{1-Q^2}, Q)$$

$$\theta_{11} = A\tan 2(P_Z, C_0 P_X + S_0 P_Y - L_0) - A\tan 2(k_{21}, k_{11})$$

$$\theta_{12} = A\tan 2(P_Z, C_0 P_X + S_0 P_Y - L_0) - A\tan 2(k_{22}, k_{12})$$

$$\theta_{31} = \theta - \theta_{11} - \theta_{21} - \theta_{31}, \theta_{32} = \theta - \theta_{12} - \theta_{22} - \theta_{32}$$

Letting:

$$Q = \frac{P_X^{\ 2} + P_Y^{\ 2} + P_Z^{\ 2} + L_0^{\ 2} - L_1^{\ 2} - L_2^{\ 2} - 2L_0\sqrt{P_X^{\ 2} + P_Y^{\ 2}}}{2L_1L_2}$$

$$k_{11} = L_1 + L_2 C_{21}, k_{21} = L_2 S_{21}, k_{12} = L_1 + L_2 C_{22}, k_{22} = L_2 S_{22}$$

Two functions were established in MATLAB, to calculate inverse and direct kinematics. These functions will be used in all further steps like workspace analysis and path-planning. Validity of equations and functions was checked by numerical examples. Constant orientation workspace analysis also performed by numerical approaches in MATLAB. The results were specially used for determining the appropriate links lengths of the 3-DOFs arm which make it possible to overcome the bent section with angles up to 100 degrees.

4. Path Planning and Simulation

Selection of an appropriate Path planner is a critical issue. In Robotics application, path planning and path tracking can be done online or offline (Figure 3). This depends on several factors include existence of obstacles, predictability level of obstacle locations and the amount of knowledge available from the robot working area before start of the robot operation [20]. We may plan paths before their execution, if we have some knowledge of the environment. Planning paths before execution allows efforts to get a shorter path time, more efficient dynamics, and absolute collision avoidance. When working in this mode a priori knowledge (i.e. Known before) is used. Priori knowledge may not be used for unpredictable or random motion as there is no detection method allowed by its definition.

For our application we should decide for using a Priori path planner or a Postieri path planner or combination of them. As stated earlier the working area of the 3DCLIMBER robot is human made 3D structures. So the geometry of the structure and thus the working area of the robot are known beforehand. Also proposed tasks of the robot are climbing and manipulating over structure. All of these tasks are performable with composition of 3 more basic tasks.

A. Climbing over the straight part of the structure. It can be done with straight line path planning algorithms.
B. Rotating around the axis of the structure. It is necessary for scanning all points of the structures, as some of the applications need.
C. Passing the bent section of the pole, and performing some complicated tasks like welding need rotation of the manipulator around Y axis.

Regarding these 3 parts and descriptions, it is possible to plan the movement of the robot in priori mode. But we should always consider errors during the operation. The most significant error is slippage of the robot. Also as the precise geometry of the structure may not bee known beforehand, it should be measured during operation with sensors and in this case a complete priori path planning is not possible because of errors and lack of the precise information about structure. To overcome this problem a higher level Priori planer should be combined with a lower level Postieri planner. This combination is called Hierarchical planner. It means to design an offline path planning plus an online path tracking (Figure 4). For instance when the robot is climbing along the straight part of the structure, the upper level Priori planner send the signal for the straight path planner module and this module send appropriate signals to all motors of the robot for going one step forward. This module also includes an online path tracking which compensate available errors.

		Collision Avoidance	
		Yes	No
Path Constraints	Yes	Off-line Collision free path planning plus, on-line path tracking.	Off-line Path Planning, plus on-line path tracking
	No	Positional Control, plus on-line obstacle detection and avoidance.	Positional Control.

Figure. 3. Chart for selection of path planning method

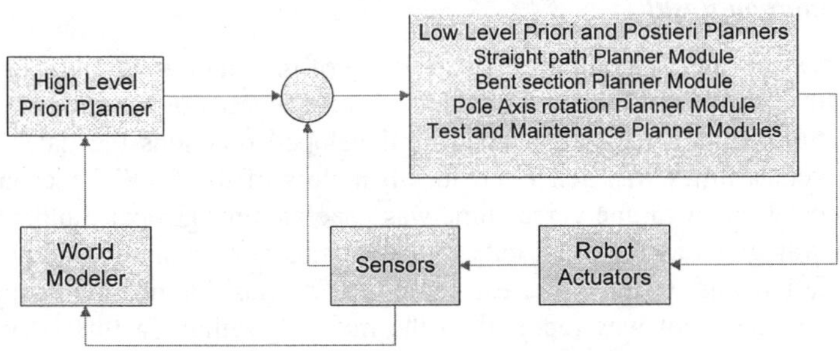

Figure 4. 3DCLIMBER path planner chart.

When the robot reaches to a specified distance from the bent section of the pole, the higher level planner receives information from the sensors about the exact distance of the manipulator from the bent section and also the angle of the bent section. Then it sends this information to the bent passing module. This module should be a Hierarchical planner. It should receive the information of geometry of the pole and plan the control of the motors during the operation of the robot.

4.1. *Straight Path Planner Module*

The climbing strategy is opening of the top gripper, moving up the manipulator, closing of the top gripper, opening of the bottom gripper, moving up the base and closing of the bottom gripper. For the robot to travel along straight path, a trajectory planner function was developed in MATLAB. This function generates appropriate data for control of motors in the 4DOF mechanism and grippers so that the robot travels along a straight path. Then this function was used for simulation of the robots model which was developed in Solidworks. Cosmos motion was used as simulation package.

Priori planner generates necessary data as input for this module. This data include start point and end point of one climbing step in the Cartesian order. The output of the module is the matrix of the position versus time for the 4DOF mechanism's motors, and torque versus time for the gripper's motors. The applied algorithm estimate n points on the straight path and by applying inverse kinematics it generates appropriate data for all of motors.

4.2. *Bent Section Passing Module*

An algorithm also developed for passing the bent section. This algorithm receives the distance of the manipulator from bent section and bent angle from world modeler and generates data for robot joints for passing the bent section without collision.

4.3. *Simulation Results*

The geometry of the pole and the start position of the robot were given as input to the path planner and the final position of the manipulator for the maximum possible thrust was calculated. Using the developed functions the database of position versus times was generated for all motors of the 4 DOF mechanism, and also database of torque versus time was generated for gripper motors. Then this data was given as the input data to the cosmos motion simulation package and simulation was started. After each step the new position of the manipulator and base of the robot was reported to the main algorithm. In this simulation process there is no difference between the desired and the resulted position, but in the real robot there would be always an error due to different sources of error and mainly slippage of the grippers. To solve this problem in the real robot laser triangulation sensors will be used in the base and manipulator of the robot and after finishing each step, the resulted error would be reported to the tracking algorithm and would be compensated. Figure 5 shows generated data from developed function for passing the bent section. X, Y and Z are position of the manipulator and Θ is the angle of manipulator in respect with base of the robot. Θ_0 to Θ_3 are angles of the 4 DOF mechanism, which are showed in both abstract and relative mode. Also curves of generated data for gripper exerting force versus time are showed. Figures 6 show some pictures from the simulation of the robot.

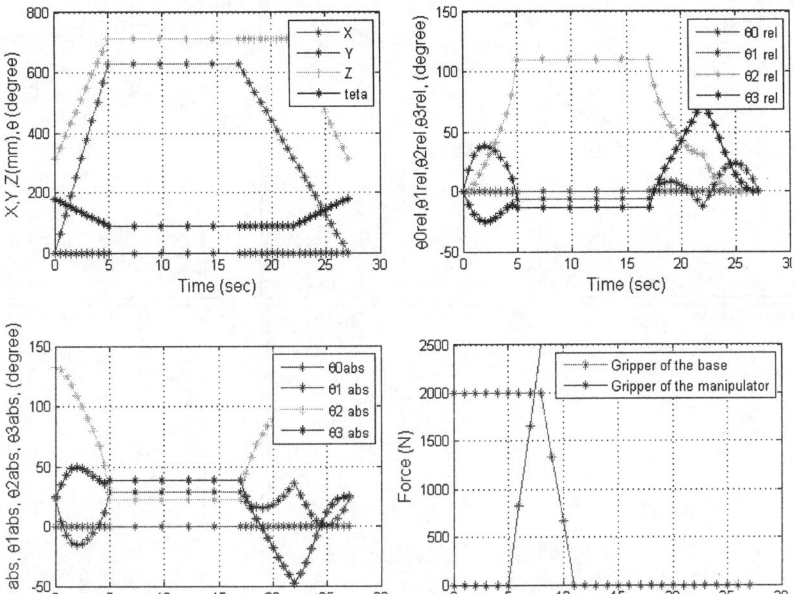

Figure.5. Generated data for passing the bent section

5. Detailed Design

A pole climbing robot was designed, based on the designed 4-DOFs arms and the obtained results from simulation. The proposed design takes advantage of novelties in the design of all parts: the climbing module, the gripping module and the rotating platform.Each gripper consists of two unique multi-fingered V-shaped bodies, a brushless motor, one right hand and one left hand ball screws and 2 linear guides. V shaped grippers has the mechanically self centering advantage. With other kinds of grippers, there would be a need for sensors and

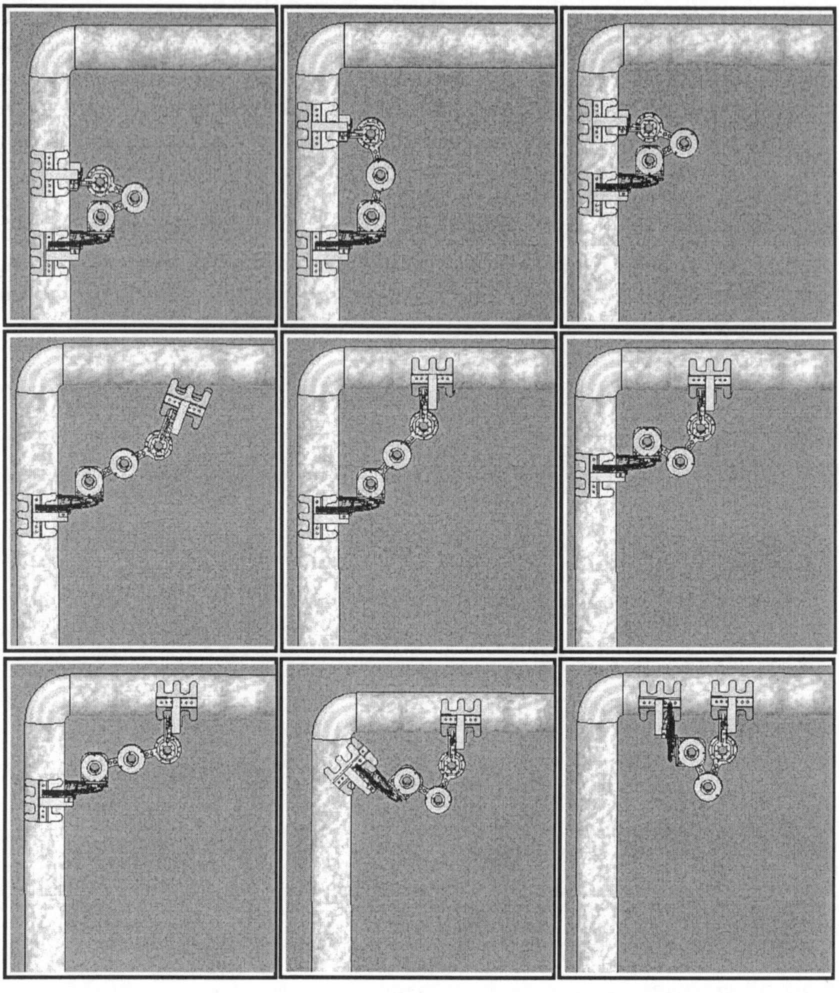

Figure 6. Simulation of the 3DCLIMBER for thrusting along pole and passing over the bent

control strategies to precisely control the position of robot, in order to perform safe gripping. Weight optimization was considered in the design of all of the robot's parts as the weight is very important factor in climbing robots. Harmonic Drive gearboxes were used in the 3-DOFs serial arm in order to reduce the weight and increase the precision and the efficiency. All non standard parts were designed and manufactured with 7075-T6 aluminum. Both grippers and rotating mechanism were assembled and tested. The 3DOF mechanism is still under tests. Figure 7 shows the grippers and rotating mechanism of the robot.

6. CONCLUSION

3DCLIMBER is a project for developing a robot able to climb and manipulate over 3D human-made structures. In this paper a short summary of the conceptual design of the robot was described. Then the path planning algorithm and the simulation results for the 3DCLIMBER model was presented. Future works include, multi-criteria optimization with genetic algorithms, assembly, control and test of the robot.

Fig. 7. Grippers and rotating mechanism of the robot.

References

1. H. Dulimarta and R. L. Tummala. Design and control of miniature climbing robots with nonholonomic constraints. In World Congress on Intelligent Control and Automation, Shanghai, P.R.China, Jun 2002.
2. A. Nagakubo and S. Hirose. Walking and running of the quadruped wall climbing robot. In IEEE Int. Conf. on Rob. And Aut., volume 2, pages 1005–1012, 1994.
3. M. Rachkov. Control of climbing robot for rough surfaces. In Int. Workshop on Robot Motion and Control, pages 101–105, 2002.
4. S. W. Ryu, J. J. Park, S. M. Ryew, and H. R. Choi. Self-contained wall-climbing robot with closed link mechanism. In IEEE/RSJ Int. Conf. on Int. Rob. And Sys, Maui, HI, 2001.
5. W. Yan, L. Shuliang, X. Dianguo, Z. Yanzheng, S. Hao, and G. Xuesban. Development and application of wall-climbing robots. In IEEE Int. Conf. on Rob. And Aut. Detroit, MI, 1999.
6. J. C. Grieco, M. Prieto, M. Armada, and P. G. de Santos. A six-legged climbing robot for high payloads. In IEEE Int. Conf. on Cont. App., Trieste, Italy, 1998.
7. S. Hirose, A. Nagabuko, and R. Toyama. Machine that can walk and climb on floors, walls, and ceilings. In ICAR, pages 753–758, Pisa, Italy, 1991.
8. D. Bevly, S. Dubowsky and C. Mavroidis. A simplified cartesian-computed torque controller for highly geared systems and its application to an

experimental climbing robot. ASME J. of Dynamic Systems, Measurement, and Control, 122(1):27–32, 2000.

9. Y. Xu, H. Brown, M. Friendman, and T. Kanade. Control system of the selfmobile space manipulator. IEEE Tr. on Cont. Sys. Tech., 2(3):207–219, 1994.
10. M. Yim, S. Homans, and K. Roufas. Climbing with snake-robots. In IFAC Workshop on Mobile Robot Technology, Jejudo, Korea, 2001.
11. H. Amano, K. Osuka, and T.-J. Tarn. Development of vertically moving robot with gripping handrails for fire fighting. In IEEE/RSJ Int. Conf. on Int. Rob. And Sys. Maui, HI, 2001.
12. C. Balaguer, A. Gim´enez, J. Pastor, V. Padr´on, and M. Abderrahim. A climbing autonomous robot for inspection applications in 3d complex environments. Robotica, 18:287–297, 2000.
13. W. Neubauer. A spider-like robot that climbs vertically in ducts or pipes. In Int. Conf. on Int. Rob. and Sys., pages 1178–1185, Munich, Germany, 1994.
14. T. Roßmann and F. Pfeiffer. Control of an eight legged pipe crawling robot. In Int. Symp. on Experimental Robotics, pages 353–346, 1997.
15. M. Almonacid, R. Saltar´en, R. Aracil, and O. Reinoso. Motion planning of a climbing parallel robot. IEEE Tr. on Rob. and Aut., 19(3):485–489, 2003.
16. Z. Ripin, T. B. Soon, A. Abdullah, and Z. Samad. Development of a low-cost modular pole climbing robot. In TENCON, volume 1, pages 196–200, Kuala Lumpur, Malaysia, 2000.
17. M. Tavakoli, M.R. Zakerzadeh, G.R. Vossoughi, S. Bagheri, "A hybrid Pole Climbing and Manipulating Robot with Minimum DOFs for Construction and Service Applications", Journal of Industrial Robot, March 2005.
18. Ali Baghani, Majid Nili Ahmadabadi, Ahad Harati, "Kinematics Modelling of a Wheel-Based Pole Climbing Robot (UT-PCR)." 2005 IEEE International Conference on Robotics and Automation, April 18-22 Barcelona.
19. M. Tavakoli, M.R. Zakerzadeh, G.R. Vossoughi, S. Bagheri, H.Salarieh, "A Novel Serial/Parallel Pole Climbing/Manipulating Robot: Design, Kinematic Analysis and Workspace Optimization with Genetic Algorithm", 21st International Symposium on Automation and Robotics in Construction, 21 to 25 September 2004, Jeju island, Korea.
20. K.S.Fu, R.C.Gonzalez, C.S.G.Lee, "Robotics; Control, Sensing, Vision, and Intelligence", NewYork: McGraw Hill, 1987.

ROBOT FOR MOTION IN TUBE*

JATSUN SERGEY

Mechatronical Department, Kursk State Technical University, 50 let Oktyabrya, 94, Kursk, 305000, Russia

MISHENKO VLADIMIR

Mechatronical Department, Kursk State Technical University, 50 let Oktyabrya, 94, Kursk, 305000, Russia

JATSUN ANDREY

Mechatronical Department, Kursk State Technical University, 50 let Oktyabrya, 94, Kursk, 305000, Russia

Mobile robots are widely utilized for various operations in environments inaccessible to a human or dangerous for him. They are utilized, for example, for inspection and repair work in nuclear power and chemical plants, operations in areas of wreckage after earthquakes or blasts, or dismantling explosive devices. Most of these robots move by means of wheels or caterpillars, some of them utilize walking mechanisms. Such robots, however, cannot enter narrow slots (for example, during rescue operations in a zone of wreckage) or move in dense media other than gases or liquids. This justifies looking for new concepts of motion to enable robots to move efficiently in environments inaccessible to robots with wheel, caterpillar, and walking propelling systems. This issue is especially topical for medical robots designed for the motion through rather narrow channels (e.g., in blood vessels or the intestines) or among muscles to reach an affected organ to perform a diagnostic or surgical operation.

1. Introduction

In the present paper, the concept of vibration-driven robots is developed. Such robots can move in various media without wheels, caterpillars or legs. The

* Work supported by RBRF project №04-01-04002, №05-08-33382 and DFG project №Zi. 540/6-1.

propulsion of the robot is provided from one hand due to vibration of internal masses inside the robot and from another hand due to interaction of the robot's body with the environment.

The vibration-driven robot is designed as a mechatronical system consisting of the mechanical, electrical, and electronic (microcomputer) components. The mechanical vibration is transmitted from the vibration exciter to the robot's body, which interacts with the environment with some force. The robot is equipped with a feedback microcontroller to maintain an efficient operation mode and provide prescribed characteristics of motion of the end-effector under the action of various disturbances. The characteristics of the exciter vibration should be tuned to a specific task to be executed by the robot. This tuning can be provided on the basis of parametric optimization in which an operating characteristic of the robot (e.g., the average velocity of the robot) is used as the objective function.

Classification of vibration – driven robots is proposed with point of view of dimension of space where the robot and his parts moves. In simplest case the robot moves in one- dimensional space (1-D). In this case motion of the robot can be provided with help of variable shape of body (this is worm like motion [1-6, 11]) or with motion of internal vibrating masses. More complicate case is the motion of robot in two - dimensional space (2-D). The motion of the robot can get with variable shape of body with open structure or close structure (snake-like motion [9]) or with mobile internal masses that can move with accordance of planar trajectory (very interesting case –is the rotational motion of masses [12]). The most complicate is motion of robot in three -dimensional (3-D) space. As it was in previous cases the motion of robot body can get with periodically variable shape of body with open or close structure or with mobile internal masses.

2. Mathematical model and design of 1-d robot with variable shape of robot body.

The 1-D motion of the robot can be provided by periodical relative motion of two parts of robot body along the line (see Fig. 8). In this case, it is necessary that the characteristic of friction force between robot body and the supporting surface should be asymmetric. The robot under consideration was designed for cleaning pipelines from solid deposits.

To determine the parameters of the cleaner robot moving in tube, we developed a mathematical model that describes the interaction of the end-effector with the technological load (due to the resistance of the solid deposits) and the vibration-driven actuator. This robot has control system that provides adequate electrical feeding of each drive (EM). Control system consists of sensor (S) microcontroller source (MC) of electrical feeding (PS).

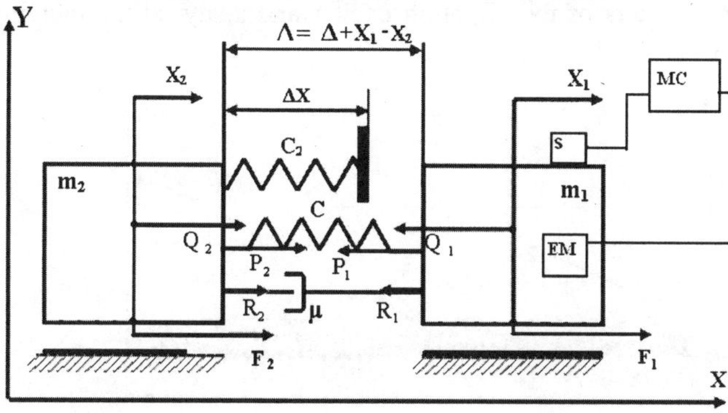

Fig. 1 Calculating scheme of mini-robot with two asymmetric friction elements.
X_1 ,X_2 – generalized coordinates of masses; C and μ – spring rate and damping coefficient; C_2 , μ - spring rate and damping coefficient of limiter; $Q_{1,2}$ -force generated by the electromagnetic exciter; $P_{1,2}$- force of elastic element; $R_{1,2}$ – force of resistance; ΔX – length of limiter; Δ –distance between masses without reference of spring deformation.

Differential equations that describe dynamics of the robot with an electromagnetic vibration actuator in not dimensional form can be introduced in a form in Eq. (1):

$$\begin{cases} \dot{\overline{\Phi}} = \varepsilon \cdot \overline{u}(\tau) - \beta \cdot (\overline{\Delta} + (\overline{X}_1 - \overline{X}_2)) \cdot \overline{\Phi} \\ \ddot{\overline{X}}_1 = -\zeta(\dot{\overline{X}}_1 - \dot{\overline{X}}_2) - \overline{P}_1 - \chi \cdot \overline{\Phi}^2 + \overline{F}_1 \\ \ddot{\overline{X}}_2 = \overline{m} \cdot \zeta \cdot (\dot{\overline{X}}_1 - \dot{\overline{X}}_2) - \overline{P}_2 + \overline{m} \cdot \chi \cdot \overline{\Phi}^2 + \overline{m} \cdot \overline{F}_2 \end{cases} \tag{1}$$

Where are:

$$\varepsilon = \frac{U_0}{\Phi_0 \cdot \omega};\ \beta = \frac{2X_0 \cdot R}{\mu_0 \cdot S \cdot \omega};\ \xi = \frac{C}{\omega^2 \cdot m_1} = \frac{\lambda_1^2}{\omega^2};\ \chi = \frac{(\Phi_0)^2}{\mu_0 \cdot S \cdot X_0 \cdot \omega^2 \cdot m_1};\ \overline{m} = \frac{m_1}{m_2};\ \overline{\Delta} = \frac{\Delta}{X_0};$$

$$\varsigma = \frac{\mu}{m_1 \cdot \omega} \varpi \cdot$$

Where are: ω – frequency of electromagnetic drive, U_0, Φ_0, X_0 – scales of electrical voltage, magnetic flux and length, Φ is the magnetic flux; I - is the current,, R- is the active electric resistance of the solenoid; U(t) is the power supply voltage, S-square of the air gap inside if electromagnet; μ_0 – electromagnetic constant.

The system of nonlinear differential equations describing the motion of the robot is solved by numerical methods. Based on the simulation results, we determined the parameters of the vibration exciter and analyzed the motion of the robot values of the friction coefficient.

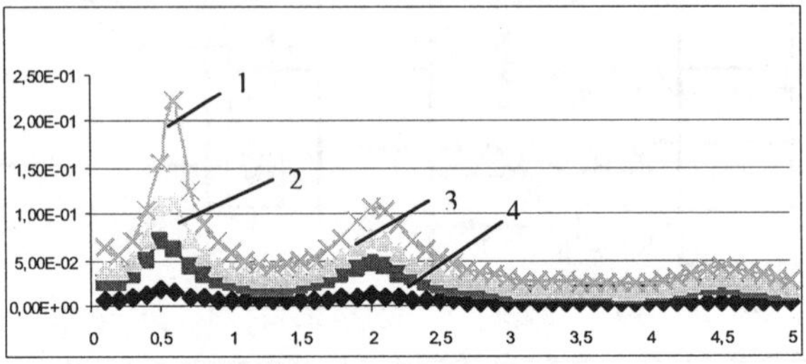

Fig. 2. The average velocity of the robot as a function of the relative frequency of the vibration exciter for different electrical feeding. 1-ε=6, 2-ε=4, 3-ε=3, 4-ε=2.

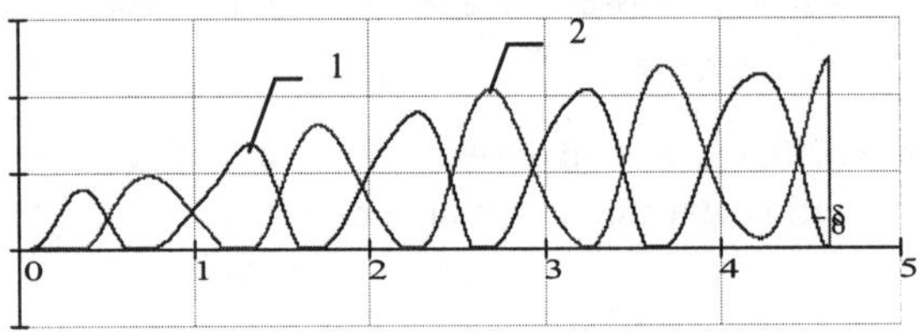

Fig. 3. Time history of velocity of robot masses for ε =2 and ξ=0,6.
1 – velocity of first mass, 2 –velocity of second mass.

Figure 3 shows the behavior of robot for different regimes of motion. Analysis of these results shows that in a field of first resonance masses move to each other and each masse has stop regimes in every step with frequency that is equal to external frequency. In a field of second resonance the masses of robot move to each other and each masse has stop regimes in every step with frequency less than external frequency in two times. And in a field of third resonance the masses of robot move to each other and each masse has stop regimes in every step with frequency less than external frequency in three times.

The some results of the calculation of the force of friction and the average velocity of the robot are illustrated in fig. 3. Analysis shows that the value of electrical voltage plays very important role for average velocity of the cleaner robot. Besides is the ratio of the natural vibration frequency of the robot to the excitation frequency of the electromagnet is very important also.

3. Description of 1-d robot with two masses and electromechanical drives

This robot consists two parts as it is shown on fig.4. The parts of the robot have three special friction elements 2 providing asymmetric friction forces. Electrical coil 1 switches direction of friction element. Electrical drive 3 displaces one part relatively another part. Due to asymmetrical friction forces acting between robot parts and tube robot moves inside in pipe. Friction element using in the robot has boll and spring. The boll moves on the inclined surface. The angle between this surface and tube surface is smaller than friction angel. It provides asymmetric of friction force dependence on direction of motion of robot body. For example, the robot, which introduced on fig. 4 moves to the left side. But if is needed to reverse direction of motion of robot control system by use of electrical coil will change inclination of the supporting surface of the bolls for each friction elements and robot will move to the right direction.

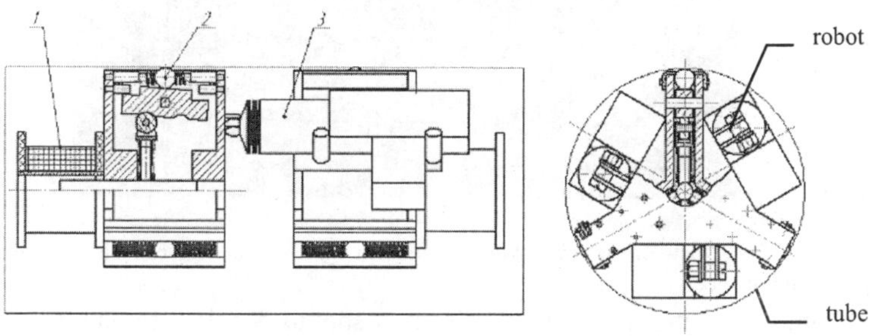

Fig. 4. Scheme of robot with two masses and electrical drives inside in tube.

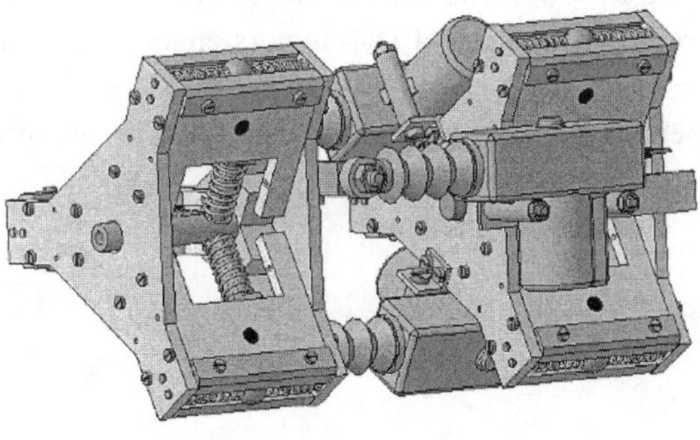

Fig. 5. The 3-D scheme of robot with two masses and electrical drives inside in tube.

Consider 3-D scheme of vibration-driven robot with two masses that move and change the shape of the body and generate friction forces is shown on fig.5. Friction elements have control system. This makes it possible to control component of the force exerted on the robot by the supporting surface. And it provides reverse motion of the robot inside in tube. In the robot three separate electrical drives is used. It is necessary in the case when robot moves inside in curve tube.

In mechatronical department of Kursk State Technical University the prototype of robot for in pipe inspection is made (see fig.6). This robot moves in tube with constant internal diameter only.

Fig.6. The general view of the robot with two masses and electrical drives inside in the glass tube (experimental device).

Control system of vibrating 1-D robot should decide different tasks main of them could be introduced in the next form:1) Trajectory planning; 2) Control velocity of motion of robot with accordance of given trajectory; 3)Optimization of velocity of robot; 4) Control of DC- motors angular velocity and phase shift angle.

In this paper control system for the 1-D robot with two masses is introduce in fig.7.

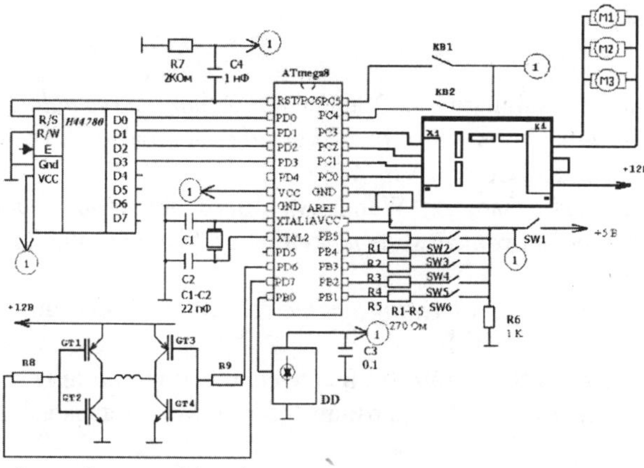

Fig. 7. Scheme of control system of the robot

Control system consists from microcontroller Atmega8, sensors, connectors LB1649 for DC-motors, remote control with infrared sensor. With accordance of signals from navigation system this control system provides change of velocity of robot.

4. Conclusions

Vibration-driven robots with a one-coordinate vibration exciter can move along a rough surface only if the friction between the robot body and the supporting surface or/and the exciter vibrations have asymmetric characteristics. Robots equipped with vibration exciters and internal masse can move even in the absence of an asymmetry in the friction or vibration characteristics. .

Robots equipped with a one-coordinate exciter and two masses has special friction elements providing asymmetric friction between robot body and internal surface of the tube and it can be utilized for cleaning pipelines from solid deposits in a case of resistance is high. The reverse in the direction of motion is provided by changing of direction of friction forces providing by friction elements.

Acknowledgements

This investigation is supported by RBRF project №04-01-04002, №05-08-33382 and DFG project №Zi. 540/6-1.

References

1. Aoshima, S.; Tsujimura, T.; Yabuta, T.: *A miniature mobile robot using piezo vibration for mobility in a thin tube,* Journal of Dynamic Systems, Measurement, and Control Vol. 115 (1993), pp. 270-278
2. Ma, J.; Lo, M.; Bao, Z.; Wang, A.: *Micro peristaltic robot simulating earthworm and its control system,* Journal of Shanghai Jiao tong University Vol. 33 No. 7, (1999)
3. Jatsun S., Safarov J. *Vibrating engine for robots*. Proceedings. CLAWAR (2000). Madrid.P.1016-1021.
4. Jatsun S., Jatsun S., Vorontsov R. *Dynamics of vibrating robot for in pipe inspection* International Symposium – SYROM. – Bucharest, (2001). 205-209.
5. Zimmermann K., Zeidis I., Steigenberger J., and Huang Jianjun. *An approach to the modelling of worm-like motion systems with a finite number of degrees of freedom*, First Steps in Technical Realization. Proc. of the 4th International Conference of Climbing and Walking Robots, Karlsruhe. (2001). P. 561- 568.
6. Yeh, R.; Hollar, S.; Pister, K.S.J.: *Design of low-power silicon articulated microrobots,* Journal of Micromechatronics, Vol. 1, Num. 3, (2001). P. 191-203
7. Bolotnik N.N., Chernousko F.L., Kostin G.V., and Pfeiffer F. *Regular motion of a tube-crawling robot in a curved tube*, Mechanics of Structures and Machines. (2002). Vol. 30. No. 4. P. 431-462.
8. Chernousko F.L. *Snake-like locomotions of multilink mechanisms*, Journal of Vibration and Control. (2003). Vol. 9. No. 1-2. P. 235-256.
9. Zimmermann K., Zeidis I., and Pivovarov M. *Dynamics of a nonlinear oscillator in consideration of non-symmetric Coulomb dry friction*, Fifth Euromech. Nonlinear Dynamics Conference. Book of Abstracts. Eindhoven Netherlands, August 7 12. (2005). P. 308
10. Zimmermann K., Zeidis I., and Steigenberger J. *Mathematical model of worm-like motion systems with finite and infinite numbers of degrees of freedom*, Theory and Practice of Robots and Manipulators. Proceedings of the 14th CISM IFToMM Symposium (RoManSy 14). (2002). P. 07 16.
11. Chernousko F., Zimmermann K., Bolotnik N., Yatsun S, Zeidis I. *Vibration –Driven Robots*, The Workshop on Adaptive and Intelligent Robots: Present and Future. Proceedings. Vol.1 The Institute for problem in mechanics RAS. Moscow. (2005), P.26-31.
12. Bolotnik N. N., Jatsun S. F., Jatsun A. S., Cherepanov A. A. *Automatically controlled vibration-driven robots*, Proceedings. International Conference on Mechatronics ICM, Budapest, (2006), P.438-441.

ROBOTRAIN AS SNAKELIKE ROBOTIC SYSTEM WITH MINIMAL NUMBER OF DOF

V.E.PAVLOVSKY, V.V.PAVLOVSKY, Jr
Keldysh Institute of Applied Mathematics of RAS
Moscow, Russia, 125047, Miusskaya sq., 4

N.V.PETROVSKAYA, V.V.EVGRAFOV
Moscow State University,
mathematics and mechanics department, Moscow, Vorob'evi Gori

The paper deals with investigating of dynamical properties and elaborating of corresponding motion planning algorithms for wheeled "robotrain", which is snakelike transport structure. In this structure there are one active robot, equipped with differential drive, and large number of trailers. The created control algorithms provide purposeful motion of the system along the given set of lines, or inside corridors, or via target points. Algorithms are verified by means of mathematical modeling. The work is supported by the RFBR grants No 02-01-00750, 04-01-00065.

1. Introduction

The problem of synthesis of control algorithms for snakelike systems of "the chain of robots" type is concerned with the need which has appeared now in view of creation of automatic multiobject transport systems. For example, such robotized systems are good for transportation and delivery of large number of scientific devices in zones where presence of the human is impossible or is difficult, and in similar cases. The great value has also creation of robots-chains as usual transport systems. Note, similar problems arise, in particular, at robotized service operations in extreme environments. So the problem of control of snakelike robots motion as chains of mobile objects is an important one and has theoretical and applied value, researches in this area are stimulated by the numerous applied problems.

In the paper there are proposed and verified by means of computer simulation the model and control algorithms for "robotrain", which is a system of one active robot and a chain of trailers. The algorithms provide purposeful movement of such chain of mobile objects along the target points set and, so forth, in the environment with obstacles (with their avoidance). The paper proposes a set of algorithms, the obtained results can be used on development of control algorithms for real transport systems with the tractor-robot and the chain of trailers providing solution of spectrum of applied problems. Note, in the system tractor-robot may be located not only as the first robot of a chain, but may be set in any position in a chain, even may be the rear object. The active robot is one equipped with differential drive and the whole chain have only two DOF, which is really minimal number of controlled parameters for snakelike system.

2. Robotrain model. The equations of motion, the solutions.

Let's note, that works in this direction actively started over the world last years. The analysis of the specified works [1-4,8] show, nevertheless, that practically all system have a large numbers of DOF, typical solution is to use drives in each joint or in each object, e.g. KORYU-II is such system. In this sense those systems are redundant in view of number of DOF, and the question is a following - what minimal DOF quantity is sufficient for organizing the snakelike robot motion. In the paper there will be below proposed a snakelike system with only two DOF, and it is possible to show that those system is quite enough to built any purposeful snakelike motion on a plane. It seems, this result is an original one.

Let's consider the snakelike robotic system "robotrain", namely, a robot with set of trailers. Let's consider following mathematical model of that chain of mobile wheel objects (carriages), see fig.1. Each carriage contains a weightless axle of length *2* with wheels of weight *m* and radius *r,* a weightless directing rod of length *2b*, on perpendicular axe, and a heavy body of mass m_i. The center of mass of the carriage coincides with the center of an axle of wheels. Carriages are coupled with each other by means of the cylindrical hinge, gap in coupling is absent. Carriages move on absolutely rough horizontal plane. The condition of rolling without slippage causes presence of nonholonomic constrains.

Let θ_1 θ_2 .., θ be angles of turn of directing rods of corresponding carriages around of a vertical axe (angles of orientation), in the further instead for angles θ_2 ..., θ_n we shall consider angles of turn of a directing rod of *i*-th carriage ($i = 1,2,\dots,n$) concerning a directing rod of the previous carriage $\psi_2,\dots,\psi_n$, where $\psi_2 = \theta_2 - \theta_1,\dots,\psi_n = \theta_n - \theta_{n-1}$, let x_1, y_1, ..., x_n, y_n - are coordinates of points C_1, ..., *n* accordingly, φ_{11} φ_{12}, ... φ_{n1}, φ_{n2} - angles of turn of wheels of corresponding carriages around their axles.

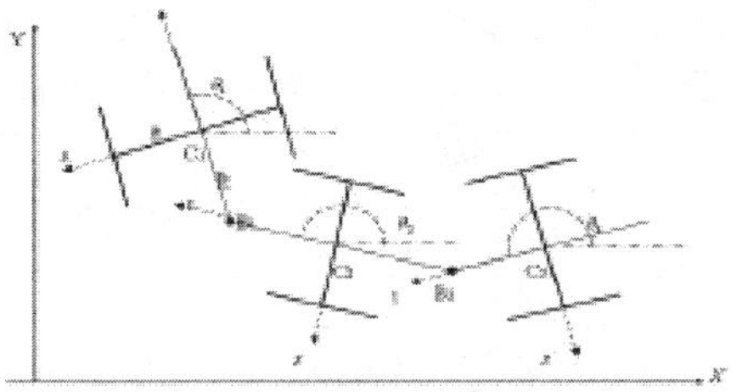

Figure 1. Projection of robotrain system onto a horizontal plane.

The equations of movement of such system with n elements are made in the form of Voronec nonholonomic equations, and in the form of Cauchy for free movement have a following appearance:

$$\dot{\varphi}_{11} = \varphi'_{11}$$

$$\dot{\varphi}_{12} = \varphi'_{12}$$

$$\dot{\varphi}'_{12} = \frac{(C_2K_4 - 2C_3K_1)}{C_2^2 - C_1C_3}\varphi'^2_{12} + \frac{(C_2K_5 - 2C_3K_2)}{C_2^2 - C_1C_3}\varphi'_{12}\varphi'_{11} + \frac{(C_2K_6 - 2C_3K_3)}{C_2^2 - C_1C_3}\varphi'^2_{11}$$

$$\dot{\varphi}'_{11} = \frac{(C_2K_1 - 2C_1K_4)}{C_2^2 - C_1C_3}\varphi'^2_{12} + \frac{(C_2K_2 - 2C_1K_5)}{C_2^2 - C_1C_3}\varphi'_{12}\varphi'_{11} + \frac{(C_2K_3 - 2C_1K_6)}{C_2^2 - C_1C_3}\varphi'^2_{11}$$

$$\dot{x}_1 = r\frac{\dot{\varphi}'_{12} + \dot{\varphi}'_{11}}{2}\cos\theta$$

$$\dot{y}_1 = r\frac{\dot{\varphi}'_{12} + \dot{\varphi}'_{11}}{2}\sin\theta_1 \qquad (1)$$

$$\dot{\theta}_1 = r\frac{\dot{\varphi}'_{12} - \dot{\varphi}'_{11}}{2}$$

$$\ldots$$

$$\dot{x}_i = \frac{r}{2a}\Bigg((a\cos\sum_{k=2}^{i}(-1)^k\psi_k - b\sin\sum_{k=2}^{i}(-1)^k\psi_k)\dot{\varphi}'_{12} +$$
$$+ (a\cos\sum_{k=2}^{i}(-1)^k\psi_k + b\sin\sum_{k=2}^{i}(-1)^k\psi_k)\dot{\varphi}'_{11}\Bigg)\cos\left(\theta_1 + \sum_{k=2}^{i}\psi_k\right)$$

$$\dot{y}_i = \frac{r}{2a}\Bigg((a\cos\sum_{k=2}^{i}(-1)^k\psi_k - b\sin\sum_{k=2}^{i}(-1)^k\psi_k)\dot{\varphi}'_{12} +$$
$$+ (a\cos\sum_{k=2}^{i}(-1)^k\psi_k + b\sin\sum_{k=2}^{i}(-1)^k\psi_k)\dot{\varphi}'_{11}\Bigg)\sin\left(\theta_1 + \sum_{k=2}^{i}\psi_k\right)$$

$$\dot{\varphi}_{i1}=\begin{cases} r\dfrac{(2ab\cos\sum\limits_{k=2}^{i}(-1)^k\psi_k+(a^2-b^2)\sin\sum\limits_{k=2}^{i}(-1)^k\psi_k)\dot{\varphi}'_{12}+(a^2+b^2)\sin\sum\limits_{k=2}^{i}(-1)^k\psi_k\dot{\varphi}'_{11}}{2ab}, & i=2p\\ r\dfrac{-(a^2+b^2)\sin\sum\limits_{k=2}^{i}(-1)^k\psi_k\dot{\varphi}'_{12}+(2ab\cos\sum\limits_{k=2}^{i}(-1)^k\psi_k-(a^2-b^2)\sin\sum\limits_{k=2}^{i}(-1)^k\psi_k)\dot{\varphi}'_{11}}{2ab}, & i=2\quad+1 \end{cases}$$

$$\dot{\varphi}_{i2}=\begin{cases} r\dfrac{-(a^2+b^2)\sin\sum\limits_{k=2}^{i}(-1)^k\psi_k\dot{\varphi}'_{12}+(2ab\cos\sum\limits_{k=2}^{i}(-1)^k\psi_k-(a^2-b^2)\sin\sum\limits_{k=2}^{i}(-1)^k\psi_k)\dot{\varphi}'_{11}}{2ab}, & i=2p\\ r\dfrac{(2ab\cos\sum\limits_{k=2}^{i}(-1)^k\psi_k+(a^2-b^2)\sin\sum\limits_{k=2}^{i}(-1)^k\psi_k)\dot{\varphi}'_{12}+(a^2+b^2)\sin\sum\limits_{k=2}^{i}(-1)^k\psi_k\dot{\varphi}'_{11}}{2ab}, & i=2p+1 \end{cases}$$

$$\dot{\psi}_i=(-1)^{i+1}\frac{1}{2ab}\Bigg[\Bigg(b\Big(\cos\sum_{k=2}^{i}(-1)^k\psi_k+\cos\sum_{k=2}^{i-1}(-1)^k\psi_k\Big)+a\Big(\sin\sum_{k=2}^{i}(-1)^k\psi_k+\sin\sum_{k=2}^{i-1}(-1)^k\psi_k\Big)\Bigg)\dot{\varphi}'_{12}-$$
$$-\Bigg(b\Big(\cos\sum_{k=2}^{i}(-1)^k\psi_k+\cos\sum_{k=2}^{i-1}(-1)^k\psi_k\Big)-a\Big(\sin\sum_{k=2}^{i}(-1)^k\psi_k+\sin\sum_{k=2}^{i-1}(-1)^k\psi_k\Big)\Bigg)\dot{\varphi}'_{11}\Bigg]$$

Just from the equations (1) it may be obtained important conclusion, that change of number of objects in system does not influence on change of number of degrees of freedom which is always equal to two for any n elements.

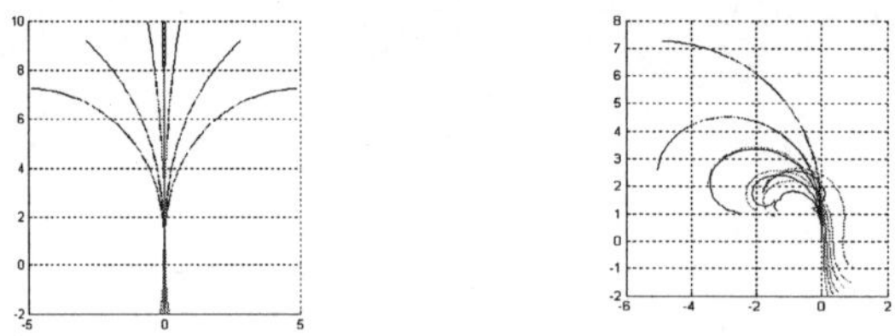

Figure 2. Examples of trajectories of ballistic movements.

Base solutions of the equations of motion (1) are obtained (see [5-7]). It is necessary to note, that such partial solutions as "straight line" and "circle" exist for any n. Also there exist dynamically smooth solution as Cornu spiral [5-7]. Other possible solutions are found by numerical methods, some their examples are resulted on fig.2.

Let's consider operated movement of a chain of the mobile system. We shall consider in this case, that there is only one active carriage (an active robot) and its wheels are set in motion by engines, and the engine is characterized by the torques developed by them (Mom_1 and Mom_2). Then the dynamic equations will become:

$$\begin{cases} \ddot{\varphi}_{12} = \dfrac{(C'_2 Mom_1 - 2C'_3 Mom_2)}{(C'^2_2 - 4C'_1 C'_3)r} + \dfrac{(C'_2 K'_4 - 2C'_3 K'_1)}{C'^2_2 - 4C'_1 C'_3}\dot{\varphi}^2_{12} + \dfrac{(C'_2 K'_5 - 2C'_3 K'_2)}{C'^2_2 - 4C'_1 C'_3}\dot{\varphi}_{12}\dot{\varphi}_{11} + \dfrac{(C'_2 K'_6 - 2C'_3 K'_3)}{C'^2_2 - 4C'_1 C'_3}\dot{\varphi}^2_{11} \\ \ddot{\varphi}_{11}, = \dfrac{(C'_2 Mom_2 - 2C'_1 Mom_1)}{(C'^2_2 - 4C'_1 C'_3)r} + \dfrac{(C'_2 K'_1 - 2C'_1 K'_4)}{C'^2_2 - 4C'_1 C'_3}\dot{\varphi}^2_{12} + \dfrac{(C'_2 K'_2 - 2C'_1 K'_5)}{C'^2_2 - 4C'_1 C'_3}\dot{\varphi}_{12}\dot{\varphi}_{11} + \dfrac{(C'_2 K'_3 - 2C'_1 K'_6)}{C'^2_2 - 4C'_1 \ '_3}\dot{\varphi}^2_{11} \end{cases} \quad (2)$$

where $C'_1, C'_2, C_3, K'_1, K'_2, K'_3, K'_4, K'_5, K'_6$ - known functions of $\psi_2, \ldots \psi_n$.

Using the received equations (1)-(2), we shall construct the alphabet of base movements of robotrain. Into a class of considered program movements enter: a straight line, a circle and a Cornu spiral [7]. It is possible to prove that this class is sufficient for constructing any dynamically smooth trajectories of robotrain. Movement on a straight line is set by a condition of equality of angular speeds of the leading carriage:

$$\dot{\varphi}_{11} = \dot{\varphi}_{12} = const = \dot{\phi}$$

Movement on a circle is set by a following relation for angular speeds of rotation of wheels of the leading carriage:

$$\dot{\varphi}_{11} = k\dot{\varphi}_{12}, \ \dot{\varphi}_{12} = const = \dot{\phi}$$

The linear law of change of speeds of rotation of wheels of the active carriage

$$\dot{\varphi}_{11} = k_1 t + \omega_{11}, \ \dot{\varphi}_{12} = k_2 t + \omega_{12}$$

sets movement on a Cornu spiral [5-7].

In a considered class of movements it is possible to obtain laws of change of torques on active wheels which satisfy dynamic equation (2) - on the basis of the assumptions made above the operating torques are defined from the equations (1) and (2) as known functions of time.

3. Method for planning of robotrain motion

For further trajectory design we will introduce new base trajectory - a double Cornu spiral which is a pair of base spirals joined in some point and ratio k_2 / k_1 will denote ratio of lengths of those spirals. Using the base trajectories (a straight line, a circle, a double Cornu spiral), we shall construct methods of planning of movement of a chain "robotrain" at travel between various points [7]. It appears, that realization of dynamically realizable transition from one point of phase space to another by means of unique program movement from class described above is impossible. The opportunity of such transition is provided at the combined movement, for example on two spirals.

The central technique is planning a trajectory as a set of pieces of a straight lines with dynamically smooth coupling. And in general it is necessary to plan such movement only for active robot, motion of the rest part of system will occur in a vicinity of that set of pieces. The deviation of the rest part of chain from the trajectory of active robot is precisely known from (1)-(2) and may be also taken into account by appropriate choice values of free parameters, as shown below. Note, pre-given set of pieces of lines may be produced, e.g., by geometrical route planning system, etc. The type of a similar trajectory is shown on fig.3. In fact just the stages of construction of resulting trajectory for active robot are shown on fig.3.

Let *I, II, III* and *IV* on fig.8 are pre-set pieces of given trajectory and let's denote them as O_1O_2, O_2O_3, O_3O_4, O_4O_5 correspondingly. In particular, movement on a trajectory, which is set by two pieces *I, II*, is broken into three sites:
- movement on a straight line setting the first piece O_1O_2, from an index point O_1 up to a point A_1 which is setting on independently on the first straight line;
- movement along two arcs of spirals (on a double Cornu spiral) from a point A_1 up to a point A_2 belongs to a straight line setting the second piece of movement O_2O_3, point C of coupling of spirals is defined by given ratio k_2 / k_1 ;
- movement on a straight line setting the second piece, from A_2 up to the target point O_3.

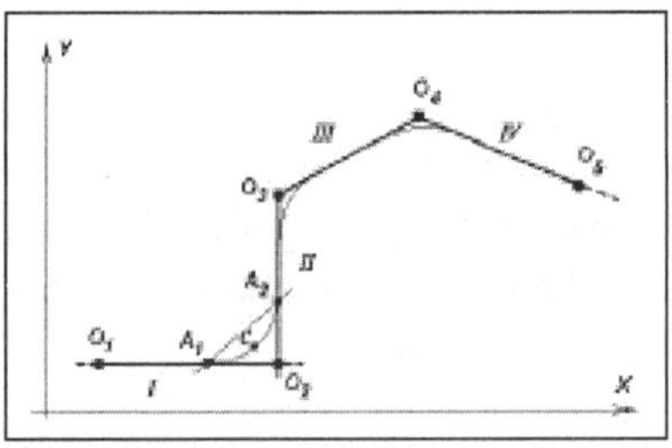

Figure 3. Stages of planning of robot movement.

And so forth for smoothing the coupling of pieces O_2O_3, O_3O_4, O_4O_5, see fig.3. Evidently, the resulting trajectory will satisfy as for dynamic equations of motion, as for equations of constrains (1)-(2). So, the resulting trajectory will be dynamically realizable. It is also necessary to note, that this technique has the following free parameters: positions of the points A_1, A_2,and C (i.e. the ratio k_2 / k_1), with using those parameters it is possible to built optimal smoothing of trajectory pieces, for example, in the sense of values of deviation resulting trajectory from given set of pre-planned lines. This means, it is possible to optimize accuracy of resulting trajectory, and so on.

With use of this feature two principles of planning of trajectory of considered system were developed: a principle of "corridors" and a principle of "reference points". First one takes into account the permissible values of deviation of system from pre-set lines. In case of a principle of "reference points" there is given a numbered set of points on a plane and the active robot has to go via all that points. The orientation of active robot is given in each point as well. It is easy to show then by constructing an auxiliary strait trajectories this problem may be reduced to a problem of planning of movement by a principle of "corridors".

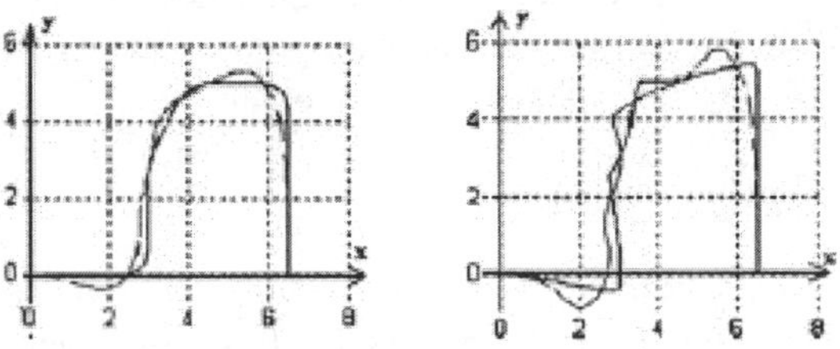

Figure 4. Examples of the planned trajectories.

Results of work of algorithms of planning of robotrain movement are shown on fig.4. The trajectory in the right figure should pass precisely through the given points, it has greater average curvatures.

To verify the results discussed above the another model was elaborated. That model was developed as multibody dynamics model, an example is shown on a figure below.

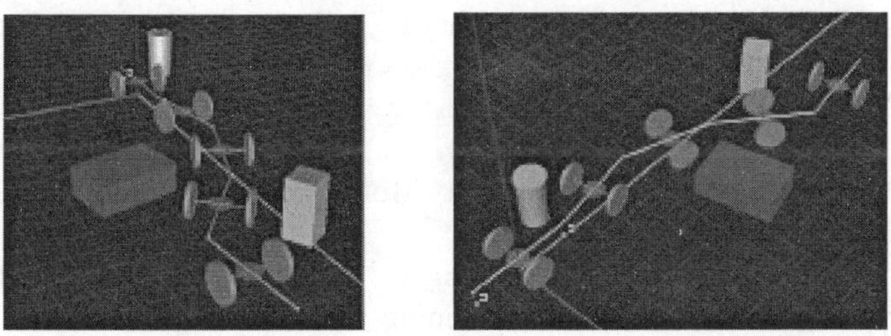

Figure 5. Simulation samples.

It is necessary to note, that simulation with using multibody model is more accurate because allow to take into account several additional details, such as more realistic contact between wheel and a surface. In fact, this model also show good functioning of the elaborated system.

4. Conclusion

Numerical experiments show good quality of elaborated control methods. From those results it is possible to conclude that proposed techniques of planning and realizing robotrain motion are quite enough for constructing and implementing motion of snakelike robot with only two DOF. Numerical researches also show that the width of "corridor" (deviations of a system from pre-set trajectories) is sufficiently small, and more, can be controlled and may be reduced due to an optimum choice of the parameters which setting curvature of turns. It has been also numerically investigated behavior of system under conditions of disturbances and it was shown the stability of the robotrain system in sufficient range of noise (disturbances) values. These properties should be considered by a control system realizing movement of real "robotrain".

REFERENCES

1. F. Lamiae, J.P. Almond. “A practical approach to feedback control for a mobile robot with trailer” in Proc.1998 IEEE International Conference on Robotics & Automation, pp. 3291-3296, 1998.

2. Y. Nakamura, Hezekiah, Yutan, W.Chung. "Design of Steering Mechanism and Control of Nonholonomic Trailer Systems" in Proc. 2000 IEEE International Conference on Robotics & Automation, pp. 247-254, 2000.

3. K.-U. Scholl, V. Kepplin, K. Berns, R. Dillmann “Controlling a Multijoint Robot for Autonomous Sewer Inspection” in Proc. 2000 IEEE International Conference on Robotics & Automation, pp. 1701-1706, 2000.

4. M.Vendittelli, G.Oriolo. “Stabilization of the general two-trailer system” in Proc. 2000 IEEE International Conference on Robotics & Automation, pp. 1817-1823, 2000.

5. V.E.Pavlovsky, N.V.Petrovskaya. Investigations of dynamics of "robotrain" chain motion. Equations of motion, specific solutions. // Preprint of KIAM, 2005, 117. In Russian.

6. V.E.Pavlovsky, V.V.Evgrafov, N.V.Petrovskaya. Investigations of dynamics of "robotrain" chain motion. Motion under control. // Preprint of KIAM, 2005, 120. In Russian.

7. V.E.Pavlovsky, N.V.Petrovskaya. Investigations of dynamics of "robotrain" chain motion. Motion planning techniques. // Preprint of KIAM, 2005, 121. In Russian.

8. Fukushima, E.F. and S. Hirose; Optimal Attitude Control for Articulated Body Mobile Robots. // Proc.of the International Symposium on Adaptive Motion of Animals and Machines, CD-ROM proceedings, Montreal, Canada, August (2000).

SERVICING SOLAR POWER PLANTS WITH WALLWALKER

RIDHA AZAIZ

RWTH Aachen, Wallwalker.Net ,
Kastanienweg 6, 52074 Aachen, Germany

More and more solar power plants are being installed worldwide in concern over climate change. The efficiency of renewable energy sources is crucial as the power output is being sold. Deposits, dust and damages lower the output of the solar plant. This is usually not recognized by the operator of the plant, as they are installed at a high altitude and inclination. For economic reasons and safety at work, inspection and cleaning of the solar plant is often neglected. This is where Wallwalker applys. Wallwalker is a light weight robot that is able to walk on the sensitive solar modules in order to inspect and clean them. The cleaning process is controlled by an imaging software. In addition damages can be detected and are reported to the operator. The efficiency of the solar power plant raises and amortises faster.

1. Introduction

1.1. *Concept of Solar Power Plants*

Solar power plants consist of several modules that are switched together. In order to gain the highest output, they are usually installed with an inclination torwards the sun.

Each module consists of many solar cells. These single cells are switched serial into a string, and the strings are again switched together. The solar module is protected by glass or lamination, embedded in an aluminium frame.

Figure 1. Solar power plant consisting of several modules on different sites.

1.2. *How deposits effect the plant*

If one solar cell of the module was shadowed, e.g. by deposits such as dust, it would produce less current than the cells that are switched before and after. The cell with the lower output would actually consume current, heat up and damage.

Therefore the manufacture of the module install diodes, in order to switch the effected cell of. But as each diode needs current itself, manufactures compromise on integrating only one diode into each string. That means that in the case of deposits on one cell, the whole string gets disconnected. The output of the effected module declines in high proportion.

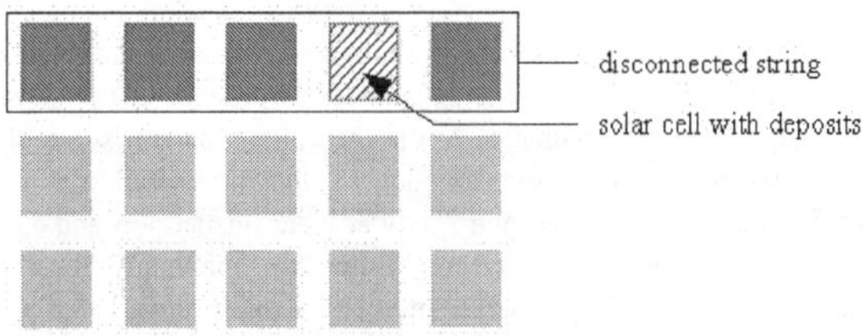

Figure 2. Solar module consisting of several cells, one with deposits.

2. Automated cleaning and inspection with Wallwalker

2.1. *Requirements of robotic servicing*

As solar modules are usually installed at high altitude and inclination, they are hard and dangerous to reach by manpower. Additional equipment and infrastructure is needed to clean them manually. The load of the human body is often too heavy for the modules. Limiting the stress during servicing by walking on the aluminium frames only remains difficult.

A criteria in automated cleaning of solar modules is light weight of the device. It is also crucial that the robot is not damaging the surface while cleaning and moving, since that would have effects on the output of the plant, too. The plant consists of several modules that have spaces and obstacles such as mounting materials in between. The robot needs to be able to recognize and to overcome such obstacles.

Picture 1 indicates that the robot should be able to move on surfaces at an inclination of up to 90 degree and a surface temperature of up to 100 degree Celsius. The robot should be able to reach every point on an rectangle shaped surface. It is also an advantage if the robot can be moved from one module-site to an other easily without additional installation.

2.2. *Characteristics of Wallwalker*

The specified requirements led to the development of Wallwalker. This is a robot that moves back and forth and takes turns in order to cover a whole rectangular surface. Wallwalker is a wall climbing robot that attaches on the surface with vacuum bulbs at an inclination of up to 90 degree.

The light weight (6kg) ensures that the surface is not damaged by the robot and the material of the vacuum bulbs is resistent towards the temperatures on the surface. The framework of the moving unit makes it possible to step over obstacles, such as mounting materials, with a width of up to 8 cm and a height of 3 cm. A supply secures the robot and carries both electricity and cleaning agent. The camera compares the surface before and after the cleaning process.

Figure 3. Wallwalker on solar modules.

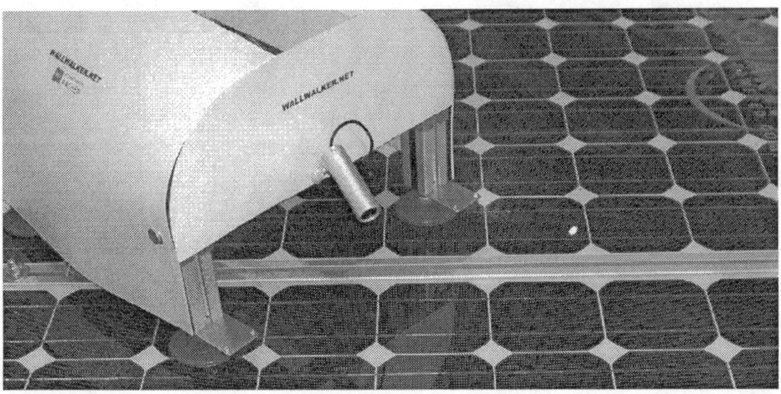

Figure 4. Front view of Wallwalker

2.3. *Locomotion*

Wallwalker moves back and forth on a so called "sliding-frame". This frame consists of two parts. One is the inner part that has its own vacuum bulbs and moves up and down. The other is the outer part that moves the inner part back and forth. First, the robot moves its inner part towards the surface and attaches the vacuum bulbs onto it. Once that happened the outer part is lifted and moves. Now the outer part can be lowered and attaches onto the surface. When the vacuum bulbs are evacuating air, the inner part is detaches and lifts. Next, that part moves and the steps are repeated.

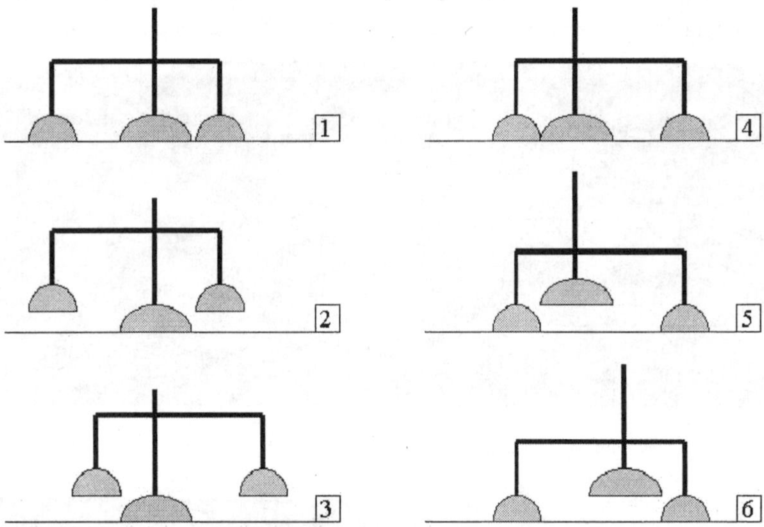

Figure 5. Simplyfied scheme of Wallwalker walking.

2.4. *Covering strategy*

Solar power plants usually have a relatively simple shaped surface. A rectangle can be covered most easily by the robot as it moves like a snake. This means Wallwalker moves forth until it reaches an edge which it can not step over. This could be the edge of the surface. There it turns 90 degrees to the right, steps its own with forth and then turns again 90 degrees. Now it reached its next track and steps forth until the next edge it can not step over (that would be the end of the surface). Wallwalker will turn left there and repeats the steps it did before.

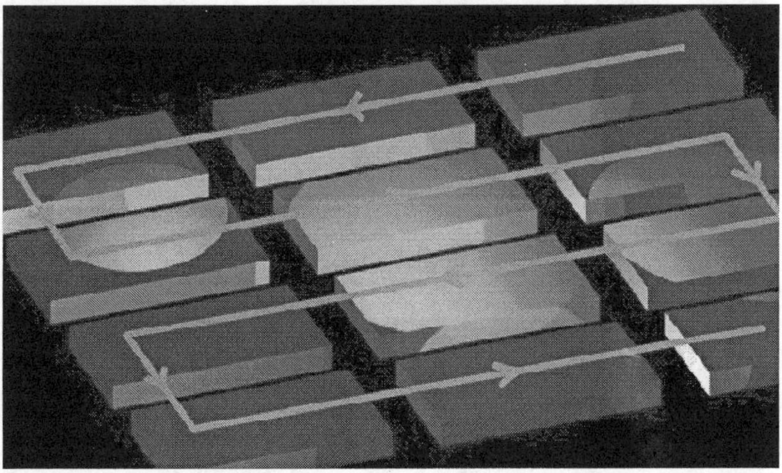

Figure 6. Wallwalker covering a large, rectangle surface.

2.5. *Services*

The efficiency loss through deposits is strongly dependend on the geographic site of the plant. In regions like Spain, California and Africa where there is less rain deposits stay on the glass longer. Rain does not clean everything. Dust and polls burn into the glass and become hard to clean off.

Nanostructures are not imposed on solar modules for several reasons. One of them is that additional surfaces on the glass will higher the reflexion of the sunbeam, lowering the efficiency of the solar modules again.

Besides this other environmental circumstances lead to corrosion, delamination and other damages. These errors are hard to identify and to keep track of in the large solar power plants. Therefore Wallwalker has a cleaning module attached and a camera. The cleaning unit consists of a rotating microfibre towel that is supplied with the cleaning agent. The camera and the imaging software compare the result of the cleanig process before and after.

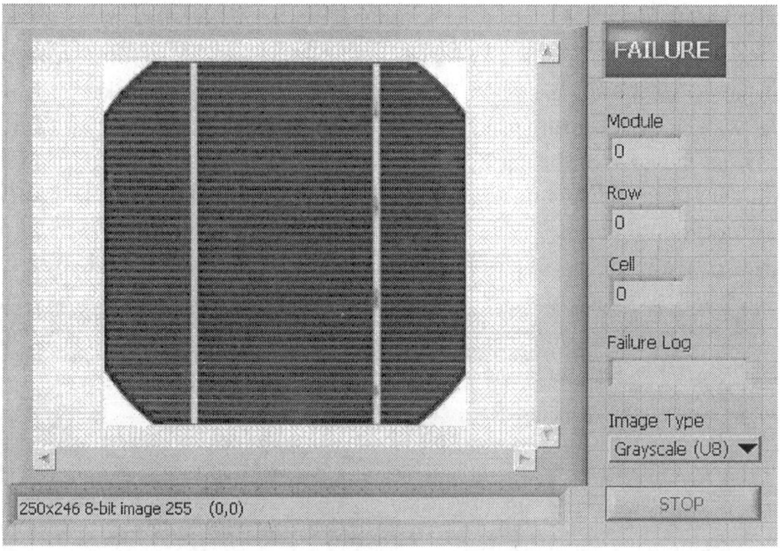

Figure 7. Imaging software with clean reference cell.

Figure 8. Solar module with deposits.

2.6. *Summary and outlook*

Wallwalker is a mobile robot that ensures the efficiency of solar power plants. With the robot surfaces of high altitude and inclination can be maintained where it has been too dangerous and expensive before. Wallwalkers' simplicity and light weight offer a unique and broad range of applications for measurement, construction and maintenance purposes.

References

1. H. Häberlein, D. Graf: Gradual Reduction of PV Yield due to Pollution, 2nd World Conference on Photovoltaic Solar Energy Conversion, 1998
2. C.Ränken, H. Häberlein: Langzeitverhalten von netzgekoppelten Photovoltaikanlagen, Ingenieurschule Burgdorf, 11/99

STABILITY AND GAIT OPTIMIZATION OF A HYBRID LEGGED-WHEELED ROVER

BYRON E. JOHNS
School of Mechanical Engineering
Human-Automation Systems Lab
Georgia Institute of Technology
Atlanta, GA 30332, USA

AYANNA M. HOWARD
School of Electrical and Computer Engineering
Human-Automation Systems Lab
Georgia Institute of Technology
Atlanta, GA 30332, USA

In this paper, we discuss the stability and gait optimization of a new kind of hybrid legged-wheeled rover. In order to control this complex robot, kinematic analysis is needed for both its wheeled and legged mobility system. With these kinematic models, we analyze the proper location of the legs during both the standing and walking phase and develop an optimization scheme to provide a stable locomotion sequence for the robot when in its legged configuration. We also compare and contrast two walking gaits associated with hexapod robots, and converge on a final gait based on this analysis for the legged mobility system of this new hybrid rover. Details of the optimization approach as well as results using the physical hardware are presented in the paper.

1. Introduction

Exploration on unknown and uncharted planetary surfaces typically involves operating in an unstructured and poorly modeled environment. To allow robust navigation capability in these natural environments, several designs are plausible for such a mobile robotic platform [1]. In order to guarantee success of robotic missions for the future, we focus on modularizing both hardware and software components to create a reconfigurable robotic explorer. This allows the deployment of rovers on a planet's surface that can ensure robust operation to deal with changes in the face of system reconfiguration, hardware failure, changes in task specifications, or alterations in the terrain environment. We have developed a new reconfigurable legged-wheeled rover known as Byrobot, which capitalizes on the positives of both wheeled and legged locomotion. Byrobot has a six-leg hexapod configuration with a 3-revolute joint mechanism on each leg as well as 4 DC motors controlling the four wheels on the robot. In this paper, we discuss our methodology for stabilizing the robot in its legged configuration as we develop an effective walking gait. Figure 1 shows CAD models and actual photos of the Byrobot platform.

Figure 1: CAD model and actual hardware photos of Byrobot

2. Stability Analysis of a Legged Mobility System

For a legged robot in remote terrain environments such as Mars, stability is a more important issue than speed. If the robot can't remain stable and stand upright on its own, then the objectives of the mission can not be achieved. A wheeled mobility system has inherent stability characteristics since the wheels are usually always in contact with the ground. However, there are areas such as cliffs, rocky, and soft terrain, which a wheeled robot may not be able to stabilize on and traverse. For a hexapod robot, at least 3 legs have to be down at all times for the robot to be in a stable configuration. At any given moment, the robot's center of mass (COM) should be within the "Polygon of Support", meaning that the center of mass is as far away from the edges of the polygon [2] that covers the area defined by the feet on the legs that are in contact with the ground. .

3. Forward and Inverse Kinematics

In order to position the robot feet for attaining proper stable polygon of support, forward and/or inverse kinematics is used [3]. Here we approximate where the center of mass (COM) is located for simplicity.

3.1. *Forward Kinematics – Geometrical Method*

The simplest method of forward kinematics calculates the foot position based on a geometric representation of the legs in the x, y, z coordinate frame, as we see in Figure 3. This method allows us to set the joint angles of the robot legs and determine the final position of the foot [4].

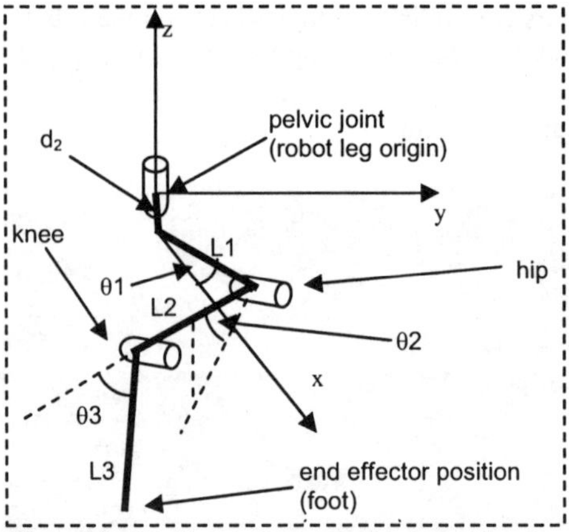

Figure 3: 3R (3-revolute joint) leg mechanism

Foot Position

$$X_3 = L_1\cos(\theta_1) + L_2\cos(\theta_2)\sin(\theta_1) + L_3\cos(\theta_2+\theta_3)\cos(\theta_1) \quad (1)$$
$$Y_3 = L_1\sin(\theta_1) + L_2\cos(\theta_2)\cos(\theta_1) + L_3\cos(\theta_2+\theta_3)\sin(\theta_1)$$
$$Z_3 = L_2\sin(\theta_2) - L_3\sin(\theta_2+\theta_3) - d_2$$

Where L_1., L_2, and L_3 are the leg segment lengths, θ_1, θ_2, and θ_3 are the pelvic, hip, and knee joints of the leg mechanism, respectively, and d_2 is the offset length in the z-direction from the leg origin down to the L_1 leg segment, all shown in Figure 3.

3.2. *Inverse Kinematics*

The inverse kinematics can be used determine the joint angles (our θ's) associated with the legs, given the position of the foot. This is important for the standing and the walking of Byrobot. A single set of three joint angles produces a single foot position; however, there is more than one set of joint angles that will result in a given foot position (as you can see by planting your foot on the floor and moving your hip and knee). Therefore, incongruous solutions, such as a knee bending backward, must be eliminated. The inverse kinematics are derived from [5], and implemented in a gait simulation for a similar 3-DOF hexapod in [6].

$$\theta_1 = \cos^{-1}\left(\frac{X_3}{\sqrt{X_3^{\,2} + Y_3^{\,2}}}\right) \quad (2)$$

$$\theta_2 = \cos^{-1}\left(\frac{L_2^2 + \left(\left(\sqrt{X_3^2+Y_3^2}\right) - L_1\right)^2 + Z_3^2 - (L_3+d_2)^2}{2*L_2*\sqrt{\left(\sqrt{X_3^2+Y_3^2} - L_1\right)^2 + Z_3^2}}\right) + \tan^{-1}\left(\frac{Z_3}{\sqrt{X_3^2+Y_3^2} - L_1}\right)$$

$$\theta_3 = \cos^{-1}\left(\frac{\left(\sqrt{X_3^2+Y_3^2} - L_1\right)^2 + Z_3^2 - L_2^2 - (d_2+L_3)^2}{2*L_2*L_3}\right) + \tan^{-1}\left(\frac{Z_3}{\sqrt{X_3^2+Y_3^2} - L_1}\right)$$

Where we use the same variables as mentioned above.

4. A Special Case – Robot on an Incline/Decline

In the case when the robot is on an incline or decline, the projected center of mass (COM) is into the surface plane, as shown in Figure 4.

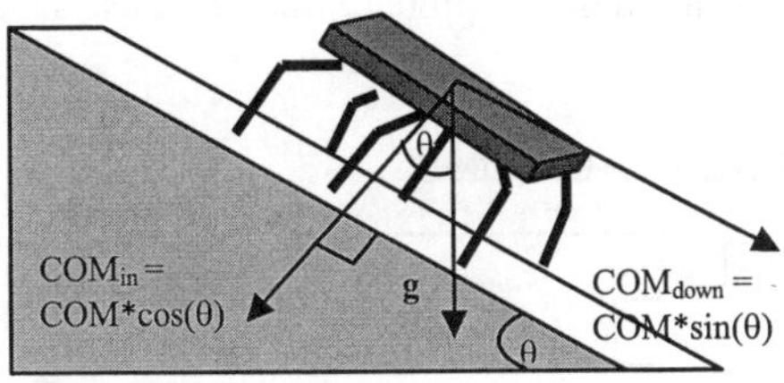

Figure 4: Byrobot projected down on an incline/decline

The COM is projected into the plane at cos(θ), where θ is the angle of the incline. In this situation, the walking gait needs to be modified so that the COM doesn't cause the robot to tip over. This modification can be done by adjusting the joint angles in the robot legs to correspond to the incline angle.

5. Walking Gait Optimization

We tested two walking gaits that are commonly used for hexapod robots [8]. Each gait depends on the load on the robot, how fast you want the robot to move, and how stable the robot will be while moving.

5.1. *The Tripod Gait*

The **Tripod Gait** is the best-known hexapod gait. A tripod consists of the front-back legs on one side and the middle leg on the opposite side. For each tripod, the legs are lifted, moved forward, and lowered in unison. Since 3 legs are on the ground at all times, this gait is both "statically" and "dynamically" stable. This gait uses two leg movements, or two "tripods" to complete one cycle, making it the fastest gait for hexapod robots.

5.2. *The Wave Gait*

In the **Wave Gait**, all legs on one side are moved forward in succession, one after the other. This is then repeated on the other side. Since only 1 leg is ever lifted at a time, with the other 5 being down, the robot is always in a highly-stable posture, making this the most stable gait and the preferred gait for robots with large loads. However, since this gait takes six leg movements to complete one cycle, it is also the slowest gait.

5.3. *Movement Scheme*

The movement scheme is easily visualized by examining Figure 5. The square wave graph next to each of the six robot legs correspond to time points. In this figure, Hi means the leg is lifted off the ground, while Lo means the legs is down on the ground.

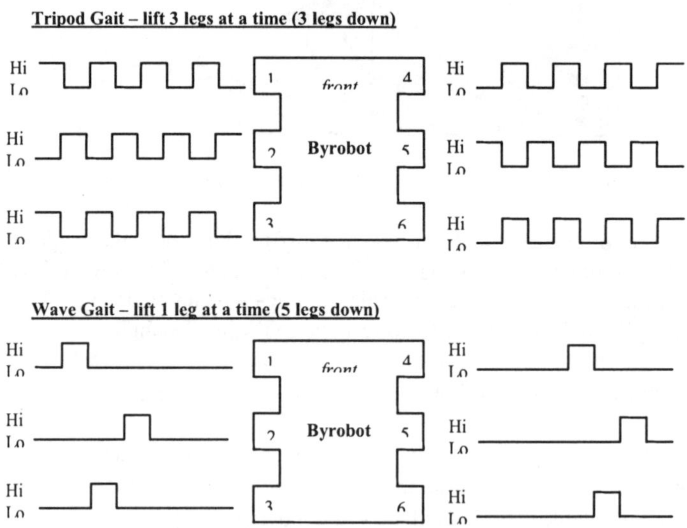

Figure 5: Different Gait movements of each leg with respect to time.

6. Results

Our goal is to stabilize our new legged-wheeled rover and test two different walking gaits for its hexapod configuration. We have shown the kinematic models for the leg mechanisms, explained how the polygon of support is needed to stabilize the robot in its legged configuration, and we have tested out the Tripod Gait and the Wave Gait. Due to the large load that is on the Byrobot (approximately 2kg), the initial tripod gait was ineffective. Byrobot simply would collapse under its own weight as soon as the three legs would lift off of the ground, despite the leg position. The Wave Gait, though a bit slow, remains stable throughout its entire walk, and therefore became our preferred walking gait. The correct servo joint angles were found using the inverse kinematics, for each step in the gait. For the Wave Gait, the movement of the six legs is in the sequence of: leg 1, leg 3, leg 2, then leg 4, leg 6, and leg 5 (robot legs labeled in Figure 5), so that there was always a stable polygon. At the end of each cycle during the gait when all six legs have moved a forward, a "forward scoot" is done to move the robot body forward. Small steps were used because they are more efficient, and we wanted to avoid interference of parts on the robot that could occur if we used larger steps. In Figure 6, we see photographs of the successful Wave Gait sequence of Byrobot on a flat surface.

Figure 6: Photographs of the Wave Gait sequence of Byrobot

In addition to the flat surface, we also tested Byrobot on an incline plane and in our Mars Sand Pit. We compared the results of both the legs and wheels in these environments. For our incline plane, we used 22° and 35° ramps. A 35° angle is the maximum angle the robot can traverse on its wheels before slipping back down. Unfortunately on the incline plane, the robot would not remain stable using the on its legged mobility system, making the wheels the preferred choice for mobility. In the sandy terrain, the wheels would lose traction, making the legs the preferred mobility system. For operation in this terrain, Mars Shoes were used on Byrobot. These shoes give more surface area to the feet so that they don't dig down into the sand. In Figure 7, we see the Mars Shoes in place on Byrobot and Byrobot standing in the shoes in the Mars Sand Pit.

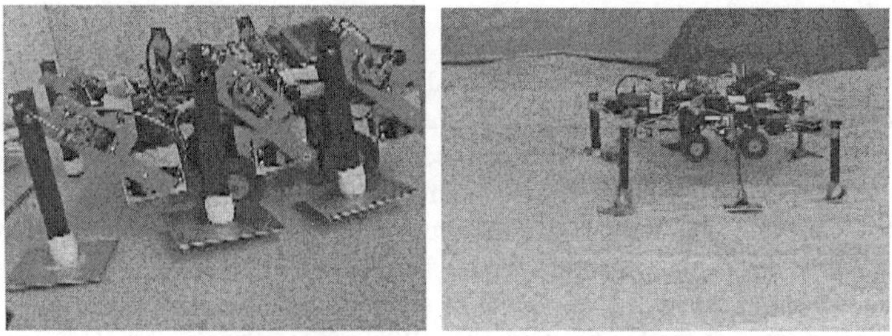

Figure 7: Mars Shoes on the feet of Byrobot

The average speed of the robot using its legs in the sandy terrain was about half of the speed of it using its legs on a flat surface. We contributed this to the lack of friction on the robot's Mars Shoes which cause its feet to slightly slip backward with each step during the walking gait. As expected, the legs move the robot much slower overall than the wheels do. Our average speed results are shown in Table 1.

Table 1: Average Speeds of Byrobot's Two Mobility Systems

Robot Surface		**Average Speed (cm/s)**
Flat Surface		
Wheels	• Full Speed • 80% Speed • 50% Speed	69.01 48.74 42.23
Legs	• Full Speed	1.16
Up an Incline Plane		
Wheels	• 22° - Full Speed • 35° - Full Speed	37.56
Legs	• Full Speed	n/a
Sandy Terrain		
Wheels	• Full Speed	n/a
Legs	• Full Speed	0.54

7. Conclusions

In this paper, we have explained how to stabilize a hexapod rover using forward and inverse kinematics to find its polygon of support. And we have shown how to find an optimal stable walking gait based on these kinematics. We tested two of the common walking gait patterns associated with hexapod robots and explained why our final Wave Gait was chosen. We will be able to improve the mobility of the robot on an incline plane in the future by adjusting the joint angles. As shown in Figure 4, the center of mass of the robot is projected down into the plane, and the legs will need to be in a different configuration depending of the angle of the incline, in order to keep the robot stable in this situation. Having a robot that can both walk and roll would be beneficial to planetary exploration since it can transition from the use of its legs or wheels based on the terrain conditions. By adding various sensors, encoders, and even a vision system, we can facilitate future autonomous capabilities of Byrobot.

References

[1]. Howard, A., Tunstel, E. “Intelligence for Space Robotics”. TSI Press, 2006

[2]. Quinn, R., Nelson, G., Bachmann, R., Kingsley, D., Offi, J., Ritzmann, Roy. “Insect Designs for Improved Robot Mobility”. 4th International Conference on Climbing and Walking Robots, 2001.

[3]. Craig J.J. Introduction to Robotics, Mechanics and Control, 3rd ed. Addison Wesley, 1986, 1989, 2005.

[4]. Sciavicco, L. and Siciliano, B. Modeling and Control of Robot Manipulators. McGraw-Hill Co., Inc. 1996, 2000.

[5]. J. Lento, Z. Huson , J. A. Haass, M. Reilley, and J. Shrestha. Development of Leg Control Mechanisms for a Radially Symmetric Octopedal Robot. The National Conference on Undergraduate Research, 2005.

[6]. G. Figliolini, V. Ripa, (2004) Kinematic Model and Absolute Gait Simulation of a Six-Legged Walking Robot. CLAWAR 2004 889-896, 2004.

[7]. Ferrell, C. "Robust and Adaptive Locomotion of an Autonomous Hexapod". Perception to Action Conference, Lausanne, Switzerland, 66-77, 1994.

[8]. NASA JPL Robotic Systems; http://www-robotics.jpl.nasa.gov

TERRAIN-ADAPTIVE LOCOMOTION OF A WHEEL-LEGGED SERVICE ROBOT USING ACTUATOR-BASED FORCE MEASUREMENTS

PETRI VIREKOSKI

Automation technology laboratory, Helsinki University of Technology, Otaniementie 17, 02015 TKK, Finland

ILKKA LEPPÄNEN

Automation technology laboratory, Helsinki University of Technology, Otaniementie 17, 02015 TKK, Finland

1. Introduction

This paper describes methods for adapting to terrain with a service robot that has four legs in a particular configuration that includes wheels at the ends of the legs. The basic motivation behind this type of configuration is to use the wheels to give more speed on terrain that allows it and use the legs more when the terrain becomes more difficult or impossible to negotiate using wheels only.

Terrain adaptation here means active utilization of the legs to gain the best possible support and propulsion in varying and rough terrain, and therefore also means trying to maintain equal ground contact with all four wheels. It acts in part the same way as the spring/dampener wheel suspension in cars and gives the following benefits: equalization of strain forces, better propulsion in soft soil, locomotion beyond the ability of wheels and also less strain to the terrain.

The terrain-adaptive locomotion is based on measuring forces and controlling positions of the legs' ankles (and thus the wheels). Furthermore the force measurements are not based on external sensors but instead the forces are calculated directly using the currents of the electric actuators (motors) and the kinematic equations of the robot.

2. The wheeled leg configuration

The robot used is the WorkPartner service robot that has been developed in the Automation Technology Laboratory of the Helsinki University of Technology. It has four legs in a particular configuration, and wheels at the ends of the legs. The configuration has been described in previous CLAWAR papers [1] and [2]. Figure 1 shows WorkPartner in rough terrain. Figure 2 shows the degrees of freedom of the leg. The main body consists of two parts, front and rear, joined by a scissors-type middle joint for steering. The front legs are connected to the front body and the rear legs to the rear body.

Figure 1 WorkPartner negotiating some obstacles in winter

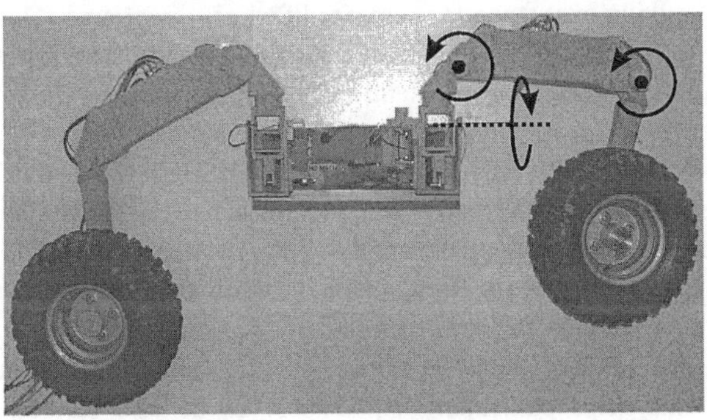

Figure 2 The degrees of freedom of the leg of WorkPartner

It should be noted here that while the legs have the "hip" joint (longitudinal axis, joins the leg to the body) for sideways ankle movement, this joint is not used for the methods described here. So only the 2nd and 3rd ("thigh" and "knee") joints are used to move the ankle in the vertical lengthwise plane.

3. Modes of Locomotion

First let us take a look at the different main modes that are determined by how the legs and wheels are used for locomotion. The three possibilities are: wheels only, hybrid mode (use of both the leg joints and wheels) and walking (use only leg joints, wheels locked). The term "rolking" (combining the words rolling and walking) will be used from here on to refer to the hybrid mode.

3.1. *Wheels-only Drive*

This basic mode uses the wheels in velocity mode to drive forward and back, and the middlejoint is then responsible for steering. This mode is similar to tractors that use a hydraulic middlejoint. The leg joints can be either completely fixed, or they can be used for adapting to the terrain. The "wheel" mode is best for mostly flat terrain, allowing greater speeds because the legs do not have to participate in moving the vehicle forward and back.

3.2. *Rolking (Rolling-walking)*

The "rolk" mode combines wheel rotation and horizontal leg motion to handle difficult terrain in such a way that the wheels can retain ground contact all the time. It is similar to walking except the legs are not lifted off the ground. This helps with a) stability problems that are related to walking where ground contact is lost periodically and b) wheel slipping problems related to normal wheel drive.

While rolking the legs can have one of two states: transfer or support. In support state the leg participates in moving (pushing) the vehicle body forward while the wheel is kept stationary. In transfer state the leg is moving forward or backward while the wheel is rolling along the ground.

When only one leg is in transfer state at a time, there are always three legs in support state pushing the vehicle forward. The transfer leg's wheel is kept in velocity mode where the velocity is exactly the same as the ankle's motion in relation to ground. This helps particularly on slippery or delicate surfaces, because there are no wheels that could slip (unless the whole vehicle slides).

Rolking can also be done with two wheels in transfer state simultaneously. This gives more speed because there are only two separate phases. Particularly the diagonal two-leg rolking has been found effective with WorkPartner in soft soil.

The rolk mode is also suitable for climbing over obstacles because only one leg is moved at a time, allowing simpler control over how the legs behave when climbing on the obstacle.

Steering is not currently possible while rolking. This is because changes to the middle joint angle would require changes to the legs' positions in order to keep the wheels from sliding sideways, and this would require rather complex adaptation for the rolking algorithm or might even be impossible to implement without sliding of the wheels. So the middle joint is kept locked while rolking.

3.3. *Walking*

Four-legged walking requires complex control of dynamics and stability. The terrain-adapting methods described later have not been developed for walking and are not applicable to it, although later in this paper we see that the "rolking" mode could lead to walking when using a particular scheme for adapting to terrain and obstacles.

4. Adapting to Terrain

When a vehicle has four wheels at fixed positions, and is moving along a terrain that is no longer flat, one wheel loses contact with ground causing essentially two problems: 1) that wheel can no longer participate in propelling the vehicle and 2) the weight of the vehicle is divided between the remaining wheels causing more strain both on the wheel suspension and the terrain. When this happens, depending on the location of the center of mass, it ultimately leads to only two diagonal wheels bearing most of the weight and leaving the other two almost without grip.

In a car this is handled by suspending the wheels so that they can move a distance vertically and putting springs and dampeners between the body and the wheels to cope with the weight and mass. In the case of our robot, the wheels are "suspended" by legs that have greater vertical reach but can also move the wheel horizontally. This gives more flexibility to handle uneven terrain, but also means that all terrain adaptation is dependent on active measuring and actuation.

The control of terrain adaptation is based on measuring forces and changing position. That is, the position of the ankle of each leg is varied vertically depending on what kind of strain is present at the ankle. For a simplified example, if a leg feels "too light", it is pushed "a bit" more down to try to regain proper load; What exactly is "too light" and how much is "a bit", are parameters of the tuning of the control.

Force control in walking robots has been studied in e.g. [3] and [4] where the goal has been to divide the support forces between the legs according to a certain principle such as to equalize the vertical forces or minimize internal forces. These have required rather complex algorithms. In a four-legged system using only the "wheel" and "rolk" locomotion modes the support force equalization can be done using simpler methods described here.

Figure 3 shows an extreme situation where one leg is driven on a large "bump" (in the "wheel" mode), and the terrain adaptation can be seen clearly.

Figure 3 An extreme case of terrain adaptation

5. Controlling Terrain Adaptation in WorkPartner

Both the "wheel" and "rolk" modes use a same scheme for controlling the legs. The horizontal and vertical coordinates of the ankle positions are controlled separately:

- The **horizontal** position is determined by the locomotion mode. When rolking the gait sequencer moves the ankles horizontally by a pre-determined sequence (each leg in turn), and in wheel mode the horizontal position is fixed.
- The **vertical** position is controlled by the terrain adaptation algorithm.

This means that the basic terrain adaptation is independent of the main locomotion mode.

Terrain adaptation can have two modes, "level" and "terrain". In "level" mode the controller attempts to keep the body perpendicular to gravitation (using two inclination sensors). In "terrain" mode the leg positions are adjusted so that the body follows the local terrain profile. The control loop can be divided into three steps: 1) initial positioning according to adaptation mode (this includes the leveling control in "level" mode), 2) equalization of the vertical forces and 3) averaging and limiting.

The forces that are used as inputs for the terrain adaptation are calculated from the electric motor currents in the "thigh" and "knee" joints. The currents are converted to motor torques and those are converted to joint torques. Using the kinematic equations of the legs the joint torques are then converted to vertical and horizontal force vectors in the robot's coordinate system. These are the actual strain forces acting on the ankle.

Step 1: Initial Positioning, Including Leveling Control

Here the initial vertical positions are set according to the adaptation mode. In "level" mode this step adjusts these positions so that the body keeps level according to gravitation and a desired height from the operator. In "terrain" mode the positions are set according to desired pitch, roll and height of the body from the operator.

Step 2: Force Equalization

Now the calculated vertical strain forces of the ankles are used to determine an adjustment to the two diagonals. This means that the legs are moved together in the two diagonal pairs (legs 1&4 and 2&3, see Figure 4) to adjust the division of forces. The division is determined by comparing the two sums of the vertical forces of each diagonal pair. In "wheel" mode the sums are kept equal. When rolking the adjustment is weighted so that the transfer leg's diagonal is lighter i.e. the sum of the two diagonal legs' vertical forces is less than the sum of the other two. This is done in order to reduce the rolling resistance of the transfer leg's wheel.

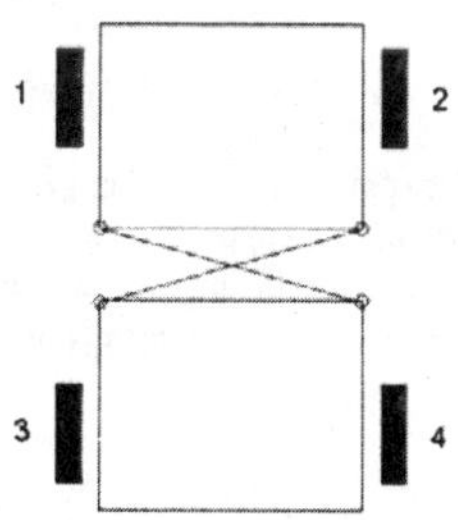

Figure 4 The numbering of the legs (wheels)

Step 3: Averaging and Limiting

When in "terrain" mode the positions are averaged between the left/right and front/rear pairs. First the mean position of the left legs is compared to that of the right legs, and both groups are moved to opposite directions by half of the difference. Then the same is done for front/rear legs. This averaging is not done in "level" mode because it would defeat the leveling. Finally the positions are limited to tested practical maximums, and set for sending to the legs.

An important point to note is that in steps 2 and 3 the legs are always moved in opposing groups of two, and to opposite directions by the same amount, so that the average height is maintained.

In step 2 the diagonal lightening during rolking can be increased even to a degree where the rolking (transfer leg) diagonal is completely without load. This has the effect of changing the rolking to walking, and has been confirmed in tests. However with the static implementation of the rolking gait this walking behavior is not completely predictable because depending on the projection of the center of mass either the rolking leg or its diagonal opposite can lift up. This has not been studied further.

Figure 5 shows a test that clearly indicates the effects of force equalization (see caption).

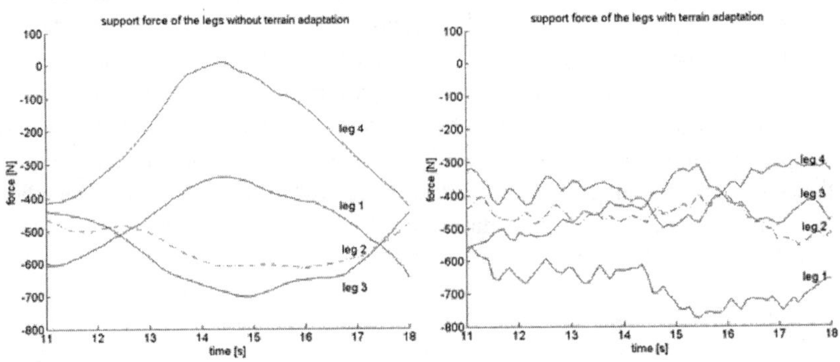

Figure 5 A test where leg 2 (front right) was driven over a 1 m long and 20 cm high smooth bump without (left) and with (right) terrain adaptation, in wheel mode. As expected, the support force variation is greatly reduced with the force equalization in terrain adaptation ("terrain" mode was used, see text for description). The ripple is caused by the relatively low frequency of the force equalization controller combined with mechanical flexibilities and play in the joints.

6. Conclusions and Future

The terrain adaptation method and algorithm described here has been widely tested in varying terrain including dirt, sand, snow and obstacles like logs on the ground. The following benefits have been confirmed:

- requires no additional sensors
- works as well in both the "wheel" and "rolk" main locomotion modes
- divides the weight equally in "wheel" mode which leads to
    - better propulsion
    - less rolling resistance in soft soil
    - less strain on both robot and terrain
- works together with the basic motion control of the legs
- gives mobility when just wheels would not suffice (e.g. deep snow)

A couple of future enhancements have been proposed and partly tested for the terrain adaptation. First, the robot could switch the locomotion mode automatically based on measuring the vehicle-terrain interaction. This has actually already been developed quite far and is working well. Second, it would be possible to control the projection of the center of mass to some extent by changing the global lengthwise and sideways position of the wheels relative to the body (this would also utilize the "hip" joints). This would give even more control to the equalization of the strain forces, particularly in inclined terrain.

A study has been started to examine the possibility of using similar active wheel suspension and terrain adaptation to improve mobility and payload handling in such machines as tractors and loaders.

References

1. Leppänen I, Salmi S, Halme A (1998) WorkPartner – HUT-Automations new hybrid walking machine. 1st International Conference on Climbing and Walk-ing Robots, Brussels
2. Halme A, Leppänen I, Salmi S (1999) Development of WorkPartner-robot – design of actuating and motion control system. 2nd International Conference on Climbing and Walking Robots, Portsmouth
3. Gorinevsky D M, Shneider A Y (1990) Force control in locomotion of legged vehicles over rigid and soft surfaces. The Int. J. of Robotics Research Vol 9. No. 2 , p. 4 – 23
4. Lehtinen H (1994) Force based motion control of a walking machine. Ph. D. dissertation, Helsinki University of Technology

Control of Quadruped Walking Robot Based on Biologically Inspired Approach

Tae Hun Kang

Phohang Institute of Intelligent Robotics, Korea
E-mail: bigxihn@postech.ac.kr

Ig Mo Koo, Young Kuk Song, GiaLoc Vo, Tran Duc Trong, Chang Min Lee and Hyouk Ryeol Choi*

School of Mechanical Engineering, Sungkyunkwan University,
Chunchun-dong 300, Jangan-gu, Suwon, Kyunggi-do, Korea,
*E-mail: kooigmo ; yksong99 ; vogialoc ; jamestran ; cmbof ; *hrchoi{@me.skku.ac.kr}*

In this research, a biological study is performed on the control of a quadruped walking robot. With a biomimetic observation of the gravity load receptor and stimulus-reaction mechanism of quadrupeds' locomotion and with the study of the stances on walking and energy efficiency, a new control method for a quadruped walking robot is derived. In particular, introduction of a new rhythmic pattern generator relieves the heavy computational burden because it does not need large computation on kinematics. The proposed controller, though it is simple, provides a useful framework for controlling a quadruped walking robot. In this paper, evaluation of the proposed method is done via a dynamic simulation. Then, it is validated by implementing in a quadruped walking robot, called AiDIN(Artificial Digitigrade for Natural Environment).

Keywords: Biomimetic; Gravity Load Controller ; Quadruped Walking Robot

1. Introduction

In spite of rapid development in robotic technologies, living creatures are superior to robots existing currently in various aspects. Thus, it is necessary to understand the principles underlying the motions and behaviors of biological subjects for innovative control of robots. Mimicking living creatures currently becomes one of the worldwide trends for robotic innovations. It is considered as one of the most adequate way of developing a robot since biological systems provide a number of useful ideas concerning control of robots. Recently, robotic researchers as well as biologists propose innovative ideas for the walking robot system. Among several ideas, mimicking

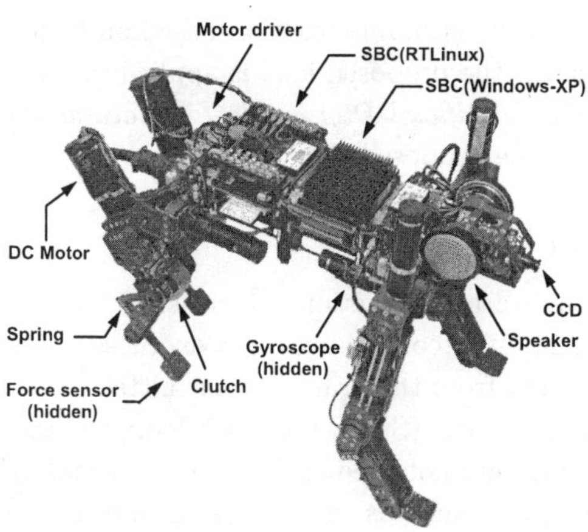

Fig. 1. AiDIN(*Artificial Digitigrade for Natural Environment*)

the rhythmic motion of animals is one of the promising ways to control the walking system. By studying on this, the locomotion of the walking robot can be close to that of the real animal. Recently, neurobiologists point out clear evidence of a neural part in the brain of animals, referred as Central Pattern Generator (or CPG), which can produce the rhythmic movement in its locomotion. Since CPG can generate rhythmic outputs even if the sensory feedback is eliminated, it became an interesting target in approaching to biomimetic control. On the way to study about this CPG, many neural models such as McCulloch and Pitts neuron, Leaky integrator neuron, and Matsuoka neuron have been proposed by investigating neurons in the real animal's brains or bodies [1,2]. As efforts of developing relevant methods to control the robots, algorithms inspired from these rhythmic patterns generators have been studied [3]. Though these neural models can illustrate the control method of animal movements, the results of applications in robotic systems are far different from real biological systems because animal's movement may not be simply mimicked by several neurons as robotic systems introduced until now [3].

On developing a quadruped walking robot, various aspects should be taken into account. In this research, mainly, the basic principles of quadrupedal locomotion controller are analyzed in terms of a biomimetic point of view. The quadrupedal locomotion controller is studied via biomimetic observation of the gravity load receptor and stimulus-reaction mechanism of animal are investigated using the stances on walking and posture control. The controller design factors is feedback information form several receptors and its stimulus-reaction based on biological sensory system analysis. Finally, the proposed idea is applied in a quadruped walking robot, called AiDIN(*Artificial Digitigrade for Natural Environment*), and the effectiveness of the proposed idea is experimentally validated.

2. Gravity Load Controller

Usually, animals move by the actions of muscles. When muscles generate arbitrary movements, the control action is executed via two loops. In the first loop, a command from the brain is transmitted to the muscle through a central nervous system(CNS). In the other loop, the muscle reacts to external stimulus which are not predictable or recognizable. The stimulus are detected with receptors such as proprioceptor and exteroceptor [4]. Even though its structure is very complex, movements can be easily simplified by the repetitive action such as flexion and extension, because these actions are directly controlled by sensory neurons and motor neurons [5]. In this section, a locomotion controller, called *Gravity Load Controller*(abbreviated as GLC afterwards) is introduced. This controller is based on a gravity load receptor and stimulus–reaction mechanism of the animal. Even though the configuration of GLC depends upon the system in hand, the basic configuration can be shared because it consists of the feedback information from several receptors and its stimulus–reaction.

As shown in Fig. 2, GLC for the quadruped walking robot consists of four parts including an oscillator, stimulator, threshold detectors, and PD controller. In addition, the proposed controller has two loops(*exceeding* and *nonexceeding*) which are separately excited according to the threshold detector. If the information from sensory feedback(it is called *load receptor* in biology) such as force sensor, touch sensor, and gyroscope exceeds the fixed threshold value, the stimulator begins to excite the extensors in the dotted rectangle. Then, *intermittent stimulus* are generated, which propel a joint to react immediately to external feedback. In the *nonexceeding* case, the extensors in the dotted rectangle are disabled, and the stimulator stimulates the flexor and extensor which are connected with δ_C called *interneuron.* δ_C is within the central nervous system that acts as a link between sensory neurons and motor neurons. Two interneurons, δ_C and δ_I in the controller are expressed with simple equations as follows.

$$\delta_C = \epsilon(S_{CG} + S_{CF}) \tag{1}$$

$$\delta_I = \epsilon(S_{IG} + S_{IF}) \tag{2}$$

where

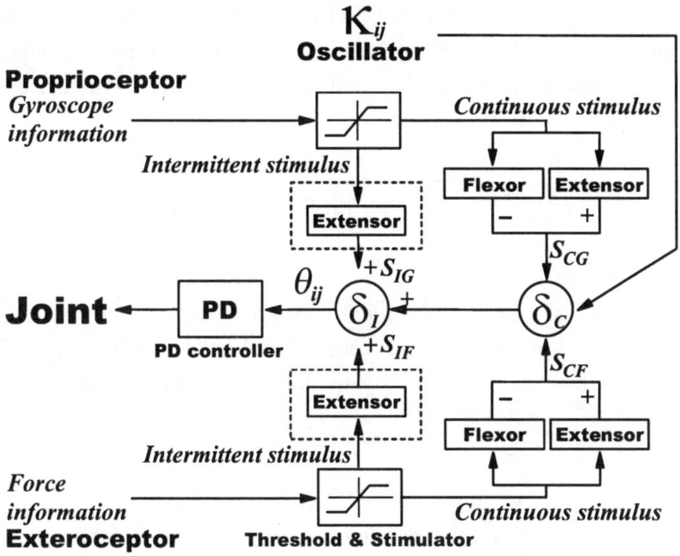

Fig. 2. The joint controller consists of four parts including an oscillator, stimulatory part, threshold detector, and PD controller

$$\epsilon = \begin{cases} 1 & \text{contact with ground} \\ -1 & \text{otherwise} \end{cases} \qquad \text{in supporting mode}$$

$$\epsilon = \begin{cases} -1 & \text{contact with ground} \\ 1 & \text{otherwise} \end{cases} \qquad \text{in swing mode}$$

	Forelimb		Hind limb	
	ST	SW	ST	SW
Walk	0.68	0.32	0.64	0.36
Trot	0.44	0.56	0.40	0.60
Run	0.27	0.73	0.26	0.74

Also, a new oscillator is proposed, which is simple and easy to implement in the locomotion controller. Although many researchers have investigated the ratio between SW and ST according to the walking speed, it is very difficult to apply this idea to the robot system. A ratio between SW and ST needs to be normalized to decide a general walking pattern for a cycle. From several references [6,7], digitigrade's walking patterns can be measured, which are normalized as shown in Table 1. From these observations, it is noted that ratio between SW and ST is decided by walking speed as well as walking types, which is applied in the design of the oscillator.

The proposed oscillator $\mathcal{K}_{ij}$(joint j of leg i) is constructed with *Bezier curve* with n control points. The oscillator $\mathcal{K}_{ij}$ can be represented as the function of time $t \in [0,1]$ such as

$$\mathcal{K}_{ij}(t) = \sum_{k=0}^{n} \frac{n!}{(n-k)!k!} \mathbf{P}_k (1-t)^{n-k} t^k \tag{3}$$

where $\mathbf{P}_k$ is a control point.

From Eq. (3) the oscillation pattern for the quadruped walking robot can be designed. The desired joint angle θ_{ij} of leg i and joint j is given by the stimulus of extensors, interneurons, and oscillator as follows.

$$\theta_{ij} = \mathcal{K}_{ij} + \delta_C + \delta_I \tag{4}$$

An oscillator $\mathcal{K}_{ij}$ in the proposed controller generates a repetitive pattern for locomotion. $\mathcal{K}_{ij}$ always generates locomotion pattern regardless of any input value as illustrated in Fig. 2. It is quite different from the other pattern generator introduced until now.

As illustrated in Fig. 3, the overall controller for the quadruped walking robot is comprised of four joint controller and the motion is synchronized with the switch for determining the walking patterns.

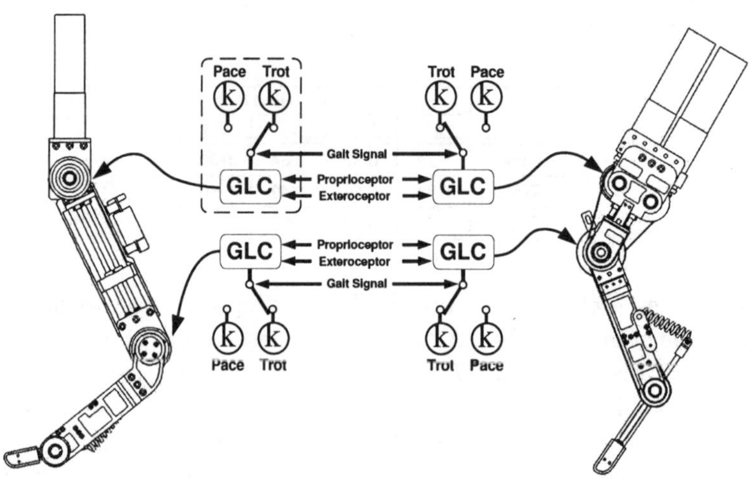

Fig. 3. Illustration of the walking controller

3. Quadruped Walking Robot: AiDIN

A quadruped walking robot, called AiDIN (*Artificial Digitigrade for Natural Environment*) illustrated in Fig. 1. AiDIN is the robot containing all the components except the battery AC 220V power is supplied via a tether cable.

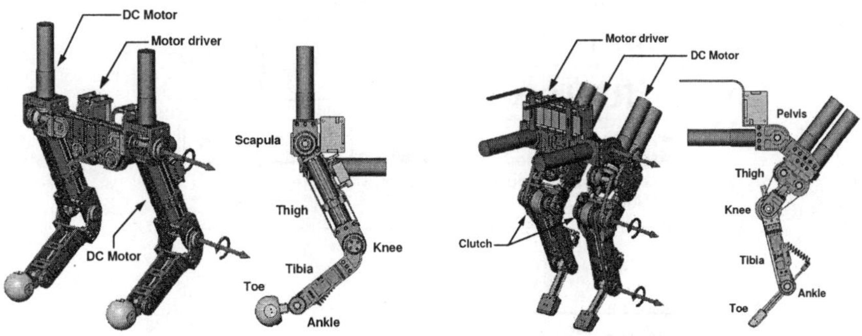

(a) Forelimb (b) Hindlimb

Fig. 4. Schematic view of limbs

As shown in Fig. 1, AiDIN has four legs with three–DOF active joints for each leg. The active joints actuated by three geared DC motors(20 watt, gear ratio of 53 : 1), respectively. In case of the forelimb, a scapular joint is mounted to change the width between supporting feet as mentioned in this research. The hind limb, in particular, is equipped with a miniature clutch in the knee. It is used to switch between a passive ankle joint and an active one. The leg spring connected between the foot and the tibia, is used as the element for storing the energy and reducing the impact on contact with the ground. Fig. 5 shows the control system layout of AiDIN. Two embedded controllers and twelve micro–controllers(PIC18f458, MicroChip) are contained in the robot. In case of embedded controllers, the first embedded controller using single board computer(Pentium–III 800*MHz* with compact flash disk, 1Mbps CAN) and *RTLinux* is ported as the operating system. The function of controller is not only control of the walking posture, but also communication management between micro–controllers based on CAN(*controller area network*). A voice speech engine and a vision library(OpenCV, Intel Image Processing Library) are involved in the other embedded controller(Pentium M. 1.1*GHz* with wireless network device). The power of the robot is supplied with a tether cable, although all the other communication is transmitted through wireless LAN(local area network).

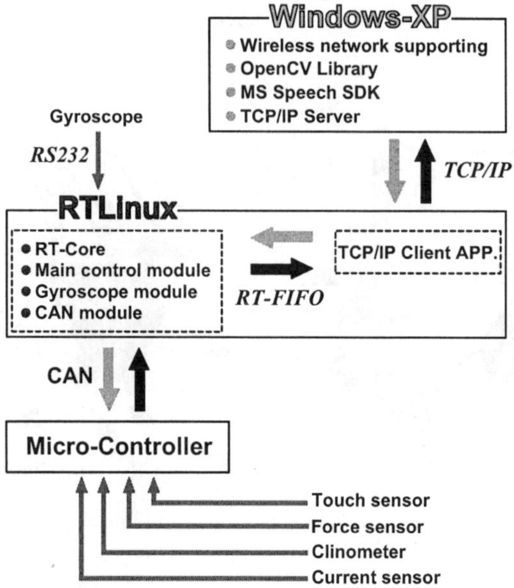

Fig. 5. System configuration of AiDIN

4. Experiments

The locomotion controller was validated via several preliminary experiments with AiDIN. In the first experiment, self–posture changeability in AiDIN was tested. The robot was set on a rolling plate as shown in Fig. 6(a). When the plate suddenly lurched laterally, GLC in AiDIN generated flexion or extension stimulus to make its posture stable as shown in Fig. 6(b). Even if a plate is slanted to the right or left, the robot sustained its balance with GLC.

In the second experiment, a comparison between *exceeding* and *nonexceeding* loops in the proposed controller were performed as illustrated in Fig. 7. In this experiments, the robot walked in a trot gait of $20cm/sec$ walking speed. In the first case, the proposed controller was not fully included, but only *nonexceeding* was applied. As shown Fig. 8(a), the body rolls with wide range of rolling angle, even could be unstable during rolling.

In this case, the robot could not walk or fall down at times. When two control loops were completely included, the robot kept balance with smaller range of rolling angle as shown in Fig. 8(b). It is noticeable that the robot can be controlled stably as well as efficiently with the proposed controller.

(a)

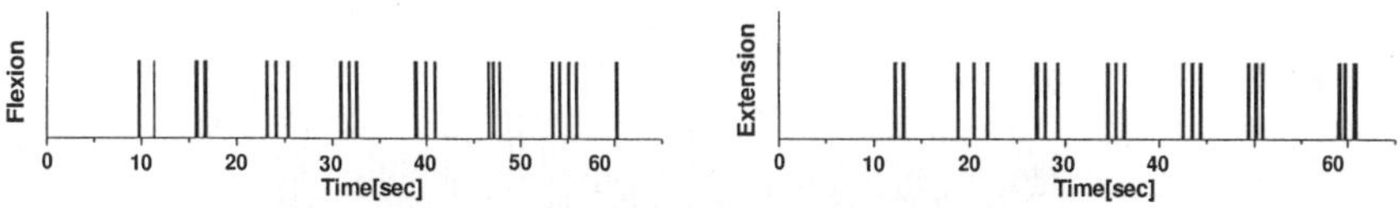

(b)

Fig. 6. Experiment to verify a self–posture changeability

5. Conclusions

To mimic living creatures has been a contemporary issue in the robotics researches for long time. It may be questionable to just copy the animal, but it can be useful if the idea is explained appropriately. In this paper, we presented a biologically inspired approach on the developing of a quadruped walking robot controller. Observation on gravity load receptor and stimulus-reaction mechanism gives us several significant tips on the control of the quadruped walking robot, which have been analyzed in terms of the robotic terminologies.

In this paper, we presented the gravity load controller with a rhythmic patterns generator included to control quadruped walking robot. Even though it is a simple model but it proves, by the experiment results, to be an effective controlling method in which the sensory feedback is nicely integrated with the oscillator to produce the rhythmic output signal. It is also shown in the simulation that this controller gives the robot a potential ability of adaptation in irregular environments.

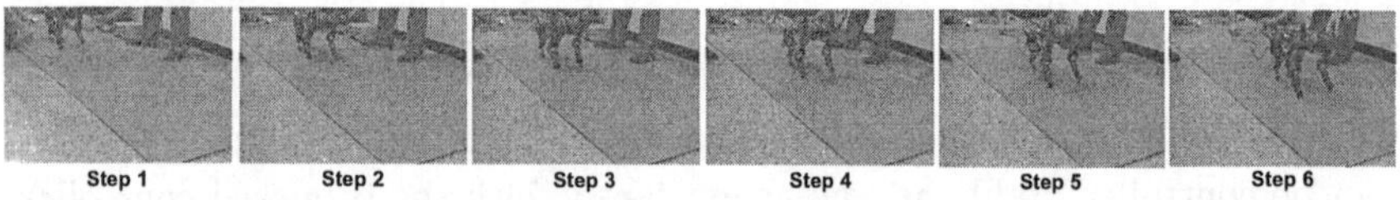

(a)

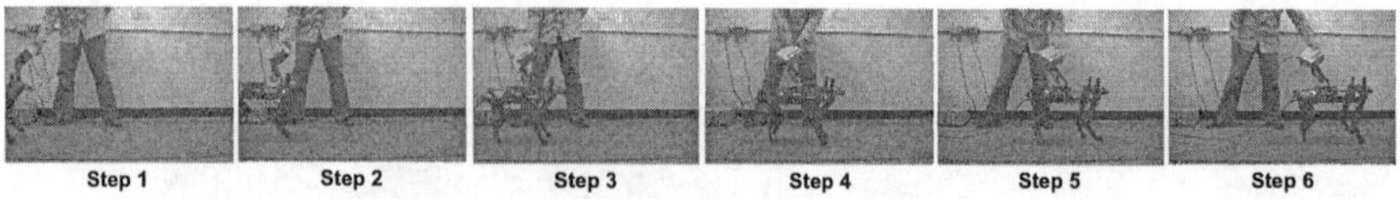

(b)

Fig. 7. Trot gait on AiDIN

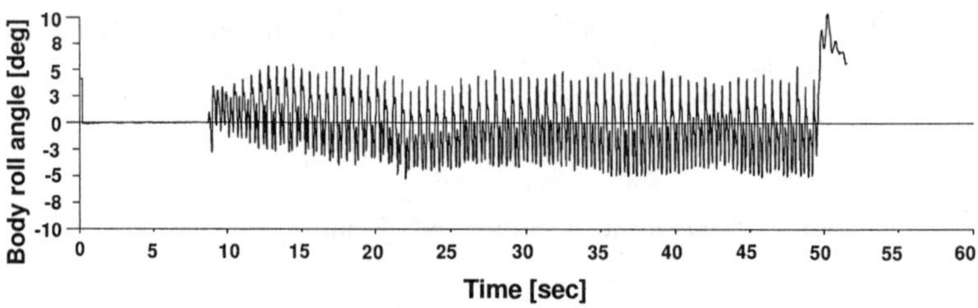

(a) Trot gait with nonfeedback

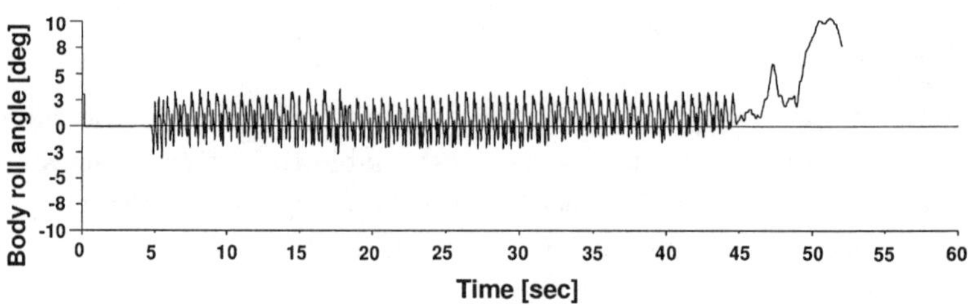

(b) Trot gait with feedback

Fig. 8. Body roll angle

Acknowledgment

This research was supported in part by the project of the dual-use technology for military and civilian missions ("Development of Quadruped Robots") of the Ministry of Commerce, Industry and Energy (MOCIE), KOREA.

References

1. K. Matsuoka, "Mechanisms of Frequency and Pattern Control in the Neural Rhythm Generators," *Biol. Cybern.*, vol. 56, pp. 345–353, 1987
2. M. A. Arbib, "The Handbook of Brain Theory and Neural Networks," *A Bradford Book, The MIT press*, 1995
3. H. Kimura and Y. Fukuoka, "Biologically Inspired Adaptive Dynamic Walking in Outdoor Environment Using a Self–contained Quadruped Robot : Tekken2", *Proc. IEEE/RSJ Int. Conf. Intell. Robots Syst.*,"pp. 986–991, 2004
4. J. Duysens, F. Clarac, and H. Cruse, "Load–Regulating Mechanisms in Giat and Posture : Comparative Aspects," *The american physiological society*, vol. 80, no. 1, pp. 83–133, 2000
5. T. A. McMahon, "Muscles, Reflexes and Locomotion," *Princeton University Press*, 1984
6. L. Robinson, "The Animal Motion Show, Volume 2," *Rhino House, http://www.rhinohouse.com*, 2003
7. G. E. Goslow, H. J. Seeherman, C. R. taylor, M. N. McCutchin, and N. C. Heglund, "Electrical Activity and Relative Length Changes of Dog Limb Muscles as a Function of Speed and Gait," *J. Exp. Biol.*, vol. 94, pp. 15–42, 1981
8. T. A. McMahon, "Using body size to understand the structural design of animals: quadrupedal locomotion," *J. Appl. Physiol.*, vol. 39, no. 4, pp. 619–627, 1975
9. A. V. Hill, "The Dimensions of Animals and Their Muscular dynamics," *Science Progress*, Vol. 38, pp. 209–230, 1950
10. , D'Arcy Wentworth Thompson, "On Growth and Form," *Cambridge University Press*, 1917
11. M. Hildebrand, "Analysis of vertebrate structure, 3rd Edition," *New York, Wiley*, 1988
12. Eliot Goldfinger, "Animal Anatomy for Artists," *Oxford University Press*
13. S. Hirose, "A Study of design and control of a quadruped walking vehicle," *Int. J. Robot. Res.*, vol. 3, no. 2, pp. 113–133, 1984
14. G. Gabrielli and T. H. von Karman, "What Price Speed ?," *Mechanical Engineering*, vol. 72, no. 19, pp. 75–78, 1950
15. D. C. Kar, "Design of Statically Stable Walking Robot: A Review," *J. Robotic Systems*, vol. 20, no. 11, pp. 671–686, 2003
16. J. A. Smith and I. Poulakakis, "Rotary Gallop in the Untethered Quadrupedal Robot Scout II," *Proc. IEEE/RSJ Int. Conf. Intell. Robots Syst.*, vol. 3, pp. 2556–2561, 2004
17. J. P. Schmiedeler, D. W. Marhefka, D. E. Orin, and K. J. Waldron, "A Study of Quadruped Gallops," *NSF Design, Service and Manufacturing Grantees and Research Conference*, pp. 1–10, 2001
18. T. M. Griffin, R. Kram, S. J. Wickler, and D. F. Hoyt, "Biomechanical and energetic determinants of the walk-trot transition in horses," *J. Exp. Biol.*, vol. 207, pp. 4215–4223, 2004

THE IMPROVEMENT OF STRUCTURAL AND REAL TIME CONTROL PERFORMANCES FOR MERO MODULAR WALKING ROBOTS

ION ION*, LUIGE VLADAREANU**, RADU MUNTEANU jr.***, MIHAI MUNTEANU***

**Department of Technology of Manufacturing, POLITEHNICA University of Bucharest,ROMANIA, **Institute of Solid Mechanics of Romanian Academy, Department of Dynamic Systems, Bucharest, ROMANIA, ***Electrical Department of Technical University Cluj-Napoca, ROMANIA*

Starting from the standard MERO walking modular robot, which is a robot on 3x6 degrees of freedom, autonomous, being able to interact with the environment, are presented the results obtained by the improvement of the structural and real time control performances. For this purpose kinematics and kinetostatics analysis are performed, and the mathematic model of the inverted kinematics is determined for controlling the main trajectory that defines legs' support and weigh centre positions. Related to this there is presented an Open Architecture system for the MERO robot position control in Cartesian coordinates through real time processing of the Jacobean matrix obtained out of the forward kinematics using the Denevit-Hartenberg method and calculating the Jacobean inverted matrix for feedback. The obtained results prove a significant reduction of the execution time for the real time control of MERO robot's position in Cartesian coordinates and increased flexibility.

1. Introduction

The improvement of performances concerning the possibility of moving the walking robots on ground with realistic configuration, in order to increase the mobility and stability in real conditions, have been realized by developing the control system in real time of the robot movement trajectory [1,2]. Finding the mathematical model allows for knowing the legs coordinates against robot's body and the body position during walking against the support points.

The mechanism within the shift system of MERO modular walking robot served as an experimental basis [2]. The mechanism is formed by 3 serially connected active groups of type RRR. The last two groups form a flat kinematics chain, functioning in vertical plane. The elements of the first group are connected to the platform and move in horizontal plane.

This kind of structure imitates quite well the shape of the human leg, is very simple and reliable, ensuring multiple possibilities of robots' maneuvers.

2. Kinematics Modeling of MERO Robot

Kinematics modeling of the legs' mechanisms related to the main system attached to the MERO modular walking robot is presented in Figure 1. There are 3 modules with 2 legs each, connected by 3 joint angles. Each leg is composed by a mechanism with 3 elements connected by 3 rotation actuators.

The Denovit-Hartenberg matrix method in homogenous coordinates is used in order to determine the legs position against the platform. Following the analysis of the Hartenberg-Denavit links system for the robot leg mechanism we have found a quite particular position. This led to decreasing the number of parameters of the transformation matrix A_i from six to four.

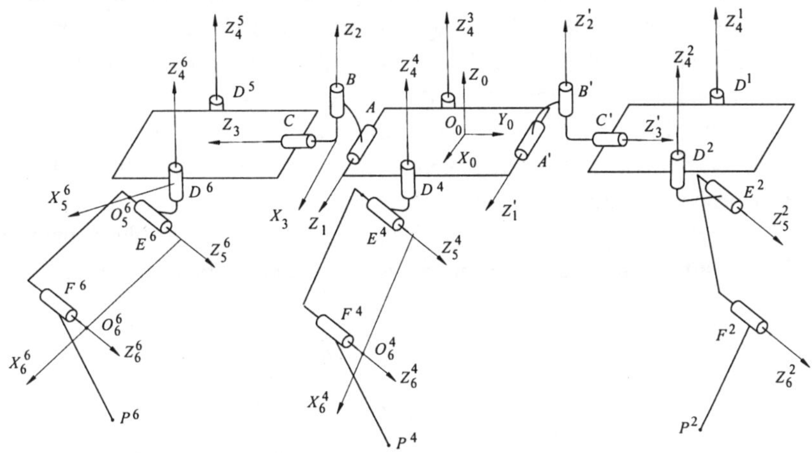

Figure 1. The Hartenberg-Danavit axis system attached to the leg of MERO modular walking robot

The work presents both the transforming connections of the a point's coordination – from system (i+1) to the system (i), that are realized by a transformation matrix A_i – as well as the transforming relations of the P_i point's coordinations from the system attached to the element (3) of the leg in the platform system. Considering as already known the legs coordinates X_{OiP}, Y_{OiP}, Z_{OiP}, the mathematic model with variable parameters of the leg joint angles θ_{ij} (j=1,6, i=4,6) have been determined by inverted kinematics analysis. Thus, knowing the trajectory that the legs should describe, the parameters necessary to the control system in real time of the robot could be established. By direct kinematics analysis are calculated X_{OiP}, Y_{OiP}, Z_{OiP}, and by inverted kinematics analysis are calculated θ_{4i}, θ_{5i}, θ_{6i}. The relative speed is calculated by derivation of movement equation and the relative accelerations by double derivation.

In the general case of the walking robots with 2n legs – the "myriapod" robots having "n" connected modules – the legs kinematics related to the environment system have been determined by the same algorithm.

Mathematic modeling of the gravity centre position, which allows the walking robots to shift on ground with difficult configuration, is presented in the Figure 2. The geometric centre position O is defined as the diagonals intersection point of the polygon formed by the points where the legs are connected to the platform, and $G_{(GxGyG,zG)}$ as the gravity centre position of robot.

Considering the positions of the walking robot's legs as X_{Pi}, Y_{Pi}, Z_{Pi} a mathematic model has been developed in order to express the kinematics characteristics of the walking robot's gravity centre. In order to determine the position of the supporting polygon against the platform, through the Denevit – Hartenberg method, where Z_{ij} (i=1,6 or 1,4, j=1,3), and m_{ij} (i=1,6, j=1,3) the leg mechanism mass, the coordinates of P_i have been transformed from the system $O_{4x4y4z4}$ into the system O_{xOyOzO}. The coordinates of P_i have been transformed from the system $O_{4x4y4z4}$ into the system O_{xOyOzO} for to determine the position of the support polygon against the platform.

The vertical projection of the G system gravity center on the supporting surface must be inside the supporting polygon in order to obtain stability.

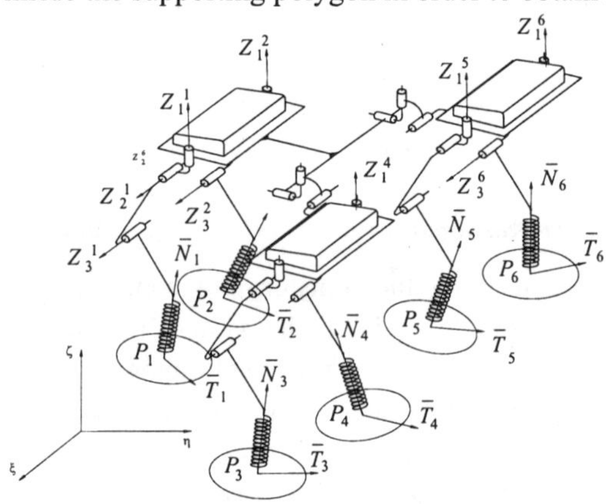

Figure 2. Mathematical modeling of the gravity centre for the MERO modular walking robot

Because the gravity centre positions of each legs mechanism element against the own system are known, the relations for the gravity robot centre coordinates, necessary for the stability control of the robot in real time have been established.

Knowing the gravity centre position, the speed $\dot{X}_G^k$ and the acceleration $\ddot{X}_G^k$ have been established by derivation.

HYRID POSITION AND FORCE CONTROL OF ROBOTS. As part of the robot control, the compliance is necessary to avoid power impact forces, to correct position error of robots or of special mechanical processes devices, and to allow tolerant relaxation of component elements. The compliance can be fulfilled either through passive compliance, as in Remote Center of Compliance (RCC) or through force control active methods [5]. Passive compliance can lower robot position capacity. The active compliance can have problems with sensibility in a rigid environment. Hybrid position and force control of robots equipped with compliant joints must take into consideration the passive compliance of the system. The generalized area where a robot works can be defined in a constraint space with six degrees of freedom (DOF), with position constrains along the normal force of this area and force constrains along the tangents. On the basis of these two constrains there is described the general scheme of hibrid position and force control in Fig.3. For simplification the coordinate transformations are not noted. Variables X_C and F_C represent the Carthesian position and the Carthesian force exerted onto the environment.
The selection matrices. Considering X_C and F_C expressed in specific frame of coordinates, its can be determinate selection matrices S_x and S_f, which are diagonal matrices with 0 and 1 diagonal elements, and which satisfy relation:

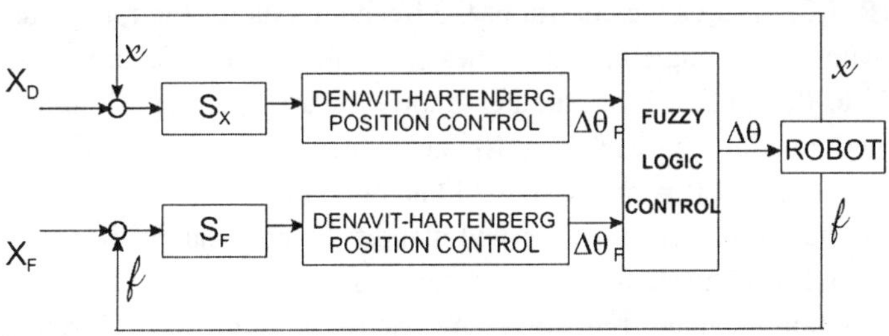

Fig.3. General structure of hibrid control.

$$S_x + S_f = I_d . \tag{1}$$

In recent approaches S_x and S_f are methodically deduced from kinematics constrains imposed by the working environment. Let A and B be two matrixes with full column rank that satisfy the equation $A^tB = 0$ and correspond to the twist and wrench spaces of constraint, then there can be determined S_x and S_f through the relations:

$$S_x = (A^t\Psi A)^{-1} A^t \Psi \ , \tag{2}$$

$$S_f = (B^t\Psi^{-1}B)^{-1} B^t\Psi^{-1} \ , \tag{3}$$

where usually Ψ is symmetrical matrix, positively defined. These theoretical concepts have been integrated into the open architecture system presented below.

3. Open Architecture System for the MERO Robot Position Control.

As a result of the studies, for the position control of the MERO walking robot (Figure 4) a real time control system with open architecture OAH has been conceived. The system has four main components: the programmable logical controller (PLC0), PC system with open architecture (PC-OPEN), multiprocessor system PLC (SM-PLC) and the position control system (**$CP^k{}_i$,** i=1-3, k=1-6). The system ensures the implementation of the algorithm for robot position control in Cartesian coordinates through real time processing of the Jacobean matrix obtained by direct kinematics of the robot, respectively of the invert Jacobean matrix for feedback. The method consists in reducing the matrix J (θ) to an upper triangulate form and finding errors in $\delta\,\theta$ joint coordinates using back-substitution. The joint angle errors $\delta\,\theta$ can be used directly as control signals for robot actuators. Joint angular errors for actuators on each robot freedom axis corresponding to the robot's new desired position generated in Cartesian coordinates are obtained by processing of Jacobean and inverted substitution.

The programmable logical controller in decentralized and distributed structure (PLC0) ensures the control of the freedom axis for the robot execution elements. The current angle motion values in absolute value $\boldsymbol{\Sigma\theta_{ci}}$ are transmitted from PLC0 to PC-OPEN. From the PC to the PLC are continuously transmitted the reference positions on each axis X_{Di} according to the trajectory generating program X_{Pi}. The PLC0 is processing 24 analogue inputs, 24 digital output, 18 analogue outputs for robot position and stability control, and other 64 digital inputs/outputs configurable for auxiliary functions: the hydraulic group control, motion limitation devices, limitation devices for homing etc. Moreover, the following analogical and digital input signals are controlled: position transducers (**$TP^k{}_i$,** i=1-3, k=1-6), proximity transducers, $2xD^k$ horizontal and $2xD^k$ on vertical for orientation, force transducers (**$T^k{}_F$,** k=1-6), vertical control transducers (X_M, Y_m) for robot stability.

The system PC-OAH (PC-OPEN Architecture) allows the introduction of new control functions based on supplementary programs. Due to the great processing speed with operating systems, which allow for programming in evolved programs, the basic functions can be implemented, using an ExTR real time multitask executive: interpolation (INP), main program (PP) of the operator interface, the movement trajectory generating program (MTG), the static stability control (SSC), robot platform control (RPT), the robot's walking shaping program (RWS), compliant function control program (CTRL C) and the prehensive control program (CTRL P). Supplementary, an interface with digital camera (IDC) could be installed, for objects recognition and the communication interface using radio modem, GSM or wireless systems. The PC-OAH system performs following operations: shapes walking stereotypes, determines the walking stability limits, ensures the horizontal position of the platform, maintains the established height of the platform and interprets the operator orders.

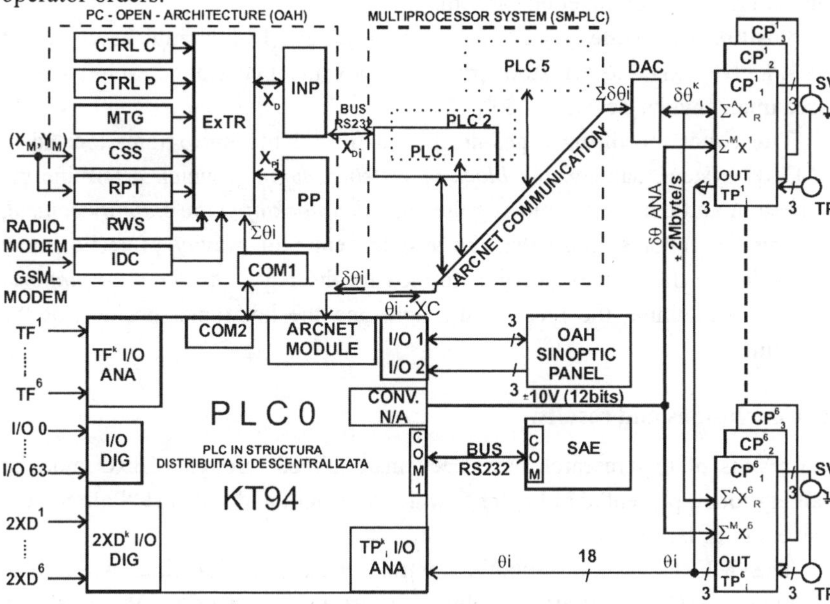

Figure 4. Real time control system with open architecture for the MERO modular walking robot

The Multiprocessor System PLC (SM-PLC) and **PLC0** has the role to send in real time, through the ARCNET fast communication network, the angular reference positions for the PIDT position regulator. For feedback, the current values θ^{K}_{ci}, (i=1-3, k=1-6), received from the position transducers (**TP^{k}_{i}**, i=1-3, k=1-6) are transmitted by PLC0 through ARCNET. Five processes for implementing the mathematical model of robot control have been identified [3]. In the active topology for (1) process, each PLC generates an ascendant data flux from PLC0 to PLC5. By calculating the transformed matrix $^{\mathbf{i}}\mathbf{A}_{\mathbf{j}}$ for the leg **k** (k=1-6), from the axis **i** to the axis **j**, we obtain the coordination matrix in axis **j**, finally resulting the coordinates for the robot environment $X^{k}_{C} = {}^{1}A^{k}_{3} = A_1*A_2*A_3$. In the active topology for the process (2), the matrix ${}^{i-1}A^{k}_{i}$ are stored for each PLC, the Cartesian coordinates X^{k}_{i} in **i** axis, by multiplying with ${}^{1}A^{k}_{j-1}$, are determined and the position variation δX^{k}_{C} is calculated. The Jacobean matrix is obtained during the process (3) by an ascending data flow, correlated with process (1). During the matricial calculation $^{1}\mathbf{A}^{k}_{j}$, the PLC0 is assigned to $J(\theta_1)$, matrix of (3x1), respectively PLC5 to $J(\theta_5)$. The active topology of process (4) brings the Jacobean matrix to triangular form by determining the pivot element and the new matrix A^{k}_{ij} elements. The active process (5) took place within a sequential process with process (4), determining the value of the angular error $\delta\theta^{\kappa}_{i}$. by inverted substitution.

Position Controllers ($\mathbf{CP}^{k}_{i}$, i=1-3, k=1-6), have 18 hydraulic servo-valves **($\mathbf{SV}^{k}_{i}$, k=1-6, i=1-3)** as actuators, providing movement control on each free axis, based upon angular deviations $\delta\theta^{\kappa}_{i}$.

The robot's walking types are issued from 3 block-programs located in PC-OAH system, namely: *the block of walking shaping*, which determine the succession and the way of moving the legs, *the block of static stability control*, granting the robot shift so that the system centre of gravity projection may remain inside the convex polygon formed by the leg's, *the block of platform control*, maintaining the prescribed height and the horizontal position of the platform.

4. Conclusions and results.

The results of this research have been materialized through MERO modular walking robot, presented in Figure 5, which was built and tested at "Politehnica" University of Bucharest.

The MERO walking robot's control system in real time with Open Architecture (OAH) ensure flexibility, short time execution, the precision targets and repeatability of the moving programs, eliminating completely the closed systems with projects meant for specified applications. Supplementary developments in order to increase the performances or new functions adding are possible only by modifying the software relating to the control modules in PC-OAH for laborious computations, respectively in the PLC multiprocessor for complex real time control. Besides, addition of new physical modules could easily be done because of the communication between the programmable logical controller (PLC) and the input/output modules made by a bus with 3 torsadate conductors.

Owing to the great computation speed of microprocessor systems and serial connection links for data transmission, the time necessary for establishing the inverted matrix is short enough to allow the robot control in real time, with no influence in performing the other programs. The results show that the time necessary to perform program for the MERO robot position control in Cartesian coordinates is 80% shorter.

Figure 5 MERO modular walking robot.

Moreover, the short time execution will ensure a faster feedback, allowing other programs to be performed in real time as well, like to the prehensive force control, objects recognition, making it possible that PC-OHA system have a human flexible and friendly interface.

References

1. Waldron, K.J., (1996) Modeling and Simulation of Human and Walking Robots Locomotion, Advanced School, Udine, Italy.
2. Ion, I., Simionescu, I., Ungureanu, M., Design and realization of the MERO modular walking robot, The Eight IFToMM International Symposium on Theory of Machines and Mechanisms, Bucharest, Romania, Aug. 28, Sept. 1, 2001pp.320-325
3. L.Vladareanu, T. Peterson – New Concepts for the Real Time Control of Robots by Open Architecture Systems, Machine Building, vol.55, ISSN 0573-7419 no.11, 2003
4. L.D.Joly, C.Andriot, V.Hayward, Mechanical Analogic in Hybrid Position/Force Control, IEEE Albuquerque, New Mexico, pg. 835-840, April 1997
5. Yoshikawa T., Zheng X.Z. - Coordinated Dynamic Hybrid Position/Force Control for Multiple Robot Manipulators Handling One Constrained *Object*, The International Journal of Robotics Research, Vol. 12, No. 3, June 1993, pp. 219-230.

Advances in Walking Robots

A BASIC VARIABLES SET BASED SCHEME OF ONLINE MOTION PLANNING FOR HUMANOID ROBOTS

JIAN WANG, TAO SHENG, HONGXU MA

College of Mechatronic Engineering and Automation, National University of Defense Technology, Changsha 410073, China.

An online motion planning scheme for humanoid robots is proposed based on basic variables. The proposed scheme transforms the problem of humanoid motion planning into the planning of a basic variables set. The chosen basic variables can reflect the motion state of humanoid robots, including Zero Moment Point (ZMP), Center of Gravity (CoG), Center of Coxa (CoC), foot position and the rotation matrices between some key frames. Then the planning flow of basic variables are shown and the dynamical and physical constraints of humanoid motion are taken into account. Finally, the efficiency and feasibility of proposed scheme are validated by computer simulation.

1. Introduction

Within the world of mobile robots, humanoid robots are of great interest these years. Originated from biped robots, humanoid robots should be able to pass obstacles easily and move on uneven terrain optionally, and they can do more besides walking with whole upper body including functional arms and fingers.

Furthermore, they have huge potential values in entertainment or other fields. Moving in human-living environment is the basic function of humanoid robots and motion control plays an important role. Currently, online motion planning has become an important approach in humanoid motion control.

In Lab of Intelligent Robots, National University of Defence Technology, biped robots have been studied since 1989. In 2000, the humanoid robot Pioneer was developed; and "Blackmann" was built in 2003. In order to achieve online motion planning for "Blackmann", an online motion planning scheme based on basic variables set is proposed. Firstly, two kinds of motion planning methods are reviewed, and requirements of real-time motion control are discussed. Then the proposed scheme is described, and algorithms of basic variables' planning are given. Finally, the effectiveness of proposed scheme is validated.

2. Related Work

Two kinds of usually used motion planning methods are reviewed firstly, and then the status quo of humanoid online motion planning is investigated.

2.1. *Motion Planning Based on Lumped-Mass Model*

A common approach for humanoid motion planning is based on lumped-mass model. Firstly, the robot is modeled as a point mass, and CoG is restricted to moving on a certain plane. Given these assumptions, the CoG motion is described by two decoupled linear ordinary differential equations of second degree[1]. Then a closed-form solution relying on ZMP is acquired. Generally, in order to ensure the feasibility of the planning results, a time evolution of the ZMP is prescribed. The advantages of the method are conceptual simplicity and the ability to plan motion online. But how to make it effective is difficult.

2.2. *Optimization Based Methods*

Another method is to solve an optimization problem. Usually most of the DOFs of the humanoid robot are optimized in joint space[2][3]. And the search space for possible solutions is so large that it is always possible to find a reasonable trajectory to fulfill the requirements. But the computation efficiency of solution search is low just because of the large search space of possible trajectories.

2.3. *Status Quo of Humanoid Online Motion Planning*

Currently, methods used for online motion planning mainly include online walking motion generation and online mixture and connection of preplanned motions. Lim[3] studied online motion planning using online mixture and connection of unit gait; Nishiwaki[4] proposed an algorithm for online gait updating based on desired ZMP; Kajita used the ultrasonic sensors to realize online gait planning of forward walking[5]; Afterward, he proposed a method of 3D online walking pattern generation[6]. Lorch[7] realized online walking generation in sagittal plane using technology of stereoscopic vision.

Compared with off-line motion planning, online motion planning cares more about feasibility and efficiency. And algorithms based on lumped mass model are chosen instead of time-consumed methods in off-line motion planning.

3. Proposed Scheme

According above analysis, lumped-model based methods are less time-consumed, but it is difficult to ensure the feasibility of results; optimized based methods can get good trajectories but cost a long time and it seems unpractical in online planning. So, lumped-model based methods are appropriate for online planning. This section describes a basic variable set based scheme of online planning. Several motion state variables called basic variables are chosen to describe humanoid motion, the task is to calculate trajectories of basic variables. Various constraints are considered to ensure the feasibility of planning results.

3.1. *Architecture of online motion planning and control*

Architecture of online motion planning and control is shown in Fig. 1. The task of humanoid motion control is divided into three parts: online motion planning, upper limbs manipulation control and posture/landing control. Here, proposed scheme mainly deals with online motion planning emphasized in Fig.1.

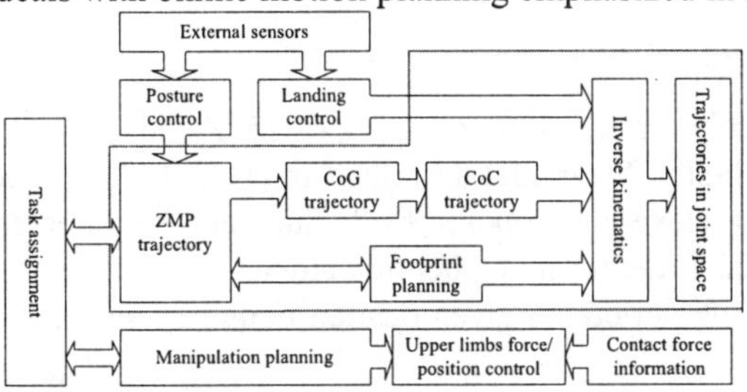

Fig. 1 Architecture of humanoid motion control

3.2. *Description of Online Motion Planning*

Generally, humanoid motion is described in joint space. But it is difficult to find a feasible motion sequence in joint space. Therefore, a basic variable set is used to describe the motion sequence instead of temporal trajectories in joint space.

The basic variable set includes:

- x_Z, y_Z, z_Z, ZMP trajectory, the criterion of humanoid walking stability;
- x_G, y_G, z_G, CoG trajectory;
- p_{LF}, p_{RF}, positions of both ankle in Cartesian frame;
- x_C, y_C, z_C, CoC in Cartesian space. CoC is the origin of the frame attached to coxa, and coxa is the part which connects both hips.
- $\theta_{rot\_left}, \theta_{rot\_right}$, rotational DoFs of both hips, variables for turning motion.
- ${}^{FSP}_{Coxa}R \in SO(3)$, rotation matrix of frame attached to coxa relative to frame attached to supporting foot;
- ${}^{O}_{FSP}R \in SO(3)$, rotation matrix of frame attached to supporting foot relative to inertial frame;
- ${}^{Coxa}_{FSW}R \in SO(3)$, rotation matrix of frame attached to swing foot relative to frame attached to coxa.

3.3. *Basic Variables Planning*

Since the motion sequence is described using basic variables, the planning methods of basic variables are proposed as forward walking for example.

3.3.1. *Humanoid Dynamic Model*

Generally, relationship between ZMP and CoG is used to describe the dynamics of humanoid robots, which is shown as follows:

$$x_Z = x_G - (z_G - z_Z)\cdot \ddot{x}_G / g \tag{1}$$

$$y_Z = y_G - (z_G - z_Z)\cdot \ddot{y}_G / g \tag{2}$$

Here, $[x_G \quad y_G \quad z_G]$ and $[x_Z \quad y_Z \quad z_Z]$ are CoG and ZMP trajectory in inertial frame respectively, g is gravity.

3.3.2. *ZMP and CoG Planning*

This section mainly discusses the planning of ZMP and CoG. Here, three points are considered: firstly, the model described in (1) and (2) is used; secondly, the idea of boundary relaxation proposed by T. Sugihara [8] is adopted; thirdly, spline functions are used as the form of CoG trajectory.

In one walking period T, given desired boundary values of ZMP and CoG in sagittal plane:

$$^{d}x \equiv \left[{}^{d}x_Z(0) \quad {}^{d}x_Z(T) \quad {}^{d}x_G(T) \quad {}^{d}\dot{x}_G(T)\right]^T \tag{3}$$

Given ZMP trajectory in sagittal (x-axis) plane:

$$x_Z(t) = f_{ZMP}(p_0, t, x_Z(0), x_Z(T)) \tag{4}$$

Here, f_{ZMP} is using the form of cubic polynomial, and p_0 is the parameters set described by ZMP boundary conditions and walking speed. Using equation (1), the solution is drawn up using spline function:

$$x_G(t) = f_{CoG}(p, t, x_Z(0), x_Z(T)) \tag{5}$$

Here, p is unknown and boundary conditions of ZMP and CoG are used to solve p Similar to [8]. Let $x \equiv [x_Z(0) \quad x_Z(T) \quad x_G(T) \quad \dot{x}_G(T)]^T$, and set $t = 0$ and $t = T$ in $\begin{cases} x_G(t) = f_{CoG}(p, t, x_Z(0), x_Z(T)) \\ \dot{x}_G(t) = df_{CoG}(p, t, x_Z(0), x_Z(T))/dt \end{cases}$ to form constraint equations:

$$g(p_0, p, t, x, x_G(0), x_G(T)) \equiv 0 \tag{6}$$

Constraint inequalities are formed using walking stability condition:

$$h(x_Z(0), x_Z(T)) < 0 \tag{7}$$

Solving (8) will obtain appropriate boundary conditions under (6) and (7):

$$\left(x - {}^{d}x\right)^T A\left(x - {}^{d}x\right)/2 \to \min \tag{8}$$

Here, $A \equiv diag\{a_i\}$, $(a_i > 0, i = 1 \sim 4)$. Using idea of penalty function, problem (8) can be replaced by (9):

$$\left(x - {}^{d}x\right)^T A\left(x - {}^{d}x\right)/2 + \sum_i C_i(\max\{h_i, 0\})^2 + \sum_j D_j g_j^{\,2} \to \min \tag{9}$$

Where C_i, D_j are weight parameters, and $C_i, D_j > 0$. Using calculated boundary conditions to compute p, ZMP and CoG trajectories are obtained.

3.3.3. *Footprint Planning*

In forward walking, only the sagittal (in $x-axis$) and vertical (in $z-axis$) parts are considered, and only the positions of ankle point are planned.

In humanoid walking, single and double support phases arise in sequence. Positions of supporting and swing foot in $t=0$ are $(x_{SP}(0), z_{SP}(0))$ and $(x_{SW}(0), z_{SW}(0))$. $x_{SW}(T) = x_{SW}(T) + 2 \cdot S$ is decided by pace S. When ZMP enters into the coming single support region, the swing leg begins to rise. When $t=T$, the swing foot reaches the desired position $x_{SW}(T)$ and $\dot{x}_{SW}(T)=0$. The time to start foot lifting T_S is decided by planned ZMP trajectory in simulation, and it is calculated using flow chart in Fig. 2, where Ω is the target single support region. Afterward, the sagittal and vertical parts of temporal trajectory of swing foot can be planned.

```
for    i = 1 : N
       if     (x_Z(i·T/N), y_Z(i·T/N)) ∈ Ω
       then   T_S = i·T/N; break
       else   i = i + 1
       end
end
```

Fig. 2 Time Decision of Swing Foot Lifting

3.3.4. *Calculation of CoC trajectory*

The positional relationship between CoG and CoC is described as follows:

$$p_G = \sum_{i=1}^{n} m_i p_i / M = p_C + \sum_{i=1}^{n} m_i \, {}^{O}_{C}R \, {}^{C}p_i / M \tag{10}$$

Where, $p_G = (x_G, y_G, z_G)$, $p_C = (x_C, y_C, z_C)$ are positions of CoG and CoC in inertial frame, ${}^{O}_{C}R \in SO(3)$ is rotation matrix of frame attached to coxa relative to inertial frame, ${}^{C}p_i = ({}^{C}x_i, {}^{C}y_i, {}^{C}z_i)$ is position of link i in frame attached to coxa, m_i is the mass of link i, and M is the mass of the robot.

Firstly, the mappings from joint angles to each link's center of mass are built and ${}^{O}_{C}R$ is calculated using ${}^{FSP}_{Coxa}R$ and ${}^{O}_{FSP}R$. Generally ${}^{O}_{FSP}R$ can be obtained beforehand, and ${}^{FSP}_{Coxa}R$ is decided by configuration of supporting leg.

$${}^{O}_{C}R(\theta, \phi) = {}^{O}_{FSP}R \cdot {}^{FSP}_{Coxa}R(\theta, \phi') \tag{11}$$

$$p_i = f_{FK\_i}(\theta, c_i) \tag{12}$$

Here $\theta = \{\theta_1, \cdots, \theta_j, \cdots \theta_n\}$ is the joint variables set, and ϕ , ϕ' , c_i are kinematical parameters sets. Then (10) can be rewritten using (11) and (12):

$$p_G = p_C + \sum_{i=1}^{n} m_i \, {}^{O}_{C}R(\theta, \phi)^C f_{FK\_i}(\theta, c_i) / M \tag{13}$$

If initial configuration of the robot is ascertained, $p_C(0)$, $\theta(0)$, $p_G(0)$ are known. Set $\Delta\theta(0) = \theta(1) - \theta(0)$, owing to $\Delta\theta \to 0$, $p_C(1)$ can be formed:

$$p_C(1) \approx p_G(1) - \sum_{i=1}^{n} m_i \, {}^O_C R(\theta(0), \phi)^C f_{FK\_i}(\theta(0), c_i)/M \tag{14}$$

Similarly, $p_C(k+1)$ can be calculated utilizing $p_G(k+1)$ and results of the k times inverse kinematics as shown in Fig.3:

$$p_C(k+1) = p_G(k+1) - \sum_{i=1}^{n} m_i \, {}^O_C R(\theta(k), \phi)^C f_{FK\_i}(\theta(k), c_i)/M \tag{15}$$

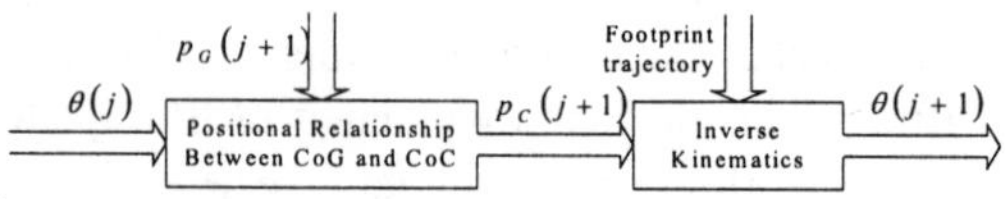

Fig. 3 CoC Calculation

3.3.5. *Inverse Kinematics*

After positions of CoC and both ankles acquired, inverse kinematics is under way using transformation matrix, Jacobian and rotation matrices ${}^{FSP}_{Coxa}R$, ${}^{O}_{FSP}R$ and ${}^{Coxa}_{FSW}R$, as seen in Fig.3. Either leg of the humanoid robots can be regarded as a series configuration with 3 links and 6 DOFs. Firstly, the transition matrices are computed using D-H coordinates, then coxa Jacobian $J_{Support}$ between joint angles of support leg and the coxa configuration and swing foot Jacobian J_{Swing} between joint angles of swing leg and the swing foot configuration are derived. Finally, RMRC [9] method is used to compute the joints angles.

3.3.6. *Considered Constraints*

During motion planning, two kinds of restrictions are considered, including stability, friction constraints and physical constraints. Stability conditions are considered in ZMP and CoG planning, so as to ensure the feasibility of walking; while in high-speed walking, friction constraints should be counted. Physical constraints considered in inverse kinematics include leg length restriction, joint angle and velocity restrictions. Leg length restriction means the distance between ankle and hip should be less than leg length. Joint angle and velocity restrictions mean the planned joint angle and angular velocity should be in a reasonable span.

4. Simulation and Analysis

Taking humanoid robot "Blackmann" as the object, forward walking simulation in corresponding platform is used to test proposed scheme.

It is supposed that pace $S = 0.2m$, walking speed $v = 0.2m/s$, step period $T = 1s$, $z_Z = 0, z_G = const$, $\theta_{rot\_left} = 0, \theta_{rot\_right} = 0$ and ${}^{FSP}_{Coxa}R$, ${}^{O}_{FSP}R$, ${}^{O}_{C}R$ are identity matrices, and the physical parameters of "Blackmann" shown in Tab.1

are used in simulation. Using aforementioned methods, ZMP, CoG and CoC trajectories as well as joint angles are obtained (seen in Fig.4 and Fig. 5). And kinematics simulation is carried out as seen in Fig.6. Then through dynamic simulation, it is shown that the recalculated ZMP (seen in Fig.4) still locate in stable margin, so results using proposed scheme are feasible in humanoid walking. In simulation environment (Pentium4 2G, 512M RAM, Windows XP), one step planning including inverse kinematics costs about 450 ms with 100 points/step. While in previous off-line motion planning [10], calculation of one walking cycle will cost tens of seconds. So, proposed scheme can be used online.

Table 1. Physical Parameters of "Blackmann"

	Mass(Kg)	Length(mm)	Width(mm)	Height(mm)
Total	63.5	\	\	1545
Head	0.5	\	\	200
Trunk	31.82	\	\	580
Arm	1.40	\	\	\
Thigh	7.06	\	\	350
Shank	3.94	\	\	340
Foot	3.16	\	\	95
Sole	\	270	175	\

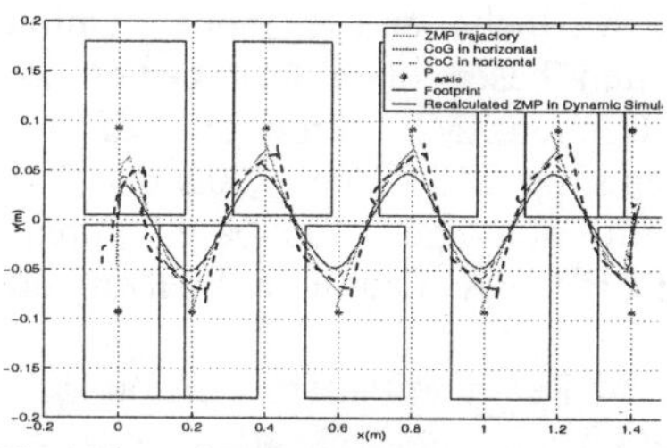

Fig.4 Planned ZMP, CoG, CoC, Footprint Trajectories and Recalculated ZMP

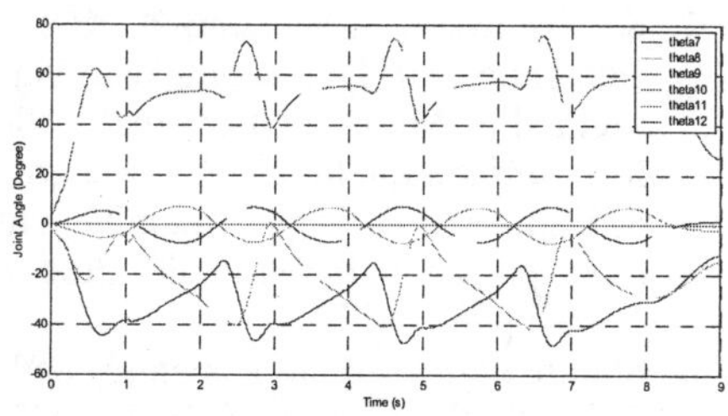

Fig.5 Joint Angles of Right Leg

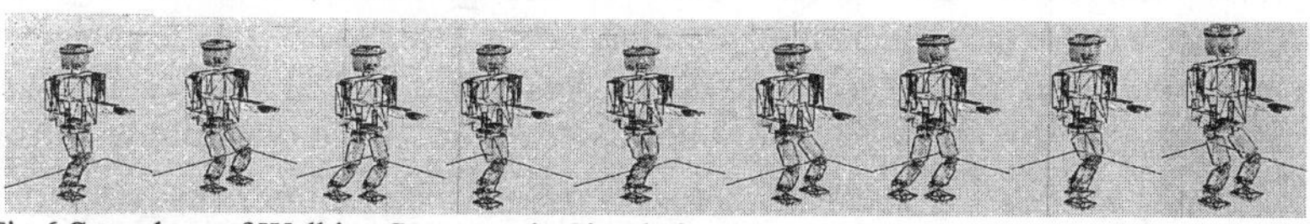

Fig.6 Snapshots of Walking Sequence in Simulation

5. Conclusion

An online motion planning scheme based on basic variables set for humanoid robots is presented. Proposed scheme transforms humanoid motion planning problem into basic variables' planning. ZMP and CoG trajectory are obtained using boundary conditions. Then positional relationship between CoG and CoC, is used to calculate CoC trajectory online. Finally, joint angles are calculated by inverse kinematics. Walking experiment using proposed scheme is not finished up to now and it can't cope with real-time motion control alone in unstructured environment, so ongoing and future work includes experimental evaluation of proposed scheme using "Blackmann".

Acknowledgments

This research is supported by grant 60475035 of the Chinese National Natural Science Foundation.

References

1. S. Kajita, F. Kanehiro, K. Kaneko, K. Fujiwara, K. Harada, K. Yokoi, and H. Hirukawa. "Biped walking pattern generation by using preview control of zero-moment point", In Proc of the IEEE ICRA, Taipei, Taiwan, sep. 2003.
2. G. Bessonnet, S. Chesse, and P. Sardain. "Optimal gait synthesis of a seven-link planar biped". *The Intl Journal of Robotics Research*, vol.23, no.10-11, pp. 1059-1073, 2004.
3. H.-O. Lim, Y. Kaneshima, and A. Takanishi. "Online Walking Pattern Generation for Biped Humanoid Robot with Trunk". Proc. of IEEE ICRA, pp. 3111-3116, 2002.
4. K. Nishiwaki, S. Kagami, Y. Kuniyoshi, M. Inaba, and H. Inoue. "Online Generation of Humanoid Walking Motion based on a Fast Generation Method of Motion Pattern that Follows Desired ZMP". Proc. of IEEE/RSJ Int. Conf. on Intelligent Robots and Systems, pp. 2684-2689, 2002.
5. S. Kajita and K. Tani. "Adaptive gait control of a biped robot based on real-time sensing of the ground". Proc. of IEEE ICRA, Lausanne, Switzerland, 1996. pp. 570–577.
6. Shuuji Kajita, Fumio Kanehiro, and etc. "A Realtime Pattern Generator for Biped Walking". Proc. Of IEEE ICRA, 2002.
7. O. Lorch, J. F. Seara, K. H. Strobl, U. D. Hanebeck, and G. Schmidt. "Perception Errors in Vision Guided Walking: Analysis, Modeling, and Filtering". Proc. of IEEE ICRA, pp. 2048–2053,2002.
8. Tomomichi Sugihara and Yoshihiko Nakamura. "A Fast Online Gait Planning with Boundary Condition Relaxation for Humanoid Robots". Proc. of the 2005 IEEE ICRA, Barcelona, Spain, April 2005.
9. Deepak Tolani, Ambarish Goswami, and Norman I. Badler. "Real-Time Inverse Kinematics Techniques for Anthropomorphic Limbs". *Graphical Models*, vol.62, pp.353–388, 2000.
10. Jian Wang, Tao Sheng, Hongxu Ma and Haili Qin. "Design and Dynamic Walking Control of Humanoid Robot Blackmann", Proc. of the 6th World Congress on Intelligent Control and Automation, Dalian, China, pp.8848-8852, 2006.

A HOPPING MOBILITY CONCEPT FOR A ROUGH TERRAIN SEARCH AND RESCUE ROBOT

SAMUEL KESNER
JEAN-SÉBASTIEN PLANTE
STEVEN DUBOWSKY

Mech. Eng. Dept., Massachusetts Institute of Technology, 77 Massachusetts Ave. Cambridge, MA 02139, USA

PENELOPE BOSTON

Earth & Env. Sc. Dept., New Mexico Inst. of Mining and Technology, 801 Leroy Place Socorro, NM 87801, USA

A new search and rescue concept based on the deployment of teams of small spherical mobile robots ("Microbots") has been proposed. In this concept, hundreds to thousands of cm-scale, sub-kilogram Microbots are released over a search site such as collapsed building rubble or caves. Microbots use hopping, bouncing, and rolling to infiltrate subterranean spaces in search of possible survivors. Key technologies enabling Microbots are the use of high energy-density micro fuel cells combined with low cost and lightweight dielectric elastomer actuators. The paper presents recent work demonstrating the feasibility of Microbots mobility in rough terrain. Experimental studies have demonstrated the possibility of using dielectric elastomer actuators to generate autonomous hops. High efficiency hydrogen fuel cells have also been used to power dielectric elastomer actuators. Simulation results show that Microbots of proper diameter and hop height can successfully traverse very rough terrains. These results suggest that teams of Microbots can effectively be used for search and rescue missions.

1. Introduction

Events such as the 2005 Pakistan earthquake and the 2001 September 11 terrorist attacks demonstrate the need for new effective search methods in rough terrain, see Figure 1. Current search methods for rough terrains are limited. Remote imaging techniques to identify subterranean features, including ground penetrating radar, ultrasonic imaging, and resistive imaging, have been developed [1,2]. However, these methods are limited in resolution and depth due to soil properties. They also cannot detect the presence of disaster survivors in difficult to reach locations. The "dog and pole" method is still the best civilian search technique.

Figure 1. Typical search and rescue sites: (a) 2005 Pakistan Earthquake (b), September 11, 2001.

A new approach for search and rescue in rough terrains based on hopping robots, called Microbots has been proposed [3]. As shown in Figure 2(a), Microbots are small spherical robots of about 10 cm in diameter. The search and rescue approach consists of deploying hundreds or thousands of Microbots over a search site. The Microbots use hopping, bouncing, and rolling to navigate rough terrains in search of survivors. Due to their small size, Microbots can diffuse inside rubble cavities to find internal passage leading to protected spaces, see Figure 2(b).

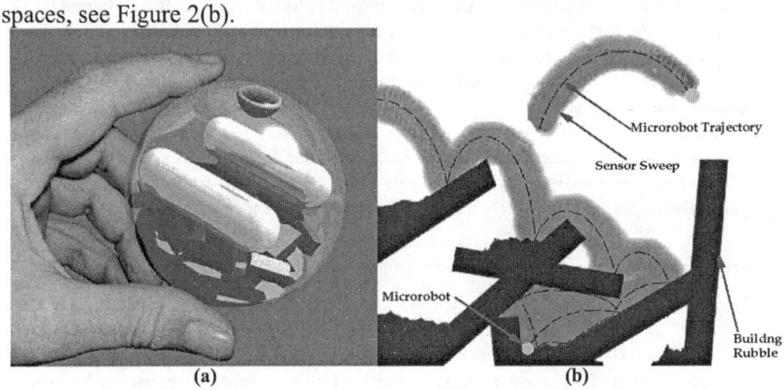

Figure 2. The Microbot concept: (a) artist representation, (b) progression in rubble.

Microbots are powered by high energy density Proton Exchange Membrane (PEM) fuel cells to assure long lasting energy supply. The mobility system is actuated by lightweight and low cost Dielectric Elastomer Actuators. Microbots are equipped with onboard miniature sensors such as cameras and chemical "sniffers" to tract and identify survivors. Their communication systems relay information between each other and a command center. Microbots components are protected by a strong plastic shell that absorbs shocks.

Microbots missions differ from conventional robotic missions that often use a single highly capable agent. Instead, Microbots missions use a very large number of low cost and simple agents, bringing a high degree of redundancy and robustness. Individual agent losses are acceptable without failing the mission objectives. Also, the low costs of Microbots make them disposable which eliminate the need for post mission recovery.

The mobility of Microbots in rough terrain is one of several important technical challenges that must be carefully understood before Microbots become a reality. Hopping robots have been proposed for space exploration and reconnaissance applications [4,5,6]. Most of this work focuses on the development of hopping mechanisms for relatively heavy robots (>1kg) and are not appropriate for lightweight Microbots. Developing a practical mobility system for small and lightweight hopping robots, especially for rough terrain environments, has not been addressed.

This paper studies the feasibility of the Microbot mobility concept for search and rescue missions using experimental validations and simulations.

An experimental Microbot prototype powered by Dielectric Elastomer Actuators has been constructed. It achieved hops of 38 cm with actuators that have less than one-half the thrust of the Microbot reference design, due to current laboratory fabrication limitations. Methods to build more powerful actuators are currently being developed. This result shows the technology to be suitable for Microbots.

Experimental miniature PEM fuel cells using hydrogen have been used to power Dielectric Elastomer Actuators. Conversion efficiencies have been measured across the energy chain and projected Microbot performance are reported here. These experiments show the concept is viable for 1000 hops missions.

Simulations of the Microbot mobility show the effect of Microbot diameter and hop height on travel distance in rough terrain. The simulations shows that a Microbot diameter of 10 cm with a projected hop height of 1 m give reasonable rough terrain mobility.

The general conclusion of this paper is that, assuming reasonable technology progress, Microbots could effectively move in rough terrains for search and rescue missions.

2. Microbot Mobility Concept

The mobility mechanism concept is illustrated schematically in Figure 3. Energy is stored in the form of hydrogen gas in a metal hydride storage vessel.

Hydrogen reacts with atmospheric oxygen in the PEM fuel cells to generate electricity. Pure oxygen could be stored onboard for anaerobic applications. A small lithium polymer battery is used to level power consumption peaks. The DEA pumps mechanical energy into a bi-stable spring over one or more actuation cycles. When a predefined energy level stored in the spring is reached, the energy is released to provide hopping power.

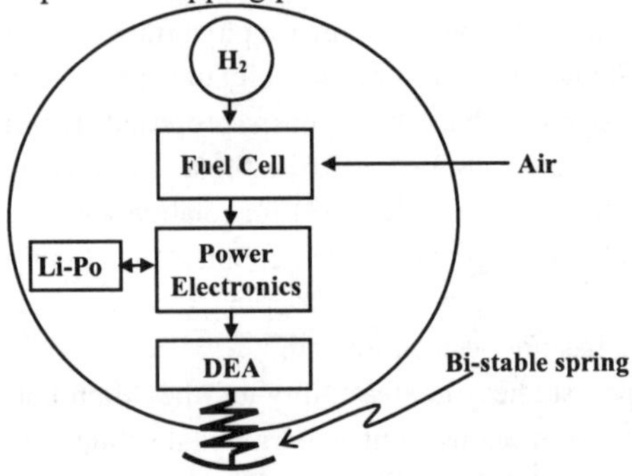

Figure 3. Schematic of the Microbot mobility concept on rough terrain.

Microbots are self-righting so that after each hop, they return to an upright position. Directionality can be provided by number of mechanisms, including small additional DEAs that tilt the Microbot prior to hopping. Directionality consumes little energy compared to hopping and is of secondary importance at this stage in the Microbot development.

The Microbot mission concept exploits the high force-to-weight and simplicity of DEAs [7,8,9,10]. These qualities make DEAs very attractive for Microbot missions since a large number of strong and lightweight actuators are needed. Another application exploiting the same characteristics of DEAs is binary actuation [11]. DEAs are also low power / high energy density devices that match well with the proposed fuel cell energy storage technology.

The preliminary design specifications of the mobility system for search and rescue missions are summarized in Table 1. These numbers are referenced throughout this paper.

Table 1. Microbot Mobility System Specifications.

Parameter	Values
Microbot Diameter	10 cm
Hop Height	1 m
Microbot Mass	100 grams
Min. Autonomy	1000 hops
Min. Hop Frequency	2 hops / minute

3. Dielectric Elastomer Actuators Powered Prototype

A simplified Microbot prototype has been built to demonstrate the feasibility of using DEAs to make a Microbot hop with an onboard energy source. The prototype is shown in Figure 4. A conical shaped DEA pumps energy into a pair of power springs. When a prescribed number of pumping cycles is reached, the stored mechanical energy is released and the Microbot hops. The transmission structure is made from carbon fiber. The power springs are made of carbon fiber strips. The strips are normally flat and are mounted in a buckled state. The combined mass of the actuator, transmission, and power springs is 18 grams.

The energy source for the prototype is a single 145 mAh Lithium-Polymer cell that weighs 5 grams. A custom electronic circuit using a pair of EMCO Inc. miniature DC/DC converters generates the 8.8 kV needed by the DEA. The Microbot prototype is shown in Figure 5. The mobility system and electronics are enclosed in a 10.5 cm diameter PETG shell. The 46 grams Microbot reaches vertical hop heights of 38 cm. Each hop requires 35 actuator pumps.

Figure 4. Mobility system prototype.

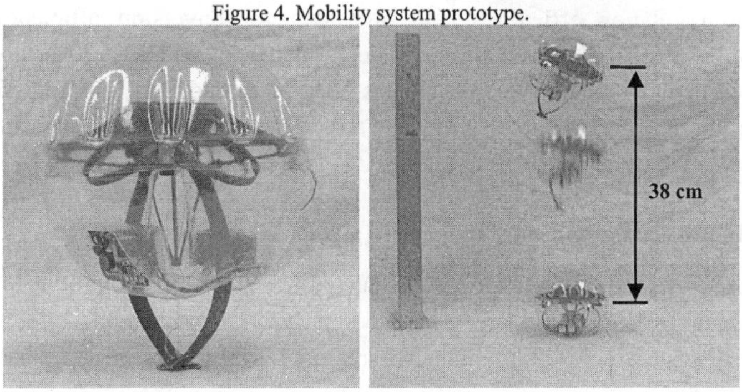

Figure 5. Autonomous Microbot prototype performing hops of 38 cm.

The Microbot prototype clearly indicates that DEAs can power lightweight hopping robots. The total mass, hop height, and pumping times of this hand-fabricated prototype are within reach of the target values of Table 1. Achieving the specifications of Table 1 appears possible with improved manufacturing techniques and further design optimization.

4. Fuel Cells Energy System

A hydrogen fuel cell energy system is proposed to power Microbots. For hopping robots, energy consumption during hopping is proportional to the system weight and hop height. An analysis of a fuel cell energy system's requirements is therefore related to the system mass via the hop height requirement and the fact that the hydrogen fuel and associated storage device has a non-negligible mass. The performance of the energy system is characterized by the relationship between the number of hops and Microbot system mass:

$$N_{hops} = \frac{\eta_T(\eta_{reg}E_{fc} - P_{elec}t)}{mgh} \tag{1}$$

where N_{hops} is the number of hops, η_T is the total energy conversion and hopping efficiencies, η_{reg} is the low voltage regulator efficiency, E_{fc} is the electric total energy generated by the fuel cell system through the conversion of hydrogen, P_{elec} is the power consumption of the onboard electronics, m is the mass of the Microbot, g is the acceleration of gravity, and h is a hop height of 1 m, and t is the length of the mission. In this case t is assumed to be 3.5 days.

An experimental fuel cell power system was constructed and used to power a Microbot prototype, see Figure 6. Values for the fuel cell efficiency, low voltage regulation efficiency, and high voltage conversion efficiency were found experimentally to be approximately 70%, 90%, and 30% respectively. A detailed explanation of the analysis has been presented [12]. Figure 7 shows a plot of the hops/mass relationship. The target of 1000 hops can be reached with a Microbot mass of about 100 grams.

5. Mobility Simulations

The experimental demonstrations conducted to date indicate that the Microbot specifications of Table 1 are realistic. Simulations studying the effect of key parameters such as Microbot diameter and hop height on performance have been performed.

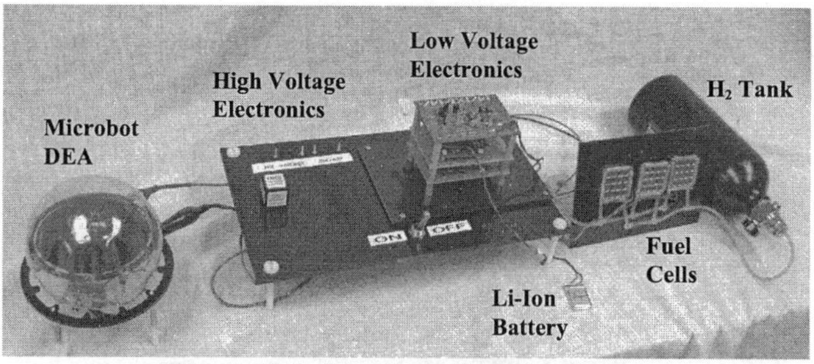

Figure 6: The hydrogen fuel cell experimental setup.

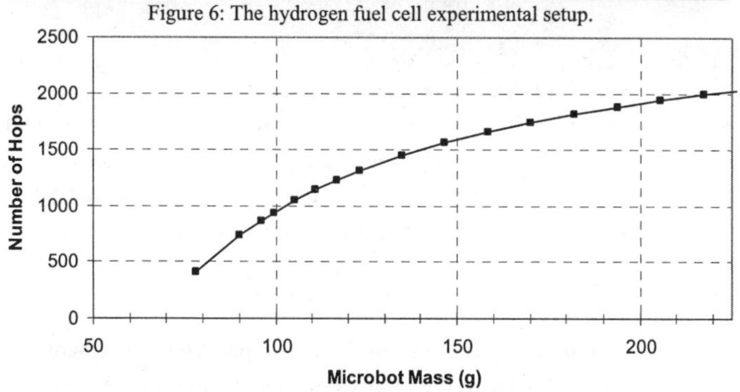

Figure 7: Number of Microbot hops as a function of system mass.

5.1. *Simulation Approach*

A tunnel with debris was selected to represent a disaster area, such as a collapsed passage in a mine, a subway tunnel after an earthquake, or the interior of a collapsed building. The simulated terrain was generated in Solidworks CAD software as an assembly of individual solid bodies, see Figure 8. The rock pile is composed of 300 rocks of different sizes randomly grouped together into a pile approximately 5x4x0.85 m. The tunnel diameter is 5 m and its length is 60 m.

The simulations are conducted with MSC Software's ADAMS dynamic simulation software. ADAMS allows the definition of mass properties, body forces, and body interaction constraints and forces. The directionality of the hop is controlled by a Simulink (Mathworks) model communicating with ADAMS.

The Microbot interacts with the environment through hopping, bouncing and rolling on the terrain. Hopping was modeled as an impulse force between the Microbot and the terrain. The hopping direction depends on the angle that the impulse is applied relative to the Microbot's body.

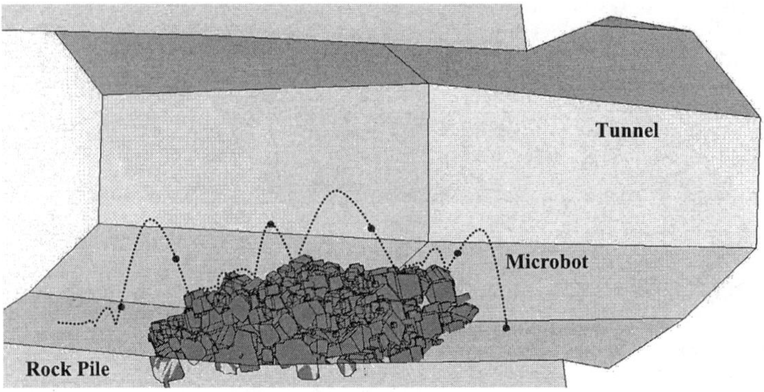

Figure 8. A Microbot traversing the simulated terrain.

The bouncing and rolling are modeled as an impact contact model with friction. The model generates a variable force between the Microbot and the terrain in a direction that resists the relative motion of the two bodies. The impact force is modeled as a nonlinear spring/damper system:

$$F_{impact} = -k(\Delta x)^{2.2} - b(\Delta \dot{x}) \tag{2}$$

where k is the spring stiffness constant, b is the position dependant damping coefficient, and Δx and $\Delta \dot{x}$ are the relative displacement and velocity.

The friction force used in the simulations is standard Coulombic friction with a velocity dependant friction coefficient, $\mu(v)$. The parameters used in the simulations were estimated from laboratory experiments in which the behavior of a Microbot on compacted dry sand and rocks was observed. Table 2 summarizes these values.

Table 2: Values used in the impact contact model.

Parameter	Value
k	240,000 N/m
b (sand)	10 N-s/m
b (rock)	0.5 N-s/m
μ static	2
μ dynamic	0.15
Stiction Transition Velocity	0.01 m/s
Friction Transition Velocity	0.1 m/s

5.2. *Results and Discussion*

A large number of simulations were run. Microbot diameters of 5, 10, and 20 cm and hop heights of 50, 100, 150, and 200 cm were used. Each combination of hop height and size was simulated with a different starting positions spread over an approximately 2 meter area. The Microbot mass is fixed to 100 grams. In each simulation, the Microbot starts approximately 2 m from the rock pile and has 14 hops to overcome the obstacle.

Figure 9(a) shows the rate of successful trials as a function of Microbot hop height. Success is defined as completely overcoming the rock pile. Failed trials were caused by three failure modes: 1) *entrapment*, when a Microbot is trapped by a group of rocks and is unable to hop out, 2) *low hop height,* when a Microbot is caught because it is unable to hop over a rock, and 3) *bouncing away* is when the Microbot hops in such a way that it bounces away from the rock pile and must start again and can not complete the task in 14 hops. The consequence of this is not a failure per se but an undesirable delay. Figure 9(b) illustrates the most common failure modes for the Microbots as a function of hop height and Microbot diameter.

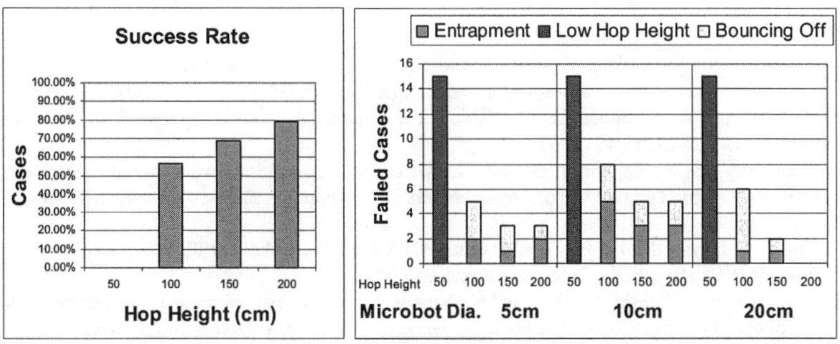

Figure 9. Simulation results: (a) The rate of successful attempts as a function of hop height, and (b), the failure modes as a function of Microbot diameter and hop height.

The results show that all trials with low hop height resulted in failure. This suggest that a hopping robot can overcome a complex obstacle only if the hop height is greater than a characteristic height of the features on which it climbs, in this case approximately 0.85 m. Here, a hop height of 1 m leads to some success. Hence, hop height should be maximized. However, increased hop height trades off with larger power consumption and mechanism weight.

The results also indicate that small Microbot size result in greater entrapment. The rock pile was randomly assembled and is not an exact model of a real pile of rubble found in disaster zones. However, it can be deduced that the maximum size Microbot should be select to minimize the chance of entrapment while still being able to fit inside the smallest openings it may need to pass through. The bouncing away failures seen in the simulations are not as much of a concern since they only retard the Microbot's progression. These failures could be improved or eliminated by effective path planning.

6. Conclusion

This paper analyzed the feasibility of the Microbot mobility system in rough terrain. An autonomous hopping DEA prototype has performed 38 cm hops in the lab. A fuel cell power system experiment and analysis indicates that a 100 grams Microbot could perform about 1000 hops. Simulations suggest that a 10 cm diameter Microbot performing hops of 1 m high could succeed in rough terrain typical of search and rescue sites. These results confirm that, with reasonable technology development, the Microbot system could become an effective tool for search and rescue missions.

References

1. A. Chamberlain, W. Sellers, C. Proctor, and R. Coard, "Cave Detection in Limestone using Ground Penetrating Radar," *Journal of Archaeological Science* **27**, 957-964 (2000).
2. W. Sellers, and A. Chamberlain, "Ultrasonic cave mapping," *Journal of the Cave Research Electronics Group* **28**, 18-19 (1997).
3. S. Dubowsky, JS. Plante, and P. Boston, "Low Cost Micro Exploration Robots for Search and Rescue in Rough Terrain", *IEEE International Workshop on Safety, Security and Rescue Robotics*, (2006).
4. P. Fiorini, S. Hayati, M. Heverly, and J. Gensler, "A Hopping Robot for Planetary Exploration," in *Proc. of IEEE Aerospace Conf.*, Snowmass, CO, 1999.
5. S. A. Stoeter, P. E. Rybski, M. Gini, and N. Papanikolopoulos, "Autonomous stair-hopping with scout robots," in *IEEE/RSJ International Conference on Intelligent Robots and Systems*, Lausanne, Switzerland, 2002, pp. 721-726.
6. G. J. Fischer and B. Spletzer, "Long range hopping mobility platform," in *SPIE Unmanned Ground Vehicle Technology Conference*, Orlando, FL, United States, 2003, pp. 83-92.
7. R. Kornbluh, R. Pelrine, Q. Pei, S. Oh, and J. Joseph, "Ultrahigh Strain Response of Field-Actuated Elastomeric Polymers," *Proc SPIE Smart Structures and Materials 2000 (EAPAD)* **3987**, 51-64 (2000).
8. R. Pelrine, R. Sommer-Larsen, R. Kornbluh, R. Heydt, G. Kofod, Q. Pei, and P. Gravesen, "Applications of Dielectric Elastomer Actuators," *Proc. SPIE Smart Structures and Materials 2001 (EAPAD)* **4329**, 335-349 (2001).
9. A. Wingert, M.D. Lichter, S. Dubowsky, and M. Hafez, "Hyper-Redundant Robot Manipulators Actuated by Optimized Binary Dielectric Polymers," *Proc. SPIE Smart Structures and Materials 2002 (EAPAD)* **4695**, 415-423 (2002).
10. JS. Plante, and S. Dubowsky, *Smart Materials and Structures* **16**, S227-S236, (2007).
11. JS. Plante, L. Devita, and S. Dubowsky, "A Road to Practical Dielectric Elastomer Actuators Based Robotics and Mechatronics: Discrete Actuation," *Proc SPIE Smart Structures and Materials 2007 (EAPAD)*, (2007).
12. S. Kesner, JS. Plante, P. Boston, T. Fabian, and S. Dubowsky, "Mobility and Power Feasibility of a Microbot Team System for Extraterrestrial Cave Exploration," *Proc. of IEEE Robotics and Automation Conf.*, Roma, Italy, 2007.

A PROPOSAL FOR BIPEDAL LOCOMOTION USING GYROSCOPIC EFFECT*

PULKIT KAPUR†

Department of Mechanical Engineering, Punjab Engineering College, Chandigarh, INDIA

RAHUL MUKHI

Department of Production Engineering Punjab Engineering College, Chandigarh, INDIA..

VINAYAK†

Project Assistant, Centre for Product Design and Development, Indian Institute of Science, Bangalore, INDIA.

In literature, there have been bipeds which involve the use of a large number of actuators and those which use gravitational force to propel i.e.; passive dynamic walkers. Study has also been carried out in context to connect the similarities between gaits of two-legged robots and their biological counterparts. The present work focuses on the idea of joining these two extremities with a novel approach to actuate two-legged systems. This paper proposes the idea on how to use gyroscopic effect, as means of propulsion, for bipedal locomotion. By modeling and simulating a two-legged system in ADAMS, the authors prove the feasibility of the proposal. The biped illustrated in the paper consists of a rotating flywheel which provides gyroscopic couple in order to move the system forward as the legs march in the frontal plane of the system. Preliminary results regarding the effect of the mass and speed of the gyroscope have been shown and the proposal is also supported by a mathematical analysis which provides a measure of the leg stroke in terms of the angle rotated by the system about the leg in support phase. The present work does not consider the start and the stop positions of the system but aims primarily at proving the concept of how gyroscopes can be used to propel a bipedal system.

1. Introduction:

Bipedal locomotion involved the movement of two legged walking systems. Due to anthropomorphic similarity of bipedal gait, and potential scope of development of gait assistive devices, there has been a lot of interest in this field.

---

* Corresponding Author:vinayak.pro@gmail.com

† Email:virtualguru@gmail.com

Research in Bipedal locomotion had previously focused on achieving the desired gait with over actuated systems, and advanced control algorithms,[3] etc. In 1990, Tad McGeer [1] showed that a simple planar mechanism with two legs could be made to walk stably down a slight slope with no other energy input or control, using the inherent dynamics of the system. This system acts like two coupled pendulums and given sufficient mass at the hip, the system will have a nominal trajectory that repeats itself. Ruina etal [6] made a 2d passive walker with knees, which could walk on level ground.

The aim of the present work is to propose a means to propel a two-legged (knee less) system on level ground using gyroscopic effect. Next, we verify whether the system that has been proposed is feasible at its most fundamental level. This is done by modeling the system in ADAMS [7] and simulating it under certain conditions and seeing whether the system works. In case the proposed system works, the third aim is to investigate certain important parameters with respect to other parameters in context to how they vary during the work period of the system.

2. Gyroscopic propelled locomotion Concept

The use of Gyroscopes in mechanical systems is not new. However, in past they have been used for providing stability, and steering capabilities. Gyrover [3] developed at Carnegie Mellon University, has an internal mechanical gyroscope that allows it to stabilize and maneuver over rough terrain, slopes etc. Also gyroscopes have been used for providing orientation estimate of a mobile robot with respect to his environment [2], for mobile hopping robots stabilized by rotating gyroscope [5]. The motion of biped is based on the mechanical principle of gyroscopic effect. A rotating flywheel, tends to presses, (rotate about an axis), at right angle to applied torque, according to fundamental equation of gyroscopic effect:

$$T = I_g \times \omega \times \omega_p \tag{1}$$

Where, T is the torque applied, I, is the moment of inertia of flywheel, and ω is its angular velocity, ω_p is the angular velocity of precession. We aim to utilize this gyroscopic couple for propelling a bipedal system (knee less) forward on level ground. This is achieved by mounting a rotating flywheel at the hip and actuating one of the legs vertically up. Then due to gyroscopic effect, the swing leg, presses about the stance leg axis, and the biped is propelled forward. A control algorithm is then developed, for controlling the on time of the gyroscope, for desired stride length.

3. Methodology:

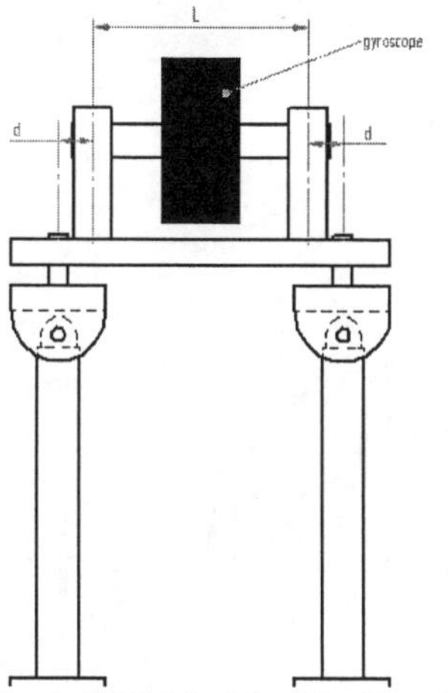

Fig.1: CAD Model of the Biped

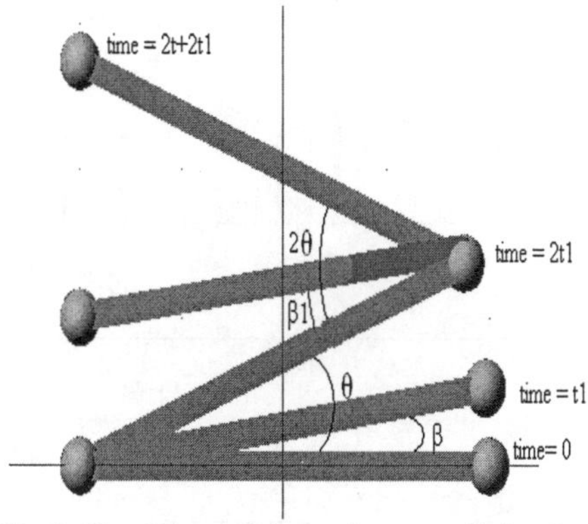

Fig.2: Top View of the trajectory of biped

3.1. *Assumptions:*

The model of the biped was constructed in ADAMS, to see if the conceptual model worked under certain conditions. The assumptions made in the model were that, the stance leg of the biped is considered fixed on the ground, for the purpose of simulation. Secondly, for our analysis we have considered the intermediate position of the gait, when it moves from an angle -θ to +θ as illustrated in Fig. 2. Also all the links and moving parts have been considered rigid in the model.

3.2. *The Control Strategy:*

The gyroscope cannot be stopped immediately, as the biped might overturn due to the inherent dynamics of the system. Hence a control strategy is needed to determine the on-time of the gyroscope. This strategy must follow a closed recursive loop, as all the energy must be returned back to the system. The following mathematical analysis provides a logical direction to our choice

The variables involved in the control algorithm are as follows:

1. I_g – Moment of inertia of gyroscope
2. I_b – Moment of inertia of body

3. ω_g – Angular velocity of gyroscope
4. ω_p – Angular velocity of precession
5. L – Distance between supports
6. d – Distance between leg axis and support

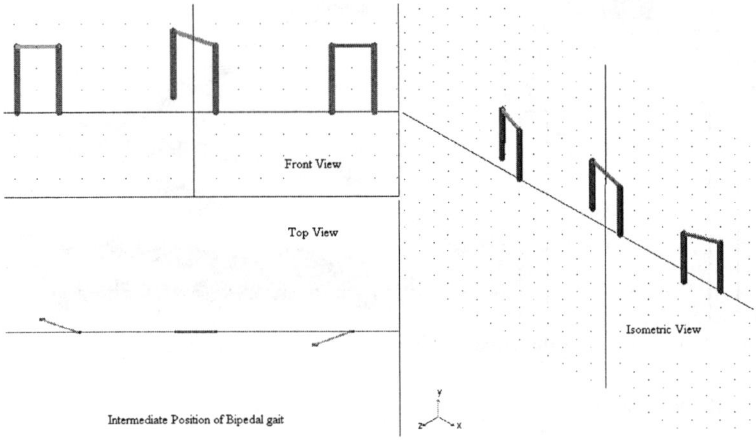

Fig. 3. Stick models showing the intermediate motion of Biped

- **When Biped starts from rest to move angle θ**

When leg is lifted up couple acts due to gyroscopic effect,

$$C = I_g \times \omega_g \times \omega_p \tag{2}$$

Force on supports, $F = \dfrac{I_g \times \omega_g \times \omega_p}{L}$ (3)

Torque on body about leg2 is given by

$$T_{body} = F \times \{(L+d) - d\}$$
$$\Rightarrow T_{body} = F \times L = I_g \times \omega_g \times \omega_p \tag{4}$$

Torque can also be represented as $I \times \alpha$, thus: $\alpha = \dfrac{I_g \times \omega_g \times \omega_p}{I_b}$ (5)

Now let body be turned about leg2 by an angular displacement β. The time taken for this amount of angular displacement is given by:

$$t_1 = \left(\frac{2\beta}{\alpha}\right)^{1/2} \tag{6}$$

And hence the angular velocity ω_{t1} attained after time 't_1' will be:

$$\omega_{t1} = \alpha \times t_1 = (2\alpha\beta)^{1/2} \tag{7}$$

The further angular displacement of leg1 is dependent on the time taken to fall on the ground. If assumed that leg1 will fall on the ground with same angular velocity ω_p, then the angular displacement is given by: =

$$\omega_{t1} \times t_1 = \alpha \times (t_1)^2 = 2\beta \tag{8}$$

The total angular displacement therefore is given by $\beta + 2\beta = 3\beta$ which further is equal to θ.

- **Intermediate Position from – θ to + θ**

For total angular displacement of 2θ, $3\beta_1 = 2\theta$

$$\Rightarrow \beta_1 = \frac{2}{3} \times \theta \tag{9}$$

$$\alpha = \frac{I_g \times \omega_g \times \omega_p}{I_b} \tag{10}$$

Time for this angular displacement, $\beta_1 = \frac{1}{2} \times \alpha \times t^2 \Rightarrow t = \left(2\frac{\beta_1}{\alpha}\right)^{\frac{1}{2}}$ (11)

$$\beta_1 = \frac{1}{2}\frac{I_g \times \omega_g \times \omega_p}{I_b} t^2 \tag{22}$$

For time t the motors should be actuated to lift the leg and same to drop the leg to the ground after gyro is turned off. Total time taken to move 2θ is 2t.

The preceding mathematical analysis provides a basic framework, which guides our logic towards the working of the system. Also, the gyroscope should not be suddenly stopped or started. This is because the biped might turn over, if the gyroscope is suddenly started or stopped, in order to conserve the angular momentum of the system. The rate of stopping the gyroscope is yet to be determined. Complete dynamic analysis and equations of motion need to be derived for that. This rate may also be calibrated from experimental study of a working model, in near future.

4. Results and Issues involved:

The model of the biped was simulated in ADAMS, and gyroscopic propulsion of the bipedal system was observed. The system was simulated for its motion under a set of parameters and the variation in certain significant variables was noted for the variation of the following parameters:

- Mass of Gyroscope
- Speed of gyroscope

Varying the above parameters, simulation results were obtained for following cases:

- Translational Velocity of center of bar
- Angle turned by the biped about the vertical axis

4.1. *Case1: Effect of the mass of Gyroscope:*

It is observed in Fig. 4 that the angle by which the biped turns is a function of the mass (inertia) of biped, as shown analytically (refer equation 12). As the mass is increased the corresponding values increase but not considerably. This shows that the mass of the gyroscope does not affect the angle significantly and also validates the parabolic nature of variation of angle with respect to time.

4.2. *Case2: Effect of the angular velocity of Gyroscope:*

Fig. 5 shows the variation of translational velocity of the centre of mass of the bar with time, with changes in angular velocity of gyroscope. It is observed that the points are more spaced out, and there is considerable effect of gyroscope speed on translational velocity.
Fig. 6 shows the variation of vertical rotation of left leg with time, upon variation in speed of gyroscope. The plots are linear in a large interval of time. This interpretation will be helpful in the design of the biped in terms of the flywheel speed.

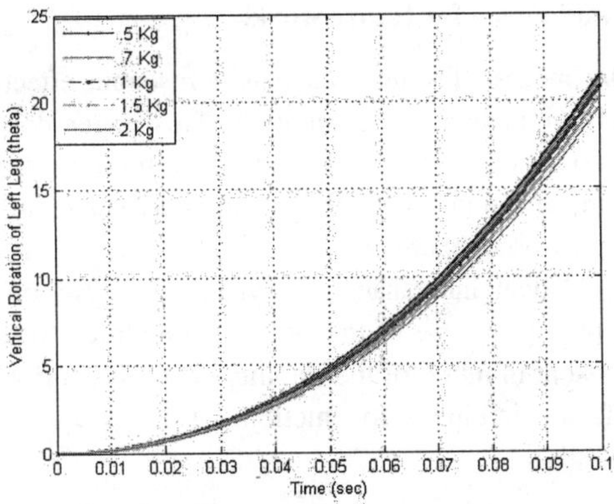

Fig.4 Variation of Vertical Rotation with time, for varying mass of Gyroscope

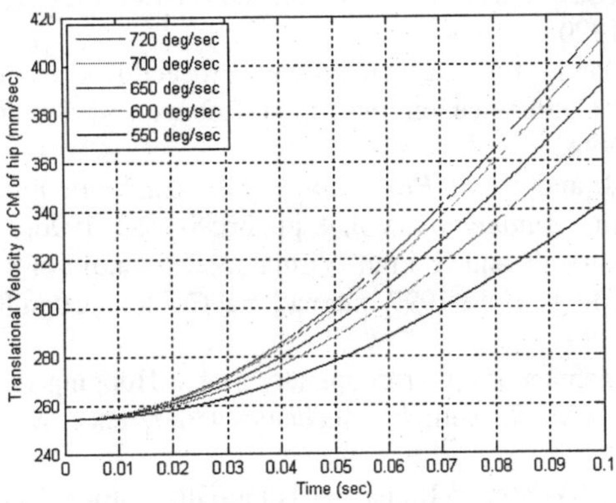

Fig. 5 Variation of Translational Velocity with time, for different speeds of Gyroscope

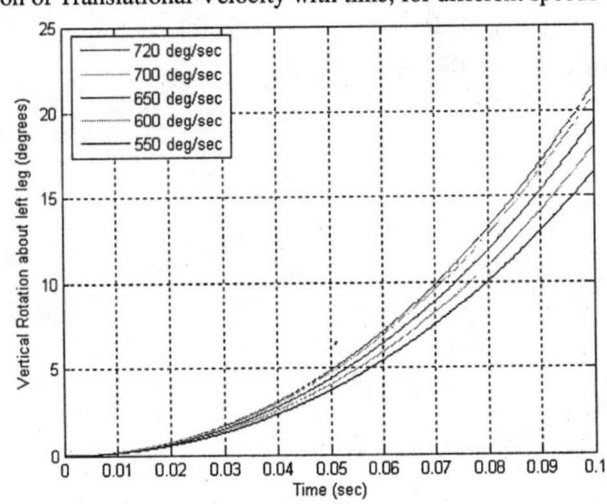

Fig.6 Variation of Vertical rotation about Left Leg with time, for different speeds of Gyroscope

5. Conclusion and Scope for future work:

It is observed by means of simulation that gyroscopic effect can be used as means of propulsion in bipedal locomotion. This paper does not attempt to provide a means to simulate anthropomorphic gait, however aim of this paper is to explore a phenomenon, and obtain an alternative means of bipedal locomotion, using gyroscopic effect.

Future work would deal into analytical dynamic analysis of the system, and experimentation on a prototype of the system. In addition to that, presently, no concern has been kept in mind for the stability of the systems which is definitely one of the main factors in bipedal locomotion.

References:

1. T. Mcgeer, IEEE International conference on Robotics and Automation, 3, 1640-1645, 1990.
2. B Barshan, and H.F. Durrant-Whyte, Proceedings of the IEEE/RSJ/GI International Conference on Intelligent Robots and Systems 1867 - 1874 vol.3, Sept 1994.
3. H.B Brown, Jr and Y. Xu, *Proceedings of the IEEE International Conference on Robotics and Automation*, Vol. 4, pp. 3658-3663, 1996.
4. J. Pratt, P. Dilworth, and G. Pratt. Virtual model control of a bipedal walking robot. Proceedings of the 1997 International Conference on Robotics and Automation, 1997.
5. M. Peck, Dynamics of a Gyroscopically Stable Hopping Robot, Proceedings of the 2001 AAS Spaceflight Mechanics Conference, Santa Barbara, CA; Feb. 2001.
6. S .H Collins, M.Wisse, A. Ruina, “A Three-Dimensional Passive-Dynamic Walking Robot with Two Legs and Knees” International Journal of Robotics Research, Vol. 20, No. 2, Pages 607-615, 2001.
7. MSC Software’s ADAMS View, 2005.

A Self-Adjusting Universal Joint Controller for Standing and Walking Legs

A. Schneider*, B. Fischer, H. Cruse and J. Schmitz

Dept. of Biological Cybernetics, University of Bielefeld,
33501 Bielefeld, P.O. Box 10 01 31, Germany
**E-mail: axel.schneider@uni-bielefeld.de*
www.uni-bielefeld.de/biologie/Kybernetik/

Recent biological studies on stick insects suggest that the character of the joint controllers (I-, P- or D-controller) depends on the compliance of the substrate (soft, intermediate or inelastic) the insect is standing on. To model these results, we propose a self-adjusting joint controller that changes its own setpoint in dependance of the substrate stiffness. The substrate stiffness is determined by means of a correlator circuit that compares injected movement commands with the actual responses of the leg joint. The negative feedback controller for standing is combined with an already published positive feedback loop for walking to form a universal controller which can be used for the control of legged robots.

Keywords: adaptive control, adaptive behavior, posture control, leg control

1. Introduction

Different biological studies provided evidence that in a walking stick insect a positive feedback effect (assistance reflex) on the single joint level is responsible for the generation of the stance movements and the coordination of all leg joints.[6] It could be shown that legs equipped with local positive velocity feedback joint controllers (LPVF controllers) are capable of walking without a central control instance.[8]

In contrast, during standing each joint angle is controlled by a negative feedback position controller (resistance reflex)[1,9] in order to maintain its posture. However, a recent study by Cruse and co-workers on the control of the knee joint of the standing stick insect changes the view on this topic drastically. It suggests that the behavior of the negative feedback joint controller depends on the compliance of the substrate the insect is standing on.[3] Figure 1(a) shows the setup of the underlying biological experiment.

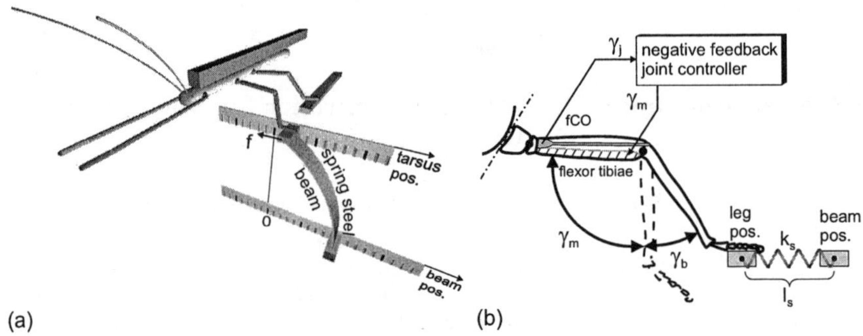

Fig. 1. (a) Design of the biological experiment. A stick insect is tethered to a balsa wood rod. The left middle leg stands on a small platform that is supported by a spring steel beam the base of which can be moved away or towards the body on a perpendicular path with respect to the body axis (lower scale). At the same time the leg position is monitored (upper scale). (b) Simplified model of the experiment.

A stick insect is tethered to a balsa wood mount. The left middle leg stands on a small platform which is mounted on a spring steel beam. The base of the beam can be moved sidewards either away or towards the insect body. Since the spring constant k_s of the spring steel is known, the force acting on the tarsus of the leg in horizontal direction can be calculated by measuring the tarsus position (p_{tarsus}) and the position (p_s) of the spring steel beam ($f = (p_s - p_{\text{tarsus}}) \cdot k_s$).

On highly elastic substrate the negative feedback control circuit behaves like an I-controller. Externally enforced deviations from the original position are compensated completely up to a spring steel stiffness of about $k_s = 0.05\,\text{N/m}$. On a substrate with intermediate elasticity, up to a spring steel stiffness of about $1\,\text{N/m}$, the animal compensates for the position only partially which resembles the behavior of a P-controller. On inelastic substrate the knee joint seems to be controlled by a D-controller. If the tarsus is forced away passively from the original position, the controller tries to compensate for this deviation. However, after a short period of time the controller seems to "give up" and the new position is "accepted". This behavior is biologically sensible as it prevents the insect from exhaustive torque generation for useless actions.

The present paper introduces a new self-regulating negative feedback joint controller that generates the described behavior of the standing animal. Moreover, an architecture for the combination of both positive feedback for walking and negative feedback for standing is proposed.

2. Simulation of the standing experiment

Figure 1(b) displays the simplified setup of the experiment shown in Fig. 1(a). The joint angle γ_j is measured (femoral chordotonal organ) and serves as the only input of the negative feedback joint controller. It is assumed that the negative feedback joint controller issues the motor command γ_m to the muscles (e.g. flexor tibiae) in order to achieve a certain angular position of the joint. If an external force acts on the tarsus, there is a difference between γ_j and γ_m which is the angle of bending γ_b ($\gamma_b = \gamma_j - \gamma_m$) due to the elasticity of the muscles and the tendons.

The angle of bending γ_b is also a measure for the amount of torque τ_γ that acts on the joint ($\tau_\gamma = \gamma_b \cdot k_\gamma$, with k_γ = spring constant of the elastic knee joint). The tarsus rests on top of the spring steel beam the base of which is moved horizontally in the experiment. The horizontal distance between the position of the spring steel beam and the position of the leg is multiplied with the spring constant k_s of the substrate in order to calculate the force that acts upon the leg. The experiment was simulated in Simulink 6.1 and SimMechanics 2.2.1 (The MathWorks Inc., Natick, MA, USA) using the kinematic data of an average stick insect as described by Ekeberg and co-workers.[4] The joint controller which was applied in the simulation is introduced in the following section.

3. A self-regulating negative feedback joint controller

In technical systems controllers are often designed as PID-controllers. Biological systems most often employ P- or PD-controllers.[2] From a technical point of view the control behavior of the knee joint controller in the standing stick insect is an apparent exception because it changes the controller characteristic and the reference value of the controller according to the elasticity of the substrate which is an environmental parameter. An apparent design approach towards a controller that regulates the knee joint as described above would be a PID-controller the parameters of which are changed at runtime according to the sensed substrate elasticity.[3] However, such an approach would ignore an important aspect of the biological results. If the base of the spring steel beam is moved and the spring steel is very inelastic, the leg is forced into the new position. The counter force of the leg decreases after some time (D-controller). But additionally, the joint controller seems to accept the new position as reference value because after a while a newly induced deviation from the latter position invokes the same actions of the controller as that from the first position.

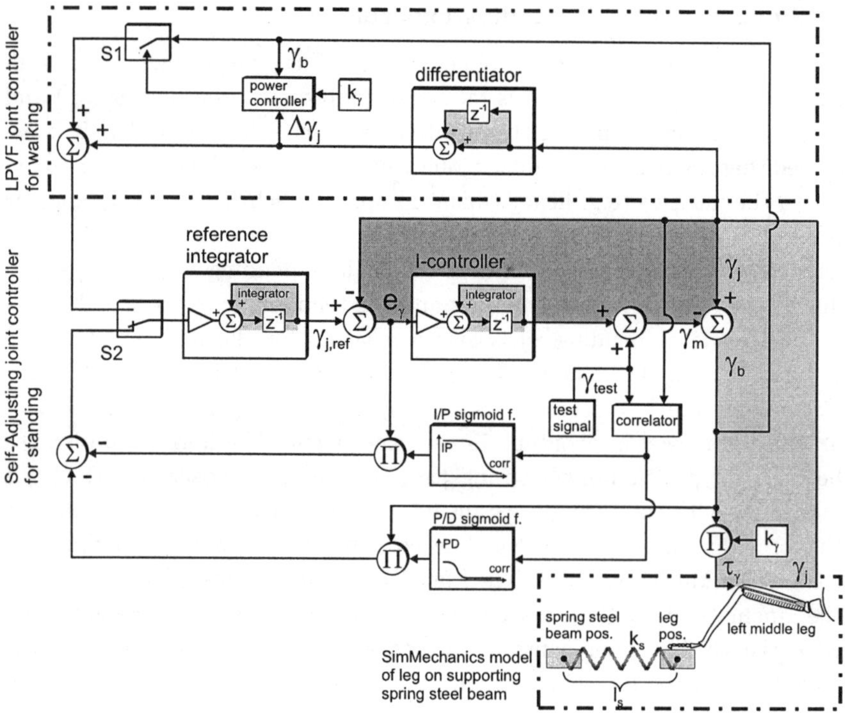

Fig. 2. Combination of the joint controller for standing (bottom) and an LPVF controller for walking (top, chain-dotted rectangle) as derived in Schneider et al.[8]

The controller introduced in this paper takes all the requirements mentioned so far into account. The block diagram is shown in Fig. 2. The core of the overall control circuit is an I-controller (center of Fig. 2) which is part of the negative feedback position control circuit that is indicated by the area shaded in dark gray. The I-controller provides the motor command γ_m and receives the difference between the actual joint angle γ_j and the reference value $\gamma_{j,\text{ref}}$ as input.

The area shaded in light gray on the right side of Fig. 2 represents the joint bending calculation, the resulting torque and the measurement of the joint angle. The joint torque τ_γ actuates the knee joint in the SimMechanics simulation of the leg. At the same time the simulation provides the actual joint angle γ_j. Taken together, the two areas shaded in dark and in light gray generate a negative feedback position control behavior of the knee joint that compensates completely for an externally enforced deviation from the desired reference position. The basic function of the circuit therefore is that of an I-controller working in the joint angle domain.

The reference value that is passed to the I-controller is produced by the reference integrator. The reference integrator acts as a memory for the current reference value $\gamma_{j,\mathrm{ref}}$ of the knee joint angle. Any input value of the reference integrator that differs from zero shifts the reference value.

In order to sense the elasticity of the substrate, a test signal γ_{test} is generated (sinus function, f=70 Hz, amp=0.1 °) and added to the motor command output of the I-controller. This test signal leads to oscillations in the joint driving torque τ_γ and thus also to periodical fluctuations in the angular position γ_j of the knee joint. However, the amplitude and the phase of the superimposed oscillations in the knee joint angle depend on the elasticity k_s of the substrate. An infinitely stiff substrate would cancel out the oscillations in the joint angle completely whereas a very soft ground would allow the joint to follow the superimposed oscillations.

The correlator in Fig. 2 receives both the test signal γ_{test} and the angular response γ_j of the knee joint. It is inspired by the Hassenstein-Reichardt correlator.[5] The output of the correlator is calibrated to deliver values between 0 ($k_s = \infty$, inelastic substrate) and 1 ($k_s = 0$, elastic substrate). In principle, the measure for the correlation is obtained by multiplication of a delayed version of the test signal γ_{test} (low-pass filter with the same time constant as the leg joint) and the actual joint movement γ_j. The absolute value of this product is filtered and represents the desired correlation value.

In order to bring the control circuit from the I-control to the P-control mode, the upper of the two sigmoid functions (I/P sigmoid function) is used. If the correlator output is 1, representing a soft substrate, the output of the sigmoid function is zero. If the correlation decreases (stiffer substrate), the output of the I/P sigmoid function increases. In this case, the product of the angular position error e_γ (input of the I-controller) and the output of the I/P sigmoid function is not zero anymore. The product is fed as input into the reference integrator to move the reference value towards the actual joint angle. In this mode, the I-controller and the reference integrator compete to make the joint angle error e_γ zero: The former by moving the actual joint and the latter by changing the reference value. As a result at a given medium elasticity of the substrate, the joint partially follows an external deflection. The behavior resembles that of a P-controller.

To shift the behavior further from the P-controller mode to the D-controller mode, the lower of the two sigmoid functions (P/D sigmoid function) is used. If the correlation is small indicating a stiff substrate, the product of the bending angle γ_b and the output of the P/D sigmoid func-

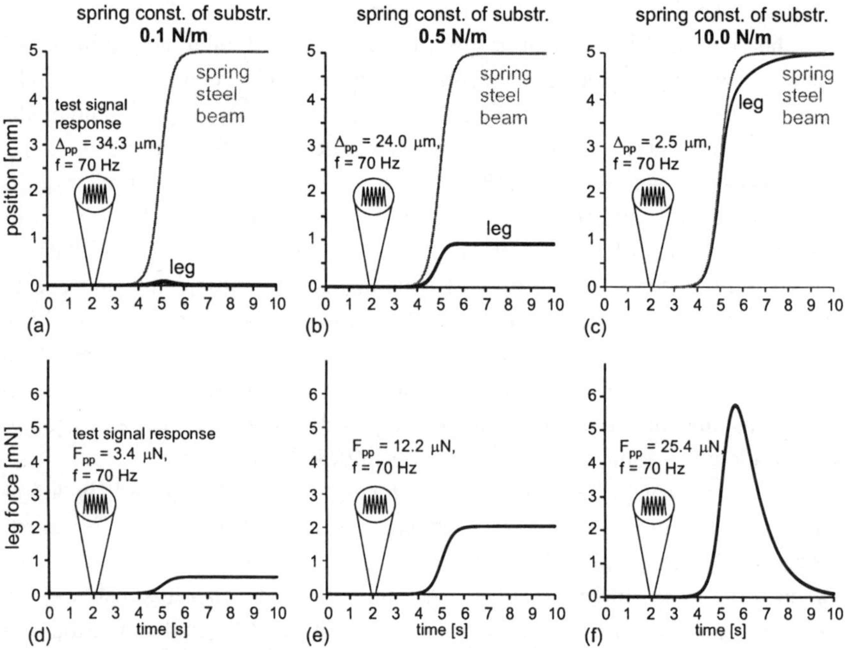

Fig. 3. (a)-(c) Position of the Spring steel beam and leg position for external deflections on three different substrates. (d)-(f) Leg forces for the same external deflections as in (a)-(c).

tion increases. Since this product is also fed as input into the reference integrator, it also shifts the reference value towards the actual joint angle. This is done as long as a joint bending (torque) is present. As a result, the controller "gives up" and does not produce any force after some time. This completes the description of the adaptive joint controller for standing. The top part of Fig. 2 is used for the control of the stance during walking (see Sect. 5).

4. Results

The controller introduced in the last section was tested on a simulation of an insect leg as described in Sect. 2 and Fig. 1(b).

Figures 3(a)-(c) depict the leg position and the position of the base of the spring steel beam during a deflection plotted over time for three different substrate elasticities, exemplarily. Figure 3(a) displays a plot of the position of the spring steel beam (gray) and the leg position (black) over time for a soft substrate with a spring constant of 0.1 N/m. The deflection function for the beam position has an amplitude of 5 mm. It can be seen that the leg has the tendency to follow the beam position at the beginning of the deflection. However, the I-controller restores the original position completely already during the course of the deflection. The I-controller domain ranges from substrate elasticities of 10^{-2} N/m to 0.2 N/m. The inset indicates the oscillations which were superimposed on the leg position by the test signal γ_{test}. The peak-to-peak amplitude of this oscillation is 34.3 μm.

Figure 3(b) shows the beam and leg position over time for a substrate with a spring constant of 0.5 N/m (intermediate elasticity) and the same deflection function for the beam position as in Fig. 3(a). The joint controller is in the P-control domain which leads to a passive movement of the leg of about 1 mm. The P-controller domain ranges from substrate elasticities of 0.2 N/m to 5 N/m. The oscillations superimposed on the leg position have a smaller peak-to-peak amplitude (24.0 μm) for the given substrate elasticity than for the softer substrate in Fig. 3(a).

Figure 3(c) presents the same plots as in Fig. 3(a) and (b) for a rather inelastic substrate (10.0 N/m). The joint controller is in the D-control mode. The leg follows the imposed deflection completely. The D-controller domain ranges from 5 N/m to 100 N/m. The superimposed oscillations are smaller than in Fig. 3(b). They have a peak-to-peak amplitude of 2.5 μm.

Figures 3(d), (e) and (f) show graphs of the leg forces during the simulated experiments plotted over time for the I-, P- and D-mode of the controller, respectively. The substrate elasticities are the same as in Fig. 3(a), (b) and (c). Note that in Fig. 3(f), which represents a simulation run in the D-domain, the controller generates a force at the onset of the deflection with a force maximum at about the end of the deflection ramp at 6 s. However, this force decreases after a few seconds due to the effect of the P/D sigmoid function. The insets in Fig. 3(d), (e) and (f) show the superimposed force oscillations which are caused by the test signal. Note that opposed to the superimposed leg oscillations in Fig. 3(a), (b) and (c) the force oscillation increase with rising substrate stiffness. This is due to the fact that the test signal generates oscillating motor commands that produce higher forces when the leg's freedom to move is decreased by a stiffer substrate.

Figure 4 depicts the leg position plotted over the leg force. For a substrate elasticity of 0.1 N/m the leg force increase up to 1 mN without changing the leg position (I-control). In the P-control mode (substrate elasticities of 0.2 N/m to $\sim$ 5 N/m) the simulation shows an almost linear relationship

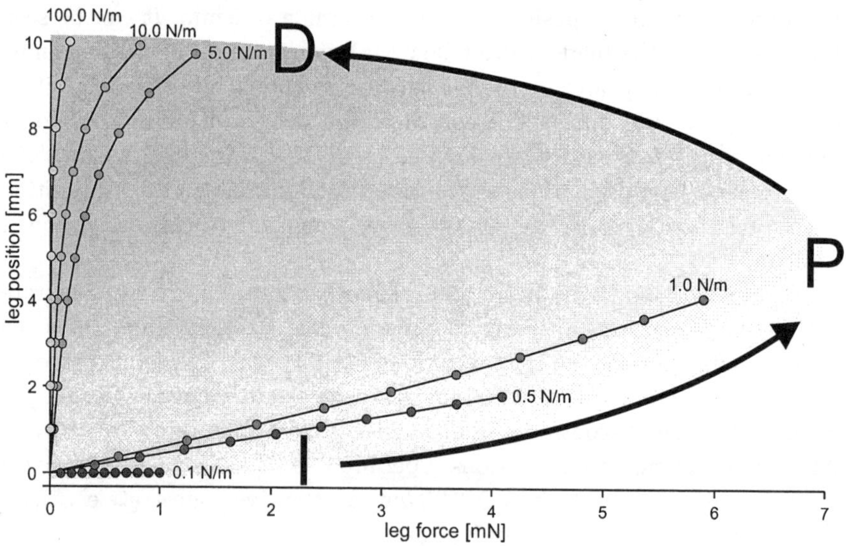

Fig. 4. Leg position plotted over leg force. Values are taken from the end of simulation cycles when the system has reached a steady-state.

between force and position which reflects the spring like behavior of a P-controller. The maximum force is achieved for a substrate elasticity of about 1 N/m. In the D-domain (substrate elasticities of $\sim 5\,\mathrm{N/m}$ to $100\,\mathrm{N/m}$ and more) the static forces decrease again until the leg follows the deflection of the base of the beam completely.

5. Combined controller

In previous publications, the control principle of **L**ocal **P**ositive **V**elocity **F**eedback (LPVF) and its derivatives have already been introduced.[7,8] The LPVF controller is used to generate and maintain movements in closed kinematic chains without a central controller. This is the case for a leg in stance during walking. This section presents a combination of both the LPVF controller for walking and the self-regulating negative feedback joint controller for standing. The block diagram of the combined controller is shown in Fig. 2.

The component that is responsible for the generation of the stance movement during walking (LPVF) can be found in the upper part of the figure (enclosed in the chain-dotted rectangle). The component of the controller that is active during standing is shown in the lower part (cf. Sect. 3). The combined controller uses the reference integrator and the I-controller (center part of Fig. 2) for both modes, i.e. standing and the generation of stance movements. The toggle switch $S2$ selects between the positive velocity feedback branch (top) and the negative position feedback branch (bottom). The two behaviors depend on the activation state of the animal and might be chosen by a central instance.

If the LPVF branch is selected to perform a stance movement, the actual joint velocity $\Delta\gamma_j$ is calculated by the differentiator (top middle) from a series of joint angle inputs γ_j. The angular velocity $\Delta\gamma_j$ is fed into the input of the reference integrator via the summation point (top left) and the switch $S2$ forming the actual positive feedback loop. Whenever a joint movement is initiated, the positive feedback loop via the differentiator maintains and continues this movement actively. However, if not coordinated, the resulting joint movement during a stance phase leads to tensions in all joints of the legs on ground which form so called closed kinematic chains. In classical, central controllers these tensions are avoided by issuing only those motor commands that lead to coordinated movements of all joints. In the LPVF approach the minimization of joint tension (and hence coordination) is achieved by an active relaxation strategy on the single joint level. The power controller (top left of Fig. 2) calculates the actual joint power which is the product of the angular joint velocity and the joint torque ($P_\gamma = \Delta\gamma_j \cdot \tau_\gamma = \Delta\gamma_j \cdot \gamma_b \cdot k_\gamma$). The switch $S1$ is toggled by the power controller depending on the degree of positive or negative mechanical power generated in the joint. If the mechanical joint power is negative, the joint torque and its corresponding joint movement point into different directions. The joint is in coasting mode which indicates unnecessary tension. Therefore, the tension is relaxed by closing the switch $S1$ which leads to a shift of the reference integrator input by the actual amount of bending (bending angle γ_b). This can be regarded as an active relaxation of the joint (active compliance). However, if the mechanical joint power is positive, the joint is in traction mode which leads to a meaningful movement of the entire walking agent. In this case, the switch $S1$ is left open and the ongoing joint movement is not changed. Instead of using a discrete switch $S1$, the active relaxation can also be achieved by a continuous mechanism that fades in the bending signal according to the amount of (negative) mechanical power that is generated in the joint.

6. Discussion

The combination of the self-regulating negative feedback joint controller and the LPVF controller results in a local controller that regulates a joint in a leg during standing and during a stance movement. The negative feedback joint controller is active during standing and is responsible for the maintenance of a certain angular position in order to keep a desired posture against external disturbances but not at any (torque-) costs.[3] Instead, the leg joint maintains a joint angle when standing on soft substrate and ceases if the leg is pulled away on a rigid substrate. The former might be the case if a leg is placed on a moving leaf, the latter if the leg is placed on a rigid branch moved by the wind. The new self-regulating negative feedback joint controller can explain the different behaviors.

In combination with the LPVF controller for walking,[8] a universal, self-adjusting joint controller for standing and walking can be built. Besides the agent's internal impulse to change between standing and walking, toggling between the two controller types might also depend on the correlation measure or on the angular acceleration for example to avoid slipping.

References

1. U. Bässler. *Neural Basis of Elementary Behavior in Stick Insects*, volume 10 of *Studies of Brain Function*. Springer, Berlin, London, New York, 1983.
2. H. Cruse. *Neural Networks as Cybernetic Systems*. Thieme, Stuttgart/ New York, 1996.
3. H. Cruse, S. Kühn, S. Park, and J. Schmitz. Adaptive control for insect leg position: Controller properties depend on substrate compliance. *J. Comp. Physiol.*, 190:983–991, 2004.
4. Ö. Ekeberg, M. Blümel, and A. Büschges. Dynamic simulation of insect walking. *Arthropod Struct. Devel.*, 33(3):287–300, 2004.
5. B. Hassenstein and W. Reichardt. Systemtheoretische Analyse der Zeit-, Reihenfolgen- und Vorzeichenauswertung bei der Bewegungsperzeption des Rüsselkäfers *Clorophanus. Z. Naturforsch.*, B11:513–524, 1956.
6. J. Schmitz, C. Bartling, D.E. Brunn, H. Cruse, J. Dean, T. Kindermann, M. Schumm, and H. Wagner. Adaptive properties of "hard-wired" neuronal systems. *Verh. Dtsch. Zool. Ges.*, 88(2):165–179, 1995.
7. A. Schneider, H. Cruse, and J. Schmitz. A biologically inspired active compliant joint using Local Positive Velocity Feedback (LPVF). *IEEE Transactions on Systems, Man, and Cybernetics - Part B: Cybernetics*, 35(6):1120–1130, 2005.
8. A. Schneider, H. Cruse, and J. Schmitz. Decentralized control of elastic limbs in closed kinematic chains. *The International Journal of Robotics Research, Special Issue on CLAWAR 2004*, 25(9):913–930, 2006.
9. G. Wendler. Laufen und Stehen der Stabheuschrecke *Carausius morosus*: Sinnesborstenfelder in den Beingelenken als Glieder von Regelkreisen. *Z. Vergl. Physiol.*, 48:198–250, 1964.

A STEP TOWARDS PNEUMATICALLY ACTUATED BIPED LOCOMOTION : A BIO INSPIRED PLATFORM FOR STIFFNESS CONTROL

G.MUSCATO, G. SPAMPINATO

Dipartimento di Ingegneria Elettrica Elettronica e dei Sistemi (D.I.E.E.S.)
Università degli studi di Catania, Viale Andrea Doria 6, 95125 Catania, ITALY
gmuscato@diees.unict.it , gspampi@diees.unict.it

This paper presents some control methodologies in the field of biped locomotion. Based on both two and three dimensional simulations, the control strategy has been implemented upon a ten degrees of freedom biped robot pneumatically actuated. The locomotion strategy has been developed using a simple but efficient force and position control algorithm, based on a different interpretation of the Virtual Model Control approach. Some energy considerations has been also carried out in order to better tune the gait parameters.

1. Introduction

The study of the walking strategy in biped robots gives more accurate insight into the strategy actuated by human beings in keeping balance during gait. Up to now only a few humanoid robots have been completely realized like the famous Honda humanoid robots [1] or the Sony Robot Qrio [2]. However, in order to investigate upon the elastic properties of human legs, some other pneumatic biped robots have also been realized, such as the Bipedal Robot LUCY [3], or the D.I.E.E.S. biped robot shown in Fig.1, which has been entirely realized in the laboratory of the University of Catania. In the present work pneumatic actuators have been adopted as an alternative solution to the most widely used electric actuators. Due to their low weight-force ratio and their high elasticity, pneumatic actuators allow the controller to adapt the stiffness of the leg during the stance phase. The D.I.E.E.S. biped robot is a 10 Dof anthropomorphic biped pneumatically actuated. The robot is 110 cm tall, and weights 24 kg (including 8 Kg payload). The design of the shape and the dimensions of the single parts are human inspired, so the dimensions of the links and the articulation movements, reproduce the corresponding biological ones [4]. Two pneumatic cylinders actuate the ankle and the hip articulations, while the knee articulation is actuated by only one. The ankle and the hip articulations are made up by a cardan joint that provides both roll and pitch degrees of freedom. Moreover, the two degrees of freedom in each articulation are strictly coupled with the two pneumatic actuators motions. In other words, a single movement of a single degree of freedom involves the extensions of both the pneumatic cylinders and vice versa. Different kind of algorithms have been implemented in order to solve the inverse kinematical problem. Some static maps, using the neural network approach, and dynamic algorithms have been proposed and widely compared in [5].

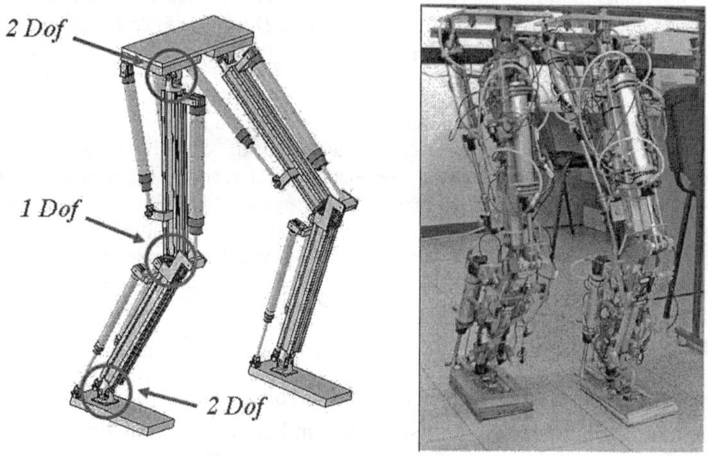

Figure 1. The DIEES Biped Robot .

2. Locomotion strategy and simulations

The locomotion strategy presented makes use of both force and position control approaches in order to actuate the stance leg and the swing leg to support the robot weight and to place the swing foot in the ahead position planned on-line by the trajectory generation algorithm. In order to generate the stance leg joint torques either in the Single Support Phase (SSP) and in the Double Support Phase (DSP), a set of virtual forces have been applied to the hip according to a different interpretation of the Virtual Model Control concept [6],[7]. In particular, a virtual spring damper couple have been applied between the hip joint and the stance foot in contact with the ground, applied to the desired COP position, as shown in Fig.2. In this way, each leg is considered as a serial manipulator, and the forces exerted by the foot, are directed in line with the robot center of mass (approximately located in the robot pelvis), preventing the body to rock and minimizing the cost of transport [8]. This approach represents a different interpretation of the VMC concept described in [6]. The implementation is quite different because no underactuated joint is required to model the foot-ground contact and no foot ground inclination measurements are needed.

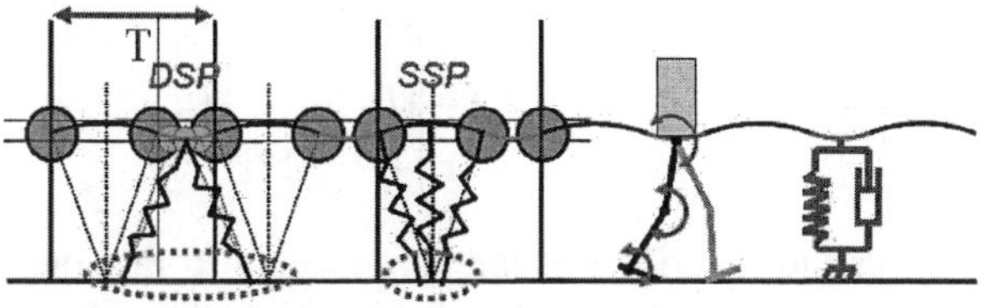

Figure 2. The two dimensional walking pattern based on the virtual spring damper components applied on each leg.

Besides, some analytical conditions have to be formalized in order to simplify the joints torques generation corresponding to the virtual forces exerted. In other words, taking into account the complete 3x3 leg jacobian matrix J_{Leg}, some conditions have been demonstrated in order to guarantee the force exerted connecting the hip joint to the desired COP position as shown in Fig.3.

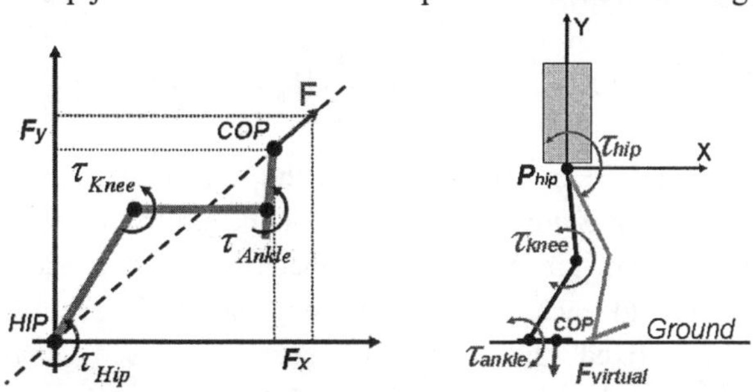

Figure 3. Two dimensional external force exerted by the stance leg to the desired COP point.

The necessary and sufficient condition (*CNS*) shown by (1) has been demonstrated, in which the terms $J_c^{1,x}$ represents the 2x2 jacobian sub-matrix containing the first and the *x-th* columns of J_{Leg}.

$$\left.\begin{array}{l}\dfrac{F_Y}{F_X}=-\dfrac{J_{11}}{J_{21}}\\ M_Z=0\end{array}\right\} \overset{CNS}{\Longleftrightarrow} \left\{\begin{array}{l}\dfrac{\tau_{Knee}}{\tau_{Ankle}}=-\dfrac{\det J_C^{1,2}}{\det J_C^{1,3}}\\ \tau_{Hip}=0\end{array}\right. \tag{1}$$

Moreover, in order to simplify the application of the condition (1), the force intensity $|F_{Ext}|$ has been directly considered. So, the torques needed to guarantee the alignment condition and satisfying the *CNS* (1) can be calculated through the relations (2).

$$\begin{cases} \tau_{Knee} = \dfrac{\det J_c^{1,2}}{\sqrt{J_{21}^2 + J_{11}^2}} \cdot |F_{Ext}| \\ \tau_{Ankle} = \dfrac{\det J_c^{1,3}}{\sqrt{J_{21}^2 + J_{11}^2}} \cdot |F_{Ext}| \end{cases} \qquad J = \begin{bmatrix} J_{11} & J_{12} & J_{13} \\ J_{21} & J_{22} & J_{23} \\ 1 & 1 & 1 \end{bmatrix} \tag{2}$$

The paper aims also to describe a theoretical method to find out some virtual components parameters according to the energy constraints and natural walking pattern. The total mechanical energy of the system can be computed as the sum of the kinetic and the potential energy terms, where the last one includes both the gravitational and the elastic energy stored into the virtual springs. For simplicity it is assumed to investigate the energy expenditure of equivalent virtual components acting in the vertical direction in order to support the robot weight during the stance phase. So, the mechanical energy stored in the system can be expressed as in (3).

$$E = \frac{1}{2} m\dot{y}_{hip}^2 + U(y_{hip}), \qquad U(y_{hip}) = \frac{K}{2}(h_0 - y_{hip})^2 + mgy_{hip} \tag{3}$$

The main constraint taken into account is the maximum energy that can be supplied to the system *Emax*. The desired *CoG* height from the ground and the elastic spring constant are also key parameters to be considered. So, three parameters related to each other by the expression (3) regulates the robot dynamics. Once fixed the maximum energy allowed *Emax*, and the uncompressed spring length h_0, the system potential energy can be calculated and represented with respect to the hip y-coordinate, as shown in Fig.4.

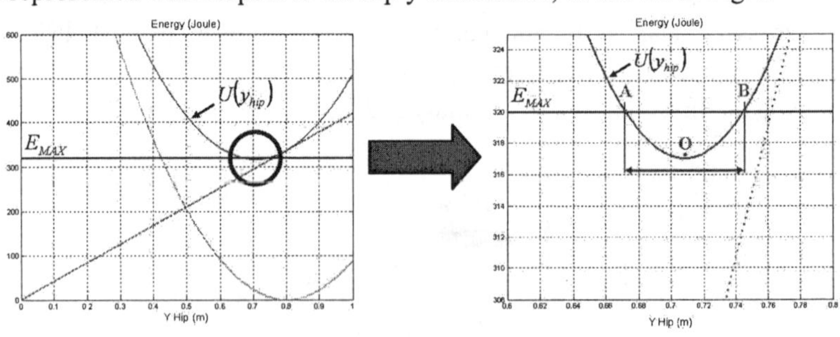

Figure 4. Gravitational and elastic potential energy with respect to the hip y-coordinate.

In the left part of the Fig.4, both elastic and gravitational potential terms are represented together with their sum $U(Y_{Hip})$. A zoom is reported in the right side. The motion of the robot hip has to remain within the segment *AB*. Indeed, outside this segment, the motion is impossible because of the potential energy exceeding the maximum energy allowed. In order to chose a proper spring elastic constant, the constraint (4) has to be satisfied.

$$\min U(y_{hip}) = U_{MIN} \leq E_{MAX} \tag{4}$$

Deriving the $U(Y_{Hip})$ expression shown by (3) with respect to Y_{hip} , the minimal potential energy point indicated by U_{min} can be found.

$$y_{\min} = h_0 - \frac{mg}{K} \qquad\qquad U_{MIN} = -\frac{(mg)^2}{2K} + mgh_0 \tag{5}$$

Finally, substituting the equation (5) into (4) and solving for *K* the desired virtual spring constraint has been found.

$$K \leq \frac{(mg)^2}{2(mgh_0 - E_{MAX})} \tag{6}$$

The energy supplied to the system E_{MAX} has been fixed to 320 J, and the uncompressed spring length h_0 to 0.8 m. The *K* parameter has also been chosen in order to maintain the energy requirements and to avoid too wide oscillations around Y_{min}. In other words, the segment *AB* length corresponding to decreasing values of *K* increases, and the steady state hip height Y_{min} decreases as shown in the left part of the Fig.5.

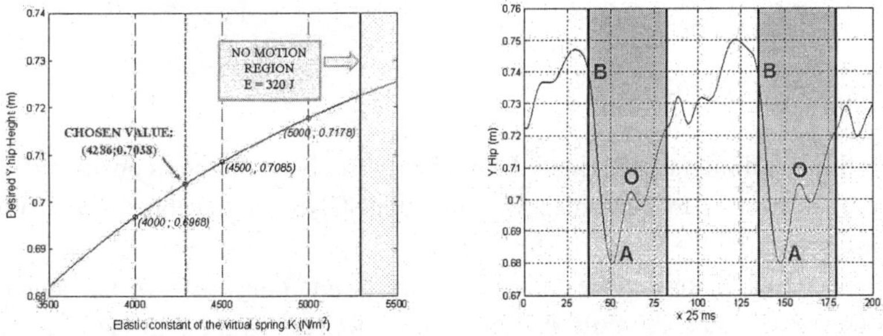

Figure 5. Minimal potential energy hip position related to the chosen value of the virtual spring constant, and related oscillations during the simulation.

Some simulations have been performed according to the virtual component criteria, where the minimal potential energy hip position has been chosen to be about 70 cm high from the ground. In the meanwhile, the swing leg, when not directly involved in balancing the robot during the Double Support Phase, is controlled in position, in order to perform parabolic trajectories, like described more in depth in [7]. Some simulation results are reported in Fig. 6.

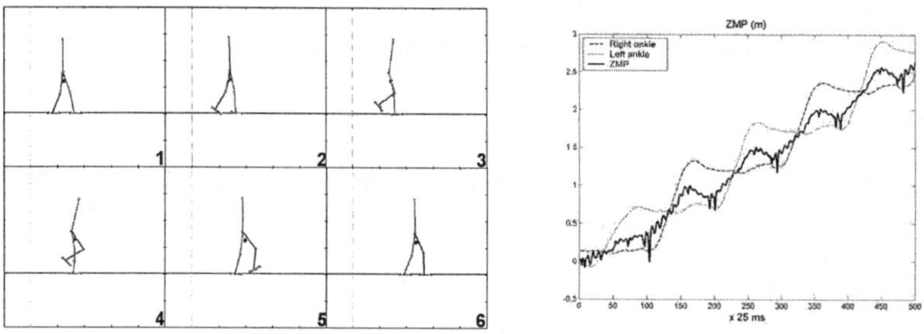

Figure 6. Two dimensional simulation results, ZMP, left and right feet positions during the gait.

After some rearrangements of the relations (1) to the three dimensional case, some equivalent equations have been found like shown by the condition (7). In other words, in order to simplify the generation of the virtual forces under the alignment condition constraint, the Jacobian matrix J_{5Dof} related to each leg is calculated during the gait, according to the robot posture and the desired *COP* position.

$$\overbrace{F_X, F_Y, F_Z}^{F_{Ext}} \quad M_X = M_Y = M_Z = 0 \quad \overset{CNS}{\Longleftrightarrow} \quad \begin{cases} \dfrac{\tau_K}{\tau_{A\_roll}} = \dfrac{\det J_C^{123}}{\det J_C^{124}} \\ \dfrac{\tau_K}{\tau_{A\_pitch}} = \dfrac{\det J_C^{123}}{\det J_C^{125}} \\ \dfrac{\tau_{A\_roll}}{\tau_{A\_pitch}} = \dfrac{\det J_C^{124}}{\det J_C^{125}} \end{cases} \quad \tau_{H\_roll} = \tau_{H\_pitch} = 0 \tag{7}$$

Again the magnitude of the exerted force $|F_{Ext}|$ has been directly considered, and the torques needed to guarantee an equivalent CNS for the three dimensional case are shown by the relations (8).
The $J_R^{\ xy}$ terms represent the 2x2 sub-matrix containing the *x-th* and the *y-th* rows of the J_{Leg} matrix, while the terms $J_c^{1,2,x}$ represents the 3x3 jacobian sub-matrix containing the first, the second and the *x-th* columns of J_{Leg}. Posing $\tau_{H\_roll}=\tau_{H\_pitch}=0$ and the other joints torques according to (8), the alignment condition shown in Fig.7 for the virtual stance forces is automatically guaranteed.

$$\begin{cases} \tau_K = \dfrac{\det J_C^{\ 123}}{D} \left| F_{Ext} \right| \\ \tau_{A\_roll} = \dfrac{\det J_C^{\ 124}}{D} \left| F_{Ext} \right| \\ \tau_{A\_pitch} = \dfrac{\det J_C^{\ 125}}{D} \left| F_{Ext} \right| \end{cases} \quad D = \sqrt{\left(\det J_R^{\ 12}\right)^2 + \left(\det J_R^{\ 13}\right)^2 + \left(\det J_R^{\ 23}\right)^2} \tag{8}$$

Also for the three dimensional case, some simulation have been carried out in order to validate the locomotion strategy based on the virtual components. Like for the two dimensional case, the swing foot is also driven directly into the three dimensional space through parabolic trajectories. A 60 cm step length locomotion pattern is reported in Fig.7.

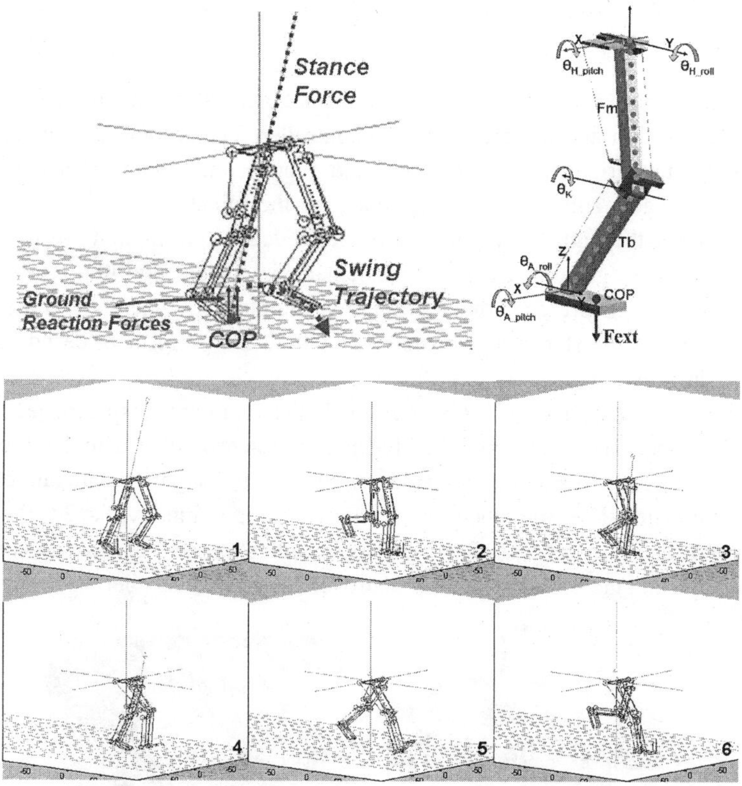

Figure 7. Three dimensional simulations based on the virtual force generation for the stance leg and parabolic trajectories for the swing leg.

3. Control architecture

As discussed in the previous section, the stance leg is actuated in order to exert the needed force to support the robot directly to the COP point, whereas a three dimensional trajectory generation has been designed for the swing leg in order to drive the swing foot directly into the operative space. So, the control architecture makes use of a three level feedback loop controller shown in Fig.8

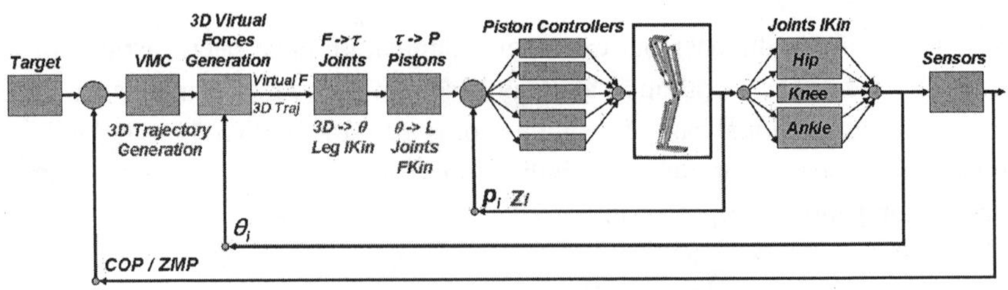

Figure 8. The three level control architecture.

The inner control loop operates on the pneumatic actuators directly performing both the position and the pressure loop controller. The second level controller works on the articulation kinematics and kinetics, in order to provide the actuators lengths and forces corresponding to the joints positions and torques generated by the high level controller. The feedback information are the joint working angles, generated from the kinematic inversion algorithm. The third level controller represents the highest level controller, where the trajectory generation and the stance leg forces computation are performed, based on the sensors information.

In order to measure the COP position inside the foot contact surface, each leg have been equipped with an inductive pressure sensor placed under the feet soles. The sensor structure is made up of two main parts, with an elastic dielectric material between them [4], as shown in Fig.9. Finally, the 2.0 B active version of the CAN protocol has been implemented at 1 Mb in order to connect the sensors and the actuators control boards to the main controller.

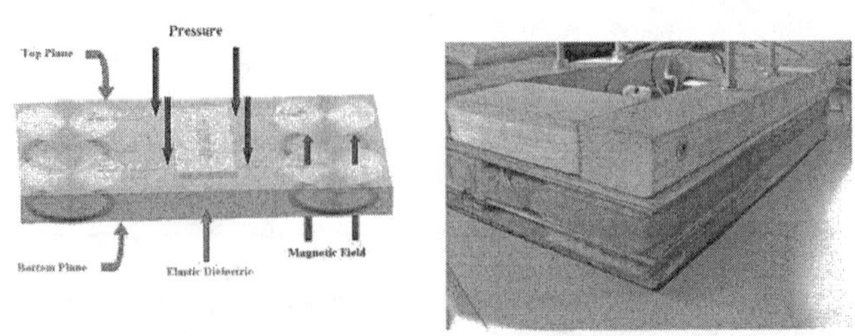

Figure 9. The foot pressure sensor structure.

4. Experimental results and conclusions

Some experimental walking tests have been performed on the real robot in order to validate the walking strategy. So, a 30 cm walking pattern locomotion test is shown in Fig.10. Apart from the virtual component parameters estimated through energetic cost considerations, the tuning process for other walking parameters like the lateral displacement or the step length are still under investigation in order to guarantee the dynamic stability conditions for robust and autonomous walking patterns.

Figure 10. Two 30 cm step length locomotion tests performed on the real robot.

References

1. J. Chestnutt, M. Lau, G. Cheung, J. Kuffner, J. Hodgins, and T. Kanade, "Footstep planning for the honda asimo humanoid," in Proceedings of the 2005 IEEE International Conference on Robotics and Automation Barcelona, Spain, 2005.
2. K.Nagasaka, Y.Kuroki, S.Suzuki, Y.Itoh,, J.Yamaguchi. "Integrated Motion Control for Walking, Jumping and Running on a Small Bipedal Entertainment Robot". Proceedings of the IEEE International Conference on Robotics and Automation, New Orleans, April, 2004.
3. B. Vanderborght, B. Verrelst, M. Van Damme, R. Van Ham, P. Beyl, D. Lefeber, "Locomotion Control Architecture for the Pneumatic Biped Lucy consisting of a Trajectory Generator and Joint Trajectory Tracking Controller," in Proceedings of the 2006 6th IEEE-RAS International Conference on Humanoid Robots, Genova, Italy, December 4-6, 2006.
4. G.Muscato, G. Spampinato, "A Pneumatic Human inspired robotic Leg: Control Architecture and Kinematical Overview", International Journal of Humanoid Robotics, Volume 3, Number 1, pp.49-66, March 2006.
5. G. Muscato, G. Spampinato, "Kinematic Model and Control Architecture for a Human Inspired Five DOF Robotic Leg", International Journal of Mechatronics, Volume 17 Number 1, pp. 45-63, February 2007.
6. J.Pratt, C-M Chew, A.Torres, P.Dilworth, G.Pratt. "Virtual Model Control: an intuitive Approch for Bipedal Locomotion". The International Journal of Robotics Research. Vol. 20, No, pp 129-143, 2, February, 2001.
7. G. Muscato, G. Spampinato, M. Costa, "Virtual Forces Based Locomotion Strategy and Energy Balance Analysis" Proceedings of the 8th International IFAC Symposium on Robot Control SYROCO 2006 Santa Cristina Convent, University of Bologna (Italy) September 6 - 8, 2006.
8. R. M. Alexander, Principles of Animal Locomotion. Princeton University Press, 2003.

AUTONOMOUS BIPEDAL WALKING GAIT ADJUSTMENT UNDER PERTURBATIONS

L. YANG C. M. CHEW and A. N. POO

Department of Mechanical Engineering, National University of Singapore, Singapore, 119260
E-mail: yanglin, mpeccm, mpepooan@nus.edu.sg
http://guppy.mpe.nus.edu.sg/ mpeccm/chewcm.shtml

This work focuses on the walking stride-frequency autonomous adjustment in response to the environment perturbations. Reinforcement learning is assigned to supervise the stride-frequency. A simple momentum estimation further assisted the adjustment. In the learning agent, a sorted action-choose table instructed the learning to find out the proper action in a straightforward way. Incorporating the real-time step-length adjustment mode, the presented gait adjustment is able to achieve different pace walking adaptive to the environment. Dynamic simulation result shows the supervision is effective.

Keywords: Q learning CMAC network real-time walking stride-frequency walking step-length

1. Introduction

The ultimate tangible goal of the research presented here is to achieve stable bipedal walking adaptive to the environment using recursive search and computations with accessible robot's dynamics.

Bipedal locomotion control is a highly nonlinear and multi-aspect control issue. Although some research work has been shown effective for giving a stable walking, walking adaptively to the environment change is still a challenge for further explorations. Among the developed methods, Central Pattern Generator (CPG) concept inspired methods are expected to be flexible with environment. Current popular CPG models such as neural oscillators[1] and Van der Pol equations[2] use nonlinear coupled equations to elaborate the leg harmonious patterns. However, so far it still has not been clearly depicted about its feedback pathways and the relationship between the oscillator models and robot dynamics.

The presented work investigated the issue about bipedal walking gait real-time adjustment under perturbations. It is based on a newly proposed motion generator Genetic Algorithm Optimized Fourier Series Formulation (GAOFSF)[3],[4] which partitions a motion control into a low-level motion generation, and a high-level motion supervision. The low-level part[34] generated optimized walking solutions and contained parameters for online adjustment.[5] For the high-level motion surveillance, a reinforcement learning (RL) algorithm is used as the core agent to supervise the walking pace to be adaptive to the enviornment. In this paper, we focus on the high-level supervision for walking under perturbations. The target robot is modelled as the same as the one modelled in a previous work report.[3]

2. Stride-frequency online adjustment

This section discusses the automatic stride-frequency adjustment using the Truncated Fourier Series (TFS) model. (1) and (2) gives the general pattern generator equation for hip and knee joint respectively.

$$\theta_{rh}, \theta_{lh} = \begin{cases} \theta_h{}^+ = \sum^n{}_{i=1} R{\cdot}A_i \text{sin} i\omega_h(t - t_h{}^+) + c_h \\ \theta_h{}^- = \sum^n{}_{i=1} R{\cdot}B_i \text{sin} i\omega_h(t - t_h{}^-) + c_h \end{cases} \tag{1}$$

$$\theta_{rk}, \theta_{lk} = \begin{cases} \theta_{k1}{}^+ = \sum^n{}_{i=1} R{\cdot}C_i \text{sin} i\omega_k(t - t_k{}^+) + c_k \\ \theta_{k2}{}^- = c_k \geq 0 \end{cases} \tag{2}$$

where $\theta_h{}^+$ and $\theta_h{}^-$ are positive and negative hip joint angles, respectively. A_i, B_i and C_i are constant parameters determined by the Genetic Algorithm, R is a scaling constant, and $t_h{}^+, t_h{}^-$and t_k are all time-shift values (refer to our previous work[3]). c_h is the defined hip offset value which indicates the place where two hip joint trajectories intersect. c_k indicates knee joint's lock phase angle value.

Stride-frequency is defined as the fundamental frequency ω_h in the TFS model. Since walking speed has been optimized even using the GAOFSF, this stride-frequency can represent the average walking speed. The main use of the stride-frequency adjustment is to maintain the kinetic energy under perturbations. The accessible range of the adjustment can be referred to our previous paper.[5]

2.1. *Modules included in the Reinforcement Learning (RL)*

Reinforcement Learning (RL) is a class of learning problem in which an agent learns to achieve a goal through trial-and-error interactions with the environment. The learning agent learns only from reward information and it is not told how to achieve the task. From failure experience, the learning agent reinforces its knowledge so that success can be attained in future trials. The environment is modelled to be a fully observable Markov decision process (MDP) that has finite state and action sets. Q-learning method recursively estimates a scalar function called optimal Q-factors ($Q^*(i,u)$) from experience obtained at every stage, where i and u denote the state and corresponding action, respectively. The experience is in the form of immediate reward sequence,$r(i_t, u_t, i_{t+1})(t = 0, 1, 2\cdots)$. ($Q^*(i,u)$) gives the expected return when the agent takes the action u in the state i and adopts an optimal policy π^* thereafter. Based on ($Q^*(i,u)$), an optimal policy π^* can easily be derived by simply taking any action u that maximizes ($Q^*(i,u)$) over the action set $U(i)$. (3) gives the single-step sample update equation for $Q(i,u)$:[6]

$$Q_{t+1} \leftarrow Q_t(i_t, u_t) + \alpha_t(i_t, u_t)[r(i_t, u_t, i_{t+1}) \\ + \gamma max_{u \in U(i_{t+1})} Q_t(i_{t+1}, u) - Q_t(i_t, u_t)] \tag{3}$$

where the subscript indicates the stage number; $r(i_t, u_t, i_{t+1})$ denotes the immediate reward received due to the action u_t taken which causes the transition from state i_t to i_{t+1}; $\alpha \in [0,1)$ denotes the step-size parameter for the update; $\gamma \in [0,1)$ denotes the discount rate. (1) updates $Q(i_t, u_t)$ based on the immediate reward $r(i_t, u_t, i_{t+1})$ and the maximum value $Q(i_{t+1}, u)$ of over all $u \in U(t_{i+1})$.

The convergence theorem for the Q-learning algorithm that uses lookup table representation for the Q-factors is: ***Theorem*** *(Convergence of Q-learning)*:For a stationary Markov decision process with finite action and state spaces, and bounded rewards $r(i_t, u_t, i_{t+1})$ If the update equation (1) is used and $\alpha_t \in [0,1)$ satisfies the following criteria: $\sum_{i=1}^{\infty} \alpha_t(i,u) = \infty$ and $\sum_{i=1}^{\infty} [\alpha_t(i,u)]^2 < \infty, \forall(i,u)$; then $Q_t(i,u) \rightarrow Q^*(i,u)$ as $t \rightarrow \infty$ with probability 1, $\forall(i,u)$.

In this paper, Cerebellar Model Articulation Controller (CMAC)[7] is used as a multi-variable function approximator for the Q-factors in the Q-learning algorithm. CMAC has the advantage of having not only local generalization, but also being low in computation. The Q-learning algorithm using CMAC as the Q-factor function approximator is summarized as:

Initialize weights of CMAC

Repeat (for each trial):

  Initialize i

  Repeat (for each step in the trial)

    Select action u under state i using policy (say ϵ) based on $\hat{Q}$.

Take action u,
Detect new state i' and reward r
if i' not a failure state
$\delta \longleftarrow r + \gamma max'_u \hat{Q}(i', u') - \hat{Q}(i, u)$
Update weights of CMAC based on δ
$i \longleftarrow i'$
else $(r = r_f)$
$\delta \longleftarrow r_f - Q(\hat{i}, u)$
Update weights of CMAC based on δ
Until failure encountered or target achieved
Until target achieved or number of trials exceed a preset limit

State variables: The input state representing and describing the motion dynamics for the learning agent is composed of: (1) trunk's CG error; (2) external force f ; (3) current stride-frequency ω_t; the output state is the normalized change of the stride-frequency $\Delta\omega_h$ with respect to different step-length scales. Such state variables contain the current average walking velocity from the current stride-frequency for a fixed step-length walking (supposing the change of stride-frequency is proportional to its step-length level), motion acceleration which depicts the dynamic trend through external force and the actual stance posture obtained from the trunk's CG error.

Reward functions: reward functions give rewards to motions which satisfied the motion objectives. However, it also punishes a failed action. The reward function is shown as (2):

$$reward = \begin{cases} k \cdot (C - |\Delta\omega_h|) & \text{for actions succeeded} \\ -k_e|E| - k_t|Tq| & \text{for actions failed} \end{cases} \tag{4}$$

k,k_e,k_t are the scales for the performances. C is a positive constant which stands for a fixed reward; E and Tq are the error of trunk's CG and the excessive ankle joint torque. For actions succeeded, there is still a difference considered among different actions. If the change of stride-frequency is small, it means the updated frequency is not deviated much from the original frequency, the action will be given more rewards. This is to encourage the walking tolerance provided the selected action is successful.

Learning tasks: In the RL supervision, the more experiences the learning agent has gone through, the more reliability the controller can ensure. The advantage is as long as the learning agent is established, it can always keep learning different tasks assigned and increase the experience unboundedly. However, it is not easy to learn enough experiences in a short time. To reduce the chance that the robot might encounter situations which have not been trained yet, 6 representative tasks were assigned to the RL. In the paper, flat-terrain walking[3] is taken as an example.

Firstly, the stride-frequency ω_h and external force f are identified into groups $\omega_s, \omega_m, \omega_b$ and f_s, f_m, f_b. The subscript s, m, b stands for small, medium, big, respectively. With the learning carried on, the following groups can be further separated into subgroups.

$$\omega_s : \omega_h \in [2.9, 4.2] \longleftrightarrow \omega_t = 3.5 \quad \omega_m : \omega_h \in [4.3, 5.6] \longleftrightarrow \omega_t = 4.9$$

$$\omega_b : \omega_h \in [5.7, 7.0] \longleftrightarrow \omega_t = 6.3 \quad f_s : f \in [10, 30) \longleftrightarrow f_t = 20$$

$$f_m : f \in [30, 50) \longleftrightarrow f_t = 40 \quad f_b : f \in [50, 70) \longleftrightarrow f_t = 60$$

The representative 6 tasks are described as:

1) $\omega_0 = \omega_s \cap f = random(f_s)$ with the objective: continuous walking at least for 4.0 seconds with the frequency increased to ω_b.

2) $\omega_0 = \omega_s \cap f = random(f_m)$ with the objective: continuous walking at least for 3.0 seconds with the frequency increased to ω_b.

3) $\omega_0 = \omega_s \cap f = random(f_b)$ with the objective: continuous walking at least for 1.2 seconds with the frequency increased to ω_b.

4) $\omega_0 = \omega_b \cap f = -random(f_s)$ with the objective: continuous walking at least for 1.5 seconds with the frequency decreased to ω_s.

5) $\omega_0 = \omega_b \cap f = -random(f_m)$ with the objective: continuous walking at least for 1.0 seconds with the frequency decreased to ω_s.

6) $\omega_0 = \omega_b \cap f = -random(f_b)$ with the objective: continuous walking at least for 0.7 seconds with the frequency decreased to ω_s.

Common constraints are: the absolute value of trunk's CG error $< 0.05m$ (trunk CG height is 0.925m); the sum of all joint torques in one discrete time step (0.01s) is less than $10000 N.m$ (one discrete-step has 100−200 time refreshes); and the stance foot has no rotation bigger than 10°. Therefore, with the torque constraint getting strict, the energy consumption will be expected reduced; with the extension of the walking time, the tolerance will be improved.

Action-choose Table: To reduce the number of iterations and choose the correct action easier, a logic instruction is established shown in Table 1.

Momentum estimation: This is to enhance the confidence for the stride-frequency online adjustment in case the current situation has not been experienced in the CMAC network. Therefore, it is only used when the current state has not been trained by the CMAC network yet. The stability issue and the allowed sudden change of the stride-frequency have

state[1]	state[2]	state[3]	action u
$+$	$+$	$\omega_h(min)$	$\uparrow$
$+$	$+$	$\omega_h(max)$	$=\uparrow$
$+$	$+$	ω_h	$\uparrow$
$-$	$+$	$\omega_h(min)$	$\uparrow$
$-$	$+$	$\omega_h(max)$	$=\downarrow$
$-$	$+$	ω_h	$\uparrow or \downarrow$
$+$	$-$	$\omega_h(min)$	$=\uparrow$
$+$	$-$	$\omega_h(max)$	$\downarrow$
$+$	$-$	ω_h	$\uparrow or \downarrow$
$-$	$-$	$\omega_h(min)$	$=\downarrow$
$-$	$-$	$\omega_h(max)$	$\downarrow$
$-$	$-$	ω_h	$=\downarrow$

been discussed in the previous work.[5] Therefore, as long as the estimated change is within the allowed range, the dynamic stability of this estimation is promised. According to the sensed or calculated external force and the stride-frequency update time-step, the input momentum given by the perturbation can be estimated by (3)to (5).

$$M = f_t \cdot \triangle t = \Sigma_{i=1}^{7} m_i \cdot \triangle v_i \tag{5}$$

$$\triangle v_1 = 0; \triangle v_2 = l_2/L \cdot \triangle v; \triangle v_3 = l_1/L \cdot \triangle v \tag{6}$$

$$\triangle v_4 = \triangle v_5 = \triangle v_6 = \triangle v_7 = \triangle v \tag{7}$$

M is the input momentum, l_1, l_2 and L are the link-length with effect of upper, lower and the whole leg respectively. i is the link number starting from the stance foot to the swing foot. The change of stride-frequency is calculated using (6):

$$\triangle \omega_h = \frac{\pi \cdot \triangle v}{S} \tag{8}$$

S is the step-length and Δv can be calculated through (3) to (5). (6) actually updates the walking to be out of the effect of the external force if the adjust is within the stable range. For the situation that the perturbation is opposite to the walking direction, stride-frequency should actually be kept to sustain the walking or call the backward motion. Nevertheless, the RL was applied to both situations even if the latter is not so important.

3. Step-length online update

The presented step-length adjustment is considered externally, not included in the learning control. It is to adjust the position error. Step-length updates itself through the scaling parameter R in the equation (1) and (2) whenever the investigated phase angle has an error to the phase defined in the transition module.[5] Actually changing the step-length is to maintain the walking pattern according to the stance posture while tuning the stride-frequency is to maintain the kinetic energy. Therefore, with the step-length and stride-frequency online adjustment, robot dynamics becomes more controllable under perturbation or transitions. The basic rule for adjusting the step-length follows the linear interpolation.[5] The further step-length adjustment mode will be presented in the future work.

4. Experiments on walking under perturbations

This part presented the learning result and dynamic simulation result of the supervisory control applied to the bipedal walking under perturbations. For the recovery of a pattern, as soon as there is no external force sensed, the robot will be assigned the recovery module which follows the linear approaching.

4.1. *Reinforcement Learning Results*

Fig. 1 and Fig. 2 show the Q-learning performances for the 6 tasks mentioned in Section 2.1. Since the basic walking pattern has already been regulated, the RL learning task is relatively simple and not time consuming yet.

4.2. *Dynamic Simulation Experiment*

The aforementioned algorithm has been implemented dynamically on a walking experiment described as: external forces $20N$, $40N$, $60N$ were applied from $t = 2s$ to $t = 5s$, $t = 6s$ to $t = 6.8s$ and $t = 9s$ to $t = 9.25s$ respectively; external forces $-20N$,$-40N$,$-60N$ were applied from $t = 11.2s$ to $t = 11.6s$, $t = 13.7s$ to $t = 14.1s$ and $t = 16s$ to $t = 16.3s$ respectively.

With the presented RL agent and the momentum estimation. The robot is able to overcome all batches of perturbation and behaves similar to human beings. Fig. 3 shows robot walking was disrupted without any motion adjustment but Fig. 4 and Fig. 5 show the successful walking and the real-time stride-frequency and step-length adjustment information under all perturbations.

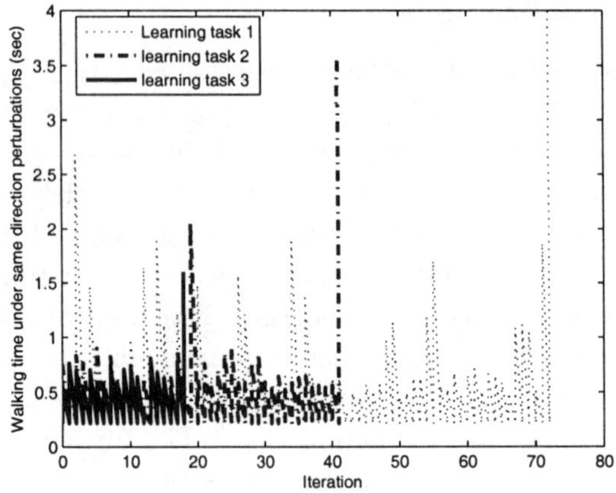

Fig. 1. Learning performance of task 1 to 3.

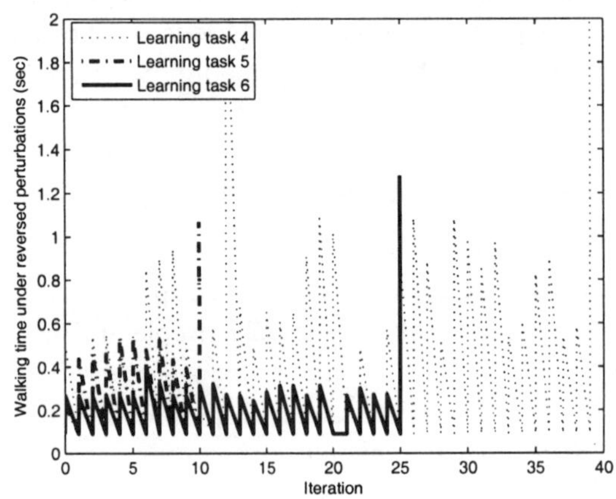

Fig. 2. Learning performance of task 4 to 6.

5. Conclusion and future works

The designed learning agent, momentum estimation and step-length update rule have been effectively applied to the motion supervision for biped's stride-frequency and step-length adjustment. As motion pattern has been regulated and optimized at low-level first using GAOFSF[3],[4] it should not take a long time to train the RL network. Concluded from the conducted

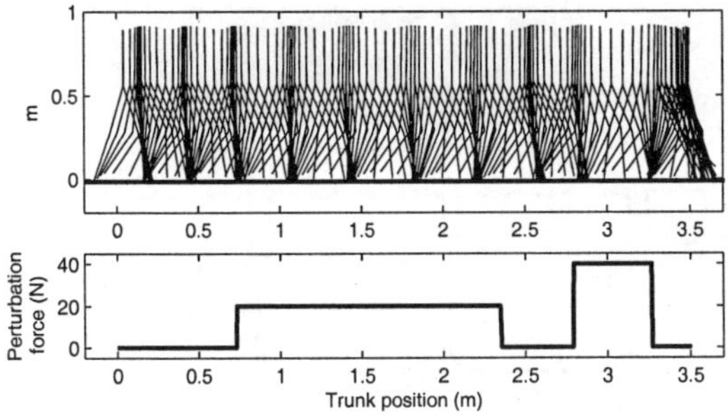

Fig. 3. Stick diagram of walking without stride-frequency or step-length adjustment (walking was disrupted finally).

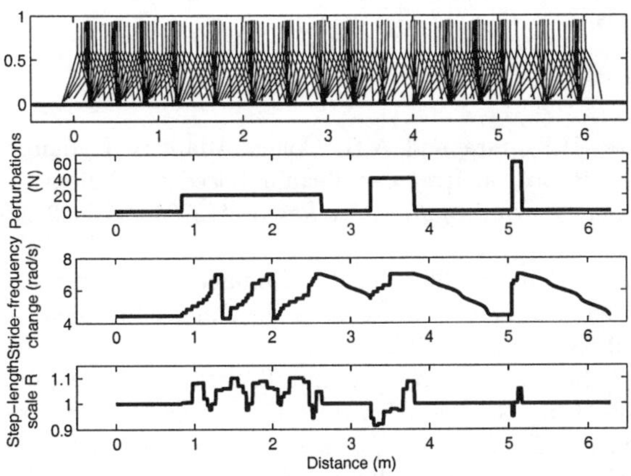

Fig. 4. Stick diagram of walking with online stride-frequency and step-length adjustment (under 1st, 2nd and 3rd batch of perturbations).

walking experiment, this supervisory control worked well and instructed the robot to get over of all batches of perturbations. Further discussions about the analysis of the performance regarding less joint torque, trunk position error and better robustness of this supervisory controller will be presented in the following report.

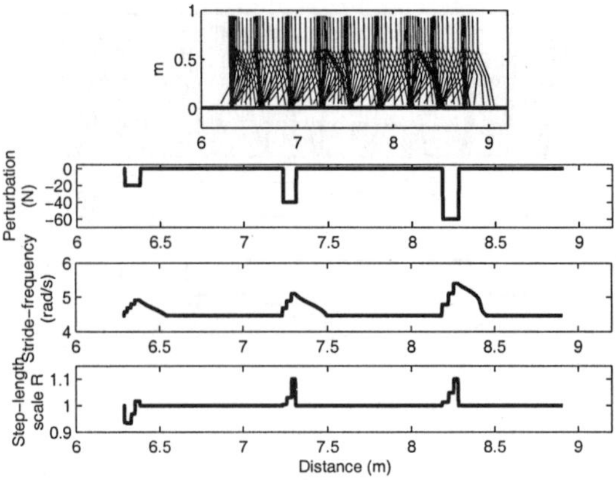

Fig. 5. Stick diagram of walking with online stride-frequency adjustment(under 4th and 5th perturbations).

References

1. Y. Fukuoka H.Kimura and A.H. Cohen Adaptive Dynamic Walking of a Quadruped Robot on Irregular Terrain Based on Biological Concepts *The International Journal of Robotics Research*(2003) **Vol** 22 No. 3-4 pp. 187-202.
2. M.S. Dutra A.C. de P. Filho and V.F. Romano Modeling of a bipedal locomotion using coupled nonlinear oscillators of Van der Pol Biological Cybernetics (2004) **Vol** 88(4), pp 286-292 .
3. L. Yang C. M. Chew A. N. Poo and T. Zielinska Adjustable Bipedal Gait Generation Using Genetic Algorithm Optimized Fourier Series Formulation *IEEE/RSJ Int. Conf. on Intelligent Robots and Systems*(2006) pp 4435-4440.
4. L. Yang C. M. Chew T. Zielinska and A. N. Poo A Uniform Biped Gait Generator with Off-line Optimization and On-line Adjustable Parameters *Robotica* (2007) accepted.
5. L. Yang C. M. Chew A. N. Poo and T. Zielinska Adjustable Bipedal Gait Generation using GA Optimized Fourier Series Formulation; Real-time Gait Adjustment *Proceedings of the Int. Conf. on the Autonomous Robots and Agents* (2006) New Zealand, pp 111-116.
6. John N. Tsitsiklis Asynchronous stochastic approximation and q-learning *Machine Learning* (1994) **Vol** 16 (3) pp 185-202.
7. Thomas W. Miller and Filson H. Glanz The university of new Hampshire implementation of the cerebellar model arithmetic computer-cmac. Unpulished reference guide to a CMAC program (1994).

DESIGN AND PROBLEMS OF A NEW LEG-WHEEL WALKING ROBOT

CRISTINA TAVOLIERI*
ERIKA OTTAVIANO
MARCO CECCARELLI

LARM: Laboratory of Robotics and Mechatronics, University of Cassino
Via di Biasio 43, Cassino, 03043, Italy

In this paper an analysis is presented for new leg mechanisms of a 1-DOF (Degree-Of-Freedom) leg-wheel walking machine. A preliminary prototype of a low-cost robot, which is capable of a straight walking with only one actuator, has been designed and built at LARM: Laboratory of Robotics and Mechatronics in Cassino. Simulation and experimental validation of the built prototype have been carried out to verify its operation. New solutions for the leg mechanism have been proposed.

1. Introduction

The walking in nature is a very flexible and complex task. For example, in generating a trajectory several parts/systems are involved: muscle as actuators, bones as linkages, nerves as sensors, and brain as a complex control system [1]. Moreover, locomotion type depends on several variables. For example, one of the most influent aspects can be considered the environment characteristic. While the legged locomotion is more adaptable in a wide range of terrain, the wheeled locomotion is faster but only on smooth surfaces. The most common walking machines are wheeled and tracked systems, but large interest is also focused on legged machines. A leg should generate an approximately straight-line trajectory of a foot point with respect to the body, as outlined for example in references [1-3]. There are also robots whose design combines wheels, legs or tracks. They can be divided in several categories: articulated legs having wheels at their end, like for example in [4]; wheels and legs in separated modules, like for example in [5]; articulated tracks, like for example in [6]; wheels and tracks in separated modules, like for example in [7].

A combined design of legs and wheels can be a good compromise for a walking machine combining the advantages of both locomotion types with the aim to achieve a successful operation of mobile robotic systems in rough, unstructured terrain, such as underground mines, forests, disaster sites, and planetary explorations, as outlined in [8].

* E-mail: tavolieri@unicas.it, ottaviano@unicas.it, ceccarelli@unicas.it.

2. A 1-DOF leg for walking machine

The leg architecture is composed by a Chebyshev mechanism and a Hart inversor, as shown in Fig.1. The Chebyshev mechanism consists of the four-bar approximate straight-line mechanism LEDCB. The lengths of the links are determined in such a way that the shape of the point B trajectory is similar to the one of a human step. Moreover, the straight part of the trajectory is quite accurate. It is an important aspect in order to avoid the body lifting during the walk. The Hart mechanism inverts and amplifies the trajectory. The amplification factor, from point B to point A, is approximately 2. Thus, the body can move horizontally by moving the feet and legs.

Several walking robots have a design with a pantograph as fundamental mechanism and even Chebyshev linkage has been used for actuator purposes like for example [9-10], specific experiences have been carried out at LARM in Cassino, as reported in [11-15].

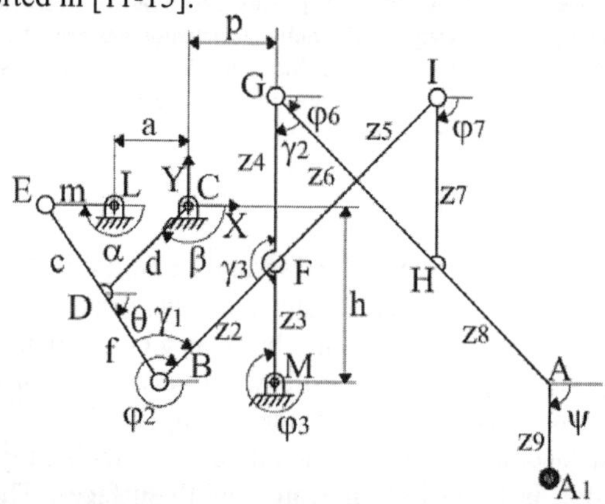

(a)

(b)

Fig.1. The new 1-DOF hybrid robot: a) kinematic scheme for the 1-DOF leg designed robot; b) built prototype.

3. The existing prototype and its operation

The mechanical design of the prototype has been developed by looking at several aspects, such as

- low costs in terms of both manufacturing and energy
- easy-operation system in terms of control,
- a compromise between flexibility for walking tasks and simplicity of the system.

The chosen material for building the robot is an aluminium alloy, because of its characteristics of lightness, easy machining and good resistance to the corrosion.

The designed robot consists of a biped module in which two legs are actuated by a DC motor. Each leg has been provided by a suitable foot, which has been rigidly installed at the leg tip, for walking in both flat and rough terrains. There are two passive wheels, in order to have always three points in contact with the ground.

Because of the use of only one actuator it is necessary to design a gear transmission system, which transmits the actuation torque to both the legs. The body of the mobile robot carries batteries for operating the robot. Furthermore, it allows the installation of suitable sensors and acquisition system on board.

The walking synchronization is obtained by moving one leg in opposite phase with respect to the other one. During the walking the prototype has always one leg and two wheels in contact with the ground. The mass distribution and body shape give that the projection of the barycentre is always in the area that is limited by the elements in contact with the ground.

In order to improve the performances of the prototype several aspects have been considered. The prototype has been tested indoor and outdoor over terrains of various nature and slope. The prototype has been tested by walking forward and backward on flat terrain. The prototype has been tested also for a slope of about 20 deg, and on outdoor various terrain, as reported in [16]. Figure 2 shows examples of the walking tests. Parameters of the mechanism showed in Fig. 1 a) are reported in Tables 1 and 2.

4. The attached problems

Main problematic aspects of the robot behaviour can be addressed to

- limited lift high,
- body shape and position of the barycentre,
- wheels and foots materials.

The built prototype has good flexibility, but one of its main limitation is the little high of the lift. The idea of using actuated legs and passive wheels was implemented in order to build a system able to climb obstacles of various size and nature.

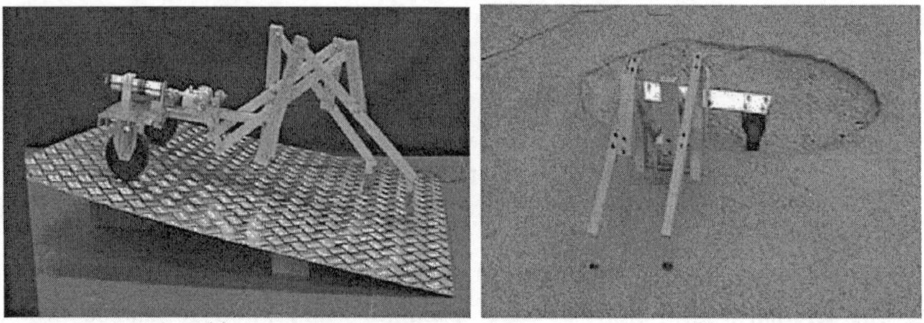

(a) (b)

Fig.2 The built prototype while: a) climbing a slope of 20 deg; b) walking over unstructured terrain.

With the aim to improve this fundamental characteristic, and, at the same time, to loose nothing in terms of easy-operability and low-cost, it is possible to change the parameters of Chebyshev links, in order to have a higher trajectory, allowing the prototype to overcome a larger obstacle variety, as reported in Fig. 3. In fact, the mechanism of the leg is the one which gives the trajectory size and shape.

The above–mentioned problem can be formulated as:

$$r < sh$$
$$Oh \geq Sh$$

where r is the wheel radius, sh is the old lift high, Sh is the height of an obstacle and Oh is the new step height. The new configuration allows the prototype to overcome obstacles of Oh high, were Oh>sh. With the proposed configuration it is possible to eliminate AA_1 link, having a linear inversion of the trajectory.

It is worth to underline that the backward walk has been obtained more efficiently than in the case in pulling walk mode. The difference in walking climbing directions can be explained by considering the position of the mass centre and friction of the floor plane.

The velocity is constant during the walking on flat terrain, and it depends on the friction between foot and ground. It is worth to underline that the backward walk has been obtained more efficiently than in the case in pulling walk mode. The difference in walking climbing directions can be explained by considering the position of the mass centre and friction of the floor plane. In fact, because of the actuator weight, the mass centre position is closer to the wheels and therefore gravity helps the walking motion. This behaviour is emphasized when the batteries are installed on board. In this case the weight of the system strongly increases, and the position of the mass centre is even more decentred because of the heaviest components lying on the passive module.

As further development of the prototype the body shape should be redesigned with a rectangular profile. In this way the batteries can be the middle of the body, improving both walking condition and stability.

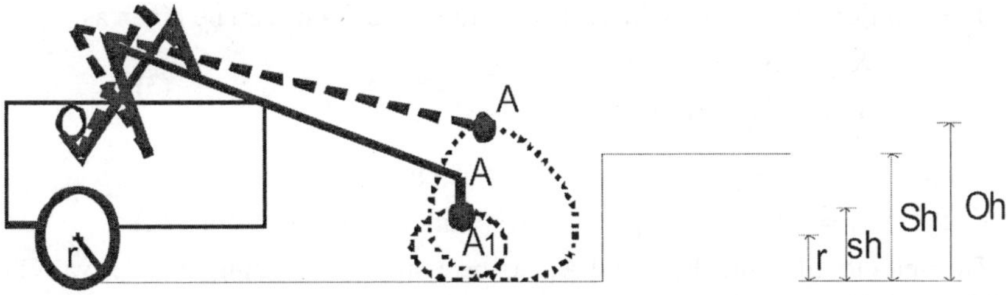

Fig.3 A scheme of the new configuration obtained by changing the Chebyshev mechanism: a higher step is obtained (dotted line represents the new solution, continuous line represents the existing leg).

From a mechanical point of view it is possible to improve the prototype walking by replacing the wheels with a version covered with rubber, in order to avoid slipping on metallic surfaces. Another important design improvement can be the replacement of the foot, with other one having spherical shape.

Table 1. Link parameters for the Chebyshev mechanism for the leg design in Fig.1 a).

Parameters (mm)	a	m	c	d	f
Built Prototype	45	15	60	60	60

Table 2. Link parameters for the Hart mechanism for the leg design in Fig.1 a).

Parameters (mm)	p	h	z_2	z_3	z_4	z_5	z_6	z_7	z_8	z_9
Built Prototype	168	104	225	150	150	225	225	150	225	100

5. A numerical simulation of the leg operation

A kinematic analysis has been developed in order to evaluate and simulate performances and operations of the leg system. A fixed reference system CXY has been considered attached at body in point C, as shown in Fig.1. The position of point B with respect to CXY frame can be evaluated as a function of the input crank angle α and kinematic parameters of the Chebyshev mechanism LEBDC in the form

$$\begin{aligned} X_B &= -a + m\cos\alpha + (c+f)\cos\theta \\ Y_B &= -m\sin\alpha - (c+f)\sin\theta \end{aligned} \tag{1}$$

The position of point A with respect to the fixed frame can be given as

$$\begin{aligned} X_A &= X_M - (z_3+z_4)\cos\varphi_3 + (z_6+z_8)\cos\varphi_6 \\ Y_A &= Y_M + (z_3+z_4)\sin\varphi_3 - (z_6+z_8)\sin\varphi_6 \end{aligned} \tag{2}$$

The position of point A_1 with respect to the fixed frame can be given as

$$\begin{aligned} X_{A1} &= X_A + z_9 \cos\psi \\ Y_{A1} &= Y_A - z_9 \sin\psi \end{aligned} \quad (3)$$

The transmission angles shown in Fig. 4 can be evaluated as

$$\gamma_1 = -\theta + \varphi_2 - \pi \,, \gamma_2 = \varphi_3 - \varphi_6 - \pi \,, \gamma_3 = \varphi_3 - \varphi_2 + \pi \quad (4)$$

The velocity of points B, A and A_1 can be evaluated by using time derivatives from Eqs. (1) to (3).

The acceleration of point A_1 can be obtained with respect to the fixed frame as

$$\begin{aligned} \ddot{X}_{A1} &= \dot{\varphi}_3^2(z_3+z_4)\cos\varphi_3 + \ddot{\varphi}_3(z_3+z_4)\sin\varphi_3 + \\ &- \dot{\varphi}_6^2(z_6+z_8)\cos\varphi_6 - \ddot{\varphi}_6(z_6+z_8)\sin\varphi_6 + \\ &- \dot{\varphi}_6^2 z_9 \cos(\varphi_6+\psi) - \ddot{\varphi}_6(z_6+z_8)\sin\psi \\ \ddot{Y}_{A1} &= -\dot{\varphi}_3^2(z_3+z_4)\sin\varphi_3 + \ddot{\varphi}_3(z_3+z_4)\cos\varphi_3 + \\ &+ \dot{\varphi}_6^2(z_6+z_8)\sin\varphi_6 - \ddot{\varphi}_6(z_6+z_8)\cos\varphi_6 + \\ &+ \dot{\varphi}_6^2 z_9 \sin(\varphi_6+\psi) - \ddot{\varphi}_6 z_9 \cos\psi \end{aligned} \quad (5)$$

Figure 5 shows numerical results for the accelerations of the leg system when the angular velocity ω of the input crank is chosen with a constant value equal to 1.0 rad/s.

By considering the obtained values of the main parameters (such as accelerations and transmission angles) and the tested behaviour of the built prototype new solution for the leg system can be considered.

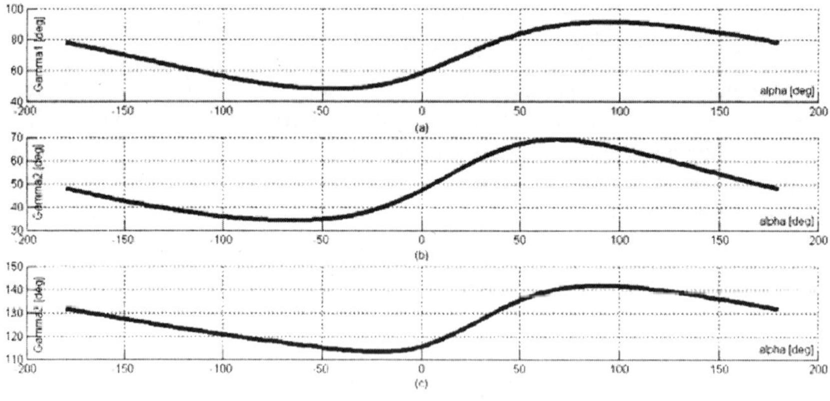

Fig. 4. Numerical simulation for the transmission angles in degrees: a) angle γ_1; b) angle γ_2; c) angle γ_3.

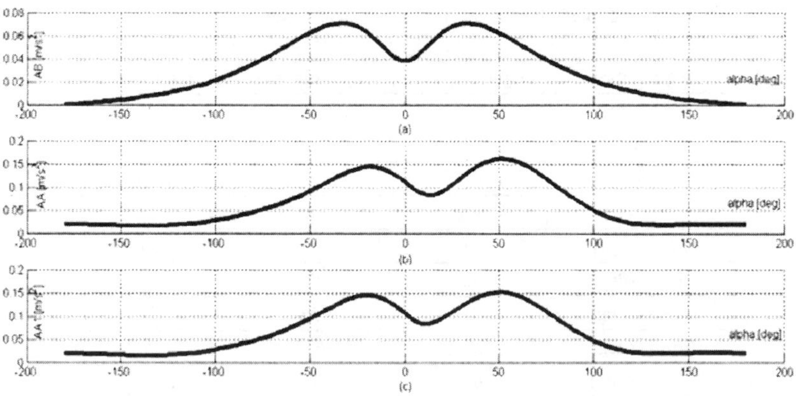

Fig. 5. Numerical simulation for the accelerations: a) point B; b) point A; c) point A_1.

6. Proposals for improving design the leg design

Design consideration and simulations have been developed in order to solve the outlined problems. One of the main issues is related with walking versatility for obtaining higher lift of the feet. A design attempt has been focused in changing the link ratios with the aim to change the size of the coupler path in the actuating Chebyshev mechanism.

By referring to the Chebyshev mechanism in the built prototype in Figures 1 and 6 a), the mechanism design is characterized by the ratios k_1 and k_2 that give

$$\begin{aligned} ED = CD = DB = k_1 LE \\ LC = k_2 LE \end{aligned} \qquad (6)$$

The convenient characteristics of the Chebyshev design can be preserved by keeping the ratios in Eq. (6) and design improvements can be studied by changing the ratio values. Among the many possibility we have considered again Chebyshev mechanisms that are listed in [17].

Thus, two alternative solutions have been selected whose main characteristics are reported in Table 3 as related with walking operation. Numerical simulation have been carried out and results are reported in Figures 6 and 7, as compared with the Chebyshev mechanism in the built prototype of Figures 1 and 2. In particular, the size of the coupler path is different to give a higher lift of the feet during the walking. Even the shape of the coupler is modified as shown in Figures 6 b) and c), which will give a modified behaviour also for the Hart mechanism that has been maintained unchanged. In fact, point M has a different location and the Hart mechanism shows different configurations. Motion characteristics are evaluated in term of transmission angles γ_1, γ_2, and γ_3 (see Fig. 1 a)) in order to synthetically show the mechanical feasibility.

Table 3. Parameters for the leg mechanisms in Fig.6.

Leg Solution	Leg in Fig. 6 a)	Leg in Fig. 6 b)	Leg in Fig. 6 c)
Lift (mm)	150	180	190
LE (mm)	20	20	15
k_1	4.0	2.17	2.94
k_2	3.0	2.8	2.83
$E\hat{D}B$ (deg)	180	180	100
Min γ_1 (deg)	48	77	37
Max γ_1 (deg)	91	153	108
Min γ_2 (deg)	34	64	35
Max γ_2 (deg)	69	117	69
Min γ_3 (deg)	113	144	115
Max γ_3 (deg)	142	166	148

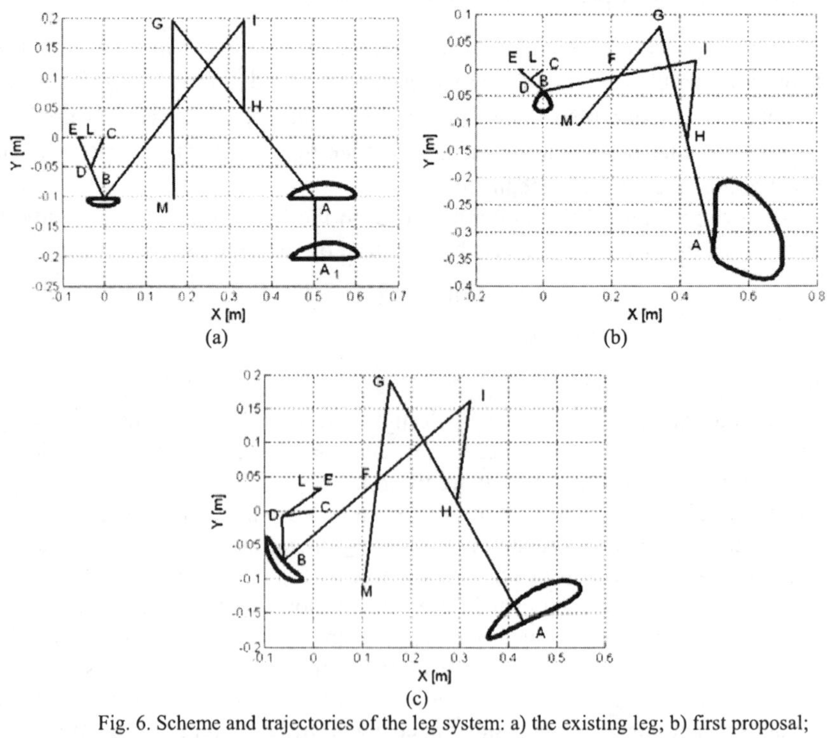

Fig. 6. Scheme and trajectories of the leg system: a) the existing leg; b) first proposal; c) second proposal.

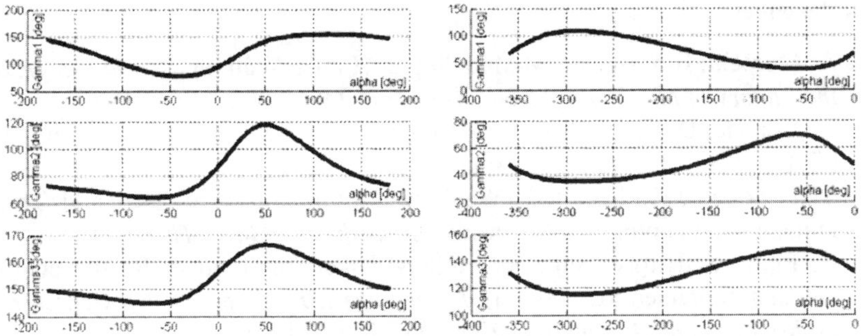

(a) (b)

Fig. 7. Numerical simulation for the transmission angles γ_1, γ_2, γ_3 in degrees: a) leg in Fig. 6 b); b) leg in Fig. 6 c).

In particular, the transmission angle γ_1 refers to the force transmission capability between the Chebyshev mechanism and Hart mechanism, as the case of the leg functionality. Plots in Figures 4 and 6 are useful to check the functionality of the proposed alternative solutions as compared with the original one. In Table 3 numerical values summarize the results of numerical simulations. One can note that although a considerable increase of the lift is obtained, in general performances are not increased accordingly. In fact, the range of the transmission angles is larger and the smooth behaviour, that is indicated by the smoothness of the plots, is even deteriorated. Nevertheless, the alternative proposed solutions can be considered feasible, since they refer to successful practical mechanisms. In addition, they could be used as initial designs in a design procedure which will use optimization techniques and proper formulation for optimality criteria for walking mechanisms, as concerning with the outlined problems in this paper.

7. Conclusion

In this paper a new prototype of leg-wheel walking robot is proposed with low-cost easy-operation features. A suitable kinematic model has been presented for a 1-DOF leg with a fully-rotative actuation and for the whole system. A kinematic analysis has been also carried out in order to evaluate and simulate performance and operations of the walking robot. A low-cost leg-wheel machine has been built at LARM in Cassino by using the proposed 1-DOF leg mechanism. Experimental tests have been conducted over several terrains with the aim of understand limits and possible improvements of the prototype. By considering the leg structure two possible new solutions have been proposed in order to have a higher lift.

References

1. A. Morecki, Biomechanical Modeling of Human Walking, *World Congr. on the Theory of Mach. and Mech.*, Vol.3, pp. 2400-2403, Milan, 1995.
2. K. Yoneda, Design of Non-Bio-Mimetic Walker with Fewer Actuators, *4th CLAWAR*, pp. 115-126, Karlsruhe, 2001.
3. S.M. Song, J.K. Lee, K.J. Waldron, Motion Study of Two and Three Dimensional Pantograph Mechanisms, *Mech. and Mach. Theory*, 1987.
4. G. Endo, S. Hirose, Study on Roller-Walker (multimode steering control and self-contained locomotion), *Procedings IEEE International Conference on Robotics and Automation,* 2000, pp. 2808-2814.
5. S. Guccione, G. Muscato, The Wheeleg Robot, *IEEE Robotics and Automation Magazine,* Vol. 10, 2003, pp. 323-330.
6. F. Michaud, D. Letourneau, M. Arsenault, Y. Bergeron, R. Cadrin, F. Gagnon, M.-A. Legault, M. Millette, J.-F. Paré, P. Lepage, Y. J. Morin, Bisson, S. Caron, Multi-Modal locomotion Robotic Platform Using Leg-Track-Wheel Articulations, *Autonomous Robots*, Vol. 18,2005, pp.137-156.
7. C. T. Chen, Y.A. Hsieh, Mobile Robot, Us Patent 6.144.180, 2000.
8. K. Iagnemma, S. Dubowsky, Mobile Robots in Rough Terrain, *Springer tracts in advanced robotics*, Vol. 12, Berlin, 2004.
9. F. Pfeiffer, T. Zielinska: Walking: Biological and Technological Aspects, *International Centre for Mechanical Sciences Courses and Lectures n. 467*, Springer Wien New York, 2004, pp. 1-29.
10. M. H. Raibert, Legged Robots That Balance. MIT Press, 1986, pp. 3-14.
11. E. Ottaviano, M. Ceccarelli, C. Tavolieri, Kinematic and Dynamic Analyses of a Pantograph-Leg for a Biped Walking Machine, *Proceeding of 7th CLAWAR*, Madrid 2004, pp. 561-568.
12. C. Tavolieri, E. Ottaviano, M. Ceccarelli, A. Di Rienzo, Analysis and Design of a 1-DOF Leg for Walking Machines, *Proceedings of RAAD'06*, 15th International Workshop on Robotics in Alpe-Adria-Danube Region, Balantonfured, 2006, CD Proceedings.
13. C. Lanni, M. Ceccarelli, E. Ottaviano, G. Figliolini, Mechanisms for Pantograph Legs: Structures and Characteristics, *Proceeding of Xth IFToMM World Congress*, Oulu, 1999, Vol.3, pp. 1196-1201.
14. C. Tavolieri, Design, Simulation and Construction of a Biped Walking Robot, Master Thesis, LARM, University of Cassino, Cassino, 2004.
15. A. Nardelli, Design and construction of a new wheel-leg mobile robot, Master Thesis, LARM, University of Cassino, Cassino, 2006 (in Italian).
16. C. Tavolieri, E. Ottaviano, M. Ceccarelli, A. Nardelli, A Design of a Nex Leg-Wheel Walking Robot, *Proceedings of MED'07*, 15th Mediterranean Conference on Control and Automation, Athens, 2007, CD Proceedings.
17. I. I. Artobolevski, Mechanisms in Modern Engineering Design, Vol. 1-5, Mir Publisher, Moscow, 1975-1980.

Stiffness and Duty Factor Models for the Design of Running Bipeds

Muhammad E. Abdallah and Kenneth J. Waldron

Mechanical Engineering Department,
Stanford University,
Stanford, CA, USA
{mesam, kwaldron}@stanford.edu

Supporting the design process for running biped robots, analytical models are presented for two aspects of running: the duty factor (DF) of the gait, and the stiffness value of the leg. For a given running speed, an optimal DF exists that minimizes the energy expenditure. Based on a model of the energetics, we present a formula for the optimal DF. This formula is validated by both human data and simulation results. In addition, a model is presented for the stiffness value of the leg as a function of the physical properties, speed, and DF. The *Gait Resonance Point* is proposed as a design target for compliant running. At this point, the gait matches the spring resonance and the stiffness value becomes independent of the DF.

Keywords: biped robot, running, duty factor, SLIP stiffness, energetics.

1. Introduction

As the field of running robotics continues to progress, a more rigorous design process is needed to meet performance criteria. This work addresses the design of a running biped robot subject to gait specifications. We present design rules and models for two aspects of running: the duty factor (DF) of the gait and the stiffness value of the leg.

The presence of compliance in the leg of a running biped is invaluable as an energy storage and thrust mechanism. It has been proven essential to the running of biological systems.[1] It has also been instrumental in the more dynamically successful robotic runners.[2–5]

Surprisingly, little theory exists for selecting the stiffness according to gait specifications. Raibert shares the effective stiffnesses for his hoppers, but nothing on the selection process.[2] Rad et al. designed their stiffness to maximize the active energy input during stance.[6] Schmeideler and Waldron selected their stiffness value according to biomimetic models.[7] Ahmadi and Buehler began with an analytical model of purely vertical hopping. They then incorporated model-specific empirical formulas to relate the stiffness to the speed.[8] None of these studies provides a generalizable design process or laws for selecting the stiffness of a general biped for any speed.

We present an analytical formula for the leg stiffness that is a function of the physical properties, speed, and DF of the system. In addition, we present the *Gait Resonance Point* (GRP) as a potential design target for compliant running. At the GRP, the gait matches the spring resonance, and the stiffness value becomes independent of the DF.

The DF is the fraction of a stride period a specific leg spends in contact with the ground. It carries significant implications for the energy consumption and ground impact forces of a gait. Despite these consequences, no documented consideration for the DF, to our knowledge, has been given for the design of any of the existing running bipeds. For a given speed, an optimal DF exists that minimizes the energy expenditure of the gait. Alexander shows the existence of this optimal DF, displaying the energy cost for a human model running at two discrete speeds.[9]

Based on a model of the energetics, we present here a formula for the optimal DF. The formula is applicable to any speed and design for the biped, and it is validated by both human data and simulation results.

2. The SLIP Model for Stiffness Analysis

A simple, yet effective model for compliant running is the Spring Loaded Inverted Pendulum (SLIP) model. It consists of a massless, spring-operated leg attached to the center of mass of the body. It has been widely used in the study of both biological and robotic systems, and its applicability has been well documented.[10–12] Fig. 1 shows the model.

While running at steady-state, the SLIP model must exhibit a symmetric contact phase.[12] This symmetry is reflected in both the configuration and velocities, as shown in Fig. 1.

The gait consists of a contact phase followed by a flight phase. The contact and flight times (t_c and t_f respectively) can be determined from the kinematics of Fig. 1.

$$t_c = \frac{2l_o \sin(\theta_o)}{v_x} \quad , \quad t_f = \frac{2v_z}{g} \tag{1}$$

v_x and v_z represent the horizontal and vertical velocities at the initial moment of contact. l_o and θ_o are the initial leg length and angle. g is the grav-

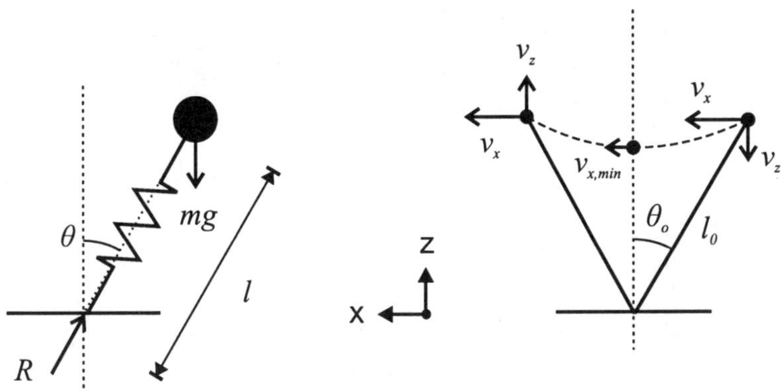

Fig. 1. The SLIP model. The free body diagram is shown to the left. The symmetry of the contact phase is shown to the right.

itational acceleration. The calculation for the contact time approximates the forward velocity as a constant v_x.

McMahon and Cheng utilized this symmetry to analyze the dynamics of running.[11] They conducted numerical integration of the SLIP equations of motion to determine the parameters that satisfy the described symmetry. They presented empirical formulas that fit the data within errors of 0.5%.

Using their empirical formulas, we can display the spring natural frequency of the system, ω_n, as a function of the speed and DF. Fig. 2 shows the results. The parameters used throughout this paper reflect an average human: a mass of 73 kg, leg-length of 0.8 m, and a touchdown angle of 0.4 radians.

We will use McMahon's data as a benchmark for our approximations. His formulas would suffice for determining the needed stiffness if it were not for two factors. First, the formulas are very complex: 24 empirical parameters are needed to compute the stiffness at any one speed. Second, his formulas are only valid for a limited range of conditions.

The next section will present an alternative analytical formula for the stiffness and compare it to McMahon's results.

3. An Analytical Stiffness Model

To derive an analytical model for the SLIP, consider the dynamical equation in the vertical direction. Fig. 1 shows the free-body diagram. k and l_o are the stiffness and initial length of the spring, and m is the mass.

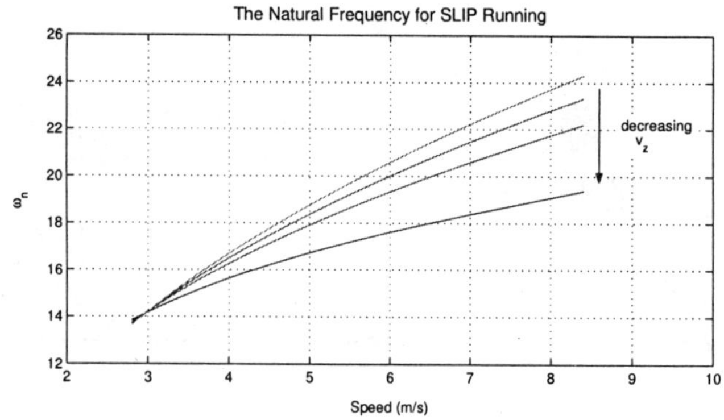

Fig. 2. The natural frequency required for steady-state running of the SLIP model, based on the empirical formulas of McMahon and Cheng.[11] Each curve represents a liftoff velocity (v_z). The arrow points in the direction of increasing DF.

$$m\ddot{z} + kz + (mg - kl_o\cos(\theta)) = 0 \tag{2}$$

Since the motion over the contact phase is known, we can replace $\cos(\theta)$ with its average value. Denoted as $\overline{c\theta}$, the geometric average is,

$$\overline{c\theta} = \frac{\sin(\theta_o)}{\theta_o}. \tag{3}$$

This results in a single-variable, linear equation of motion. This equation can be solved through boundary values based on the symmetry of the contact phase.

We will introduce the *contact frequency*, ω_c, as the frequency of the contact phase. Its period is twice the contact time.

$$\omega_c = \frac{\pi}{t_c} \tag{4}$$

The solution to this system gives us our final equation relating the spring frequency to the system parameters. We will refer to it as the *SLIP Governing Equation*. For brevity, $\cos(\theta_o)$ is expressed in shorthand as $c\theta_o$.

$$\left(g - \omega_n^2 l_o(\overline{c\theta} - c\theta_o)\right)\tan\left(\frac{\pi}{2}\frac{\omega_n}{\omega_c}\right) + v_z\omega_n = 0 \tag{5}$$

The stiffness a biped needs to run is a function of both physical properties and gait characteristics. It is a function of the mass, leg length, and initial leg angle of the system. It is also a function of the desired speed and

DF of the gait. The SLIP Governing Equation provides an analytical tool for determining the stiffness value given those parameters.

The equation predicts the spring frequency with good accuracy. Compared to McMahon's data, it produced errors less than 7%. A side-by-side comparison is shown in Fig. 3.

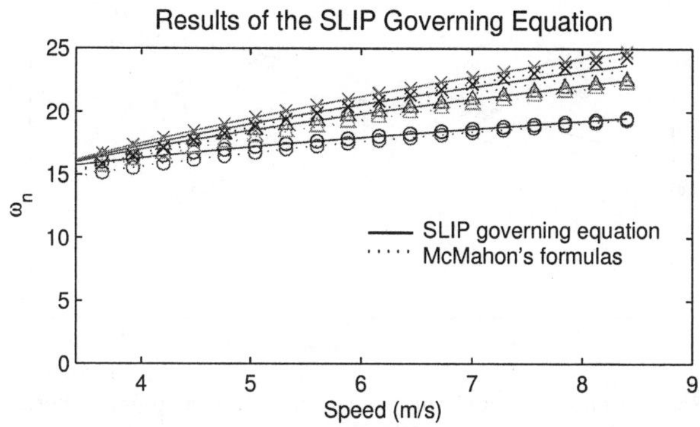

Fig. 3. The *SLIP Governing Equation* accurately predicts the spring frequency needed. The plots compare results from the governing equation with McMahon's empirical data at different DF's. The errors are all less than 7%.

4. The Gait Resonance Point

The SLIP stiffness curves display an interesting behavior: the stiffness values tend to converge at a select speed. This phenomenon is exhibited in Fig. 2. Although the stiffness is normally a function of both the speed and the DF, at this speed the value is effectively constant, independent of the DF. We call this point the *Gait Resonance Point* (GRP) for reasons that will become apparent.

This behavior consistently demonstrated itself in McMahon's formulas, even with changes to the parameters. The SLIP Governing Equation (5) allows us to solve for the point. The GRP represents the solution of the equation at its singularity, where the *tan* term approaches infinity and its coefficient approaches zero. This results in the following two conditions.

$$\omega_n = \omega_c \tag{6}$$

$$v_x = \frac{2}{\pi}\sqrt{gl_0(1 + c\theta_o)} \tag{7}$$

According to the first condition, the solution occurs when the contact frequency matches the spring natural frequency—hence naming it the *Gait Resonance Point.* These simple design rules predict the GRP with good accuracy. Fig. 4 shows the analytical GRP alongside the frequency curves.

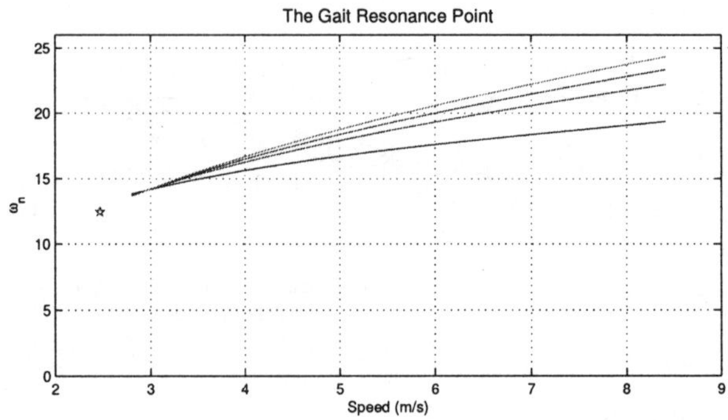

Fig. 4. The *Gait Resonance Point* targets the intersection point, where the stiffness becomes effectively independent of the DF. Shown as the star, the design rules predict the point with good accuracy. These frequency curves are the same curves from Fig. 2.

We propose the GRP as a potential design target for compliant running due to two advantages. First, the stiffness is effectively independent of the DF; hence, we can implement different DF's without changing the stiffness. Second, the stiffness is governed by simple design rules, where the spring frequency equals the contact frequency.

If the GRP does not meet the desired design specifications, one can return to the SLIP Governing Equation to determine the needed stiffness. In this case, a DF needs to be selected. The next sections turn to the issue of the DF. First, the energy cost is defined. Then, an optimal DF model is presented.

5. The Energy Cost of Running

5.1. *Defining the Energy Cost*

The kinetic energy (KE) of a multi-body system can be partitioned into two segments. The first is defined as the *external work*, and it consists of the KE of the center of mass (CM) in a ground reference frame. The second is defined as the *internal work*, and it consists of the KE of the limbs in the CM reference frame.[13,14]

This framework allows for a simple characterization of the energy expenditure. By contrasting the top-of-flight and the bottom-of-contact instants, the external work itself reveals two components. First, it must provide the change in height, i.e. the hopping height energy. Second, it must provide the change in KE as the system accelerates from its minimum horizontal speed at the nadir to its maximum horizontal speed at the apex.

Similarly, the internal work can be characterized by two components. First, it must provide the energy to swing the non-support leg, work conducted primarily by the respective hip. Second, it must provide the work conducted by all the other joints, primarily to maintain posture.

Of these four total components, two dominate the energy cost: the hopping height energy and the swing energy. Each scales by the square of the DF. We will thus consider these two only.

These two functions capture the tradeoff inherent in the DF. For example, a smaller DF translates into a longer flight time. This long flight time entails a large hopping height energy. At the same time, it increases the stride period and thus reduces the swing energy required.

5.2. *Deriving the Energy Cost*

For the following energy calculations, we will consider the net absolute work performed over a step. A step includes a contact phase and subsequent flight phase, i.e. half the stride.

Given a change in height of h, the net absolute work performed for the hopping height is $2mgh$. Neglecting the change in height during contact, the hopping height energy as a function of v_z follows.

$$E_{hh} = mv_z^2 \tag{8}$$

The swing energy, E_{sw}, is derived elsewhere.[15] The resulting energy is,

$$E_{sw} = \theta_o^2 m_l l_s^2 (\omega^2 - \omega_n^2). \tag{9}$$

l_s and m_l are the effective length and mass of the swing leg. ω is the stride frequency and ω_n here is the pendulum natural frequency. The total energy cost, E_{tot}, is the sum of both components.

The DF, β, is defined as the ratio of a leg's contact time to the stride period. Assuming symmetric right and left steps,

$$\beta = \frac{t_c}{2(t_c + t_f)}. \tag{10}$$

Given (10) and (1), we can solve for the total energy cost as a function of the DF. This gives us our final energy cost formula.

$$E_{tot} = c_1 \left(\frac{1}{2\beta} - 1 \right)^2 + c_2\beta^2 - c_3 \tag{11}$$

where c_i are constant functions of the parameters.

$$c_1 = m \left(\frac{g l_o \sin(\theta_o)}{v_x} \right)^2 \tag{12}$$

$$c_2 = \theta_0^2 m_l l_s^2 \left(\frac{\pi v_x}{l_o \sin(\theta_o)} \right)^2 \tag{13}$$

$$c_3 = \theta_0^2 m_l l_s g. \tag{14}$$

This formula quantifies the energy tradeoff associated with the DF. Fig. 5 displays the energy cost for an average human running at 4 m/s.

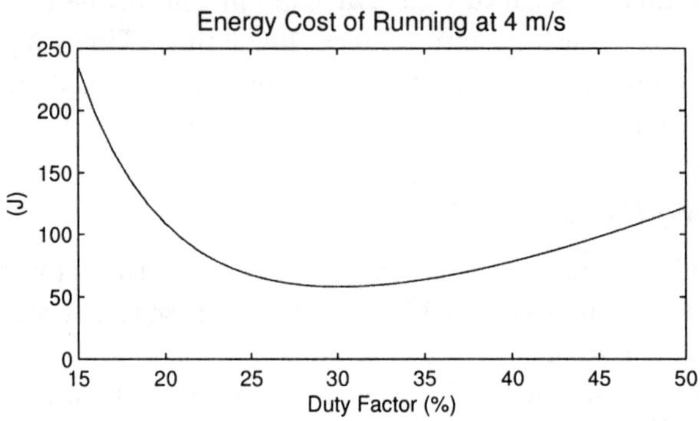

Fig. 5. Plot of the energy cost formula, shown for an average human running at 4 m/s. The energy cost at higher DF's is dominated by the swing energy; while at lower DF's it is dominated by the hopping height energy. This tradeoff results in a global minima that depends on both the running speed and physical parameters.

6. The Optimal DF

To solve for the minima of the energy cost, we will optimize (11). This gives us our final characteristic equation, the *Optimal DF Formula.*

$$\left(\frac{4c_2}{c_1} \right) \beta^4 + 2\beta - 1 = 0 \tag{15}$$

This formula was validated in two ways. First, it was compared to data of human running. As a widely accepted premise, biological systems adapt their gaits to minimize energy consumption.[9,16] Minetti experimentally determined the DF of humans running at a range of speeds, using an average sample size of over 30 runners.[14]

Second, the results were validated in simulation. Using the dynamic model described in Appendix A, the biped was tested at two select speeds. For each speed, a range of DF's was applied and the energy consumption tabulated. The model approximated an average human running at speeds of 3.5 m/s and 4.0 m/s respectively.

Both the human data and the simulation model exhibited close correspondence with the Optimal DF Formula. Results of the formula nearly fall within a standard deviation of the average human DF's. This close correspondence occurs despite neglecting the error between mechanical and metabolic work, and not knowing the actual physical properties of the runners. Fig. 6 displays the combined results, comparing the DF's from the formula with the human data and simulation results. The Optimal DF Formula is computed here with the same average human parameters used throughout.

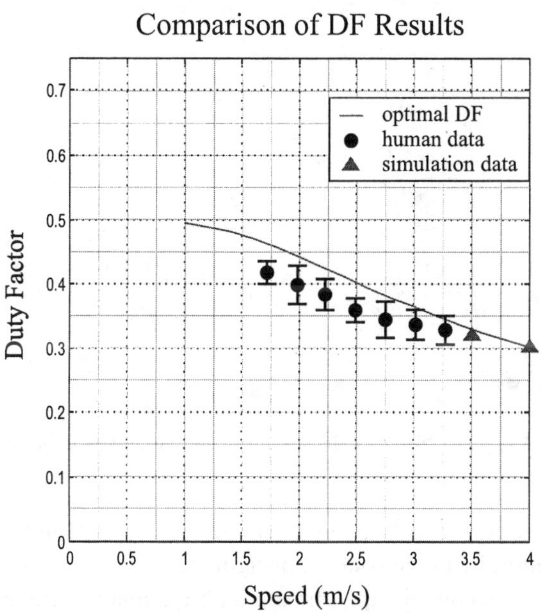

Fig. 6. The Optimal DF Formula closely corresponds to both simulation results and natural human DF's. The experimental human data is provided by Minetti and is displayed as a mean ± S.D.[14]

7. Conclusion

This work answers the following basic question: *I want to design a biped robot to run at a target speed. What DF should I design it for? And what stiffness value does it need?*

In designing a running biped, selecting a stiffness to meet a desired gait specification is not trivial. It depends on the speed, configuration, and DF in highly-nonlinear systems. The DF itself warrants careful consideration, given its effect on the energy consumption and impact forces of the gait.

It is one thing to design a compliant robot that can run stably; it is another thing to design it for set gait specifications. This work presents formulas and design laws that supplement the design process for a running biped robot.

Appendix A. The Simulation Model

The simulation system modeled a telescoping legged biped, shown in Fig. A1. The model consists of three rigid bodies, representing the torso and two legs, connected by a revolute joint at the hip. The feet are point-masses connected by prismatic joints to the legs.

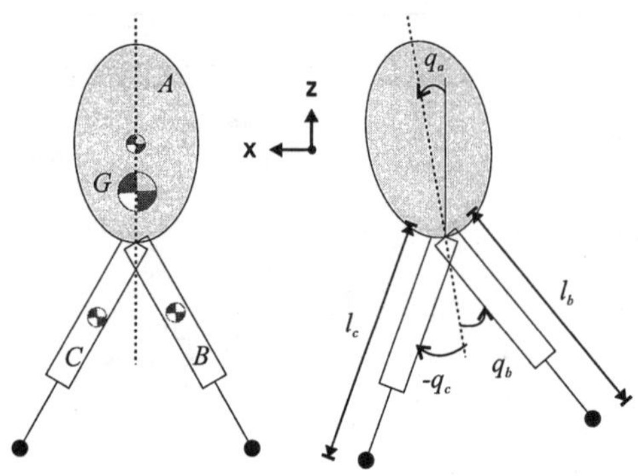

Fig. A1. The biped model used for the dynamic simulation.

A fully dynamic simulation was created. The foot-ground contact was modeled as a rigid, inelastic contact using motion constraints. Friction was assumed sufficient to avoid slip. The model implemented the control strategy of Abdallah and Waldron for running.[17] It's parameters modeled an average human as shown in Table A1. A detailed description of the modeling and experiment results is available elsewhere.[15]

Table A1. Model Parameters

	distance from hip to CM (m)	mass (kg)	inertia (kg m^2)
torso	0.34	48.3	8.12
leg	0.384	11.5	1.03
foot	0.8 (nominal)	1.0	0

References

1. R. Alexander, *Elastic Mechanisms in Animal Movement* (Cambridge University Press, Cambridge, 1988).
2. M. Raibert, *Legged Robots that Balance* (MIT Press, Cambridge, MA, 1986).
3. J. Nichol, L. Palmer and K. Waldron, Design of a leg system for quadraped gallop, in *Proceedings of the 11th Congress in Mechanism and Machine Science*, ed. T. Huang (China Machinery Press, Tianjin, China, August 2003).
4. M. Ahmadi and M. Buehler, The ARL Monopod II running robot: control and energetics, in *IEEE International Conference on Robotics and Automation*, (Detroit, MI, 1999).
5. B. Brown and G. Zeglin, The bow leg hopping robot, in *IEEE International Conference on Robotics and Automation*, (Leuven, Belgium, 1998).
6. H. Rad, P. Gregorio and M. Buehler, Design, modeling and control of a hopping robot, in *IEEE/RSJ Conference on Intelligent Systems and Robots*, (Yokohama, Japan, 1993).
7. J. P. Schmiedeler and K. J. Waldron, Leg stiffness and articulated leg design for dynamic locomotion, in *Design Engineering Techinal Conferences and Computers and Information in Engineering Conference*, (Montreal, Canada, 2002).
8. M. Ahmadi and M. Buehler, *IEEE Transactions on Robotics and Automation* **13**, 96(February 1997).
9. R. Alexander, *Phil. Trans. of the Royal Society of London* **338**, 189 (1992).
10. R. Blickhan, *Journal of Biomechanics* **22**, 1217 (1989).
11. T. A. McMahon and G. C. Cheng, *Journal of Biomechanics* **23**, 65 (1990).
12. W. Schwind, Spring loaded inverted pendulum running: a plant model, PhD thesis, University of Michigan, (Ann Arbor, MI, 1998).
13. G. Cavagna and M. Kaneko, *Journal Physiology* **268**, 467 (1977).
14. A. E. Minetti, *Journal of Biomechanics* **31**, 463 (1998).
15. M. Abdallah, Mechanics motivated control and design of biped running, PhD thesis, Stanford University, (Stanford, CA, 2007).
16. D. Hoyt and R. Taylor, *Nature* **292**, 239(July 1981).
17. M. Abdallah and K. Waldron, A physical model and control strategy for biped running, in *IEEE International Conference on Robotics and Automation*, (Rome, Italy, 2007).

Constraint Based Trajectory Simplification of Full Body Trajectories for a Walking Robot

Hanns W. Tappeiner, Alfred A. Rizzi

The Robotics Institute, Carnegie Mellon University, Pittsburgh, PA 15213, USA
E-mail: hanns@ri.cmu.edu
http://www.ri.cmu.edu/people/tappeiner_hanns.html

Keywords: Walking; Legged Robots; Trajectory Simplification;

1. Introduction

In this paper we present an algorithm and experimental results on a real robot which solves two related problems: First we simplify a robot trajectory created by a human driver by reducing the amount of control points, according to a set of constraints. Second - as a side effect - we eliminate local small scale driver errors. The goal is to generate trajectories which are closer to what the driver "intended" to do and can be used as good examples for robot learning strategies.

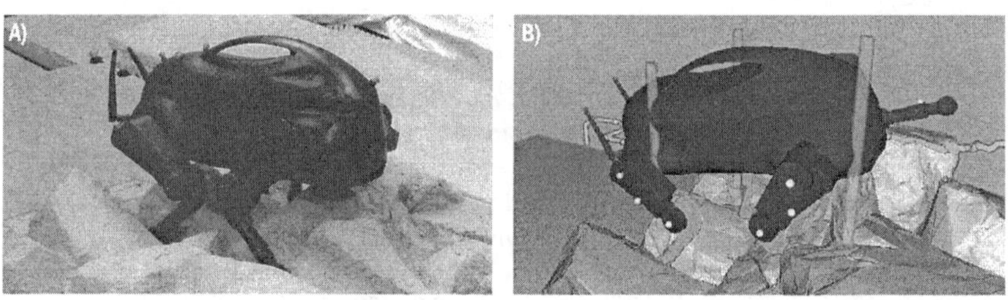

Fig. 1. A) Robot LittleDog climbing over a terrain board. B) An accurate 3D model of both LittleDog and terrain is used to perform full body collision detection, COM checks etc. The arrows mark collision points, the orange line is the robots COM position trajectory over the terrain.

In our experiments with controlling a four legged walking robot over

rough terrain, it turns out that a skilled human driver, equipped with a good control device like a joystick or gamepad, can outperform our used autonomous control techniques. Even though controlling a 12dof, four legged robot with a joystick is not straightforward, it seems that the driver developes a level of understanding of the terrain and the corresponding control actions of the robot which is more effective than our autonomous control techniques.
It seems reasonable to consider using examples from a human driver to learn autonomous behavior for the robot. One problem is that a driver does not create "clean" trajectories but often introduces irrelevant motion or even makes mistakes, especially at the small scale: for example the operator may try to place a leg several times before actually ending up in a good spot. Additionally it is difficult for an operator to control multiple axes at the same time with a joystick, and the result is somewhat jerky motion of the robot. Even though the overall trajectory is probably very good and captures the idea the driver had to solve the problem, it will contain a lot of unneccessary noise at the smaller scale. We would like to remove that "noise" and produce trajectories which are closer to what the driver intended to do rather then what he actually commanded.

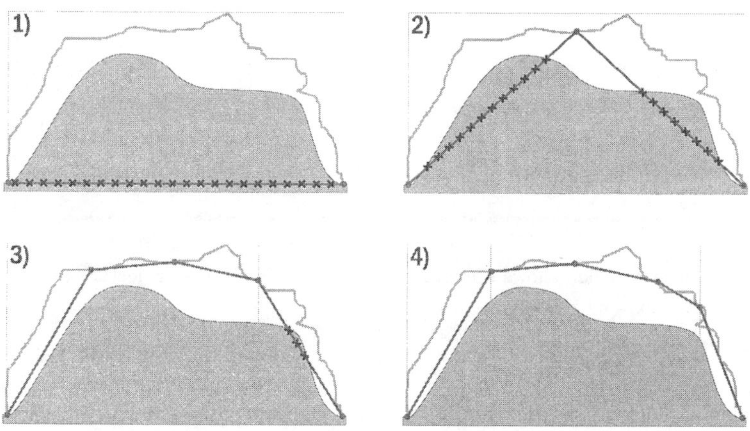

Fig. 2. Example trajectory and the corresponding simplified trajectory. The red x indicate constraint violations, in this case simply unwanted collisions with the terrain.

Secondary, our recordings of trajectories generated by a human driver contain control points at every timestep, which means a large amount of data. It is possible that there are only very few key control points in the motion which need to be there and everything else is much less important. Reducing the number of control points may make trajectory learning more effective.

2. The basic algoritm

Figure 2 shows four subsequent snapshots of the algorithm simplifying a trajectory. Figure 3 the result of applying the algorithm to a real robot trajectory. We start with a trajectory, containing control points at every timestep. In addition, we define a set of constraints. A robot state s_n at a certain point in time t_n is valid if all the constraints are fullfilled. We could remove contol points from the original trajectory and interpolate between the remaining points until we violate some of the constraints.

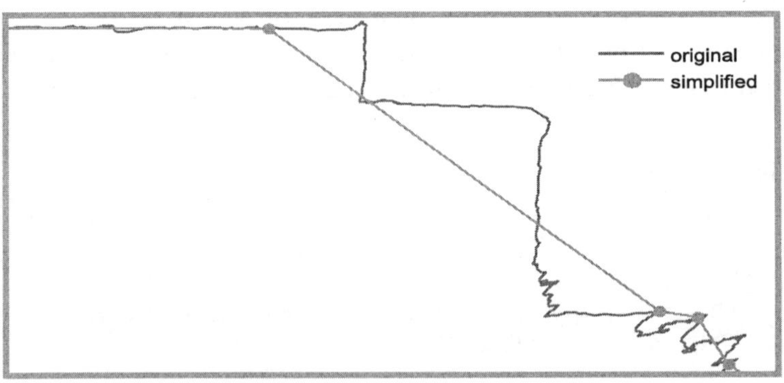

Fig. 3. Closeup which shows operator introduced noise in the trajectory and the corresponding simplified version.

In this version, we do the inverse. We start with only two control points in the trajectory, the first and the last, interpolate between them and check the constraints. If we violate some of them, we add an additional point from the original trajectory in between and check the constraints again. We keep doing this until the whole trajectory is valid. As we will see later, the real application requires a set of more complex constraints (approximate collisions, valid center of mass) than 2D collisions to generate valid trajectories.

3. Test Results

In our experiments, a full body state of the robot is defined by 19 parameters: Position (x,y,z) and orientation (quaternion) of the body in the world and the 12 joint angles of the legs.

$$S = \{x, y, z, \theta_x, \theta_y, \theta_z, \theta_w, \phi_1, ..., \phi_{12}\}$$

The algorithm adds control points S_n to the simplified trajectory and interpolates between them until no constraints are violated. There are two main groups of constraints: collisions with the terrain and keeping the center of mass (COM) above the support triangle. The algorithm has to keep existing wanted collisions (like footsteps), but not introduce unwanted collisions. The simplified trajectory is valid if:

(1) It produces the same collisions (time τ and position ω) with the environment as the original: $||\tau_o - \tau_s|| < \epsilon_t$, $||\omega_o - \omega_s|| < \epsilon_p$
(2) It has a COM above the support triangle of the stance feet at any time the original has one.

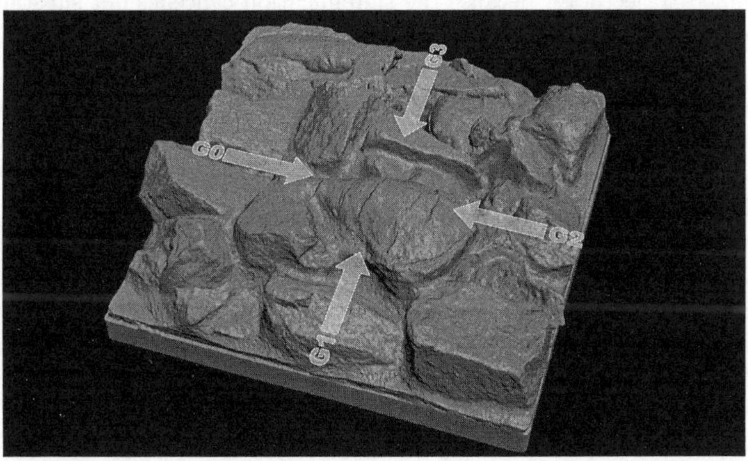

Fig. 4. 3D model of TerrainG with the four tested walking directions G0, G1, G2 and G3.

Depending on the complexity of the terrain the algorithm is able to reduce the amount of control points from about 92% (complex rock terrain) to 99.7% (flat ground). Figure 4 shows one of the terrain boards from the

LittleDog program, TerrainG. For the two directions G0 and G2 a human operator created a trajectory that reliably crossed the terrain at different speeds. Those trajectories have been simplified according to the algorithm described in this paper, and original and simplified version have been compared in terms of behavior on the real robot and number of remaining control points. Table 1 shows the result of this comparison.

Table 1. Comparison of original and simplified trajectory properties on TerrainG.

Trajectory	Control Points	Fastest run [sec]	Average joint speed [rad/sec]
Original G0	21240	21 (10x)	0.0758
Simplified G0	301 (1.4%)	15 (14x)	0.0624
Original G2	27417	28 (10x)	0.0825
Simplified G2	417 (1.5%)	19 (14.7x)	0.0702

The amount of control points is usually reduced by about 97% and the maximum overall speed is higher for the simplified trajectory because of smoother body and leg motions. Also, even though the maximum forward speed is higher, the average joint speed during the runs is lower in the simplified trajectories, because the simplification cancelles out substancial parts of unnecessary motions (parts of noise and driver mistakes).

Figure 5 shows views of a full body trajectory for each leg and robot body (z coordinate over time) on a rock board. The original trajectory contains 12780 control points, the corresponding simplified trajectory 47 control points, which is 0.37% of the original. This plot shows 2D views of the high dimentional trajectory.

Videos of the simplification process, the robot crossing TerrainG using original and simplified trajectories etc. can be found online at:
`http://www.cs.cmu.edu/~hanns/clawar07/index.htm`

4. Summary

In this paper we developed a simple but effective algorithm to reduce the amount of control points in legged robot trajectories according to a set of defined constraints. The effect is that the simplified trajectories are cleaner in terms of noise and small scale driver errors than the originals and are likely to be closer to what the driver intended to do. We think that such trajectories will be more usefull as examples for learning algorithms and be better trajectories for trajectory libraries [2]. The results of the simplifica-

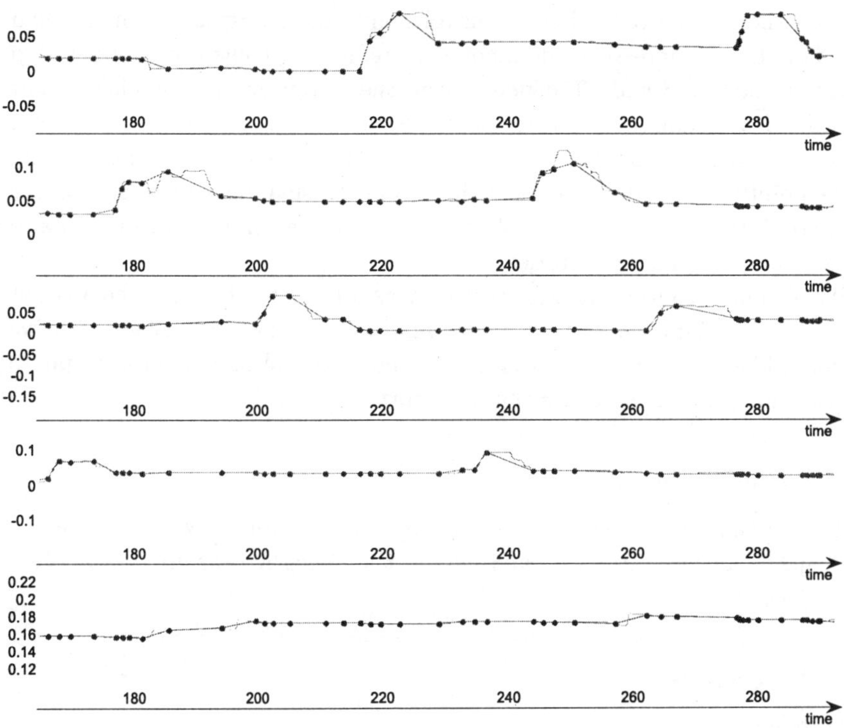

Fig. 5. Z coordinate over time views of a trajectory for all four legs (FL, FR, HL, HR) and body, both original and simplified.

tion have been tested on the real robot as well as in simulation and have been compared to original trajectories on the robot. In all experiments, the described algorithm produced trajectories which contain only a fraction of the control points of the original trajectories. The simplified versions are not only executable on the robot, but allow the robot to move at a faster overall speed with smoother motions.

5. Future work

There are three main improvements to the algorithm which we think would be very useful. The first one is that the actual constraints don't take into account robot dynamics. This turns out to be ok at most of the speeds we are walking at so far, but as the robot becomes faster, dynamics plays a more important role and should be part of the constraints and the point selection algorithm.

Even though the reduction of control points by almost always more than 95% is quite impressive, the amount of remaining control points per step varies between 3 and 30, depending on the terrain etc (there where a few outliers with about 90 control points per step). In some cases a different point selection algorithm would produce better results, e.g. by not picking the point in the middle, but near the actual collision. Instead of using linear interpolation, we plan to fit splines trough the remaining control points to get even smoother trajectories.
We also plan to use this algorithm only as a first step to guess the amount of control points needed to generate valid trajectories. In a second step, we would like to optimize the positions of those control points, for example to maximize the robustness of the trajectory.

6. Acknowledgements

The LittleDog project is supported by the Defense Advanced Research Projects Agency (DARPA) and part of the DARPA Learning Locomotion Project.

7. References

References

1. Footstep Planning for the Honda ASIMO Humanoid - J. Chestnutt, M. Lau, K.M. Cheung, J. Kuffner, J.K. Hodgins, and T. Kanade (Proceedings of the IEEE International Conference on Robotics and Automation, April, 2005).
2. Policies based on trajectory libraries. Martin Stolle and Christopher G. Atkeson (Proceedings of the International Conference on Robotics and Automation (ICRA), 2006).
3. A concept for manipulator trajectory planning - F. Pfeiffer and R. Jahnanni (Proceedings of the IEEE International Conference on Robotics and Automation, pages 1399-1405, 1986).
4. An integrated method for planning smooth collisionfree trajectories for robot arms - J. Zhang (Proceedings of Third Annual Conference on AI, Simulation and Planning in High Autonomy Systems, Perth, pages 46–53, 1992).
5. Case Studies in Trajectory optimization: Trains, Planes, and other Pastimes - Robert J. Vanderbei (Operations Research and Financial Engineering Princeton University, 2000).

FOOT PLANNING MOTION OF HUMANOID ROBOT RH-1 USING LAG ALGORITHM*

MARIO ARBULU, LUIS CABAS, DMITRY KAYNOV, PAVEL STAROVEROV, CARLOS BALAGUER†

Department of Systems and Automation Engineering, University Carlos III of Madrid, Av. de la Universidad 30
Leganés, Madrid 28911, Spain

Real-time gait generation algorithm for humanoid robot is presented in this paper. The 3D foot motion planning for the humanoid global motion is developed in order to walk in any surface as plane, ramp, climbing stairs. Furthermore, it is possible continuous change the step length and orientation in real time.. The proposed algorithm takes into account physical robot constraints as joint angles, angular velocity and torques. Torques are computed by Lagrange method under Lie groups. Some results are shown and discussed..

1. Introduction

The gait pattern generation of humanoid robots has several solutions and it is a no closed problem, so we propose in this work a feasible solution of foot planning for being implemented in real time. Two reference trajectories should be introduced to the humanoid: The COM trajectory and swing foot trajectory. The COM trajectory [5], [6] must maintain the robot stability while it is walking. The swing foot trajectory [1], [3] should change the humanoid position avoiding support leg collision, obstacle collision, maintain stability limit (ZMP in stable zone [2], [4]) and adapt on landing surface, reduce the reaction force on the floor when it is landing, and reduce the joints angular velocity in particular of knee joint. In this work, foot planning motion of Rh-1 humanoid robot are described in order to get stable walking, and decide in real time the foot placement, so the "Local Axis Gait" (LAG) algorithm is proposed. The performance of this algorithm is described as following: at first, goal foot position and orientation should be decided; it is composed by 3 parameters at local suitable axis: step length ($\boldsymbol{L_x}$), lateral foot motion ($\boldsymbol{L_y}$) and final foot orientation ($\boldsymbol{L_\theta}$). After that, a 5th order polynomial interpolator is proposed for going from actual foot position

* This work is supported by CICYT (Comisión Interministerial de Ciencia y Tecnología).

to goal foot point at local axis, which allows us to control the position, velocity and acceleration of the foot motion. Human like foot motion is obtained and smooth joints motion could reduce the mechanical efforts. For avoiding support leg collision a preview motion range study of swing foot had been done. Furthermore, for avoiding obstacle collision, maintain stability limit and adapt on landing surface, robot sensors decide the goal foot position and orientation. The present work is divided as following: at first, in section II, an overview of foot motion planning is developed. In section III, motion interpolators are developed, and planning foot position and orientation are introduced. In the section IV, dynamics analysis is detailed in order to obtain the torque limits on each robot joint by using Lagrange formulation, Lie groups and screw theory. In section V, some simulation on Rh-1 simulator are shown and discussed. Finally in section VI the conclusions of this work are presented.

2. Overview of foot motion planning

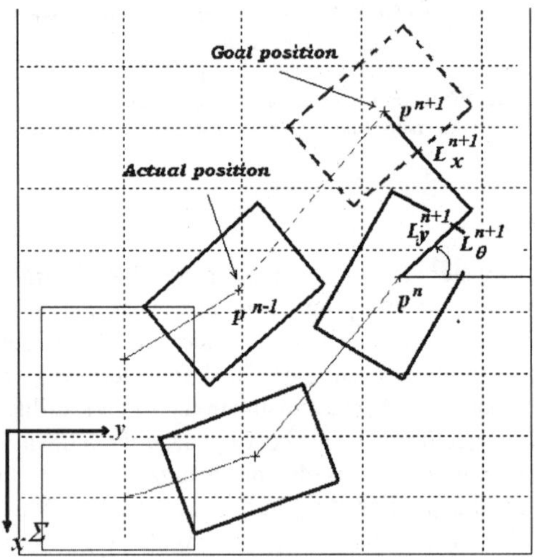

Figure 1. Foot motion planning, at local axis in p^n (support foot) for doing ***n-th*** step

The foot motion planning for doing the *n-th* step should be modeled, as shown in Fig. 1. The swing foot moves from actual configuration (position and orientation *(n-1)-th* frame) to the goal configuration (position and orientation *(n+1)-th* frame).

The input parameters are the swing foot goal configuration (position and orientation), defined in appropriate local axis as: $(Lx, Ly, L_\theta)^{n+1}$. It is possible to identify for the *n-th* step the parameters and frames as following:

$\sum$: World frame

$\sum^{n}, \sum^{n-1}, \sum^{n+1}$: *n, n-1, n+1* local configuration frame, support foot

p^{n}, p^{n-1}, p^{n+1} : *n, n-1, n+1* position of support foot local frame

$L_x^{n+1}, L_y^{n+1}, L_\theta^{n+1}$: goal swing foot motion and orientation

Goal foot configuration (position, $\boldsymbol{p}^{n+1}$, and orientation, $\boldsymbol{L}_\theta^{n+1}$) are the input parameters for doing the next step. It could been obtained by humanoid sensors or external command. Those input parameters could be generalized in order to compute the *n-th* step in real time as:

$$\begin{bmatrix} p_x^{(n)} \\ p_y^{(n)} \end{bmatrix} = \begin{bmatrix} p_x^{(n-1)} \\ p_y^{(n-1)} \end{bmatrix} + \begin{bmatrix} \cos\left(L_\theta{}^{(n)}\right) & -\sin\left(L_\theta{}^{(n)}\right) \\ \sin\left(L_\theta{}^{(n)}\right) & \cos\left(L_\theta{}^{(n)}\right) \end{bmatrix} \begin{bmatrix} L_x^{(n)} \\ -(-1)^{(n)}.L_y^{(n)} \end{bmatrix}. \tag{1}$$

As see in (1), goal foot configuration depends of support foot position, because the support foot is the local axis of gait input parameters.

3. Swing foot trajectory

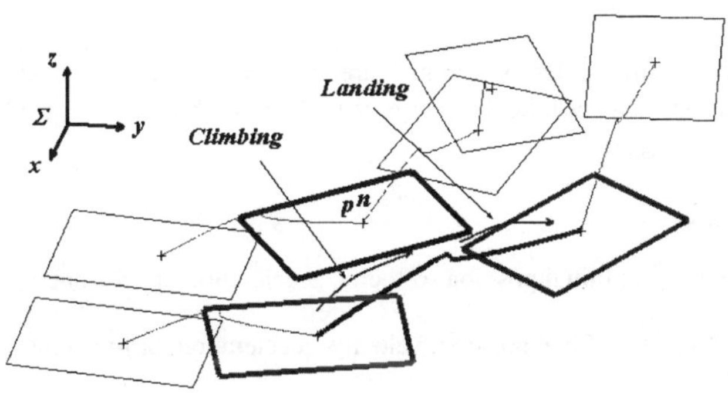

Figure 2. . Swing foot motion planning changing walking direction in real time. The ***n-th*** step trajectory in black bold line, and swing foot is represented by the red bold rectangles, while the support foot is represented by the blue bold rectangle

Polynomials have been used for approximation because they can be evaluated, differentiated, and integrated easily and infinitely many steps using just basic arithmetic operations of addition, subtraction and multiplication (2). A polynomial of order $\boldsymbol{n}$ or degree $\boldsymbol{n}$ is a function of the form:

$$\phi(t) = w_1 + w_2.t + ... + w_n.t^{n-1} = \sum_{k=1}^{n} w_k.t^{k-1} \quad (2)$$

Where $\phi(t) \in P^n$, being $\boldsymbol{P^n}$ the set or linear space of all polynomials of order $\boldsymbol{n}$.

The swing foot motion is planned by two fifth order interpolators (3), one for climbing the foot ($\boldsymbol{\phi_1(t)}$) and one for landing the foot ($\boldsymbol{\phi_2(t)}$), on each axis motion and orientation motion.

Those ones allow to control the foot position, velocity and acceleration, so it is minimized the foot landing reacting force, the knee angular velocity, smooth and natural joints motion is obtained.

$$\phi(t) = \begin{cases} \phi_1\left(t, x_i^1, \dot{x}_i^1, \ddot{x}_i^1, x_f^1, \dot{x}_f^1, \ddot{x}_f^1\right) \quad \forall \quad 0 \le t < \frac{T}{2} \\ \phi_1(t) = w_6^1.t^5 + w_5^1.t^4 + w_4^1.t^3 + w_3^1.t^2 + w_2^1.t + w_1^1 \\ \\ \phi_2\left(t, x_i^2, \dot{x}_i^2, \ddot{x}_i^2, x_f^2, \dot{x}_f^2, \ddot{x}_f^2\right)_2 \quad \forall \quad \frac{T}{2} \le t \le T \\ \phi_2(t) = w_6^2.t^5 + w_5^2.t^4 + w_4^2.t^3 + w_3^2.t^2 + w_2^2.t + w_1^2 \end{cases} \quad (3)$$

Where the fifth order polynomial are characterized by the next boundary conditions on each zone; i.e. for *j-th* zone (*j*=1for climbing, *j*=2 for landing), as it have seen in Fig. 2:

T : Step time

$(x_i^j, \dot{x}_i^j, \ddot{x}_i^j)$: Initial position, velocity, acceleration of *j-th* zone

$(x_f^j, \dot{x}_f^j, \ddot{x}_f^j)$: Final position, velocity, acceleration, of *j-th* zone

So the polynomial coefficients could be developed as following from boundary conditions described above:

$$w_1^j = x_i^j \quad (4)$$

$$w_2^j = \dot{x}_i^j \quad (5)$$

$$w_3^j = \frac{\ddot{x}_i^j}{2} \tag{6}$$

$$w_4^j = 10.\left(\frac{x_f^j - x_i^j}{T^3} - \frac{\dot{x}_i^j}{T^2} - \frac{\ddot{x}_i^j}{2.T}\right) - 4.\left(\frac{\dot{x}_f^j - \dot{x}_i^j}{T^2} - \frac{\ddot{x}_i^j}{T}\right) + \frac{1}{2}.\left(\frac{\ddot{x}_f^j - \ddot{x}_i^j}{T}\right) \tag{7}$$

$$w_5^j = -15.\left(\frac{x_f^j - x_i^j}{T^4} - \frac{\dot{x}_i^j}{T^3} - \frac{\ddot{x}_i^j}{2.T^2}\right) + 7\left(\frac{\dot{x}_f^j - \dot{x}_i^j}{T^3} - \frac{\ddot{x}_i^j}{T^2}\right) - \left(\frac{\ddot{x}_f^j - \ddot{x}_i^j}{T^2}\right) \tag{8}$$

$$w_6^j = 6.\left(\frac{x_f^j - x_i^j}{T^5} - \frac{\dot{x}_i^j}{T^4} - \frac{\ddot{x}_i^j}{T^3}\right) - 3.\left(\frac{\dot{x}_f^j - \dot{x}_i^j}{T^4} - \frac{\ddot{x}_i^j}{T^3}\right) + \frac{1}{2}\left(\frac{\ddot{x}_f^j - \ddot{x}_i^j}{T^3}\right) \tag{9}$$

The continuity at the ending of climbing zone and at the starting the landing one, should satisfied the next conditions:

$$\begin{cases} \phi_1(T/2) = \phi_2(T/2) \\ \dot{\phi}_1(T/2) = \dot{\phi}_2(T/2) \\ \ddot{\phi}_1(T/2) = \ddot{\phi}_2(T/2) \end{cases} \tag{10}$$

So, the a set of six splines for swing foot motion are obtained, three for spatial motion and three for spatial orientation. Furthermore, natural and smooth foot motion could be developed at low computational cost for being applying in real time.

4. Dynamic Modelling

4.1. *Lagrange formulation*

In order to compute the joint torques and dynamics constraints, dynamic model should be proposed, so Lagrange formulation under lie groups and screw theory had been developed [7]; because that give us a natural description of jacobian manipulator.

The Lagrange formulation could be expressed as following (Lagrangian):

$$L(\theta, \dot{\theta}) = K(\theta, \dot{\theta}) - V(\theta) \tag{11}$$

Being the motion equation by Γ torques the next:

$$\frac{d}{dt}.\left(\frac{\partial L}{\partial \dot{\theta}}\right)-\frac{\partial L}{\partial \theta}=\Gamma \tag{12}$$

Which could be expressed as following:

$$Mm(\theta).\ddot{\theta}+N(\theta,\dot{\theta})=\Gamma \tag{13}$$

So, $Mm(\theta)\in \Re^{nxn}$ is the inertial manipulator matrix and it is defined by the operational jacobian manipulator J_{sli} and the inertial generalized matrix M_i as (14):

$$Mm(\theta)=\sum_{i}^{n} J_{sli}^{T}(\theta).M_{i}.J_{sli}(\theta) \tag{14}$$

The Rh-1 humanoid robot has 21 degrees of freedom, for this fact $Mm(\theta)\in \Re^{21x21}$.

The $N(\theta,\dot{\theta})$ term give us the joints torque due to the potential forces, and it is defined as (15):

$$N(\theta,\dot{\theta})=\frac{\partial V}{\partial \theta} \tag{15}$$

From (15), $V(\theta)=\sum_{i=1}^{n} m_i.g.h_i(\theta)$, where m_i is the mass of the *i-th* link, g is the gravity acceleration and $h_i(\theta)$ the COM height of the *i-th* link.

5. Simulation results

5.1. *Rh-1 humanoid robot platform*

Several simulation tests have been done in Rh-1 simulator software, developed under VRML environment, before to test in Rh-1 humanoid robot platform (Fig. 3), which specifications are described in Table I.

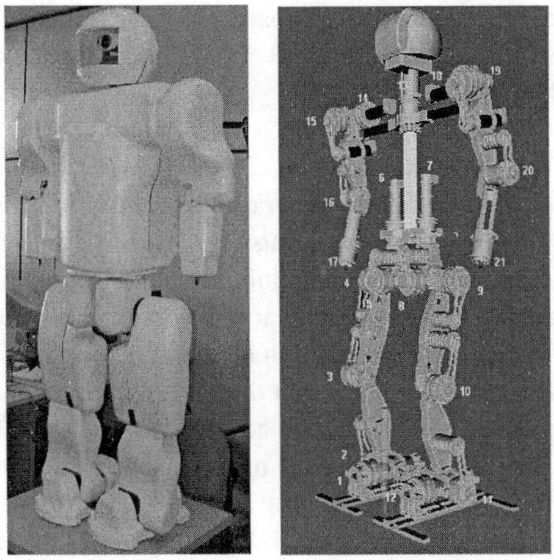

(a) (b)

Fig. 3. Rh-1 humanoid robot (a) and its mechanical configuration (b)

Table 1. Rh-1 humanoid robot specifications

Link	Number of DOF	Total	Joint	Human angular range	Rh-1 angular range
Head	-	-	*Head*	-50 ° to 50° (Roll)	-
				-50 ° to 60° (Pitch)	-
				-70 ° to 70° (Yaw)	-
Waist	1 (Yaw)	1	*Waist*	-50° to 50° (Roll)	- (Roll)
				-30° to 45° (Pitch)	- (Pitch)
				-40° to 40° (Yaw)	-45° to 45° (Yaw)
Arm	4	4 x 2	*Arm*		
	1 (Pitch)		*Shoulder*	-90° to 0° (Roll)	-95° to 0° (Roll)
Shoulder	1 (Roll)			-180° to 50° (Pitch)	-180° to 60° (Pitch)
				-90° to 90° (Yaw)	-
	1 (Pitch)		*Elbow*	-145° to 90° (Pitch)	-135° to 0° (Pitch)
Elbow				-90° to 90° (Yaw)	-
Wrist	1 (Roll)		*Wrist*	-55° to 25° (Roll)	-90° to 90° (Roll)
				-70° to 90° (Pitch)	-
Leg	6	6 x 2	*Leg*		
Hip	1 (Yaw)		*Hip*	-45° to 20° (Roll)	-20° to 20° (Roll)
	1 (Roll)			-125° to 15° (Pitch)	-80° to 80° (Pitch)
	1 (Pitch)			-45° to 45° (Yaw)	-15° to 15° (Yaw)
Knee	1 (Pitch)	1 x 2	*Knee*	0° to 130° (Pitch)	0° to 100° (Pitch)
Ankle	1 (Roll)	2 x 2	*Ankle*	-20° to 30° (Roll)	-20° to 20° (Roll)
	1 (Pitch)			-20° to 45° (Pitch)	-25° to 25° (Pitch)

5.2. *Gait patterns and constraints*

Some gait patterns should be seen in Fig. 4, taking into account the angular joint range limits as see in Table 1. The proposed gait generation algorithm allows to change directions while the humanoid is walking, continuous step length changing and walking in any surface.

Angular velocity constraints have been satisfied as actuator limits, thus the smooth gait patterns could be validated; i.e the actuator knee angular velocity limit is 8000 RPM, the gait pattern generated is suitable for this requirement as we can see in Fig. 5. The next two joints (Fig. 6) drive the sagital and frontal ankle support leg motion, 2 and 1. The joints actuator torque limits are defined by the reduction gear and motor nominal torque. So, the limit nominal torque for actuators 11 is 200 N.m, as it is shown in the chart that each joint never overlap the nominal torque limit; it is notice that the knee joint have the highest torque, the gait pattern generated should have in account to reduce it. The 1 joint torque limit is 17.6 Nm, the chart shown 12 N.m as maximum, it is near the nominal one, because the frontal ankle joint has less reduction respect to other ones. Furthermore, this joint support overall humanoid robot mass during single support stage and it drives the frontal balance in order to maintain the ZMP inside the convex hull.

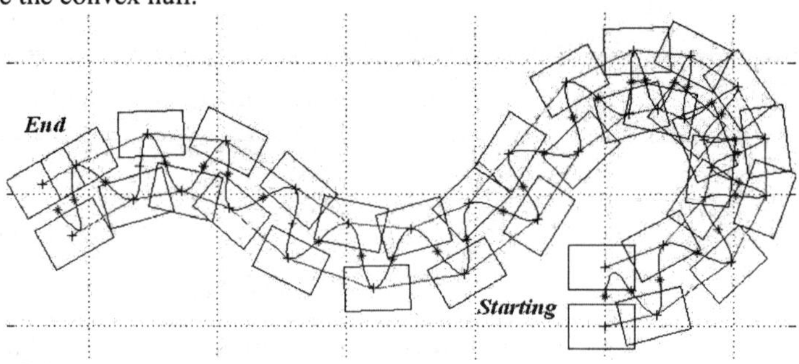

Fig. 4. Gait patterns generated for continuous humanoid walking. It is possible to generate any direction continuous walking

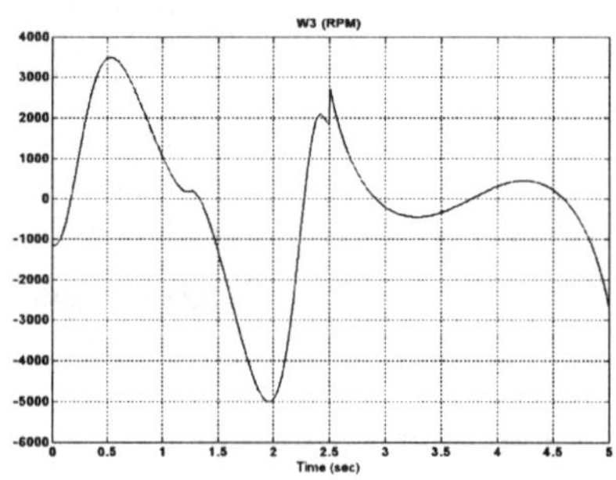

Fig. 5 The angular velocity requirement by knee joint, its actuator run up 8000 RPM

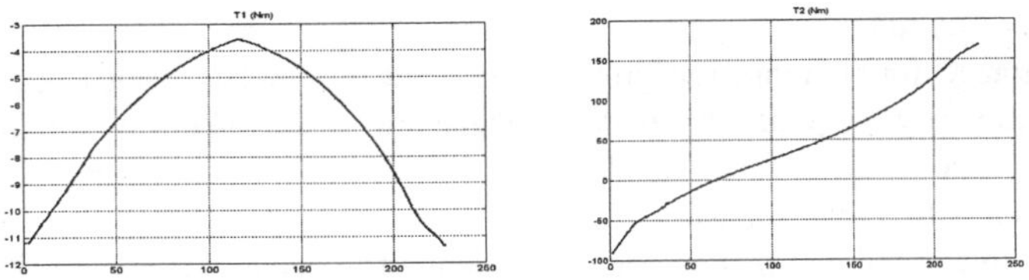

Fig. 6. Ankle joints torque of the support foot in the *n-th* step

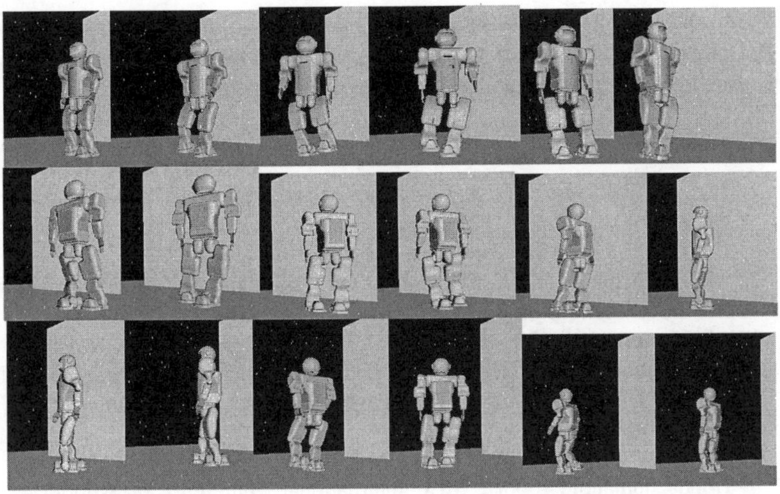

Fig. 7. Snapshots of VRML Rh-1 simulation, for smooth naturally and continuous walking as Figure 4.

5.3. *VRML Rh-1 simulations*

Snapshots for doing a circle and curve path on flat surface have shown in Fig. 7. Natural and smooth change direction could be done by the proposed algorithm, thus natural walking is obtained as human one. Furthermore, faster walking motion should be developed until 1 Km/h at any direction on flat surface; in another kind of surface slower walking motion by physical constraints and stability one would be obtained.

6. Conclusion

Foot motion planning had been developed for applying in real time, by satisfying the stability criterion on bipedal walking robots, joint angle, angular velocity and torque conditions.

Lagrange formulation, Lie groups and screw theory had been employed in order to verify the joint torques in the humanoid robot, while it is walking, thus dynamics constraints are obtained.

Simulation results on Rh-1 simulator environment had been shown, and it allows to test the gait patterns before to apply it in actual humanoid robot. The next step is to test the above algorithm in actual Rh-1 humanoid robot platform.

Acknowledgments

We would like to thank the members of the Robotics Lab of University Carlos III of Madrid for their cooperation an suggestions. We also express acknowledgment to the Humanoid team for their support.

References

1. H.Lim, Y.Kaneshima, A.Takanishi, et al., "Online Walkinge Pattern Generation for Biped Humanoid Robot with Trunk," Proc. of IEEE Int. Conf. on Robotics and Automation, pp.3111-pp.3116, 2002
2. M. Vukobratovic and A. Frank, "One the Gait Stability of Biped Machines", IEEE Transactions on Automatic Control, Dec. 1970.
3. M. Morisawa, K. Harada, S. Kajita, F. Kanehiro, K. Kaneko, K. Fujiwara, S. Nakaoka and H. Hirukawa, "A Biped Pattern Generation Allowing Immediate Modifications of Foot Pacement in Real-time", Humanoids '06, Dec. 2006.
4. K. Hirai, M. Hirose, Y. Hikawa and T. Takanaka, "The development of Honda humanoid robot", IEEE International Conference on Robotics and Automation (ICRA 1998) Leuven (Belgium)
5. Kajita, S., Kanehiro, F., Kaneko, K., Fujiwara, K., Harada, K., Yokoi, K. and Hirukawa, H., "Biped Walking Pattern Generation by using Preview Control of Zero-Moment Point," Proc. of ICRA2003, pp.1620-1626, 2003.
6. M.Arbulú; L.M.Cabas; P.Staroverov; D.Kaynov; C.Pérez; C.Balaguer. "On-line walking patterns generation for Rh-1 Humanoid Robot using a simple three-dimensional inverted pendulum model". 9th Internacional Conference on Climbing and Walking Robots (Clawar 2006). Brussels. Belgium. Sep, 2006.
7. R.M. Murray, Z. Li, and S.S. Sastry, "A Mathematical Introduction to Robot Manipulation", CRC Press, 1993

GAIN PROPERTY FOR BIPED WALKING VIA LEG LENGTH VARIATION

Tetsuya Kinugasa†‡, Shoichi Miwa‡, Yannick Aoustin† and Christine Chevallereau†

†Institute de Recherche en Communications et en Cybernetique de Nantes (IRCCyN)
BP 92101, 1 rue de la Noë, 44321 Nantes cedex 03, FRANCE
‡Okayama University of Science
1-1, Ridai-cho, Okayama, 700-0005, JAPAN
{Tetsuya.Kinugasa, Yannick.Aoustin, Christine.Chevalereau}@irccyn.ec-nantes.fr, miwa@mech.ous.ac.jp

The purpose of this paper is to show a gain property of frequency response analysis for a biped based on passive dynamic walking mechanism. The walking method is composed of a cyclic length variation of legs and a body constraint control. Some simulated results show that the walking gain property represents an efficiency in the sense of consumed energy. It was found from the result that walking frequencies only depended on frequencies of the leg length variation, and there is a resonance frequency which can be called a walking resonance frequency. Finally, our biped called Prototype Biped Emu-IV (PBEmu-IV) is introduced, and some experimented results are shown in order to verify the effectiveness of the frequency analysis.

Keywords: frequency analysis; passive dynamics based walking; variable length leg.

1. Introduction

It has been reported that Passive Dynamic Walking (PDW)[1] by McGeer is extremely natural and efficient. There was no reference trajectory and walking is not constrained by ZMP. Almost of all basic ideas for actuated walking based on PDW were given in his paper (e.g.[2]). Now many researchers are trying to give improvements and to realize such kind of walking by experiments with stability analysis (e.g. in the literature[3,4]). We are focus on walking by cyclic variation of leg length which is one of the methods of McGeer. However the given analysis is only for stability of a cyclic motion.

The objective of this paper is, therefore, to analyze the frequency response of biped walking by the sinusoidal leg length variation. The gain property is defined and it gives the same efficiency of walking as consumed energy. Finally, some simulations and experiments are shown to verify a validity of the analysis.

2. The Biped Modeling

2.1. *Dynamic model of single support phase*

The biped appeared in Figure 1 is composed of a body and two symmetric legs which consist of prismatic actuated knees and spherical feet. Feet radius are chosen so that the biped can be originally stable for its upright position. The hips are rotational actuated joints with a little viscosity friction. We assume that there is no slip between the spherical foot and the ground. The equation of motion is as follows:

$$D_e(q_e)\ddot{q}_e + \{h_e(q_e, \dot{q}_e) + C\}\dot{q}_e + G_e(q_e) = B_e\Gamma + D_{R1}(q_e)R_1, \quad (1)$$

$$\ddot{x}_H = -\{(l_1 - R)\ddot{q}_1 - 2\dot{q}_1\dot{l}_1\}\cos q_1 - (\ddot{l}_1 + (l_1 - R)\dot{q}_1^2)\sin q_1 - R\ddot{q}_1, \quad (2)$$

$$\ddot{y}_H = -\{(l_1 - R)\ddot{q}_1 + 2\dot{q}_1\dot{l}_1\}\sin q_1 + (\ddot{l}_1 - (l_1 - R)\dot{q}_1^2)\cos q_1. \quad (3)$$

$q_e = [q_1, q_2, q_3, l_1, l_2, x_H, y_H]'$, and x_H and y_H are the hip position. The

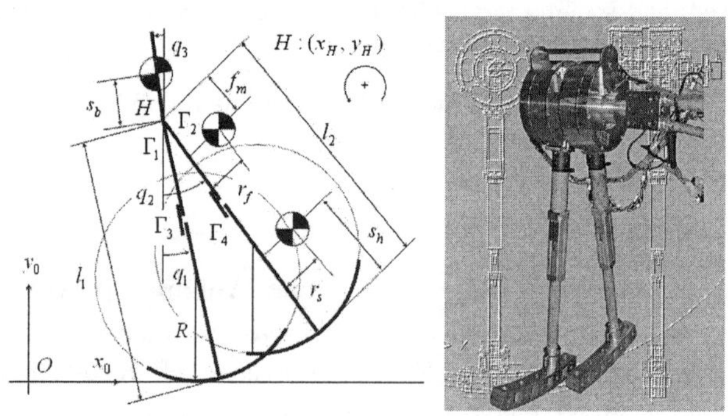

Fig. 1. Biped model and Experimental apparatus PBEmu-IV

Table 1. Physical parameters for the dynamic model

m_s	1[kg]	I_s	0.05[kgm^2]	r_s	0[m]	s_h	0.4[m]	L	0.78[m]
m_f	1.5[kg]	I_f	0.05[kgm^2]	r_f	0.15[m]	f_m	0.3[m]	R	0.78[m]
m_b	15[kg]	I_b	3[kgm^2]	d	0.01[P$_a$s]	s_b	0[m]		

counterclockwise rotation is positive. $D_e, h_e, C \in \Re^{7\times7}, G_e \in \Re^7$ are an inertial matrix, a Coriolis and centrifugal effect matrix, a viscosity effect matrix and a gravity vector, respectively. $B_e, D_{R1} \in \Re^{7\times 2}$ are a coefficient matrix for the inputs and a Jacobian matrix connecting to the stance leg tip and joints, respectively. $\Gamma = [\Gamma_1, \Gamma_2, \Gamma_3, \Gamma_4]'$ consists of relative torques of the first hip, the second hip, the first and the second knees. $R_1 = [R_{x1}, R_{y1}]'$ represents ground reaction force vector at the stance leg tip.

2.2. *Impact model*

We assumed that the impact is inelastic and instantaneous without slip. After the impact, the biped is sure to return the single support phase. The following impact equation is given:

$$D_e(q_e)(\dot{q}_e^+ - \dot{q}_e^-) = D_{R2}(q_e)I_{m2}. \tag{4}$$

The impact occurs when the leg tip of the swing leg contacts to the ground. q_e in equation 4 is the state vector at the impact. $\dot{q}_e^-$ and $\dot{q}_e^+$ are the angular velocities just before and just after the impact. D_{R2} is a Jacobian matrix connecting to the swing leg tip and joints. $I_{m2} = [I_{mx2}, I_{my2}]'$ is an impulse vector at the contact point of the swing leg.

3. Control method and Walking Gain

3.1. *Sinusoidal leg length variation and body control*

Sinusoidal functions with respect to time are used for reference trajectories of leg length which are controlled by PD law with the gravity compensation. Let K_ω, ω, and ϕ_ω be reference amplitude, angular frequency and phase angle, respectively, then the reference trajectory is described as follows:

$$l_{id} = K_\omega \sin(\omega t - \phi_w + \pi i) + L, \quad i = 0, 1, \tag{5}$$

$$\dot{l}_{id} = K_\omega \omega \cos(\omega t - \phi_w + \pi i), \quad i = 0, 1. \tag{6}$$

The PD law is also used for attitude control of the body and realized by the torque at the hip joint of the stance leg. The control laws are written in the following equations:

$$\Gamma_1 = K_{pb}q_3 + K_{db}\dot{q}_3 + G_{cb}(q_e), \tag{7}$$

$$\Gamma_2 = 0, \tag{8}$$

$$\Gamma_3 = -K_{pk}(l_1 - l_{1d}) - K_{dk}(\dot{l}_1 - \dot{l}_{1d}) + G_{c1}(q_e), \tag{9}$$

$$\Gamma_4 = -K_{pk}(l_2 - l_{2d}) - K_{dk}(\dot{l}_2 - \dot{l}_{2d}) + G_{c2}(q_e), \tag{10}$$

where K_{p*}, K_{d*} and G_{c*} are proportional gains, differential gains and the gravity compensations, respectively.

3.2. *Definition of walking gain*

All variables are defined when the motion is cyclic. Let ω be a angular frequency for the sinusoidal input of the legs. Let one step be a motion from just after the impact to just after the next impact, and n be the step number. Let T_w be a walking cycle[s] which is a duration for two steps. Let ω_w be a walking angular frequency [rad/s] given by $\omega_w = 2\pi/T_w$ All variables with the subscript ω are functions with respect to the angular frequency of the input, ω. A relative acute angle between the both legs at the double support phase is called step angle which is notated by α_{ω_w}[rad]. If we can assume the linear relationship, $\omega = \omega_w$, then the walking by the leg length variation can be treated as a frequency response between the leg length variation (the input) and the step angle (the output). Therefore a walking gain W_ω is defined by the following equation such that a ratio between the amplitude of the length variation:

$$W_\omega = |\alpha_{\omega_w}/K_\omega|, \tag{11}$$

4. Simulation

The control method can generate some walking patterns for the biped given in section II through numerical simulations. The result for a cyclic motion is illustrated in figure.2. The parameters of the biped are shown in table 1. m_* and I_* are mass and inertia, r_*, s_h, f_m and s_b identify the COG position of each link as shown in figure 1. Note that the COG of the both legs are forward, therefore, $r_f > 0$ and $r_s = 0$ in the table. Subscripts s, f and b mean a shank, a thigh and a body, respectively. R and L are a radius of feet and a default length of the legs, respectively. The amplitude and the angular frequency for the leg length variation are $K_w = 10$[mm] and $w = 1.25\pi$[rad/s] in this case. The average velocity converged to 0.41[rad/s].

4.1. *Walking gain*

In the simulations, the radius of the feet is $R = 0.78$. It is the same length as the initial length of the legs. Roughly speaking, the walking cycle, T_w is correspond to the cycle of the leg length variation for any ω, because the impact occurs when the both legs are the same as 0.78 [m] again at $t = n \cdot T_w/2$. Therefore the phase property are not important. The gain property is only analyzed as a frequency response. The walking gain and the step angles are presented in figure 3. From the figure, a resonance point is found. It appears when the velocity of the swing leg was correspond to the

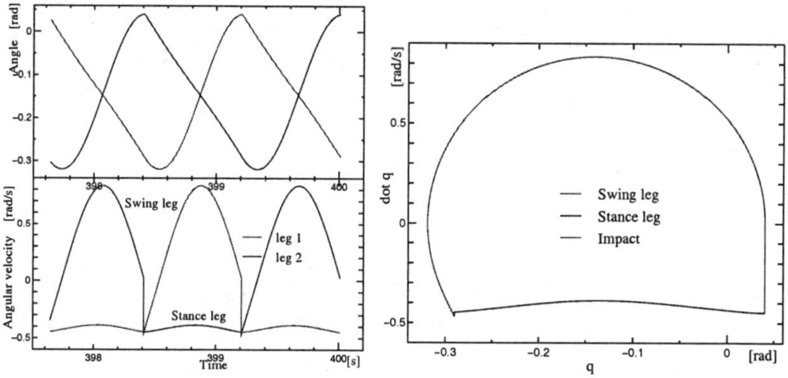

Fig. 2. A cyclic step of walking when $K_w = 10$[mm], $w = 1.25\pi$[rad/s]. Left figure shows angles and velocities of the stance and the swing leg. Right shows the phase diagram.

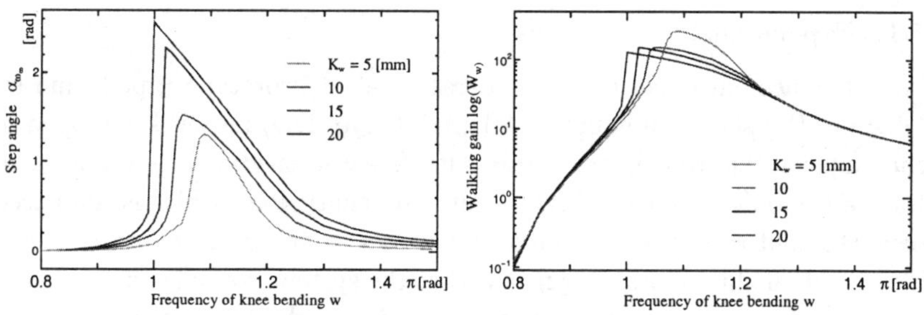

Fig. 3. Walking gain and step angle

velocity of the stance leg. Therefore, the resonance frequency is smaller than the characteristic frequency of the legs. Though the frequency shifts to left with respect to the amplitude K_w, the gain characteristics with respect to K_w correspond except the region around the walking resonance frequency, in the left of figure 3. Therefore the step angle can be regarded as a linear function with respect to the amplitude of the leg length variation. (Remark: since the feet length are not limited, the step angles around the resonance frequency are too large.) There are appropriate frequencies which excite biped walking for a desired step angle. Figure 4 shows energy consumption with respect to the input frequency. Figures 3 and 4 are compared, and the higher the walking gain is, the more effective walking is in the sense of consumed energy. Thus high gain frequencies are also appropriate for effective walking.

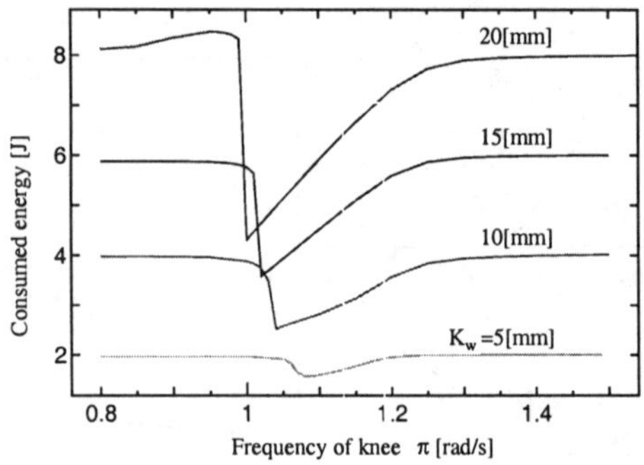

Fig. 4. Consumed energy

5. Experiment

5.1. *Experimental apparatus*

We introduce our experimental apparatus called Prototype Biped Emu IV (PBEmu-IV) shown in figure 1. A Direct Drive (DD) motor is equipped in each hip joint so that it can rotate with few viscosity under no actuation. Each knee joint is developed by a kind of prismatic joint which is composed of a DC motor and a ball screw. Each foot is spherical whose center is located at the hip joints. It is meant that the system is stable and it is easy to keep standing upright position. In addition, the COG of the body is located at the hip joint. The property makes PBEmu-IV walk more easily. A parallel link is attached to the apparatus so that the lateral motion can be constrained, the absolute position can be detected and the biped can walk on a circle which radius is about 2[m]. An information of foot impact is detected by a touch sensor embedded in each ankle which moves upside down for 5[mm]. A sampling cycle for control is 1[ms] and data are logged each 100[ms]. The configuration for the apparatus are shown in Table 2.

Table 2. Configuration of PBEmu-IV

	Leg	foot	Body
Mass	2.0[kg]	1.0[kg]	16.1[kg]
Length	0.75-0.80[m]	0.25[m]	0.2[m]
radius		0.80[m]	0.1[m]

5.2. *Experimental results of biped walking*

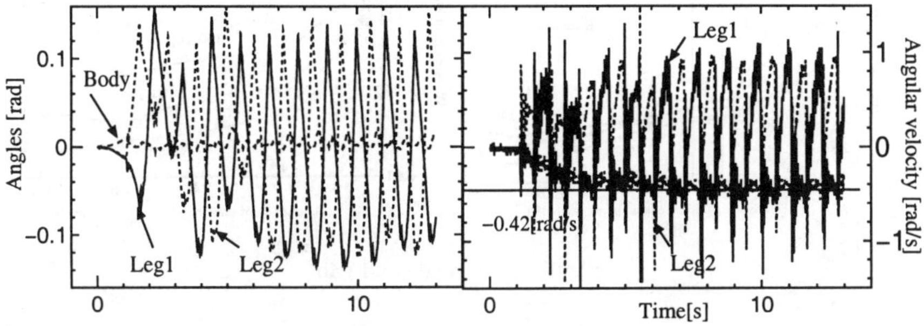

Fig. 5. Left figure: Relative angles of legs and absolute angle of the body. Right figure: Angular velocities of the legs. Walking is stable in 5 steps (3.5[s]) and the average velocity of the stance leg is converge to about 0.42[rad/s].

PBEmu-IV can walk using the leg length variation with the body control when $K_w = 18$[mm] and $w = 1.8\pi$[rad/s]. The biped is standing upright at the beginning, next the both knees are expanded 20[mm], and lastly the legs start varying and the biped starts walking at the same time. The experimental results are shown in figure 5. The figure presents angles and angular velocity of the legs and the body. The origin is the position where the biped stands upright. The walking gait is stabilized in five steps (3.5[s]) and the average velocity of the stance leg is converge to 0.42[rad/s]. The body is kept around the upright position. The step angles have some variation which is caused by undulation of the floor. It should be remarked that the convergence speed is much faster than the simulations. It is also different from the analysis of McGeer that an eigenvalue of speed mode approaches to unity with respect to feet radius increase.

5.3. *Frequency response analysis for walking*

Next, a frequency property of the biped walking is analyzed for the experimental results. As the result of the experiments, w_w were constants with respect to K_w, and vary with respect to angular frequency of the knee w. The relationship could be described $w = w_w$. Therefore the previous assumption is validated. The step angle a_{w_w} and the walking gain W_w denoted by the equation 11 are shown in figure 6. The figure is described in $1.0\pi \leq w \leq 1.8\pi$ and $6 \leq K_w \leq 18$. a_{w_w} and W_w increase when w decreases and they are the largest when $w = 1.0\pi$. Thus, the most effective

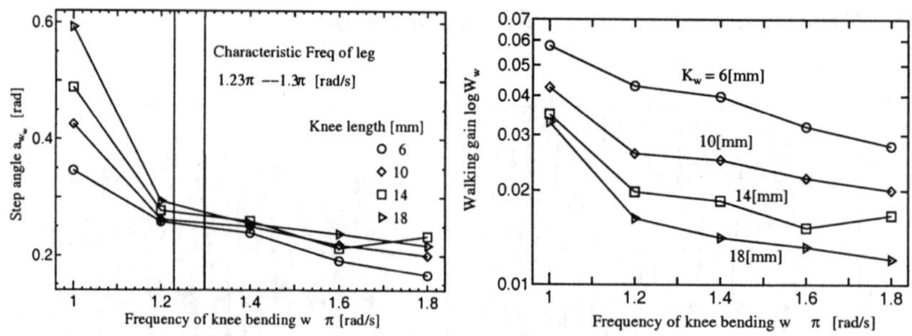

Fig. 6. step angle and gain.

frequency for our experimental apparatus is $w = 1.0\pi$ in the sense of the walking gain. In the case, the velocity of the swing leg at the impact was close to the stance leg. The large K_w is, the larger a_{w_w} is. Graphs of W_w with respect to K_w did not correspond each other, which is a little different from the simulated results. We can also expect the existence of a walking resonance frequency from the experimental results. However the step angle around the resonance frequency is too large to walk according to the simulation. On the outside range of w and K_w, indeed, the biped could not walk. It is caused that the length of the feet are limited in the experiments.

6. Conclusion

For a planer biped, the frequency property was revealed for biped walking by leg length variation. Some simulations gave the gain property and showed a walking resonance frequency as a resonance point of the walking gain. In addition, some experimental results were presented and validity of the frequency analysis was verified. The frequency property showed appropriate frequencies for biped walking in the sense of energy consumption.

References

1. T. McGeer, *Passive Dynamic Walking*, IJRR, Vol. 9, pp. 62–82, 1990
2. McGeer: "Stability and control of two-dimensional biped walking", Technical report CSS-IS TR 88-01. Simon Fraser University, Centre for Systems Science, Burnaby, BC, CANADA, 1998.
3. S. Collins, et al. *Efficient Bipedal Robots Based on Passive-Dynamic Walkers*, Science, no. 5712, pp. 1082–1084, 2005.
4. M. Wisse et. al., "How to Keep From Falling Forward: Elementary Swing Leg Action for Passive Dynamic Walkers", *IEEE Transactions on robotics*, Vol.21, No.2, pp.394–401, 2005.

Leg Control for Changing Locomotion between Leg-type and Wheel-type Designs based on Effective Use of Total Power

Tokuji OKADA[1], B. Wagner TANAKA[2], Toshimi SHIMIZU[2]
[1]Department of Biocybernetics, Niigata University
[2]Graduate School of Niigata University
Ikarashi 2-8050, Niigata-Shi, Japan 950-2181
[1]E-mail: okada@eng.niigata-u.ac.jp
http://okada.eng.niigata-u.ac.jp

This paper proposes a method for determining joint control variables of a mobile robot having the ability to move using a leg-type or wheel-type design when necessary. Specifically, we discuss the transient state of changing the types based on minimization of total energy cost and distributing necessary power into motors for actuating hip and knee joints so that the energy cost of each type becomes equal. In both cases, we calculate the control variables of the two joints uniquely by using an electrically calculated ratio of the hip joint motor capacity to the knee joint motor capacity that are found in the robot's hardware design specification. Demonstrative simulation clarifies that the type change is carried out continuously and smoothly without any need for mechanical adjustment and manual assistance in a stationary state while it stands in the same position and in a dynamic state while it moves in the same direction of locomotion.

Keywords: Legged Locomotion; Wheeled Locomotion; Minimization of Power Consumption; Energy Cost Function; Motor Capacity.

1. Introduction

Leg-type robots have high mobility on irregular terrain with adaptability in climbing up and down stairs, and walking over obstacles. On the other hand, the robot cannot move fast like a car even on a flat surface. The legged robot proposed by Y. Zhou,[1] is a simple model of a rigid-link binary walking robot that moves forward in a straight line with a wave gait. If it has a hybrid mechanism with legs and wheels, the robot can move on different terrain while steering itself. For example, the robot[2] can select either of wheel-type or leg-type electro-mechanically. Other robots having legs and wheels at its hinged leg ends[3] can move on flooring and rugged terrain. The wheelchair robot proposed by A. Gonzales *et al*,[4] is able to

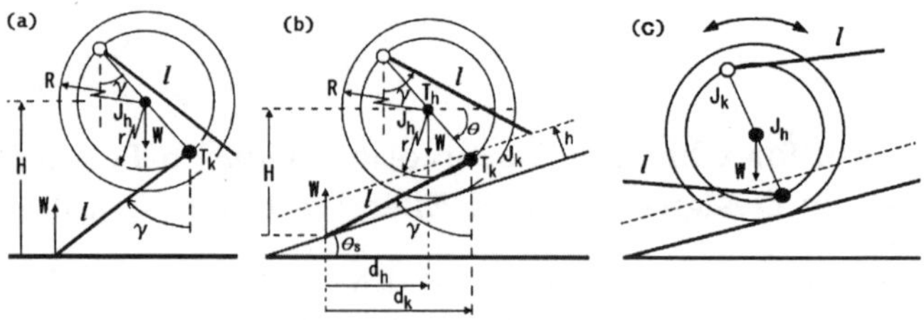

Fig. 1. Definition of parameters and symbols. Configurations in (a), (b) and (c) are seen in the movements of leg-type, change-round and wheel-type, respectively.

climb up a staircase.

This paper proposes two methods for changing the locomotion between the leg-type and the wheel-type design without human assistance or mechanical operations using jacks, screwdrivers, monkey wrenches, levers and so on[5].[6] The first is to simply minimize the total power consumption. The second is to share the necessary power between the hip and knee joints. In both cases, we take into consideration the ratio of the energy cost for producing the hip joint torque to the cost for producing the knee joint torque. Actually, this ratio is estimated using motor capacities found in the robot's hardware design specifications. For instance, such a robot driving hip and knee joints by motor capacities of 30[W] and 20[W] makes a ratio of 1.5. This value becomes greater when the motor at the hip joint is more powerful than that at the knee joint.

We show that the joint control variables are determined uniquely by applying the above methods and that the robot can change locomotion type at the same position smoothly without bumping, swinging or sliding on the ground. In the demonstrative simulation, we verify that the two methods are advantageous in making the robot barrier free from obstacles on the ground and in giving it high mobility in a real environment.

2. Definition of parameters for the motion analysis

Our PEOPLER is composed of 4 hip joints and 2 legs pivoted at each hip joint rim with the same mechanism.[6] Also, it is designed symmetrically on the right and left side, and controlled with synchronism in motion of the front and rear structures. Thus, one hip joint side view in Fig.1 is enough to understand the robot's motion. The left and right diagrams depict side views of the leg-type and wheel-type design, respectively. The middle diagram shows an instantaneous case belonging to the two types on a common slope since it has two landing contacts. We call this case, *change round configuration.* Parameters and symbols in the figure are helpful to analyze and extract equations for determining the joint control variables.

c :	ratio of the J_h's motor capacity to the J_k's motor capacity
c' :	modified value of c for unusual motor use
d_h :	horizontal distance of the J_k from the ground contact
d_k :	horizontal distance of the J_h from the ground contact
h :	height of the obstacle
l :	leg length of the robot
r :	radius of the knee joint rim
H :	height of the hip joint, i.e. height of the robot body
$J_.$:	joint (J_h; hip joint, J_k; knee joint)
R :	radius of a wheel (a circular profile surrounding the two knee joints)
T_h :	torque of the hip joint
T_k :	torque of the knee joint
W :	axle load
γ :	leg posture angle (a pair of γ have a bisymmetrical direction)
ζ :	ratio of l to R
θ :	arm angle from the robot's front direction
θ_s :	inclination angle of the road surface

3. Leg motion analysis of the locomotion type change

In order to always maintain the robot in a standing position without disturbance from the scattered obstacles on the ground while it moves according to the leg-type, we consider kinematics related to the joint control variables of θ and γ with a torque of T_h and T_k. When referring to Fig.1, we have the following equations.

$$T_h = Wd_h = W(r\cos\theta - l\sin\gamma) \tag{1}$$

$$T_k = Wd_k = Wl\sin\gamma \tag{2}$$

$$H = r\sin\theta + l\cos\gamma, \tag{3}$$

where in general H varies in the following range.

$$R \leq H \leq r + l \tag{4}$$

3.1. *Power distribution for using input energy effectively*

Motor capacity is an important specification in distributing electric power effectively without heat loss. Evidently, the capacity shows maximal energy consumption. Using more than the specified amount will shorten the motor's life. Since power is consumed with the two motors, energy cost of a

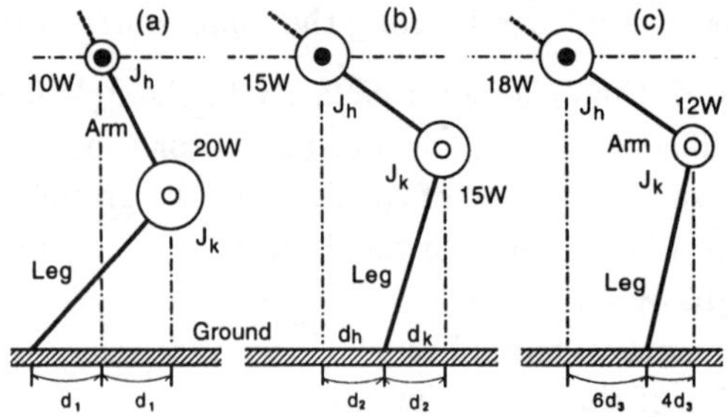

Fig. 2. Reasonable standing configuration when the load factors of the two motor operations are proportional to motor capacities. d_i stands for a scale unit

small capacity motor is high when it operates with heavy loads. On the other hand, the energy cost of a large capacity motor is low when it operates for the same load. Therefore, we define c as the ratio of motor capacity of the hip joint to the knee joint for using power effectively.

This idea is clarified by the designs using three different values of c to generate driving torque suitably for motor equipment. Let suppose that c_1, c_2 and c_3 take the values of 0.5, 1.0 and 1.5, respectively. Also, suppose that the total motor capacity, say 30[W], is available to drive the two joints. Then the configuration to keep the same hip joint height is illustrated in Fig.2. The configurations in (a), (b) and (c) are generated under c_1, c_2 and c_3, respectively. In (a), capacities of the hip and knee joint motors are 10[W] and 20[W], respectively. That is, the knee joint can use much power than the hip joint. In (b), the two motor powers are the same. In (c), they are 18[W] and 12[W] and the hip joint can use 1.5 times more power than the knee joint.

3.2. *Method for minimizing total energy cost*

We define the energy cost of each joint for driving is reversibly proportional to the motor capacity but proportional to the necessary torque that the motor has to produce using reduction gear mechanisms. According to this definition the joint using a large amount of motor capacity is less expensive and the joint using a small amount of motor capacity is more expensive for producing the same joint torque. Hence, we evaluate total energy cost spent on the two joints by E_a, then we have

$$E_a = (T_h)^2 + (cT_k)^2 \tag{5}$$

To make the value of E_a minimal it follows that

$$\partial E_a / \partial \theta = 0 \tag{6}$$

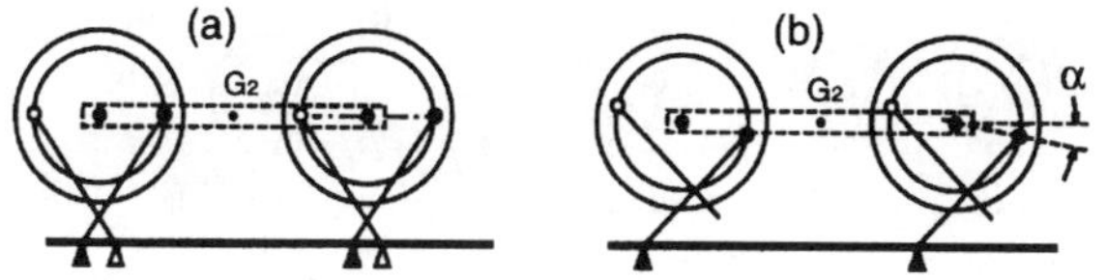

Fig. 3. Solution to the robot becoming stuck using angular shift α.

Using Eqs.(1-2), one can get the 6th order equation of unknown parameter sinθ. Solving the equation is not simple, but graphical tracing and exploration gives the angle θ through sinθ. Then γ is calculated from Eq.(3).

Also, the six order equation produces the critical values of H and γ when θ becomes zero. In fact,

$$H = \sqrt{l^2 - \{r/A\}^2} \tag{7}$$

$$\gamma = \tan^{-1}\{r/\sqrt{(Al)^2 - r^2}\ \} \tag{8}$$

One end of the two legs must be free from sliding on the ground. But this approach is not practical and we thought about shifting θ a little bit from zero so that one of the two legs ends separates from the ground. This shift, say α, prevents the legs from becoming immovable. Therefore, we assign the shift α heuristically to θ for determining γ in the lower part.

$$\gamma = cos^{-1}(H - r\sin\alpha)/l \tag{9}$$

Figure 3 shows the stuck robot configurations on the left and solved robot configurations on the right by using the shift $\alpha(=\pi/9[\text{rad}])$. A special robotic sensor can sense an obstacle on the ground and surmise whether it is possible to maneuver around the obstacle by checking the height h defined by

$$h = \eta R, \tag{10}$$

where η $(0 < \eta < 1)$ means the coefficient of the self rolling up power of the robot. Of course, the swing phase leg in the change round configuration is also needed to avoid the obstacle without pushing on the obstacle, however, if the robot cannot avoid the obstacle it is better to step on it than go around it. To make this possible the value of θ in Fig.1(b) is assigned using the height h to have

$$\theta{=}\sin^{-1}\left\{\frac{R}{r} - \frac{(2R-h) - \tan\theta_s\sqrt{4l^2\cos^2\theta_s - (2R-h)^2}}{2r}\right\} - \theta_s \tag{11}$$

Also, the γ is assigned by Eq.(3).

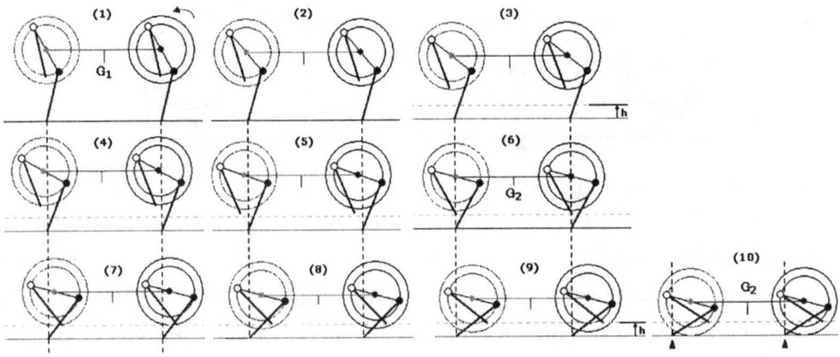

Fig. 4. Sequence of leg configurations based on minimizing total energy cost. Two neighboring configurations differ in the hip joint height by 5[mm].

3.3. *Method for bisecting motor power inversely to the energy cost*

In such a case when the condition of the motor changes caused by such things as overheating, excessive friction, frame weakness and etc., each motor's load factor might be changed. In general, it is meaningful to distribute enough power to each of the two motors so that $T_h = c'\, T_k$ is satisfied irrespective of the total energy cost. Unbalanced energy cost error, say E_b, between the two joints is evaluated by

$$E_b = \{\ (T_h)^2 - (c'\ T_k)^2\ \}^2 \tag{12}$$

Similar to Eq.(6), the value of θ making E_b minimal is calculated by $\partial E_b/\partial\theta = 0$. However, it is difficult to solve the equation, since T_h and T_k change their signs, in general. Therefore, we simplified the equation as follows by assuming that T_h and T_k are positive. Therefore,

$$T_h\ -\ c'\ T_k\ =\ 0 \tag{13}$$

Insert Eqs.(1-2) into $\partial E_b/\partial\theta = 0$ to obtain

$$r^2(B-1)\sin^2\theta - 2BHr\sin\theta + B(H^2 - l^2) + r^2 = 0\ , \tag{14}$$

where $B = (c'+1)^2$. Hence, we find

$$\sin\theta = \{BH\ \pm\ \sqrt{B^2l^2 + B(H^2 - l^2 - r^2) + r^2}\}/\{r(B-1)\} \tag{15}$$

Introducing sinθ into Eq.(3) yields γ. This method is useful to assign the modified ratio of c' when needed in unusual cases.

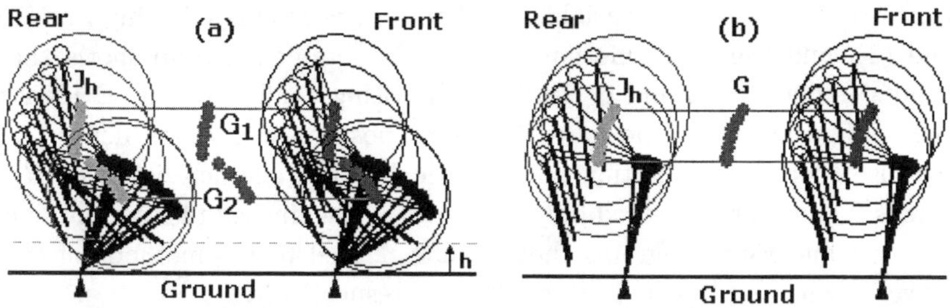

Fig. 5. Overwritten leg motion in the period while the robot changes locomotion type. Results of (a) and (b) are based on the total energy cost minimization method under c=1 and bisected power sharing method when H decreases with a small difference in ΔH=3[mm], respectively.

4. Simulated motion results of the locomotion type change

We simulate robot motion over a transient period while the robot changes locomotion from one leg type to the other type. Using overwritten configurations and sequential snapshots we verify the motion based on the methods explained in Section 3. Particularly, we demonstrate variation of the robot configurations generated under the influences of c or c'

4.1. *Minimization method of the total energy cost*

Aforemensioned 6th order equation is utilized to determine the control variables to minimize total energy consumption at higher positions of the hip joint from the ground. On the other hand, at lower positions of the hip joint, the control variable θ becomes α, and γ is calculated from Eq.(9). In Fig.4, the robot motion sequence is illustrated separately under c=0.5, η=0.4, ζ=1.5 and $\alpha=\pi/12$[rad]. The sequence from (10) to (1) stands for the locomotion change from the wheel-type to the leg-type design, reversionary. In Fig.5 all of the transient configurations are overwritten. In the left diagram, the robot's center in the higher and lower positions is distinguished by the trajectories of G_1 and G_2, respectively. In Fig.4, the robot centers of G_1 and G_2 appear in the configurations from (1) to (5) and from (6) to (10), respectively. We notice that the mode is changed at the same contact with the ground without any disturbance from the obstacle having the height h. Therefore the robot makes a smooth locomotion change without bumping or sliding. And one leg end does not slip on the ground.

4.2. *Bisected power sharing method*

In our power sharing control strategy, two motors' load factors are assigned, therefore power for driving the hip and knee joints are given independently.

In this case, the control variables of θ and γ are calculated using Eqs.(15) and (3), and the robot motion sequence is obtained. Fig.5b shows these results under c'=1.0 and ζ=1.5 by overwriting configurations. Evidently, the configurations in the lower hip joint position cannot be determined because there is no solution mathematically. The characteristics of H/R versus θ and γ are obtained but not shown in the limited pages in number. The left and right diagrams show results related to the methods of total power saving and bisected power sharing, respectively.

5. Conclusion

We proposed the two methods to use motor power effectively for driving the hip and knee joints in the locomotion change between the leg-type and wheel-type designs. In order to evaluate energy cost for driving the two joint motors, we defined the ratio between the two motor capacities and showed two methods for calculating the two joint control variables using not only the ratio but also the rate of leg length to wheel radius. The first is to minimize total energy cost effectively. The second is to bisect necessary motor power so that each energy cost becomes equal. In unusual cases, the ratio is modified gradually so that the energy cost is changed appropriately. In the simulation of locomotion, we made it possible to visualize the robot's motion on a display while it changes locomotion from the leg-type to wheel-type design and vice versa. This simulation resulted in showing that the locomotion change is carried out on a flat road without disturbance from scattered obstacles. A demonstration using an experimental model and also extension of the mode change on a slope are parts of our future work.

References

1. Y.Zhou: "On the planar stability of rigid-link binary walking robots", in *Robotica*, Vol.21, no.3, pp.667-675, 2003.
2. S.Hirose and H.Takeuchi:"Study on roller-walk", Proc. IEEE Int. Conf. on Robotics and Automation, pp.3265-3270, 1996.
3. K.Sonehara, T.Yamamoto and et.al.: "Development of Breadboard Model for Locomotion of 3-Lewg Wheeled Type Robot", Tech. Rep. of IHI, Vol.44, no.3, pp.241-247, 2004.
4. A.Gonzalez, R.Morales, V.Feliu and P.Pintado:"Improving the mechanical design of a new staircase climing wheelchair", Proc. of CLAWAR 2006, pp.14-18, 2006.
5. T.Okada, W.T.Botelho, T.Shimizu:"Compatible use of a legged robot as a wheeled robot and its demonstrative simulation", Proc. of CLAWAR 2006, pp.34-44, 2006.
6. T.Okada, W.T.Botelho, T.Shimizu:"Development of a rotating four-legged robot, PEOPLER for walking on irregular terrain", Proc. of CLAWAR 2003, pp.593-600, 2003.

MOVEMENT SIMULATION FOR MERO MODULAR WALKING ROBOT

ION ION[1], ION SIMIONESCU[2], ADRIAN CURAJ[3] ALEXANDRU MARIN[4]
[1]Technology of Manufacturing Department [2]Mechanism and Robot Theory

Department,[3]Automatic Control and System Engineering Department, [4]Hidraulic Department, POLITEHNICA University of Bucharest, Romania

E-mail: ioni51@ yahoo.com

Modern methods of drawing up machines and tools necessarily include a simulation stage of their functioning, which means to use a mathematical pattern of the real, original system. These activate the functioning simulation that encompasses several rules and specifications whose enactment generates behavior data and the instructions operating on the pattern's description variables. The veracity and validity of a real system depend on the compliance degree. The real system and the model are different by the fact that whereas for the former the manner to generate conduct data can be completely unknown, for the latter they mean a group of rules or specifications, whose enactment puts out the conduct data, namely instructions operating of the model's description variables. In the paper, a robust model of the direct and inverse analysis of the open kinematic chains of the walking robot legs is constructed.

1. Introduction

Scientists of all times have been permanently mesmerized and have studied the simplest but the most important movement, namely the mechanical movement of humans and animals. Humankind is so much anthropomorphism addicted that it is almost impossible for it to conceive or imagine automatic systems, even having artificial intelligence, and that are not anthropomorphic.

Since long ago, many research teams worldwide have been focused on goals such as creating an autonomous walking robot equipped with functions like handling objects, locomotion, perceiving, navigation, learning, judgment, information storage and intelligent control, and that can carry out tasks like altering the multitude of the parts belonging to a dynamic universe.

The access of man to dangerous areas where his safety is jeopardized made the scientific research approach topics of various purposes and conceive devices that through their performances aim at covering different fields. The architecture of these systems is quite different and depends on their purpose and destination.

For example, walking robots protect the environment better because their contact to the ground is discrete, which substantially diminishes the surface to be crushed, the robot's weight can be optimally distributed on the contact surface through controlling the forces. The variation of the distance from the ground allowed the robot to step over young trees or other vegetation growing in the area it moves on.

The walk is defined by the manner the waking robot moves between two points, under specific circumstances. To achieve and guide a walking robot requires thorough knowledge about all walking possibilities because choosing the number of legs and their structure depends very much on the selected walk. The selection of the walk type depends on several elements such as:

shape and consistency of the ground the robot walks on,

walk stability driving and controlling the movements of the elements of the walking systems,

speed and mobility movement requires.

It is very difficult to select the type of walk, mainly during real walking. Therefore, it is necessary that the ground surface to be defined before selecting the walk.

The walking robot's steps are a sequel of movements of the legs, coordinated with a succession of movements of the body for the purpose of moving the robot from one place to another

2. The establishing the mathematical pattern of the movement system

Theoretical research of robot movement assumes nine important stages:

1. the establishing of the mathematical model of the movement of the kinematic and dynamic system;
2. the process and construe the results obtained through simulation with a view to determining the system's conduct;
3. the structural and geometrical optimization of the movement systems compounds;
4. the movement of compounds system;
5. the establish the functional block scheme to calculate the compounds;
6. the establishing of the test methodology to perform the system's functions that is need and the identity of the functions generators walking requires;

7. the design of the guidance system's software;
8. the determination of the initial parameters and data characteristics of the system's structure and state;
9. the conduct of the simulation program.

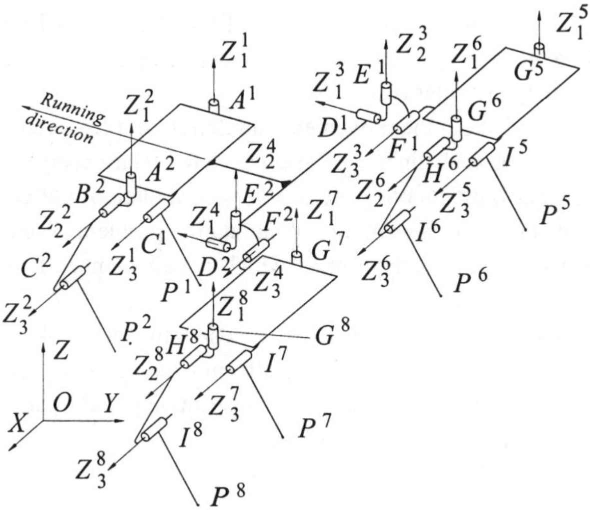

Figure1 The Denavit –Hartenberg axis system attached of modular walking robot; it is suggested support of technological equipments to work in farming and forestry.

2.1. *Movement simulation by Denavit – Hartenberg formalism*

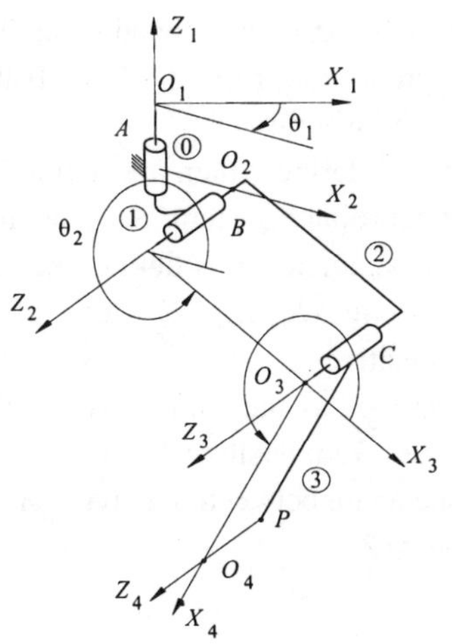

Figure 2. The Denavit – Hartenberg coordinate axes systems for a leg mechanism

Let us a modular walking robot consist of three modules. [3]. Each module has two 3-DOF legs, symmetrically arranged on the platform axis (fig.1).

The legs on the right – onto the movement direction are superscript marked with $2i$, $i=\overline{1,6}$ whereas the legs on the left with $2i-1$. Each platform of the rear modules is connected to the platform of first module by a 3-DOF kinematic chain with two links and three rotational pairs. The axes if these pairs are concurrent and perpendicular two by two.

In order to carried out the movement simulation of a leg, a coordinate axes system is attached to each link, with the Denavit – Hartenberg rule [2]. This formalism may not only simplify the problem formulation, but can also yield considerable advantage in the solution of simulation problem. The pairs of each leg are numbered consecutively from *A* which is pair number 1 to *C* which is pair number 3.

The Denavit - Hartenberg systems attached to each link are subscript numbered as the pairs respectively. The platform is designed as link number (**0**) and the remaining links are numbered consecutively. All pairs of the leg mechanism are rotational and actuated ones.

2.1.1 Denavit - Hartenberg systems attached to the leg links of the walking robot

The characteristic axis Z_i of each pair should be defined. The positive sense of each of these axes is defined arbitrarily. If the axes Z_i and Z_{i-1} are skew with respect to each other, then there is one common perpendicular between them. The perpendicular is designed as the X_i axis. If the Z_i and Z_{i-1} axes are parallel, the X_i axis may chosen as any common perpendicular. The positive direction of the X_i axis is designed as proceeding from Z_{i-1} to Z_i. If the Z_{i-1} and Z_{i-1} intersect, the positive sense of X_i axis is arbitrarily.

When the X_i axes are all defined, then are define both the Y_i axes and the origin of each right hand coordinate system. So, a coordinate system defined is attached to each link. The parameter a_i is defined as the distance from O_iZ_i to $O_{i+1}Z_{i+1}$ axes, measured along $O_{i+1}X_{i+1}$. Because of the orientation of the $O_{i+1}X_{i+1}$.axis, a_i is always positive.

The parameter α_i is defined as the angle between the positive O_iZ_i and the positive $O_{i+1}Z_{i+1}$ axes, as seen from positive $O_{i+1}X_{i+1}$.

The parameter $\theta_{\underline{i}}$ is the angle between positive O_iX_i and the positive $O_{i+1}X_{i+1}$ axes, as seen from positive O_iZ_i.

The parameter s_i is defined as the distance from O_iX_i to $O_{i+1}X_{i+1}$ axes, measured along the O_iZ_i axis.
Under this definition, the Denavit – Hartenberg transformation matrix $\mathbf{A}_i^j$ has the well-known form:

$$\mathbf{A}_i^j = \begin{Vmatrix} 1 & 0 & 0 & 0 \\ a_i^j \cos\theta_i^j & \cos\theta_i^j & -\cos\alpha_i^j \sin\theta_i^j & \sin\alpha_i^j \sin\theta_i^j \\ a_i^j \sin\theta_i^j & \sin\theta_i^j & \cos\alpha_i^j \cos\theta_i^j & -\sin\alpha_i^j \cos\theta_i^j \\ s_i^j & 0 & \sin\alpha_i^j & \cos\alpha_i^j \end{Vmatrix} \quad (1)$$

To mould the walking robot's moves is assumed that:

1. The kinematical length of the binary link (**1**) is null and it is connected to the platform (**0**), by pair A and to the link (**2**) by pair B; the axis of pairs A and B are perpendicular.
2. the binary link (**2**) is connected to the link (**1**) by the pair B and to link (**3**) by the pair C; the axis of pairs B and C are parallel.

The $\mathbf{A}_3^j$ matrix performed the coordinate transformation of a point belonging to link (**3**) from $O_4^jX_4^jY_4^jZ_4^j$ system to $O_3^jX_3^jY_3^jZ_3^j$ system attached to link (**2**). In a similar manner, the coordinates of lower end point P belonging to link (**3**) from $O_4^jX_4^jY_4^jZ_4^j$ system to $O_1^jX_1^jY_1^jZ_1^j$ system attached to the platform (**0**) is performed by the equation

$$\begin{Vmatrix} 1 \\ X_{1P}^j \\ Y_{1P}^j \\ Z_{1P}^j \end{Vmatrix} = \prod_{k=1}^{3} \mathbf{A}_k^j \begin{Vmatrix} 1 \\ X_{4P}^j \\ Y_{4P}^j \\ Z_{4P}^j \end{Vmatrix}, j = 1, 2. \quad (2)$$

This matrix equation described the geometrical model of the leg 1 and 2 of the walking robot. The goal of the direct kinematic analysis is to calculate the position, velocity and acceleration of the end point P, in terms of the pair variables θ_i^j, $i = \overline{1,3}$. In inverse kinematic analysis, matrix equation (2) is solved with respect to the pair variables θ_i^j, $i = \overline{1,3}$.

The positions of the point P and the positions of the platform with respect to the reference coordinate axes system $OXYZ$ fastened to the ground are considered as known. Therefore, the position of the point P with respect to the platform coordinate axes system are known.

The movement of the legs of the rear modules are controlled by the following equations:

$$\begin{Vmatrix} 1 \\ X_{1P}^5 \\ Y_{1P}^5 \\ Z_{1P}^5 \end{Vmatrix} = \mathbf{A}_1^3\mathbf{A}_2^3\mathbf{A}_3^3\mathbf{A}_1^5\mathbf{A}_2^5\mathbf{A}_3^5 \begin{Vmatrix} 1 \\ X_{4P}^5 \\ Y_{4P}^5 \\ Z_{4P}^5 \end{Vmatrix}, \begin{Vmatrix} 1 \\ X_{1P}^6 \\ Y_{1P}^6 \\ Z_{1P}^6 \end{Vmatrix} = \mathbf{A}_1^3\mathbf{A}_2^3\mathbf{A}_3^3\mathbf{A}_1^6\mathbf{A}_2^6\mathbf{A}_3^6 \begin{Vmatrix} 1 \\ X_{4P}^6 \\ Y_{4P}^6 \\ Z_{4P}^6 \end{Vmatrix},$$

$$\begin{Vmatrix} 1 \\ X_{1P}^7 \\ Y_{1P}^7 \\ Z_{1P}^7 \end{Vmatrix} = \mathbf{A}_1^4\mathbf{A}_2^4\mathbf{A}_3^4\mathbf{A}_1^7\mathbf{A}_2^7\mathbf{A}_3^7 \begin{Vmatrix} 1 \\ X_{4P}^7 \\ Y_{4P}^7 \\ Z_{4P}^7 \end{Vmatrix}, \begin{Vmatrix} 1 \\ X_{1P}^8 \\ Y_{1P}^8 \\ Z_{1P}^8 \end{Vmatrix} = \mathbf{A}_1^4\mathbf{A}_2^4\mathbf{A}_3^4\mathbf{A}_1^8\mathbf{A}_2^8\mathbf{A}_3^8 \begin{Vmatrix} 1 \\ X_{4P}^8 \\ Y_{4P}^8 \\ Z_{4P}^8 \end{Vmatrix}. \quad (3)$$

Each of these matrix equations is equivalent with three nonlinear equations and has six unknowns, namely variables of the pairs. In the inverse kinematic analysis three out of six unknowns must be imposed from independent conditions.

Table 1.

Pairs	A^2	B^2	C^2
Coordinate systems	$O_1^2X_1^2Y_1^2Z_1^2$	$O_2^2X_2^2Y_2^2Z_2^2$	$O_3^2X_3^2Y_3^2Z_3^2$
Transformation matrix	$\mathbf{A}_1^2$	$\mathbf{A}_2^2$	$\mathbf{A}_3^2$
a [mm]	0	350	350
α	-π/2	0	-π/2
θ	θ_1^2	θ_2^2	θ_3^2
s [mm]	150	0	0

Table 2

Pairs	D^4	E^4	F^4	G^8	H^8	I^8
Coordinate systems	$O_1^4X_1^4Y_1^4Z_1^4$	$O_2^4X_2^4Y_2^4Z_2^4$	$O_3^4X_3^4Y_3^4Z_3^4$	$O_1^8X_1^8Y_1^8Z_1^8$	$O_2^8X_2^8Y_2^8Z_2^8$	$O_3^8X_3^8Y_3^8Z$
Trans. matrix	$\mathbf{A}_1^4$	$\mathbf{A}_2^4$	$\mathbf{A}_3^4$	$\mathbf{A}_1^8$	$\mathbf{A}_2^8$	$\mathbf{A}_3^8$
a [mm]	0	0	670	0	350	350
α	π/2	π/2	π/2	π/2	0	0
θ	θ_1^4	θ_2^4	θ_3^4	θ_1^8	θ_3^8	θ_3^8
s [mm]	200	0	0	380	0	0

Fig. 3 shown the schemes of the sequence of the Denavit – Hartenberg coordinate systems attached to the links of the left leg of the first module (*a*) and to the links of the leg of the left rear module (*b*).

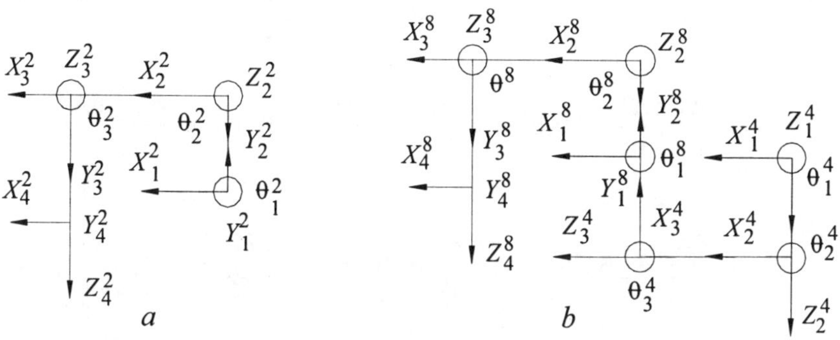

Figure 3. The Denavit – Hartenberg axes systems attached to the links of the left leg of the first module (*a*) and of the rear module (*b*)

In the movement simulation program, the parameters of the Denavit – Hartenberg transformation matrices have the values in Tables 1 and 2.

2.1.2 *First and second – time derivative of the pair variables*

Through the repeated differentiated of the equation (2) with respect to the time, yields:

$$\begin{Vmatrix} 0 \\ \dot{X}_{1P}^i \\ \dot{Y}_{1P}^i \\ \dot{Z}_{1P}^i \end{Vmatrix} = \left(\frac{\partial \mathbf{A}_1^i}{\partial \theta_1^i} \mathbf{A}_2^i \mathbf{A}_3^i \dot{\theta}_1^i + \mathbf{A}_1^i \frac{\partial \mathbf{A}_2^i}{\partial \theta_2^i} \mathbf{A}_3^i \dot{\theta}_2^i + \mathbf{A}_1^i \mathbf{A}_2^i \frac{\partial \mathbf{A}_3^i}{\partial \theta_3^i} \dot{\theta}_3^i \right) \begin{Vmatrix} 1 \\ X_{4P}^i \\ Y_{4P}^i \\ Z_{4P}^i \end{Vmatrix}, i = 1, 2;$$

$$\begin{Vmatrix} 0 \\ \ddot{X}_{1P}^i \\ \ddot{Y}_{1P}^i \\ \ddot{Z}_{1P}^i \end{Vmatrix} = \left(\frac{\partial^2 \mathbf{A}_1^i}{\partial \left(\theta_1^i\right)^2} \mathbf{A}_2^i \mathbf{A}_3^i \left(\dot{\theta}_1^i\right)^2 + \mathbf{A}_1^i \frac{\partial^2 \mathbf{A}_2^i}{\partial \left(\theta_2^i\right)^2} \mathbf{A}_3^i \left(\dot{\theta}_2^i\right)^2 + \mathbf{A}_1^i \mathbf{A}_2^i \frac{\partial^2 \mathbf{A}_3^i}{\partial \left(\theta_3^i\right)^2} \left(\dot{\theta}_3^i\right)^2 + \right.$$

$$+ 2 \frac{\partial \mathbf{A}_1^i}{\partial \theta_1^i} \frac{\partial \mathbf{A}_2^i}{\partial \theta_2^i} \mathbf{A}_3^i \dot{\theta}_1^i \dot{\theta}_2^i + 2 \frac{\partial \mathbf{A}_1^i}{\partial \theta_1^i} \mathbf{A}_2^i \frac{\partial \mathbf{A}_3^i}{\partial \theta_3^i} \dot{\theta}_1^i \dot{\theta}_3^i + 2 \mathbf{A}_1^i \frac{\partial \mathbf{A}_2^i}{\partial \theta_2^i} \frac{\partial \mathbf{A}_3^i}{\partial \theta_3^i} \dot{\theta}_2^i \dot{\theta}_3^i +$$

$$\left. + \frac{\partial \mathbf{A}_1^i}{\partial \theta_1^i} \mathbf{A}_2^i \mathbf{A}_3^i \ddot{\theta}_1^i + \mathbf{A}_1^i \frac{\partial \mathbf{A}_2^i}{\partial \theta_2^i} \mathbf{A}_3^i \ddot{\theta}_2^i + \mathbf{A}_1^i \mathbf{A}_2^i \frac{\partial \mathbf{A}_3^i}{\partial \theta_3^i} \ddot{\theta}_3^i \right) \begin{Vmatrix} 1 \\ X_{4P}^i \\ Y_{4P}^i \\ Z_{4P}^i \end{Vmatrix}, i = 1, 2. \tag{4}$$

In the inverse analysis, these equations are solved with respect to the first and second – time derivative respectively of the pair variables.

The velocity and the acceleration components of the point P on the axes of the coordinate system attached to the platform (**0**) are considered as known. In a similar manner are differentiated the equations (3).

In these matrix equations, the variables θ_k^{2i} $\dot{\theta}_k^{2i}$ and $\ddot{\theta}_k^{2i}$, $k = \overline{1,6}$ of the pairs and their derivatives with respect to the time are imposed by the walking simulation program.

Because of these variables' particular values, the Denavit – Hartenberg transformation matrices have the following simpler particular forms:

$$\mathbf{A}_1^{2i} = \begin{Vmatrix} 1 & 0 & 0 & 0 \\ 0 & \cos\theta_1^{2i} & -\sin\theta_1^{2i} & 0 \\ 0 & \sin\theta_1^{2i} & \cos\theta_1^{2i} & 0 \\ s_1^1 & 0 & 0 & 1 \end{Vmatrix}$$

$$\mathbf{A}_2^{2i} = \begin{Vmatrix} 1 & 0 & 0 & 0 \\ a_2^{2i}\cos\theta_2^{2i} & \cos\theta_2^{2i} & -\sin\theta_2^{2i} & 0 \\ a_2^{2i}\sin\theta_2^{2i} & \sin\theta_2^{2i} & \cos\theta_2^{2i} & 0 \\ s_2^1 & 0 & 0 & 1 \end{Vmatrix}$$

$$\mathbf{A}_3^{2i} = \begin{Vmatrix} 1 & 0 & 0 & 0 \\ a_3^{2i}\cos\theta_3^{2i} & \cos\theta_3^{2i} & -\sin\theta_3^{2i} & 0 \\ a_3^{2i}\sin\theta_3^{2i} & \sin\theta_3^{2i} & \cos\theta_3^{2i} & 0 \\ s_3^{2i} & 0 & 0 & 1 \end{Vmatrix}. \tag{5}$$

Keeping stable is a special problem that occurs while the robot walks, when one or more legs are in the transfer phase. When all the legs are in the support phase, it is obvious that the protection of the center of gravity is within the support polygon.

If one or more legs are in the transfer phase, the geometry of the support polygon changes and it occurs the risk that the protection of the center of gravity moves outside the support polygon.

Solutions to such situations depend on how the modular walking robot is configured. In fig. 4 are shown some sequence of the computer simulation of the gait of the modular walking robot

Figure. 4 Computer graphics simulation the gaits of the modular walking robot

Conclusions

The movement simulation of the walking robots may be idealized into a mathematical model for the purpose of kinematic analysis. The techniques of idealization can play the decisive role in easiness, precision and time of calculus for the problem solving. The Denavit – Hartenberg method is numerically robust, the solutions are either exact in the sense that is possible to refine them up to an arbitrary accuracy. A modular walking robot could have one or more modules. The motions of the legs must be coordinated so that the conditions of the gait stability of the system to be ensured.

References

1 A.P. Bessonov, N.V. Umnov, *The Analysis of Gaits in six-legged Robots according to their Static Stability*, Proc. Symp. Theory and Practice of Robots and Manipulators, Udine, Italy, 1973

2. Denavit J., Hartenberg R.S., *A Kinematic Notation for Lower Pair Mechanisms Based on Matrices*, Journal of Applied Mechanics, Tr. ASME, 1955, Vol. 77
3. I.Ion, I. Simionescu,. A. Curaj, MERO Modular Walking Robot Support of Technological equipments, The 8thInternational Conference on Climbing and Walking Robots, September 12-14, 2005 London, UK
4. I.Ion, I. Simionescu,. A. Curaj, *Mobil Mechatronic System With Applications in Agriculture and Sylviculture*.The 2th IFAC International Conference on Mechatronic Systems , December 8-12, 2002-Berkeley – USA pp941-946.Pergamon Press
5. R.B. McGhee, G.I. Iswandi, *Adaptive Locomotion of a Multi-legged Robot Over Unarranged*, IEEE Trans. On Systems, Man, and Cybernetics, Vol. SMC-9, No. 4, April 1979, pp. 176-182
6. S.M. Song, and J.K. Waldron, *An Analytical Approach for Gait Study and its Application on Wave Gait,* International Journal of Robotics Research, Vol. 6, No. 2, 1987, pp. 60-71
7. J.K. Waldron, *Modeling and Simulation of Human and Walking Robots Locomation* ,Advanced Scholl Udine ,Italy 1996

Trajectory Generator for Rhythmic Motion Control of Robot using Neural Oscillators

Weiwei Huang, Chee-Meng Chew, Geok-Soon Hong and Nithya Gnanassegarane
Mechanical Engineering, National University of Singapore, Singapore

This paper presents a structure of coupled neural oscillator which can be used as central pattern generator(CPG), for humanoid robot locomotion. Parameters tuning is always a big challenge for oscillators. In this article, some properties of neural oscillator, like limit cycle requirement, amplitude and frequency are studied. The objective is to develop a trajectory generator which has simple parameters' tuning. This architecture is suitable in various fields where rhythmic motion is required. We tested the oscillator structure on a simulated biped robot. The result demonstrates a stable walking behavior in both forward and backward straight walking and different walking speeds.

Keywords: Neural Oscillator; CPG; Limit Cycle; Trajectory Generator; Biped Locomotion.

1. Introduction

Exploiting the nature of human walking is always an open question for human locomotion researchers. Neurophysiological studies of animal locomotion have revealed that the basic rhythm of movement is controlled by rhythm generating networks in the nervous system, which are called central pattern generator(CPG).[1] A CPG is a network of neuron oscillators coupled by mutual inhibition which can be used to control the rhythmic movement of a walking robot. The goal of the CPG research is to generate the desired motion for the joints. Many interesting works based on neural oscillator have been done to explain such automatic oscillatory activities.[2–6] In these research works, the oscillators' parameter tuning is the key challenge, especially for complex CPG network with multiple neurons. Some researchers have addressed this problem by tuning the parameters mathematically.[7] Another approach is to use optimization tools like GA[8,9] to search for the desired parameters. Other researchers focus on deriving the learning algorithms in order to adjust nonlinear oscillators automatically.[10] However, tuning the oscillator parameters is still a problem in the CPG research.

In this paper, we present the analysis which enable us to understand and tune the oscillator with ease. The details are organized as follows: In Section 2, limit cycle behavior of neural oscillator is discussed and mathematical formulas for oscillator frequency and amplitude are given. In Section 3, the oscillator structure is designed to generate the desired trajectory. In Section 4, we empirically verify the robustness of the proposed structure by numerical simulation.

2. Limit Cycle Behavior

In this section, we will study the limit cycle behavior of the neural oscillator model proposed by Matsuoka.[3] A new approach based on Poincaré-Bendixson theorem and piecewise linear analysis is proposed. We introduce the approximate formula of amplitude and frequency to help tuning the oscillator parameters.

2.1. *Stability Analysis*

The model of the two neurons Matsuoka's oscillator is:[3]

$$\tau_1 \dot{u}_1 = c - u_1 - \beta v_1 - a[u_2]^+ - \sum h_j [g_j]^+ \tag{1}$$

$$\tau_2 \dot{v}_1 = [u_1]^+ - v_1 \tag{2}$$

$$\tau_1 \dot{u}_2 = c - u_2 - \beta v_2 - a[u_1]^+ + \sum h_j [g_j]^- \tag{3}$$

$$\tau_2 \dot{v}_2 = [u_2]^+ - v_2 \tag{4}$$

$$[u]^+ = \max(0, u) \qquad [u]^- = \min(0, u) \tag{5}$$

$$Y = [u_1]^+ - [u_2]^+ \tag{6}$$

where $u_{1(2)}$ represents the inner state of the neuron; $v_{1(2)}$ is the degree of neural adaptation; c is the constant stimuli; $\tau_{1(2)}$ is the time constants; β is the parameter that indicates the effect of adaptation; a represents the strength of inhibition connection between neurons; g_j is the feedback from environment and Y is the oscillator output.

Poincaré-Bendixson Theorem is one of the important results for limit cycle behavior analysis of nonlinear dynamics. In particular, it can be used to establish the existence of limit cycle in the state space.

Poincaré-Bendixson Theorem:[11] Suppose that

(1) R is a closed, bounded subset of the phase plane;

(2) $\dot{x} = f(x)$ is a continuously differentiable vector field on an open set containing R;

(3) R does not contain any fixed points; (Note:Fixed point is a point that satisfies $f(x^*) = 0$);

(4) There exists a trajectory C that is confined in R; (i.e. a trajectory that starts in R will stay in R for all future time)

Then either C is a closed orbit or it spirals toward a closed orbit as $t \to \infty$. That is, R contains a limit cycle.

Based on the theorem, we will prove the existence of limit cycle in neural oscillators step by step. For better understanding, we begin our proof with condition (2) and condition (1) will be proved later. Here, we assume that there is no sensor feedback to the oscillator. That is , $g_j = 0$.

Satisfying of condition 2: The oscillator model is not continuously differentiable, because $[u]^+ = max(u, 0)$ cannot be continuously differentiable. In order to meet the smoothness requirement, we used a smoothed version of the function by first replacing the $max(u, 0)$ function with the function $\frac{|u|+u}{2}$, and then by replacing the absolute value with the function $f(u) = \frac{2u}{\pi} \arctan(ku)$. When $k \to \infty$, the vector fields of $f(u)$ and $|u|$ are the same. Since the function $f(u)$ is continuously differentiable, $[u]^+$ can be considered as continuously differentiable.

Satisfying of condition 3: It is difficult to check whether there are fixed points or not in equations (1)-(6). To overcome this, we separate the u_1, u_2 region into four subset quadrants $\{\{u_1 \geq 0, u_2 \geq 0\}, \{u_1 \geq 0, u_2 < 0\}, \{u_1 < 0, u_2 \geq 0\}, \{u_1 < 0, u_2 < 0\}\}$. For each respective quadrant, the equations (1)-(4) can be transformed into a linear equation:

$$\tau_1\tau_2\ddot{Y} + (\tau_1 + \tau_2 - a\tau_2)\dot{Y} + (1 + \beta - a)Y = 0 \tag{7}$$

$$\tau_1\tau_2\ddot{Y} + (\tau_1 + \tau_2)\dot{Y} + (1 + \beta)Y - c = 0 \tag{8}$$

$$\tau_1\tau_2\ddot{Y} + (\tau_1 + \tau_2)\dot{Y} + (1 + \beta)Y + c = 0 \tag{9}$$

$$Y = 0 \tag{10}$$

where $Y = [u_1]^+ - [u_2]^+$

Equation (7) has a fixed point at $\mathbf{u}^* = (0, 0)$. It can be proved that when $1 + \frac{\tau_1}{\tau_2} - a < 0$, equation (7) has a negative damping and the fixed point is a repeller. Therefore we can modify the region with the punctured region R as shown in Fig.1 which does not include (0,0) point.[11]

Equation (8) has a fixed point at $\mathbf{u}^* = (\frac{c}{1+\beta}, c(1 - \frac{a}{1+\beta}))$. The requirement for $\mathbf{u}^*$ to be a fixed point is $1 - \frac{a}{1+\beta} < 0$, because it has to be within the quadrants $\{u_1 \geq 0, u_2 < 0\}$. If $1 - \frac{a}{1+\beta} > 0$ that is $a < 1 + \beta$, then $\mathbf{u}^*$

is not in the region. This implies that $\mathbf{u}^*$ cannot be a fixed point. We can also give a similar proof for Equation (9).

Here we have proved that when $a - 1 - \frac{\tau_1}{\tau_2} > 0$ and $a - 1 - \beta < 0$ are satisfied, then the output Y has no fixed point. If the output has no fixed point, this implies that the state variables also have no fixed point. On the other hand, if the state variables have fixed points, the output will converge and maintain a constant value.

Satisfying of condition 4: To ensure that a confined trajectory exists, we construct a trapping region R, which is a closed connected set such that the vector field points inward everywhere on the boundary of R as shown in Fig.1. Then all trajectories in R are said to be confined. Matsuoka has proved that the oscillator state is bounded,[3] that is $u_{1(2)} \leq |u_{1(2)}(0)| + c + \beta|v_{1(2)}(0)|$ where $|u_{1(2)}(0)|$ and $|v_{1(2)}(0)|$ are initial condition. We construct a larger boundary for R in order to show the vector fields clearly, where $bound = c + \beta v_{max} + a u_{max}$, v_{max} and u_{max} are the largest value of $v_{1(2)}$ and $u_{1(2)}$.

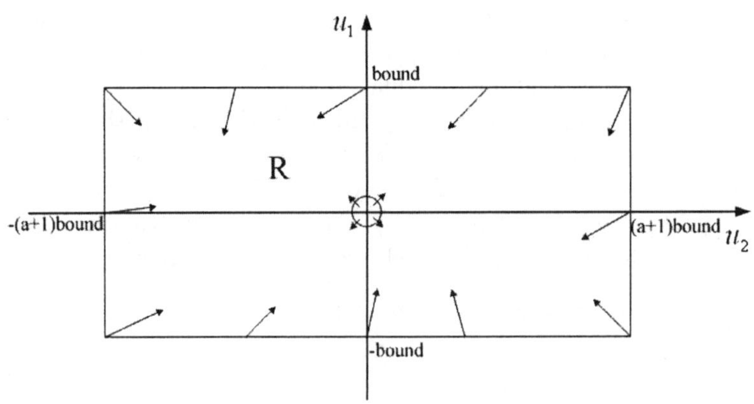

Fig. 1. Trapping region R

Satisfying of condition 1: The state region R shown in Fig.1 is a bounded subset of the plane.

Therefore the neural oscillator model described by equations (1)-(6) has a unique limit cycle in the bounded region R if

$$a - 1 - \frac{\tau_1}{\tau_2} > 0 \tag{11}$$

$$a - 1 - \beta < 0 \tag{12}$$

The trajectory is unique in region R based on the Existence and Uniqueness Theorem.[11]

2.2. *Frequency and Amplitude*

Frequency and amplitude are two important properties of a periodic signal. We found that when $|a-1-\frac{\tau_1}{\tau_2}|$ is small, two neurons are always in the active state, that is in the region $\{u_1 \geq 0, u_2 \geq 0\}$. Generally, we can approximate the oscillator frequency F by using the damped natural frequency of the linear system in equation (7) as follows:

$$F = \frac{\omega_n\sqrt{1-\zeta^2}}{2\pi} = \frac{1}{2\pi\tau_2}\sqrt{\frac{1+\beta-a}{b} - \frac{(a-1-b)^2}{4b^2}} \tag{13}$$

where $b = \frac{\tau_1}{\tau_2}$, $\omega_n^2 = \frac{1+\beta-a}{\tau_1\tau_2}$, $2\zeta\omega = \frac{\tau_1+\tau_2-a\tau_2}{\tau_1\tau_2}$.The amplitude is given by:[12]

$$U_{Amp} = \frac{2c}{1+\beta+a} \tag{14}$$

There are several advantages in deriving these expressions. *First:* with the knowledge of both frequency and amplitude, we can know the shape of the oscillator output. *Second:* we can set the parameters of oscillator based on the desired frequency and amplitude. *Third:* from the formula we found that if we fixed the value of a, b and β, the value of τ_2 is directly proportional to the frequency and c is directly proportional to the amplitude. Therefore changing the performance of the oscillator becomes easy.

3. Trajectory Generation

In this section we will present the neural oscillator structure that we have used to generate periodic trajectory. First we present the structure of the pattern generator which is made of coupled neural oscillators and then we discuss those properties of the structure that make it suitable for rhythmic motion control.

3.1. *Structure of the Trajectory Generator*

In the previous section, we found that both amplitude and frequency could be modified by changing certain parameters. When $|a-1-\frac{\tau_1}{\tau_2}|$ is small, the shape of the oscillator is similar to the sine wave. According to Fourier series, an arbitrary periodic function can be decomposed into a weighted sum of sinusoidal components. It is plausible to have the idea of using several oscillators to generate any desired trajectory.

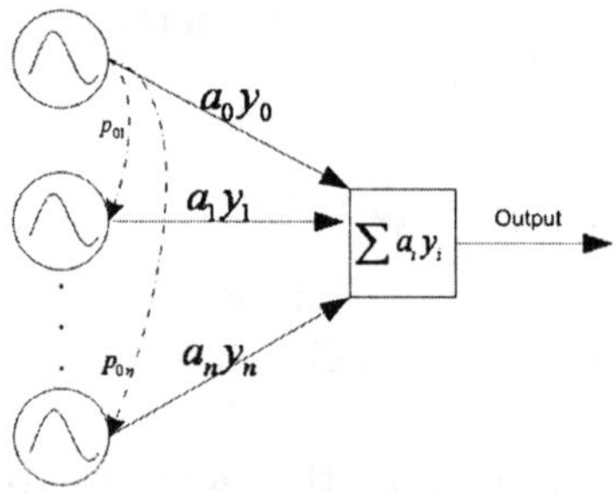

Fig. 2. Network structure of the oscillators

Fig.2 shows the structure of this network. We associate to each oscillator a phase adjustment for the difference in phase of each oscillator with respect to the first oscillator in the network, thus enabling us to correct any phase error between the oscillators. The equations describing this structure are as follow

$$\tau_{i1}\dot{u}_{i1} = c - u_{i1} - \beta v_{i1} - a[u_{i2}]^{+} - [p_i]^{+} - \sum h_{ij}[g_{ij}]^{+} \tag{15}$$

$$\tau_{i2}\dot{v}_{i1} = [u_{i1}]^{+} - v_{i1} \tag{16}$$

$$\tau_{i1}\dot{u}_{i2} = c - u_{i2} - \beta v_{i2} - a[u_{i1}]^{+} + [p_i]^{-} + \sum h_{ij}[g_{ij}]^{-} \tag{17}$$

$$\tau_{i2}\dot{v}_{i2} = [u_{i2}]^{+} - v_{i2} \tag{18}$$

$$Y = \sum a_i([u_{i1}]^{+} - [u_{i2}]^{+}) \qquad i = 1...n \tag{19}$$

Note: p_i is the difference in phase of ith oscillator with respect to the first oscillator.

3.2. *Property Analysis*

In this section, we present several numerical experiments to verify the property of the neural oscillator structure. Firstly, we set the structure to generate a knee trajectory which is obtained from the normal human walking experiment. After Fourier analysis, we found that the trajectory is mainly composed of three different frequencies components. Therefore we assign three oscillators to generate this trajectory. Fig.3(a) shows the comparison between the reference and the generated knee joint trajectory. If we want to reduce the number of oscillators to simplify the structure, we can choose the main frequency components. Because of the intrinsic properties of stability, the structure can produce trajectories that are stable to perturbations(Fig.3(b)). When we consider the perturbation as a feedback to the oscillators, then the oscillators tend to adjust the output automatically.

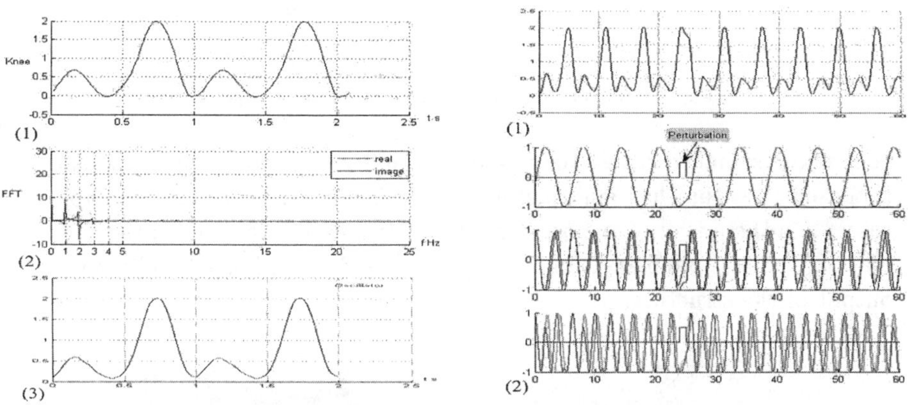

(a) (1),(2)The reference knee joint trajectory and its FFT; (3)generated knee trajectory

(b) (1)Stability against perturbation given through g_{ij}, (2)performance of each oscillator

Fig. 3. Result of trajectory generation and stability against perturbation

Another important aspect of this neural oscillator structure is that it allows easy modulation of the amplitude and the frequency of the trajectory. Since the frequency and amplitude are linearly related to τ_2 and c, simple modulation of these values can cause a large variation of stable trajectories. Some properties are shown in Fig.4. Because of the properties of coupled oscillators, modulation of these parameters is always smooth and thus interesting for robot trajectory generation.

4. Empirical Verification

In this section, we will test the structure on a biped robot in dynamic simulation environment. We assign the oscillators to the task space of a 3D robot, which has a significant reduction in the number of parameters compared to joint space. It was demonstrated that robust steady walking can be achieved with a proper arrangement of oscillators.[5]

As shown in Fig.5, this structure is composed of three oscillators which control X,Y,Z direction movement. The three oscillators are coupled together by the phase adjustment. Y direction oscillator generates the refer-

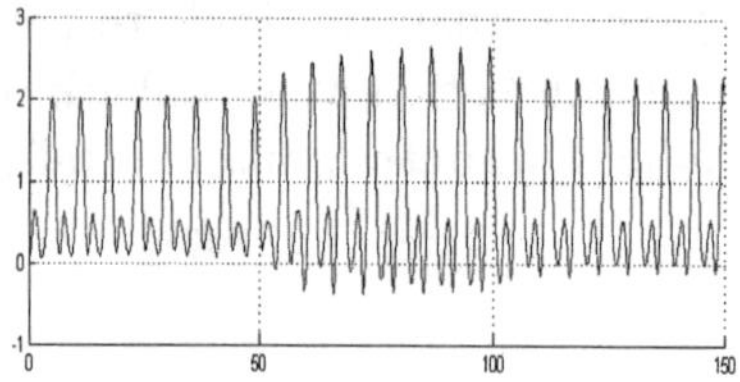

(a) Modulation of amplitude

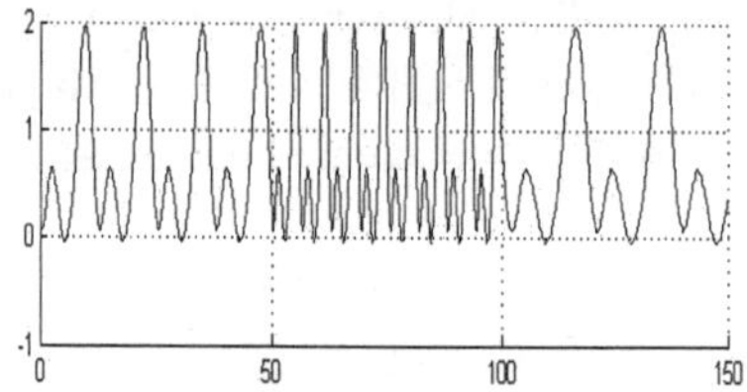

(b) Modulation of frequency

Fig. 4. Smooth response of the parameter change

ence phase signal to adjust the phase of X and Z direction oscillators. The proposed structure has multiple oscillators. However, we choose a simple case when there is only one oscillator to control each direction. In the simulation, roll and pitch angle of the robot are used as feedback to adjust the output of the structure.

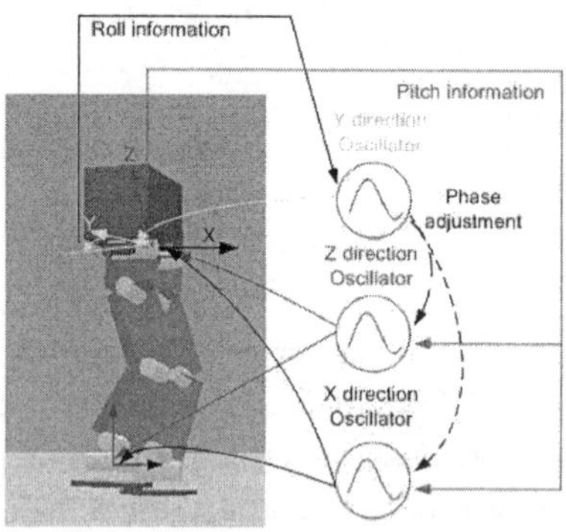

(a) Proposed control architecture using neural oscillators

(b) NUSBIP-II

Fig. 5. Simulation model

The simulation is carried out in Yobotics environment which is a Java package for simulating fully dynamic systems. The structure of the robot model is the scale down model of NUSBIP-II(Fig.5(b)). The body height is 36cm; the foot length is 8cm and the whole weight is 5.84kg. In Fig.6 both forward and backward walking are demonstrated. Backward walking was achieved by changing the sign of the X direction oscillator. The Oscillator output and sensor feedbacks are given in Fig.7(a). In Fig.7(b), the phase diagram of the Y direction motion are plotted. The trajectory starts from the origin and eventually converges to a constant oscillatory swinging motion.

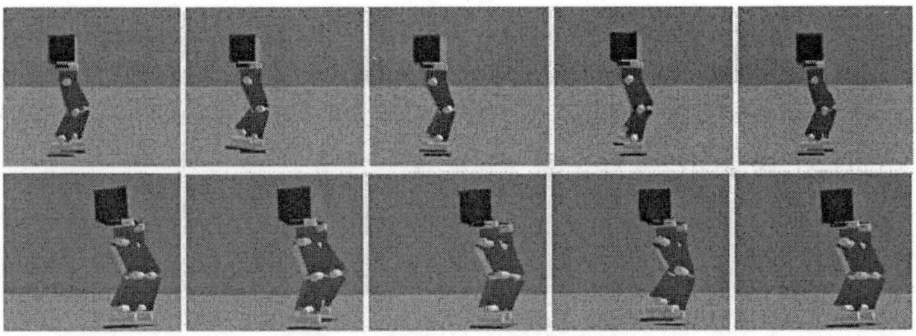

Fig. 6. Snapshots of forward walking (Step length=0.02m, Velocity=0.08m/s) and backward walking (Step length=0.012m, Velocity=0.048m/s)

The simulation shows a robust walking behavior. The forward walking speed is 0.08m/s and backward speed is 0.048m/s. Since the size of the robot is small, the walking speed is acceptable compared to the previous work. In the simulation, Y direction oscillator is able to adjust the phase of the other two oscillators. This implies that our structure also shows the ability of entrainment property. This is our preliminary work and for simplicity, we have generated trajectories with only one frequency component. In the future, we plan to demonstrate a more complex oscillator structure and the robustness of the structure will be studied.

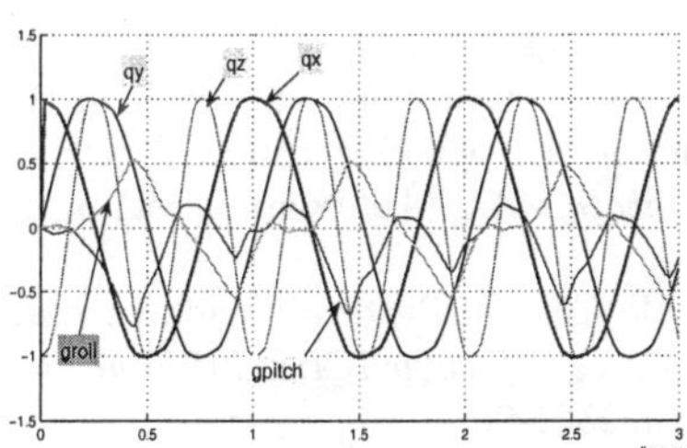

(a) q_x, q_y, q_z entrainment and roll, pitch feedback in the straight walk

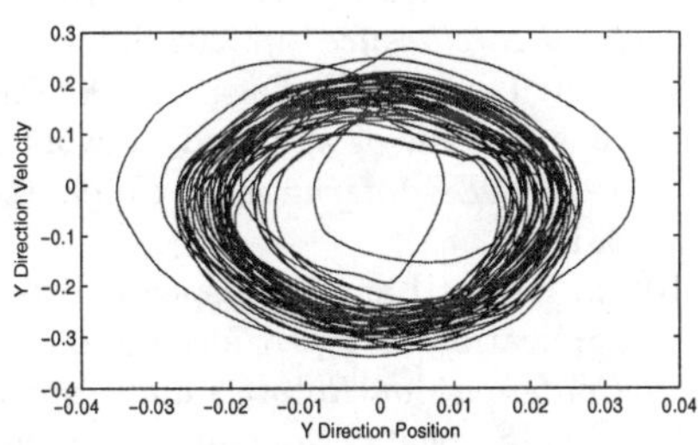

(b) The limit cycle behavior of frontal plane

Fig. 7. Simulation results

5. Conclusion

In conclusion, this paper presents a new way of analyzing the neural oscillator. A simple trajectory generator structure is designed. We analyze the properties of this structure through numerical and dynamic simulations. The robot shows a robust walking behavior with three coupled neural oscillators. In the future, we may consider implementing the structure to achieve more complex walking behaviors. We will also explore more efficient feedback pathway to achieve robust walking.

References

1. S. Grillner, Neurobiological bases of rhythmic motor acts in vertebrates, *Science*,**vol. 228** 1985, pp. 143-149.
2. K. Feng, C.M. Chew, G.S. Hong, T. Zielinska, Bipedal locomotion control using a four-compartmental central pattern generator, *Procs. of 2005 IEEE International Conference on Mechatronics and Automation*,**Vol. 3**, pp. 1515-1520.
3. K. Matsuoka, Sustained oscillations generated by mutually inhibiting neurons with adaptation, it Biological Cybernetics, **vol. 52**, 1985, pp. 345-353.
4. G. Taga, A model of the neuro-musculo-skeletal system for human locomotion I. emergence of basic gait, *Biological Cybernetics*, **vol. 73**, 1995, pp. 97-111.
5. G.E. J Nakanishi, J. Morimoto and G. Cheng, Experimental Studies of a Neural Oscillator for Biped Locomotion with QRIO, *Procs. of 2005 IEEE International Conference on Robotics and Automation*, pp.596-602.
6. M. Williamson, Neural control of rhythmic arm movements, *Neural Networks*, **vol.11**, 1998, pp.1379-1394.
7. A. Arsenio, Neural oscillator networks for rhythmic control of animats, *From Animals to Animats* MIT-Press, 2000.
8. D. Garis, Evolution of a time dependent neural network module which teaches a pair of stick legs to walk, *Procs. of 1990 9th European Conference on Artificial Intelligence*, pp. 204-6.
9. M.A. Lewis, A.H. Fagg, AS Solidum, Genetic programming approach to the construction of a neural network for control of a walking robot, *Procs. of 1992 IEEE International Conference on Robotics And Automation*, **vol.3**, pp 2618-23.
10. L. Righetti and AJ Ijspeert, Programmable Central Pattern Generators: an application to biped locomotion control, *Procs. of 2006 IEEE International Conference on Robotics and Automation*, pp.1585-90.
11. S.H. Strogatz, Nonlinear Dynamics and Chaos with application to physics Biology, Chemistry, and Engineering, (Addison-Wesley Publishing Company), 2000.
12. W. Huang, C.M. Chew, G.S. Hong, Neural Oscillator Analysis for Rhythmic Control of Robot Locomotion, *Procs. of 2007 IEEE/RSJ International Conference on Intelligent Robots and Systems*, submitted.

Observer-based control of a walking biped robot: stability analysis

V. Lebastard, Y. Aoustin, and F. Plestan

IRCCyN, Ecole Centrale de Nantes, CNRS, Université de Nantes - Nantes - France
{Vincent.Lebastard, Yannick.Aoustin,
Franck.Plestan}@irccyn.ec-nantes.fr

Abstract

The considered problem in this paper is the estimation the absolute orientation of a dynamical stable five-link biped during a walking gait. The solution is founded on a nonlinear observer based on the super twisting approach, such that the estimation error converges in finite time and only uses the measurements of the joint variables. This observer is coupled to a nonlinear control law, the stability of the cyclic walking gait under the observer-based controller being also proved.

1 Introduction

The interest to take into account the imbalance phases in the walking gait of a biped is real in order to make its locomotion more fluid and faster. However, such improvements induce the use of the whole state of the biped in the design of reference trajectories and the control law computations. For example, accurate sensors are necessary to measure the joint variables, the absolute orientation of the biped in imbalance phases. Unfortunately, a precise measurement of the absolute orientation of a walking robot in these imbalance phases is, by a technical point-of-view, quite difficult to get. For the joint variable measurement, usually a gearbox reducer leads to a good accuracy. But to estimate the orientation of a walking biped in imbalance phases with for example accelerometers, gyrometers, inertials units many others problems have to be taken into account in the measurement results such as the offset errors, the impacts effects, the temperature effects, the bandwidth limits of the sensor versus the dynamics of the walking biped, the robustness, the price... For these latter reasons, there is a real interest to explore the possibilities of observers in order to estimate absolute angular positions and velocities from only the knowledge of the relative angular variables.
Very few works have been done for the design of such observers, the previous works on observers design being done especially for the estimation of velocities

(for noiseless differentiation) by supposing that all the angular variables are measured [8]. First results on the design of observer/controller using only the measurement of joint link angular variables (relative angles) for three-links and five-links biped robots with no actuator in the ankles, in both cases of its stabilization in a vertical position and its walking, have been proposed by the authors and are based on Kalman filter [1], high gain observers, (high order) sliding mode observers [6, 5]. This latter class allows to establish the finite-time convergence of the estimation error and, through the high order of sliding mode, removes the chattering effect while preserving robustness and improving the accuracy.
The paper proposes a new version of second order sliding mode observer, based on *super twisting* approach [7], in order to estimate the absolute orientation of a five-link biped without feet during a walking gait. The imbalance phases have been taken into account when the biped is in single support. This algorithm is using only measurements of the joint variables, and not yet its time derivatives as previous works, which implies less noise in the controller. Furthermore, this paper presents the proof of stability of observer-based control, which has been still done only for a 3 links-biped robot: this result is an extension of [6].

2 Model of a planar five-link biped robot

The considered biped is walking on a rigid and horizontal surface. The robot is modeled as a planar biped. It consists of a torso, hips, and two legs with knees but no actuated ankles (see Figure 1). The walking cycle takes place in the sagittal plane and consists of successive of single support phases and impacts. The complete model of the biped robot consists of two parts: the differential equations describing the dynamics of the robot during the swing phase, and an impulse model of the contact event (the impact between the swing leg and the ground is modeled as a contact between two rigid bodies). The dynamic model of the biped in single support phase between successive impacts is derived from the Lagrange formalism

$$D(q_{rel})\ddot{q} + H(q,\dot{q})\dot{q} + G(q) = B\Gamma \tag{1}$$

with $q = [q_{31}\ q_{41}\ q_{32}\ q_{42}\ q_1]^T$, $q_{rel} := [q_{31}\ q_{32}\ q_{41}\ q_{42}]^T$, the vector joint angles vector, and $\Gamma = [\Gamma_1\ \Gamma_2\ \Gamma_3\ \Gamma_4]^T$ (see Figure 1)[1], D the inertia matrix, H the centrifugal and Coriolis effects matrix, G the gravity matrix. The torques u are applied between the torso and the stance leg, the torso and the swing leg, at the knee of the stance leg, and at the knee of the swing leg, respectively. Then, the model can be written in state space form by defining [3]

$$\dot{x} = \begin{bmatrix} \dot{q} \\ D^{-1}(-H\dot{q} - G + B\Gamma) \end{bmatrix} = f(x) + g(q_{rel}) \cdot \Gamma \tag{2}$$

[1] Leg 1 is the stance one, leg 2 the swing one.

with $x = [q^T\ \dot{q}^T]^T$. The state space is taken such that $x \in \mathcal{X} \subset I\!R^{10} = \{x = [q^T\ \dot{q}^T]^T \mid \dot{q} \in \mathcal{N},\ q \in \mathcal{M}\}$, where $\mathcal{N} = \{\dot{q} \in I\!R^5 \mid |\dot{q}| < \dot{q}_M < \infty\}$ and $\mathcal{M} = (-\pi, \pi)^5$. The walking surface is taken as $x \in \mathcal{S} = \{x \in \mathcal{X} \mid z_2(q) = 0,\ \dot{z}_2(q) < 0\}$ with $z_2(q)$ the altitude of the swing leg tip and its time derivative $\dot{z}_2(q)$. An impact occurs when the swing leg touches the walking surface, also called the ground. The impact between the swing leg and the ground is modeled as a contact between two rigid bodies. Under the same hypotheses and results of [4], the final result gives an expression for $x^+ := (q, \dot{q}^+)$ (the state value just after the impact) in terms of $x^- := (q, \dot{q}^-)$ (the state value just before the impact), which is expressed as $x^+ = \Delta(x^-)$. The overall biped robot model can be expressed as a nonlinear system with impulse effects [3]

$$\dot{x} = f(x) + g(q_{rel})\Gamma, \quad \text{for } x^- \notin \mathcal{S}, \qquad x^+ = \Delta(x^-), \quad \text{for } x^- \in \mathcal{S}. \tag{3}$$

3 Controller and observer design

3.1 Controller design

The control for the walking gait [4] consists in tracking the four joints reference angles q_{31d}, q_{41d}, q_{32d} and q_{42d} of the biped. During the single-support phase, the degree of under-actuation equals one, as only four outputs can be driven. Then, the robot gets a walking motion if the controller drives to zero the output vector $h(x)$ defined as (with θ defined in Figure 1)

$$h(x) := [q_{31} - q_{31d}(\theta)\ \ q_{32} - q_{32d}(\theta)\ \ q_{41} - q_{41d}(\theta)\ \ q_{42} - q_{42d}(\theta)]^T .$$

The control strategy consists in decoupling the system and in forcing the system to evolve by arbitrarily stated dynamics through the use of a finite time convergent control law [6].

3.2 Analysis of observability

Consider the dynamical system (2), with y the vector composed of the measured variables $y := [y_1\ y_2\ y_3\ y_4]^T = q_{rel}$

$$\dot{x} = f(x) + g(y)\Gamma, \quad y = Cx \tag{4}$$

In the biped context, this model describes the swing motion and is studied over one step, *i.e.* for $t \in [T_I^i, T_I^{i+1}[$, with T_I^i (resp. T_I^{i+1}) the initial (resp. final) impact time of the step i. As $g(y)\Gamma$, the *input-output injection* term of (4), is fully known, an observer for (4) can be designed by the following way. With abuse of notation, consider the next nonlinear system, which is the part of (4) without the input-output injection term $g(y)\Gamma$,

$$\dot{x} = f(x), \quad y = Cx \tag{5}$$

The observer displayed in Subsection 3.3 is designed from this model (5).

Proposition 1 ([6]). *There exist observability indices* $[k_{31}\ k_{32}\ k_{41}\ k_{42}]^T$ *and* $\mathcal{T} \subset \mathcal{X}$ *such that system (5) is observable for* $x \in \mathcal{T}$. ■

From Proposition 1, system (5) is locally equivalent to the canonical form,

$$\dot{z} = Az + \varphi(z), \quad y = C_z z \tag{6}$$

with (A, C_z) Brunovsky form, under state transformation $z = \Phi(x) = \left[y_1 \ \cdots \ y_1^{(k_1-1)} \ \cdots \ y_4 \ \cdots \ y_4^{(k_4-1)}\right]^T$.

3.3 Step-by-step observer

In this section, the observer is based on triangular form one [2] and its main property is the finite time convergence to zero of the estimation error. System (6) is still on (particular) triangular form. A such observer of (6) can be written as [2]

$$\dot{\hat{z}} = A\hat{z} + \varphi(\hat{z}) + E(t)\chi(\cdot) \tag{7}$$

with $\hat{z} \in \mathcal{Z}$ the estimated vector of z. Knowing that the principle of this class of observers consists in forcing, each in turn, estimated state variables to corresponding real ones, in finite time, this latter property is based on an adequate choice of $E(t)$ and $\chi(\cdot)$, *i.e.* $E(t)$ and χ are defined such that the estimation error $e = \hat{z} - z$ converges to zero in finite time. The originality in the following is the use of the *super twisting algorithm* [7] in order to ensure the finite time algorithm: this allows to not use time derivative of estimation error in the computation of the observer.

Finite time convergence observer

In a sake of clarity, and without loss of generality, only the observer design for a third order system is fully displayed in the sequel, *i.e.* $z \in \mathcal{Z} \in I\!R^3$. Then, in this case, system (6) reads as

$$\dot{z}_1 = z_2, \ \dot{z}_2 = z_3, \ \dot{z}_3 = f_3(z), \ y = z_1 \tag{8}$$

with $z = [z_1\ z_2\ z_3]^T$. Then, an observer for (8) reads as

$$\dot{\hat{z}}_1 = \hat{z}_2 + E_1(t)\chi_1(\cdot), \ \dot{\hat{z}}_2 = \hat{z}_3 + E_2(t)\chi_2(\cdot), \ \dot{\hat{z}}_3 = f_3(\hat{z}) + E_3(t)\chi_3(\cdot) \tag{9}$$

with $E_i(t)$ and χ_i $(1 \leq i \leq 3)$ defined such that each estimation error $e_i = \hat{z}_i - z_i$ converges to zero in finite time. To ensure a finite time convergence, the function $\chi_i(\cdot)$ is based on the *super twisting algorithm* [7] and reads as

$$\chi_i = \varsigma_i + \lambda_i \ |S_i|^{1/2} \ \text{sign}(S_i), \ \dot{\varsigma}_i = \alpha_i \ \text{sign}(S_i) \tag{10}$$

with

$$S_i = \begin{cases} y - Cz \text{ for } i = 1 \\ \tilde{z}_i - \hat{z}_i \ \text{ for } i > 1 \end{cases} \tag{11}$$

and $\tilde{z}_j = \hat{z}_j + E_{j-1}(t)\chi_{j-1}$ for $j \in \{2, 3\}$.

Determination of E_i and sketch of proof of finite time convergence

From (8)-(9), the dynamics of estimation error, $e_i = \hat{z}_i - z_i$, reads as

$$\dot{e}_1 = e_2 + E_1(t)\chi_1(\cdot),\ \dot{e}_2 = e_3 + E_2(t)\chi_2(\cdot),\ \dot{e}_3 = f_3(\hat{z}) - f_3(z) + E_3(t)\chi_3(\cdot).$$

- **Step 1.** Suppose that $e_1(0) \neq 0$, and observer (9) is initialized such that $E_1 = 1$ and $E_2 = E_3 = 0$. The error dynamics reads as

$$\dot{e}_1 = e_2 + \chi_1,\ \dot{e}_2 = e_3,\ \dot{e}_3 = f_3(\hat{z}) - f_3(z) \tag{12}$$

  As χ_1 is based on the *super twisting algorithm* with appropriate gain value, e_1 reaches zero in finite time at $t = t_1$. Then, $\forall\, t \geq t_1$, $e_1(t) = \dot{e}_1(t) = 0$, *i.e.*

$$e_1 = 0, \quad \dot{e}_1 = e_2 + \chi_1 = \hat{z}_2 - z_2 + \chi_1 = 0 \tag{13}$$

  From (13), one gets $\hat{z}_2 + \chi_1 = z_2$ and $z_2 = \tilde{z}_2$.
- **Steps 2 and 3.** The proof takes the same way than step 1.

Loss of observability and observation algorithm

As mentioned in [6], during the swing phase, along the *nominal* trajectories, for each observability indices possibilities, there is loss of observability. This problem has been solved through the use of two observers structures based on different observability indices [6].

3.4 Proof of stability

In this section, the objective is to prove the stability of the walking of the biped controlled by previously presented controller with finite-time observer. The stability can be proven on the basis of a restriction of the Poincaré's map to an one-dimensional manifold [4]. In [6], an original extension of [4] to observer-based controlled systems has been proposed for a three-links biped robot. In the sequel, an extension is made for five-links biped robots. As the "real" state vector is not fully measured, the "real" zeros dynamics and impact surface manifolds can not be used in the stability proof, the idea is then to suppose that the estimated state is on "estimated" zeros dynamics and impact surface manifolds. The finite-time convergence of the observer and controller, at the end of the first step, ensures that the estimated manifolds are the same than "real" ones. Then, it is possible to use the standard reduced Poincaré's approach to establish, over the second step, the stability.

Notations. The impact time at the end of the ith step is noted T_I^i and is taken as the time origin for the $(i+1)^{th}$ step. Let T_O^i denote the convergence (towards 0) time of the estimation error over the ith step and T_C^i the convergence time of the controller over the ith step. The observer and the controller have been tuned such that $T_O^i \leq T_C^i < T_I^i$, *i.e.* the observer converges faster than the controller over the step i. As the observer converges to real state in finite time, it is obvious that $T_O^i = 0$ for $i > 1$: then, from T_O^i, one has $\hat{x} = x$.

Numerical procedure. Let $\hat{Z}$ denote the "estimated" zero dynamics manifold, $\hat{Z} = \{\hat{x} \in \hat{\mathcal{X}} \mid h(\hat{x}) = 0,\ \dot{h}(\hat{x}) = 0\}$, $\hat{\mathcal{S}}$ the "estimated" impact surface manifold, $\hat{\mathcal{S}} = \{\hat{x} \in \hat{\mathcal{X}} \mid z_2(\hat{x}) = 0,\ \dot{z}_2(\hat{x}) < 0\}$, $\hat{\mathcal{X}} := \{\hat{x} := [\hat{q}^T, \dot{\hat{q}}^T]^T\}$, $\hat{q}$ and $\dot{\hat{q}}$ being the estimated values of q and $\dot{q}$. The conditions required to define the restricted Poincaré map are

1. $\hat{\mathcal{S}} \cap \hat{Z}$ is a smooth sub-manifold of $\hat{\mathcal{X}}$. It is equivalent to the fact that the map $\left[h(\hat{x})^T\ \dot{h}(\hat{x})^T\ z_2(\hat{x})\right]^T$ has constant rank equal to 9 on $\hat{\mathcal{S}} \cap \hat{Z}$, which is obvious to prove. If $(\dot{\hat{x}}, \hat{x}) \in \hat{\mathcal{S}} \cap \hat{Z}$, $\hat{x}$ equals a constant, denoted $\hat{x}_0$. Let $\gamma := [h(\hat{x})^T\ \hat{\theta}(\hat{x})]^T$ which has full rank at $\hat{x}_0$ (for definition of θ, see Figure 1). On $\hat{Z}$, one has $\dot{h}(\hat{x}) = 0$

$$\begin{bmatrix} 0 \\ \dot{\hat{\theta}} \end{bmatrix} = \frac{\partial \gamma}{\partial \hat{q}} \dot{\hat{q}} \tag{14}$$

Proposition 2. *Let us define p, a diffeomorphism from $I\!R \to \hat{S} \cap \hat{Z}$ to complete equation (14) with the configuration vector $\hat{q}_0$ of the impact such as*

$$p(\dot{\hat{\theta}}) = \begin{bmatrix} \left[\frac{\partial \gamma(\hat{q}_0)}{\partial \hat{x}}\right]^{-1} \dot{\hat{\theta}} \\ \hat{q}_0 \end{bmatrix} \tag{15}$$

At each configuration $\hat{q}_0$ is associated an unique angle $\hat{\theta}_0$.

2. The cross section for the Poincaré map will be taken to be $\hat{S}$, the "estimated" impact surface. Define $\lambda : I\!R \to I\!R$ computed by the following manner
   - Let $\dot{\hat{\theta}}_0^+ = \dot{\hat{\theta}}^+(T_I^0)$ (resp. $\hat{\theta}_0$) denote the initial estimated angular velocity (resp. position) just after the *initial* impact. Compute $\hat{x}^+(T_I^0) = [\hat{q}_0^T\ \dot{\hat{q}}_0^T]^T := p(\dot{\hat{\theta}}_0^+)$, the estimated state vector of the biped after the impact. State the real state after the impact as (given that q_{31}, q_{41}, q_{32}, q_{42} and the corresponding velocities are measured)

$$\begin{aligned} x^+(T_I^0) &:= \Big[q_{31}^d(\hat{\theta}_0)\ q_{41}^d(\hat{\theta}_0)\ q_{32}^d(\hat{\theta}_0)\ q_{42}^d(\hat{\theta}_0)\ q_1(T_I^0) \\ &\qquad \dot{q}_{31}(T_I^0)\ \dot{q}_{41}(T_I^0)\ \dot{q}_{32}(T_I^0)\ \dot{q}_{42}(T_I^0)\ \dot{q}_1(T_I^0)\Big]^T \\ \hat{x}^+(T_I^0) &:= \Big[q_{31}^d(\hat{\theta}_0)\ q_{41}^d(\hat{\theta}_0)\ q_{32}^d(\hat{\theta}_0)\ q_{42}^d(\hat{\theta}_0)\ \hat{q}_1(\hat{\theta}_0) \\ &\qquad \dot{q}_{31}^d(\hat{\theta}_0, \dot{\hat{\theta}}_0^+)\ \dot{q}_{41}^d(\hat{\theta}_0, \dot{\hat{\theta}}_0^+)\ \dot{q}_{32}^d(\hat{\theta}_0, \dot{\hat{\theta}}_0^+)\ \dot{q}_{42}^d(\hat{\theta}_0, \dot{\hat{\theta}}_0^+)\ \dot{\hat{q}}_1(T_I^0)\Big]^T \end{aligned}$$

   - Use $x^+(T_I^0)$ as the initial condition in (3) controlled by a control law which uses $\hat{x}$. Simulate until one of the following happens :

a) There exists a time T_I^1 for which $\hat{z}_2 = 0$, $\dot{\hat{z}}_2 < 0$ and $T_O^1 \leq T_C^1 < T_I^1 < \infty$ (recall that, for $t \geq T_O^1$, $\hat{x}(t) = x(t)$): apply again the impact model to $\hat{x}^-(T_I^1)$. Note that, fotThen, $\hat{x}^+(T_I^1) = \Delta(\hat{x}^-(T_I^1))$. At this time, the real and estimated state variables have same values, viewed that the observer has finite-time convergence of estimation error, and that the observer gains have been tuned such that $T_O^1 < T_I^1$. Use $\hat{x}^+(T_I^1)$ as the initial condition in (3). If there exists a time T_I^2 for which $\hat{z}_2 = z_2 = 0$, $\dot{\hat{z}}_2 = \dot{z}_2 < 0$ such that $T_O^2 \leq T_C^2 < T_I^2 < \infty$, then apply the impact effect $\hat{x}^+(T_I^2) = \Delta(\hat{x}^-(T_I^2))$. One gets $\lambda\left[\dot{\hat{\theta}}^+(T_I^1)\right] := \dot{\hat{\theta}}^+(T_I^2) = p^{-1}\left[\hat{x}^+(T_I^2)\right]$; else $\lambda\left[\dot{\hat{\theta}}^+(T_I^1)\right]$ is undefined at this point.

b) There does not exist a $T_I^2 > 0$ such that $\hat{z}_2 = 0$, $\dot{\hat{z}}_2 < 0$; in this case, it is also true that $\lambda\left[\dot{\hat{\theta}}^+(T_I^1)\right]$ is undefined at this point.

4 Simulations

The parameters of the five-links biped prototype "Rabbit" [3] are used to design the five-links biped parameters. The masses and lengths of the links (Indices 31, 41, 32, 42, 1: swing leg (femur, tibia), stance leg (femur, tibia), torso, resp.) are $m_{31} = m_{32} = 3.2\,kg$, $m_{41} = m_{42} = 6.8\,kg$, $m_1 = 17.0528\,kg$, $l_{31} = l_{32} = l_{41} = l_{42} = 0.4\,m$, $l_1 = 0.625\,m$. The distances between the joint and the mass center of each link are $s_{31} = s_{32} = 0.127\,m$, $s_{41} = s_{42} = 0.163\,m$, $s_1 = 0.1434\,m$. The inertia moments around the mass center of each link are $I_{31} = I_{32} = 0.0484\,kg{\cdot}m^2$, $I_{41} = I_{42} = 0.0693\,kg{\cdot}m^2$, $I_1 = 1.8694\,kg{\cdot}m^2$. The inertia of the rotor for each DC motor is $I = 3.32\,10^{-4}\,kg{\cdot}m^2$. The ratio N of the gearbox reducers equals 50. Value U of the applied torques equals $150\,N\cdot m$.

In order to determine if the closed-loop system is stable, function λ is evaluated for $\dot{\hat{\theta}}_0^+ \in [0.5, 4]$, $q_1 \in [\pi/12 - \pi/30 \text{ rad}, \pi/12 + \pi/30 \text{ rad}]$ and $\dot{q}_1 \in [0.02 \text{ rad/s}, 0.22 \text{ rad/s}]$. Figure 2 displays function λ : λ is undefined for $\dot{\hat{\theta}}_1^+(T_I^1)$ less than 0.825 rad/s and more than 3.75 rad/s. A fixed point appears at approximately $\dot{\hat{\theta}}_1^+(T_I^1) = 1.665\,rad/s$, and corresponds to an asymptotically stable walking cycle. Figure 3 displays $\dot{\hat{\theta}}^+(T_I^0)$ with respect to the initial "real" kinetic energy $E_c = E_c(q_{31}, q_{32}, q_{41}, q_{42}, q_1, \dot{q}_{31}, \dot{q}_{32}, \dot{q}_{41}, \dot{q}_{42}, \dot{q}_1)$ and its initial "estimated" value $\hat{E}_c = E_c(\hat{q}_{31}, \hat{q}_{32}, \hat{q}_{41}, \hat{q}_{42}, \hat{q}_1, \hat{q}_{31}, \hat{q}_{32}, \hat{q}_{41}, \hat{q}_{42}, \hat{q}_1)$ andthe initial estimation errors on unmeasured variables q_1 and $\dot{q}_1$. This representation allows to know the possible initial estimation errors on all the variables which are admissible in order to guarantee the convergence to the limit cycle through the kinetic energy. For each point of this 3D-area, one ensures that each estimated state variable converges to the corresponding one, the control outputs reach zero before the end of each step, and the biped robot converges to a stable limit cycle. In terms of this 3D-area, the final stable walking cycle corresponds to the red star point. Figure 4 displays the walking motion of the biped robot as a series of stick figures over three steps on the stable limit cycle.

5 Conclusion

The first contribution of this paper is the design of an original finite time observer, based on second order sliding mode approach, which estimates the absolute orientation of a dynamical stable five -ink biped. This task is important and vital in order to control its locomotion with high accuracy. The stability of the nonlinear association control/observer is proved for cyclic walking gait, which is a hard task and an originality in the litterature. The next step will consist in experimentally testing the strategy on prototype Rabbit [3]. An other extension of this work consists in estimating the orientation of humanoid robot in 3D when its feet sole is partially in contact with the ground, or when the ankles torques are dramatically limited.

References

1. Y. Aoustin, G. Garcia, and A. Janot. Estimation of the absolute orientation of a two-link biped using discrete observers. In *Proc. Mechatronics and Robotics Conf. MECHROB*, pages 1315–1320, Aachen, Germany, 2004.
2. T. Boukhobza and J.P. Barbot. High order sliding modes observer. In *Proc. IEEE Conf. on Decision and Control CDC*, pages 1912–1917, Tampa, Florida, USA, 1998.
3. C. Chevallereau, G. Abba, Y. Aoustin, F. Plestan, E.R. Westervelt, C. Canudas de Wit, and J.W. Grizzle. Rabbit: a testbed for advanced control theory. *IEEE Control Systems Magazine*, 23(5):57–79, 2003.
4. J.W. Grizzle, G. Abba, and F. Plestan. Asymptotically stable walking for biped robots : analysis via systems with impulse effects. *IEEE Transactions on Automatic Control*, 46(1):51–64, 2001.
5. V. Lebastard, Y. Aoustin, and F. Plestan. Absolute orientation estimation for observer-based control of a five-link walking biped robot. *Robot Motion and Control: Recent Developments, Lecture Notes in Control and Information Sciences*, 335, 2006.
6. V. Lebastard, Y. Aoustin, and F. Plestan. Observer-based control of a walking biped robot without orientation measurement. *Robotica*, 24(3):385–400, 2006.
7. A. Levant. Sliding order and sliding accuracy in sliding mode control. *Int. J. of Control*, 58(6):1247–1263, 1993.
8. P. Micheau, M.A. Roux, and P. Bourassa. Self-tuned trajectory control of a biped walking robot. In *Proc. Int. Conf. on Climbing and Walking Robot CLAWAR*, pages 527–534, Catania, Italy, 2003.

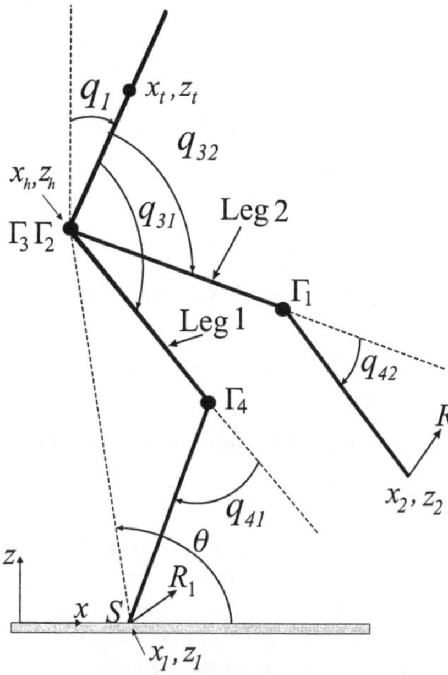

Fig. 1. Schematic of biped robot: absolute and relative angles - Cartesian coordinates.

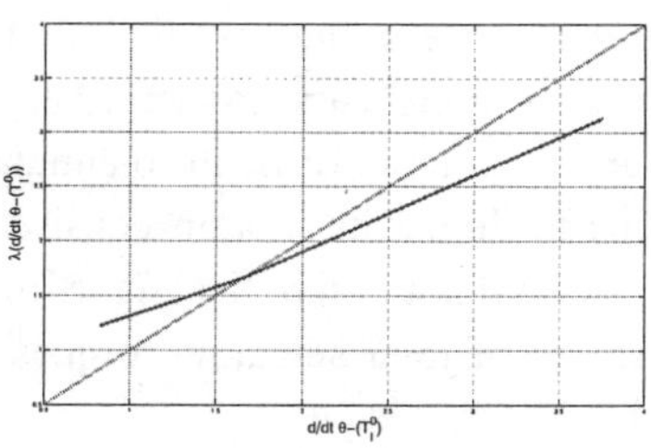

Fig. 2. Function λ (solid line) and identity function (dotted line) versus $\hat{\theta}_1^+(T_I^1)$. This graph establishes the existence of an asymptotically stable walking motion.

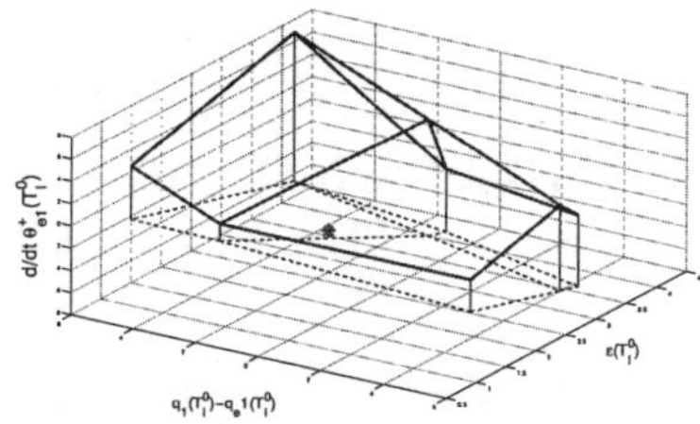

Fig. 3. $\dot{\hat{\theta}}_1^+(T_I^0)$ versus initial kinetic energy estimation error ϵ and initial estimation error on q_1. All the initial conditions in the 3D-box allow tha convergence to the stable limit cycle characterized by the red star.

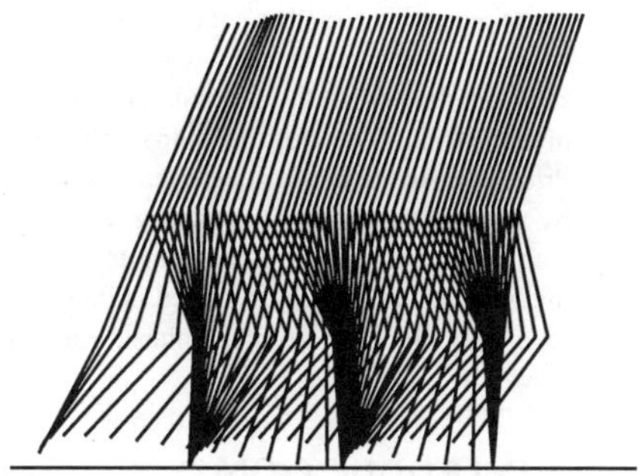

Fig. 4. Plot of walking as a sequence of stick figures.

OPTIMIZATION OF HUMANOID ROBOT MOTION DURING THE ELEVATION OF AN OBJECT

HAMED AJABI NAEINI

Biomedical Department, Azad University Sciences and Researches Branch, Tehran ,Iran

MOSTAFA ROSTAMI

Biomedical Department, Azad University Sciences and Researches Branch, Tehran ,Iran

This research is aimed at generation optimal motions of two dimensional humanoid robot while robot must elevates an object to hypothetical height with knee bending and without knee bending from the ground. The approach is demonstrated by using a 4 and 5 link planar biped. The optimization problem is dealt with using Pontryagin's Maximum Principle. As the biped is essentially submitted to gravity forces, the motion is generated by minimizing the joint actuating torques and time. After implementation optimization technique, a final two point boundary value problem is solved by applying a shooting method. The approach presented is illustrated by various numerical simulations applying to a four and five segment planar robot which has four or five active joints during elevation.

1. INTRODUCTION

In the field of biped locomotion,our reference model is human gait.Although the kinematics organization of human motion is quite easy to describe, the underlying dynamics is complex to analyze and even more difficult to simulate. If we want to move a robot, a simple means to define its motion consists in introducing a kinematic model of elevation. It would not ensure the quality of its dynamics. Synchronization with restrained values of joint actuating torques, as well as moderate energy consumption is needed.

The prevailing idea is based on the minimization performance criterion with a dynamic content involving essentially the joint actuating torques, or the energy consumption.

To deal with the optimization problem, which is stated within the frame of the optimal control theory. Accordingly, the implemented optimization technique is the Pontryogin Maximum Principle (PMP). Kinematic constraints are taken into account by means of an augmented integral criterion, and are dealt with using a penalty method.In this way, the constrained optimization problem is transformed into a sequence of unconstrained problems which are solved as two-point boundry value problems.

2. KINEMATICS AND DYNAMICS OF A PLANAR BIPED

The planar multibody system more free joint are considered, and all of them are powered. A complete Hamiltonian dynamic model is formulated on the basis of a set of relative joint coordinates.

2.1. *Kinematic model*

We describe a sagittal model of an anthropomorphic humanod. This model is made up of 4 and 5 limbs. In accordance with the schematic representation in figure 1, we define the dimension and complementary notations :

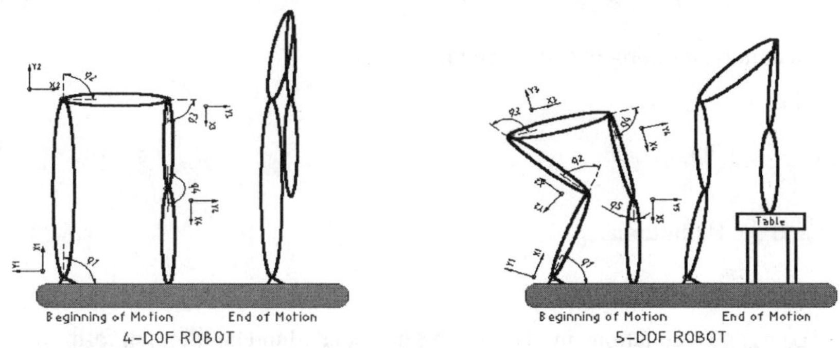

Figure 1. 4&5-link planar humanoid at first and end of elevation

$$q = (q_1,\dots\dots,q_n)^T, vector \quad of \quad joint \quad coordinates$$
$$\dot{q} = (\dot{q}_1,\dots\dots,\dot{q}_n)^T, vector \quad of \quad joint \quad velocities$$
$$\ddot{q} = (\ddot{q}_1,\dots\dots,\ddot{q}_n)^T, vector \quad of \quad joint \quad accelerations$$

2.2. *Dynamic model*

As we intend to use the Pontryogin Maximum Principle (or PMP)for solving the dynamic optimization problem stated in what follows, let us recall that the implementation of the PMP requires theb formulation of the dynamic model in state space form. As indicated in reference [1, 2] the Hamiltonian dynamic model not only best fulfills this requirement but, as well, strengthens the robustness of algorithms used to solve the optimization problem. We present the outlines of the formulation we need.

Firstly, introducing the Langrangian of the mechanical system

$$T(q,\dot{q}) = \frac{1}{2}\dot{q}^T A(q)q \tag{1}$$

Where V stands for the gravity potential, and T is the kinetic energy defined as

$$L(\mathbf{q},\dot{\mathbf{q}}) = T(\mathbf{q},\dot{\mathbf{q}}) - V(\mathbf{q}) \tag{2}$$

M being the (n× n) mass matrix of the kinematic chain, Lagrange's equations of motion may be derived as

$$\frac{d}{dt}\left(\frac{\partial L}{\partial \dot{\mathbf{q}}_i}\right) - \frac{\partial L}{\partial q_i} = Q_i^{\ d} + Q_i^{\ a} \quad ,i = 1,\ldots,n \tag{3}$$

Secondly, defining the conjugate momenta

$$p_i = \frac{\partial L}{\partial \dot{q}_i} \qquad i = 1,\ldots,n \tag{4}$$

and the Hamiltonian

$$H(\mathbf{p},\mathbf{q}) = \mathbf{p}^T\dot{\mathbf{q}} - L(\mathbf{q},\dot{\mathbf{q}}) \tag{5}$$

Lagrange's equations in (4) may be reformulated in Hamiltonian form

$$\begin{cases} \dot{q}_i = \dfrac{\partial H}{\partial p_i} \\ \dot{p}_i = -\dfrac{\partial H}{\partial q_i} + Q_i^a + Q_i^d \end{cases} \tag{6}$$

Or

$$\begin{cases} \dot{\mathbf{q}} = \mathbf{A}^{-1}\mathbf{p} \\ \dot{\mathbf{p}} = -1/2\,\mathbf{p}^{\mathrm{T}}\mathbf{A}_{,i}^{-1}\mathbf{p} - \mathbf{V}_{,i} + Q_i^a + Q_i^d \end{cases} \quad \text{where:} \quad \begin{matrix} p = A\dot{q} & A_{,i}^{-1} = \partial A^{-1}/\partial q_i \\ \dot{q} = A^{-1}p & V_{,i} = \partial V/\partial q_i \end{matrix} \tag{7}$$

With this formulation, Hamiltonian equations are ideally structured for applying the Pontryogin Maximum Principle.

The double set of vectorial equations (7) can be recast as the 2n-order differential vector equation

$$\dot{x}(t) = F(x(t)) + Bu(t) = F(x(t), u(t)) \tag{8}$$

Where F is a nonlinear function in X, and B is the constant (2n× n) matrix .

Let us mention that a complementary transformation remains to be achieved in order to perfect formulation (8). In consist in rescaling all the variables of the problem to homogenize their order of magnitude. To that end, one can introduce of the following reference quantities:L,M,T,I,Q, respectively length, mass, time, moment of inertia and torque of reference which can be define and linked as

$$\overline{M} = \frac{1}{n}(m_1 + \ldots + m_n) \qquad \overline{I} = \overline{ML}^2 \qquad \overline{T} = \sqrt{\frac{\overline{I}}{\overline{Q}}} \qquad \overline{Q} = \frac{1}{n}(Q_1^{a,\max} + \ldots + Q_n^{a,\max})$$

3. RESULTS

Simulated motion of robot from initial to end of elevation and relative joint coordinates and angular velocities and anglular acceleration of joints during lifting of anobject in following figures being represented :

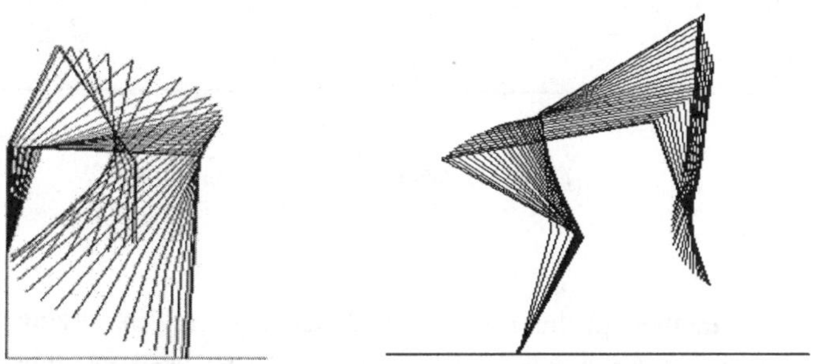

Figure 2.Optimal motion with and without knee flextion

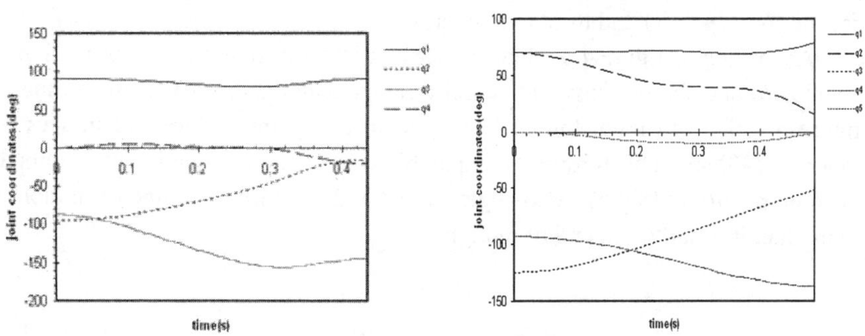

Figure 3.Evolution of joint angle

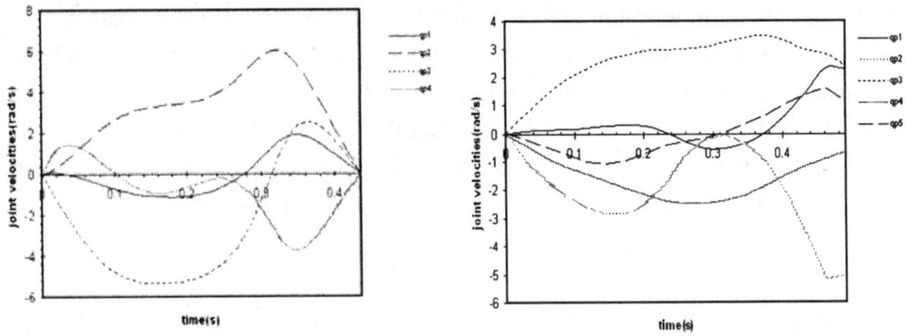

Figure 4. Evolution of joint angular velocities

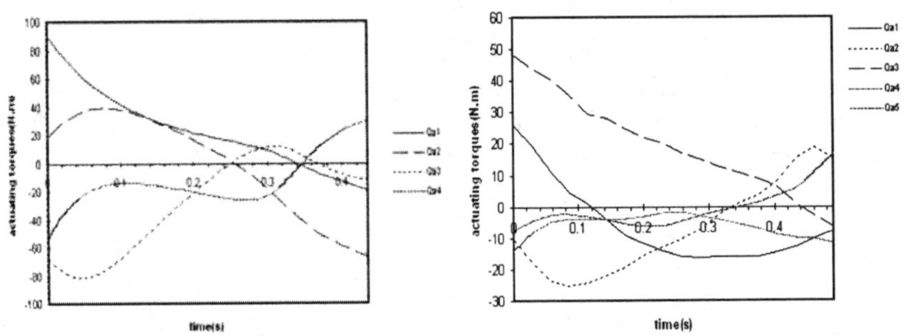

Figure 5. Evolution of joint actuating torques

Used method in this research made a lot of way in solving. Modeling and optimal control problems , and its result can be used in other biomechanical researches.

According to figure 3 , the maximum range of joint angle is seemed in second joint (pelvis) and third (shoulder) .

According to figure 4 angular velocity of joints in first and end of motion is equal zero and shapeof ang ular velocity can consider a sine wave. According to figure 5 , the maximum torque is created in first joint (ankle) and in start of motion , (90Nm). This torque is comparable with created torque in pelvis joint at bending position (48Nm). Also angle of first and third joints in first of motion is more than its quantity in end of motion.

References

1. M. Rostami ,G. Bessonet,"Sagittal gait of a biped robot during the single support phase.",Robotica,2001
2. B.Colbert and F.Multon, "Biomechanical Simulation of Human Lifting" , Bennes University, France, 2001
3. L.L.Menegaldo, A.D.T.Fleury, H.I.Weber, "Biomechanical Modeling and Optimal Control of Human Posture", Journal of biomechanics, 36:1701-1712, 2003.
4. W.Park, B.J.Martin, S.Choe, D.B Chaffin, M.P.Reed, "Representing and Identifying Alternative Movement Techniques For Goal-Directed Manual Tasks", Journal of biomechanics, 38:519-527, 2005.
5. M.Parnianpour and etc., "A Computational Method For Simulation of Trunk Motion", Biomechanics Research Laboratory,Yale University School of Medicine, Jun 2000
6. William H.Press, Saul A. Teukolsky, William T.Vetterling & Brian P.Flannery, "Numerical Recipes in FORTRAN: the Art of Scientific Computing", Press Syndicate of the University of Cambridge, 2th edition, 1986
7. H. Asada and E. Slotine , "Robot Analysis and Control",NewYork ,John Wiley,1986

Postural stability control for Robot-Human cooperation for sit-to-stand assistance.

V. Pasqui*, L. Saintbauzel and P. Bidaud

Université Pierre et Marie Curie-Paris6, FRE 2507, ISIR
18 Route du Panorama 92 265 Fontenay-aux-Roses
**E-mail: pasqui@robot.jussieu.fr*
http://lrp6.robot.jussieu.fr/

J. Graefenstein

Leibniz University of Hanover
E-mail: j.graefenstein@web.de

This article presents a fuzzy controller, for a robotic device, to ensure stability of the user during the assisted sit-to-stand transfer. The first problem to be addressed is the postural analysis of the chair rising. Experiments, with healthy subjects, were performed with this aim in view. Analysis of external forces shows that sit-to-stand transfer can be subdivided into several phases. The observation of the Center of Pressure and of the horizontal component of the handle force yields rules to observe the stability of the patient and consequently adjust the robotic interface motion to the human volontary movement. These rules are used in the fuzzy control implementation. The controller is validated on experiments with healthy subjects and diseaded patients.

Keywords: Assistive device; robotic interface; human centered robotic; postural stability; sit-to-stand; fuzzy control.

1. Introduction

The aim of the work presented in this paper is to realise a robotic interface for equilibrium assistance during Sit-To-Stand (STS) transfer. Here, it is supposed that interactive robotic devices, as human-centered robotics, is more comfortable and more efficient that traditionnal technical devices [1].
Robotics technologies have been investigated in the last few years to prevent falls by a postural control of patients and to promote safe mobility [2], [3], [4], [5], [6], [7], [8], [9]. But these robotic devices have no postural correction to restore equilibrium.
Based on an analysis of the most common walking troubles associated with aging or cerebellar syndrome we have designed and developed a robotic device (in Figure 1) [10] to help the patient to sit-down, stand-up and walk [11]. The assistive device handles guide the patient to rise from a chair or to sit down,

following trajectories which are based on parameters reflecting personal strategies [12].

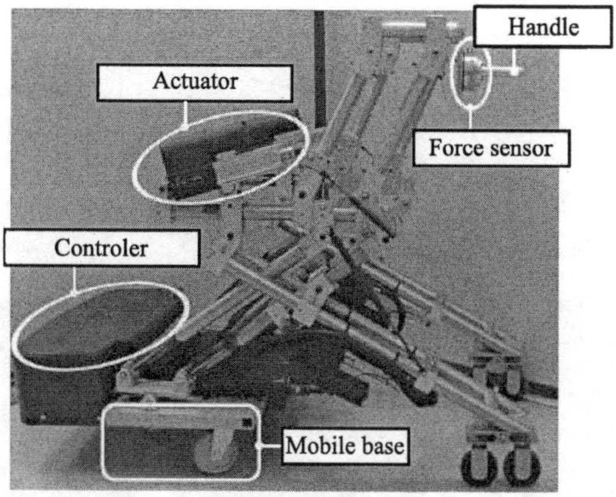

Fig. 1. Robotic Interface prototype

We propose here to detail how an adapted control can give interactive ability to this robotic interface.

2. Method

By interactivity, we intend the capacity to interpret the postural movements detected by the sensors to trigger the movement or to maintain the postural equilibrium.
To observe the postural state, experimental dynamical analysis of the stand-up have been done in our laboratory (Figure 2) . Results show different phases of chair rising, that are matching with physiological litterature [13]. Each phase depends on interaction forces between human and handle : $\vec{F_h} = (F_{hx}, F_{hy})$; or human and ground : $\vec{F_g} = (F_{gx}, F_{gy})$ and their time variations. Reaction force between human and ground is computed at Center of Pressure (CoP) which position may be used as a stability criteria [14].
The observation of the CoP position and direction of the force $\vec{F_h}$ yields simple rules to identify unstability cases or desired movement to trigger (i.e. beginning of the STS). Fuzzy controller is used in intention detection to control neural prostheses [15] or orthosis using FES [16].
Fuzzy controller seems to be a good way for interactivity, then we have extended the role of the fuzzy controller from the detection of voluntary movement to the detection of the unstability.
The fuzzy control has to fulfill two tasks, thats defined two output:

- **output 1**: recognition of the current phase,
- **output 2** : determination of proper reaction to ensure stability of subject.

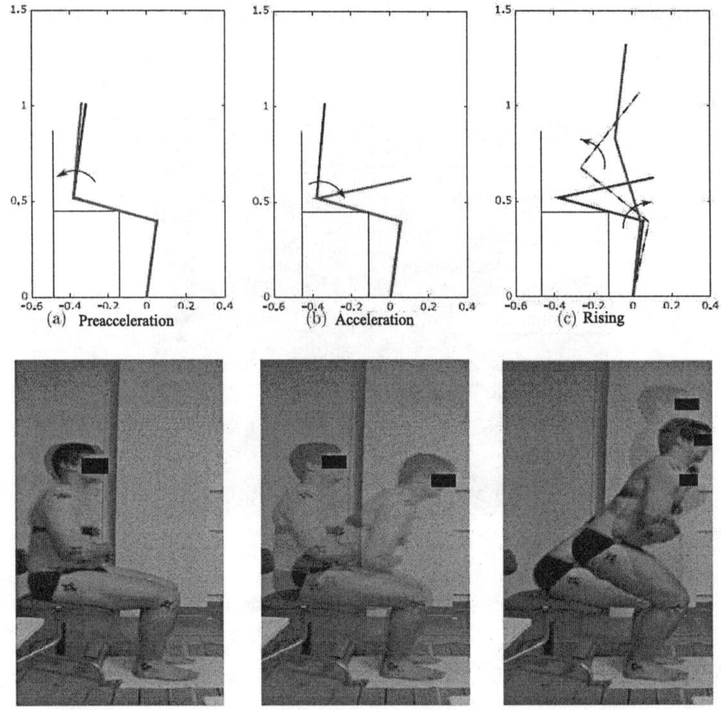

Fig. 2. Different sit-to-stand phases analysis

The following fuzzy sets were defined for the output 1 (in Figure 3) : seated, returned, preaccelaration (preac), acceleration (acc), start rising (startr), rise. The detection of the phases of the STS is obtained analysing the value of the $\vec{F}_h, \vec{F}_g$ and the time variation of $\vec{F}_h$.

"Returned" identifies the case when subject aborts stand-up and returns to the seated position.

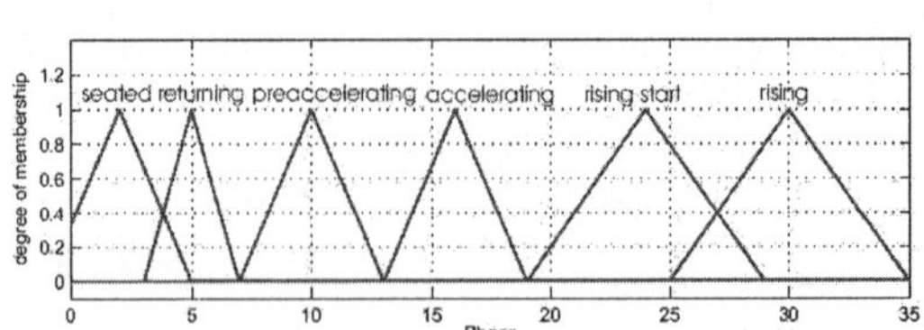

Fig. 3. Membership functions for output 1

The membership functions for the output2, that determines the movement, if proper phase is detected, are shown in Figure 4.The following fuzzy sets were defined:
instable (inst) : object underlies high unbalance. Quick reaction is required.
stabilize (stab): object indicates desire of stabilization.
no move (nm): no movement is necessary in the horizontal direction.
adjust (adj) : object desires another position of the handles.

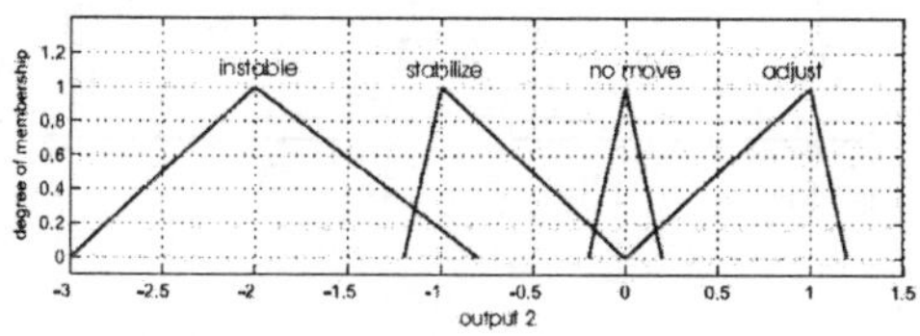

Fig. 4. Membership functions for output 2

If we denote H for high, Z for zero, L for low, EL for extremly low and EH for extremly high, we have for exemple :

IF F_{gy}=*EL AND* F_{hx}=*L AND* $\frac{dF_{hy}}{dt}$=*H THEN the human is RISING.*
IF F_{hx}=*H AND CoP*=*L THEN the human posture is stable.*

The rulebase for the fuzzy control is presented in the table figure 5.

F_{gx}	F_{gy}	$\frac{dF_{gx}}{dt}$	$\frac{dF_{gy}}{dt}$	F_{hx}	F_{hy}	$\frac{dF_{hx}}{dt}$	$\frac{dF_{hy}}{dt}$	CoP	$\frac{d\,\text{CoP}}{dt}$	out 1	out 2
Z	Z	–	–	Z	Z	–	–	–	–	seated	nm
H	–	H	–	H	0	H	–	–	–	preacc	nm
L	–	EH	L	–	–	–	H	–	–	accel	-
–	EL	–	–	L	–	–	H	–	–	rise	-
H	EL	Z	Z	L	–	Z	Z	–	–	rise	-
L	EL	H	EL	L	H	–	–	–	–	startr	-
EH	EH	H	H	H	H	–	–	–	–	return	-
–	–	–	–	H	–	–	–	L	–	-	stab
–	–	–	–	L	–	–	–	L	–	-	adj
–	–	–	–	H	–	–	–	L	–	-	stab
–	–	–	–	H	–	–	–	H	–	-	adj
–	–	–	–	Z	–	–	–	Z	–	-	nm
–	–	–	–	Z	–	–	–	–	EL	-	inst
–	–	–	–	Z	–	–	–	–	EH	-	inst

Fig. 5. Rulebase for the fuzzy control

Detection of unstable posture is illustrated in figure 6, where both patient and robot are modelised by a 3 links model each. The difference between these two models is in the interaction with ground. We assume that robotic interface cannot loose contact with ground while patient could if he is unstable.

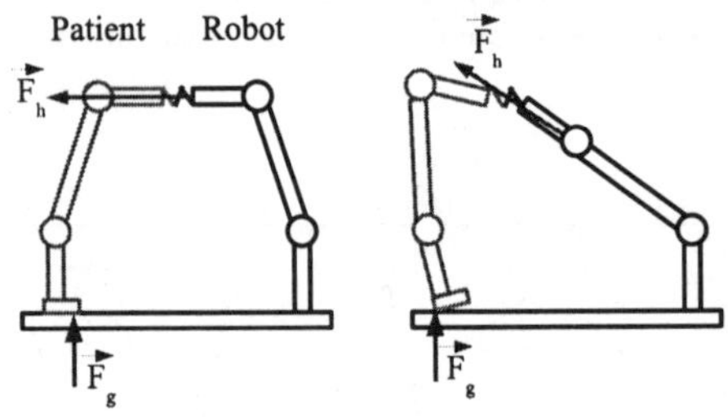

Fig. 6. Interaction between patient and robotic interface

If a subject, under perturbations, is verge on to loose his balance, he quickly shifts the load within the foot support area in the opposite direction of fall direction. If the impending fall directed forwards, the CoP will rapidly move in the same direction. An according reaction could be observed for a fall backwards.

3. Application

The complete structure of the controller is shown in figure 7.
The preprocessing block receives forces measurement, applies a filter and calculates the position of CoP and its time derivatives. These outputs are processed by the fuzzy logic block to identify patient posture state. Then, the corresponding control mode is selected between those:

- **Normal** : tracking trajectory,
- **Impedance** : Impedance control according to the efforts of interactions measured,
- **Stabilization** : modification of the tracking trajectory to stabilize the patient,
- **Return** : the interface returns to the initial position,

The movement is triggered by the preacceleration phase.
For cases identified as a patient aborting movement, the robot returns to the initial position. If postural unstability is detected the device motion in vertical direction is stopped and a new desired position is computed that guarantee patient stability.

4. Results

The presented prototype is currently in Bellan Hospital for a rehabilitation protocol validation. Many diseaded patients, with cerebellar syndrome, have tested the device. This kind of pathology imposes a ballast walking-aid to filter

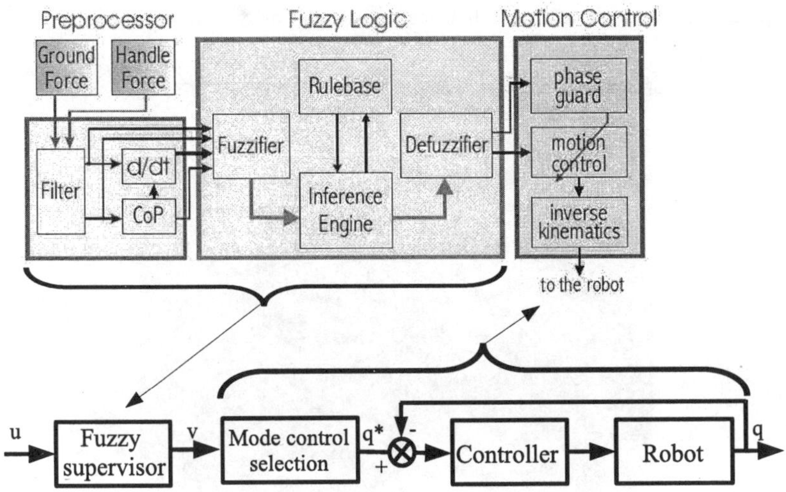

Fig. 7. Control structure

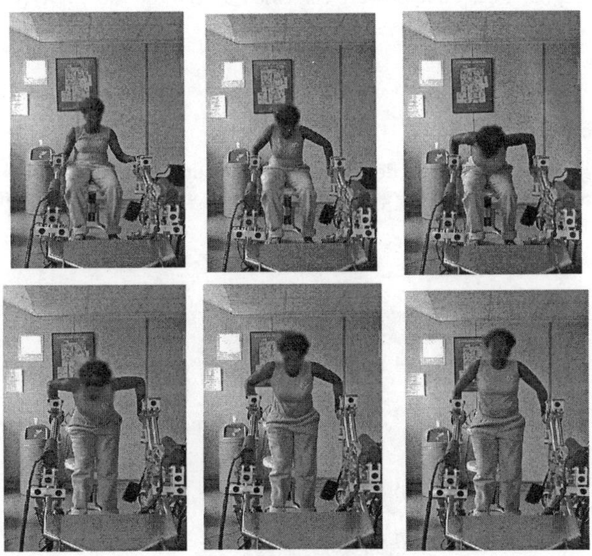

Fig. 8. Prototype in Bellan Hospital

shaking. The robotic interface is more comfortable, less tiresome and easy to drive. In addition to assist in position change and walking, this device can detect onset of fall. Then the robotic interface will response by changing handle position, producing a force to balance the patient, as it is shown in Figure 9.

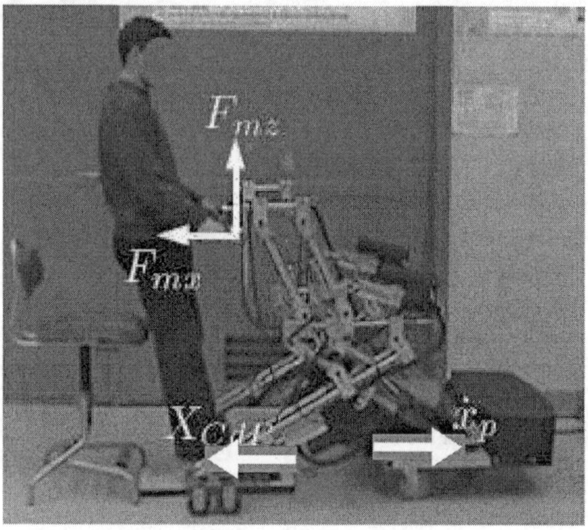

Fig. 9. Unstable posture corrected by the robotic interface

5. Conclusion

Fuzzy logic is very usefull for detection of movement intention and unstable postures. Provided with a fuzzy supervisor, the robotic interface becomes highly interactive.

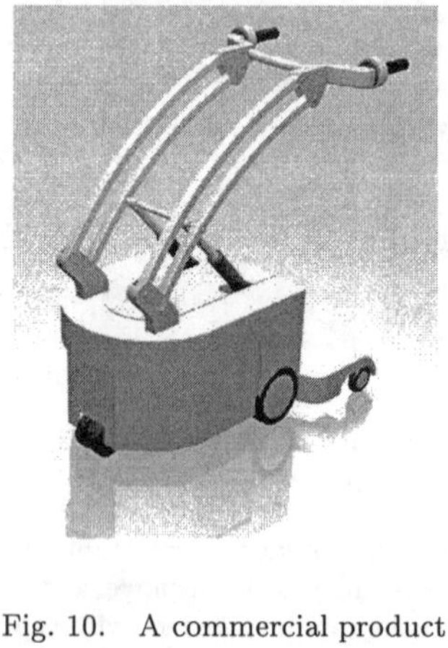

Fig. 10. A commercial product

Now, objectives are to develop rehabilitation protocols with collaboration with the medical team of Bellan Hospital. A second objective is to produce a commercial product with ROBOSOFT society. Such a product would be designed as in Figure 10.

References

1. R. reiner, L. Lunenburger and G. Colombo, human-centered robotics applied to gait training and assement., in *Journal of rehabilitation research and development*, (5)2006.
2. S. C. Lee, K. Oh K. and J. Lee, A system for gait rehabilitation : Mobile manipulator approach., in *Proc. IEEE Int. Conference on Robotics and Automation, (ICRA'02)*, (Washington, USA, 2002).
3. K. Nagai, I. Nakanishi and H. Hanafusa, Assistance of self-transfer of patients using a power-assisting device., in *Proc. IEEE Int. Conference on Robotics and Automation*, (Taipei, Taiwan, 2003).
4. Korean eldercare robots are coming. (2006), `http://robotgossip.blogspot.com/2006/07/korean-eldercare-robots-are-com%ing.html`.
5. M. Spenko, H.-Y. Yu and S. Dubowsky, A robotic personal aid for the mobility and health monitoring for the elderly., in *IEEE Trans on Neural Systems and Rehabilitation Engr*, 2007.
6. N. Rea, G. Lacey, C. Lambe and R. Dahyot, Multimodal periodicity analysis for illicit content detection in videos., in *The 3rd European Conference on Visual Media Production (CVMP 2006)*, 2006.
7. B. Graf, M. Hans and D. S. Rolf, Care-o-bot ii development of a next generation robotic home assistant., in *Autonomous Robots*, 2004.
8. J. Pineau, M. Montemerlo, M. Pollack, N. Roy and S. Thrun, Towards robotic assistants in nursing homes : Challenges and results., in *Robotics and Autonomous Systems*, (3-4)
9. H. Park, H. Hong, H. Kwon and M. Chung, A nursing robot system for the elderly and the disabled., in *International Journal of Human-friendly Welfare Robotic Systems*, (4)
10. P. Médéric, V. Pasqui, F. Plumet and P. Bidaud, Sit to stand transfer assisting by an intelligent walking-aid, in *Proc. 7th International Conference on Climbing and Walking Robots, (CLAWAR'04)*, (Madrid, Espagne, 2004).
11. P. Médéric, V. Pasqui, F. Plumet and P. Bidaud, Elderly people sit to stand transfer experimental analysis, in *Proc. 8th International Conference on Climbing and Walking Robots, (CLAWAR'05)*, (London, UK, 2005).
12. V. Pasqui and P. Bidaud, Bio-mimetic trajectory generation for guided arm movement during assisted sit-to-stand transfer, in *Proc. 9th International Conference on Climbing and Walking Robots, (CLAWAR'06)*, (Bruxel, Belgium, 2006).
13. R. Aissaoui and J. Dansereau, Biomechanical analysis and modelling of sit to stand task: a literature review, in *Proceedings of IEEE International Conference on Systems, Man, and Cybernetics*, 1999.
14. P. Sardain and G. Bessonet, Forces acting on a biped robot. center of

pressure-zero moment point, in *IEEE Transactions on Systems, Man and Cybernetics*, (5)2004.
15. P. C. Sweeney, G. M. Lyons and P. H. Veltink, Finite state control of functional electrical stimulation for the rehabilitation of gait, in *Medical and Biological Engineering and Computing*, (2)2006.
16. S. Hussein and G. M.H., Intention detection using a neuro-fuzzy emg classifier, in *IEEE Engineering in medecine and biology*, 2002.

RESEARCH ON UNDERACTUATED DYNAMICAL WALKING OF 3D BIPED ROBOT

SHENG TAO, MA HONG-XU

College of Mechatronic Engineering and Automation, National University of Defense Technology, Changsha, Hunan Province, China

A new underactuated 3D biped robot and its control strategy are presented. Its hybrid dynamic model is developed and gait is planned using method of time-invariant. By the feedback controller and finite-time stable control strategy, the robot realizes stable dynamical walking. Simulation results show that a stable limit cycle of dynamical walking is achieved, and proposed control strategy is feasible. Based on analyzing the relationship between walking velocity and robot configuration, a velocity control strategy is presented and validated by simulation.

1. Introduction

Biped robot is always a research focus because it has the same appearance as people and can do things instead of people. After many years' development, this domain has got a lot of achievements. There are many famous projects such as Honda P series biped robot, HRP series of Japan Trade and Industry (METI) [1], SDR series and QRIO biped robot of Sony. Leg Laboratory of MIT is one of the pioneers in this area, Flamingo, Turkey, M2 were developed since 1980s.

Despite the nearly half-century's efforts, no biped robots have made their way into sectors where their utility exceeds their novelty. It is conjectured that one main factor contributing to so slow development of usable biped robot is the low energy efficiency. It costs ten times or more energy than people when walking the same length. In order to improve the energy efficiency, the research in passive walking appeared and originated by McGeer in the late 1980s. In his seminal work [2], McGeer built a four-link planar passive walker and performed a detailed parameter variation and stability analysis. With his success, plenty of researches on passive walking were carried out [3], [4].

Passive walking is stable in theoretic, but in practice, vast work should be done to adjust its parameters before realizing stable walking. Furthermore, the robot should walk on a slope, so it has little actual significance. Based on the passive walking, Grizzle begins to study underactuated planar biped robot. The robot has no sole, contacts with ground by point feet. It maintains dynamic balance by changing the contact states of the two feet. Grizzle in [5] proved that the planar biped robot has a hybrid zero dynamics and gave theorems to estimate the stability. But the lateral stability was not dealing with in his work. Rabbit is a prototype of five links planar biped robot [6], which is the most successful one at present. Underactuated planar biped robot can only walk around an origin because it needs a no-actuated rotating bar to keep its lateral balance. Catherine E in [7] and J. Maxwell in [8] proved respectively that the lateral stability of human need active control. Steven H [9] and D. Kuo [10] respectively designed their 3D passive walking robot to perform 3D passive dynamic walking. But up to now, there are little reports about 3D underactuated biped robot.

Based on the work of Grizzle, we design a 3D biped robot. The robot uses feet with lateral distributing sole to realize lateral moment control. So the lateral motion is controllable completely and the sagittal is underactuated. It can realize 3D dynamic walking. The remainder of the paper is structured as follows: Section II develops the dynamical model, which is not continuous and includes single support phase and double support phase. Section III develops the time-invariant strategy to plan the gait of the robot. Section IV introduces the feedback controller and uses finite-time stable control strategy to control the walking. Simulations are done in Section V to validate the performance of the control strategy and the stability of the robot. At last, a strategy is introduced in section VI to control the walking velocity.

2. Dynamical Model

The robot we consider is shown in Fig. 1 It consists of 7 links: two identical femurs, tibiae, feet and a torso. Each knee joint is a revolute joint with one DOF (Degree Of Freedom). Each hip consists of revolute joint with 2 DOF. The sole is a cylindrical bar, distributing in lateral plane. Ankle lies in the center of the sole, and provides lateral moment with the robot.

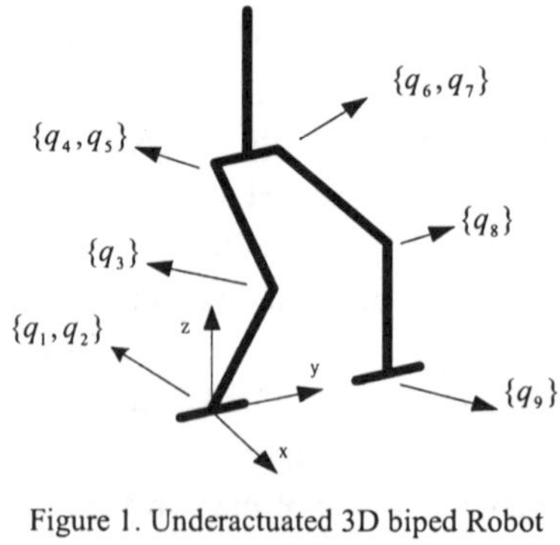

Figure 1. Underactuated 3D biped Robot

We make several assumptions to the robot as in [5], [11]:

1. Walking takes place on a flat surface;
2. The walking is symmetric in the steady state;
3. The double support is instantaneous and involves the impact of the swing leg with the ground;
4. The impacts are plastic and there is no rebound or slippage;
5. The impulsive forces may result in an instantaneous change in the velocities, but there is no instantaneous change in the configuration;
6. In the single support phase, the stance leg acts as a pivot in sagittal plane.

2.1. *Single Phase Dynamical Model*

A biped robot in the single support phase is an open chain manipulator and made up of tandem joints. A set of generalized coordinates of the robot is $q = [q_1 \cdots q_9]^T$, q_1 is the sagittal DOF of the ankle, which is an absolute angle between the robot and the ground. $q_2 \cdots q_9$ are the relative coordinates of the robot: q_4 and q_5 are the sagittal and lateral joints of the standing hip; q_6 and q_7 is the lateral and sagittal joints of the swing hip; q_9 is the lateral joint of the swing foot.

Letting $x := [q^T, \dot{q}^T]^T$, applying the Euler-Lagrange's equations, the dynamical equations of the robot in single support phase can be written as:

$$\dot{x} = \begin{bmatrix} \dot{q} \\ D^{-1}(q)(-C(q,\dot{q})\dot{q} - G(q)) \end{bmatrix} + \begin{bmatrix} 0 \\ D^{-1}(q)Bu \end{bmatrix} := f(x) + g(x)u \tag{1}$$

$D(q)$ is the inertia matrix and always positive definite. $C(q,\dot{q})$ is the Coriolis matrix and $G(q)$ is the gravity terms. The matrix B is a constant matrix used to indicate whether a joint is actuated. Our robot has 8 active joints, and q_1 is unactuated, then $B = [0_{1\times 8}\ I_8]^T$. $u = [u_1 \cdots u_8]^T$ are the applied torque at the joints.

2.2. *Double Phase Dynamical Model*

The double support phase is the end of the current step and the start of next step. As the assumptions 3~5, the double support phase is assumed to be instantaneous and consists of two distinct processes: impact and relabeling. In former a plastic impact takes place between the swing foot and the ground, and the later is responsible for changing the roles of the two legs so that they can share the same dynamical model.

The result of the impact and relabeling of the states can be expressed as:

$$x^+ = \Delta(x^-) \tag{2}$$

where "-" indicates the state before the impact and "+" after it. The detailed process can refer the literature [11, 14].

2.3. *The Complete Dynamical Model*

The complete dynamical model of biped robot is:

$$\begin{aligned} \dot{x} &= f(x) + g(x)u \qquad && x \notin S \\ x^{+} &== \Delta(x^{-}) && x \in S \end{aligned} \tag{3}$$

where S is the switching surface, by the assumptions, it is the level ground.

3. Gait Planning

By far, the most popular algorithms are time dependent in planning the gait of biped robot. The central idea of those algorithms is: planning the trajectories $q(t)$ of each joint, and tracking them by controller to realize walking control. But for underactuated biped robot, we can't control the time of walking directly and the synchronization of time is not realizable, so the robot can't walk stably under those algorithms.

Time-invariant algorithm is employed in this paper to plan the gait of biped robot. Time-invariant algorithm based on the definition of the reference trajectory for M outputs (where M is the number of actuators), not as a function of time, but as a function of a configuration variable θ. θ is monotone and independent of the M outputs. We call those outputs as virtual constraints, as in [15]. To a dynamic system of N DOF, defining M virtual constraints, it can be reduced to a $N-M$ dimension zero dynamics system. We can analyze the stability of the whole system by studying the zero dynamics system.

During walking, lots of variable can be selected as θ. In this paper, we select the angle between the hip and the standing ankle in sagittal plane, as in Fig.2.When the femur and the tibiae have the same length:

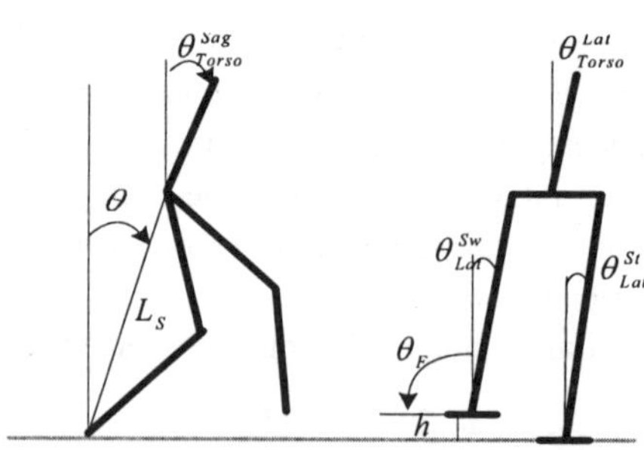

Figure 2. The left one is the pose looking from side, and the right is from back.

$$\theta = q_1 + q_3 / 2 \tag{4}$$

When walking, one observes that the torso is maintained at a nearly fixed angle, the hips remain roughly centered between the feet, and the end of the swing leg traces an approximately parabolic trajectory[6]. The hip sways left and right slightly and the two legs are always parallel. All of these characters can be expressed by some parameters, such as the angle between the torso and the uprightness (θ_{Torso}^{Sag}), the length of standing leg (L_S), the maximum of θ (θ^{Max}), the height of the swing foot (h) and the swing angle of the hip (θ_{Lat}^{St}). As mentioned before, the biped robot has 8 actuators, and 8 independent outputs are need to restrict the walking. We define them as $y_1 \cdots y_8$. They are:

1. Keeping the torso upright in the lateral plane;
2. θ_{Lat}^{St} is a parabola trajectory;
3. The swing leg parallels the standing leg in lateral plane;
4. The swing foot is level in lateral plane;
5. θ_{Torso}^{Sag} keeps a fixed angle;
6. h is parabola trajectory;
7. L_S keeps a fixed length.
8. The center of the hip is in the middle if the two feet in sagittal plane.

where 1~4 are the lateral constraints and 5~8 are the sagittal constraints. They can express the walking pose of the robot basically. All of these can be expressed as the output:

$$y = [y_1 \quad \cdots \quad y_8]^T = H(\theta) \tag{5}$$

The dynamical model of biped robot is Eq.(1) and Eq.(2), its output is Eq.(5). Then the task of the controller is to settle the output in finite time so that the robot can impact the ground with desired configuration. All of those are geometrical constraints to the configuration, not to velocities. So the walking velocity is not control directly, it would differ from the desired one. The strategy to control the walking velocity will be studied in the later section.

4. Control and Stability

4.1. *Feedback Controller Design*

The period of a step is finite, so controller should settle the system in finite time. The control objective is driving the outputs to zero. Since the outputs only depend on the generalized positions, and the dynamic model is second order, the relative degree of each output component is at least two. Using standard Lie derivative notation direct calculation yields:

$$d^2y/dt^2 = L_f^2 H(q,\dot{q}) + L_f L_g H(q)u \tag{6}$$

$L_f L_g H(q)$ is decoupling matrix and always reversible. When the input is:

$$u(x) = (L_g L_f h(x))^{-1}(v - L_f^2 h(x)) \tag{7}$$

The nonlinear system can be linearized as a double integral system. The feedback functions used here come from [13]:

$$v = \Psi(h, L_f h) = \frac{1}{\varepsilon}\begin{bmatrix} \psi_1(y_1, \varepsilon \cdot \dot{y}_1) \\ \vdots \\ \psi_8(y_8, \varepsilon \cdot \dot{y}_8) \end{bmatrix} \tag{8}$$

$$\psi_i = -sign(\phi_i(y_i, \varepsilon \cdot \dot{y}_i))|\phi_i(y_i, \varepsilon \cdot \dot{y}_i)|^{\frac{\alpha}{2-\alpha}} - sign(\varepsilon \cdot \dot{y}_i)|\varepsilon \cdot \dot{y}_i|^{\alpha} \tag{9}$$

with $\varphi_i = y_i + (1/2 - \alpha) sign(\varepsilon \cdot \dot{y}_i)|\varepsilon \cdot \dot{y}_i|^{2-\alpha}$, $0 < \alpha < 1$, $\varepsilon > 0$ are the parameters to control the settling time。

4.2. *Stability of the Controller*

Westervelt and Grzzie have proved that the hybrid zero dynamics is existent and the limit cycle is stable for planar biped robot in paper [14]. If the ankles can offer enough moment to settle the lateral motion, the lateral motion is controllable completely and the hybrid zero dynamics of the 3D biped robot is the same as the planar one. Then the stability of 3D biped robot includes the stability of periodic motion and the stability of lateral motion.

Stability of periodic motion has been proved in detail in [14]. Here we only give the conditions to verify the stability of lateral motion. The ground reaction force applied to the supporting leg is unilateral, so the moment that the standing ankle can offer is limit. In order to keep the robot lateral stable and not tip outside, the lateral FRI (Foot Rotation Index) point should in the foot area:

$$y_{FRI} < y_{Foot}^{Max} \tag{10}$$

where y_{Foot}^{Max} is the maximal lateral length of the foot.

5. Simulation

Considering the biped robot model with the following parameters (Table1), we use MATLAB to simulate the walking process and validate the control strategy. In order to simplify the simulation, we develop the dynamic model by lumped mass.

Table 1. Parameters of the robot

	torso	hip	femur	tibiae	foot
Mass (kg)	10	0	2.5	2.5	1
Length (m)	0.4	0.15	0.3	0.3	0.10
CoM (m)	0.2	0	0.15	0.15	0.05

The control parameters are $\alpha = 0.9$ and $\varepsilon = 0.03$. The walking process can be seen in Fig.3 and the walking parameters as Table2:

Table 2. Walking parameters

δ^2_{zero}	$V_{zero}(\theta^-)$	V^{MAX}_{zero}	θ^{Max}	v
0.9872	-64.273	15.103	0.393	1.073

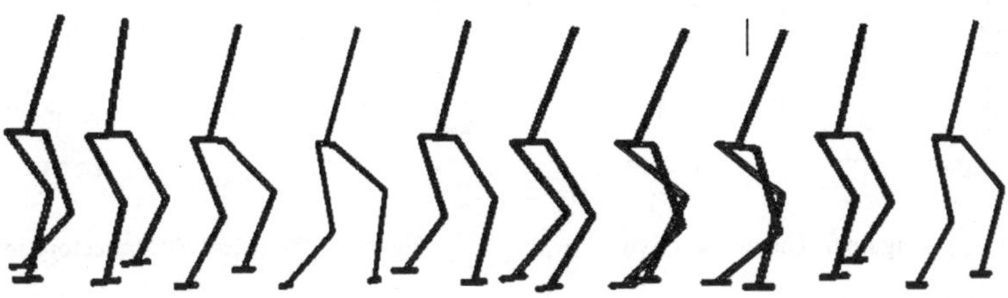

Figure 3. The walking process

Fig.4~7 are some characteristic curve during this walking process. Fig.4~5 indicate that the motions of the joints are all convergence to stable limit cycles. Fig.6 indicates that the outputs are complete settled. And Fig.7 indicates that the lateral motions are stable and the FRI point is in the range of the sole.

6. Walking Speed Control

Walking velocity V is effected by step length L and step period T. The gait is planned by time-invariant strategy, and the walking gesture is control by geometrical constraints. There is no control to the velocity, so the step period can't be controlled directly and the walking velocity would not be the one we want.

When restricted by $N-1$ restrictions, the dynamical model can be reduced to one dimension zero dynamic. In practice, the reduced model is an inverted pendulum with alterable length and alterable moment of inertia. The period of inverted pendulum is determined by the length, moment of inertia and range of the swing. Based on them, we control the walking velocity by adjust the gesture of the robot in walking.

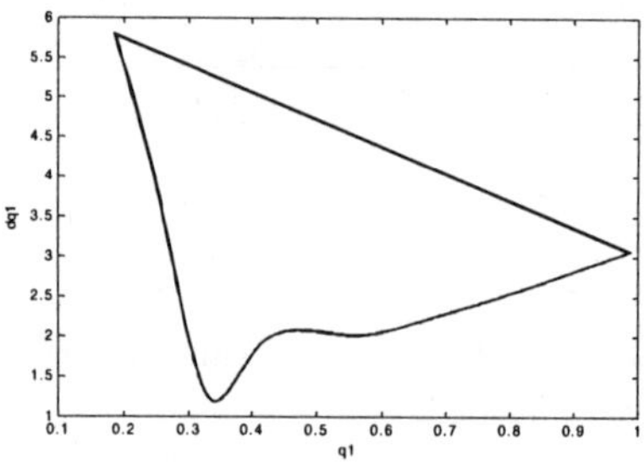

Figure 4. Phase plots of q_1

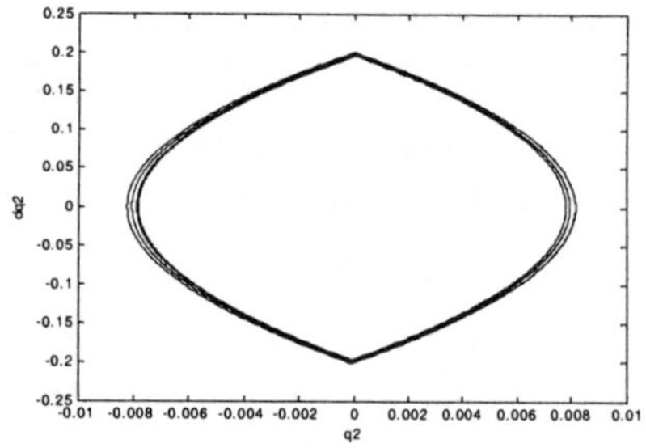

Figure 5. Phase plots of q_2

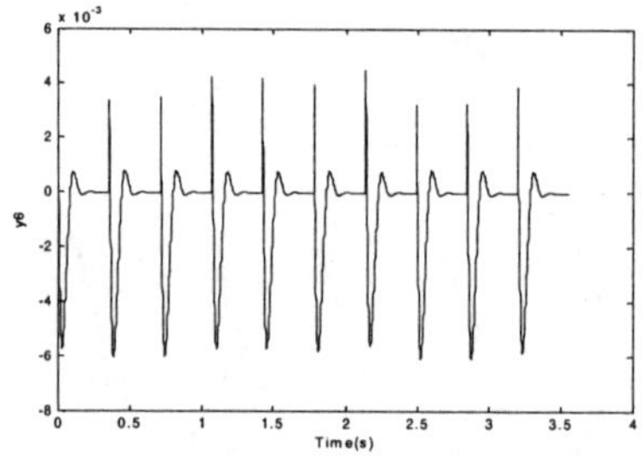

Figure 6. The output of y6

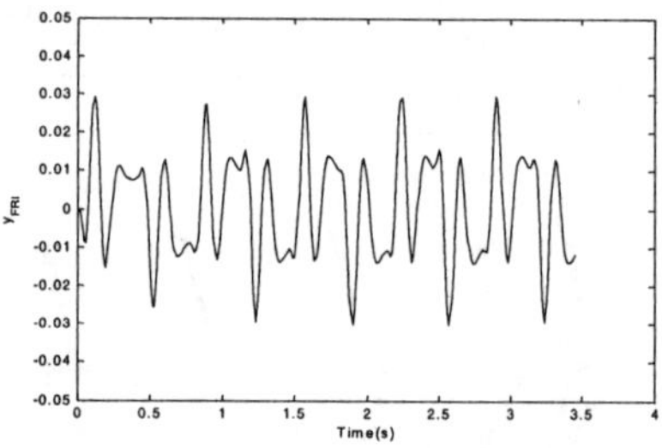

Figure 7. The lateral FRI trajectory(m)

To our robot, the length of the pendulum is determined by L_S, moment of inertia is by θ_{Torso}^{Sag}, and the range of swing by θ^{Max}. We can control the walking velocity by adjusting those parameters.

The fallowing rules can be summarized from experiments:

1. L_S and θ^{Max} fixed, larger θ_{Torso}^{Sag} implies shorter T and larger V.
2. θ_{Torso}^{Sag} and L_S fixed, larger θ^{Max} implies longer T and less V.
3. θ_{Torso}^{Sag} and θ^{Max} fixed, shorter L_S implies shorter T and larger V.

From those rules, we can't get the certain relationship between velocity and the gesture parameters, but the trend. Using them we can control the velocity by adjusting it. From those rules, we can't get the certain relationship between velocity and the gesture parameters, but can get the trend of them. Using them we can control the velocity by adjusting it.

Here we validate our strategy by simulation. We control the walking velocity by adjusting θ_{Torso}^{Sag}, where L_S=0.5 and θ^{Max}=0.393. P controller is used here:

$$\theta_{Torso}^{N} = \theta_{Torso}^{N-1} + P\cdot(V^{N-1} - V_d) \tag{11}$$

Where θ_{Torso}^{N} is the current torso angle, θ_{Torso}^{N-1} is the last one, V^{N-1} is the last velocity, and V_d is referenced velocity, P is controller parameter.

Using the controller (11) and $P=-0.38$, we control the velocity to 1m/s first, then accelerate it to 1.3m/s, and fix it to 1.15m/s at last. Fig.8 is the angle of torso and Fig.9 is the walking velocity of the process.

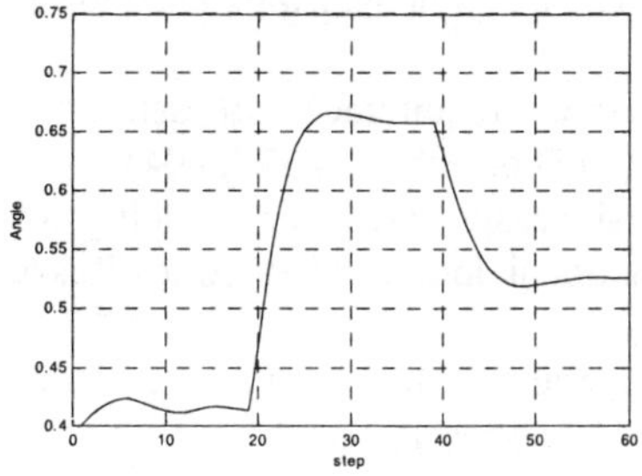

Figure 8. Trajectory of θ_{Torso}^{Sag}

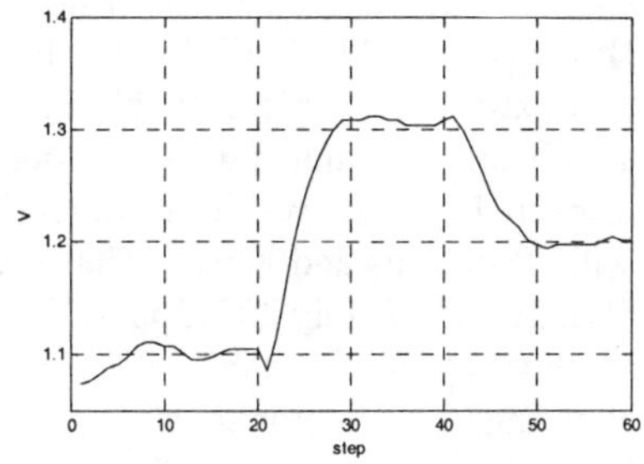

Figure 9. Walking velocity of the robot (m/s)

7. Conclusion

This paper presents a new underactuated 3D Biped Robot. The robot has two cylindrical sole distributing in lateral plane, which can offer lateral moment to stabilize the lateral motion. Using the time-invariant algorithm and feedback controller, the robot realizes 3D dynamical walking. Simulation results indicate that stable 3D dynamical walking has been achieved. And by adjusting the gesture of walking, it realizes walking velocity control. Simulation results indicate that stable 3D dynamical walking has been achieved.

Acknowledgments

This research is supported by National Natural Science Foundation of China (No. 60475035).

References

1. Kenji Kaneko, et al, Humanoid Robot HRP-2, Proc, International Conference on Robotics & Automation, 2004.
2. T. McGeer, Passive dynamic walking, International Journal of Robotics Research 9 (1990), no. 2, 62–82.
3. K. Osuka and K. Kirihara, Motion Analysis and Experiments of Passive Walking Robot QUARTET II, Proc. of the 2000 IEEE Int. Conf. on Robotics & Automation, pp.3052–3056, 2000.

4. M. Garcia, et al, Efficiency, speed, and scaling of two dimensional passive-dynamic walking, Dynamics and Stability of Systems 15 (2000), no. 2.
5. Grizzle J W, Abba G, Plestan F. Asymptotically stable walking for biped robots: analysis via systems with impulse effects, IEEE Transactions on Automatic Control, 2001, 46 (1) : 51 - 64.
6. C. Chevallereau, et al, RABBIT: A Test bed for Advanced Control Theory, PAPER NUMBER CSM-02-038 Revision 08, June, 2003.
7. Catherine E. Active control of lateral balance in human walking, Journal of Biomechanics 33 (2000) 1433~1440.
8. J. Maxwell, et al. Mechanical and metabolic requirements for active lateral stabilization in human walking. Journal of Biomechanics 37 (2004).
9. Steven H. et al. A Three-Dimensional Passive Dynamic Walking Robot with Two Legs and Knees. The International Journal of Robotics Research Vol. 20, No. 7, July 2001, pp. 607-615.
10. Arthur D. Kuo. Stabilization of Lateral Motion in Passive Dynamic Walking. The International Journal of Robotics Research Vol. 18, No. 9, September 1999, pp. 917-930.
11. Guobiao Song and Milos Zefran, Underactuated Dynamic Three Dimensional Bipedal Walking. Proc of the 2006 IEEE International Conference on Robotics and Automation Orlando, Florida - May 2006.
12. Jessy W. Grizzle, et al. Stable Walking of a 7-DOF Biped Robot. IEEE Transactions on Robotics and Automation, vol. 19, NO. 4, August 2003.
13. S. P. Bhat and D. S. Bernstein, Continuous finite-time stabilization of the translational and rotational double integrators, IEEE Trans. Automat. Contr., vol. 43, pp. 678–682, May 1998.
14. E.R. Westervelt, J.W. Grizzle, and D. Koditschek. Hybrid zero dynamics of planar biped walkers, IEEE Transactions on Automatic Control, 2002.
15. C. Chevallereau, A. Formalsky, and D. Djoudi, Tracking of a joint path for the walking of an underactuated biped, Robotica, vol. 22, pp. 15-28, 2004.

ROTOPOD: A NOVEL APPROACH TO EFFICIENT LEGGED LOCOMOTION*

DAMIAN M. LYONS†

Robotics and Computer Vision Lab.
Fordham University. Bronx NY 10458 USA

A number of attempts have been made to integrate the efficiency of wheeled locomotion with the terrain versatility of legged locomotion, e.g., Univ. Michigan's Rhex platform and Case Western's Whegs. Those platforms cast legs as rotating spokes placed traditionally at the corners of a rectangular platform. In this paper, we present an alternate approach, with three legs radiating down from a central hub. The energy to move the platform is generated by a rotating reaction mass mounted at the hub and, at rest, rotating parallel to the ground plane.

Our approach is to construct a platform whose *natural, uncontrolled* motion is energy efficient and useful, and which requires only a small amount of control to produce effective locomotion. We describe the platform and analyze the characteristics of its uncontrolled motion. Several strategies to produce directed motion will be presented and evaluated. Simulation models and a prototype platform have been built and will be discussed.

1. Introduction

Our objective is to develop a legged robot platform that can perform ground surveillance and reconnaissance activities in an effective and energy-efficient manner. Legged platforms have a mobility advantage over wheeled and tracked vehicles on rough terrain: They can step into depressions, onto rises or over obstacles that could defeat a wheeled vehicle. However, conventional legged platforms offer mobility at a high cost in energy use: Unlike a wheel or track, a leg must *lift* as well as *propel* the robot platform and hence will use more energy. Wheels, therefore, have an advantage of energy-efficiency over legs. Researchers have attempted to combine legs and wheels within a single platform, resulting in many different platform designs. Halme et al.'s WorkPartner [4] has wheels at the end of its legs. Birch et al.'s cricket-inspired robot [1] has wheels in the front and legs in the back. Saranli et al.'s RHex [9] has four legs that are single-spokes – legs – mounted on axles, and Quinn et al.'s Whegs robots [7] have six legs that are each 3 spokes on axles. There is evidence from insect and associated robotic studies that the reason animals exploit legged locomotion in an energy efficient manner has as much to do with the use of a tuned physical mechanism as with control or intelligent planning [8, 10]. Our design goal is therefore, to design a mechanism that *naturally* produces

* This work was supported by STIR P-49411-CI-II Grant from the US Army Office of research.

(1) an energy efficient style of locomotion, and

(2) a motion pattern appropriate for ground reconnaissance and surveillance applications.

In section 2 we introduce a mechanism designed according to this strategy, the *rotopod,* a three-legged mechanism that walks by rotating its body around its leg endpoints and its natural motion is captured as a hypertrochoid. In section 3 we describe how the natural motion can be controlled. In section 4 we look at the energy efficiency of the mechanism in terms of its specific resistance. We conclude in section 5.

2. The Rotopod mechanism

Our approach is inspired by the observation that the RHex and Whegs leg designs are essentially one or more wheel spokes (Fig. 1(a)). An alternate design in which legs can be considered analogous to wheels is shown in Fig. 1(b). In this case, the legs are spokes rigidly mounted to the frame, as opposed to mounted on an axle as in Fig. 1(a); however, if the platform itself can be made to tilt and rotate, then the legs will raise and fall to follow the circumference of a 'virtual' wheel as shown. This mechanism design has the potential to fulfill our design goal because:

(1) it exploits the spoke/leg approach already shown to be successful, and

(2) the natural 'rolling' motion of this kind of mechanism can produces convoluted paths that exhibit a high-degree of area coverage.

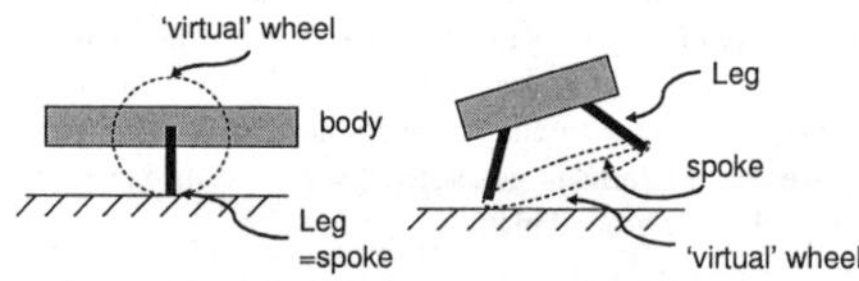

Figure 1: Legs as Analogs of Wheels

Leg motion is effected in Fig. 1(a) by rotating the leg axles. However, in Fig. 1(b) the legs are rigidly attached, and the body must be rotated. If the mechanism is extended to include a rotating reaction mass mounted on the body, then as the mass rotates it can be used to produce the desired body motions.

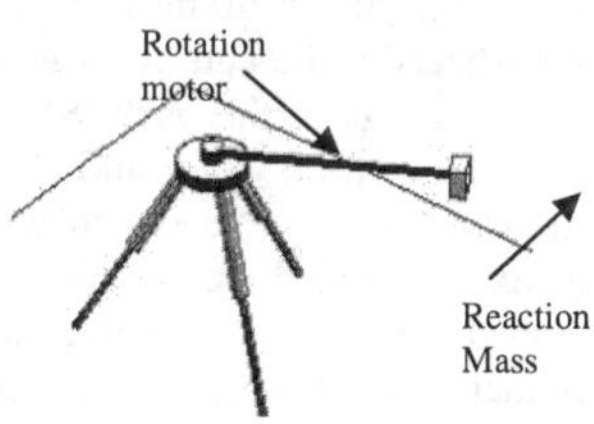

Figure 2: Rotopod Mechanism.

Figure 3: Mechanism Stepping

Our approach is to investigate whether the natural motion of a tripedal structure surmounted by a rotating reaction mass (Fig. 2.) can be harnessed to provide a basis for energy-efficient locomotion for surveillance and reconnaissance activities. We call such a mechanism a *rotopod.*

The natural motion of such a platform is to rotate an amount around each of its legs in turn. The raised legs can be used to *step over* obstacles that might center-ground a wheeled vehicle (Fig. 3) giving this platform the mobility to work on a wide range of terrains. When two legs are raised from the ground, the platform rotates around the third, moving and rotating its center. The amount of rotation and height of the leg stepping is a function of the leg lengths and masses and the rotational velocity of the reaction mass.

The resultant epicycloid-like path for the platform has advantages over straight line motion when the platform must cover an area for search or reconnaissance purposes. The rotating reaction mass is the principal consumer of energy of in the system; if effective search locomotion behavior can be obtained with simple control of the natural motion, a highly energy efficient system will result.

2.1. *Description of the Mechanism*

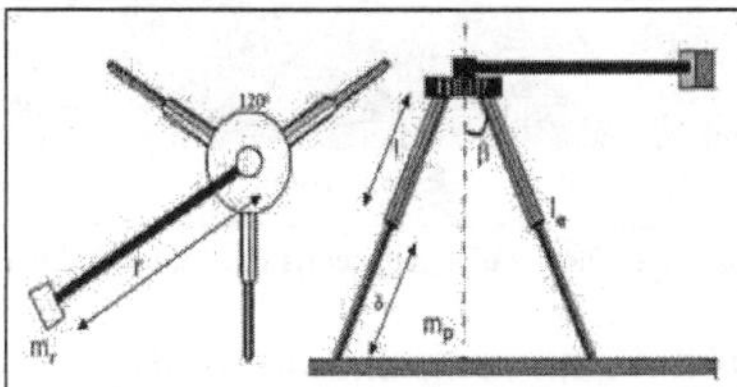

Figure 4: Top (a) and Side (b) Views

Figure 5: Prototype mechanism

The mechanism concept introduced above raises a number of design parameters. For study purposes we select the following values for these (see Fig. 4):

(1) Number of Legs: 3. **Rational**: the minimum number for static stability.

(2) Position of legs on body: evenly distributed around central axis (120° separation) subtending β=45° with vertical midline. **Rational**: each leg should behave interchangeably. The value of β is chosen to enhance static stability.

(3) Degrees of freedom per leg: 1 translational degree. **Rational**: This is the minimum and most direct configuration to change leg length.

(4) Location of reaction mass joint: At apex of legs. **Rational**: maximize the lever arm that can be applied by the reaction mass.

(5) Platform lengths and masses: Reaction mass arm limited to about 1 meter max, and total platform weight limited to less than 20 Kg. **Rational**: safety and internal testing

purposes, though in fact a small reconnaissance drone might be of a roughly similar size.

An Open Dynamics Engine (ODE) simulation of the 3D kinematics and dynamics of the full mechanism has been build and used to evaluate motion strategies as well as energy use. This simulation takes reaction mass motor velocity and saturation torque and leg linear actuator forces and velocities as input and produces a 3D simulation (e.g., Fig. 3) and various kinematic and dynamic measurements as output.

A small prototype (~10cm height) was constructed in [5] that used an alternative rotational, instead of translational, leg degree of freedom. This ultimately proved limiting. A larger prototype (~0.45m, Fig. 5) has now been build and is being evaluated. This prototype and the ODE simulation have the same degrees of freedom, and similar lengths and masses.

2.2. *Natural Motion of the Mechanism*

If the mechanism is placed on a level surface, legs fully extended, and the reaction mass is rotated slowly (ω <*0.2rps* for our model) the mechanism remains stationary. If the rotational velocity is slowly increased, then a small chattering motion of the leg endpoints first results. Chattering may result in a small rotation of the platform around its center point *in the same direction* as the reaction mass rotation. As the velocity is further increased (ω ~ *1 rps*) the platform tilts in the direction of the reaction mass, and the chattering turns into the rotation of the platform around each leg endpoint *in the opposite direction* to the reaction mass, as the reaction mass passes over that leg. It is this final mode of motion that we are interested in, and we refer to as the natural motion of the platform, though the chattering motion may be of some value for small 'backwards' motions.

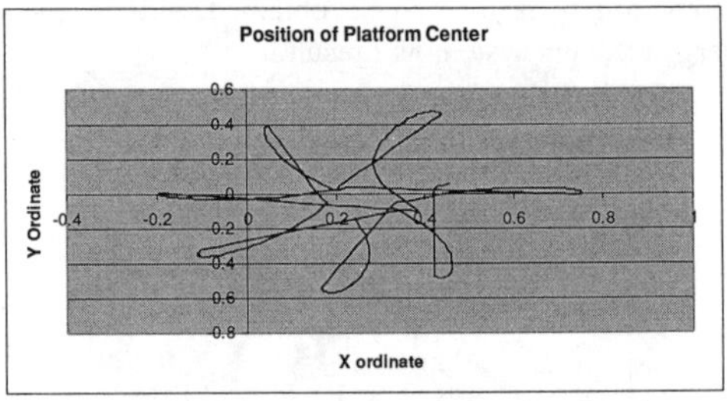

Figure 6: Plot of platform center during natural motion

Fig. 6 shows a plot of the *(x,y)* or plan view of the motion of the center of the platform during natural motion. Note that it is a series of 'loops' which are caused by the tilting of the platform and the effect of the reaction mass approaching and then leaving the position of the leg endpoint. This pattern of motion belongs to the family of curves called *hypertrochoids* generated by rotating a circle along the inside of the perimeter of a second larger circle. The standard equations for a hypertrochoid are:

$$y = (a-b)\sin t - h\sin\left(\frac{a-b}{b}t\right),\quad x = (a-b)\cos t + h\cos\left(\frac{a-b}{b}t\right) \qquad (1)$$

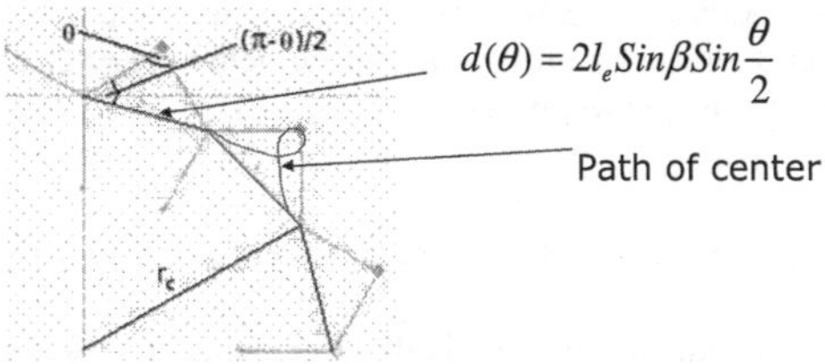

Figure 7: Kinematic simulation of mechanism rotating same angle around three successive endpoints.

We can relate the rotopod motion to these equations. Let l_e be the length of the leg projection onto the *xy* plane, let β be the leg/midline angle, and let α be the max angle that the platform tilts when the reaction mass passes over a leg. Fig. 7 shows the plan view of the mechanism rotating an amount θ around each of the three leg endpoints in turn. As it rotates, the mechanism center follows a curve with radius r_c. The distance moved on each rotation can be calculated as $d(\theta)$ as shown, and the relationship with the hypertrochoid curve established as:

$$r_c = a-(b+h) \quad 2\pi b = d(\theta) = 2l_e Sin\beta Sin\frac{\theta}{2}$$
$$a-(b-h) = l_e Sin\beta - l_e Sin(\beta+\alpha) \tag{2}$$

3. Control of Motion

The epicyloid-like path has good ground coverage properties. However, to be useful, the motion must be controlled to cover a desired area. The natural motion of the platform is fixed

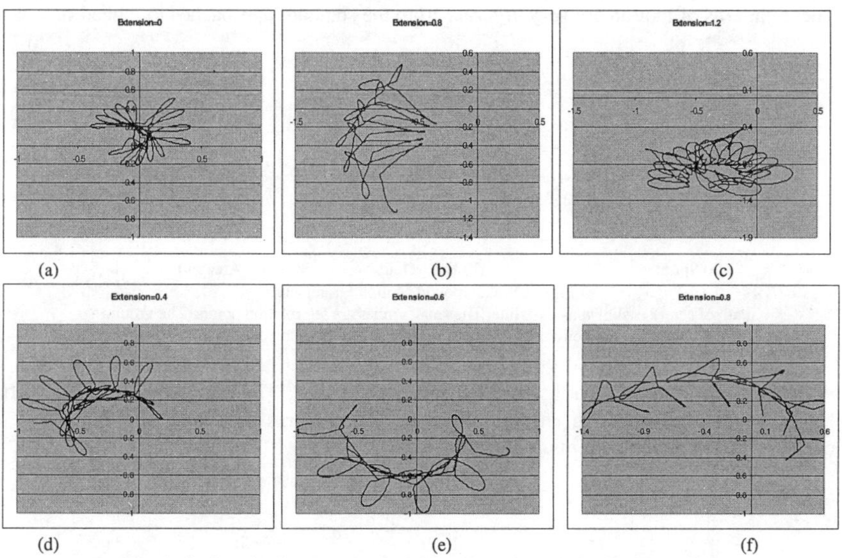

Figure 8: Changing Leg Lengths, (a-c) 2 Leg Extensions; (d-f) 1 Leg Extensions.

by the leg lengths, masses and rotational velocity, and is, simplifying somewhat, a closed circle on the ground plain. There are several choices of control parameters include the rotational velocity, moving the reaction mass along the reaction arm (with a linear actuator) or changing the leg lengths. We selected changing the leg length because of its energy-efficiency:

(1) leg lengths can be changed while a leg is in the air, and hence require a minimal injection of energy into the system.

(2) Once a leg length has changed, the linear actuator model requires no injection of energy to maintain that length

Fig. 8 shows the effect of modifying leg length, while leaving all other parameters the same: the radius of the circle of rotation increases in direction proportion. Extending two legs allows for smaller circles, while extending just one leg generates bigger circles. These graphs were generated from the ODE simulation. These results mean that the rotopod can be driven along a curve; however, a practical limitation is that the radius of curvature must always have the same sign.

The closed circular paths in Fig 8. can be used to build compound motions. There are three compound motions we have studied, spiral search, epicycloids and area fill. Fig. 9 shows examples of each of these. Each has the advantage that it has strong area coverage properties, and is hence a useful mode of locomotion for searching, reconnaissance and surveillance. The spiral [2] and area fill patterns are appropriate when looking at a limited geographical area, while the epicycloid is a traveling pattern and useful for traversing large areas.

These patterns of motion are *very different* from the standard approaches to legged or wheeled

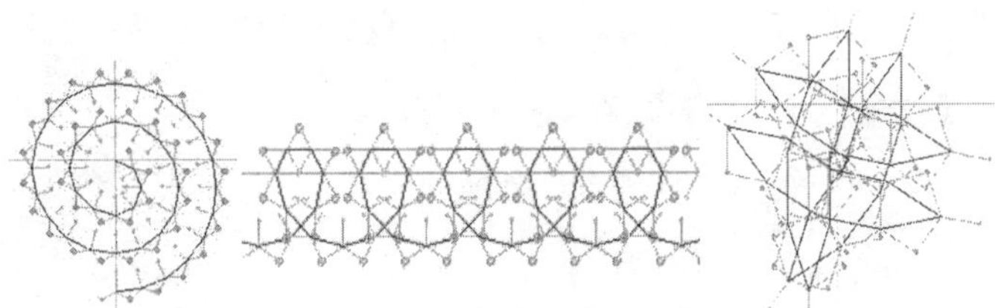

(a) Spiral (b) Epicycloid (c) Area Fill

Figure 9: Compound Motion Strategies.
Path of center is shown as dark line. The small circles are leg rotation points. The ground projections of the three legs are shown in light color.

vehicles [6, 13], and its worth recalling how we got here: The rotopod is designed to have legged capabilities but high energy efficiency. The patterns shown in Fig. 9 are therefore energy-efficient search patterns.

4. Energy efficiency

The specific resistance is a common measure of energy efficiency for wheeled and legged vehicles (though by no means the only one [12]). It is defined as

$$\varepsilon = \frac{power}{weight \times velocity}$$

The most efficient legged locomotion (passive-dynamic walking [3]) has ε between *0.05 and 0.08* and the less efficient legged approaches (e.g., Asimo biped) can be as great as *1.8.*

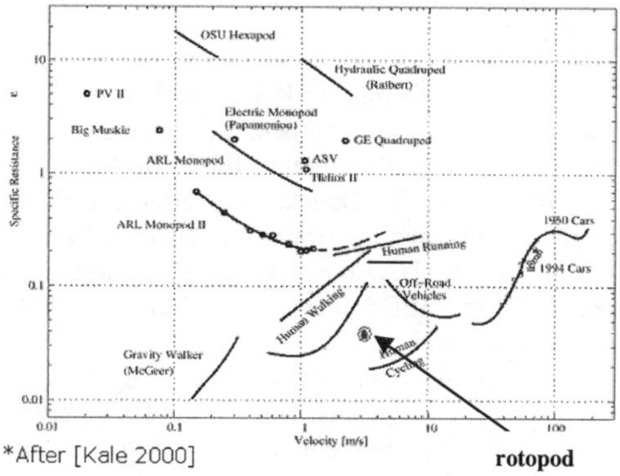

Figure 10: Potential Efficiency of Rotopod with respect to other locomotion

We used ODE to measure the mechanical power expended by the rotopod in natural motion (e.g., Fig 6). Typical speed for the robot was about 1.29 m/s or 2.8 mph – about walking speed, and the specific resistance measured 0.03. Note this *impressively low figure* should be taken with some caution: It's the result of ODE measurements, it's the only the physical cost of motion, and it includes motion along the entire perimeter of the hypertrochoid (e.g. 8m in Fig. 6. for a curve of average radius 0.2m). We have not make measurements on the physical prototypes on the electrical cost of motion, which is what will determine longevity given current battery technology. However, this low number, which should be regarded as a best case, does indicate that the basic idea of the rotopod as energy efficient legged locomotion is a strong one as shown by the comparisons in Fig. 10.

5. Conclusions

We have described a robot mechanism, the rotopod, that exploits a novel model of legged locomotion. The natural, uncontrolled motion of the platform has good area coverage properties and promising energy efficiency. By varying the lengths of the legs of the mechanism, this natural motion can be controlled and compound motion patterns such as a spiral (for search), epicycloid (for directed motion) or area fill produced. These results were obtained using an ODE simulation.

A physical prototype has been constructed that stands approx. 0.45m high and uses one Firgelli model CYCJ linear actuator per leg and a Korebot/Koremotor embedded computer running Linux is used as a controller, with wireless 802.11B connection. Preliminary results duplicating those in Fig. 8 have been produced. Compound motions and energy efficiency are now being evaluated.

Beyond this, the next steps involve adding visual sensing to the prototype by mounting a camera on one or more legs. The rotating camera can in theory be used to produce a stereo from motion depth map to aid in foot placement.

References

1. Birch, M., Quinn, R., Hahm, G., Phillips, S., Drennan, B., Fife, A., Beer, R., Yu, X., Garverick, S., Laksanacharoen, S., Pollack, A., Ritzmann, R., A Miniature Hybrid Robot Propelled by Legs. Proceedings IROS 01, Maui, Hawaii, Oct. 29. – Nov. 3, 2001.
2. Burlington, S., Dudek, G., Spiral Search as an Efficient Mobile Robotic Search Technique. Proceedings of the 16th National Conf. on AI, Orlando Fl. 1999.
3. Collins, S. H., Ruina, A. (2005) A bipedal walking robot with efficient and human-like gait. Proc. IEEE International Conference on Robotics and Automation, Barcelona, Spain, April 2005
4. Halme, A., Leppanen, I., Montonen, M., Ylonen, S., Robot Motion by Simultaneous Wheel and Leg Propulsion. 4th Int. Conf. Climbing & Walking Robots (CLAWAR), 2001.
5. Lyons, D., and Pamnany, K., Rotational Legged Locomotion. Proceedings of ICAR 2005, Seattle WA.
6. Mei, Y., Lu, UH., Hu, C., and Lee, CSG. Energy-Efficient Motion Planning for Mobile Robots. Int. Conf. on Robotics and Automation, New Orleans LA, 2004.
7. Quinn, R. D., Nelson, G.M., Bachmann, R.J., Kingsley, D.A., Offi, J., and Ritzmann, R. E., (2001). Insect Designs for Improved Robot Mobility. Proc. of Climbing and Walking Robots Conference (CLAWAR01), Karlsruhe, Germany, pp. 69-76.
8. Quinn, R., Offi, J., Kingsley, D., and Ritzman, R., Improved Mobility Through Abstracted Biological Principles. Proc. Int. Conf. on Int. Robots and Systems (IROS), Lausanne Switzerland, Oct. 2002.
9. Saranli, U., Buehler M., and Koditschek, D.E., Design, Modeling and Preliminary Control of a Compliant Hexapod Robot, IEEE Int. Conf. Robotics and Automation, San Francisco, California, April 2000.
10. Saranli,U., Buehlerand M., Koditschek D., Hex: A Simple and Highly Mobile Hexapod Robot. Int. Jour. of Robotics Research, Vol. 20, No. 7, July, 2001, pp. 616 - 631.
11. Schroer, R.T., Boggess, M.J., Bachmann, R.J., Quinn, R.D., and Ritzmann, R.E. Comparing Cockroach and Whegs Robot Body Motions, IEEE Conference on Robotics and Automation, New Orleans LA, 2004.
12. Silva, M.F., Machado, JAT, and Lopes, AM. Energy Analysis of Multi-Legged Locomotion Systems. 4th Int. Conf. on Climbing and Walking Robots (CLAWAR), 2001.
13. Wong, S.C., Middleton, L., MacDonald, B.A., Performance Metrics for Robot Coverage Tasks. Proc. Australasian Conf. on Robotics and Automation. Auckland, New Zealand Nov. 2000.

THE DESIGN OF A HUMANOIDAL BIPED FOR THE RESEARCH ON THE GAIT PATTERN GENERATORS

P. KRYCZKA

Faculty of Power and Aeronautical Engineering
Warsaw University of Technology
ul. Nowowiejska 24, 00-665 Warsaw, Poland
E-mail: P.Kryczka@onet.eu

C. M. CHEW

Department of Mechanical Engineering
National University of Singapore
10 Kent Ridge Crescent, Singapore 119260
E-mail: mpeccm@nus.edu.sg

It was proven that it is possible to build a gait pattern generators, producing the human-like leg joints trajectories. For this purpose the van der Pol equations, as well as the neural networks can be applied. A robot described in this paper will serve as a platform for research on implementation of the mentioned gait pattern generators in the bipedal antropomorphic structures. The paper contains the detailed design process description, especially the dynamic analysis process. The mechanical design methods, as well as the designed preload mechanism and the foot compliance are also presented.

Keywords: Biped, CPG, CAD/CAE, Compliant foot, Dynamic analysis, Preload mechanism

1. Introduction

The research on the vertebrae locomotion system, conducted in the 20th century, led to the conclusion that the rhythmic movements, such as locomotion, are generated by the Central Nervous System (CNS).[1] It was first suggested[2] and then experimentally proven[3] that human beings are also equipped with this kind of system, called the Central Pattern Generator (CPG). Since 80s the idea of the neural systems generating a set of rhythmic joints trajectories, started being used in robotics. Until now, it was successfully implemented in various legged robots.

The research being conducted at Warsaw University of Technology aims at developing CPGs[4–6] for legged robots. The CPG based on the information from the sensors, should be able to produce a set of legs joints trajectories fulfilling the stability criteria.[7] As a first step of the research, a set of the human gait data was recorded. Based on the recorded data the two kinds of the gait pattern generators were developed. The first one utilizing the van der Pol coupled oscillators and the second one the models of neural networks.[8] Both approaches proved to be efficient in generating the gait patterns similar to that of the human being. The current works focus on building an atropomorphic biped for tests on the implementation of the created oscillators.

Over the last two decades, a number of advanced humanoidal bipeds were developed, among the others Asimo,[9] HRP-3,[10] H7,[11] Qrio,[12] Robian,[13] Wabian.[14] All of these robots have the rigid structure made of metal or composites. Usually, each joint (DOF) is actuated by the single motor, unlike in humans, where each joint is actuated by muscle groups. As it is known from the biomechanics, as well as from our own experience, a human motion system is very compliant. Our muscles and tendons tend to elongate under the load, thus making our body more adaptable to the impact loads (e.g. occurring during the foot initial contact). A progress, worth mentioning, towards the more human like structure made the scientists from The University of Tokyo building a Kotaro[15] robot. However, from the robots mentioned above (except Kotaro) only Robian and H7 are equipped with the compliant feet.[16] Moreover, the compliance in these two robots' feet is located in their frontal part, imitating human being's fingers. This solution improves the 'toe off' phase of the gait cycle, however it doesn't influence the 'initial contact' phase, which is the source of the disturbing impact. The solution implemented in the described design influences both the 'initial contact' and the 'toe off' phases of the gait cycle.

The very fast development of the personal computers and their computational power, in past two decades, drastically improved the design process. Nowadays, most of the mechanical designers (if not all) use the computer-aided design (CAD) and engineering (CAE) software as their basic tools at work. The CAD software lets them for the fast and very accurate modelling of the structures, whereas CAE software lets them to check the structural properties (e.g. the stress distribution under complicated load conditions) of the design and improve its strength and geometrical properties. The design described in this paper was also performed with use of the CAD/CAE software, which will be described in more details in the following chapters.

2. The project formulation

The designed robot is meant to serve as a research platform for CPGs based motion synthesis mentioned in the introduction. The robot will comprise only of the lower extremities and the torso. Its mechanical structure will consist of 12 active DOFs and 2 passive DOFs. The active DOFs will be directly driven by series of the Robotis' *Smart Actuators*. The two passive DOFs will be located in the foot. Justification of their implementation will be presented in the further part of the paper.

Unlike in the professional design sequence we first had the servos and then had to suit the biped's structure to their dynamical capabilities. In order to do that the dynamic analysis had to be performed in the initial and the final stage of the design process.

3. Dynamic analysis

The main purpose of the analysis was to set the length of the links, so that the used servos are able to actuate the structure. The used DX–113 and DX–116 Robotis actuators are capable of generating 1 Nm and 2.8 Nm torque, respectively. By solving a set of inverse dynamic problems for different links lengths and the gait parameters the joints actuating torque with respect to the links lengths were found.

The inverse dynamic problem was solved only for the single support phase, considering robot as a serial kinematic chain and using the Newton-Euler formulation.[17] The analysis comprised of the following stages:

- choosing the kinematic model — the kinematic model was limited to five links (Fig. 2) with the assumption that the pelvis is perpendicular to the sagittal plane and that only hip pitch, knee pitch and ankle pitch joints take part in the motion. The five values of the links lengths were tested $l = \{100, 120, 130, 140, 150\}[mm]$, where l - length of the thigh and shin links (further referred to as the links length). The length of the pelvis link was constant in all tests and was equal $l_p = 100\ mm$.
- choosing the gait parameters — the parameters, such as the gait speed, step length and swing phase time were those of human being,[18] but appropriately scaled to suit the size of the robot. The overall size of the robot was assumed to be 500 mm for all data sets and accordingly the speed $V = 0.076\ \frac{m}{s}$, the step length $d = 172\ mm$ and the swing phase duration $t = 1.4\ s$.
- solving the inverse kinematic problem — the two equations describ-

ing the simple geometric relationships (Fig. 1) were used:

$$\begin{cases} x = r_1 cos\alpha + r_2 cos(\alpha + \beta) \\ y = -r_1 sin\alpha - r_2 sin(\alpha + \beta) \end{cases} \tag{1}$$

For the given pelvis and foot trajectories the set of non-linear equations (1) was solved numerically with respect to α & β, with use of the Newton-Raphson method implemented in Matlab environment. The obtained results were the hip pitch (α) and knee pitch (β) angles with respect to time. Finally, from the evaluated data angular velocities and angular accelerations were calculated.

- creating a mechanical model and extracting mass properties — the simplified mechanical model was created in CAD environment. Because of the biped's sagittal symmetry only one leg and a pelvis were modelled. From the created model the mass properties (links' masses, their centre location and moments of inertia) required for the dynamic analysis were extracted.
- solving the inverse dynamic problem — calculation of the torque in the particular joints was performed in the two, forward and backward sub-stages, according to the Newton-Euler formulation.[17] In the forward sub-stage (from the foot in the support phase to the foot in the swing phase), the linear and angular accelerations of the mass centre were evaluated. Then the forces and moments acting on the mass centre of each link were calculated, with use of the following formulae:[17]

$$^{i+1}F_{i+1} = m_{i+1}\, ^{i+1}v_{C_{i+1}}$$

$$^{i+1}N_{i+1} = \, ^{C_{i+1}}I_{i+1}\, ^{i+1}\dot{\omega}_{i+1} + \, ^{i+1}\omega_{i+1} \times \, ^{C_{i+1}}I_{i+1}\, ^{i+1}\omega_{i+1}$$

In the backward sub-stage, beginning from the end of the kinematic chain (the foot in the swing phase) forces, moments and finally torque in each join were evaluated, using the following formulae:

$$^{i}f_i = \, ^{i}_{i+1}R^{i+1}f_{i+1} + \, ^{i}F_i$$

$$^{i}n_i = \, ^{i}N_i + \, ^{i}_{i+1}R^{i+1}n_{i+1} + \, ^{i}P_{C_1} \times \, ^{i}F_i + \, ^{i}P_{i+1} \times \, ^{i}_{i+1}R^{i+1}f_{i+1}$$

$$\tau_i = \, ^{i}n_i^T\, ^{i}\hat{Z}_i$$

Because of the space limitation, only four graphs of the joint torque for links length $l = 100\ mm$ are presented on figure 3. The analysis described above was preformed with many simplifications. One of them was the trajectory of the swing leg's foot, in these tests assumed to form a semicircle. The trajectory of the human's foot of the leg being in the swing phase, during the normal gait, is much lower, with its beginning and the end almost tangential to the support surface. Another simplification was the omission of the influence of 'toe off' and 'initial contact' phases on the swing phase. The foot accelerates from the zero velocity and then stops, at the beginning and the end of the motion respectively. Due to these factors the obtained joint torque did't fully resemble that of the human being. Nevertheless, the obtained results were authoritative and sufficient for determining the maximum torque occurring during the normal gait. Taking into account the obtained values and the additional structural losses (e.g. friction) the links length was set to $l = 100mm$.

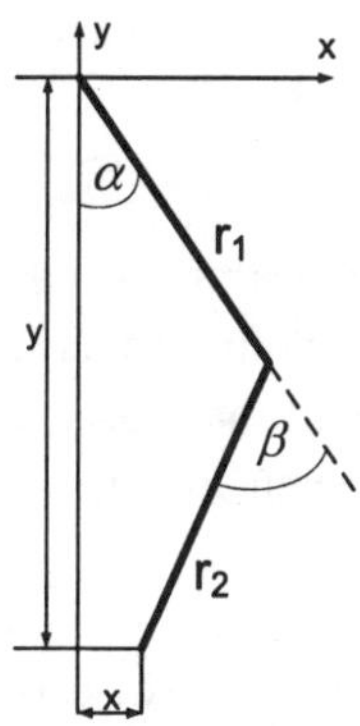

Fig. 1. The leg representation.

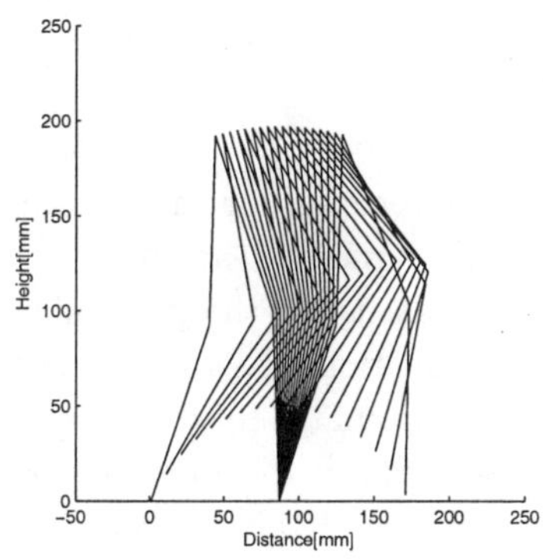

Fig. 2. The stick diagram of the analysed gait phase.

After completing the robot design, the dynamic properties of the structure were verified using the *Adams* environment. The *Adams* is a motion simulation software for analysing the complex behaviour of mechanical structures. The analysis was performed in the following stages:

- exporting the 3D model of the robot from CAD environment and importing it into *Adams* — from the CAD assembly model, created during the design process, the individual links were exported as *parasolids* (a geometrical model in form of the mathematical definitions). The exported links were then imported into *Adams*.
- assigning the mass properties to the imported model — as with the parasolid only the geometrical properties of the model were

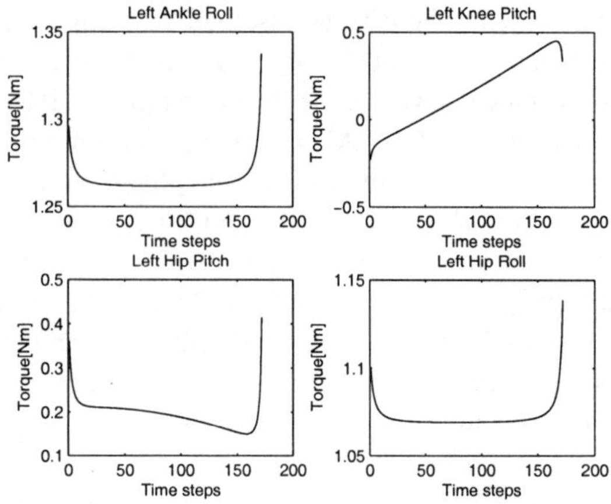

Fig. 3. Torque in Ankle Roll, Knee Pitch, Hip Pitch, Hip Roll joints obtained from the analytical dynamic analysis of the leg being in the support phase.

imported, for the dynamic analysis the mass properties had to be assigned to each link. In this case, it was enough to assign only the density to the particular solids. The resultant mass, mass centre location and the moments of inertia of each link were calculated automatically by *Adams.*

- defining the motion parameters of the model — one of the last steps of this analysis was to assign the motion trajectory to each of the joints. In order to compare the results of the two methods, the motion assigned to the joints was the same as the one from the theoretical analysis, conducted at the initial stage of the design.
- obtaining torque in the joints

The results of the dynamic analysis conducted in Adams, for the same joints as for the initial analysis, are presented on figure 4. This analysis was conducted with the same simplifications as the initial one. The magnitudes of torque, except the ankle roll joint, were similar to that conducted before designing the full structure. During the design the ankle roll axis was shifted towards the sagittal plane (compared to the model used during the initial analysis), which resulted in decreasing the maximum torque in this joint. The signs of some of the results were opposite compared to that from the initial analysis, this was caused by the different orientation of the local coordinate systems in both analysis. The analysis of the structure in the

Adams environment confirmed the initial analytical analysis.

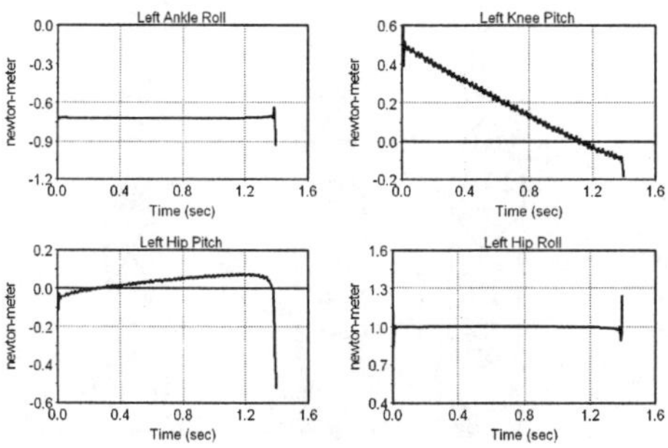

Fig. 4. Torque in Ankle Roll, Knee Pitch, Hip Pitch, Hip Roll joints obtained from the Adams' dynamic analysis of the leg being in the support phase.

Fig. 5. Results of the FEM analysis of the Ankle Pitch - Ankle Roll link. The bright colour denotes the stress concentration areas.

4. Mechanical Design

The robot structure was designed in *Modelling* module of *Unigraphics NX4* — a powerful integrated CAD/CAM/CAE software. As the actuators used in the design had relatively low torque, making the structure as light as possible was of the primary importance. In this connection, the *Advanced Simulation* module (utilizing the Finite Elements Method – FEM, for the structural analysis) was iteratively used to tailor the structure's mass and strength properties. An example of results of the FEM analysis of the ankle pitch - ankle roll link is shown on figure 5.

The mechanical structure will comprise of the elements cut by laser in aluminium rectangular pipes. The crucial elements, such as hip-knee and knee-ankle (Fig. 6) links will form a semi-3D truss, that is known from its high weight to strength and rigidity ratio.

4.1. *Preload mechanism*

From the full range of the dynamic analysis results it was visible, that regardless the length of the links, the load in the ankle roll and the hip roll joints exceeded the DX–113 actuator's capabilities. Thus, in order to

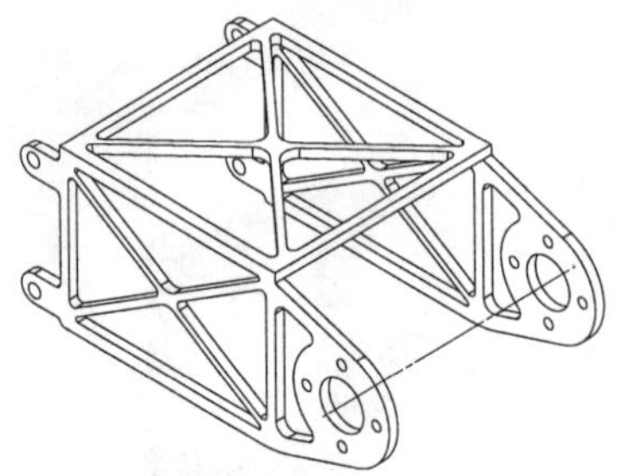

Fig. 6. The isometric view of the knee-ankle joint.

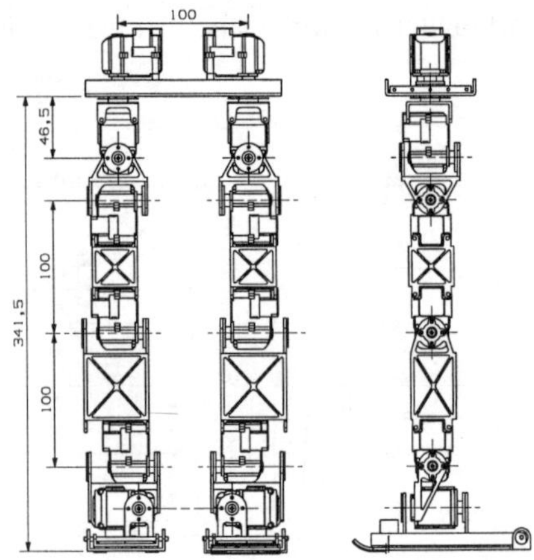

Fig. 7. The front and side view of the assembly (without the trunk).

decrease the maximum absolute value of the actuator's load, the pre-load mechanism was designed.

The pre-load mechanism principle of operation is based on shifting the range of the asymmetrical load (2), to make it symmetrical (3) and thus decrease the absolute maximum load.

$$0 < \tau < \tau_{max} \tag{2}$$

$$-\frac{\tau_{max}}{2} < \tau < \frac{\tau_{max}}{2} \tag{3}$$

Let's take the right hip roll joint as an example. The torque τ in the joint, during the motion, ranges from 0 Nm to 1.15 Nm, where 1.15 Nm exceeds the capability of the DX–113 actuator. By applying the pre-load, by the torsional spring acting in the opposite direction to the structural load (Fig. 8), the maximum torque acting on the joint is half of its original value.

4.2. *Foot compliance*

The robot's foot (Fig. 9) comprises of two plates connected in frontal part by one rotating joint. The rear upper and lower parts of the foot are connected by a convertible spring. The motion range of the joint is limited, so that the spring does not fall out in the swing phase of the leg.

The basic principle of operation is that when the leg begins the 'initial contact' phase, the rear lower part of the foot touches the ground first. From that moment the spring is being compressed, until the moment when the lower part of the foot touches the upper part. Due to this transitional state between the beginning of the 'initial contact' and the 'full support' phase, there should not be the impact. During the 'toe off' phase, the compressed spring begins to release the stored energy and additionally "actuates" the gait.

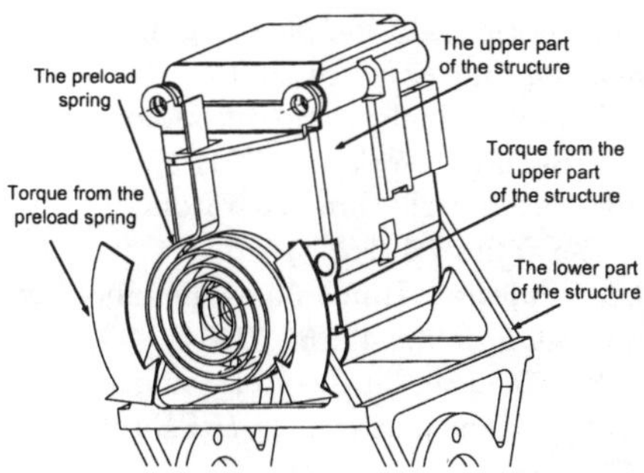

Fig. 8. The preload mechanism of the right hip roll joint. The upper part of the structure is not visible.

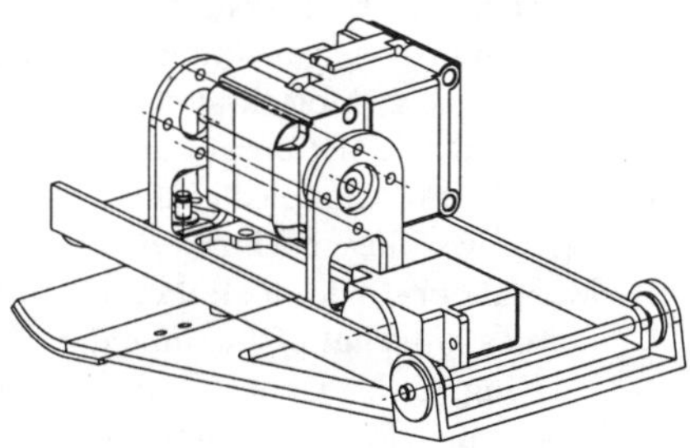

Fig. 9. The isometric view of the compliant foot.

5. Summary

This paper presents the results of the mechanical model analysis and design of the robot. It contains a proposed solution to lighten the actuator load by using a pre-load mechanism. It is expected that the resultant biped robot will serve as a tool for tests and development of the gait pattern generators and the influence of the foot passive compliance on the gait stability and smoothness.

6. Acknowledgements

The authors gratefully acknowledge the support. This work has been supported by National University of Singapore, and Warsaw University of Technology (WUT) Research Program. The participation in CLAWAR event was financially supported by Rector of Warsaw University of Technology, Dean of Faculty of Power and Aeronautical Engineering and Organizers of CLAWAR Conference from Nanyang Technological University. This research is supervised by T.Zielinska (WUT).

References

1. C. L. Vaughan, Theories of bipedal walking: an odyssey, *Journal of Biomechanics* **36** (2003).
2. S. Grillner, Neurobiological bases of rhythmic motor acts in vertebrates, *Science* **228**, 143 (1985).
3. M. R. Dimitrijevic and M. M. Pinter, Gait after spinal cord injury and the central pattern generator for locomotion, *Spinal Cord* **37**, 531 (1999).
4. T. Zielinska, Biological inspirations in robotics: Motion planning, in *Proc. of 4th Asian Conf. on Industrial Automation and Robotics*, 2005.
5. T. Zielinska, Motion synthesis in walking: Biological and technological aspects, in *CISM Courses and Lectures*, (Springer Verlang, 2004).
6. T. Zielinska, Coupled oscillators utilized as gait rhythm generators of a two-legged walking machine, *Biological Cybernetics* **74**, 263 (1996).
7. M. Vukobratovic, V. Potkonjak and S. Tzafestas, Human and humanoid dynamics, *Journal of Intelligent and Robotic Systems* **41**, 65 (2004).
8. K. Feng, C. M. Chew, G.-S. Hong and T. Zielinska, Bipedal locomotion control using a four-compartmental central pattern generator, in *IEEE Int. Conf. on Mechatronics and Automation*, 2005.
9. http://asimo.honda.com/.
10. http://www.kawada.co.jp/mechs/hrp-3p/index.html in Japanese.
11. K. Nishiwaki, S. Kagami, J. J. Kuffner, M. Inaba and H. Inoue, Humanoid JSK-H7: Research platform for autonomous behavior and whole body motion, in *Proc. of 3rd IARP International Workshop on Humanoid and Human Friendly Robotics*, 2002.
12. http://www.sony.net/sonyinfo/qrio/technology/index_nf.html.
13. F. B. Ouezdou, A. Konno, R. Sellaouti, F. Gravez, Mohamed, B. Bruneau and O. Bruneau, ROBIAN biped project – a tool for the analysis of the human-being locomotion system, in *Proc. of 5th International Conf. on Climbing and Walking Robots*, 2002.
14. http://www.takanishi.mech.waseda.ac.jp/research/wabian/index.htm.
15. http://www.jsk.t.u-tokyo.ac.jp/~ikuo/kotaro/index-e.html.
16. F. B. Ouezdou, S. Alfayad and B. Almasri, Comparison of several kinds of feet for humanoid robot, in *Proc. of 5th IEEE-RAS Int. Conf. on Humanoid Robots*, 2005.
17. J. J. Craig, *Introduction to Robotics: Machanics and Control* (Addison Wesley Longman, 1989).
18. Z. Bejek, R. Parczai, r. Illys and R. M. Kiss, The influence of walking speed on gait parameters in healthy people and in patients with osteoarthritis, *Knee Surgery, Sports Traumatology, Arthroscopy* **14**, 612 (2006).

THINKING ABOUT BOUNDING AND GALLOPING USING SIMPLE MODELS

KENNETH J. WALDRON[1], J. ESTREMERA[2], PAUL J. CSONKA[1], S.P.N. SINGH[3]
[1]*Stanford University,* [2]*Instituto de Automática Industrial-CSIC, Madrid,* [3]*University of Western Australia*

In this paper simple models are used to develop insights into the mechanics of bounding and galloping. The focus is on the entire stride cycle, rather than the stance phase. The models suggest reasons why galloping mammals have their centers of mass closer to the shoulders than the hips, The galloping model also reveals a hidden symmetry.

Introduction

Here we attempt to understand the important mechanical features of bounding and galloping using only arguments based on simple models. Despite their obvious limitations, simple models can lead to deep insights. Based on those insights, accurate modeling in simulation can then produce reliable quantitative results.

Front-Rear Force Distribution

It is notable that almost all mammals, and certainly all cursorial mammals, are built with their center of mass closer to the shoulders than to the hips. Schmiedeler *et al* (2002) located the center of mass of a large dog at 65% of the shoulder-hip distance ahead of the hips. This is fairly consistent with Jayes and Alexander (1978) who found that dogs of various breeds support 61% of their weight on the front legs. This is a result of the head and neck being cantilevered forward of the shoulders in running posture, and also the forward position of the rib cage and the muscles anchored to it. The reason for this persistent feature remains mysterious. Most likely there are several effects pushing evolution in this direction.

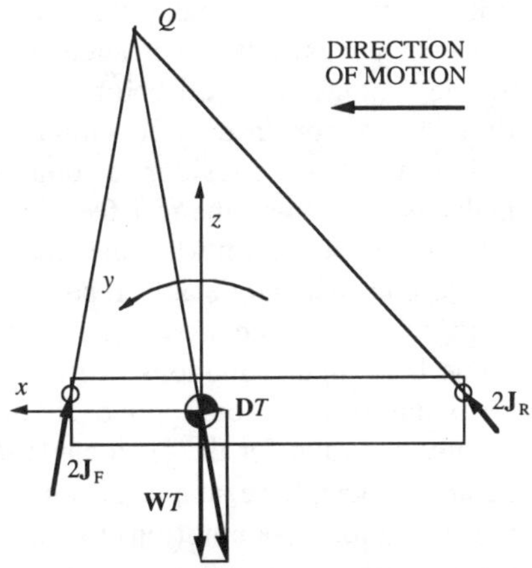

Figure 1: Impulsive equilibrium over the duration of a stride for a bound. **W** is the weight of the system, and **D** is the drag. T is the stride duration. $2J_F$ is the net impulse imparted by the front feet, and $2J_R$ is that imparted by the rear feet.

Experimental evidence (Pandy *et al*, 1988; Alexander and Jayes, 1983; Gambaryan, 1974) indicates that when running most animals place their fore feet well ahead of the vertical projection of the shoulder. The shoulder pole vaults over the leg thereby converting some horizontal kinetic energy into vertical kinetic energy and helping to generate the vertical component of the leg thrust. The net impulse from the front leg winds up having a small rearward

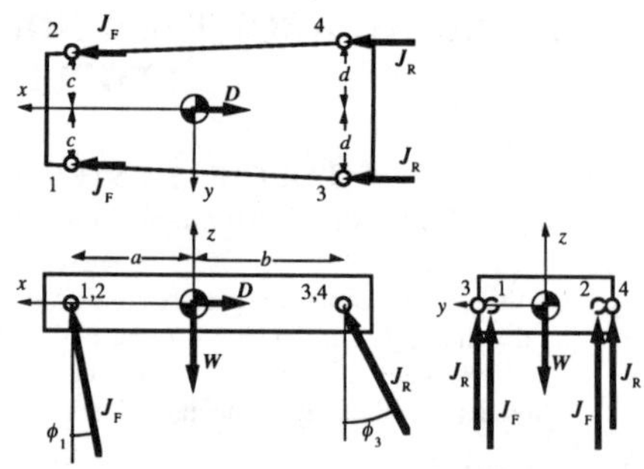

Figure 2: Free body diagram. The center of mass is assumed to be coplanar with the shoulder and hip joints. The shoulder and hip joints are assumed to be pairs of intersecting revolutes that are, respectively, parallel to the x axis of the body and orthogonal to the plane of the leg. The body reference frame is centered on the center of mass and aligned as shown. The principal axes of inertia are assumed to coincide with the x, y, z axes.

component in the horizontal direction, as shown in Figure 1. Thus, the front legs brake the system a little, and provide the largest component of the vertical impulse.

In contrast, the rear legs are placed only a little ahead of the vertical projections of the hips, or possibly even further back, and generate an impulse that is strongly inclined in the forward direction. The horizontal component overcomes the braking effect of the front legs, and provides enough forward thrust to overcome the drag on the system. The vertical component is smaller than that at the front legs, but is sufficient to provide moment equilibrium, as ensured by the concurrency condition.

Let us first consider a bound. Here the front feet touch down simultaneously, as do the rear feet. External forces, other than gravity, can only be applied to the system when the feet are on the ground.

We consider only stably repetitive motion in which the system returns to exactly the same state after one complete stride, or after one motion cycle. We assume that angular motions of the body are small. Consequently small angle approximations can be applied about all three of the attitude axes. This is certainly not valid for the pitch angle, θ_y, in a deep bound, as would be used to leap an obstacle, for example. However, for relatively shallow bounds like those used for rapid locomotion by some small animals it is a reasonable approximation. Certainly it is reasonable when galloping, which has the effect of minimizing pitch rotations.

We also assume that the forces imparted to the body by the legs act approximately along the leg axes from the point of contact with the ground to the hip/shoulder center. There is much experimental evidence to back up this assumption (Pandy *et al*, 1988; Alexander and Jayes, 1983).

The mass of the legs is neglected and the principal moments of inertia of the system about its mass center are assumed to be constant. These are the least defensible of the simplifying assumptions made. As was shown by Schmiedeler *et al.* (2002) the leg masses, and variations in the positions of the legs, result in significant variations in the pitch moment of inertia, I_y. Nevertheless, in the interests of transparency we choose to regard I_y as constant.

We choose to work in the impulse-momentum domain allowing us to consider an impulse balance over the entire stride period, T. We also work in a reference frame fixed to the vehicle body. This does not imply that we regard the stance period as being instantaneous. Rather, we integrate the contact force during stance with respect to time. This is analogous to replacing a spatially distributed force system by its resultant.

If the net impulse imparted to the system by the front feet is $2\mathbf{J}_F$, and that imparted by the rear feet is $2\mathbf{J}_R$, **W** and **D** are the weight of the system and the drag respectively, the system becomes a three-force system. Therefore the lines of action of the front and rear impulses must be concurrent with that of the impulse of the resultant of the weight and drag as shown in Figure 1.

The equivalent dynamic equilibrium equations are:

$$2\mathbf{J}_{Fx} + 2\mathbf{J}_{Rx} = \mathbf{D}T \quad (1)$$

$$2\mathbf{J}_{Fz} + 2\mathbf{J}_{Rz} = \mathbf{W}T \quad (2)$$

$$a\mathbf{J}_{Fz} = b\mathbf{J}_{Rz} \quad (3)$$

The effect of the near vertical thrust of the front legs is to pitch the system back, when viewed in the world reference frame. This allows the rear legs to contact the ground in a more nearly vertical position than they would have done without the effect of the front legs. That, in turn, increases the horizontal component that can be generated by the rear feet without slipping. The rear legs are free to extend to their limit, in order to generate as much thrust as possible, without any risk of interference with the fronts,

Impact Energy

It is notable that animals can run over diverse surfaces without any noticeable effect on their gait, and at very nearly the same speed. At first glance this seems anomalous since the coefficient of restitution between the foot and the ground must vary considerably, and unpredictably. The reason for this capability is very simple when one considers the way the foot works.

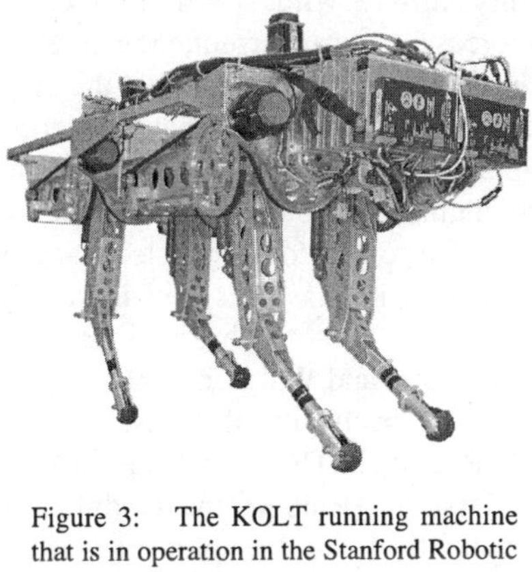

Figure 3: The KOLT running machine that is in operation in the Stanford Robotic Locomotion Laboratory.

For simplicity we consider a system having only one leg, and consisting of a small mass, m, representing the foot, connected to a large mass, M, representing the rest of the system by a spring. Initially the foot is at rest relative to the rest of the system with the spring at its natural length. The whole assemblage is descending with velocity v_z. When the foot strikes the ground it does not rebound because the mass, M, compresses the spring, and the spring force holds the foot down. Eventually M is propelled upwards again by the spring, and energy is added by an actuator assisting the spring.

The energy that the foot had when it descended is lost, and must be replaced by the actuator. Because m is very much smaller than M this energy loss can be accepted. In return, the system can run without adjustment, and with very nearly the same energetic cost, on most surfaces.

The energy that is lost in the impact of the foot with the ground appears to the system as an increase in the external drag, **D**. The way this works is somewhat analogous to the generation of rolling resistance at a wheel. It was discussed by Schmiedeler and Waldron (1999).

There is a considerable literature on internal energy storage during running, and the amount of energy stored as strain energy, as opposed to gravitational potential energy during stance. However, our objective here is to consider the entire gait cycle, so we will not attempt a more detailed discussion of stance. It appears that the loss of energy due to the vertical velocity of the foot at impact is the principal energy drain on the system. Horizontal kinetic energy varies proportionately relatively little when running fast, and horizontal kinetic energy is easily converted into gravitational potential energy with minimal loss by the mechanism of pole vaulting over the leg, discussed above, and the converse is equally easily and efficiently accomplished by the body falling forward over a rearward thrusting leg. Impact losses in the horizontal direction can be avoided by swinging the leg so that the forward velocity of the foot when it touches down is zero, or at least very small.

Observation of animals running shows that their stride frequency increases only slowly with speed. In fact, one might hypothesize that the observed increase in stride frequency is a result of the necessary decrease in stance duration with speed, and that the effective duration of each ballistic trajectory is constant. This implies that the vertical velocity at lift-off is for a given foot is approximately constant, and hence the height of each hop is also constant. Since the number of footfalls per second is approximately constant, the rate of energy loss due to the foot impacts is approximately constant. The increase in energy cost with increasing speed is largely due to increasing external drag. This is consistent with our observations running the KOLT machine shown on Figure 3 (Estremera and Waldron, 2006, 2007).

The mechanics of the system during stance is often considered to follow the SLIP model. This has been extensively studied and its properties are well understood (Raibert and Hodgins, 1991; Saranli *et al.*, 1998). The principal implication for the present purpose is that the upward vertical velocity component immediately after stance is exactly equal in magnitude to the downward vertical velocity immediately before stance. We will show that this condition is excessively restrictive, and that the slip model is not, in general, consistent with a bound, and cannot be consistent with a three-dimensional gallop.

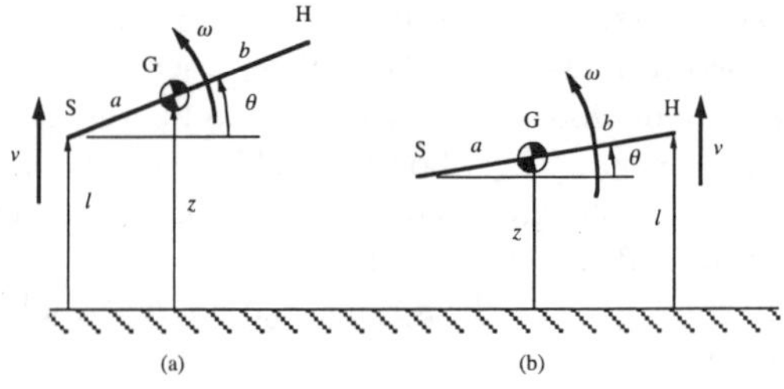

Figure 4: Coordinate senses during (a) front foot stance and (b) rear foot stance.

Bounding

Let the vertical velocity of the center of mass of the system immediately before front foot impact be u_F; and that immediately after the impact be u'_F. The vertical velocity immediately before rear foot impact is u_R, and that immediately after is u'_R. If the vertical impulse delivered by a front foot is J_{zF}, and that delivered by a rear foot is J_{zR} then applying impulse-momentum across the impacts gives

$$u'_F = u_F + \frac{2J_{zF}}{M}, \quad u'_R = u_R + \frac{2J_{zR}}{M}, \quad \omega_{yFR} = \omega_{yRF} - \frac{2aJ_{zF}}{I_y} \tag{4}$$

where ω_{yFR} is the pitch angular velocity after the front foot stance and before the rear foot stance, and ω_{yRF} is the corresponding angular velocity after the rear foot stance.

Looking at the velocity changes during the two ballistic trajectories (hops)

$$u_R = u'_F - g\tau_{FR}, \quad u_F = u'_R - g\tau_{RF} \tag{5}$$

Here τ_{FR} is the time period from the front foot stance to the rear foot stance, and τ_{RF} is that between the rear stance and the front. Obviously

$$\tau_{FR} + \tau_{RF} = T \tag{6}$$

where T is the stride period.

Applying the condition that the vertical height of the center of mass at the end of the stride must be identical with that at the beginning

$$\left(u'_F + u_R\right)\tau_{FR} + \left(u'_R + u_F\right)\tau_{RF} = 0 \tag{7}$$

M is the total mass of the system and I_y is its pitch moment of inertia about G. J_{Fz} is the vertical component of the impulse delivered by each front foot.

The condition that the system must return to the same pitch angle after a complete stride gives

$$\omega_{yFR}\tau_{FR} + \omega_{yRF}\tau_{RF} = 0 \tag{8}$$

Combining Equations 4 through 8 gives

$$\begin{aligned} u'_F &= \frac{g}{2(a+b)}\left(2a\tau_{FR} - T(a-b)\right), \quad u'_R = \frac{g}{2(a+b)}\left(T(b+a) - 2b\tau_{FR}\right), \\ u_F &= \frac{g}{2(a+b)}\left(2a\tau_{FR} - T(a+b)\right), \quad u_R = \frac{g}{2(a+b)}\left(T(b-a) - 2b\tau_{FR}\right) \end{aligned} \tag{9}$$

Similarly, Equations 4 and 8 can be combined to give

$$\omega_{yFR} = \frac{\kappa g T}{a+b}(p-1), \quad \omega_{yRF} = \frac{\kappa g T}{a+b}p \tag{10}$$

where

$$\kappa = \frac{Mab}{I_y} \tag{11}$$

Energy Loss

The velocity, v_F, with which the front feet strike the ground may be obtained from Equations 9 with reference to Figure 4:
The energy lost in each stride due to impact with the ground is minimized if

$$\tau_{FR} = T\left(2b^2 m_R(1-\kappa) + (am_F - bm_R)(a+b)\right) \Big/ \left(2\left(a^2 m_F + b^2 m_R\right)(1-\kappa)\right) \tag{12}$$

as the value of τ_{FR} that minimizes the energy consumption. The corresponding value of U_L is given by

$$U_{\mathrm{L}} = \frac{g^2T^2 m_{\mathrm{F}} m_{\mathrm{R}}}{4\left(a^2 m_{\mathrm{F}} + b^2 m_{\mathrm{R}}\right)}\left(a + b - \frac{2ab(1-\kappa)}{a+b}\right)^2 \tag{13}$$

This might be compared with the corresponding values for a trot for which τ/T is constrained to be 0.5 and there is no body rotation

$$U_{\mathrm{L}} = g^2T^2\left(m_{\mathrm{F}} + m_{\mathrm{R}}\right)/16 \tag{14}$$

The impact energy loss for the bound is always greater than that for the trot. There is always a term which is a function of κ that increases the energy loss for the bound. κ expresses the importance of body pitch rotation.

So far we have ignored motion in the longitudinal direction and analyzed only dynamic equilibrium in the vertical direction, and rotation about the pitch axis. However, there is an interchange of horizontal and vertical kinetic energy resulting from the inclination of the impulses indicated in Figure 1. Usually there is a rearward component of the impulse from the front feet that causes a reduction in the horizontal velocity. There is a greater forward component of the impulse from the rear feet to meet the requirements of Equation 1. Thus there is an increase in horizontal velocity after rear foot stance.

We assume that the leg is swung so that the longitudinal foot velocity relative to the ground is close to zero when the foot is placed. This means that there is no "horizontal impact" causing energy to be lost from the system in the same way as the vertical impact.

After a complete stride, the total dynamic energy of the system returns to the same value. Thus a positive value of $U_{z\mathrm{F}}$ appears as a corresponding loss of horizontal kinetic energy at front foot stance, and the same amount of kinetic energy is returned to horizontal motion at rear foot stance. The work done by the horizontal component of the impulse is divided between making up the decrease in kinetic energy due to transfer to vertical energy, and making up the energy lost to drag.

Galloping

We follow the suggestion of Minetti (1998) and regard the gallop as a bound in which the front and rear leg pairs "skip". The advantage of splitting out the legs at each stance becomes obvious if we recall that the impulse impelling each hop is halved. Based on the discussion above, to a first approximation this means that the energy going into vertical motion in each hop is one eighth of that for the hops in a bound, taking into account that two feet impact at the end of each hop in a bound. However, since there are four

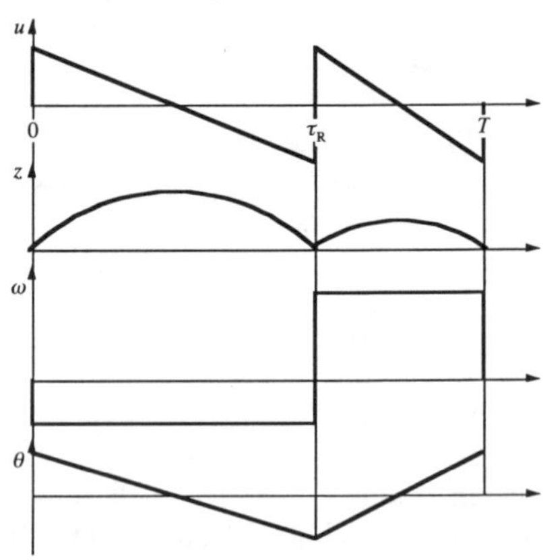

Figure 5: Variations in vertical velocity and vertical position of center of mass, and pitch angular velocity and pitch angle through a bound stride. The intervals used in the text are $\tau_{\mathrm{FR}} = \tau_{\mathrm{R}}$, $\tau_{\mathrm{RF}} = T - \tau_{\mathrm{R}}$.

hops rather than two, the energy loss in a gallop stride should be approximately one quarter of that in a bound.

We can regard a gallop as being an asymmetric bound in which the front and rear leg pairs "skip". That is, we start with a bound and examine the effects of separating the front feet, and the back feet. Instead of placing the front feet together, we place foot 1 ahead of foot 2. For a transverse gallop, we would then place foot 3 ahead of foot 4. The hops FR, and RF of the bound correspond to the flight phases, 23 and 41 of the gallop, while the skips are the hops 12 and 34.

The stride cycle impulse-momentum relationships are still satisfied. The separation of the two front foot impulses does not affect Equations 1, 2 and 3 above. Likewise, the relationship between the angular velocities $\omega_{y23} = \omega_{yFR}$ and $\omega_{y41} = \omega_{yRF}$ remain the same. The change in angular velocity occurs in two steps instead of one.

We apply a similar analytical approach to that used for the bound to a transverse gallop to get, after considerable algebraic development that cannot be presented here because of space limitations:

$$\begin{aligned} u_1' &= \frac{g}{2}\tau_{12}, u_2 = -u_1', u_2' = \frac{g}{2}\left(-\tau_{12} + \frac{bT}{a+b}\right), u_3 = -u_4', \\ u_3' &= \frac{g}{2}\tau_{34}, u_4 = -u_3', u_4' = \frac{g}{2}\left(-\tau_{34} + \frac{aT}{a+b}\right), u_1 = -u_2' \end{aligned} \tag{15}$$

and

$$\begin{aligned} &\omega_{y12} = 0, \quad \omega_{y23} = -\frac{\kappa g T}{2(a+b)}, \quad \omega_{y34} = 0, \quad \omega_{y41} = \frac{\kappa g T}{2(a+b)} \\ &\omega_{x12} = \frac{\lambda g(T - \tau_{12} - \tau_{34})}{2(c+d)}, \omega_{x23} = -\frac{\lambda g(\tau_{12} - \tau_{34})}{2(c+d)}, \omega_{x34} = \omega_{x12}, \omega_{x41} = \omega_{x23} \end{aligned} \tag{16}$$

In order to get the above result, it is necessary to assume that the left and right legs in each pair operate at the same working length, and to apply a constraint that the roll angle, and angular velocity, ω_x, at the end of a complete stride are the same as at the beginning.

On the face of it, this solution, with the flight phases being of equal duration, is not consistent with the apparent asymmetry of the gait. It is a basic feature of the gallop that the gathered flight phase (23) is always present, but the spread flight phase (41) is only present in a fast gallop, and then appears to be much shorter than the gathered flight phase.

There is a partial explanation if we recall that the stances are of finite duration and that the impulses are inclined as illustrated in Figure 1. As may be seen in Figure 8 the front foot touches down far in front of the shoulder, and lifts off a short distance behind the shoulder. The rear foot touches down almost directly beneath the hip, and lifts off after the leg has extended far to the rear of the hip. The effect is to shorten, or erase the period when all feet are physically out of contact with the ground when going from a rear foot stance to a front foot stance (41), and to emphasize the flight period when going from a front foot stance to a rear foot stance (23). Thus, the apparent asymmetry between the gathered flight phase (23), and the spread flight phase (41) is not necessarily inconsistent with equal values of τ_{23} and τ_{41}. Nevertheless, this result does call in question the assumption that the legs in each pair deliver identical impulses.

Following a development similar to that used for the bound, we can identify the footfall intervals that minimize impact energy loss:

$$\tau_{12} = \frac{a\kappa + b(1-\lambda)}{2(a+b)(1-\lambda)}T, \tau_{34} = \frac{a(1-\lambda)+b\kappa}{2(a+b)(1-\lambda)}T, \tau_{23} = \tau_{41} = \frac{1-\kappa-\lambda}{4(1-\lambda)}T \quad (17)$$

The corresponding minimal value of the impact energy loss per stride is

$$U_L = g^2T^2\left(m_F\left(a(\kappa+\lambda)+b\right)^2 + m_R\left(a+b(\kappa+\lambda)\right)^2\right)\Big/16(a+b)^2 \qquad (18)$$

It may be observed from Equations 41 that the optimal durations τ_{23} and τ_{41} of the flight phases are functions solely of the inertial constants $\kappa = Mab/I_y$ and $\lambda = Mcd/I_x$. This is important since animals do not increase speed by increasing stride frequency. Rather, they increase speed by increasing the distance covered during each stride particularly by increasing the durations of the flight phases. The stride period, T, declines only slowly with increasing speed. Although we do not model the stance phase in this paper, we hypothesize that the decrease in T is due to the decreasing durations of the stance phases. The duration of a stance phase depends on the angle through which the leg moves relative to the body while the foot is on the ground. It appears that this does not change much with changes in horizontal velocity. Hence, the duration of the stance phase is inversely proportional to horizontal velocity. If the durations of the flight phases remain constant, due to maintenance of constant upward velocities of the center of mass at lift-off, the overall stride duration will decline slowly due to the shortened durations of the stances.

Equation 18 may be compared to the corresponding expressions for the bound and trot given by Equations 13 and 14 above. The energetic advantage of the gallop over the trot depends on the value of $\kappa + \lambda$. The smaller the value of $\kappa + \lambda$, the more attractive it is to gallop from the energetic perspective, as long as the inter-stance intervals are close to the optimal values given by Equations 17. The significance of Equations 17 and 18 is that the system can increase the proportion of the stride period devoted to the flight phases, but only at the cost of increasing impact losses.

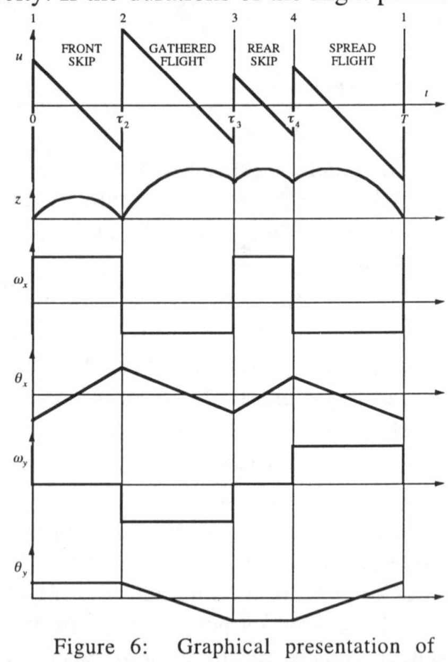

Figure 6: Graphical presentation of variations of velocity and position variables through the transverse gallop stride.

Conclusions

The models presented here have obvious shortcomings. For example, the actual duration of each stance period is significant, typically declining from about $0.2T$ to $0.1T$ with increasing speed (Gambaryan, 1974). This means that the stances of the front and rear leg pairs actually overlap, creating double support periods that require higher fidelity models than those presented here.

We have also almost ignored the velocity in the longitudinal direction, which is, of course, the whole point of running. Likewise, we have not discussed the effect of splitting the front and rear stances on motion about the yaw axis. There is a significant effect, but it is easily corrected by a slight displacement of the center of the front track relative to that of the rear track.

Acknowledgement

The authors acknowledge the support of the National Science Foundation, Grant Nos. IIS-0208664 and IIS-0535226, and the Secretaria de Estado de Universidades of the Spanish Ministry of Education and Science.

References

Alexander, R. McN. and Jayes, A. S., 1983, "A dynamic similarity hypothesis for the gaits of quadrupedal mammals," *J. Zoology (London)*, **201**:135-152.

Estremera, J. and Waldron, K.J., 2006, "Leg Thrust Control for Stabilization of Dynamic Gaits in a Quadruped Robot, " Proc. ROMANSY, Warsaw, Poland.

Estremera, J. and Waldron, K.J., 2007, "Thrust Control, Stabilization and Energetics of a Quadruped Running Robot", *IJRR*, under review.

Gambaryan, P.P., 1974, How Mammals Run: Anatomical Adaptions, Wiley.

Jayes, A.S. and Alexander, R.M., 1978, "Mechanics of Locomotion of Dogs and Sheep", *J. Zoology, London*, **185**, pp. 289-308.

Minetti, A.E., 1998, "The Biomechanics of Skipping Gaits: a Third Locomotion Paradigm?", *Proc. Royal Society, B*, **265**, pp. 1227-1235.

Pandy, M. G., Kumar, V., Waldron, K. J. & Berme, N. 1988, "The Dynamics of Quadrupedal Locomotion," *J. Biomechanical Engineering*, **110**, 3, pp. 230-237.

Raibert, M.H. and Hodgins, J. 1991, "Animation of Dynamic Legged Locomotion," *Computer Graphics*, **25**, pp. 349-358.

Saranli, U., Schwind, J. Koditschek, D., 1998, "Toward the Control of a Multi-Jointed, Monopod Runner," *Proc. ICRA*, Leuven, Belgium.

Schmiedeler, J.P., Siston, R., and Waldron, K.J., 2002, "The Significance of Leg Mass in Modeling Quadrupedal Running Gaits," ROMANSY 14, ed. G. Bianchi, J.C. Guinot, C. Rzymkowski, Springer, pp481-488.

Schmiedeler, J.P. and Waldron, K.J., "The Effect of Drag on Gait Selection in Dynamic Quadrupedal Locomotion," *IJRR*, **18**, 12, 1999, pp. 1224-1234.

Waldron, K.J., 2007, "A Coordination Scheme for an Asymmetrically Running Quadruped", to appear Proc. IFToMM World Congress, Besançon, June.

USING OPTIMIZATION TECHNIQUES FOR THE DESIGN AND CONTROL OF FAST BIPEDS

T. LUKSCH* and K. BERNS

Robotics Research Lab, University of Kaiserslautern,
Kaiserslautern, Germany
**E-mail: t.luksch@informatik.uni-kl.de*

K. MOMBAUR* and G. SCHULTZ

IWR, University of Heidelberg
Heidelberg, Germany
**E-mail: katja.mombaur@iwr.uni-heidelberg.de*

Fast, dynamic or energy efficient locomotion of bipeds is still an unsolved problem in robotics. Nature seems to have solved many of the arising difficulties in thousands of years of evolution, optimizing both mechanics and control along the way. This paper proposes to use techniques from numerical optimization and optimal control combined with behavior-based control concepts to address some of the problems when designing and controlling two-legged robots. Results from the optimization process will affect both the mechanical construction and the control strategies.

Keywords: Optimal Control, Behaviour-based Control, Bipeds.

1. Introduction

Speed, efficiency and stability of biological bipedal walking and running is still unparalleled by contemporary bipedal robots, despite all the technological progress made in past decades. The observation of biological motions leads to the conjecture that nature does apply optimization – not only when changing the properties of populations in the course of evolution, but also in the development of characteristic motion patterns of each individual. One key feature of biological motion is that it fits naturally to the system itself and to its kinematical and dynamic quantities, and that no un-natural motion is imposed that would result in inefficiency or increased need for stability control. Another important characteristic of nature is the exploitation of elasticity in the passive tissue and in the actuators themselves in order to enhance efficiency and stability of the system.

The goal of the research presented here is to apply the same kind of optimization that nature applies in the process of designing, building and controlling a bipedal running robot. A methodology, based on dynamical multibody system models and efficient numerical optimization techniques has been developed that assists during all stages of the robot's construction and behaviour-based control design – starting at the earliest prototype design and component selection, and ending only when the robot has acquired the capability to perform the desired motions efficiently. In this paper, all steps of the methodology are described and it is shown how it has been applied to design a leg prototype to be used in a running biped. To control the robot, a behavior-based reflex network is designed based on results from the optimization and knowledge on neuro-motor control in nature. Optimization is also used to increase the exploitation of elasticity in the system by adequately designing springs for the running motion. In current research, this same methodology is to be applied in the design and construction of the full biped robot.

Other projects focusing on dynamic locomotion of bipeds include the research on the robot Lucy, where elasticity is included in form of pneumatic muscles [1,2]. Another project working on dynamic walking and running is the Rabbit robot. Questions examined include stability of posture or robustness against external forces [3]. The benefits of self-stabilizing systems are exploited in the field of passive-dynamic walking. With robots like Denise or Meta from the University of Delft it is examined how actuation can be included to this concept [4,5].

2. Methodology: Optimization-based design and control of robots and components

Inspired by nature, a methodology for designing and controlling robots and robot components based on dynamic models of the robotics system and efficient state-of-the-art optimization techniques has been developed. The proposed methodology requires a frequent interaction and exchange of results between the computational and the experimental side. The following steps are taken:

(1) Specify a preliminary mechanical design, including dynamical properties of material, include ranges of characteristics of motors and other components. Define the desired form of locomotion (i.e. walking, running, hopping, type of ground etc.)

(2) Create a dynamic model for this initial construction and the desired locomotion, include all realistic constraints in the model
(3) Optimize this model with design and control parameters as free variables
(4) Modify design according to results from optimization
(5) Iterate 2-4 if necessary
(6) Integrate optimal control results into a generalizing behavior-based control scheme
(7) Build the prototype and implement the control
(8) Validate by comparing the results from the prototype with those from the simulated model

3. Experimental Setup

As described in the previous section, the first step of the proposed methodology is to develop an initial design for the robot. To verify the feasibility of the methodology in the framework of this initial study, instead of directly investigating a full biped, a single, jumping leg is started with. Jumping on the spot is an extremely dynamic, periodic movement putting high demands on the mechatronical design and the control, this way providing meaningful evidence on the quality of the proposed method.

The experimental setup of the prototype leg on the treadmill is shown in figure 1(b). The leg has an active hip and knee joint with one rotary degree of freedom each. Being fixed to a linear vertical slider, it can perform vertical motions in the hip. Each of the joints is powered by a non-retardant drive composed of a d/c motor with low ratio gear allowing for a freely moving joint and thus for less mechanical stress and more time to react to disturbances.

The knee joint is supported by a spring mounted in parallel to the drive with fixed equilibrium position. This newly designed pneumatic, rotatory spring (see fig. 1(a)) has the advantage that it can produce high torques and the stiffness can be adjusted by changing the working volume using switchable separations including valves (for more details, see [6]). A custom-made load cell in the lower limb allows measuring of the force in leg length direction (z-axis) and the torques around the x- and y-axis during impact and stance phase. The overall weight of the prototype sums up at about 12kg. This initial construction forms the starting point for the subsequent optimization process. Several of the mechanical design parameters are defined as free variables for the optimization.

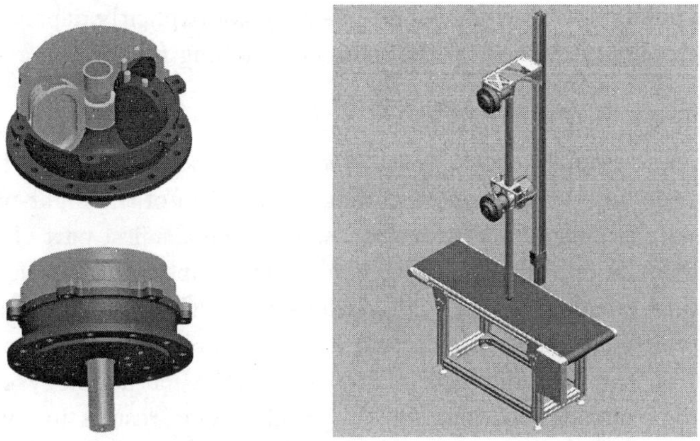

Fig. 1. (a) Design of the pneumatic spring showing the moving piston and one separation – (b) The initial single leg prototype mounted on a linear slider of the test rig.

4. Mathematical Model, Optimal Control Problem Formulation and Solution

Setting up a mathematical model of the leg is step (2) of the above methodology. A rigid multibody system model of the prototype leg has been established including all free design parameters p and free input quantities $u(t)$ driving the system (in this case the torques produced at hip and knee). The model of this leg performing a running – or rather hopping – motion on this leg results in a hybrid dynamical system involving both continuous phases (flight and contact) and discrete model phases (discontinuous touchdown event).

Choosing the three coordinates hip height y, absolute angle of upper leg α_u, and relative angle of lower leg ϕ_l, we obtain a set of differential equations of form

$$\dot{x}(t) = f_j(t, x(t), u(t), p) \quad \text{for} \quad t \in [\tau_{liftoff}, \tau_{touchdown}] \tag{1}$$

for the flight phase with $x^T = (y, \alpha_u, \phi_l, \dot{y}, \dot{\alpha}_u, \dot{\phi}_l)^T$ and differential algebraic equations for the contact phase

$$\dot{x}(t) = f_j(t, x(t), u(t), p) \tag{2}$$

$$0 = g(t, x(t), u(t), p) \quad \text{for} \quad t \in [\tau_{touchdown}, \tau_{liftoff}] \tag{3}$$

The touchdown discontinuity is described by

$$x(\tau^+_{touchdown}) = J(x(\tau^-_{touchdown}), p). \tag{4}$$

Starting points τ_j of new phases do generally not explicitly depend on time, but are implicitly defined by the roots of switching functions

$$s_j(t, x(\tau_j), p) = 0. \tag{5}$$

The periodic hopping cycle considered in the model starts in the flight phase at the highest point, goes through the rest of of the flight phase, the touchdown discontinuity, contact phase, and then the first part of the flight phase until the cycle is completed. Periodicity constraints are formulated on all state variables $x(0) = x(T)$, where the cycle time T and all phase times are free variables. Realistic constraints are imposed on all design parameters, controls and states. Step (3) of the methodology consist in determining optimal motions for this model. This results in solving the following multi-phase optimal control problem:

$$\min_{x,u,T} \int_0^T \phi(x(t), u(t), p)\, dt + \Phi(T, y(T), p) \tag{6}$$

$$\text{s. t.} \quad \text{model eqns. (1)/(2) and (4)}$$
$$+ \text{ addl. equality and inequality constraints}$$

Problems of this form can efficiently be solved by optimization techniques based on the direct boundary value problem approach using multiple shooting [7–9]. This technique involves a discretization of both state and control variables transforming the optimal control problem into a nonlinear programming problem (NLP) with simultaneous simulation for the evaluation of objective function and constraints. For an application of these methods to hybrid gait problems and required extensions if stability is used in the objective function, see [10].

5. Optimization Results for Prototype Leg

The goals of the optimization runs performed for the prototype leg were

- to determine if the original leg design and the initial choice of components would be adequate to preform the desired hopping motion (with a desired hopping height of 20-30 cm, and at least 10 cm)
- if this was not the case, to establish a better setup, parameters and actuation patterns
- to identify which modifications are most crucial in improving the performance of the leg
- to evaluate the use and the best design of the pneumatic rotary spring in order to exploit spring elasticity to produce more efficient hopping.

Since the prototype leg only moves on the the test rig, stability has not been considered in the optimization at this point. The optimal control problem formulations aimed at finding periodic hopping motions that maximize jumping height under different conditions. In a first series of optimization runs we have determined the dependency of the maximum hopping height on the average torque limits and the peak torque limits in the knee and the hip for a first set of design parameters of the prototype leg. The results in figure 2 show that an average torque limit of 20-30 Nm and a peak torque limit of 40 - 50 Nm are required. For the choice of a maximum average

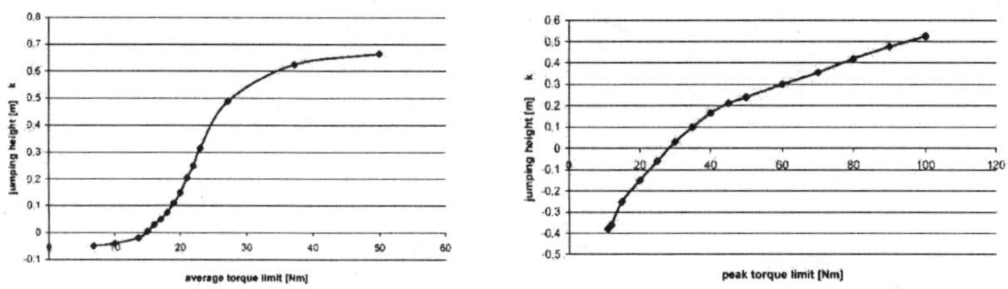

Fig. 2. Dependency of maximum hopping height on limits of (a) average torques and (b) peak torques

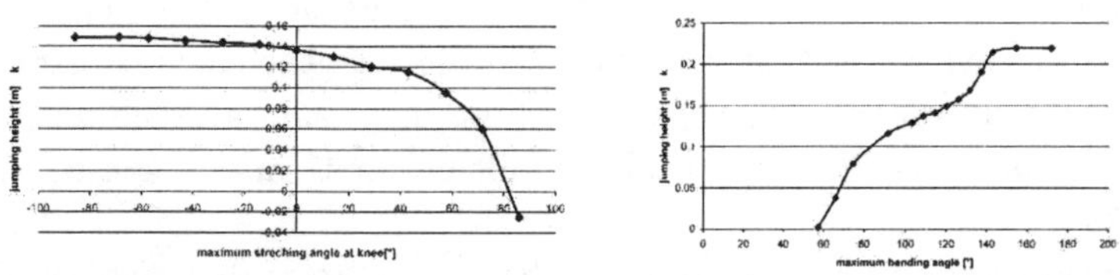

Fig. 3. Dependency of maximum hopping height on the maximum (a) stretching and (b) bending angle at the knee

torque of 20 Nm we investigated the influence of several parameters on the maximum hopping height:

- the maximum stretching angle of the leg: the more the leg can be stretched, the higher is the jump, but the possibility to overstretch the leg at the knee only leads to slight increases of the hopping height, see figure 3
- the maximum bending angle of the leg: this has a significant influence, see figure 3, should be at least 100 deg, more than 160 deg is not useful

- the maximum pressure in the pneumatic spring has a major influence on the hopping height (see figure 4). In this case, three different average torques have been investigated, and we found that with a maximum pressure of 30 bar, even an average torque of 13.6 Nm was sufficient to produce a jumping height of 25 cm. Pressures of .... in the spring are realistic.

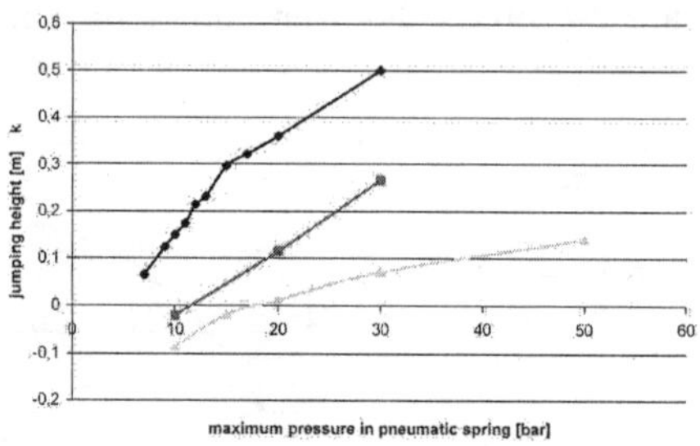

Fig. 4. Dependency of maximum hopping height on the maximum pressure in the pneumatic spring

In addition, we have investigated the effect of a treadmill on the hopping height. Not too surprisingly, having to produce an additional forward or backward motion while joint torques are subject to the same constraints, results in a reduced hopping height. With an maximum average torque of 20 Nm, the prototype leg only is able to lift off for treadmill speeds up to $\approx 1.2m/s$

After all these initial computations that served to evaluate the original prototype design some necessary modifications have become obvious (compare step (4) of the methodology) such as the introduction of gears with higher ratios in hip and knee in order to be able to produce the high torques required for the desired motions. These resulted in additional masses that were taken into account in a new set of computations performed subsequently. With limits of average torques at $27.2Nm$ and $17.0Nm$ for knee and hip, and limits of peak torques at a factor 2.5 respectively a maximization of jumping height resulted in the state variable and control histories shown in figure 5. The jumping height was $13.5cm$. The resulting parameters of the optimally designed pneumatic spring, in which it best supports the computed target motion, were $1.88bar$ initial pressure, a compartment partition at 185 deg, and a spring zero position at -12.5 deg.

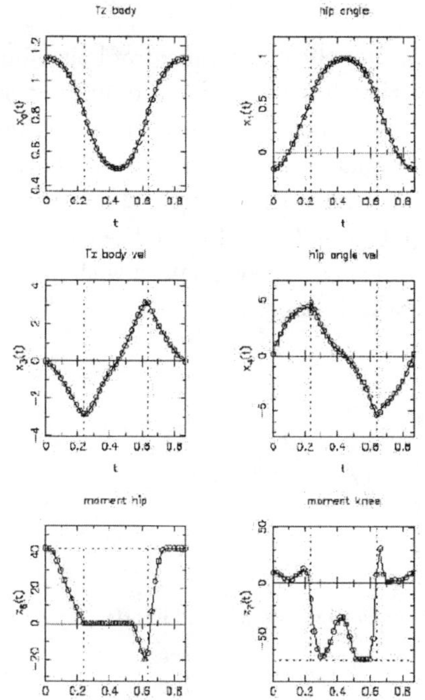

Fig. 5. (a) Position and velocity variables and torques of optimal motion

6. Behavior-based Control of Prototype and Experimental Results

After optimizing the initial construction in several iterations, a suitable control must be conceived (step (6) of the methodology). Beside the optimal mechanical parameters, the optimization process also provides optimal control trajectories for the motion aimed at in the undisturbed case. Those trajectories can directly be used as open-loop control strategy in hope the real robot does not differ to much from the modeled one. This way a one-on-one comparison between the optimal control results and real prototype is possible.

The second strategy is to derive a reflex control. The optimal control trajectories are analyzed to find generalized reflexes that produce similar control output in the undisturbed case, but also give reasonable results with arising external disturbances. The selection of reflexes is also influence by knowledge on neural control from biology. This control is realized as behaviour-based network.

In previous works, a framework for the development of behaviour-based robot control has been designed that has been successfully used on many different machines (see e.g. [11]). It consists of a network of reflexes or behaviors. Besides the sensor input $\vec{e}$ that is transformed to the control values $\vec{u}$ by the transfer function F, each behavior has a special motivation input ι and two meta information outputs a and r that state the current activity and target rating and are used for behavior-coordination with fusion nodes. The basic behavior module is shown in figure 6.

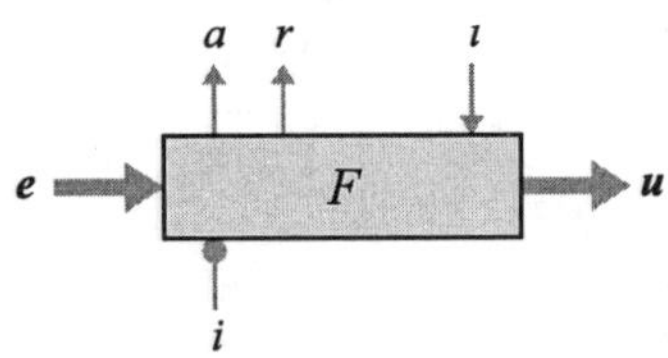

Fig. 6. The basic behavior module with its interface

Figure 7 shows the resulting reflex network. The *set posture* reflex guarantees as slightly flexed leg stance for stable standing and secure landing after the flight phase. During ground contact the hip torque is reduced by the *relax hip* reflex similar to what can be seen in the optimal control trajectories. This also reduces mechanical restraint. To achieve enough acceleration to take off, the knee joint has to be bend sufficiently, which is done by the *prepare jump* reflex after touching the ground. Finally, the *push off* reflex produces full torque in the knee and hip joint as soon as the knee is bended far enough to produce maximum acceleration. The control flow, the fusion nodes and the inhibiting connections can be taken from figure 7.

As the finishing of the real prototype leg is still pending, the behavior-based control is tested in a dynamics simulation environment. The activities of the reflexes during one jumping cycle can be seen in figure 8. Despite no explicit cyclic behavior like an oscillator being present, the leg shows periodic jumping with nearly exactly the same cycle duration as was calculated as optimal by the optimization process. The control shows robust behavior to disturbances, e.g. sudden changes to the ground height.

Figure 9 shows the resulting joint angles and the upper body position. The ground contact force is also plotted to show the points in time when the leg is hitting the ground, is slightly bouncing back, resting on the

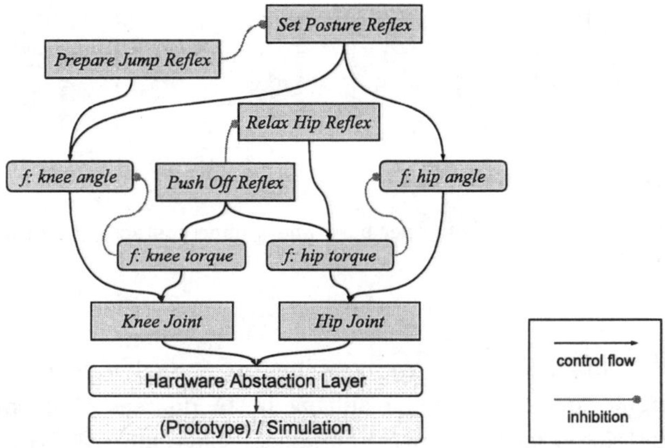

Fig. 7. The reflex network producing robust jumping of the prototype leg. Edges ending in a filled dot represent inhibiting connections. The rounded boxes denoted with an f are fusion nodes (e.g. *f: knee torque*).

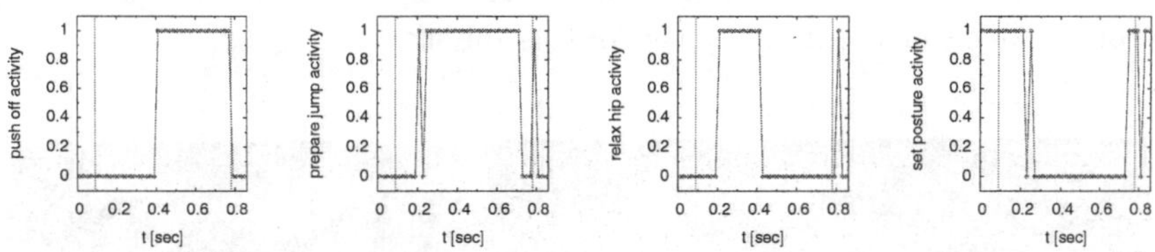

Fig. 8. Activity of the reflexes during one jump cycle

ground and finally lifting off. The resulting trajectories are similar to those predicted by the optimization calculations. The pneumatic spring always produces a certain torque as the equilibrium position is set to an negative angle and the spring is pre-stretched. The slider position shows a lift-off height of about 5cm.

7. Extensions of Methodology to Bipedal Running

Our ultimate goal is to apply the described methodology to construct a bipedal running biped. The use of the prototype leg in this context was twofold:

- to test the methodology of optimization-based design
- to manufacture a first component that might serve as a basis for the future bipedal robot.

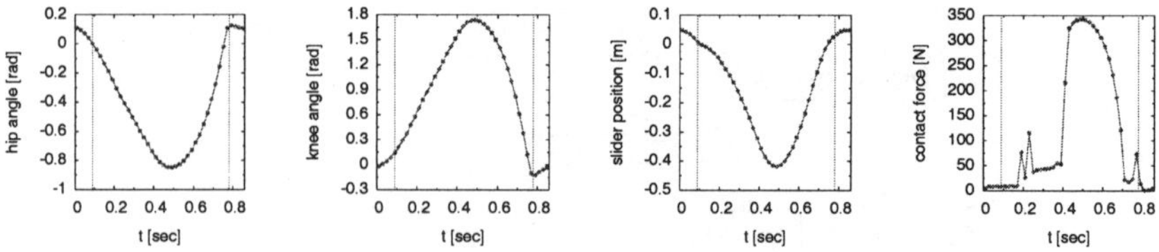

Fig. 9. Joint angles, position of upper body and ground contact force during one jump cycle

The target configuration of the running biped has nine bodies (trunk, shanks, thighs, feet and arms) resulting in 11 degrees of freedom if motions are restricted to the sagittal plane. We have already established the bipedal model and produced first optimization results for energy-efficient running at different speeds, using human-like geometries and mass distributions. One of these optimized motions is shown in figure 10.

In our computations, all model parameter ranges as well as torque limits etc. will be adapted to the technical restrictions of robotics components as soon as we enter the design phase of this robot.

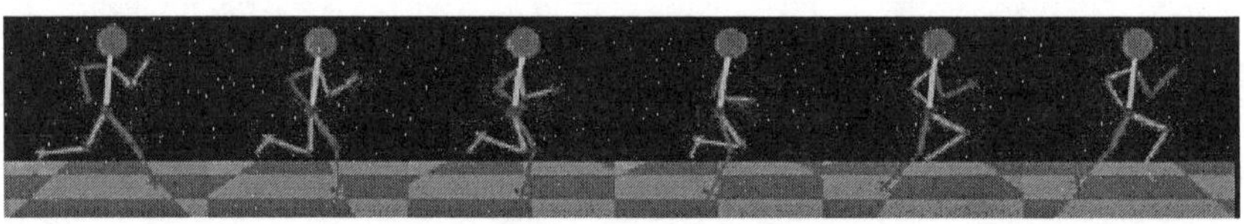

Fig. 10. Optimization of running biped

For the computations and the design of the bipedal running robot, stability and robustness of the motion play a crucial roles, in addition to the goal of moving efficiently at certain speeds. We therefore also include open-loop stability and robustness as optimization criteria for the biped computations.

Our latest biped optimizations have shown that it is possible to achieve stable running without feedback for human-like model parameters, and we therefor expect that we will also be able to heavily exploit natural stability by correctly designing the biped. Detailed descriptions of the bipedal robot model and the optimization results are given in two other publications currently in preparation.

8. Outlook

After validating the methodology with the prototype leg, the concept will be continually extended to two-legged systems. The design of a reflex network for a biped robot has already been started. The development and control of biped prototypes with increasing complexity including elastic elements will be the next steps.

9. Acknowledgements

The research project has been supported by the German Ministry of Education and Research (BMBF) within the framework *"Bionik – Innovationen aus der Natur"*.

References

1. B. Verrelst, R. Van Ham, B. Vanderborght, F. Daerden and D. Lefeber, *Autonomous Robots* **18**, pp. 201 (2005).
2. J. Vermeulen, D. Lefeber, B. Verrelst and B. Vanderborght, Trajectory planning for the walking biped lucy, in *7th International Conference on Climbing and Walking Robots*, 2004.
3. C. Sabourin, O. Bruneau and G. Buche, Experimental validation of a robust control strategy for the robot rabbit, in *IEEE Conference on Robotics and Automation (ICRA)*, 2005.
4. M. Wisse, D. G. E. Hobbelen, R. J. J. Rotteveel, S. O. Anderson and G. J. Zeglin, Ankle springs instead of arc-shaped feet for passive dynamic walkers, in *IEEE International conference on Humanoid Robots*, 2006.
5. S. Collins, A. Ruina, R. Tedrake and M. Wisse, *Science* **307**, pp. 1082 (2005).
6. T. Luksch, K. Berns and F. Flörchinger, Actuation system and control concept for a running biped, in *Fast Motions in Biomechanics and Robotics*, eds. M. Diehl and K. Mombaue (Springer Verlag, 2005).
7. H. G. Bock and K.-J. Plitt, A multiple shooting algorithm for direct solution of optimal control problems, in *Proceedings of the 9th IFAC World Congress, Budapest*, 1984.
8. D. B. Leineweber, I. Bauer, H. G. Bock and J. P. Schlöder, *Comput. Chem. Engng* **27**, 157 (2003).
9. D. B. Leineweber, A. Schäfer, H. G. Bock and J. P. Schlöder, *Comput. Chem. Engng* **27**, 167 (2003).
10. K. Mombaur, H. Bock, J. Schlöder and R. Longman, Open-loop stable solution of periodic optimal control problems, in *ZAMM*, 2005.
11. J. Albiez, T. Luksch, K. Berns and R. Dillmann, *The International Journal on Robotics Research, Sage Publications* **vol. 22**, pp. 203 (2003).

Using Virtual Model Control and Genetic Algorithm to Obtain Stable Bipedal Walking Gait Through Optimizing the Ankle Torque

Van-Huan Dau, Chee-Meng Chew and Aun-Neow Poo

Department of Mechanical Engineering, National University of Singapore,
10 Kent Ridge Crescent, 119260, Singapore
E-mail: dvhuan, mpeccm, mpepooan@nus.edu.sg
www.nus.edu.sg

This paper presents a new approach of controlling the biped robot to achieve stable dynamic walking by optimizing the stance-leg ankle behavior. In this study, the motion control in the sagittal plane of the robot is formulated using a control framework known as Virtual Model Control. Genetic Algorithm is used to search for optimal gains of the controllers such that stable walking gait with smooth velocity profile can be achieved. The stability of the biped is checked using the zero-moment-point criterion. Simulation results show that the obtained walking gait is stable and its forward velocity profile is very smooth.

Keywords: Bipedal walking, Virtual Model Control, Genetic Algorithms

1. Introduction

Stability is probably the most important issue in bipedal walking. Although many approaches have been introduced to tackle this problem, this is still an open question. One of the key characteristics of biped robot that make it difficult to be controlled is that bipeds are non-linear, high dimensional systems. Another important factor that causes most of the difficulties in maintaining stability is the *underactuation* between the feet and the ground. Human beings can walk stably in different kind of terrains because they can flexibly control the ankle joints and thus effectively control the contact between the feet and the ground. For robots, however, controlling the ankle joint to obtain stable walking is much more challenging. M. Guihard et al. (2002) introduced the notion of hip and ankle strategies coming from physiological observation on the human, in the control and design of a biped robot. K. Young Yi (1997) designed a biped which has a compliant ankle joint. The compliant joint gives a flexibility of joint and compliance when touching the ground thus making good contact between the feet and the ground. However, no research so far focuses on optimizing the ankle behavior to achieve a smooth, human-like walking and to regulate the walking speed of the biped.

In this paper, we propose a method in which Virtual Model Control (VMC)[1,2] is used as the control framework and ankle torque of the stance leg is formulated and optimized such that stable walking gait could be achieved with a smooth velocity profile and the ankle torque is as low as possible. Genetic Algorithm (GA)[4] is used as a searching engine to search for the optimal value of the key parameters which optimize the ankle torque. Zero-moment-point (ZMP)[5] has been widely used as a stability criterion in bipedal walking, hence it will be used to verify the feasibility of the biped's walking gait in this study.

This paper is organized as follows. Section two briefly describes NUSBIP, the biped being simulated. Section three shows the implementation of Virtual Model Control and explain the importance of controlling the stance ankle to achieve stable walking. In section four, the application of GA to search for the optimal value of ankle gains will be presented. Section five is devoted to walking algorithm and ZMP computation. Simulation results and discussion will be shown in section six. Finally, some conclusions will be made in section seven.

2. The Biped Robot

The robot considered in this study is a 7-link biped robot named NUSBIP, which was built and developed in our lab (see Fig. 1). The total weight is about 22.2 kg. The biped has 12 DOFs. Each leg has six active DOFs of which three DOFs at the hip (pitch, yaw, roll), one at the knee (pitch) and two at the ankle joint (pitch, roll). Each DOF is driven by a DC motor and integrated with an angular position sensor to measure the relative angle between two consecutive links. Each of the feet is equipped with four force sensors (two at the toe and two at the heel) to sense the contact forces between the feet and the ground.

3. Virtual Model Control (VMC) Implementation

Virtual model control is a motion control framework using virtual components to create desired joint torques. These virtual components such as springs, dampers, dashpots, masses, or any imaginable components are

Fig. 1. Picture of NUSBIP

placed at strategic positions to achieve certain task. Virtual forces generated by virtual components will act on the robot as if they really exist. Then these forces will be transformed into joint torques using the relation $\tau = J^T F$ to control the robot as desired. For more details on VMC, please refer to Ref. 1. Fig. 2 shows the mounting of virtual components on the body, the stance and the swing leg.

For the stance leg, the torques of the ankle, knee and hip joints $[\tau_a \ \tau_k \ \tau_h]^T$ can be derived from the virtual forces $[f_x \ f_z \ f_\theta]^T$ which are generated by the virtual components, by means of Jacobian:

$$\begin{bmatrix} \tau_a \\ \tau_k \\ \tau_h \end{bmatrix} = - \begin{bmatrix} L_1 c_a + L_2 c_{a+k} & L_1 s_a + L_2 s_{a+k} & 1 \\ L_2 c_{a+k} & L_2 s_{a+k} & 1 \\ 0 & 0 & 1 \end{bmatrix} \begin{bmatrix} f_x \\ f_z \\ f_\theta \end{bmatrix} = \begin{bmatrix} A & B & -1 \\ C & D & -1 \\ 0 & 0 & -1 \end{bmatrix} \begin{bmatrix} f_x \\ f_z \\ f_\theta \end{bmatrix} \tag{1}$$

Where f_x, f_z and f_θ are respectively the horizontal, vertical and rotational virtual forces applied on the body of the biped, $c_a = \cos(\varphi_a)$; $c_{a+k} = \cos(\varphi_a + \varphi_k)$; $s_a = \sin(\varphi_a)$; $s_{a+k} = \sin(\varphi_a + \varphi_k)$.

The effect of the virtual force f_x is to pull the biped to move forward while f_z and f_θ to maintain the body's height and pitch angle, respectively. Ideally, if the stance foot is fixed to the ground all the virtual forces can be set independently to achieve a desired gait. However, in humanoid robot the contact between the foot and the ground is unilateral and under-actuated. As a result, if the resultant ankle torque derived from the prescribed virtual forces is too high, the foot may be rotated, and this may lead to instability. In addition, when the motion of the swing foot is specified to meet the step length and speed requirement and the trunk is kept upright as in human

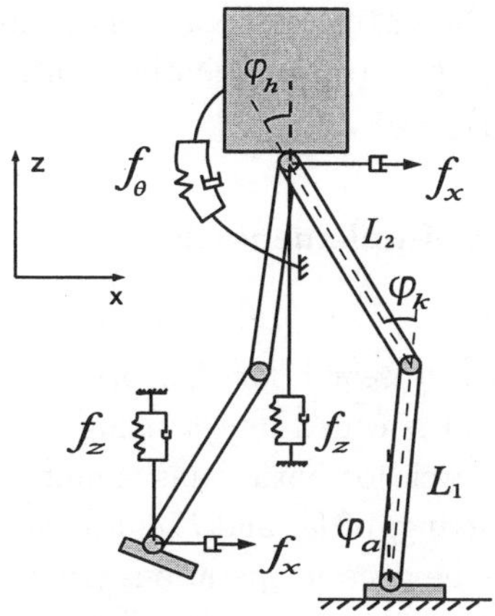

Fig. 2. The placement of virtual components on the biped.

walking, to obtain stable walking the ankle torque must be constrained. As the ankle torque is constrained, the virtual forces cannot be set arbitrarily. The ankle torque is formulated as follows:

$$\tau_a = K_p(v_{desired} - v) + K_d(\dot{v}_{desired} - \dot{v}) \tag{2}$$

Where v is current forward velocity of the body, $v_{desired}$ is the desired velocity. The optimal value of gains K_p and K_d will be searched by GA. Since ankle torque is constrained and f_z, f_θ are set independently to satisfy the desired hip's height and hip's pitch respectively, f_x must be solved from (1). Once f_x is known, the knee torque τ_k and hip torque τ_h are determined as in equations (3) and (4):

$$\tau_k = \frac{C}{A}\tau_a + (D - \frac{BC}{A})f_z + (\frac{C}{A} - 1)f_\theta \tag{3}$$

$$\tau_h = -f_\theta \tag{4}$$

For the knee torque in (3) to be determined, the denominator $A = -L_1 c_a - L_2 c_{a+k}$ must be non-zero. A equals to zero when $\varphi_k + 2\varphi_a = 180^o$ (for the case of $L_1 = L_2$). This condition only happens when the hip is at the same level as the ankle joint. However, this condition may not happen during walking, so we are sure that A is always different from zero.

For the swing leg, the torque of the joints is easier to control because all the virtual forces can be set independently. The Jacobian for the swing leg is derived from the swing foot to the hip. The joint torques are determined using the same relation $\tau' = (J')^T F$.

The parameters of the virtual components are spring stiffness (k_x, k_z), damping coefficient (b_x, b_z), torsional spring stiffness (k_α) and torsional damping coefficient (b_α).

4. Genetic Algorithm Implementation

4.1. *GA's parameters*

In this study, GA is used to search for the optimal value of ankle gains K_p and K_d in (2). Each gain is coded into a string of ten-bit binary number. However, we need to search for both gains simultaneously. In order to do that, we combined the strings of K_p and K_d into one single string of twenty-bit binary number. All the genetic operations (reproduction, crossover and mutation) will be performed on this string. The probability of crossover is selected to be 80%. The mutation probability is 2%.

4.2. *Fitness function*

The objective that we want to achieve is reflected through the fitness value. In this study, the interested parameters are the ankle gains of the ankle torques. These parameters will be searched by GA to find the optimal values satisfying the requirement that the horizontal velocity profile of the body is smooth, the average velocity is close to desired velocity and the ankle torque is minimized. The performance index is as followed:

$$P = w_1 \frac{1}{n} \sum_{i=1}^{n} (V_{\text{max}}(i) - V_{\text{min}}(i)) + w_2 \left| V_{\text{avg}} - V_{\text{desired}} \right| + w_3 r(t) \quad (5)$$

Where

$$r(t) = \begin{cases} \sum\limits_{i=1}^{m} \left| \frac{\tau_i}{\tau_{\text{ref}}} \right| & \text{if } |\tau_i| \geq \tau_{\text{ref}} \\ 0 & \text{Otherwise} \end{cases} \quad (6)$$

Where n is the number of walking steps, m is the number of integration steps and i denotes ith step. V_{max} and V_{min} are respectively the maximum and minimum speed within that step. V_{avg} is average speed of the biped which is equal to distance of walking divided by duration of walking. V_{desired} is desired speed of the biped. w_1, w_2 and w_3 are the weighting factors. τ_{ref} is the reference torque value.

The performance index must be minimized. However, as a convention, in GA the fitness function is always being maximized. Therefore, the fitness function is $F = \frac{1}{P}$.

5. Walking Gait

5.1. *Walking Gait Generation*

In this approach, thanks to the use of virtual model control, the walking gait is generated using simple, intuitive strategies instead of planning foot and hip trajectories. The advantage of this method is that it is simple, intuitive. The strategy adopted for the swing leg is as follows:

(i) Lift the swing foot to a specified height H_s. This job can be achieved easily by setting the equilibrium point for the vertical spring-damper.

(ii) Swing the leg forward with a desired speed to a specific horizontal end-position (this end-position is determined such that desired step length could be achieved). In this case, the spring setpoint is moving with the same desired speed.

(iii) Land down the swing foot while maintaining the foot angle such that the foot is parallel to the ground. This can be done by damping the swing foot in x direction.

5.2. *Stability Verification*

In this method, the ZMP[5] criterion is used to check for stability of the biped. It is noted that no desired ZMP trajectory is proposed in this approach and ZMP is only used to verify that the achieved walking gait is feasible. For the computation of ZMP position please refer to Ref. 3. The biped is stable if the computed ZMP is within the support polygon.

6. Simulation Results And Discussions

The specifications of the simulated biped are taken from a real biped, which was named NUSBIP and developed in our lab (Fig. 1). Table 1 summarizes its specifications and desired walking parameters. The simulation is done in Yobotics, a dynamic simulation software which allows the running of batches of simulation. Velocity profile of the biped when the ankle gains

Table 1. Specifications of the simulated biped.

Parameters	Value	Parameters	Value	Parameters	Value
Body mass	10.45 kg	Thigh length L_2	32 cm	Body height	65 cm
Thigh mass	2.68 kg	Shank length L_1	32 cm	Step length	35 cm
Shank mass	1.68 kg	Foot length	23 cm	Desired speed	0.75 m/s
Foot mass	1.52 kg	Foot thickness	5.5 cm	Desired acceleration	0 m/s^2

are optimal values ($K_p = 13.13, K_d = 1.06$) is shown in Fig. 3. From the figure we can see that the velocity profile is smooth and the average velocity is very close to 0.75 m/s which is the desired velocity of the biped. Fig. 4 shows the torque applied at the right ankle. It is observed that the maximum ankle torque is about 20 Nm. This maximum torque only appears at some time instants, most of the time the ankle toque is below 10 Nm which is quite small.

To assure the effectiveness of the proposed method we have to prove that the achieved walking gait is stable. In view of this, we use ZMP as a criterion to check for stability. The biped is stable if the computed ZMP falls within the foot support polygon. Since we consider only the sagittal plane, the support polygon becomes the upper and lower bounds for the ZMP. In single support phase, the lower bound is the heel and the upper bound is the toe of the supporting foot, respectively. In double support phase, the lower bound is the heel of the back foot and the upper bound is the toe of the front foot. It is shown in Fig. 5 that the ZMP trajectory is always within the lower and upper bound. Therefore, the resulting walking motion is stable. This result convinces that the stance-ankle is very important to achieve a stable walking motion especially when the swing leg motion has been chosen to obtain certain desired tasks. Fig. 6 shows the stick diagram of the biped walking on level ground, the solid line represents the right leg and dashed line represents the left leg.

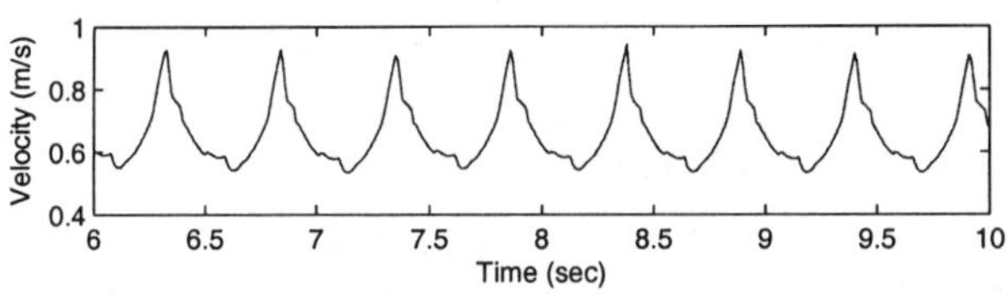

Fig. 3. Velocity profile of the biped when the ankle gains are optimal.

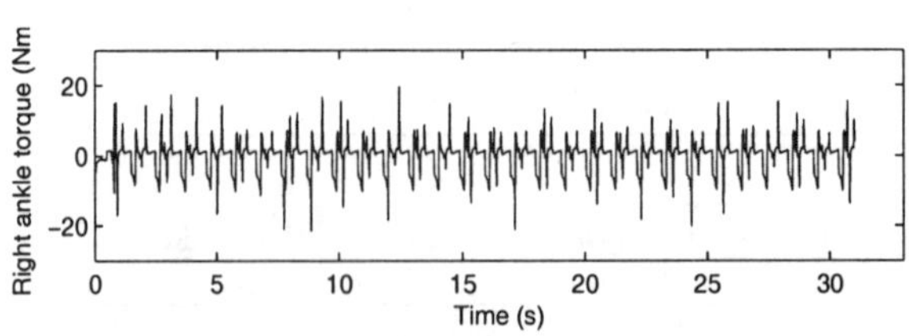

Fig. 4. The torques applied at the right ankle

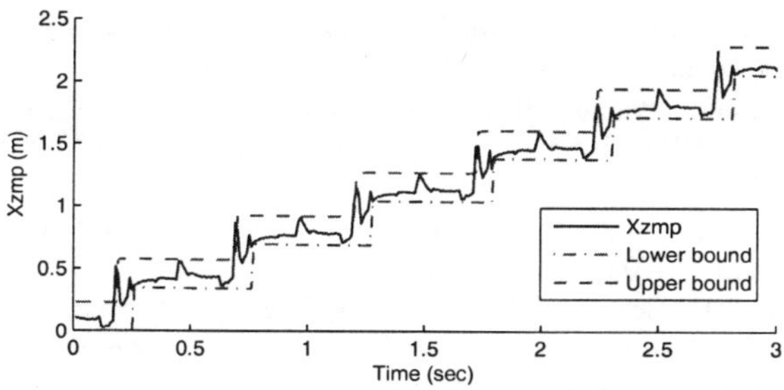

Fig. 5. ZMP trajectory of the biped

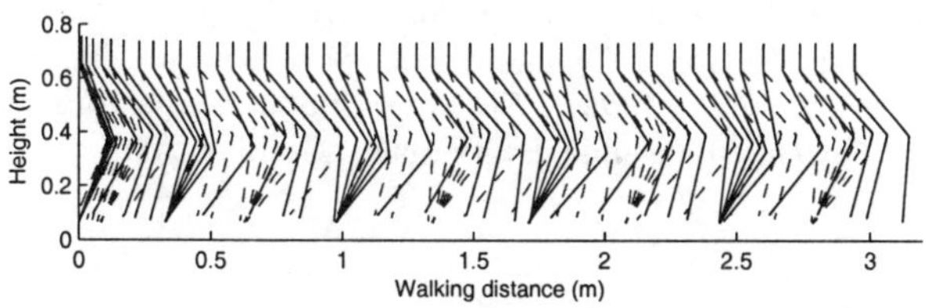

Fig. 6. The stick diagram of the biped depicted at 0.1s apart

7. Conclusion

In this paper, we have presented a method in which VMC is used as a control framework and GA is used to search for optimal value of ankle gains. The simulation results show that the optimal values of the ankle gains result in a smooth and stable motion. The average speed is also close to desired speed. Our study shows that when the motion of the swing leg is pre-defined, appropriately controlling the stance ankle torque is an effective way to achieve a stable and smooth walking motion. In this study, the generation of the walking gait is very simple and intuitive thanks to the use of Virtual Model Control.

References

1. J. Pratt, CM Chew, A. Torres, P. DilWorth, F. Pratt, Virtual Model Control: An Intuitive Approach For Bipedal Locomotion, *The International Journal Of Robotics Research*, **Vol. 20** No. 2 February 2001, pp 129-143.

2. CM Chew, G. A. Pratt, Dynamic bipedal walking assisted by learning, *Robotica* , (2000), **Vol. 20**, pp. 477-491.
3. C. Qiang Huang, Kazuhito Yokoi et al., Planning Walking Patterns for a Biped Robot, *IEEE Transactions on Robotics and Automation*, **Vol. 17**, No. 3, June 2001.
4. D.E. Goldberg, *Genetic Algorithm in Search Optimization and Machine Learning* (Addison-Wesley, MA), 1989.
5. Vukobratovic, B. Borovac, D. Surla, and D. Stokic. *Scinetific Fundamentals of Robotics 7. Biped Locomotion: Dynamics Stability, Control and Application* (Springer- Verlag, New York), 1990
6. Keon Young Yi, Walking of a Biped Robot with Compliant Ankle Joints, *Procs. of the 1997 IEEE International Conference on Intelligent Robots and Systems*, September 7-11, 1997, World Trade Center Atria, Grenoble, France.
7. Keon Young Yi and Yuan F. Zheng, Biped Locomotion by Reduced Ankle Power, *Proceedings of the 1996 IEEE International Conference on Robotics and Automation*, Minneapolis, Minnesota - April 1996.
8. M. Guihard and P. Gorce, Dynamics control of bipeds using ankle and hip strategies, *Procs. of the 2002 IEEE International Conference on Intelligent Robots and Systems*, Lausanne, Switzerland, October 2002.

WALKER SYSTEM WITH ASSISTANCE DEVICE FOR STANDING-UP

D. CHUGO*, W. MATSUOKA* and K. TAKASE*

Graduate School of Information Systems, The University of Electro-Communications, Chofu, Tokyo 182-8585, Japan
**E-mail: chugo, matsuoka, takase@is.uec.ac.jp*
http://www.taka.is.uec.ac.jp/

This paper proposes a walker system with power assistance device for standing up motion. Our system focuses on family use for aged person who needs nursing in their daily life. Our key ideas are two topics. The first topic is new assistance manipulator mechanism with four parallel linkages. Our proposed manipulator mechanism requires only smaller actuators and realizes rigid structure with lighter linkages comparing with general manipulator. Thus, we can design our assistance system compactly with low-cost using our mechanism. The second topic is the combination of force and position control. According to the patient's posture during standing up, our control system selects more appropriate control method from them. We use the reference of standing-up motion which is based on the typical standing up motion by nursing specialist for realizing the natural assistance. The performance of our proposed assistance system is verified by computer simulations and experiments using our prototype.

Keywords: Force assistance; Active Walker; Standing up motion; Force control; Position control,

1. Introduction

In Japan, the population ratio of senior citizen who is 65 years old or more exceeds 20[%] at January 2004 and rapid aging in Japanese society will advance in the future.[1] In aging society, many elderly people cannot perform normal daily household, work related and recreational activities because of decrease in force generating capacity of their body. Today, the 23.5[%] of elderly person who does not stay at the hospital cannot perform daily life without nursing by other people.[2] For their independent life, they need assistance system which enable them to perform daily life easily in their home even if their physical strength reduces.

Standing up motion is the most serious and important operation in daily life for elderly person who doesn't have enough physical strength.[3,4] In typical bad case, elderly person who doesn't have enough physical strength will cannot operate standing up motion and will falls into the wheelchair life or bedridden life. Furthermore, if once elderly person falls into such life, the decrease of physical strength will be promoted because he will not use his own physical strength.[5] Therefore, force assistance system is required to use part of the remaining strength of the patient for standing-up motion in order not to reduce their muscular strength.

In previous works, many researchers developed power assistance devices for standing up motion. However, these devices are large scale and it is difficult for family use.[6,7] Furthermore, these devices assist all necessary power for standing up and they do not discuss the using the remaining physical strength of patients. Therefore, there is a risk of promoting the decrease of their physical strength.

In this paper, we develop a walker system with power assistance device for standing up motion. Our system is based on a walker which is popular device for aged person in normal daily life and realizes the standing up motion using the support pad which is actuated by the manipulator with three degrees of freedom. For using the remaining physical strength, our system uses the motion pattern which is based on the typical standing up motion by nursing specialist as control reference.[8]

Our key ideas are two topics. The first is new assistance manipulator mechanism with four parallel linkages which enables the system to be rigid and compact. The second is the combination of force and position control. Using our control scheme, the patients can stand up with fewer loads and can use their own remaining physical strength during the motion. We verify the performance of our proposed assistance system through computer simulations and experiments using our prototype.

2. System Configuration

2.1. *Assistance Mechanism*

Fig.1 shows our proposed assistance system. The system consists of support pad with three degrees of freedom and the walker system. (Fig.1(a)) The support pad are actuated by our new assistance manipulator mechanism with four parallel linkages.

Fig.1(b) shows the frame-kinematic model of our assistance manipulator. The position of the support pad (2DOF, red lines in Fig.1(a)) is coordinated by Actuator 1 and Actuator 2 which are equipped on O point.

Actuator 1 drives Link1 (α) and Actuator 2 drives Link 2 (β). Using four parallel linkages mechanism, two actuators can generate the position of support pad. The inclination of the support pad (1DOF, blue lines in Fig.1(a)) is coordinated by Actuator 3 which is equipped on P point.

The advantages of our proposed mechanism are two topics. The first is that two main actuators (Actuator 1 and 2) are required to keep only own weight. In general manipulator, the actuator of lower part supports not only weight of linkages but also actuators of upper part. Therefore, the actuators of lower part are required high output traction and tend to be heavy. On the other hand, using our mechanism, main actuators are mounted on the walker body (O point) and they are required to keep only weight of linkages. As the result, we can use smaller actuators for our assistance manipulator.

The second is parallel linkage mechanism. The parallel linkage mechanism is strong for a twist load and we can use lighter linkages for our proposed manipulator comparing with the general one. Using our proposed mechanism, we can use smaller actuators and lighter linkages, and our system realizes compact design with low cost.

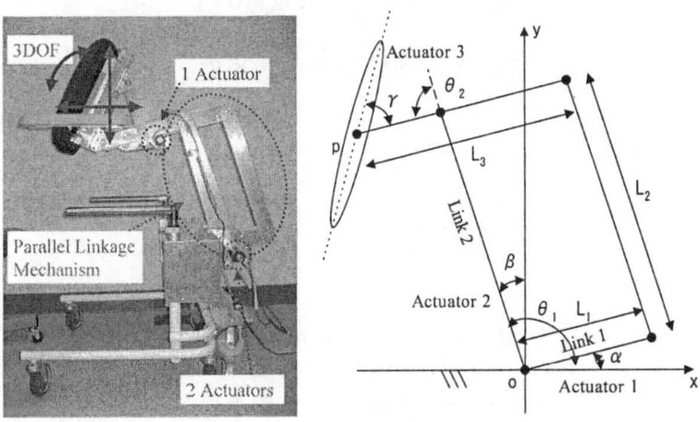

(a) Prototype system (b) Parallel Linkage Mechanism

Fig. 1. Our assistance system

2.2. *Kinematics*

In this section, we derive the inverse kinematics of our proposed linkage mechanism. Using our proposed mechanism, the position of P point (x_p, y_p) is derived as follows;

The first, we set angular values and length of linkages as in Fig.1(b).

$$\theta_1 = \frac{\pi}{2} + \beta, \qquad \theta_2 = \frac{\pi}{2} + \alpha - \beta \tag{1}$$

$$l_1 = L_2, \qquad l_2 = L_3 - L_1 \tag{2}$$

Now, we consider the geometric relationships among the position of P point and these angular values, we can derive (3) and (4). From (3) and (4), θ_2 is (5)

$$x_p = l_1 \cos\theta_1 + l_2 \cos(\theta_1 + \theta_2) \tag{3}$$

$$y_p = l_1 \sin\theta_1 + l_2 \sin(\theta_1 + \theta_2) \tag{4}$$

$$\theta_2 = \arccos\left(\frac{x_p^2 + y_p^2 - l_1^2 - l_2^2}{2l_1 l_2}\right) \tag{5}$$

We set k_1 and k_2 as (6), (3) and (4) are expressed as (7) and (8).

$$k_1 = l_1 + l_2 \cos\theta_2, \qquad k_2 = l_2 \sin\theta_2 \tag{6}$$

$$x_p = k_1 \cos\theta_1 - k_2 \sin\theta_1 \tag{7}$$

$$y_p = k_1 \sin\theta_1 + k_2 \cos\theta_1 \tag{8}$$

Furthermore, we set r and γ as (9), k_1 and k_2 are expressed as (10). Using (10), (7) and (8) are expressed as (11) and (12).

$$r = \sqrt{x_p^2 + y_p^2}, \qquad \tan\gamma = \frac{k_2}{k_1} \tag{9}$$

$$k_1 = r\cos\gamma, \qquad k_2 = r\sin\gamma \tag{10}$$

$$x_p = r\cos(\gamma + \theta_1) \tag{11}$$

$$y_p = r\sin(\gamma + \theta_1) \tag{12}$$

From (11) and (12), we can derive (13).

$$\tan(\gamma + \theta_1) = \frac{y_p}{x_p} \tag{13}$$

Thus, θ_1 is (14). Using (5) and (14), we can control our manipulator.

$$\theta_1 = \arctan\left(\frac{y_p}{x_p}\right) - \arctan\left(\frac{l_2 \sin\theta_2}{l_1 + l_2 \cos\theta_2}\right) \tag{14}$$

3. Motion Control

3.1. *Motion by Nursing Specialist*

In previous study, a lot of standing up motions for assistance are proposed. Kamiya[9] proposed the standing up motion which uses remaining physical strength of the patients maximum based on her experience as nursing specialist. In our previous work, we analyze this standing up motion and find that Kamiya scheme is effective to enable standing up motion with smaller load.[8]

We assume the standing up motion is symmetrical and we discuss the motion as movement of the linkages model on 2D plane.[10] We measure the angular values among the linkages, which reflect the relationship of body segments. The angular value is derived using the body landmark as shown in Fig.2(a).

From measuring results, the trunk needs to incline to forward direction during lifting up from chair as shown in Fig.2(b). Y-axis shows the angular value (Pelvis and trunk, knee, ankle) and X-axis shows the movement pattern ($\hat{s}$)[11] which means the ratio of standing up operation as (15).

$$\hat{s} = \frac{t}{t_s} \tag{15}$$

where t_s is required time to the standing up operation and t is present time.

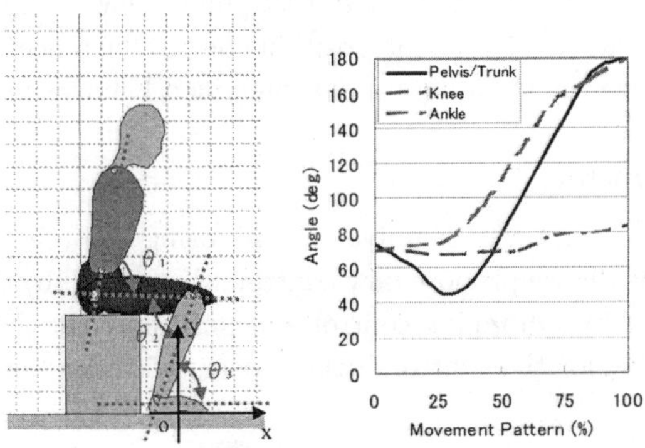

(a) Coordination (b) Angular Value

θ_1, θ_2 and θ_3 show the angles of pelvis/trunk, knee and ankle, respectively.

Fig. 2. Standing-up motion with Kamiya scheme

3.2. *Derivation of Control Reference*

In this section, we derive the control reference of our assistance system which can realize the standing up motion proposed by Kamiya using a computer simulation. Fig.3(a) shows the simulation setup. The parameters are chosen from a standard body data of adult male[12].[13] We use the Working Model 2D as a physical simulator and MATLAB as a controller. Both applications are linked by Dynamic Data Exchange function.

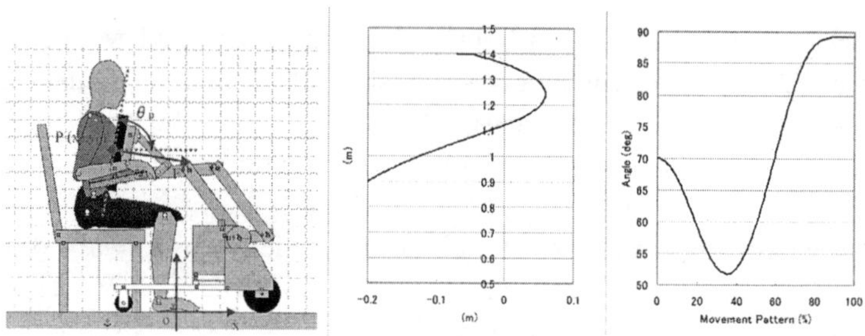

(a) Simulation Setup (b) Position Reference (c) Angle Reference

Fig. 3. Derived control references

From these simulation results, Fig.3(b) shows position references of the support pad and Fig.3(c) shows reference angle. The coordination is defined as shown in Fig.3(a). In Fig.3(b), the start point is lower and the end point is upper. Using these tracks as the position control reference, our assistance system can realize the standing up motion which Kamiya proposes.

3.3. *Force Control*

For realizing the motion of Kamiya scheme using the remaining physical strength of patients, we propose new control scheme as (16). Proposed control scheme combines dumping control and position control. The dumping control is suitable for the control of the objects with contact.[14]

$$\begin{cases} v_i = v_i^{ref} - B\left(F - F_0\right) - K\left(x_i - x_i^{ref}\right) & \cdots \quad (if\ F > F_0) \\ v_i = v_i^{ref} - K\left(x_i - x_i^{ref}\right) & \cdots \quad (if\ F \leq F_0) \end{cases} \tag{16}$$

where F is the applied force on the support pad as shown in Fig.3(a) F_0 is the threshold which selects force or position control. v_i^{ref} is the velocity reference and x_i^{ref} is the position reference from track references in

Fig.3(b) and (c). v_i is actucal control reference. i is identification number of actuators. $(i = 1, \cdots, 3)$ B and K are constants.

Using this control scheme, when the patient does not use own physical strength, the applied force on the support pad will be large and system will assist him more slowly. Therefore, the patient can use own physical strength easily.

4. Experiment

4.1. *Computer Simulation*

We verify the performance of our control scheme by the computer simulation. In this experiment, the human model stands up with Kamiya motion and our assistance system assists him using our proposed control scheme. Furthermore, we compare the result by our proposed scheme with the result using only the position control reference.

We use the control references as shown in Fig.3(b) and (c) which are derived from standing up motion with Kamiya scheme in previous section. The simulation parameters are chosen from Table 1. The coordination is defined in Fig.3(a). Furthermore, we set F_0=15[N] as a threshold which is derived experimentally.

Fig.4 shows the simulation results. Fig.4 is standing up motion using our proposed assistance control. Allows in Fig.4 show the applied assistance force to the patient. The support pad applies the force vertically to the chest and the hand. Using our proposed control scheme, we verified that our assistance system realizes the natural standing up motion.

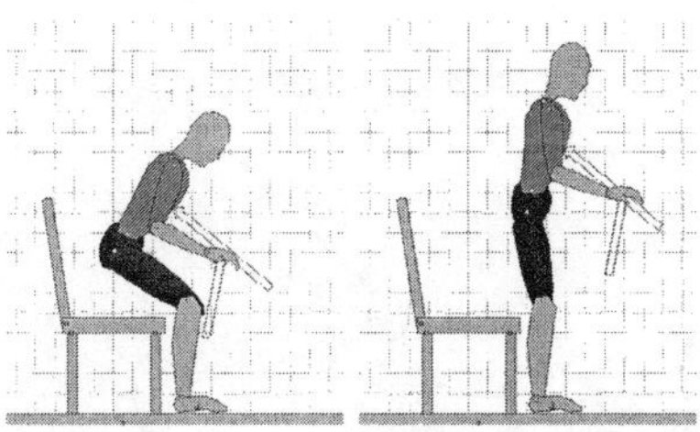

Fig. 4. Simulation results

Table 2 shows the maximum output and output power of each joint. From these results, our proposed control scheme reduces maximum output into 0.5[Nm/kg] comparing with the result by the position control. In general, if the applied load to each joint is heavier than 0.5[Nm/kg], it is difficult to stand up for the elderly person. Therefore, the patient can adapt this motion using his remaining physical strength.

Furthermore, with our proposed control scheme, the workload of each joint is maintained comparing with the result by the position control. During standing up motion, the patient is required to use his 92[%] of physical strength comparing with physical strength without the force assistance control. This means our assistance system can use part of his remaining strength in order not to reduce muscular strength. Therefore, we can verify that the system selects more appropriate control method and our proposed control scheme is effective.

Table 1. Simulation results

		Pelvis/Trunk	Knee	Ankle
Only Position Control	Peak(Nm/kg)	0.48	0.59	0.38
	Workload(Ws)	29.0	39.3	31.0
Proposed Control	Peak(Nm/kg)	0.49	0.50	0.38
	Workload(Ws)	26.9	36.3	28.8

4.2. *Experiment*

Here, we verify the performance of our prototype system by the experiment. In this experiment, our prototype system assists the patient with control references shown in Fig.3(b) and (c). As the result of the experiment, our system can assist the patient as shown in Fig.5. The height of the patient is 170[cm] and the system lifts him at 30[sec].

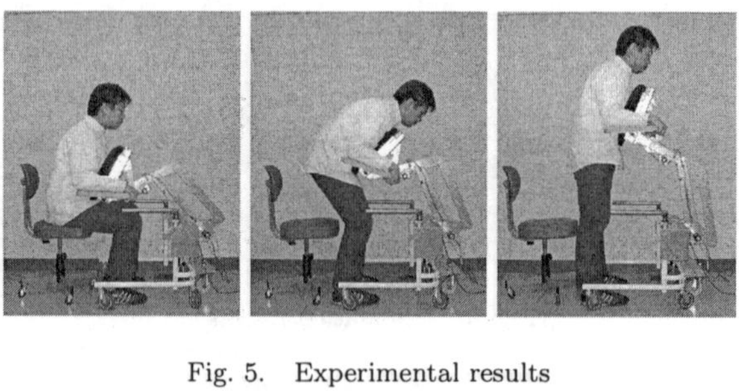

Fig. 5. Experimental results

Fig.6 shows the tracks of the position of the patient's waist, knee and ankle joint and their control references. From Fig.6, both tracks are al-

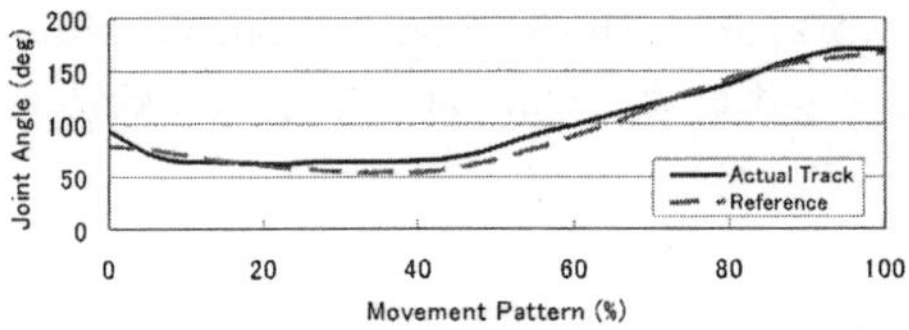

(a) Pelvis / Trunk

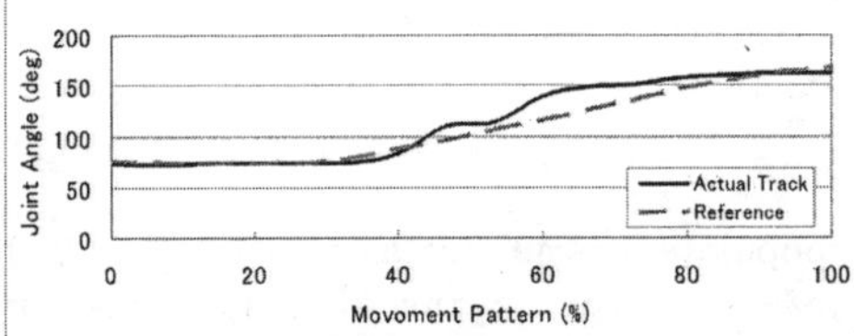

(b) Knee

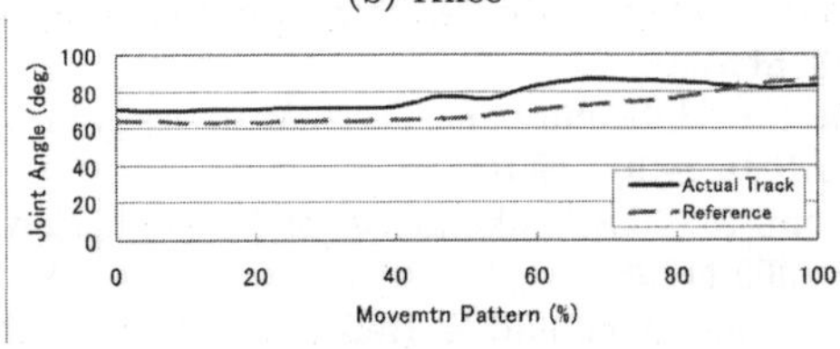

(c) Ankle

Fig. 6. Angular value of each joint

most same line and this means our assistance system realizes the natural standing-up motion by nursing specialist.

5. Conclusion

In this paper, we develop the novel assistance system for the standing up motion. Our system focuses on family use and system can assist the elderly person using part of their remaining strength, in order not to reduce muscular strength. In order to fulfill this condition, we propose new assistance manipulator mechanism with parallel linkages. Our developed mechanism enables the assistance system to be rigid and compact with low costs. Furthermore, we design the novel control scheme which combines the dumping and the position control. According to the posture of the patient during standing up motion, our control system can select more appropriate control method from them. Our assistance system realizes the natural standing up motion by nursing specialist and it is effective to assist the aged person to stand up without reducing their muscular strength.

In our future work, we will develop the active walker function, which enables to assist the patients during standing, walking and seating motion continuously.

Acknowledgments

This study is supported by *France Bed Medical Home Care Research Subsidy Foundation.*

References

1. Japan current population estimates as of october 1, 2004 (2004), `http://www.stat.go.jp/english/data/jinsui/2004np/index.htm`.
2. Japan annual reports on health and welfare 2001 social security and national life (2001), `http://www.mhlw.go.jp/toukei/saikin/hw/k-tyosa/k-tyosa01/4-3.html`.
3. N. B. Alexander, A. B. Schultz and D. N. Warwick, *J. of Geometry: MEDICAL SCIENCES* **46**, M91 (1991).
4. M. A. Hughes and M. L. Schenkman, *J. of Rehabilitation Research and Development* **133**, 409 (1996).
5. M. Hirvensalo, T. Rantanen and E. Heikkinen, *J. of the American Geriatric Society* **48**, 493 (2000).
6. K. Nagai, I. Nakanishi and H. Hanabusa, Assistance of self-transfer of patients using a power-assisting device, in *Proc. IEEE Int. Conf. on Robotics and Automation (ICRA'03)*, (Taipei, Taiwan, 2003).
7. A. Funakubo, H. Tanishiro and Y. Fukui, *J. of the Society of Instrument and Control Engineers* **40**, 391 (2001).
8. D. Chugo, E. Okada, K. Kawabata, H. Kaetsu, H. Asama, N. Miyake and K. Kosuge, Force assistance control for standing-up motion, in *Proc. IEEE/RAS-EMBS Int. Conf. on Biomedical Robotics and Biomechatoronics (BioRob'06)*, (Pisa, Italy, 2006).
9. K. Kamiya, Development and evaluation of life support technology in nursing, in *Proc. 7th RACE Symp., Research into Intelligent Artifacts for the Center of Engineering*, (Chiba, Japan, 2005).
10. S. Nuzik, R. Lamb, A. Vansant and S. Hirt, *J. of Physical Therapy* **66**, 1708 (1986).
11. P. O. R. R. W. M. M. Schenkman, R. A. Berger and W. A. Hodge, *J. of Physical Therapy* **70**, 638 (1990).
12. Human body properties database (digital human research center of aist) (2006), `http://www.stat.go.jp/english/data/jinsui/2004np/index.htm`.
13. K. Omori, Y. Yamazaki, H. Yokoyama, U. Aoki, M. Kasahara and K. Hiraki, *Sogo Rehabilitation* **30**, 167 (2001).
14. T. Sugihara, K. Kawabata, H. Kaetsu, H. Asama, K. Kosuge and T. Mishima, Development of a reasonable force sensor for a standing-up and sitting motion support system, in *Proc. of Robotics and Mechatronics Conf. 2004*, (Nagoya, Japan, 2004).

Advances in Humanoid Soccer Robots

A Distributed Embedded Control Architecture for Humanoid Soccer Robots.

Carlos Antonio Acosta Calderon, Changjiu Zhou, Pik Kong Yue,
Mike Wong, and Mohan Rajesh Elara

*Advanced Robotics and Intelligent Control Centre,
School of Electrical and Electronic Engineering,
Singapore Polytechnic, 139651, Singapore
Email: {CarlosAcosta, ZhouC}@sp.edu.sg
www.robo-erectus.org*

This paper presents an embedded control architecture developed for a soccer-playing humanoid robot, namely Robo-Erectus Junior. Robo-Erectus is a project developed in the Advanced Robotics and Intelligent Control Centre of Singapore Polytechnic that will participate in the KidSize category in the Humanoid League of RoboCup 2007. The Robo-Erectus project has reached a new stage with this latest version of the robot. The new mechanical, electronic design, and embedded control architecture are described in this paper. The limitations of size force us to implement this architecture in an embedded network. The features of this new version are described in detail focusing on the modules of gait generation, vision, behaviour control, and communication.

Keywords: Behaviour Control, Humanoid Robot, Soccer Robots

1. Introduction

The Robo-Erectus project is developed in the Advanced Robotics and Intelligent Control Centre of Singapore Polytechnic. Robo-Erectus is one of the foremost leading soccer-playing humanoid robots in the RoboCup Humanoid League.[1] Robo-Erectus has collected several awards since its first participation in the Humanoid League of RoboCup in 2002. Some of the awards include the 2nd place in the Humanoid Walk competition at the RoboCup 2002, the 1st place in the Humanoid Free Performance competition at the RoboCup 2003. In 2004, Robo-Erectus won the 2nd place in Humanoid Walk, Penalty Kick, and Free Performance.

The competitive level of the humanoid teams in RoboCup has increased and the challenges in the competition are becoming tougher every year. For the Robo-Erectus project to keep to as one of the leading teams is imperative to have a robust and reliable control system that allows the robot to exploit its full capabilities. This paper provides the inside of the architecture for the Robo-Erectus Junior which controls not just a robot but display the emergent behaviour of a team.

The development of Robo-Erectus has gone through many stages either in the design of its mechanical structure, electronic control system and gait movement control. The latest version of Robo-Erectus, namely Robo-Erectus Junior, has been designed to cope with the complexity of a soccer game. Robo-Erectus is able to perceive different colours and to track them. The robot also contains a dedicated processor used to control its behaviour, wireless communication with the control PC and teammates, and a subsystem to control sensors and actuators. For more detailed information about the Robo-Erectus humanoid soccer robots, please refer to the team's website www.robo-erectus.org.

This paper describes the current state of the project developed for the 2007 RoboCup competition. The rest of the paper is organized as follows. In the next Section, details of the mechanical and electronics design are presented. Section 3 describes the software developed: the image processing, the hierarchy of the control system for the robot behaviour, and the infrastructure needed to support a team of soccer playing robots, respectively. Finally, some concluding remarks are presented in Section 4.

2. Hardware Design

The Robo-Erectus project aims to develop a low-cost fully-autonomous humanoid platform that could be used for competition, education, and research purposes.[2] The previous generations of humanoid soccer robots, namely RE40I, RE40II, RE40III and RE70; have provided us with a robust knowledge about the hardware and software design. This Section provides an inside to the latest development namely Robo-Erectus Junior. Appendix A presents a compendium of the specifications of the Robo-Erectus Junior.

2.1. *Mechanical Design*

Figure 1(a) shows the design of the humanoid robot Robo-Erectus Junior. The skeleton of the robot is constructed with aluminum braces. The head and arms of the robot are made of plastic. Despite its simplicity, the mechanical design of the robot is robust and lighter weight in comparison with its predecessors. Its human-like body has a height of 50cm and weight of just 3.2kg, including batteries.

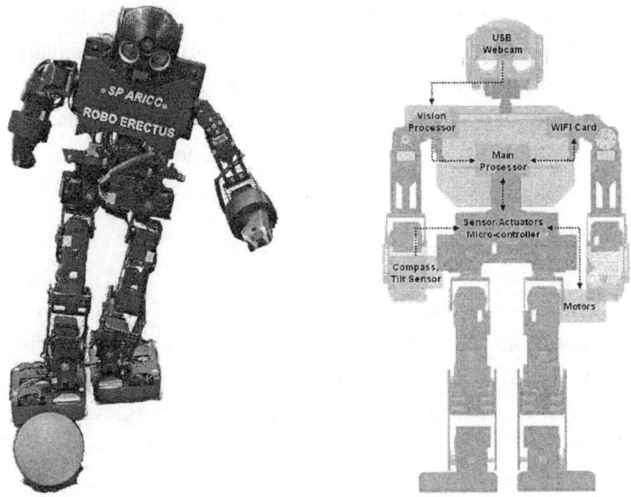

(a) (b)

Fig. 1. Robo-Erectus Junior, the latest generation of the family Robo-Erectus.

Robo-Erectus Junior has a total of 24 degrees of freedom. Further details of the body parts and their associated degrees of freedom can be found in Appendix A. Each degree of freedom uses as actuator a *Dinamixel DX-117 Digital Servomotor.* These servomotors have a typical torque of $28.89 kg \cdot cm$ and a speed of $0.172 sec/60°$. Each smart actuator has a micro-controller in charge of receiving commands and monitoring the performance of the actual motor. An RS485 serial network connects all the servomotors to a host processor which, sends positions and receives the current data (angular positions, speed, voltage, and temperature) of each actuator.

2.2. *Electronic Design*

Figure 1(b) shows the network of three processors and the main devices that controls the Robo-Erectus. The use of processors for particular tasks improves the performance of the system. Each processor is dedicated to different tasks:

(1) The main processor coordinates all behaviors of the robot, collects the sensor information and sends the data to the motors, and communicates by WIFI with the other robots and the main PC.
(2) The vision processor processes the image obtained from the USB camera. This processor is connected with the main processor by RS232.

(3) The Sensor/Actuator dual micro-controller receives the motor commands from the main processor. These commands are validated and finally sent to the servomotors by RS485. The motor feedback is collected and sent back to the main processor every 16.6ms. This system is also responsible to collect the sensors' values to be sent to the main processor.

Robo-Erectus has four main sensors: a *camera* to capture images, a *tilt* sensor to recognize whether it is standing or falling down, a *compass* to detect its own orientation, and finally a couple of *ultra–sonic* sensors to measure distance from front and rear objects. As mentioned earlier, the servomotors send back the feedback data i.e. angular positions, speed, voltage, and temperature. To communicate with its teammates, Robo-Erectus uses a wireless network. The WIFI interface is connected to the main processor. The connection of the main building blocks can be seen in Fig. 1(b).

Finally, Robo-Erectus is powered by two high-current Lithiumpolymer rechargeable batteries, which are located in each foot. Each battery cell has a weight of only 110g providing 12V which is able to operate about 20 minutes.

3. Control Architecture

The control architecture for each robot is distributed in a network of three processors. The main processor runs Linux as operative system. Due to the limitations of the system the footprint of the *Embedded Linux* is very small, but yet powerful to permit to take all the advantages of this operative system, such as threading, networking, connectivity, and so forth. The team behaviour in a game emerges form interaction of the robots and the role selection. The functionality of Robo-Erectus is divided in several software modules, which are described in detail below.

3.1. *Motion Control*

Our team has studied different approaches to motion control, including kinematics, dynamics, fuzzy logic, neural networks and genetic algorithms.[3–6] How to generate a dynamically stable gait for the humanoid soccer robots with consideration of various constraints is still an important research topic in this area.

Our latest approach employs the *Estimation of Distribution Algorithm EDA* for gait optimization.[7,8] This approach speeds up the searching in a highly dimensional coupling space constructed by the permutation of the optimization parameters to establish a periodic orbit in biped locomotion. Based on the maximum entropy principle, we also developed a *Factorized Distribution Algorithm FDA* based gait optimization method to better understand how information are transferred between these parameters so that we may progress toward better understanding human locomotion and extend the results to design of humanoid robots.[6,9] We used the EDA and FDA have been successfully used to generate biped gaits that satisfy a criterion. The gaits have been efficiently used to drive the humanoid robot Robo-Erectus Junior.

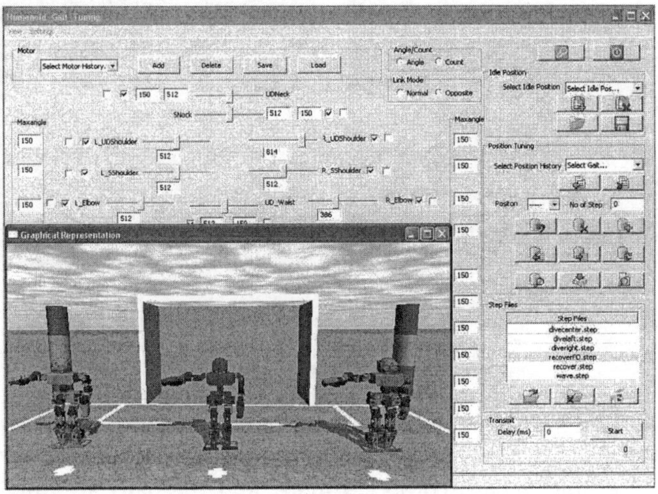

Fig. 2. The suit of in-house development tools.

These gaits and other behaviours are programmed by using a IDE software, which controls from a single servo position to a full complex behaviour. This IDE allows that any changes to the gait can be implemented and tested in short turnaround time (See Fig. 2). In addition, the data of each joint can be monitored and analyzed in real time. In order to be able to design behaviors without access to the real hardware, we implemented a physics-based simulator for the robots. This simulation is based on the Open Dynamics Engine.[10]

3.2. *Image Processing*

The vision system provides the main source of information about the environment for the robot. The system consists of a 70° camera mounted on a pan-tilt system that allows the robot to scan 240° wide. The computation of the vision system occurs in a dedicated embedded system, only the output is transmitted to the main system.

The main process on the image is *colour segmentation* based on the HSI colour space. Each image's line is scanned pixel by pixel. During the scan, the pixels are classified by color. A look-up-table is employed to classify the pixels by colours described by an HSI range for each colour.

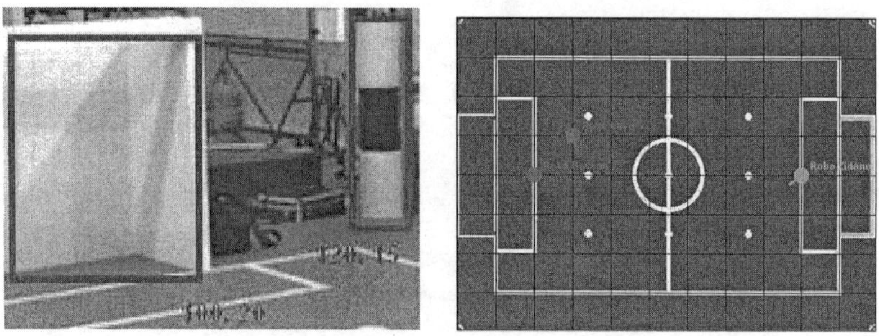

Fig. 3. The image processing

In a multistage process we discard insignificant colored pixels and detect colored objects (See Fig. 3). Recognition algorithms for the most important features (i. e. lines, goals, corners, robots, and the ball) are implemented avoiding a wrong identification. In addition, previous observations of these objects provide confidence for their identification.

Finally the system estimates the position of each object in an egocentric frame (distance from the robot and its orientation). This egocentric map is merged along with sensory data and teammate information into a allocentric map, which provides a robust world representation.

The gaze control uses a *fovea* as main tracking area. When the object of interest is found inside the fovea the pan and tilt motor will not move. On the contrary, if the object of interest is outside the fovea, but still in the image, the pan and tilt motor will move to compensate the distance from the center of the fovea. If the object is not found on the image, it is reported and the motor will not move.

A system helps to turn the vision system when the robot is used in different light conditions. This system uses a remote connection via wireless network to turn the thresholds that determine a particular colour.

3.3. *Behaviour Control*

The module provides the control and strategy for the autonomous mode of the robot when playing. A framework of *hierarchical reactive behaviours* is the core of this control module.[11] This structure restricts interactions between the system variables and thus reduces the complexity. The control of the behaviours happens in three layers: skill, reactive, and planning layer (See Fig. 4(a)).

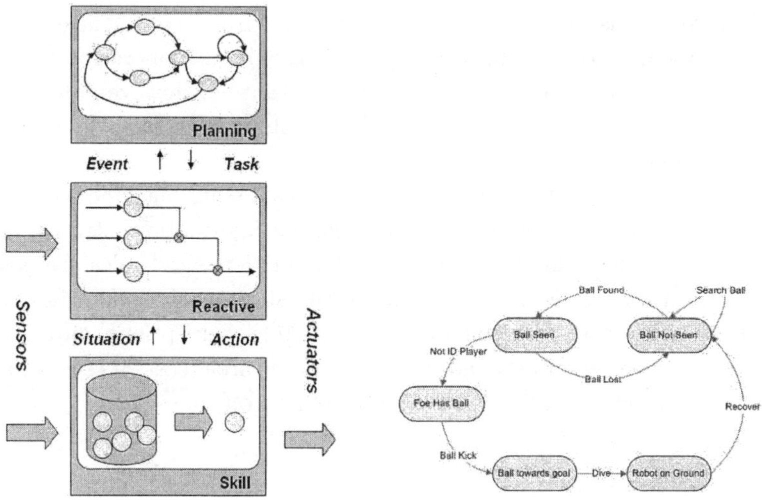

(a) (b)

Fig. 4. The behaviour control has been implemented as a hierarchical reactive framework, which is organized in three layers.

The skill layer controls the servos, monitors targets, actual positions, and motor duties. The skill layer receives actions from the reactive layer and convert them into motor commands. After performing the motor commands a feedback is sent back to the reactive layer.

The reactive layer implements the robot behaviours like walking, kicking, getting-up, and so forth. This layer determines the parameters for the behaviour and these parameters can be adapted on time. This makes possible to correct deviations in the actual data and to account for changes in the environment by using the sensor feedback. Each of these behaviours consists of several actions, which are sent to the skill layer. The selection of the behaviours depends on the desire task of the planning layer. The behaviours in this layer are implemented as a subsumption architecture to enable the robot to satisfy the task while it can navigate safely in the environment.

The planning layer used the behaviours of the reactive layer to implement some soccer skills like approaching the ball, dribbling, attacking and defensive behaviors. The planning layer guide the robot to coordinate its efforts with the teammates to score goals and defend their goal. The behaviours at the planning layer are abstract goals. These abstract goals are passed to the reactive layer to be sent to the actuators.

Each robot has the same hierarchical reactive framework. The emergent team behaviour is the product of communication and cooperation between players. In the game each robot has a particular role. However, the robots can dynamically switch theirs roles. The robots change theirs roles based on a particular strategy.

Each role is defined in the planning layer as a state-machine with transitions triggered by a combination of sensorial information, strategy, and messages received from the teammates. The implemented roles are *goalie, defender,* and *attacker*; Fig. 4(b) shows the state-machine for the goalie role. Each robot takes decisions based on its role, perceptions, and the information it receives from teammates. The robot broadcasts its state to the teammates whom use this information to decide on role.

3.4. *Communication*

Robo-Erectus is equipped with wireless network adapters. The robots communicate with each other to negotiate roles and to share perceptions. The wireless communication is also used to transmit information to an external computer for recording and visualization. A monitor system was developed to follow the state of the robots during a game. This monitor system also broadcasts to the players information about the game, e.g. kickoff, finish, penalty, free-kick. The information received from each robot can be recorded for analysis purposes.

4. Conclusion

In this paper, we introduced the state-of-art of the Robo-Erectus project. Robo-Erectus Junior is an autonomous humanoid robot with a network of three embedded systems, 24 degrees of freedom, and several kinds of sensors that serves as a platform of education, edutainment, and research issues. The latest version of the Robo-Erectus holds several advantages in contrast with the previous generations, e.g. faster, robust control, gait improvement, vision. The new features prepare the Robo-Erectus for the 2007 RoboCup competition. New development tools were conceived from the gained experience of the previous versions of the Robo-Erectus. In addition, the improvement in the robot platform allows a more robust and efficient performance of the robot in the autonomous mode. Research with this platform has lead to develop a new approach to better control of humanoid robots in particular for RoboCup.

Appendix A. Mechanical Specifications of Robo-Erectus Junior

1. Physical Specifications.

Weight	Dimensions			Speed	
	Height	Width	Depth	Walking	Running
3.2kg	480mm	270mm	150mm	2mts/min	—

2. List of Degrees of Freedom.

Joint	Roll	Pitch	Yaw
Head		✓	✓
Body		✓	✓
Shoulder	✓	✓	✓
Elbow		✓	
Hip	✓	✓	✓
Knee		✓	
Ankle	✓	✓	

3. Sensors of the Robo-Erectus Junior.

Sensor	Details
Camera	320x240 Resolution 30fps.
Compass	1° heading accuracy.
Tilt	Two dimensions.
Sonar	Distance range 5cm–250cm.

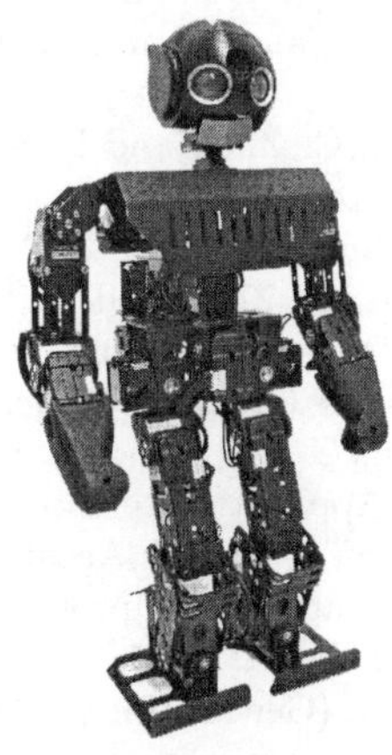

4. Specifications of the boards of the Robo-Erectus Junior.

Features	Main Processor	Vision	Sensor/Actuator
Processor	Intel ARM XScale	Intel ARM XScale	Dual PIC18F8720
Speed	400Mhz	400Mhz	25Mhz
Memory	16MB	32MB	8KB
Storage	16MB	16MB	256KB
Interface	RS232, WIFI	RS232, USB	RS232, RS485

Acknowledgements

The authors would like to thank staff and students at the Advanced Robotics and Intelligent Control Centre (ARICC) and higher management of Singapore Polytechnic for their support in the development of our humanoid robots. In particular to Chin Keong Ang, Weiming Yuan, Edric Lee Lai Fatt, Guohua Yu, Weijie Ye, Yang Bo and Aung Aung Kyaw for their contributions to this project.

References

1. H. Kitano and H. Asada, The RoboCup humanoid challenge as the millennium challenge for advanced robotics, *Advanced Robotics* **13**, (2000).
2. C. Zhou and P. Yue, Robo-Erectus: A Low Cost Autonomous Humanoid Soccer Robot, *Advanced Robotics* **18**, (2004).
3. Z. Tang, C. Zhou and Z. Sun, Gait synthesizing for humanoid penalty kicking, in *Proc. 3rd International DCDIS Conference on Engineering Applications and Computational Algorithms*, (Ontario, Canada, 2003).
4. C. Zhou and Q. Meng, Dynamic balance of a biped robot using fuzzy reinforcement learning agents, *Fuzzy Sets and Systems* **134**, (2003).
5. L. Hu, C. Zhou and Z. Sun, Biped gait optimization using estimation of distribution algorithm, in *Proc. of IEEE-RAS Int. Conf. on Humanoid Robots*, 2005.
6. L. Hu, C. Zhou and Z. Sun, Biped gait optimization using spline function based probability model, in *Proc. of IEEE Int. Conf. on Robotics and Automation*, (Orlando, USA, 2006).
7. L. Hu, C. Zhou and Z. Sun, Estimating probability distribution with Q-learning for biped gait generation and optimization, in *Proc. of IEEE Int. Conf. on Intelligent Robots and Systems*, 2006.
8. L. Hu, C. Zhou and Z. Sun, Hybrid Estimation of Distribution Algorithm with Application to Biped Gait Generation, *Information Sciences* (2006).
9. C. Zhou, L. Hu, C. Acosta and P. Yue, Humanoid soccer gait generation and optimization using probability distribution mode, in *Workshop on Humanoid Soccer Robots, Proc. on the IEEE International Conference on Humanoid Robots*, (Genoa, Italy, 2006).
10. Open dynamics engine, www.ode.org url.
11. S. Behnke, B. Frotschl, R. Rojas, P. Ackers, W. Lindstrot, M. de Melo, A. Schebesch, M. Simon, M. Sprengel and O. Tenchio, Using hierarchical dynamical systems to control reactive behavior, in *RoboCup*, 1999.

Design of a Humanoid Soccer Robot: Wukong

Qing Tang, Rong Xiong, Jian Chu, Xinfeng Du

State Key Lab. of Industrial Control Technology
Zhejiang University
Zheda Road No.38, Hangzhou 310027
Zhejiang Province, P.R. China
tangq@iipc.zju.edu.cn
http://www.nlict.zju.edu.cn/humanoid/index.html

This paper presents a complete robot system for humanoid soccer robot, Wukong. 22 DOFs and 2 cameras are designed for the robot to take part in the soccer game. Specific hardware architecture is introduced. Robot locomotion, robot vision and robot decision are all achieved on the robot.

Keywords: Humanoid Soccer Robot; Biped Locomotion; Robot Vision; Robot World Model; Wukong

1. Introduction

Recent years, the humanoid robot research has drawn hundreds of researchers' interests and some of their robots have been acting as the receptionist or can deliver drinks for people. The most exiting humanoid robots include the Asimo made by Honda,[1] Wabian developed in Waseda University,[2] HRP-3P from AIST,[3] KIST's KHR-3[4] and some other research groups' robots. They are leading the research of humanoid robots in robot design and locomotion.

At the same time, motivated by the rapid development of the artificial intelligence research and robotics, the RoboCup Federation organizes international robotic soccer competitions since 1997. The long-term goal of the RoboCup Federation is to develop a team of humanoid soccer robots that wins against the FIFA world Champion.[5] After 10 years' research, the most attractive match has focused on the humanoid soccer league which started from 2002. This paper present the design of a robot which is capable to attend the humanoid soccer competition. The rest of the paper contains the summarized descriptions of each components of the robot. Mechanical specifications, hardware specifications, and software specifications are described in Section 2, 3, and 4 respectively.

2. Mechanical Design of the Robot

In order to design a humanoid soccer robot to play soccer, we should enable our robot Wukong capable of accomplishing the following movements in physical characters.

- walk straight
- turn around
- Kick the ball
- catch the ball
- get up from the ground

Table 1 shows the specification of our robot Wukong. Wukong is driven by 22 servo motors: 6 per leg, 3 in each arm, 2 in the trunk and 2 in the head. The six leg-servos allow for flexible leg movements. Three orthogonal servos constitute the 3-DOF hip joint. Two orthogonal servos form the 2-DOF ankle joint. One servo drives the knee joint. The pitch and role trunk joints are specifically designed for humanoid soccer robot. It is much easier for the robot to get up having the pitch DOF. And the role joint enable the robot to walk fast. The head is manipulated by 2 motors, so the camera can turn in quite wider direction as human's head. Figure 1(a) shows the distribution of the DOFs of Wukong.

Table 1: General specifications of the robot

Parameters	Value
Height[mm]	550
Width[mm]	30
Weight[kg]	3.2
Dof	22

Networked smart motor DX-117 is selected as the actuator, because it offers a high torque with less weight. The motor is controlled with an in-built microchip. The communication is through RS485 protocol with a transfer speed up to 1MHz. We connect the motors one by one with wires and control the motors by their unique ID. With this type of motor we designed our humanoid soccer robot Wukong which can be seen in fig 1(b).

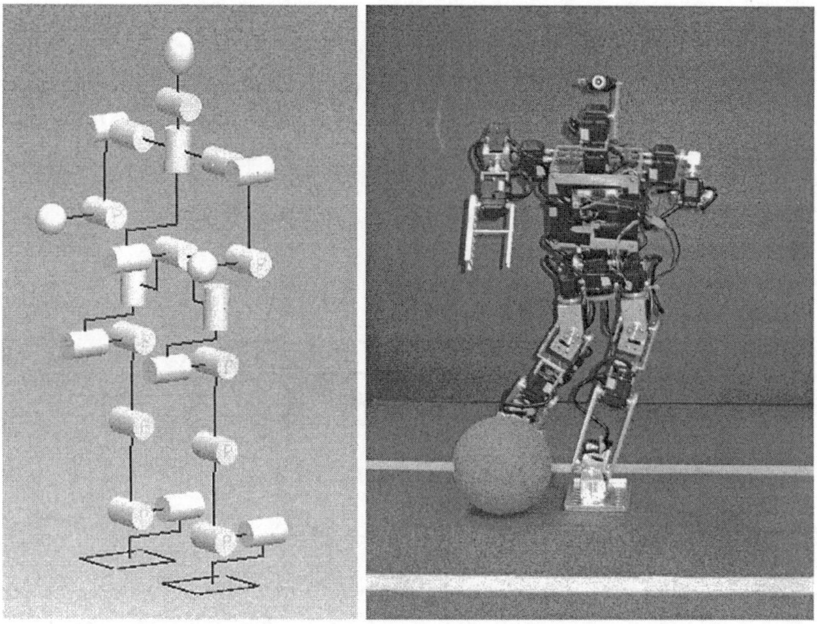

(a) The DOF distribution (b) Our robot Wukong

Fig. 1: Robot's Mechanics

3. Hardware Architecture

To motivate the 22 pieces of motors move simultaneously, we choose ATMega128 as its microcontroller. ATMega128 is an 8-bit processor worked at 16MHz with 128K Bytes of In-System Reprogrammable Flash and 4K bytes EEPROM. The Baud Rate of the serial port can run up to 1MHz. The main controller sets the motors' position one by one and broadcast an instruction to motivate all the motors together every cycle during robot's locomotion.

Vision is provided by two stereo analog cameras using IEEE 1394 interface for output and power. The cameras are 640x480 progressive scan CCD with a standard miniature lenses. The frame rate is 32Hz for 640x480 with automatic control of exposure, gain and black level. The image processor DSP DM642 is used to take image data from the camera. Image threshold segmentation, shape description and target identification would be done on this processor. The reason we choose DSP as our main processor is that DSP has a good capability of processing image data rapidly compared to the other processors such as ARM and PDA. Some basic algorithms are implemented by hardware. On the other hand, DSP DM642 can connected to the Ethernet based on IEEE 802.11b protocol, which makes it very convenient to communicate between robot teammates and easy for the humanoid soccer robots to receive the referees' signals. The overall hardware architecture of Wukong can be seen in fig. 2.

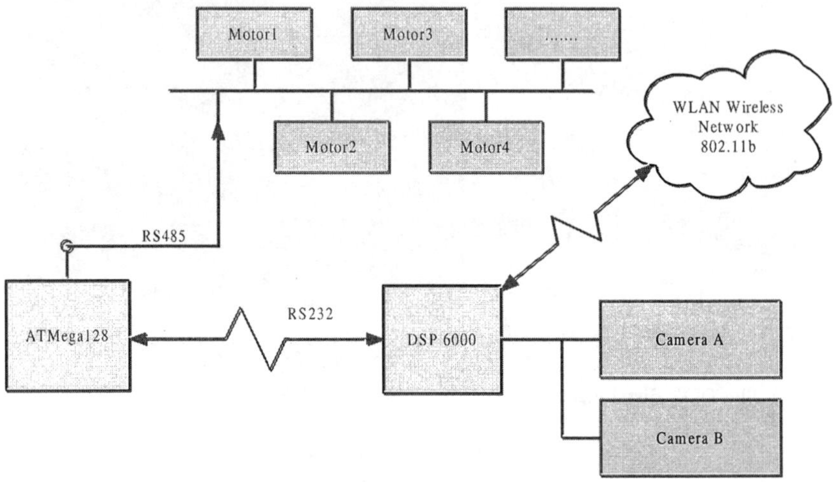

Fig. 2: Hardware Architecture

4. Software Architecture

What makes a man a good soccer player, not only how strong, quick and agile he is that matters, but also he must distinguish and locate the target clearly and accurately as well as make the right decision according to the situation of the playground and teammates. So software architecture should be divided into three parts: gait planning, image processing and decision making.

4.1. *Gait Planning*

To generate the agile movement for the robot, ZMP(Zero Moment Point) Criterion[6,7] proposed by Vukobratovic is considered. The robot is simplified by link model and its kinetic and kinematic movement are described using conventional Denavit-Hartenberg coordinate. Under these hypothesis, the robot's gait is generated in the following steps:

- Preset the lower limbs' trajectory
- Preset the ZMP's trajectory
- Optimize the hip's position towards the minimum ZMP domain based on the robot's structural parameters and the preset limbs' trajectory.
- Plan the upper limbs' trajectory to compensate the robot's dynamic balance.
- Change the trajectory data from cartesian coordinate into joints coordinate.

After getting the trajectory of the left ankle $(x_{aL}(t), y_{aL}(t))$, the trajectory of the right ankle $(x_{aR}(t), y_{aR}(t))$, and the movement of the ZMP point $x_{zmp-ref}(t)$, the hip trajectory $(x(t), y(t))$ is obtained considered the geometrical limitation and the optimization of the ZMP criterion. The problem could be described as follows:

$$\begin{cases} \min \left\| f(x, y, x_{aL}, y_{aL}, x_{aR}, y_{aR}) - x_{ZMP-ref} \right\|^2 \\ \qquad (x, y) = g(x_{aL}, y_{aL}, x_{aR}, y_{aR}) \\ \qquad x_1 \leq x \leq x_2, y_1 \leq y \leq y_2 \end{cases}$$

Function f returns the ZMP position of the robot under given hip point and function g constitutes the geometrical limitation.

Computation of the hip's position is an optimization problem of functional extreme value. If each generated hip position would lead the actual ZMP approaching to the objective ZMP, we could admit that the generated hip trajectory would also lead the actual ZMP trajectory approaching to objective ZMP trajectory. So the optimization problem of functional extreme value could be converted into objective function optimization problem which is easier in computing. The method could be expressed in the following way.

$$\begin{cases} f(\vec{x}(t_1)) \rightarrow x_{zmp-ref}(t_1) \\ f(\vec{x}(t_2)) \rightarrow x_{zmp-ref}(t_2) \\ \qquad \vdots \\ f(\vec{x}(t_n)) \rightarrow x_{zmp-ref}(t_n) \end{cases} \Rightarrow f(\vec{x}(t)) \rightarrow x_{zmp-ref}(t)$$

$t_1, t_2, \cdots t_n$ stands for the sample time intervals of the planning gait.

The hip trajectory results in a stable gait. Robot achieved the basic gait of walking, turning, kicking the ball, catching the ball and standing up using the method. To verify the robot motion, we choose Adams which is a dynamical mechanical simulation software as our simulator. First we create a model based on our robot's structure. Then Adams takes the joints angle's trajectory as the input of the model, considers the robot's inertial effect and the counterforce of the ground, and outputs the stability of the robot. It is convenient and useful to have the gait tested using this software, as well as it avoids some potential damage to the robot.

4.2. *Image Processing*

Cameras worked as the robot's eyes are the main feedback of the humanoid soccer robot. It is an essential capability for the robot to estimate its pose on the field robustly and accurately. The main difficulties are that the robot has a constrained field of view, that the role and pitch angle of the camera changes continuously and can only be roughly estimated, and that the colors of the landmarks are usually affected by the lighting and the pose of the camera. Current work has been focused on solving the last problem with self-adaptive threshold segmentation. In this method, the threshold values are updated every sample time, processor records the history thresholds and associated weights for each target. The current thresholds are pre-calculated based on the historical records as follows:

$$Threshold_{current} = \frac{\sum (HistoryThreshold\,[i] * Weight\,[i])}{\sum Weight\,[i]}$$

With these pre-thresholds and shape description, targets in the color image are identified. A new threshold value and associated weight of targets are calculated according to the reliability of identification. The weight of the threshold decrease autonomously and both of them would be discarded after fixed times. Much better results have been attained using the adaptive threshold for segmentation. Then we can specify the distance of the target according the position and orientation of the cameras. More specific research has been carrying on.

4.3. *Decision Making*

Decision making is made based on what the robot learned about the environment. It covers updating the robot's world model and path planning.

The robot's world model contains all the targets' position and the robot itself's pose. It is constructed according to the fusion of perception and locomotion of the robot. The multi data fusion change the data from polar coordinate into cartesian coordinate. Because the robot is not equipped with omni-camera, limited vision is obtained, the most accurate and trusted landmarks are selected based on the robot's world model and the image process results. Using this landmarks, robot could relocate itself and other objectives and update its world model every frame of image.

Attack phrase and defensive phrase are defined in the world model. Each robot's moving path is calculated to optimize the shortest time considering the locomotion. Robots would communicate with each other using wireless network to balance their strategies.

5. Experiment

Experiment is carried out on Wukong. Maximum walking speed of 7cm/s is achieved and all the movements listed above are all realized. Fig 3 shows the identification of the ball on the playground. Fig 4 shows our humanoid soccer robot Wukong is kicking the ball.

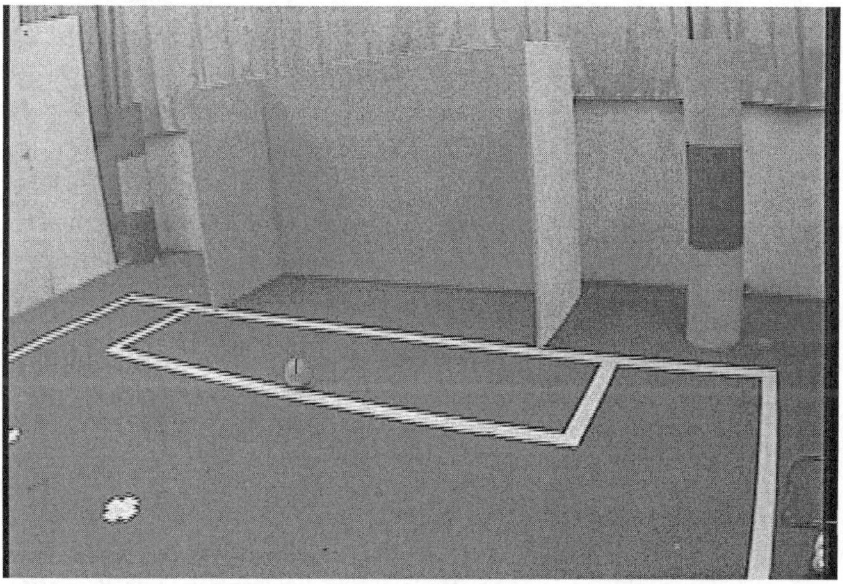

Fig. 3: Identification of a ball on the playground.
A black line indicates the radius of the ball.

Much more videos of our robot could be found on the web site:

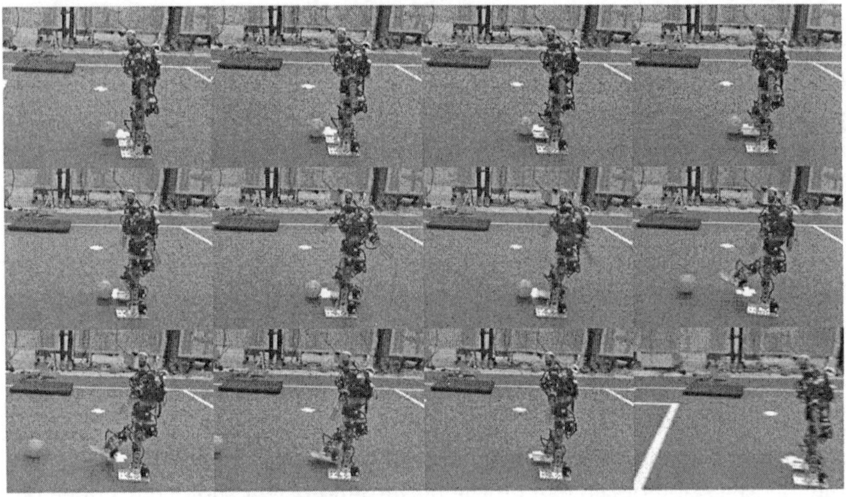

Fig. 4: Wukong is kicking the ball

http://www.nlict.zju.edu.cn/humanoid.

6. Conclusion

This paper presented the development of the humanoid soccer robot Wukong. It was proved that Wukong provided with the current mechanical, hardware and software control system can walk, turn around, kick the ball, catch the ball, and get up from the ground autonomously and is qualified as a soccer player for RoboCup 2007.

In the future, we would intend to provide Wukong with more complicated movements and much better understanding of the world. Furthermore, the analysis and improvement of the hardware and software architecture will also be continued.

References

1. Y. Sakagami, R. Watanabe, C. Aoyama, S. Matsunaga, N. Higaki and K. Fujimura, The intelligent asimo: system overview and integration, in *Proceedings of 2002 IEEE/RSJ International Conference on Intelligent Robots and System*, 2002.
2. Y. Ogura, H. Aikawa, K. Shimomura, H. Kondo, A. Morishima, H.-o. Lim and A. Takanishi, Development of a new humanoid robot wabian-2, in *the 2006 IEEE International Conference on Robotics and Automation*, (Orlando, Florida, 2006).

3. K. Akachi, K. Kaneko, N. Kanehira, S. Ota, G. Miyamori, M. Hirata, S. Kajita and F. Kanehiro, Development of humanoid robot hrp-3p2005.
4. P. Ill-Woo, K. Jung-Yup, P. Seo-Wook and O. Jun-Ho, Development of humanoid robot platform khr-2 (kaist humanoid robot-2)2004.
5. H. Kitano and M. Asada, *The RoboCup Humanoid Challenge as the Millennium Challenge for Advanced Robotics*, *Advnaced Robotics* **13**, 723 (2001).
6. M. Vukobratovic and D. Juricic, *Contribution to the synthesis of biped gait*, *IEEE Transactions on BioMedical Engineering* **16**, 1 (1969).
7. M. Vukobratovic and B. Borovac, *Zero-moment point - Thirty five years of its life*, *International Journal of Humanoid Robotics* **Vol. 1**, 157 (2004).

FORMULATION OF DESIRED ZERO MOMENT POINT TRAJECTORY USING STATISTICAL METHOD

LINGYUN HU, CHANGJIU ZHOU, BI WU, TIANWU YANG

School of Electrical and Electronic Engineering, Singapore Polytechnic
Dover 500 Road, 139651, Singapore
E-mail: ZhouCJ@sp.edu.sg
http://www.robo-erectus.org/

This article explains how to construct desired Zero Moment Point (ZMP) region by statistical method. Gait samples, generated by humanoid soccer robot named Robo-Erectus Senior, are converted to unified ones with standard step length. They are evaluated by ZMP margin and energy consumption. The estimated values are also used as weights in the calculation of means and covariance. Thereby, ZMP quality in term of dynamical stability and energy efficiency can be evaluated by covariance. The less the covariance is, the better the ZMP is. For given covariance, desired ZMP region can be constructed easily and quickly for further optimization and learning.

Keywords: humanoid robot; desired ZMP;

1. Introduction

Intuitional evaluation on optimum performance of biped walking is maximum walking speed and minimum power consumption. To be more specific,it can be quantified by the following criterions.

(1) **Balance** Biped walking is supposed to be stable and maintains balance during the locomotion;
(2) **Efficiency** Dynamical load of each robot joint should be balanced and harmonious;
(3) **Anthropomorphosis** Gaits are not only feasible but also hominine. Anthropomorphic characteristics, like the flexion of knees should be considered in gait control methods.

Well balanced gaits with minimum energy cost and hominine characters are called desired gaits. They have wide application in biped gait generation and optimization as the reference ones [1, 2].

Since biped robot tends to fall in nature because of the unilateral and underactuated foot joint [3], criterion (1) is the basis to generate realizable gaits for biped mechanism. Indicators of the degree of dynamical balance has been profoundly interwind [2, 3, 4]. For dynamical stability, the ZMP relative position with respect to the support footprint is popularly used in evaluation and control [2].

Research in biological motion pattern has shown that biological information, like structural information and dynamic cues, is encoded in biped gaits [5]. It was reported that, discrimination of biological motion mediated by gender is supposed to be determined mainly by hip width and shoulder width [6]. In other words, candidate has personal gaits carrying individual information for recognition. Like human, biped robots have specified limb and mass distributions, which make the individual difference in their personal gaits. Thereby, desired ZMP, whose corresponding biped gait fits the designated robot best, has personal characteristics.

Fast motion pattern generation techniques that follows the desired trajectory in form of ZMP trajectory have been proposed for different digital and physical robots [7, 1]. Moreover, many works have been devoted to define the optimal desired gait of biped robot based on minimizing the integral of ZMP displacement. It always yielded a nonlinear programming problems, that can be solved by artificial intelligent methods [8] or optimization algorithms [9].

However, those methods were all absence of generation and analysis of desired ZMP trajectories. Generally, to realize walking with large stability margin, the middle line in the stable region is chosen as the desired ZMP trajectory [10]. It is discontinuous when shifting from double support phase to single support phase, unless such a shift is constrained to happen only when ZMP is exactly in the center of the foot about to become the single support one. Actually, in the absence of noise, any trajectory that stays within stable region would be equally viable in term of dynamical stability. In addition, gaits with such kind of ZMP trajectories are infeasible in energy cost of impulsive control. The interested aim of this paper is to construct the desired ZMP trajectory region using statistical method.

Besides ZMP margin, the second criterion on energy cost is also considered in ZMP trajectory evaluation and desired ZMP region construction. For human walking, a minimum expenditure curve with appropriate combinations of speed and step length has been reported [11]. Similar to that, mechanical energy expenditure can be evaluated by the time integral of the instantaneous power and minimized by optimization and control methods.

Among the three criterions, the last one is the hardest to achieve for anthropomorphosis can not be estimated by equations and principles easily. Some artificial intelligent methods have been proposed to present the human walking characters.

These three criterions can also be used to estimate control methods for biped mechanism locomotion.

The first two criterions are combined with statistical method in this paper to evaluate gaits in term of desired ZMP trajectories. Gaits are firstly converted to ones with the same length of pace, maintaining their individual speed. Then ZMP trajectories are rated from the view of stability and energy efficiency point. Using statistical method, desired ZMP trajectory region with confidence interval will finally be given.

The rest of this paper is organized as follows. Section 2 introduces the ZMP and energy consumption concept for gait estimation. Based on it, desired ZMP region construction is proposed in Section 3. Related experiment result is shown in Section 4, followed by the conclusion in Section 5.

2. Formulation of desired ZMP trajectory

2.1. *ZMP concept*

ZMP is defined as the point on the ground where the total moment generated due to gravity and inertia equals zero [12]. It is originally proposed by Vukobratoivc and Juricic in 1969, and has been widely applied, frequently cited [13].

Early work related with ZMP criterion minimized the deviation between the ZMP and the center of shape of the supporting area by treating the position of the body as free variables [9]. Based on the off-line algorithm, a fast online motion pattern generation technique is adopted in humanoid robot H7 to track the desired ZMP when carrying objects [7]. Besides being used as the optimization objective, ZMP is also applied in constraints design [4]. A convex optimization algorithm for whole body stabilization is introduced to find the optimal acceleration profile subject to ZMP constraints [14].

ZMP position can be determined in case of real walking mechanisms with pressure or force sensors. Dynamical model is required if only partial counter forces can be measured by pressure or force sensors. Compared with it, 6-axis force/moment sensor gets practical counter force more directly and conveniently. It fits for walking on rough terrain or stairs when both feet are not contacting in the same horizontal plane. Its practical application in humanoid robot includes WABIAN-2R [15], Honda [12] and so forth. Moreover, 6-axis force sensing footwear has been developed for natural walking analysis in humanoid robot H7 [16].

When sensor is unavailable, ZMP position along x-axis and y-axis can be calculated by Equations (1) and (2) respectively.

$$x_{\text{zmp}} = \frac{\sum_{i=1}^{N_q}(m_i(\ddot{z}_i + g)x_i - m_i\ddot{x}_i z_i - (I_i\ddot{\theta}_i)_y)}{\sum_{i=1}^{N_q}(m_i(\ddot{z}_i + g))}, \tag{1}$$

$$y_{\text{zmp}} = \frac{\sum_{i=1}^{N_q}(m_i(\ddot{z}_i + g)y_i - m_i\ddot{y}_i z_i + (I_i\ddot{\theta}_i)_x)}{\sum_{i=1}^{N_q}(m_i(\ddot{z}_i + g))}, \tag{2}$$

where m_i is the mass of link i and g is the gravity acceleration. Position of link i is described by $\mathbf{p} = (x_i, y_i, z_i)$. And correspondingly, accelerations of link i in x, y and z direction are represented by $\ddot{x}_i$, $\ddot{y}_i$ and $\ddot{z}_i$. $(I_i)_x$ and $(I_i)_y$ are the inertial components. $(\ddot{\theta}_i)_x$ and $(\ddot{\theta}_i)_y$ are the absolute angular velocity component around x and y axis at the center of gravity of link i.

2.2. *Energy consumption calculation*

By assuming that power regeneration is unavailable by motor doing negative work, power can be computer by the product of the motor torque and the angular velocity as $E_{ei} = \tau_i\dot{q}_i$, where τ_i and $\dot{q}_i$ are torque and angular velocity of joint i, $i = 1, 2, ..., N_q$. Let T be the time for one gait cycle. We get the average mechanical power

$$E_e = \frac{1}{N_q}\sum_{1}^{N_q}\frac{1}{T}\int_0^T |E_{ei}|dt \tag{3}$$

to measure the power consumption.

3. Formulation of desired ZMP trajectory using statistical method

From the calculation equations of ZMP position (Equations (1) and (2)), it can be found that ZMP position changes with position $\mathbf{p}$ and acceleration $\ddot{\mathbf{p}}$. That is, the desired ZMP trajectory p_{dzmp} satisfies

$$p_{\text{dzmp}} = f(\mathbf{p}, \ddot{\mathbf{p}}).$$

Studies on optimal forward velocities and step length for both normal adults and biped robot was presented by experimental statistical method [17]. For example, desired stride length in human walking has the following relationship with leg length L_h as $L_g = (1.0 \backsim 1.2) \times L_h$.[18] It is reported that, with given walking speed, there is an optimal step length that acquires minimal power consumption in both human and robot walking. Such a walking speed at average pace size is called "natural walking speed" [19].

Similar to that, normal region of pace length and the desired ZMP trajectory $\mathbf{p}_{\text{dzmp}}$ for given robots can be formulated by statistical methods.

Before statistical operation, gait lengthes are converted to standard ones $L_g = 1.0 * L_h$. Correspondingly, ZMPs are recalculated with individual velocities.

Let

$$E_d = \frac{1}{T}\int_0^T |\mathbf{p}_{\text{zmp}} - \mathbf{p}_b| dt \tag{4}$$

calculate the average ZMP margin, where $\mathbf{p}_{\text{zmp}}$ and $\mathbf{p}_b$ are ZMP position and stable region boundary.

With Equations (3) and (4), gait i can be estimated by

$$f_i = \beta_e \aleph(1/E_e) + \beta_d \aleph(E_d). \tag{5}$$

Where $\aleph$ is the normalization operator, β_e and β_d are weighting parameters, $\beta_e + \beta_d = 1$, $i = 1, 2, .., N_g$, N_g is the number of gaits. For simplicity, integration in Equations (3) and (4) is implemented by summarization on the N_s unified samples in one gait cycle.

ZMP trajectories are rated by ZMP margin and energy consumption. Mean $U = [U_x, U_y]$ and covariance $V = [V_x, V_y]$ of ZMP probability distribution are given by

$$U = \frac{f_i}{\sum_{i=1}^{N_g} f_i} \mathbf{p}_{\text{zmp}}$$

$$V^2 = \frac{\sum_{i=1}^{N_g} (\mathbf{p}_{\text{zmp}} - \mu)^2}{N_g}$$

4. Experimental Results

Fig.1 is the biped robot Robo-Erectus Senior. After many stages of design optimizations for its mechanical structure, electronics and control systems, the final physical parameters of Robo-Erectus Senior are roughly 150cm in height and 30kg in weight. Each leg consists of a hip, a knee and an ankle joint, formed by three, one and two motors respectively. These joints are identical in structure and are composed of a servo motor, a harmonic drive gear and an incremental encoder. The energy supply, the power circuits for the joint-drives as well as the micro controller for the low level robot control are on-board of the robot. Robo-Erectus Senior is equipped with a monocular USB camera with miniature lenses.

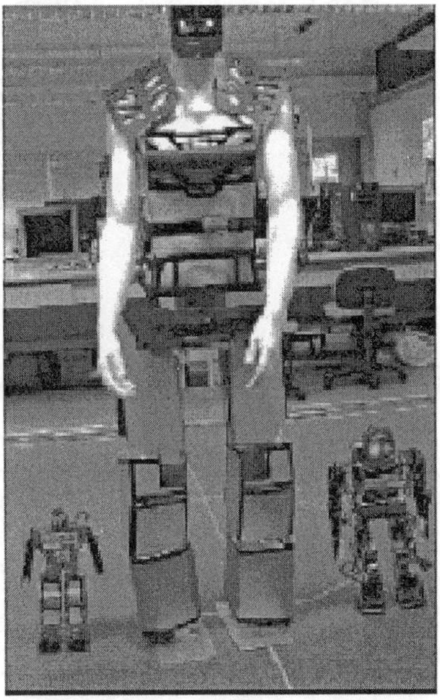

Fig. 1. Robo-Erectus Senior (middle one) and Robo-Erectus Junior (left and right ones)

$N_g = 10$ experiments have been taken on this platform using following parameters, $\beta_e = \beta_d = 0.5$, $L_h = 0.7$, $N_s = 100$. Joint torques and ZMP trajectories are calculated with feedback parameters.

Fig.2 is the N_g ZMP trajectories (see blue $*$ mark) and their weighted average results. For comparison, the proposed weighting method (see Equation (5) and red $\circ$ mark) and $f_i = \frac{1}{N_g}$ (see purple $\triangleright$ mark) are applied on ZMP statistic. It can be found that, statistical ZMP result using average weights comes to be more compact because that stable gaits with little energy consumption act as those with much energy consumption in this statistic method. While the proposed weighting method emphasizes function of those gaits which are both stable and energy effective.

Let t_i be the unified samples in one gait cycle, $i = 1, 2, ..., N_s$. Take t_1 as an example, Fig.3 is the covariance distribution of $V_x(t_{50})$ and $V_y(t_{50})$. According to the definition of gait estimation as shown in Equation (5), the larger the covariance is, the more unstable and energy consuming the gait is. In words, gaits with smaller covariance tend to be more stable and

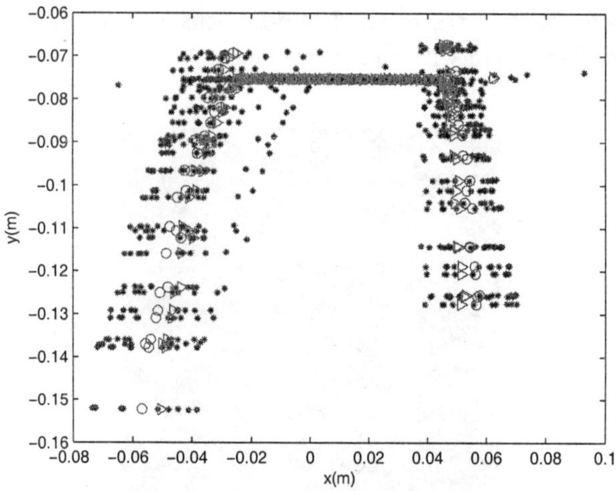

Fig. 2. ZMP trajectories and statistical results

energy efficient.

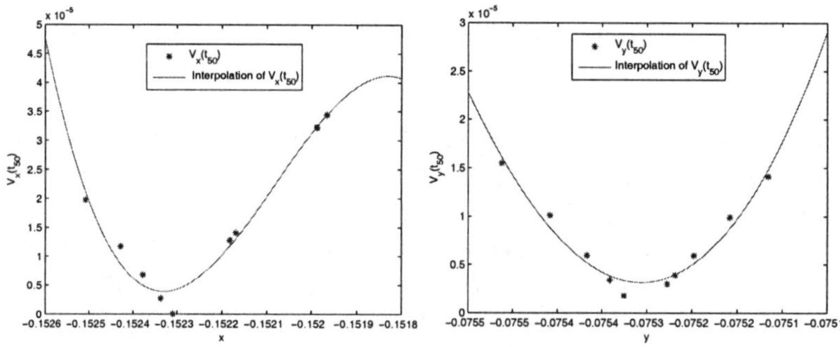

Fig. 3. Covariances of x and y coordinates at the 50th sample moment

Based on it, Fig.4 shows $V(t_{50})$ in the region of support foot as $-L_b < x < L_a$ and $-W_f < y < 0$, where L_a, L_b and W_f are length of frontal foot, heel and width of foot.

Thereby, for given $V(t)$, e.g. $V(t_{50}) = 0.006$, desired ZMP region can be found with the help of the statistic result, which is the cone-shaped area under the partition plate in Fig.4 for the given example.

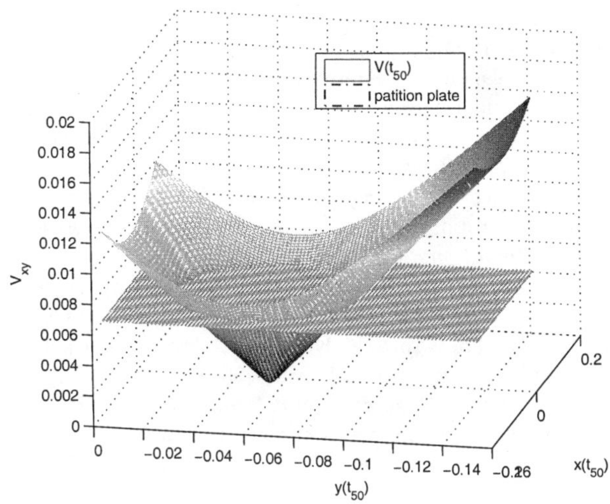

Fig. 4. Covariances at the 50th sample moment

5. Conclusion

This paper proposed an experimental construction method of desired ZMP region based on statistics. Gaits are converted to ones with standard pace length. Dynamical stability in term of ZMP criterion and energy consumption are estimated on gait samples. Taking the weighted average of gaits as their means, covariance can be calculated to measure how good the position is. From the statistical view of point, region with covariance under the given value can be treated as the desired area. This result will be used to generate desired ZMP trajectory for gait optimization and analysis in our future work.

References

1. S. Kagami, T. Kitagawa, K. Nishiwaki, T. Sugihara, M. Inaba and H. Inoue, *Autonomous Robots* **12**, 71 (2002).
2. M. Vukobratovic and B. Borovac, *Int. Journal of Humanoid Robotics* **1**, 157 (2004).
3. A. Goswami, *The International Journal of Robotics Research* **18**, 523 (1999).
4. M. Morisawa, S. Kajita, K. Kaneko, K. Harada, F. K. andK. Fujiwara and H. Hirukawa, Pattern generation of biped walking constrained on parametric surface, in *Proceedings of the IEEE International Conference on Robotics and Automation*, 2005.
5. N. F. Troje, *Journal of Vision* **2**, 371 (2002).
6. G. Mather and L. Murdoch, Gender discrimination in biological motion dis-

plays based on dynamic cues, in *Proc. of the Royal Society of London*, B: Biological Sciences1994.
7. K. Nishiwaki, S. Kagami, Y. Kuniyoshi, M. Inaba and H. Inoue, Online generation of humanoid walking motion based on a fast generation method of motion pattern that follows desired zmp, in *IEEE/RSJ International Conference on Intelligent Robots and System*, 2002.
8. Y. Zhang, Q. Wang, W. Qiang and P. Fu, A new method of desired gait synthesis in biped robot, in *Proc. of the 3rd World Congress on Intelligent Control and Automation*, 2000.
9. C.-L. Shih, Y. Zhu and W. A. Gruver, Optimization of the biped robot trajectory, in *Proc. of IEEE Int. Conf. on Systems, Man, and Cybernetics*, 1991.
10. Q. Huang, K. Yokoi, S. Kajita, K. Kaneko, H. Arai, N. Koyachi and K. Tanie, *IEEE Trans. on Rob. and Aut.* **17**, 280 (2001).
11. H. Elftman, *Journal of Bone and Joint Surgery* **48A**, 363 (1966).
12. K. Hirai, Current and future perspective of Honda humanoid robot, in *Proc. of IEEE Int. Conf. on Intelligent Robots and Systems*, 1997.
13. M. Vukobratovic and D. Juricic, *IEEE Trans. on Bio-Medical Engineering* **16**, 1 (1969).
14. J. Park and F. C. Park, *Advances in Robot Kinematics* (Springer Netherlands, 2006), ch. A convex optimization algorithm for stabilizing whole-body motions of humanoid robots, pp. 157–166.
15. *http://www.takanishi.mech.waseda.ac.jp/research/wabian/index.htm.*
16. Y. Takahashi, S. Kagami, Y. Ehara, M. Mochimaru, M. T. M and H. M. H., Six-axis force sensing footwear for natural walking analysis, in *IEEE International Conference on Systems, Man and Cybernetics*, 2004.
17. M. Hardt, O. von Stryk, D. Wollherr and M. Buss, Development and control of autonomous, biped locomotion using efficient modeling, simulation, and optimization techniques, in *Proc. of IEEE Int. Conf. on Robotics and Automation*, (Taiwan, 2003).
18. E. A. B. John, P. D'ANTONIO, J. PARDO and V. L. David, *Journal of motor behavior* **34**, 309 (2002).
19. L. Roussel, C. Canudas-De-Wit and A. Goswami, Generation of energy optimal complete gait cycles for biped robots, in *Proc. of IEEE Int. Conf. on Robotics and Automation*, 1998.

LOCOMOTION CONTROL SCHEME FOR FAST WALKING HUMANOID SOCCER ROBOT

WEERAYUT SAWASDEE

Institute of Field Robotics (FIBO), King Mongkut's University of Technology Thonburi, Bangkok 10140, Thailand

PASAN KULVANIT

Department of Science Service, Ministry of Science and Technology, Bangkok 10400, Thailand

THAVIDA MANEEWARN

Institute of Field Robotics (FIBO), King Mongkut's University of Technology Thonburi, Bangkok 10140, Thailand

In the game of soccer, the ball approaching speed of the humanoid robot is vital. Fast moving robot gains advantage over the other players. For a limited computing resource and actuator specifications our kid size humanoid robot can walk fast with robustness to external disturbance in the form of stepping foot placement impact. The inverted pendulum model is used for walking pattern generator. The step size and sway distance are adapted online using feedback information from gyro sensors. The impact reduction scheme based on foot force feedback information is implemented to damp out excessive impact force at heel strike. The locomotion scheme is implemented successfully with our humanoid soccer robot competing in the Robocup 2007 Humanoid Soccer League.

1. Introduction

Most biped robot systems nowadays has very little problem stabilizing the walking motion on an ordinary flat surface [1, 2, 3, 4]. The next step may be finding the way to increase the walking speed or achieving the running motion. Fast walking speed or running motion helps the robot reaching the destination faster and renders the humanoid robot more useful than a normally slow pace robot. In humanoid soccer match, it is obvious that the robot with fast ball-approaching speed can reach the ball faster than its opponent and be able to shoot or snatch the ball giving the team more chance to score the goal or defending its own goal.

The problem of walking fast is the effect of the inertia force to the dynamic stability condition is more pronounced. The foot placement scheme to match the ground reaction force and the planned Zero Moment Point (ZMP) must be realized and the compensation must be computed in real-time. The formal way of calculating ZMP [8] may not satisfy the real-time requirement here due to computing hardware and time limitation.

The other problem stemming from the fast walking robot is the impact created by the heel strike event. This impact force can cause the humanoid robot to deviate from planned trajectory or even fall down completely. The experiment results from [2] and the data from real human walking from [9] show that the impact force at heel strike increase proportionally with the walking speed. In normal human walking with the speed of 2 to 4 Kilometer per hour, the impact force is approximately 1.2 to 1.4 times the body weight. If the walking speed increases to 8 Kilometer per hour, the impact force is approximately 1.8 times the body weight. Human body is structurally build to deal with such impact force. The soft tissue around the foot and the tendon structure of the foot help damping out such impact disturbances. For the bipedal walking machine with mechanical body equipped with multiple DC servo motors and no passive energy storage devices, there is no help from compliancy. The introduction of compliance into the rigid system is needed. Park [7] suggests the approach of using the software to help damping out the impact force. [2, 6] use simple compliance in the form of flexible pad and/or springy foot mechanism attaching to each foot of the humanoid robot in order to partially damped out the vibration created by the impact.

This work shows the small size and compact humanoid robot system that can walk stably fast on the nominal flat surface with ability to treat the impact force that may de-stabilize the walking locomotion. There are three main controllers that help stabilizing the walking pattern. First is the regulation of the body uprightness using the feedback information from gyro sensor. Second is the adaptation of the step size based on the concept of the ZMP. The observed value of the angular velocity is used here to predict the direction of the robot CM's acceleration and adjust the foot placement position. Third is the controller that deal with the impact force via the vertical foot force feedback. The compliance at the knee is adapted with software using foot force information. All three controllers work collectively to adapt the pre-planned walking pattern in real time. Note that all schemes are executed with a low-power consumption and low cost microcontroller. Comparing to the more complicated humanoid robot systems such as [1, 2, 3] the resulting walking performance is quite good. The robot is used in the Robocup 2007 Humanoid Soccer League Robot competition [5] with satisfying results.

The structure of this paper is as follows. Section 2 presents the biped robot system used to obtain the experiment results. Section 3 discusses the locomotion control and the impact reduction scheme. Section 4 shows the results of implementation on the biped robot. We then conclude this paper with section 5.

2. Humanoid robot system

The actual biped robot design and build at the Institute of Field Robotics (FIBO) is used in the experiment process. The general specifications are displayed in Table I. The robot is originally built to compete in the Robocup Humanoid league competition.

Table 1 Biped robot system specifications

Name: *Jeed*	Specifications
Height	45 CM
Weight	2.7 KG (total)
Walking Speed	18 m/min (max)
Number of DOF	22
Actuator	DC servomotor
Structural Material	Aluminum Alloy

There are 2-axis accelerometer [+/-2g], 2 rate gyros [+/-100 deg/sec], 8 force-sensing resistors, and a camera. The accelerometer tells the robot if there is any longitudinal and/or transversal tilt. The two rate gyros, installing at the lower trunk, measure angular velocity at longitudinal and transversal axis. The angular velocity information will be used to adapt the attitude of the body during walking and calculating the instantaneous acceleration of the CM. The force sensing resistors are used to sense the vertical ground reaction force. Force information will be used to adapt the knee joint impact absorption characteristic. The camera is used to track the ball and other objects of interest, which is crucial for navigation decision-making software.

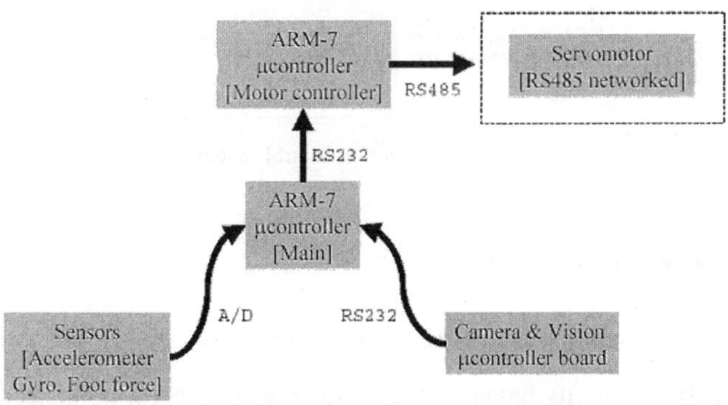

Figure 1. System overview of the biped robot

Figure 1 shows the overview of the biped robot system. The main computer for the biped robot is ARM-7 [60MHz] RISC microprocessor. Its major function is to make decision based on the information obtained from the camera system and sensors. The main computer, then, issues commands to the motor locomotion controller (the other ARM-7 processor) to generate walking motion. The inverse kinematics of the robot legs is stored in ARM-7 motor controller. The kinematics is calculated in real time.

3. Locomotion control and impact reduction scheme

The small size and compact humanoid robot system can walk stably fast with the ability to treat the impact force that may de-stabilize the walking locomotion. There are three sets of the control scheme which are designed to help stabilizing the walking pattern at before, during and after the foot placement as shown in figure 2.

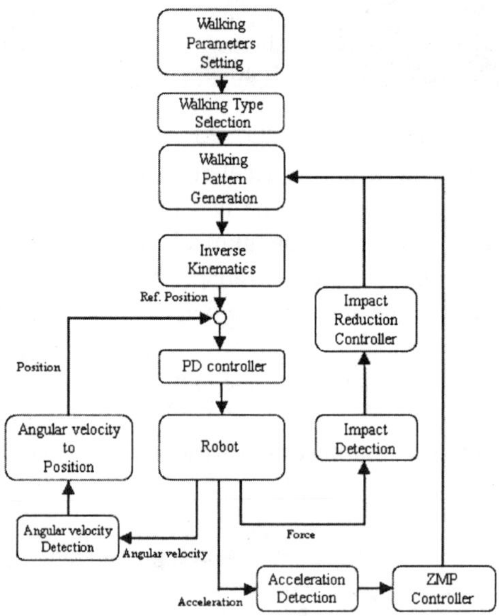

Figure 2. Locomotion control scheme

3.1. *Before foot placement*

3.1.1 Controlling balance before foot placement

When the robot loses its balance, the angular velocity can be sensed from the gyro sensors. The measured angular velocity is fed back to the system so that the robot can lean its body to compensate for the angular rotation which can help balancing the robot before the foot placement. The angular position of four motors attached to the robot's ankle are adjusted directly from the sensed angular velocity. When the inverse kinematics is used to calculate joint positions from the predefined gait which is used as the reference trajectory, the PD controller is used to adjust the position command which is the input for these motors at the ankle as shown in Eq.(1)

$$CMD = CMD \pm ((Kp(Err_i)) + ((Err_{i-1} - Err_i)Kd)) \tag{1}$$

CMD is the command angular position

Kp, Kd is the PD gain

Err_i is the angular velocity error at this time step

Err_{i-1} is the angular velocity error at the previous time step

3.1.2. Controlling the foot placement

When the robot walks, inertial force and gravitational force affect the walking acceleration. These forces can be referred to as the total inertial force. When the foot touches the ground, the robot receives the reaction force from the ground. The point of intersection between the ground reaction force vector and the total inertial force vector is called the zero moment point because it has zero moment. The position on the ground that the reaction force passes through is called the ground reaction point. When the walking pattern is generated, the target total inertial force can be calculated from the robot model. When the robot can perfectly balance during the walk, the target total inertial force and the actual ground reaction force is projected to the ground at the same position. When the robot walks on an uneven ground , these two positions will be separated which results in lost of balance and causing the robot to fall. The force that causes the fall can be shown in term of the mismatch ZMP and the actual ground reaction force as shown in figure 3.

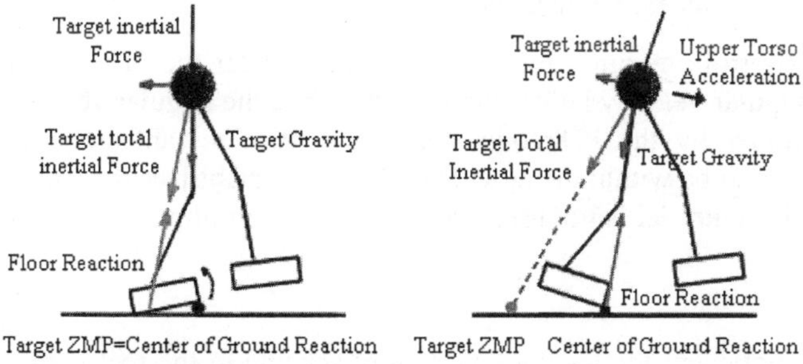

figure 3. the diagram shows the robot losing balance

When the robot is starting to fall forward, the velocity and acceleration increase in the forward direction. The ZMP control system can help the robot's body position to move toward the direction of acceleration by adjust the foot placement position. The ZMP control system is designed to adjust the walking step size based on the relationship with the predicted direction of the body acceleration. The direction of the body acceleration is predicted from the sensed angular velocity of the robot's CM. The value and direction of the sensed angular velocity of the robot's CM (in two axes) are scaled and then added to the planned foot placement position in x and y axis. The adjusted foot placement compared to the planned foot placement are shown in figure 4.

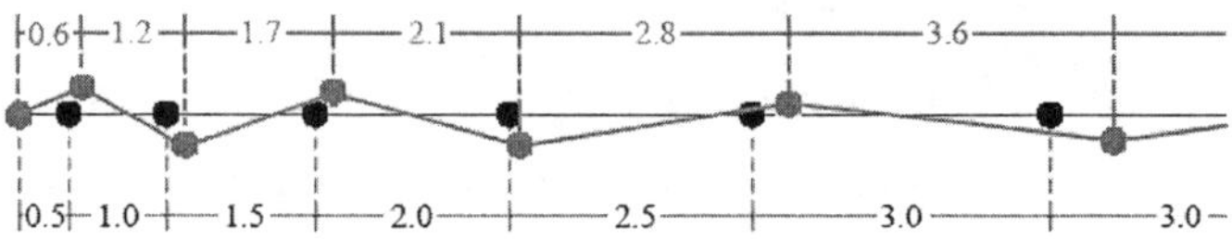

Figure 4. the adjusted foot placement from ZMP calculation

3.2. *During foot placement*

The impact reduction control is used to reduce the high impact force occurred during foot placement. The force that is acting to the foot when the foot touches the ground can be measured by force sensors that attached to the four corners of the robot's foot. The impact force can be identified from the average force from all four sensors. When the value of force and the average change of force is higher than the specified threshold in the positive direction, the impact state is established. In the impact state, this average change of force is used for the knee bending adjustment. The knee bending adjustment depends on the amount of impact force. The difference between the average change of force and the specified threshold is scaled (in the P-control fashion) and this amount is then taken out from the height of the robot which is the input for the knee bending inverse kinematics equation. The command input for the motors attached to the leg are then adjusted accordingly.

3.3. *After foot placement*

The control algorithm for after impact is similar the PD controller that control the angular velocity before the impact. When the angular velocity after impact is controlled by the PD controller, the controller can successfully reduce the velocity to be within the acceptable level and stabilizes the walking cycle. When the PD controller is not used, the robot tends to falling forward after the impact.

4. Results

We implement the locomotion control scheme to the real humanoid robot. The robot walks at the nominal speed approximately 0.2 m/s. We detect the acceleration at the foot via vertical accelerometer as shown in figure 5. The angular velocity in the side sway direction of the CM is measured from the side sway gyro sensor. We can see that when the impact reduction scheme is off, there are many sub-peaks occur in the vertical acceleration profile at the foot. When the impact reduction scheme is on, the sub-peaks are partially eliminated and only left with the major peaks, which represent the initial impact at heel strike. Note that the unit of the time axis is second and the unit of the vertical axis is volt since we map the force quantity to voltage.

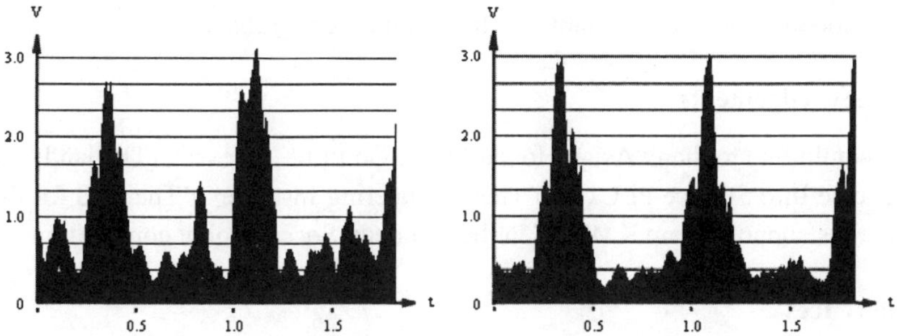

Figure 5. Vertical acceleration at the foot during stepping motion. Left figure is the acceleration for the case without the impact reduction scheme. Right figure is the acceleration profile for the case with the impact reduction scheme.

We compare the result of the side sway behavior of the humanoid robot between the case when gyro feedback control is turned on (Figure 6. Right Figure) and the case when gyro feedback control is turned off (Figure 6. Left Figure). The graph of the gyro reading of the side sway behavior is periodic. This means that the controller is acting to stabilize the swaying motion. The left figure shows no constructive pattern, which means that the swaying motion is not rhythmic and the robot is now prone to falling down. Note that the unit of the time axis is second and the unit of the vertical axis is volt since we map the angular velocity to voltage.

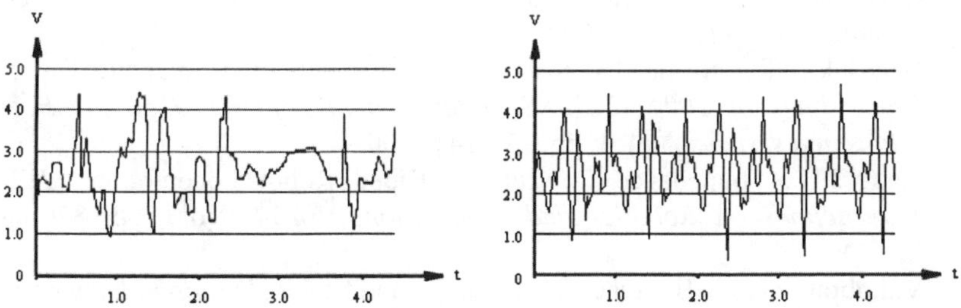

Figure 6. Angular velocity data read from side sway gyro sensor. The right figure is the case when the gyro feedback control is on. The left figure is the case when the gyro feedback control is off.

5. Conclusions

Fast and simple controllers are introduced to the humanoid robot locomotion control scheme. The controllers can be divided into three main parts. First is the preview control to generate the foot placements for the fast walking motion. Second is the gyro feedback control to help regulating the trunk of the robot to be upright during walking. Last controller is the impact reduction control scheme to help damping out the excessive impact force at the heel-strike event. All three controllers work together to stabilize the bipedal walking as a whole and help increasing the robustness of the walk. The result is the fast and stable biped locomotion that is suitable for humanoid soccer robots.

Acknowledgements

We would like to thank Asian Honda Motor Co., Ltd., Chevron (Thailand) Ltd., Advance Info Service PLC., and The Engineering Institute of Thailand for their generous support Team KMUTT in the Robocup Soccer Robot competition.

References

1. Hirai, K., Hirose, M., Haikawa, Y., and Takenaka, T., "The development of Honda Humanoid Robot", *Proceedings of the 1998 IEEE International Conference on Robotics & Automation,* Vol.2, pp. 1321-1326 (1998).
2. Honda Motor Co., Ltd., Public Relations Division, "ASIMO, The Honda Humanoid Robot", *Technical Information* (2002).
3. Kagami, S. et al., "Online 3D Vision, Motion Planning and Bipedal Locomotion Control Coupling System of Humanoid Robot: H7", *Proceedings of the 2002 IEEE/RSJ International Conference on Intelligent Robots and Systems*, Vol. 3, pp.2557-2562 (2002).
4. Kajita, S. et al., "The 3D Linear Inverted Pendulum Mode: A simple modeling for a biped walking pattern generation", *Proceedings of the 2001 IEEE/RSJ International Conference on Intelligent Robots and Systems*, Vol. 1, pp. 239-246 (2001).
5. Kulvanit, P., et al, "Team description paper: Team KMUTT", *Robocup Humanoid League 2007* (2007).
6. Nagasaki, T., et al, "A Running Experiment of Humanoid Biped", *Proceedings of 2004 IEEE/RSJ International conference on intelligent robots and systems*, Vol. 1, pp. 136-141 (2004).
7. Park, J.H., "Impedance Control for Biped Robot Locomotion", *IEEE Transactions on Robotics and Automation*, Vol.17, No.6, pp. 870-882 (2001).
8. Vukobratović, M, Boravac, B., Surla, D., and Stokic, D., *Biped Locomotion Scientific Fundamentals of Robotics 7*, Springer-Verlag (1989).
9. Zatsiorsky, V.M., *Kinetics of Human Motion*, Human Kinetics (2002).

OPTIMUM PERFORMANCE OF THE FAST WALKING HUMANOID SOCCER ROBOT: EXPERIMENTAL STUDY

PASAN KULVANIT

Department of Science Service, Ministry of Science and Technology, Bangkok, 10400, Thailand

BANTOON SRISUWAN

Institute of Field Robotics, King Mongkut's University of Technology Thonburi, Bangkok, 10140, Thailand

DJITT LAOWATTANA

Institute of Field Robotics, King Mongkut's University of Technology Thonburi, Bangkok, 10140, Thailand

The optimum performance of the humanoid robot is needed in the game of robot soccer. In order to gain advantage over the opponent, the robot must output the fastest possible walking speed at the minimal expense of on-board power source. We implement the gyro feedback control to the inverted pendulum model to regulate the attitude of the trunk during walking. Through experimental study with the real humanoid robot, we have found the optimum step size the humanoid robot should take in order to walk fast at minimum power consumption. Our robot is able to achieve the optimum speed of 0.18 m/s.

1. Introduction

A humanoid robot is composed of various sub-systems working together to achieve a controlled bipedal walking locomotion. Such a complex system is difficult to design exactly to the required specifications. Only estimated performance can be predicted. The optimal performance can only be obtained via experiment and testing. The needed performance for humanoid soccer robot is maximum walking speed and minimum power consumption. Walking pattern with lengthy double support phase such as in [3, 10] are not sufficient for soccer playing robot. [9] realize a natural walk using angular momentum information. [2] state that although the system of inverted pendulum is unstable in the sense of Lyapunov, we can still design a stable trajectory for the gait by switching the support leg at the mirror location of the angular position of the starting point of the center of mass (CM)'s trajectory. The condition for leg switching follows the locomotion continuity condition where the terminal CM's velocity of previous cycle equals the CM's initial velocity of the present cycle. [4] comment on the phase plane trajectory of the constant hip height walk that a kind of energy is conserved and called it orbital energy. The behavior of the trajectory is much depended on the orbital energy. In addition, the constant horizontal hip velocity minimizes "jerk" in the trunk area. Minimization of jerk and the hip ripple increases walking reliability due to decreasing in a risk of encountering unexpected external disturbances.

In this paper, we use a constant hip height, gravity compensated inverted pendulum model to construct a gait pattern. The energy and angular momentum of such gait is conserved. The gait also has self-stabilized trajectory in nature. The gyro feedback control is used to adapt the gait pattern in real time in case when the robot encounters any disturbances. The experiment is used to find the optimal operating point, i.e., natural dynamic, for the robot.

The structure of this paper is as follows. Section 2 introduces the humanoid robot system used for this work. Section 3 explains the inverted pendulum model used to implement fast dynamic gait and the adaptation of gait pattern via gyro stabilization control. Section 4 presents the findings from experiment. Section 5 concludes the paper.

2. Humanoid robot system

The actual humanoid robot, which is designed and built at the Institute of Field Robotics (FIBO) is used for the experiment study. The robot competes in the Robocup Soccer Humanoid League 2007. Table 1 shows the overall specifications of the robot.

Table 1: Biped robot system specifications.

Name: *Jeed*	Specifications
Height	45 CM
Weight	2.7 KG (total)
Walking Speed	18 m/min (max)
Number of DOF	22
Actuator	DC servomotor
Structural Material	Aluminum Alloy

There are 2-axis accelerometer [+/-2g], 2 rate gyros [+/-100 deg/sec], 8 force-sensing resistors, and a camera. The accelerometer tells the robot if there is any sagittal and/or transversal tilt. The two rate gyros, installing at the lower trunk, measure angular velocity at sagittal and transversal axis. The force sensing resistors are used to sense the vertical ground reaction force. The camera is used to track the ball and other objects of interest in the soccer field.

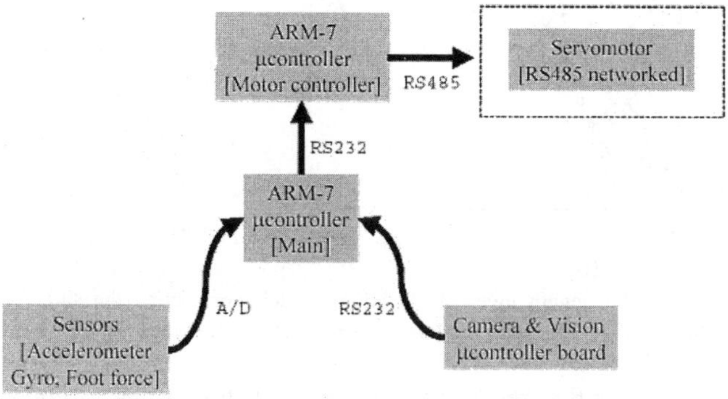

Figure 1. System overview of the biped robot

As shown in Figure 1, the main computer for the biped robot is ARM-7 [60MHz] RISC microprocessor. Its major function is to make decision based on the information obtained from the camera system and sensors. The main computer, then, issues commands to the motor locomotion controller to generate walking motion. The inverse kinematics of the robot legs is stored in ARM-7 motor controller. The kinematics is calculated in real time.

3. Fast dynamic gait

In our fast dynamic gait, we minimize time spent on an exchange of support phase and keep the biped robot stable by relying on the high gain control at the single support phase. By switching the leg fast enough, the change of support phase is squeezed into a very short time in the gait cycle such that it produces infinitesimal effect on locomotion's stabilization process. The inverted pendulum model is the model of choice. It can be readily described by the four basic walking parameters - Step length/ Side sway distance/ Trunk bend angle/ Step time interval.

3.1. *Two dimensional Inverted pendulum model*

We, first, pay attention to the single support phase of the bipedal walking, where the walking cycle occupy the largest percentage of the total walking. As shown in figure 2(a), we represent the bipedal walking during the single support phase as a variable length inverted pendulum. Assume that all biped robot's mass is concentrated at the top most part of the pendulum. Let's assume further that there is external torque acting at the ankle. In real robot, this is the torque that is produced from the power ankle joint.

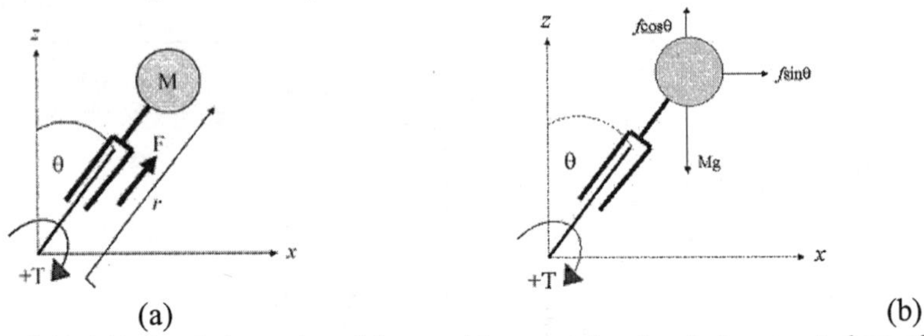

Figure 2: a) Variable length inverted pendulum model representing the single support phase of bipedal walking b) A free body diagram corresponded to the variable length inverted pendulum model

Figure 2(b) shows the free body diagram of all forces and torque that act on the CM of the inverted pendulum. The equations of motion are:

Rotational dynamic

$$r^2\ddot{\theta} + 2r\dot{r}\dot{\theta} - gr\sin\theta = \frac{Tq}{M} \tag{1}$$

Linear dynamic

$$\ddot{r} - \dot{\theta}^2 r + g\cos\theta = \frac{f}{M} \tag{2}$$

Since the vertical component of the force acting on the CM are $f\cos\theta$ and Mg; therefore in order to cancel out gravity effect, f must equal $Mg/\cos\theta$. We can, then, achieve a constant hip height trajectory. Figure 3 shows the phase plane plot of the inverted pendulum motion with constant hip height under the influence of several sets of initial conditions. Note that the ankle is not active. We use a set of initial condition that consists of initial angular position (θ) and angular velocity ($\dot{\theta}$). There are nine sets of initial conditions. Some of them manage to travel from negative angular position to the positive region. Some of them cannot make it to the positive side. The phase plane plot shows that the system of inverted pendulum is unstable since the trajectory of the system neither goes to the equilibrium point nor constitutes a limit cycle. However, the stable trajectory that satisfies the locomotion continuity condition can be achieved. As shown in Figure 3, the 4th trajectory from the top has equal value of $\dot{\theta}$ at the beginning and at the end of the trajectory. We will use inverted pendulum model as a basis to generate a walking pattern for the humanoid robot.

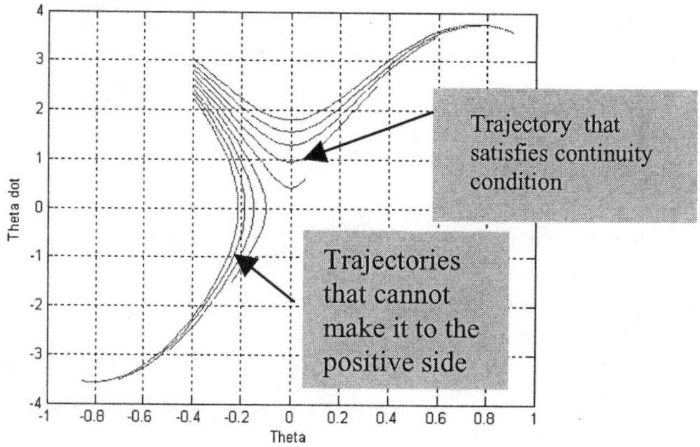

Figure 3. Phase plane plot for the inverted pendulum model mass in motion at constant hip height (i.e., gravity compensated motion)

3.2. *Gyro feedback control scheme*

The walking pattern based on gravity compensated inverted pendulum model alone is not sufficient to stabilize the walking robot. We need adaptation scheme to compensate the error that happens due to external disturbance. We use two gyro sensors to feedback the attitude information in 2 axes- sagittal and transversal plane. This information tells the walking controller the states of angular velocity at each sampling moment. We sample the gyro sensor data every 10 ms. The control loop for bipedal locomotion used with the real system is shown in Figure 4. Assuming that this controller is controlling the attitude of the biped robot in the sagittal plane. In this case the angular velocity of the ankle's joint in the sagittal plane is regulated based on the feedback information from the sagittal plane's gyro sensor. We use the ankle torque to control the whole body attitude of the biped during walking. The control equation is as follows.

$$\omega_{ankle\_cmd} = K_p(\overline{\omega}_{gyro} - \omega_{kinematics}) + \omega_{kinematics} \tag{3}$$

Where

ω_{ankle_cmd} : Angular velocity command for the ankle joint's motors

K_p : Proportional gain

$\overline{\omega}_{gyro}$: Angular velocity reading from gyro sensor

$\omega_{kinematics}$: Joint angular velocity calculated from the inverse kinematics

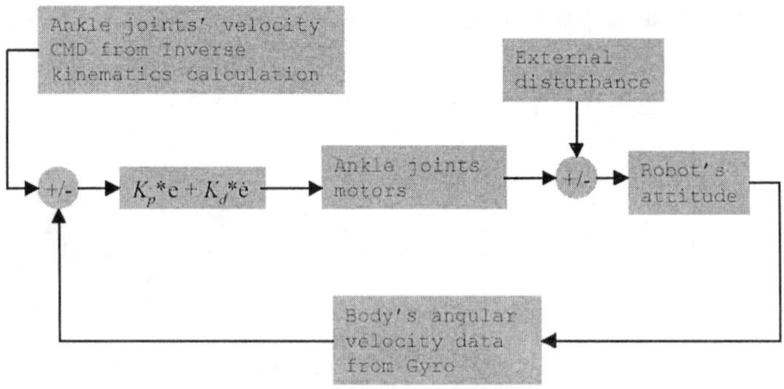

Figure 4: Bipedal motion control loop used to stabilize the fast gait

The walk control scheme of [10] involved the following torque control equation:

$$T_{ref} = K_L(L_{ref} - L_{robot}) + \overline{T}_A \tag{4}$$

Eq. (4) is the feedback control scheme used to regulate the angular momentum of the robot (L_{robot}) to some reference value (L_{ref}) by way of adapting the ankle torque ($\overline{T}_A$). If we let $K_p = K_L / \Delta t$ and we know that $T = I\omega$, then eq. (3) is equivalent to eq. (4) and the angular momentum of the robot with the gyro feedback control scheme is conserved.

4. Experiment results

We operate the robot at three different speeds- 0.12, 0.16, and 0.18 m/s. For each walking speed we vary the step size and tune other walking parameters to stabilize the walk. The total power consumption is recorded via load current sensor. The walking speed is the average walking speed.

Figure 5 shows three sets of data at three different CM's speeds. Note the parabolic nature of the plot. The trend of each data set is that, at the same walking speed, the energy consumption takes parabolic characteristic such that at the shortest step length and longest step length, the energy consumption is at maximum. There is an optimal step length that the biped robot can use to acquire minimal power consumption. It should be noted that the biped that walk with higher speed tends to consume more power than the biped robot walking at slower speed. Some of the data points that distributed out of the parabolic range are the results of some discrepancy during the practice of the experiment, and assume to be ignored. It can be concluded that for the biped robot walking with the same speed, there is an optimal step length that leads to optimal power consumption. For our robot used in the experiment, such step length is approximately 0.046 m. This is also true for human locomotion. Such a walking speed at average human step size is called "natural walking speed" [8] or "self-selected speed". This natural walking speed is the walking speed that minimizes the human's power consumption.

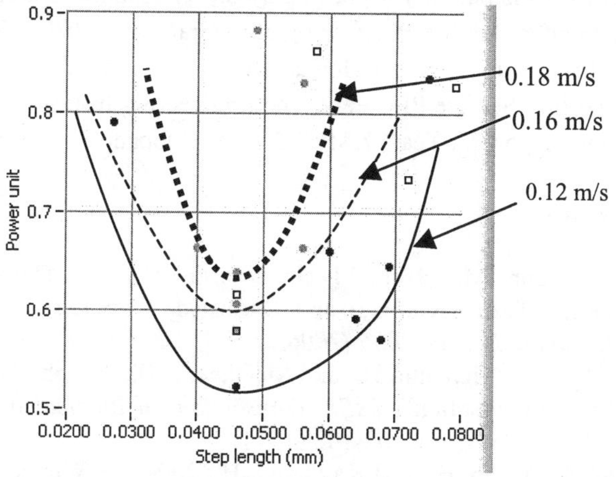

Figure 5. For the biped walk with the same CM's speed, there is an optimal step length that minimizes the power consumption used during the walk. Grey dots represent the walk at 0.18 m/s. Squares represent the walk at 0.16 m/s. Solid dots represent the walk at 0.12 m/s

We select 0.18 m/s as our operating speed. Comparison via Froude number, $Fr = v^2/(g*L_{leg})$, where v is walking speed and L_{leg} is leg length [5], shows that our robot (Fr = 0.018) performs equally well when compared with WABIAN-2R [7] (Fr = 0.021) and performs better than HRP-2LR [6] (Fr = 0.0043) or Hajime Robot [1] (Fr = 0.0136). It should be noted that the maximum speed of our robot is 0.304 m/s with higher power consumption than the one at the optimum walking speed.

5. Conclusions

An experimental study of the humanoid robot to achieve an optimal performance is the focus of this work. The interested objectives are walking speed and power consumption. The basic gait pattern for our robot is the gravity compensated inverted pendulum model. The implementation of the gyro feedback controller is for added locomotion stability robustness. The result obtained is the step length of 0.046 m that always yield minimum power consumption at any walking speed. This result is specific to the particular robot. It is crucial to find this "natural parameter" in order to operate the biped robot at the optimum level.

Acknowledgments

This research is partially supported by the Thailand Research Fund through the Royal Golden Jubilee Ph.D. Program (Grant No. PHD/1.M.KT.44/C.1). We would like to thank Asian Honda Motor Co., Ltd., Chevron (Thailand) Ltd., Advance Info Service PLC., and The Engineering Institute of Thailand for their generous support for Team KMUTT in the Robocup Soccer Robot competition.

References

1. Friedmann, M., et al., "Team description paper: Darmstadt Dribblers & Hajime Team (KidSize) and Darmstadt Dribblers (TeenSize)", *Robocup Humanoid League 2006 (2006).*
2. Gubina, F., Hemami, H., and McGhee, R.B., "On the Dynamic Stability of Biped Locomotion", *IEEE Transaction on Biomedical Engineering*, Vol. BME-21, No.2, pp. 102-108 (1974).
3. Ito, S., Asano, H., and Kawasaki, H., 2004, "A Weight Shift by Control of Center of Pressure of Ground Reaction Forces in Biped Double Support Phase", *Journal of the Robotics Society of Japan*, Vol. 22, No.4, pp113-120.
4. Kajita, S., Tani, K., and Kobayashi, A., "Dynamic Walk Control of a Biped Robot along the Potential Energy Conserving Orbit", *International Workshop on Intelligent Robots and Systems*, Vol. 2, pp789-794 (1990).
5. McNeil, A.R., "Gaits of Mammals and Turtles", *Journal of the Robotics Society of Japan,* Vol.11 No.2, pp 314-319 (1993).
6. Nagasaki, T., et al., "A Running Experiment of Humanoid Biped", *Proceedings of 2004 IEEE/RSJ International conference on intelligent robots and systems*, Vol. 1, pp. 136-141 (2004).
7. Ogura, Y., et al., "Evaluation of Various Walking Patterns of Biped Humanoid Robot," *In proceedings of the 2005 IEEE International conference on Robotics and Automation*, pp. 603-608 (2005).
8. Roussel, L., Canudas-de-wit, C., and Goswami, A., "Generation of Energy Optimal Complete Gait Cycles for Biped Robots", *Proceedings of the 1998 IEEE International Conference on Robotics and Automation*, Vol. 3, pp. 2036-2041 (1998).
9. Sano, A. and Furusho, J., "Realization of Natural Dynamic Walking Using the Angular Momentum Information", *Proceedings of 1990 IEEE International Conference on Robotics and Automation*, Vol. 3., pp. 1476-1481 (1990).
10. Takanishi, A., et al., I., 1989, "Dynamic Biped Walking Stabilized with Optimal Trunk and Waist Motion", *IEEE/RSJ International Workshop on Intelligent Robots and Systems*, pp. 187-192.

Supporting Technologies

A BIOLOGICALLY INSPIRED ARCHITECTURE FOR CONTROL OF GRASPING MOVEMENTS OF AN ANTHROPOMORPHIC GRIPPER

S. VARONA MOYA[a], J. MOLINA VILAPLANA[a], A. LINARES BARRANCO[b], J. FELIÚ BATTLE[a], J. LÓPEZ CORONADO[a]

[a]*Department of Systems Engineering and Automatics, Polytechnic University of Cartagena, Campus Muralla del Mar, C/Dr. Fleming S/N. 30202, Cartagena, Murcia, Spain*

[b]*Department of Computer Technology and Architecture, University of Sevilla, Avda. Reina Mercedes S/N. 41012 Sevilla, Spain*

In this paper we introduce a research plan for the development of an architecture for control of grasping movements in a manipulation system. It consists in an anthropomorphic robotic gripper with tactile and propioceptive sensing capability plus a stereoscopic visual guidance system, electronic control boards designed for implementing a powerful and neurologically inspired inter-chip communication technique called AER and a HYPBF neural networks model that will emulate the performance of the most important brain areas involved in motor behaviour control for efficient object manipulation.

1. Introduction

How does the brain control our hand when we use it to grasp something? As with any other complex system, the initial approach to answering this question starts with a *black box* model where inputs and outputs are mapped in a convenient way. We know several things about this box: it is located in the central nervous system, inputs to the box must be visual, tactile, propioceptive and task-related flows of information, the output is a prehension behaviour measured in terms of static gestures and dynamic forces during task execution and the end-effector gripper can be a real human hand, a prosthesis or a robotic mechanism. Also, knowledge obtained from previous grasping operations must be stored and recovered inside that black box.

Following MacKenzie [1], several disciplines have tried to model this system, such as control engineering, robotics, mathematics and computer science. Regarding the engineering approach, human motor control has been identified as a slow adapting, low resolution, sensor-driven non-linear system. However, it has not been possible, based on these assumptions, to develop systems capable of matching the skills that humans and other animals show when manipulating things and, above all, speaking of cognitive functions (Arbib [2])).

It is clear that, at present, there is not a simple solution to this problem. Even now that we can dispose of high computational power, it is still necessary to solve certain questions such as efficient coordination of hand joints regarding synergic movements, development of bio-inspired sensors, design of an appropriate control of end-effector forces in order to achieve an efficient and stable grasp in a highly anticipatory manner, also capable of reacting to unexpected events like object slipping, ways of extracting information about the object and the task and, finally, the implementation details in order to make simulations and obtain results.

With the aim of structuring and guiding research work, Marr [3] suggested three levels in the analysis of the performance of any system that process a big amount of information in a complex way. These are the task level, the representation or algorithmic level and the implementation level. Computational models implement the ideas present in conceptual models, letting us test them and obtaining useful data that can suggest modifications or even new conceptual models. Expert systems of artificial intelligence, non-linear equations and, more recently, neural networks can be found in this level.

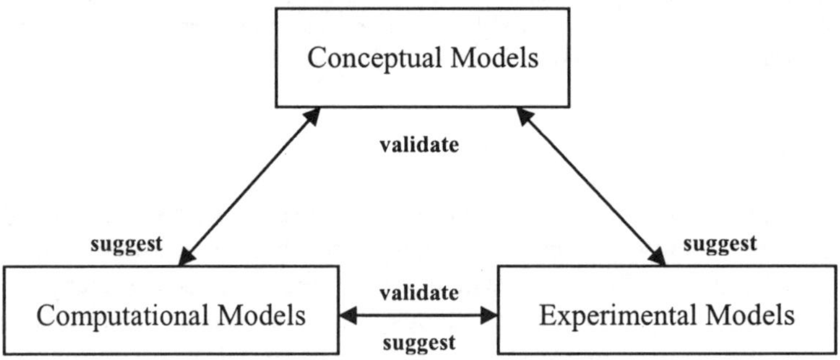

Figure 1. Marr's [3] triangle strategy.

Nowadays, neurorobotics has emerged as a fresh research field in which the method of cognitive neuroscience can be applied to robotics in order to present a bio-inspired solution to the grasping problem. This is one of the research lines of the NEUROCOR (NEUrotechnology, COntrol and Robotics) group of the Polytechnic University of Cartagena (UPCT). Our aim is to develop biologically inspired control systems that achieve the closest human-like performance of robots in a wide area, applying the functional principles of the highly efficient central nervous system, gaining robustness and flexibility.

In cognitive neuroscience, clinical data and brain imaging are used to form a high level conceptual model of the involvement of various brain regions in the studied behaviour –grasping motor behaviour, in our case-. As Arbib [2] details in his brain theory, we may model the brain either functionally, analyzing some behaviour in terms of interacting schemas, or structurally, through the interaction of anatomically defined units, such as brain regions, or substructures of these regions, such as layers or columns, seeking an explanation in terms of neural networks, elements of Marr's computational level, since the neuron may be considered the basic unit of function as well as of structure. However, schemas may be implemented not only in neural networks, but also in conventional computer programs or special-purpose devices. All these approaches rest on effective design of VLSI chips or other computing materials. Using neural networks in the computational level has the advantage of allowing the researcher to take direct inspiration in the performance of the nervous system and exploit, at the same time, the benefits of working with simple yet powerful computational units that offer a biologically plausible solution.

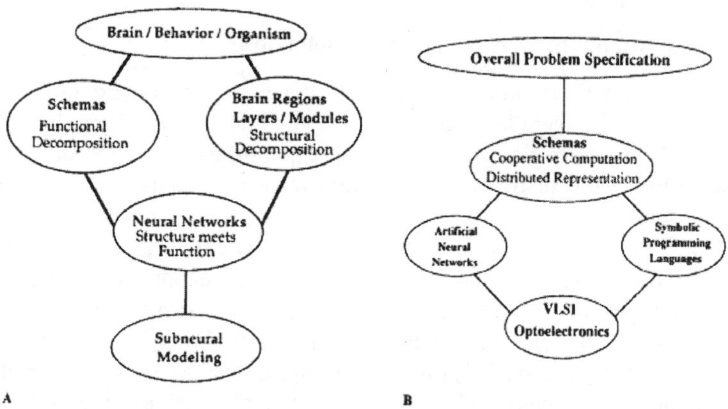

Figure 2. Arbib's [2] views of level of analysis of brain and behaviour (A) and a distributed technological system (B), highlighting the role of schemas as an intermediate level of functional analysis in each case

2. The cortical motor system

Behavorial, cell recording and anatomical data from both human and monkey have been used to anatomically and functionally identify cortical areas that control motor functions. As a brief overview, two of the most important areas are the rostral inferior premotor cortex (referred to as F5) and the anterior intra-parietal cortex (AIP). It is suggested that F5 contains a storage ("vocabulary") of motor actions (Rizzolatti and Luppino [4]), whose words are represented by a population of neurons that codify an action at different degrees of abstraction. It is thought that when an appropriate stimulus is shown, F5 neurons automatically code a potential action, which is the representation of that action. A sector of F5, F5ab, receives a major parietal input from area AIP (Luppino *et al.* [5]). The study of the responses of area AIP neurons indicates that they code three-dimensional objects in visual terms. Taken together, the circuit formed by AIP and F5 plays a role in visuomotor transformations for object grasping and manipulation (Rizzolatti and Luppino [4]). Various attempts have been made to explain how AIP and F5 neurons perform that transformation. Common to these proposals is the idea that some neurons of AIP code the object's intrinsic properties and, then, either directly or via other AIP elements, send this information to specific sets of F5 neurons. F5 transforms the received information into patterns of hand movements that are appropriate to the size and shape of the objects to be grasped. Finally, the F5 grasping pattern recruits specific F1 neurons to execute the grasp (Rizzolatti and Luppino [6]).

3. Computational models for grasping behaviour

Several computational models have been developed following this trend of research. Not all of them have gone down to the level of modelling those areas, but they all rely on exploiting the learning capabilities of neural networks as a way to emulate the distributed information process that is presumed to take place in them and they have all been inspired at some extent by physiological evidence. Uno's model, for instance, tries to solve the grasping problem by modelling a multilayered neural network that learns mappings between objects and grasping gestures, storing those bindings as an abstract representation in an internal layer (Uno [7]). Other models do work with the mentioned brain areas, like those of Sakata *et al.* [8], Moussa [9] and above all Fagg and Arbib [10] and Rizzolatti and Luppino [11] who proposed an alternative model to Fagg and Arbib's model.

In addition to these proposals, NEUROCOR is developing a model of radial basis functions (RBF) based neural networks that performs the dynamic visuomotor transformation made in F5 and AIP, following the interaction model presented by Murata et al. [13], as seen in figure 3, based on the idea that parietal areas of the cortex can correlate different types of sensorial information or different neural coding in a single space of representation through a process that Burnod and his partners (Baraduc *et al.* [14,15]) called *matching process.* From this point of view, the correlation or matching between two different sensorial types or different codings is made in a *matching unit*. We propose that AIP can be understood as one of these *matching units*. Indeed, the closed loop of the circuit below is a mechanism that allows a continuous comparison between the internal representation regarding the grasp target (the object) and another representation regarding the present motor action, generating at every moment the necessary motor commands until the task is completed.

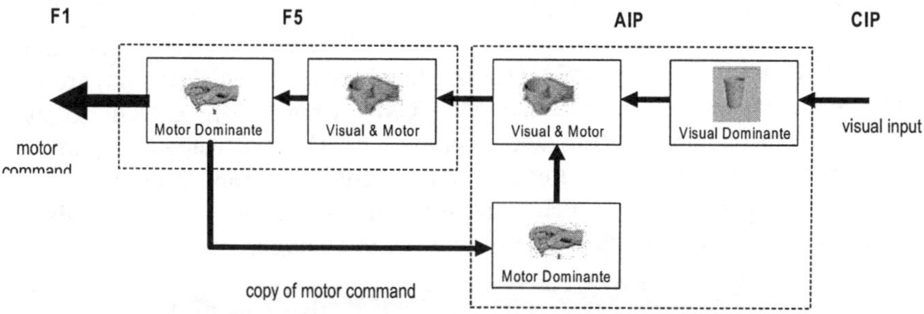

Figure 3. Model of interactions between different cortical areas by Murata *et al.* [13] for control of the prehension movement.

4. A new control architecture

Based on this neuro-inspired control policy, NEUROCOR is developing and implementing neural networks to control a manipulation system composed of an anthropomorphic robotic gripper and a stereoscopic visual guidance system (Coronado *et al.* [16], Taddeucci [17]).

The goal in our approach to efficient object manipulation is to implement a multisensorial neuro-inspired control architecture. It is new in the sense that visual, tactile and propioceptive information will be integrated and handled together; also, because it will be highly modular and also completely platform independent. Its control core will be a group of interconnected HYPBF (hyperplane basic functions) neural networks that will be functionally equivalent to certain brain areas (AIP, F5, F1 and, in the future, the areas that constitute the mirror system). Some of them will constitute a *matching unit* for the visuomotor correspondences made in AIP. It will transform the simultaneous inputs from the actual motor program held by AIP motor dominant neurons and from AIP visual dominant neurons that extract the object's visual properties to a common coding, in order to compare them and compute a difference vector, an operation that is presumably performed by AIP visuomotor dominant neurons, that will be used to adjust the actual motor program prepared by F5 and executed by F1 to control the grasping operation. Through an unsupervised reinforcement learning process the system will tune itself as a neurocontroller capable of executing complex manipulation tasks, guided by sensorial information, thus emulating human grasping behaviour. Its real performance will be tested in our robotic gripper designed in our laboratory using the AER technique, described later

5. The anthropomorphic UCPT-NEUROCOR hand

The biomechanical design of the anthropomorphic UPCT-NEUROCOR hand was inspired by the skills of the human hand. It has three fingers and an opponent thumb, with 4 DOF each of them. They are mounted on a rigid palm. Each of them has 3 independent joints (MCP, PIP y DIP for the fingers and CMC, MCP and IP for the thumb). DIP/IP (thumb) and PIP/MCP (thumb) are 1-DOF joints (flexion/extension) and MCP/CMC (thumb) are 2-DOF joints (flexion/extension and abduction/adduction). Each DOF is actuated by a pair of DC Maxon motors and a couple of polystyrene tendons, enrouted through a set of pulleys through the finger, that stand for artificial muscles, as they emulate the agonist/antagonist muscle actuation made by our nervous system

Figure 4. The NEUROCOR-UPCT anthropomorphic hand. Detail of the fingers, with propioceptive position Hall effect sensors, tactile FSR sensors and the pulley enrouted tendons that move the phalanxes in an agonist/antagonist way

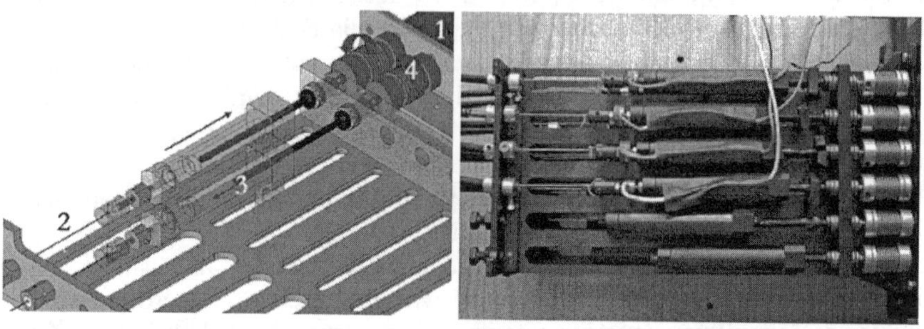

Figure 5. The actuation system for the NEUROCOR-UPCT anthropomorphic hand. Each DC motor (1) moves one polystyrene tendon (2) attached to a horizontal shaft (3) by means of a planetary gearhead (4). There is an encoder adjoined to each motor to control number of turns.

To measure joint position, speed and direction Hall effect sensors, integrated in each joint, have been used. Tactile information is read with FSR sensors placed on the fingertips and on the inner face of medial phalanxes. The hardware control boards receive commands from a dedicated PC and send backwards the requested information. through a PCMCIA communication bus. The neural networks control software is programmed in *MatLab*.

6. AER in control hardware

NEUROCOR works with other groups in a new research field concerning bio-inspired hardware systems that exploit the inter-chip communication technique known as "Address-Event-Representation" (AER). The AER concept, coined in 1991 by Sivilotti [18] in Caltech, is being adpoted and developed by a growing community of researchers interested in the design of VLSI circuits dedicated to emulate real neural systems (Abusland [19]).

At first sight, AER is just a TDM asynchronous communication technique between chips that emulates the high connectivity found between adjacent layers of neurons by means of a high speed digital bus with few wires. The idea is to "place" a group of neurons in one chip, another one in another chip, and use AER to simulate the massive connectivity between both groups. Moreover, AER is capable of complex operations, as a chip working like an emitter connected to several receivers or vice versa, such as translations or rotations that can be made while digital signals are being transmitted.

Among the most ambitious projects regarding AER, we have to mention the works of IMSE research group, trying to develop microchips capable of performing real-time convolutions that would have an enormous potential for artificial vision systems both in a conventional manner (Pitas [20]) and in a bio-inspired way (Sheperd [21]).

7. Conclusion

In conclusion, the innovative aspect of our research work in NEUROCOR consists in the development of AER-based hardware and bio-inspired neural networks software to implement a real platform for performing real-time object manipulation tasks with a robotic gripper in a human-like manner, merging visual, tactile and propioceptive information in the same control architecture and controlling the planning and execution of the hand movements emulating the performance of the central nervous system.

References

1. C. L. MacKenzie and T. Iberall, *Advances in Psychology: The Grasping Hand,* **104** (1994).

2. M.A. Arbib, *The handbook of Brain Theory and Neural Networks,* 13, The MIT Press (2003).
3. D. Marr, *Vision*, San Francisco: W.H. Freeman.
4. G. Rizzolatti and G. Luppino, *Neuron* **31,** 892 (2001).
5. G. Luppino, A. Murata, P. Govoni and M. Matelli, *Exp. Brain Res.* **128,** 181-187 (1999).
6. G. Rizzolatti & G. Luppino, *The handbook of Brain Theory and Neural Networks*, 501-504, The MIT Press (2003).
7. Y. Uno. N. Fukumura, R. Suzuki and M. Kawato, *Neural Networks* **8,** 839-851 (1995).
8. H. Sakata. M. Taira, S. Mine and A. Murata, *Exp. Brain Res.* **22,** 185-198 (1992).
9. M.A. Moussa, *IEEE Transactions on Neural Networks* **15,** 629-638 (2004).
10. A.H. Fagg and M.A. Arbib, *Neural Networks* **11,** 1277-1303 (1998).
11. G. Rizzolatti and G. Luppino, *Neuron* **31**, 889-901 (2001).
12. J. Molina Vilaplana, *A biologically inspired neural architecture for learning and control of grasping movement in anthropomorphic robotic platforms*, Ph.D. Thesis, Polytechnic University of Cartagena (2006).
13. J. Molina Vilaplana, *A biologically inspired neural architecture for learning and control of grasping movement in anthropomorphic robotic platforms*, Ph.D. Thesis, Polytechnic University of Cartagena (2006).
14. P. Baraduc. E. Guigon and Y. Burnod, *Cerebral Cortex* **11,** 906-917 (2001).
15. A. Murata, L. Fadiga, L. Fogassi, V. Gallese, V. Raos and G. Rizzolatti, *Journal of Neurophysiology* **78,** 2226-2230 (1997).
16. J. López Coronado, J.L. Pedreño Molina, A. Guerrero González and P. Gorce, *Robótica* **20,** 23-31 (2002).
17. D. Taddeucci, P. Gorce, Y. Burnod, J. López Coronado, J.L. Pedreño Molina, A. Guerrero González, C. Laschi and P. Darío, *First IEEE-RAS International Conference on Humanoid Robots*, The Massachusetts Institute of Technology, Cambridge. MA (2000)
18. M. Sivilotti, *Wiring Considerations in analog VLSI Systems with Application to Field-Programmable Networks*, Ph.D. Thesis, California Institute of Technology, Pasadena CA (1991).
19. A. Abusland, T.S. Lande and M. Hovin, *IEEE International Symposium on Circuits and Systems, ISCAS'*96 **vol.III**, 401-404 (1996).
20. I. Pitas, *Digital Image Processing Algorithms and Applications,* John Wiley and Sons (2000).
21. G. M. Sheperd, *The Synaptic Organization of the Brain,* Oxford University Press, 3rd Edition (1990).

A CONCURRENT PLANNING ALGORITHM FOR DUAL-ARM SYSTEMS *

JEN-HUI CHUANG AND TING-WEI CHAN

Depart. of Computer Science, National Chiao Tung University
1001 Ta Hsueh Road, Hsinchu, Taiwan 30010, R.O.C.

CHIEN-CHOU LIN†

Depart. of Computer Science and Information Engineering, Shu-Te University
No. 59, Hunshan Rd., Yenchau, Kaohsiung, Taiwan 824, R.O.C.

An algorithm for path planning of object held by multiple manipulators is presented in this paper. The proposed path planning algorithm is composed with two planners: path planner of the object and motion planner of manipulators. Both planners use a generalized potential model to evaluate the repulsion between object/manipulators and obstacles, so as to avoid collision. By adding virtual obstacle planes to roughly divide the workspace into several subspaces, a concurrent approach of path planning is also proposed. With each subspace designated to a manipulator, collisions between manipulators can be avoided. Therefore, paths of all manipulators can be planned at the same time.

1. Introduction

While using only one robot in a workspace limits the classes of tasks that can be performed, more and more multi-robot systems have been considered in recent decades [1..6]. For many applications, manipulators are holding an object and, together, form a closed linkage, named closed kinematic chains. When planning the manipulator motion and the object path, the manipulators should be connected to the object so that the closure constraint is satisfied. Thus, the complexity of motion and path planning for multi-manipulator systems increases significantly from single manipulator systems. In [1], the motion and path planning algorithms of closed kinematic chain manipulator systems are divided into three basic approaches: exact motion planning [1, 2], probabilistic motion planning [3, 4] and mechanism singularity analysis.

This paper proposes a concurrent planning algorithm for chain robot systems. The proposed algorithm is composed of two stages of path planning

* This work is supported by National Science Council of Taiwan under grant no. NSC 94-2213-E-366-011 & NSC 95-2221-E-366 -010.

† Corresponding author, email: jacoblin@mail.stu.edu.tw.

algorithms: (i) global planning of the object, and (ii) motion planning of the manipulators. Both algorithms use the generalized potential model [7] to plan collision-free paths. While (i) is for planning the safest path of the object, (ii) is for planning safest configurations of manipulators whose end-effectors are attached to the object.

The remainder of this paper is organized as follows. In Section 2, we propose the concurrent planning algorithm with some implementation details. Simulations considering more realistic situations are presented in Section 3 to show the effectiveness of the proposed algorithm. Section 4 concludes this paper and outlines some possible directions for future works.

2. Concurrently Planning Algorithm

The scenario of this work involves two manipulators trying to move an object to its destination while avoiding obstacles (including other manipulators). Since the end-effectors of both manipulators will be attached to the object, the path planning of manipulators is thus reduced to simpler motion planning with known end-effector trajectories. Therefore, the proposed algorithm is composed of two planners: (i) the global planner for the moving object and (ii) the manipulator motion planner. Similar to the path planning of a rigid object, the former will find a collision-free path for the object from its start to its goal location, while the latter is to plan a sequence of manipulator configurations to hold the object along the planned object path. The generalized potential model (reviewed next) is used in both planners to avoid collision and to adjust manipulator configurations for minimum potential (minimum risk of collision).
The generalized potential field proposed in [7] is used to model the 3-D environment by representing object/manipulators by charged sample points, and obstacles by charged polygons, respectively. In [7] the potential function is assumed to be inversely proportional to the distance between two point charges to the power of an integer (m). In particular, it is shown that the repulsive force exerted on a point charge due to a 3-D polygon can be obtained analytically by evaluating the gradient of the potential function (for m=3)

$$\int_S \frac{dS}{R^3} \triangleq \Phi(x,y,z) = \sum_i \left[\phi\left(x_2^i, y_2^i, z\right) - \phi\left(x_1^i, y_1^i, z\right)\right] + \frac{\alpha}{z} \tag{1}$$

with (see [10] for details)

$$\phi(x,y,z) = \frac{1}{z}\tan^{-1}\frac{xz}{y\sqrt{x^2+y^2+z^2}}, \tag{2}$$

where the x^i-y^i-z coordinate system is determined by the right-hand rule for each edge i of the polygon such that z is measured along the normal direction of S, and x^i is measured along edge i of the polygon, respectively, and α is a constant. The discussion for repulsive torque is similar and is omitted for brevity.

2.1. *Global Path planning*

For a rough description of the object path, the proposed approach uses one or more guide planes (GPs) [8] as final or intermediate goals in the 3-D workspace. Similar to the approaches in [8], the GPs provide the object a general direction to move forward. A collision-free traversal of a given sequence of GPs by the object is regarded as a global solution of the path planning problem of cooperative manipulators. In this paper, it is assumed the sequence of GPs is given in advance.

Assuming that a guide plane GP_i is given as an intermediate goal, the basic path planning procedure for moving the object onto GP_i is described below:

Algorithm of Global Path Planning

Step 1 Translate the object with distance δ_i along the direction of attraction of the GP_i. Find the smallest such that corresponds to a feasible and collision-free translation. If the object can't move anymore, go to Step 5; otherwise, add an intermediate guild plane perpendicular to the direction of attraction.

Step 2 Slide the object without changing its orientation on the GP to minimize the potential. Considering the forces exerted on the object, let F be the repulsive force exerted on the object due to the repulsion between the object and the obstacles. To determine the direction in which the object should slide on GP, and thus to reduce the repulsive potential, the projection of the resultant force exerted on the object is calculated.

Step 3 Adjust the object orientation for the minimum potential configuration with the fixed position. In Step 3, the object is adjusted in orientation while fixed in position. The direction in which the object should rotate is determined by the repulsive torque experienced by the object with respect to its centroid.

Step 4 If the object reaches GP_i, the planning is completed. Otherwise, go to Step 1.

Step 5 Exit and report that goal is unreachable.

For path planning involving multiple GPs, the above algorithm will be executed for each of them sequentially. It is assumed that the planning for a GP starts as the planning of the previous GP is accomplished. The path planning ends as the object reach the goal, which is usually a (goal) GP in the path planning problems.

2.2. *Motion Planning Algorithm of Manipulators*

The planning of manipulators can simply be regarded as to finding the safest configuration for each robot with fixed-position end-effector and base. The proposed concurrently planning algorithm cuts a chain manipulator into two manipulators and plans two paths of manipulators independently in two separated subspaces, which are formed by auxiliary obstacle planes.

2.3. *Free Space Decomposition*

As shown in Fig. 1(a), there are five auxiliary obstacle planes to roughly separate the free space into two subspaces. The sizes, orientations and positions of these five auxiliary planes is depending on the direction of the object. To make the explanation more clear, the x-y-z coordinate system of the scene is depicted as Fig. 1(a). The first auxiliary plane, *midpoly* shown in Fig. 1(b), is formed at the middle of the object. And, other four planes, the *sidepoly1*, *sidepoly2*, *sidepoly3* and *sidepoly4*, are formed above, below, and on two sides of the object, respectively. While *midpoly* is attached to the object, *sidepoly1* and *sidepoly2* are parallel to *x-z* plane. Fig. 3(c) shows the different orientations of object through rotation with respect to the z–axis, and with various sizes of *sidepoly1* and *sidepoly2*. *sidepoly1* and *sidepoly2* are in fact the extensions of the ends of *midpoly* to the walls of workspace. It is easy to see that the workspace is thus roughly divided into two subspaces.

Similar to *sidepoly1* and *sidepoly2*, *sidepoly3* and *sidepoly4* are parallel to *y-z* plane, and their sizes are also adjustable when the object rotates respective to *z*-axis. Similarly, *sidepoly3* and *sidepoly4* are the extensions of the ends of *midpoly* to the walls of workspace.

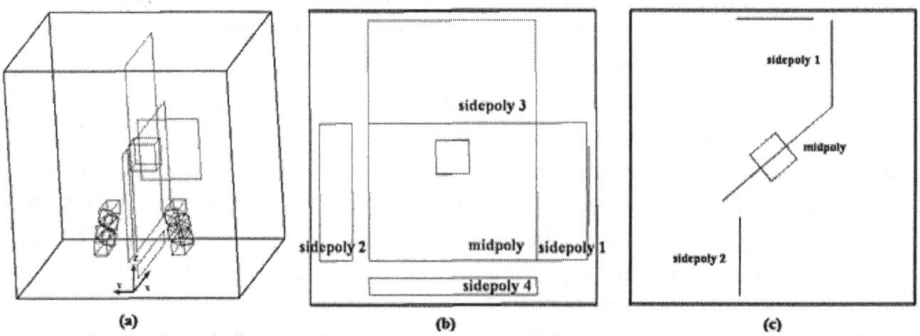

Fig. 1 (a) A dual-manipulator system has two 7-link manipulators, Master and Slave, their goals. (b) Five auxiliary obstacle planes. (c) An example of the space decomposition where the object rotates an arbitrary angle and the positions of the five auxiliary obstacle planes vary as well.

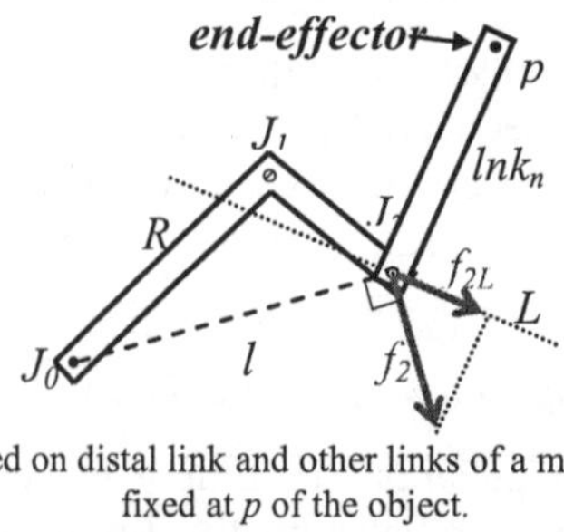

Fig. 2 The forces exerted on distal link and other links of a manipulator with its end-effector fixed at *p* of the object.

2.4. *Configuration Optimization*

As shown in Fig. 2, the direction to which the distal link should rotate is determined by the total repulsive torque experienced by the distal link with pivot *p*. Let f_2 be the repulsive force from obstacles exerted on J_{n-1} (n=3 in Fig. 2). The resultant torque experienced by lnk_n with respect to *p* is equal to

$$\tau^* = \tau_n + f_{2L} \cdot l_n \quad (3)$$

where l_n is the length of lnk_n and f_{2L} is the projection of f_2 onto a plane perpendicular to lnk_n. To find the minimum potential orientation of lnk_n, binary searches are performed using the projection of τ^*_n along three orthogonal axes of rotation, one axis at a time. After the optimal orientation of lnk_n is found, the orientations of the rest links are adjusted recursively for connectivity and for minimum potential using τ^*_{n-1}, τ^*_{n-2}...., etc.

3. Simulation Results

In this section, we will carry out computer simulation of path planning algorithms regarding multiple robot arms holding an object. The simulating platform is based on Windows XP running on a Pentium4 2.8Ghz, 512MB RAM PC.

In this example we place a narrow passage between two closed spaces. The object is in the right space initially. We hope to move it to the left space by two five-link manipulators, as shown in Fig. 3(a). Two GPs (*bot1* and *bot2*) are used in this example. We first use *bot1* to guide the object and the robot arms go through the narrow passage, then use *bot2* to pull them to the goal position. Fig. 3(b) and Fig. 3(c) show the partial trajectories of the dual-arm systems.

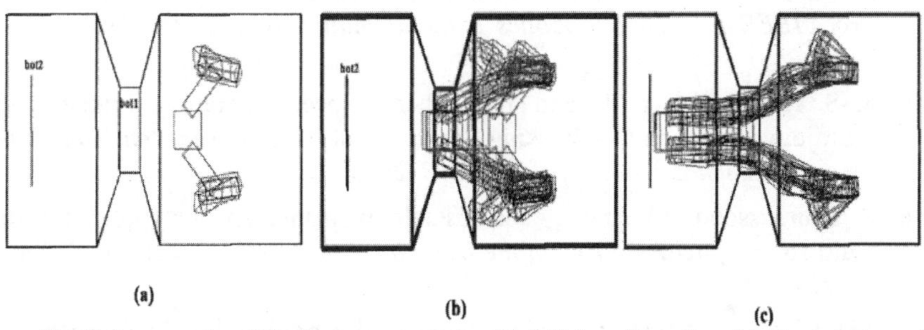

Fig. 3 An example within the narrow passage. (a) Initial configuration. (b) The trajectory of robots from start to bot1. (c) The trajectory of robots from bot1 to bot2.

4. Conclusions

The main purpose of this paper is to route the path for multiple robot arms holding an object in a three dimensional workspace. The proposed algorithm is based upon the generalized potential field model which produces repulsive force between obstacles and the robot arms along with the object, assuring that the arms and the object will not collide with the obstacles when moving along the planned path in the work space. The concept of virtual spacing plane is introduced to appropriately divide the workspace into subspaces, each for a robot arm, in order to resolve the issue regarding robot arms colliding each other and to make parallel routing possible instead of establishing a process queue for multiple robot arms. According to computer simulation, the paths we obtain are moderately smooth and continuous.

In our frame of reference, we only bring out simulations and tests aiming at the applications of multiple robot arms but did not take in some of the actual mechanical concerns of robot arm operations such as acceleration, energy consumption, and the load of the arms and joints. Nonetheless, the path we obtained is still valuable as a start or a reference for further detail modifications for actual applications. Our routing solutions then become a bridge between theory and reality.

References

1. J. C. Trinkle and R. J. Milgram, “Complete path planning for closed kinematic chains with spherical joints,” *Intl. Journal of Robotics Research*, vol. 21, no. 9, pp. 773-789, Sep. 2002.
2. G. F. Liu and J. C. Trinkle, “Complete path planning for planar closed chains among point obstacles,” *in Proc. Robotics: Science and Systems*, 2005.
3. J. H. Yakey, S. M. LaValle, and L. E. Kavraki, “Randomized path planning for linkages with closed kinematic chains,” *IEEE Trans. on Robotics and Automation*, vol. 17, no. 6, pp. 951-958, 2001.
4. J. Cortés, T. Siméon, and J. P. Laumond, “A randomized loop generator for planning the motions of closed kinematic chains using PRM methods,” *in Proc. IEEE Intl. conf. Robotics and Automation*, vol. 2, pp. 2141-2146, 2002.
5. K.-S. Hwang, M.-Y. Ju and Y.-J. Chen, “Speed Alteration Strategy for Multijoint Robots in Co-Working Environment,” *IEEE Trans. on Industrial Electronics, vol. 50, no. 2, pp. 385- 393, 2003.*
6. P. Tournassoud, “A strategy for obstacle avoidance and its application to multi-robot systems,” *Proc. of the IEEE Int. Conf. on Rob. & Auto. (ICRA), pp. 1224--1229, 1986.*
7. J-H. Chuang, “Potential-based modeling of three-dimensional workspace of the obstacle avoidance,” *IEEE Trans. on Rob. & Auto.*, vol. 14, no. 5, pp. 778-785, 1998.
8. C.-C. Lin, J.-H. Chuang, “Potential-based path planning for robot manipulators in 3-D workspace,” *IEEE Int. Conf. on Rob. & Auto. (ICRA), pp. 3353-3358, 2003.*

A modular approach for controlling mobile robots

K. Regenstein, T. Kerscher, C. Birkenhofer T. Asfour, J.M. Zöllner, R. Dillmann

Interactive Diagnosis- and Servicesystems (IDS), Research Center for Information Technologies (FZI),

Karlsruhe, 76131, Germany

**E-mail: regenstein@fzi.de, kerscher@fzi.de, birkenhofer@fzi.de, asfour@fzi.de, zoellner@fzi.de, dillmann@fzi.de*

www.fzi.de/ids

A variety of different robots was built at our institute. As these robots differ as well in size, shape and in actuation principle it would be very time consuming and inefficient to tailor a computer and hardware architecture especially to the specific robot. In this paper it will be described how common aspects in robot control can be identified and how modular hardware components can be derived from a modular software framework and a respective computer architecture. A decentralized computer architecture based on embedded PC systems connected to local controller modules via CAN Bus was developed. The requirements and restrictions that led to the development of these controller modules and their associated power amplifier boards will be described

Keywords: Computer Architecture; Modular Control Concept; Hardware/Software Co-Design

1. Introduction

In a large number of robotic systems a decentralized architecture is used.[1–4] At the Research Center for Information Technologies (FZI) different kind of robots - like humanoid robots, four- or six-legged walking machines, mobile platforms and snakelike sewer inspection robots - are developed. For these robots we designed a computer architecture based on embedded PCs and distributed controller modules connected with each other via one or more CAN-Busses.[5] Though the requirements concerning the distributed components are quite different we wanted to implement a persistent design that could be used in all robots with only small amount of adaptation. The main issues for the controller modules used in our robots are space requirement, power consumption, several inputs for sensor value acquisition and communication interfaces (i.e. CAN-Bus). As none of the available of-the-shelf products suited all these needs we decided to build a controller module and an associated power amplifier ourselves.

2. Computer Architecture

The mechatronical construction of a robot can roughly be divided into mechanical aspects and into aspects of setting up the electronic and computer system. In this section we will describe how the electronic system of our robots is set up and how a computer architecture suiting the needs in these robotic systems was designed.

We started by identifying the concepts of how a robot should accomplish given tasks and thus proposing a control architecture that then was the basis for developing the computer architecture. We chose a hierarchically organized control system for the robots with the three following levels:[6]

- The task planning level specifies the subtasks for the multiple subsystems of the robot. Those could be derived from the task description autonomously or interactively by a human operator
- The task coordination level generates in sequence/parallel primitive actions for the execution level in order to achieve the given task goal. The subtasks are established by the task planning level. The execution of the subtasks in an appropriate schedule can be modified/reorganized by an operator using an interactive user interface
- The task execution level is characterized by control theory to execute specified sensory-motor control commands. This level uses task specific local models of the environment and objects, which represent the active scene

According to the control architecture the computer architecture is structured into three levels as well. Choosing suitable devices for these three levels yielded that the requirements of the task planning and task coordination level could be met with industrial PCs and PC/104 systems. As cabling in the robot is a major issue it is desirable to reduce cabling efforts as much as possible. Because of this we decided to use a decentralized system for sensory-motor control. By placing controller modules close to the motors and sensors cabling can be reduced to one common power supply and a bus connection. Wires for supplying the motors and connecting the sensors to the controller can be kept short and have not to be passed through moving joints. To fulfil the requirements of the task execution level we designed the so called Universal Controller Module (UCoM) that in combination with our

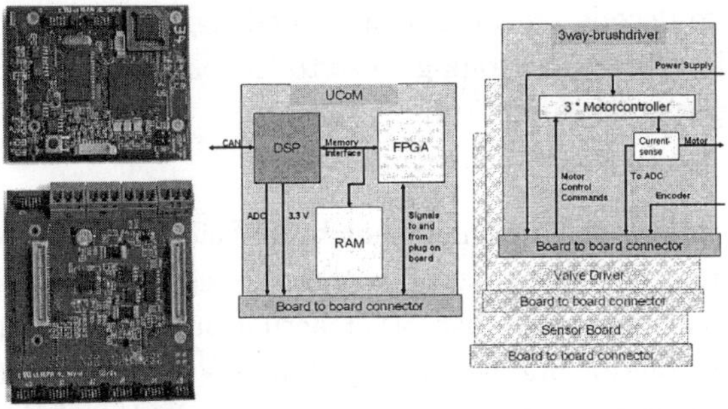

Fig. 1. The Universal Controller Module (UCoM) (upper left) and the 3way-brushdriver (lower left); Schematic overview of data flow on the UCoM and to the piggyback board

motor controller 3way-brushdriver (Fig. 1) is responsible for the sensory-motor control of the robot. The features of the combination of UCoM and 3way-brushdriver will be described in section 3 in more detail.

As software framework we use the Modular Controller Architecture (MCA2)[7] that was developed at our institute and is available under GPL online here.[8] The idea behind MCA2 is to structure the software into reusable modules with simple interfaces. Each module has the three data channels: control data, sensor data and parameters. Via these data channels information is exchanged between the modules. A number of modules can be combined into a group which has the same interface as one module.

3. Controller modules on sensory-motor level

Following the modular strategy that is realised in the software framework MCA2 we wanted to reach this modularity in the computer architecture as well. To achieve this goal not only for one of our robots but spanning all different robots we have to choose a common controller unit. This is important to reduce programming efforts as well. For example you can implement a PID-controller only once and as you use the same hardware that can run the same software in different robots you only have to adapt the PID parameters for the chosen joint. As already mentioned the main issues for the controller modules are space requirement, power consumption, several inputs for sensor value acquisition and communication interface. Especially for motor control the mandatory requirements for the controller module were

- Suitable to control three brushed DC motors at 24 V at up to 5 A
- Achieve cycle times as low as 1 ms

- Able to decode six quadrature coded signals
- Small outline, positioning close to the actuator possible
- Low power consumption
- Interface to access CAN-Bus

In some applications the 5 A might not be sufficient but this was the maximum current that could be realized without exceeding the space limitation. Besides electrical motors we use other actuation principles in our robots. For example in one of our walking machines - Airbug[9] - fluidic muscles were used and there is still ongoing research evaluating fluidic muscles as actuation. For this kind of actuation a valve driver is needed. Furthermore in some robots like LAURON[10] we need extended sensor input like posture information from gyroscopes and acceleration sensors. To avoid building a special controller module for each of these applications we decided to split the controller module into one part that actually contains the controller and one part that contains the power amplifier, the valve driver or sensor acquisition electronics. As mentioned above we named the part with the actual controller Universal Controller Module (UCoM) as it will be universally used in our robots together with the respective piggyback board.

3.1. *Universal Controller Module - UCoM*

The choice for a suitable microcontroller/DSP for the UCoM could be narrowed down rather quickly as nearly no available controller had all the required features. The Freescale "DSP56F803 16-bit Hybrid Controller"[11] came closest to our needs. This controller is a DSP featuring a set of peripherals usually only known from microcontrollers.

Though this hybrid controller nearly matches the requirements it still misses some essential features. On the one hand side it does not have enough general purpose IOs to control three motors on the other hand it only has two quadrature timers capable of decoding quadrature coded signals. To extend the DSP's flexibility we decided to put an FPGA next to it. As suitable FPGA we chose the Altera EPF10k30A. With this FPGA we can equip the UCoM with a high number of general purpose IOs. We gain a high flexibility concerning routing and assignment of pins to the piggyback board. Through the FPGA we can reassign most of the signals so that it suits the used piggyback board. The FPGA is also used to preprocess data that is exchanged between the UCoM and the piggyback board.

By this approach we can disburden the DSP from tasks that are done in hardware more efficiently. For example we implemented six decoders for quadrature encoded signals. The communication between DSP and FPGA takes place via the external memory interface. As FPGA and the external RAM that we integrated on the UCoM share the external address range we implemented an address decoder into the FPGA. This address decoder deasserts the chipselect for the lowest 64 addresses of the external address range and receives the sent data. For all other addresses the data is routed to the external RAM so that nearly the whole external memory range is available to the DSP and only the lowest 64 addresses are used as FPGA-registers. The DSP always initiates communication with the FPGA by writing to or reading from a FPGA-register. These registers are used to exchange data between the two devices. So to get the value of a quadrature coded signal all the DSP has to do is access the respective external RAM address.

The UCoM combines the Freescale "DSP56F803 16-bit Hybrid Controller", an external RAM and an Altera EPF10k30A on one board. It is interfaced to the desired piggyback board via two 60-pin 0.8 mm pitch board-to-board connectors. Via this connector the UCoM is supplied with a 5 V power supply. From this 5 V we generate the 3.3 V that are needed on the UCoM. We do not directly feed 3.3 V to the UCoM to avoid problems with voltage drop or disturbances that must be expected due to the wiring close to the motor power wires. The 3.3 V generated on the UCoM are then fed back to the board-to-board connector to be available on the piggyback board.

3.2. *Motorcontrol Board*

As the main actuation principle in our robots are electronic motors the first piggyback board we developed was a power amplifier able to drive three brushed DC motors. We named this piggyback board 3way-brushdriver. On the 3way-brushdriver we integrated three H-Bridges which are driven by a 3-phase brushless DC motor controller chip each. This motor controller chip can be configured to drive brushless or brushed motors. To drive brushless motors the hall-inputs to the driver must be connected to the hall-sensors of the motor. To drive brushed DC motors the hall-inputs are simply tied to ground. We chose this motor controller chip so that it can be interfaced by software in the same manner if we need to design a piggyback board for brushless DC motors in the future. In both branches of the H-Bridge we integrated a shunt via which we can measure the motor current for each motor. As the motors are driven by PWM signals we had to use an OP-AMP with a high gain-bandwidth-product to amplify the signals for use in

Fig. 2. Robots in which the presented computer architecture is actually used: ARMAR III (left) and LAURON IV (right)

the AD-Converter.

As we wanted to keep the UCoM as small as possible we decided to put all interface connectors except the ones that are directly wired to the DSP like CAN-Bus, serial communication interface and JTAG to the piggyback board. Thus the piggyback board is responsible for supplying the UCoM with the 5 V input voltage. The 3way-brushdriver has a connector for input of the aforementioned 5 V, 24 V as power supply to the motors and a common Ground.

Further connectors on the motor control board are six small connectors for the quadrature coded encoder signals. Each of these connectors has six pins, two of which carry supply voltage of 5 V and ground for the encoder, two are the quadrature channels A and B. The remaining two are an index signal and a pseudo absolute code. A schematic overview of the dataflow on the UCoM and to the piggyback board can be found in Fig. 1.

3.3. *Software Components on DSP*

To extend the modularity down to the code for the DSP each UCoM is treated as one module in the MCA2 software framework. Thus the UCoMs can be seamlessly integrated into the software running on the linux PC via a MCA-driver interfacing to the CAN-Bus. A number of different basic application programs for the UCoM were developed so that they can be used in most of the robots with only small adaptations. In the humanoid robot ARMAR-III[12] we only use one generic program for all different joints. This generic program can be adapted via a configuration file that is evaluated at system startup. Other basic application programs that are implemented are: direct_pwm, a p_controller, a pid_controller and a Speed/Position Controller. Further control programs including for example time discrete and torque control algorithms are in development.

To download these programs to the UCoM a bootloader that accepts data via CAN-Bus is used. This is very convenient if programming is done frequently in the development phase.

3.4. *Functions on FPGA*

The FPGA can be seen as an extension to the DSP. It equips the DSP with a large number of general purpose IOs and processes data that is exchanged between the DSP and the piggyback board. As stated above we use the memory interface to communicate between DSP and FPGA. The function blocks in the FPGA are programmed in VHDL. There are some blocks that are common to all designs independent of the plug on board: An Address Decoder, a version supervision sytem, Initialization and a Watchdog.

For the introduced 3way-brushdriver the following function blocks were already implemented into the FPGA: A Quadrature decoder, Pseudo Absolute Decoder, Serial Synchronous Interface and a Motor Control register. Further modules easily can be integrated into the FPGA as they are needed.

4. Summary and Outlook

In this paper we presented a modular concept to control robots which we applied to our robots ARMAR III and LAURON IV (depicted in Fig. 2). This concept includes a control architecture from which the used computer architecture was derived, a modular software framework and the development of a hardware architecture that can be mapped into the computer architecture. The focus of this paper was on setting up a modular system that can be used in a variety of robots so that not only software components can be reused but that also the hardware is interchangeable. It was laid out how this goal was achieved and especially the development of the UCoM and the 3way-brushdriver were described. Some examples of our robots in which these control devices are successfully used were presented above. In ongoing research we will implement this concept in robots that are going to be built. Most likely further piggyback boards will be designed for that purpose.

5. Acknowledgement

This work is partially supported by the German Research Foundation (DFG) within the Collaborative Research Centre SFB 588 (Humanoid

Robots - Learning and Cooperating Multimodal Robots).

References

1. J.-Y. Kim, I.-W. Park, J. Lee, M.-S. Kim, B. kyu Cho, and J.-H. Oh, "System Design and Dynamic Walking of Humanoid Robot KHR-2," in *Proceedings of the 2005 IEEE International Conference on Robotics and Automation, Barcelona, Spain, April 2005*, 18-22 April 2005, pp. 1431–1436.
2. T. Sugihara, K. Yamamoto, and Y. Nakamura, "Architectural design of miniature anthropomorphic robots towards high-mobility," in *Intelligent Robots and Systems, 2005. (IROS 2005). 2005 IEEE/RSJ International Conference on*, 2-6 Aug. 2005, pp. 2869–2874.
3. J. Butterfass, G. Hirzinger, S. Knoch, and H. Liu, "DLR's multisensory articulated hand. I. Hard- and software architecture," in *Robotics and Automation, 1998. Proceedings. 1998 IEEE International Conference on*, vol. 3, 16-20 May 1998, pp. 2081–2086vol.3.
4. R. Bischoff and V. Graefe, "HERMES - a versatile personal robotic assistant," *Proceedings of the IEEE*, vol. 92, no. 11, pp. 1759–1779, Nov. 2004.
5. K. Regenstein and R. Dillmann, "Design of an open hardware architecture for the humanoid robot ARMAR," in *Humanoids 2003, International Conference on Humanoid Robots, Conference Documentation, October 1 - 3, 2003, Karlsruhe and Munich, Germany*, October 2003, p. 3.
6. T. Asfour, D. Ly, K. Regenstein, and R. Dillmann, "Coordinated Task Execution for Humanoid Robots," in *Experimental Robotics IX*. Springer Berlin / Heidelberg, 2006, vol. 21, pp. 259–267.
7. K. U. Scholl, J. Albiez, and B. Gassmann, "MCA - An Expandable Modular Controller Architecture," in *3rd Real-Time Linux Workshop, 2001, Milano, Italy*, 2001.
8. FZI, "Modular Controller Architecture Version 2." [Online]. Available: http://www.mca2.org/
9. K. Berns, J. Albiez, V. Kepplin, and C. Hillenbrand, "Airbug - Insect-like MachineActuated By Fluidic Muscle," in *Proceedings of CLAWAR 2001, Int. Conference on Climbing and Walking Robots*, 2001.
10. B. Gassmann, T. Bär, J. Zöllner, and R. Dillmann, "Navigation of Walking Robots: Adaptation to the Terrain," in *Proceedings of CLAWAR 2006, 9th International Conference on Climbing and Walking Robots (CLAWAR)*, 2006.
11. Freescale, *DSP56F803 Data Sheet.* [Online]. Available: http://www.freescale.com/files/dsp/doc/data_sheet/DSP56F803.pdf
12. T. Asfour, K. Regenstein, P. Azad, J. Schröder, A. Bierbaum, N. Vahrenkamp, and R. Dillmann, "ARMAR-III: An Integrated Humanoid Plattfrom for Sensory-Motor Control," in *Proceedings of IEEE-RAS International Conference on Humanoid Robots*, 2006.

A SELF ORGANIZING NETWORK MODEL FOR CLAWAR SYSTEMS COMMUNICATION COEVOLUTION

FABIO P. BONSIGNORIO

RTD, Heron Robots s.r.l., V.R.Ceccardi, 1/18
Genova, I-16121, Italy

This paper shows a model of robot-robot communication that is based on the concept of 'network embodied cognition'. The communication between intelligent clawar agents is seen as an activity made possible by the self-organisation of coevolving 'situated' and 'embodied' cognitive processes, physically distributed among the inter-communicating agents, motivated and initiated by physical 'finalized' interactions with the environment.

The benefits and trade-offs of this approach for managing swarms of CLAWAR systems are compared and discussed against alternative solutions.

1. Introduction

This paper exposes a model for the evolution of a communication pragmatics between networks of CLAWAR autonomous agents, between them and in perspective with humans.

In nature there are many kinds of loosely coupled networks of intelligent agents, largely varying in terms of quantity of agents and cognitive and adaptive capacity (i.e. of computational needs) of each agent.

In the natural domain the most widely used method of 'intelligence', computation and 'cognition' are 'embodied' biological neural networks.

A number of empirical and theoretical researches, [2,3,4] are investigating, on one side on the aspects and implication of 'embodiment', particularly interesting in the walking machine domain,[15], on the other side on the 'emergence' of cognition from network interaction of physical agents, [7,8,9].

We propose here a model of cognition were the model of the environment 'emerges' from the collective interaction of a networked robot system with its environment.

We assume (for simplicity) that the 'cognitive network ' can be accessed by all the agents which were/are coevolving it and in fact share (constitute) it.

In this conceptual framework 'models of the environment' are shared very naturally, while is needed a concept of 'self'.

This is actually local to the agent and evolved by means of evaluation of the interaction of the specific agent with the environment.
By means of random action generation the nodes which have no consequences on the individual agent are labeled by that agent as 'non self'.
This mimic the basic behavior of the human and mammals immune system (which can be regarded by itself as a 'cognitive system'.

2. The cognition model

The self organizing coevolutive model of communication, and cognition, proposed here, consider the building of the model of the environment by a number of networked physical agents, a 'swarm' , 'situated' in a an evironment and interacting with it, as a collective learning process represented by the growth of a network of nodes mapping, for simplicity, the already identified sensory motor correlations.

The random activity of the network of moving and interacting CLAWAR agents allow the system to identify the regular patterns in the variables (sensors and actuator propioceptors data flow) and to connect them in a model of the environment.

After that the first node has been created, new nodes (highly correlated groups of sensors and actuators) are attached preferentially to the previous one, according to their 'fitness'.

The resulting representation is shared (again as a simplification) between all the agents.

In the following paragraph some information are given about the modeling of network growth process, the fitness function which drive the preferential attachment of new nodes and the 'self' model of each agent.

2.1. *Network growth model*

In this section we review the recent fast progress in statistical physics of evolving networks, see [21], with the aim of grounding a model of communication for physical automous agent typically represented by CLAWAR.

So far researcher interest has focused mainly on the structural properties of random complex networks in communications, biology, social sciences and

economics.
A number of giant artificial networks of such a kind came into existence recently. This opens a wide field for the study of their topology, evolution, and complex processes occurring in them.
Such networks possess a rich set of scaling properties. A number of them are scale-free and show striking resilience against random breakdowns. In spite of large sizes of these networks, the distances between most their vertices are short, a feature known as the "small world" effect.
It is known that growing networks self-organize into scale-free structures through the mechanism of preferential linking.
Numerous networks, e.g., collaboration networks, public relations nets, citations of scientific papers, some industrial networks, transportation networks, nets of relations between enterprises and agents in financial markets, telephone call graphs, many biological networks, food and ecological webs, metabolic networks in cell etc., can be modeled in this way.
It is thought that evolving self organizing networks can (could) model the collective knowledge of a network of intelligent (artificial) autonomous agents. An 'object/process' in the environment is modeled by a node of the network with many links generated by a fitness process within a coevoluted learning process.
In our model the probability that a new node will connect to a node i already present in the network is a function of the connectivity k_i and on the fitness η_i of that node, such that:

$$\prod_i = \frac{\eta_j k_i}{\sum_j \eta_j k_j} \tag{1}$$

A node i will increase its connectivity ki at a rate that is proportional to the probability that a new node will attach to it, giving:

$$\frac{\partial k_i}{\partial t} = m \frac{\eta_i k_i}{\sum_j k_j \eta_j} \tag{2}$$

The factor m accounts for the fact that each new node adds m links to the system.
Thus the connectivity distribution, i.e. the probability that a node has k links, is

given by the integral:

$$P(k) = \int_0^{\eta_{max}} d\eta \, \frac{\partial P(k_\eta(t) > k)}{\partial t} \propto \int d\eta \, \rho(\eta) \left(\frac{m}{k}\right)^{\frac{C}{\eta}+1} \tag{3}$$

We define a proper η_i function which is basically a performance index of the effectiveness of sensory motor coordination.

2.2. *The fitness function*

In our model the fitness function of the node i, η_i ,controls the growth process of the cognitive network.
It is given by a function of the mutual information, between the sensors and the actuators connected to that node. The mutual information between two given variables is given by equation (4), where X and Y two random variables:

$$I(X,Y) = -\sum_i \sum_j P_{XY}(i,j) \log \frac{P_X(i)P_Y(j)}{P_{XY}(i,j)} \tag{4}$$

If X and Y are statistically independent eq. 4 gives $I(X,Y)=0$
It is possible that a more sophisticated performance index, including for instance 'transfer entropy' [10] could be helpful, as mutual information doesn't give any indication about the direction of information flow from X to Y, or the opposite, but only a measure of correlation between the two variables.
It is interesting to notice that entropy based performance indexes have been proposed to guide the search within evolutionary algorithm as in [11]

It is important to notice that nodes are not coinciding with the physical CLAWAR agents.

They can be sub processes of a single robot system, for instance the coordination process between the vision sensors and the front legs, or the one between the infrared and the sound sensors.

2.3. *The 'self' model*

As told, in the model of cognition described in this paper the sharing of the model of the environment is very natural, while of the contrary, a model of 'self'

must be defined explicitly.
Each agent is able distinguish its 'self' from the 'environment' with a parallel selection process inspired by the mammals immune system.
The key concept is that any physical agent will activate at random all the network node actuators and will identify which of them will trigger a change in the environment measuring the transfer entropy between the actuators of the node and the related sensors.
The 'self' of the robot will be the list of nodes for which it has experimented consequences in the environment.
This is similar to the 'self' identification process in some models of the immune system, [12].

3. A toy system representation

A simplified model of what we are discussing is described in this paragraph.
In Table 1 is represent the cartesian product between the sensor measure (streams) $\boldsymbol{S_i}$ and the actuator propioceptor values $\boldsymbol{A_{j,,}}$ of all the CLAWAR agents of the 'swarm'.
The nodes are created when the fitness function ,representing the mutual information between the nodes, is above a 'relevance' threshold.
The nodes are created as the system evolve in time, as the individual agent 'explore' randomly the environment. Any new node attach preferentially to the already existing according to their 'fitness '.

Table 1. A simplified represenataion of the node mapping in a example case.

	S_1	S_2	S_3	S_4	S_5	S_6
A_1						N_3
A_2		N_1		N_4		
A_3			N_6		N_2	
A_4				N_5		

In figure 1 is represented the resulting connected graph of the nodes.
Table 1 and figure 1 represent the 'model of the world' shared by the networked robot systems.

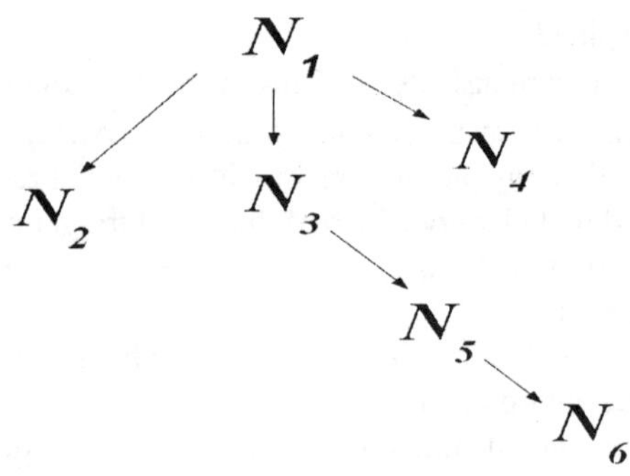

Figure 1. Node connection graph

In figure 2. is depicted the representation of the 'self' of an agent.

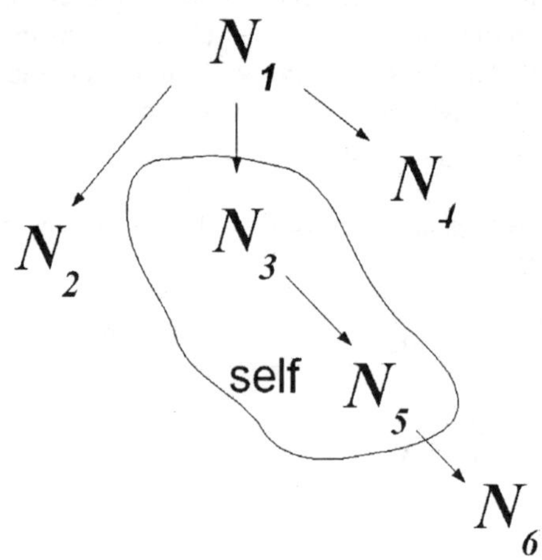

Figure 2. Schematic representation of an agent self.

As in the 'real' environment' we should have 'many' sensors and actuators belonging to many CLAWAR systems the network will grow in accordance to the equations recalled in paragraph 2.1 .

4. Computing Backbone Architecure

As it should emerge from the previous discussion, the architecture of this cognition system is inherently highly parallel.

We think that the advantages of this kind of algorithm should be emphasized by the appropriate distribution of the modules constituting the system over a highly parallelized multiserver backbone computing facility.

We describe below how we plan to implement the cognition system on an advanced grid based distributed architecture, named RoboGrid [22].

The n-tiered structure of the RobotGrid middleware is depicted in Fig. 3. At the lower layer we have the infrastructure resources, we have the grid services layer, which implements the virtualization of the resource of the bottom level, according to the SOA (Service Oriented Architecture) paradigm on demand. Above we have the application server level, where the (e.g. enterprise java) beans implementing the planning and control logic reside.

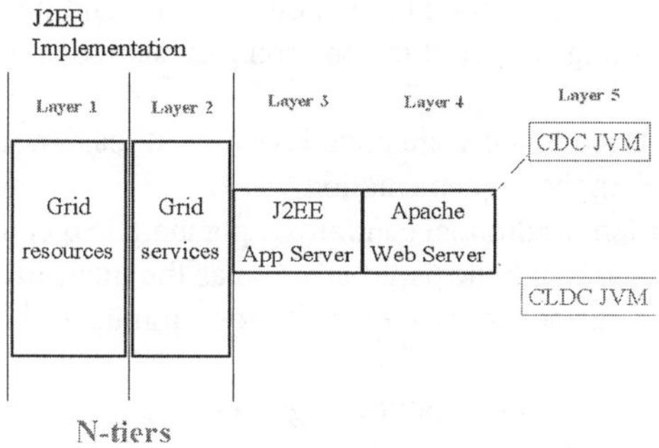

Figure 3. The n tiers

We show in figure 4 the flow of the sense/reaction scheme of a single robot. The 'applet' running on the robot onboard virtual machine sense e.g. a movement of a 'focus object' in the 'environment'.

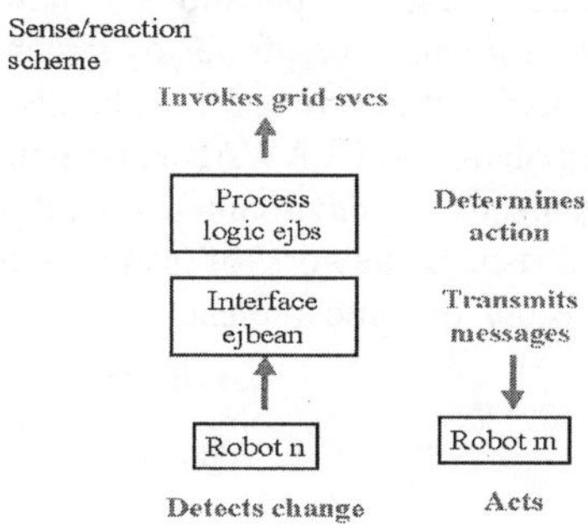

Figure 4. The sense/reaction scheme

A front end unit of service (an interface 'enterprise java bean') is invoked with a string of parameters , the bean activate the necessary actions instantiating the necessary transient grid services (adding resources where necessary) and some where another bean dispatch a message that trough an on board applet act in the environment (e.g. moving one or more robot on the field from one position to another) .

All sensor and actuator information and the network representation of the environment are shared.
Only the 'self' information is stored in the single agent.

5. Discussion

The approach described here has something in common with evolutionary programming and particle swarm optimization and 'swarm intelligence' in general as it is an emergent coordination process, on the other end it is 'deeper' as it generates and 'emerging' model of the environment shared among the networked robot systems.
This occurs at the price of a certain complexity, or at least, not immediately intuitive 'simplicity', in the implementation.
In comparison to more traditional centralized planning and control schemes based on the so called 'symbolic paradigm' , it has the main advantage in the fact that doesn' t need, in principle, any explicit programming of knowledge of the environment.
It also seems a good model for 'natural cognition' processes if the model is extended to take care of 'embodiment'.

6. Futher Research

From an experimental perspective it would be useful to benchmark in a practical situation against other more classical approaches in order to verify if it could enable more more robust and autonomous applications of networked CLAWAR.
From a theoretical standpoint a formalization of the 'toy system' in section 3 could help to develop a quantitative framework for 'embodied' and 'emergent' cognition in robot systems in general and CLAWAR in particular.
An application to the self organization of environment cognition in single robot systems (including humanoids) seen as 'network' of (semi) independent sensory motor units could also be, possibly, of some interest.

7. Conclusions

We have described a model of self organizing communication and cognition schema that allow the development of new architecture networked robot systems, which could, in principle, show more autonomy than current one, on the other side this model could help formalize a (semi) quantitative theory of the emergence of cognition in 'embodied' agents.

References

1. C.E. Shannon, “The Mathematical Theory of Communication”, *Bell Sys. Tech. J.* 27,379,623 (1948).

2. R.A Brooks, “Robust Layered Control System for A Mobile Robot” *IEEE Journal of Robotics and Automation*, (1986)

3. R. Pfeifer, “Cheap designs: exploiting the dynamics of the system-environment interaction. Three case studies on navigation.”, In: Conference on Prerational Intelligence --- Phenomonology of Compexity Emerging in Systems of Agents Interacting Using Simple Rules. Center for Interdisciplinary Research, University of Bielefeld, 81-91, (1993).

4. R. Pfeifer, F. Iida, “Embodied artificial intelligence: Trends and challenges. Embodied artificial intelligence” Iida et al. (Eds), LNCS/AI Vol. 3139, 1-26, Springer, (2004).

5. H. Touchette, S. Lloyd, “Information-theoretic approach to the study of control systems “, *Physica* **A 331**, 140-172, (2004).

6. P. Bak, C. Tang, K. Wiesenfeld, “Self-Organized Criticality: An Explanation of 1/f Noise” *Phys. Rev. Letter.* 59,4,381-384, (1987).

7. G. Gomez , M. Lungarella, D. Tarapore, “Information-theoretic approach to embodied category learning” Proc. of 10th Int. Conf. on Artificial Life and Robotics, pp.332-337,(2005).

8. D. Philipona, J.K. O' Regan, J.-P, Nadal, O.J.-M.D. Coenen, “Perception of the structure of the physical world using unknown multimodal sensors and effectors”, in *Advances in Neural Information Processing Systems*, (2004).

9. L. Olsson, C.L. Nehaiv, D. Polani, “Information Trade-Offs and the Evolution of Sensory Layouts” In Proc. Artificial Life IX, (2004).

10. M.Lungarella, O. Sporns, “Mapping information flow in sensorimotor networks”, PLOS Computational Biology, 2/10,1301-1312, (2006).

11.S.Liu, M.Mernik, B.R. Bryant, “Entropy-driven parameter control for evolutionary algorithms”, *Informatica*, **31**, 42-50, (2007).

12. D. Morpurgo, R. Serenità, P. Seiden, F. Celada, “Modelling

thymic functions in a cellular automaton", *International Immunology*, 7/4, 505-516, Oxford, (1995).

13. P. Gacs, "The Boltzmann Entropy and Randomness Tests" In: Proc. 2nd IEEE Workshop on Physics and Computation (PhysComp'94), 209-216, (1994).

14. P. Gruenwald, P. Vitanyi, "Shannon Information and Kolmogorov Complexity", IEEE Transactions on Information Theory, (2004).

15. M. Garcia, A. Chatterjee, A. Ruina, M.Coleman, "The Simplest Walking Model: Stability, Complexity, and Scaling," *Transactions of the ASME, Journal of Biomechanical Engineering,* 120, 281-8, (1998).

16. S. Lloyd, "Use of mutual information to decrease entropy: Implication for the second law of thermodynamics", *Phys. Rev.* **A, 39**,10, 5378-5386,(1989).

17. N. Wiener, "Cybernetics: or Control and Communication in the Animal and the Machine", MIT Press, Cambridge, MA, (1948).

18. J.C Maxwell,"Theory of Heat",Appleton, London, (1871).

19. J. Kennedy , R. Eberhart,"Particle swarm optimization",Proc.of the IEEE Intl. Conf. On Neural Network, Washington DC, USA,1942-1948, vol.4 (1995).

20. M. M Millonas, "Swarms, Phase transitions, and Collective Intelligence" In C.G. Langton (Ed.), Artificial Life III, pp. 417-445, Santa Fe Institute Studies in the Sciences of the Complexity, Vol. XVII, Addison-Wesley , Reading, Massachussetts,(1994).

21.R.Albert, A.L. Barabasi, "Statistical physics of complex networks", Rev. Mod. Phys. **74**,47-97,(2002).

22. F.P. Bonsignorio, "A grid based distributed multiagent enabling system for intelligent autonomous robot swarms", ISR 2005, http://www.isr2005.com/,Tokyo, (2005).

23.F.P. Bonsignorio, "Preliminary Considerations for a Quantitative Theory of Networked Embodied Intelligence", in Special issue AI50, *LNAI*, Springer, Heidelberg, (2007) (to appear)

AN APPROACH TO GLOBAL LOCALIZATION PROBLEM USING MEAN SHIFT ALGORITHM

GIOVANNI MUSCATO, SALVATORE SESSA
D.I.E.E.S. University of Catania, Viale A. Doria 6
Catania 97100, ITALY
gmuscato@diees.unict.it, ssessa@diees.unict.it

This paper describes a global localization algorithm (GLA) for a mobile robot in indoor environments, based on a particle filter. This algorithm uses data obtained by two kinds of sensors: encoder and scanner laser. Given a map of the environment, where the robot moves, a GLA tries to localize the robot on the map by using its sensor data. The map of the environment is preliminary built mixing laser readings from well-known poses of the robot. The mean shift algorithm (MSA) processes the map and obtains a list of features, which is the synthetic map of the environment used in the GLA. The MSA is also applied for each sampling step in order to calculate the importance factor of the particles. The trials have been performed by using a dynamic simulator of a differential drive robot and the 3Morduc mobile robot.

1. Introduction

Robot localization is the problem of estimating the robot pose relative to a fixed external reference frame. A Global Localization Algorithm (GLA) tries to localize the robot on the map by using its sensor data and the map of the environment. In other words, the GLA answers to the question "Where is the robot?" placing a robot somewhere in a known environment [1]. Grid-based techniques were among the firsts GLAs implemented; the map is divided into cells and their values represent the probability of the robot to be located in it. Unfortunately, the computational requirements are proportional to the square of the dimension of the map. Simmons et al. have applied this method in [2] to localize a robot in an office environment and Burgard et al. in [3] to localize a robot in a museum. A solution to these computational problems is the map matching technique that can be applied directly with the raw data of the sensors, as described in [4] by Konolidge and Chou, or by identifying suitable landmarks of the environment, as described by Thrun in [5]. Motivated by the famous condensation algorithm used in computer vision, Dellaert et al. in [6] and Fox et al. in [7], were the first to develop a particle filter for mobile robot localization. Jensfelt et al. in [8] applied Montecarlo localization to features based maps.

In the implemented algorithm, the map of the environment is a list of features obtained by using the Mean Shift Algorithm (MSA) [9]. The implemented algorithm relies on a particle filter; each sample contains the estimated pose of the robot (position and heading). In the measure phase, when a scanner laser measure is ready, the MSA computes the local map of the features. The importance factor of the samples can be updated by evaluating the distance of the recognized features (in the local map) from their positions in the global map. The proposed algorithm has been tested by using a dynamic simulator of a differential drive robot based on Open Dynamic Engine library [10]. Moreover several experimental trials have been performed on the 3Morduc, one of the mobile robotic platforms available in our laboratory (Fig.1).

Figure 1. 3Morduc mobile robotic platform.

In order to increase the interoperability between the simulator and the real environment, the software architecture of the robot is fully compatible with the exchanged commands between the GLA application and the simulator software [11].

2. The clustering algorithm

The MSA is a classical way to extract features from images in computer vision. The algorithm shown is an adaptation of that described in [9] to scanner laser measurements with, in additions, some rules to extract clusters. These clusters represent the features extracted from the environment, used to build the global map during the initialization phase and the local maps in the execution phase. The main target of the MSA algorithm is that of finding all the local maxima of the multivariate kernel density estimate function. That function is obtained as sum of all the kernel function each centered in a laser scanner measure point and its local maxima represent all the point in which the scanner laser measurements are more dense as explained below. The algorithm is iterative and has been demonstrated in [9] that it is convergent.

Given a scanner reading $\{x_i\}_{i=1...n}$, the multivariate kernel density estimate is obtained as follows:

$$\hat{f} = \frac{1}{nh^2} \sum_{i=1}^{n} K\left(\frac{x - x_i}{h}\right) \tag{1}$$

Where $K(x)$ is a kernel function with windows radius h. The optimum kernel function yielding minimum mean integrated square error is the *Epanechnikov kernel*:

$$K_E(x) = \begin{cases} \frac{2}{\pi h^2}(1 - x^T x) & x^T x < 1 \\ 0 & otherwise \end{cases} \tag{2}$$

The use of a differentiable kernel allows computing the estimate of the density gradient as the gradient of the kernel density estimate:

$$\hat{\nabla} f(x) \equiv \nabla \hat{f}(x) = \frac{1}{nh^2} \sum_{i=1}^{n} \nabla K\left(\frac{x - x_i}{h}\right) = \hat{f}(x) \frac{4}{h^2} M_h(x) \tag{3}$$

Where $M_h(x)$ is the sample mean shift vector:

$$M_h(x) \equiv \frac{1}{n_x} \sum_{x_i \in S_h(x)} (x_i - x) \tag{5}$$

Where $S_h(x)$ is a circumference of radius h centered in x and containing n_x points. The mean shift vector has the direction of the gradient of the density estimate at x and points towards the direction of the maximum increase in the density. As a result, the mean shift algorithm is executed by iterating the two phases: computation of the mean shift vector $M_h(x)$ and translation of the window $S_h(x)$ by $M_h(x)$

The iteration process ends when the components of $M_h(x)$ are less than a threshold. To improve the independence of the feature determination by the distance of the scanning point, the h radius is expressed as:

$$h_i = kd_i \tag{4}$$

Where k is the *feature compression parameter* and influences the final number of features found for each laser scan. Fig.2 shows the results of the feature extraction algorithm applied to a square room of 4 x 4 meters by using three different values of k. If k is too small (Fig.2a) all the scanning points are evaluated as features, on the contrary, if k is too big (Fig.2c) the algorithm will found only one features in the barycenter of the scan. For this reason, the tuning of this parameter is fundamental to obtain good feature extraction results (Fig.2b). After this pre-elaboration phase, the points (the n centers of the windows) are clustered by considering their distances. The barycenter of the cluster is the representative point of the set, and corresponds to a feature of the environment. Each feature is characterized by a weight and a variance parameter. The weight parameter represents the importance of a feature in the localization problem, with respect to the other features. It is computed by counting the cluster elements. The variance parameter, instead, represents the precision of the estimation in the feature location referred to the fixed frame. It is calculated as the maximum distance between the barycenter of the cluster and the windows centers that belong to the cluster itself.

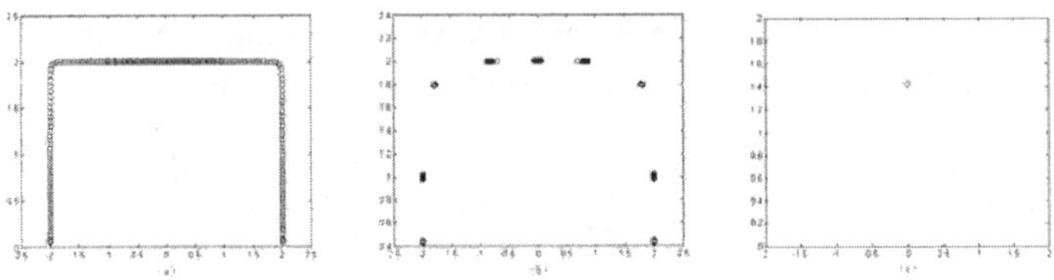

Figure 2. Centers of the windows with three values of k parameter a:$\pi/100$ b: $\pi/20$ and c:π.

3. The implemented algorithm

The particle filter technique is well suited to deal with the GLA. It does not make strong assumptions on the probability density of the robot pose after the sensors readings. As a result, a particle filter can be used to represent complex multimodal beliefs due to multiple possible poses.

The application of this algorithm requires an initialization phase, during which a map of the environment is built by means of a features list. In order to do that, the robot is placed on a set of well-known poses in a global reference system and the clustering algorithm is applied to the corresponding scanner readings, as described in section 2. The segmentation algorithm creates a list of features that are points in the global reference system characterized by a weight (number of scanning point linked to the feature) and a variance. These two parameters are used to estimate the importance factor of the particles. The particles filter algorithm is initialized by placing the samples randomly in the map with the same initial heading of the robot. For each laser reading, corresponding to the measure phase of the particle algorithm, the clustering process calculates the local map of the features. For each sample, the local map is roto-translated in order to compare it to the global map, assuming that the robot is located in the sample pose. It is then possible to compute the importance factor of each particle, as described in the following section. This parameter leads the re-sampling phase of the particle filter.

4. Estimation of the importance factor

The importance factor is the posterior probability of the robot to be in the particle pose after the sensors readings. After the measure of the scanner laser, the features of the local map are computed in the global reference frame for each particle. In this way, the global list of features can be compared to the features obtained by a scanner reading in the same reference frame. The first step to obtain the importance factor is to compute the minimum distance of the N_f features from the feature list (made of N_{fl} elements), for each particle k.

$$d_{ki} = \min \arg(\left\| z_i - z_j \right\| \; j = 1 \ldots N_{fl}, i = 1 \ldots N_f \tag{4}$$

Where z_i is the i-th feature obtained from the scan and z_j is the j-th feature of the feature list. The importance weight of the k-th particle now can be computed as follows:

$$w_k = \prod_i N_i \exp\left(-\frac{d_{ki}^2}{\sigma_i^2} \right) \tag{4}$$

Where N_i is the number of scanning points that belong to the cluster i-th on the feature list and σ_i is the variance of the i-th cluster.

5. The software architecture of the system

The first step of the software development was to give the possibility to distribute the calculus in many computers, so a client-server architecture has been chosen.

The other target to satisfy was the interoperability between the simulator and the real environment. As a result the “Simulator Server” and the “3Morduc Server” provides the same interface to the “Localization Client”.

Fig. 3 shows a block diagram of the whole system architecture [11].

The next subsection describes:

•Simulator

•3Morduc Server

•Localization client

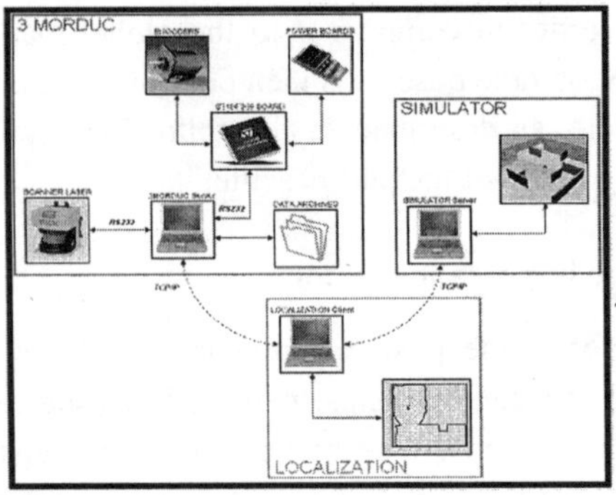

Figure 3. The system architecture

6. Simulator

The environment simulator has been developed using GLscene components for Delphi 7 that integrates Open Dynamic Engine [10].
The geometric dimensions of the simulated robot (wheelbase, radius of the wheels, etc.) and the weight are the same of the 3Morduc platform.
The simulator models the measures coming from the scanner laser; in order to obtain the intersections between rays and walls, a geometric algorithm has been used. The scanner laser noise has been modeled using the classical beam model of range finders, thus by adding hit, short, max and random noise [1].
Hit noise has a narrow Gaussian distribution characterized by a mean equal to zero and a standard deviation σ_{hit}.
Short noise simulates the presence of unexpected objects and the probability of range measurement in such situations and is described by an exponential distribution.
Max noise simulates failures, i.e. when a ray is lost. This problem is modeled by using a point-mass distribution centered at the maximum range of the laser.
Random noise, finally, models the unexplainable measurements as a uniform distribution.
The simulator models also the encoders and examines the effect of two kinds of noise sources. First a Gaussian noise, due to the encoder quantization, has been added to the measures, with mean $\mu = 0$ and standard deviation $\sigma = \dfrac{\pi}{2N_e}$, where N_e is the number of encoder pulse per turn. Moreover, it has been considered the wheel-ground interaction modeled as a contact joint [11] (see fig. 4).

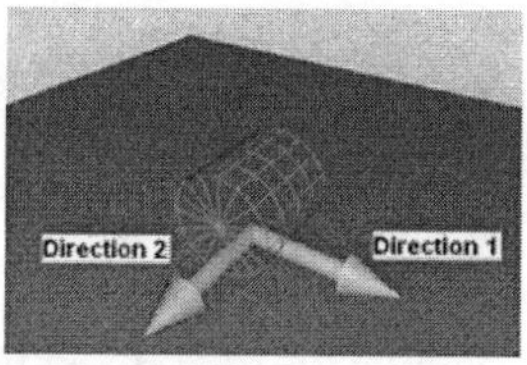

Figure 4. Contact joint

This interaction prevents cylinder and plane inter-penetrating at the contact point and forces the bodies to have an "outgoing" velocity in the normal direction of the contact. Contact joints typically have a lifetime of one time step. They are created when a collision is detected and then immediately deleted.
Contact joints can simulate friction at the contact by applying forces in the two friction directions, perpendicular to the normal one.
The main parameters to characterize the contact are:
•mu: Coulomb friction coefficient ($mu \in \left[0,+\infty\right[$); zero results in a frictionless contact, and ∞ in a contact that never slips.
•mu2: Optional Coulomb friction coefficient for friction direction 2 ($mu2 \in \left[0,+\infty\right[$).
•Bounce: Restitution parameter ($Bounce \in \left[0,1\right]$); zero means that the surfaces are not bouncy at all, 1 represents maximum bounciness.
•Bounce vel : The minimum incoming velocity necessary for bounce (in m/s). Incoming velocities below this will effectively have a bounce parameter of zero.
•Soft erp and Soft cfm: normal "softness" of the contact. Fig. 5 shows a screenshot of the simulator.

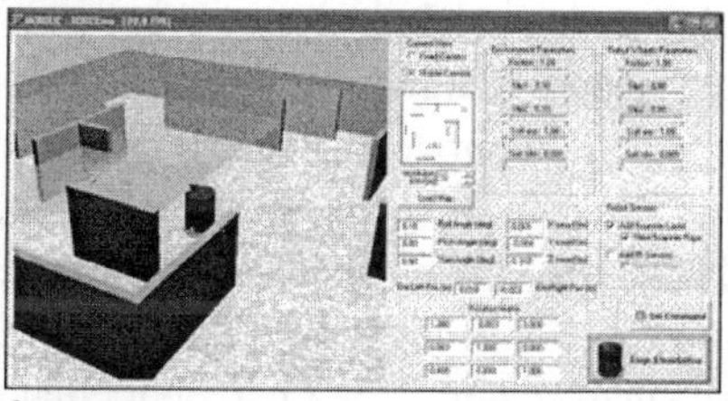

Figure 5. Simulator screenshot

7. The 3Morduc Server

The 3Morduc Server provides to the client the same command interface of the Simulator application.

It implements the RS-232 communication protocol with scanner laser to setting its parameters and obtains the measures.

The 3Morduc server also performs the communication protocol with the ST10F269 motor control board to tuning the control parameter, set the speed of the robot and get the encoders measures.

Finally, it adds a timestamp to the measures to permit, on the client, the synchronization of the data.

8. The Localization Client

The Localization client requests the sensors measures to the server and allows the user to choose between Simulator data or 3Morduc real measures.

Another important task of the client is the data synchronization using timestamp, to have scanner laser measures consistent with the encoders data.

The localization client execute the GLA and allows the user to set the number of the particles and the type of estimation of the importance factor.

The implemented weight estimation algorithms are:

- Standard - uses the minimum distance of the laser data.
- Mean Shift – uses the algorithm previously described.

The user has to load the map of the environment, an image that represents the walls and the obstacles with black pixels on white background, before launching the GLA.

9. Test and results

The first test in simulation mode has been prepared in a simple room of $11m^2$. Initially the clustering algorithm is used to obtain a list of features representing the global map. The number of particles used in the GLA is 10000. They have been initialized randomly in the map space as shown in Fig.6a with an initial heading of $-\pi$. The robot goes straight at constant speed and the scanner laser is acquired at 4 Hz. The standard method, that uses only the minimum distance of the laser data to estimate the importance factor, does not localize the robot after 23s (Fig.6a, Fig.6b and Fig.6c). The Mean shift method localizes the robot pose in 5s by using only 20 scanner laser readings, the particles tends to condensate in the correct position as shown in Fig.7.a, Fig.7.b and Fig.7.c. The particles are concentrated in a very small region with a maximum diameter smaller than one centimeter with a distance from the real position of 5mm.

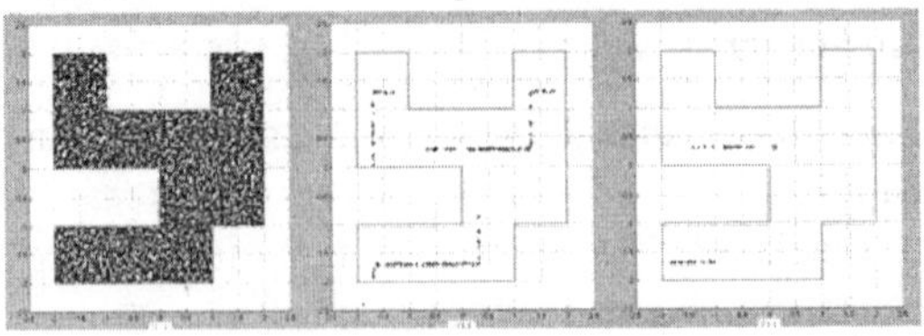

Figure 6. Particles distribution in classical GLA for three different time (a) t=0 s, (b) t=5 s and (c) t=23 s.

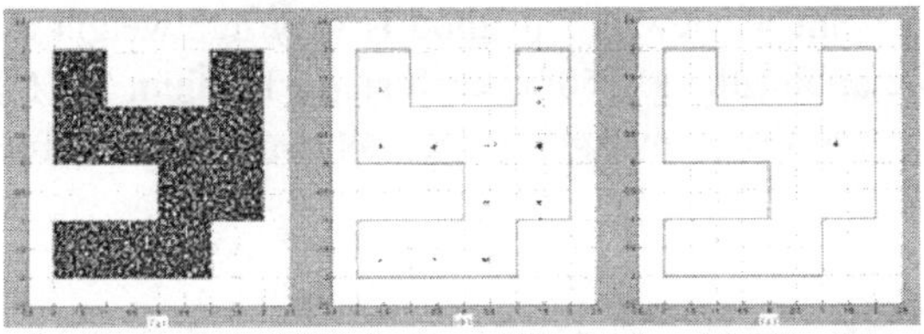

Figure 7. Particles distribution using MSA for three different time (a) t=0 s, (b) t=2 s and (c) t=5 s.

The second test was performed using the real 3Morduc platform and has been prepared in a room of about $24m^2$. The environment map has been reconstructed by using six scanner data rows obtained when the robot was in well-known pose. The number of samples in this case was 5000, uniformly distributed in the room with initial heading of $-\pi/2$ as shown in Fig.8a. The scanner laser data was acquired at 4 Hz sample frequency and the encoders measures at 10 Hz. The GLA localizes the correct pose after 6 s, only 24 scanner laser data were needed to condensate the samples in a small zone with 7 cm of diameter (Fig.8a, Fig.8b and Fig.8c).

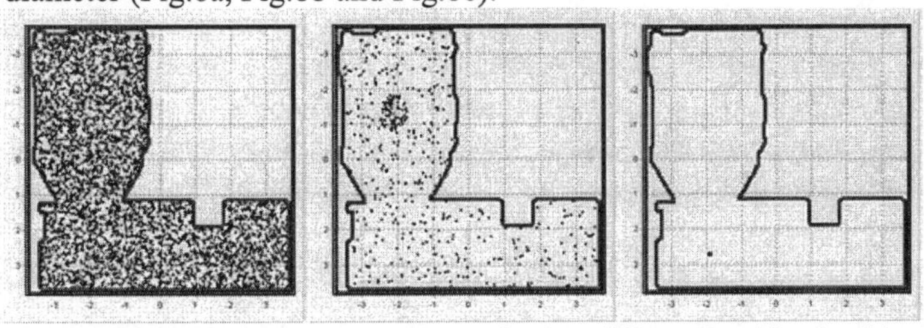

Figure 8. Particles distribution in a real environment for three different time (a) t=0 s, (b) t=3 s and (c) t=6 s.

10. Conclusion

This paper presents a GLA based on features extraction and particle filters, applied to a differential drive robot. The environment map is a list of features created using an adaptation of the MSA to the scanner laser readings. The features are not geometric entities (corners, walls, objects, etc.) and most of all they are very robust to noise. Moreover, two parameters are assigned to each feature. The weight represents the relative importance of a feature with respect to the other ones, while the variance is a measure of the feature precision in the map. In this way, the map quality can be evaluated by considering the variance of its features. The importance factor of the particles used for re-sampling is computed by comparing the feature position on the local map to the features on the global one. The local map is computed from the scanner laser measures using the same segmentation algorithm: mean shift and clustering. From a computational point of view this method is very lightweight and so it can be used in real-time applications (150ms each step - Pentium 3GHz) . Finally, the algorithm has a very fast convergence rate, as demonstrated from the reported results.

Acknowledgments

This work has been partially supported by STMicroelectronics of Catania, Automation and Robotics Team.

References

1. S. Thrun, W. Burgard, D. Fox, "Probabilistic robotics", Ed. The MIT Press, 2005.
2. R.G. Simmons, D. Apfelbaum, W. Burgard, D. Fox, M. Moors, S. Thrun, and H. Younes. 2000 , "Coordination for multi-robot exploration and mapping". Proc. of the National Conference on Artificial Intelligence (AAAI), 2000.
3. W. Burgard, A.B. Cremers, D. Fox, D. Hähnel, G. Lakemeyer, D. Schultz, W. Steiner and S. Thrun, "Experiences with an interactive museum tour guide robot", Artificial intelligence 1999, 114:3-55.
4. K. Konolidge and K. Chou, "Markov localization using correlation", In proceedings of the international Joint Conference on Artificial Intelligence (IJCAI), 1999.
5. S. Thrun, "Bayesian landmark learning for mobile robot localization", Machine Learning 33,1998
6. F. Dellaert, D. Fox, W. Burgard and S. Thrun, "Montecarlo localization for mobile robots", Proceedings of the international Conference on Robotics and Automation (ICRA), 1999.
7. D. Fox, W. Burgard, F. Dellaert and S. Thrun, "Montecarlo localization: Efficient position estimation for mobile robots", Proceedings of the National Conference on artificial intelligence (AAAI), 1999
8. P. Jensfelt and H.I. Christensen, "Pose tracking using laser scanning and minimalistic environment model", IEEE Transaction on robotics and Automation 17 pp.138-147, 2001.
9. D. Comaniciu and Peter Meer, "Mean shift analysis and applications", IEEE International conference Computer Vision (ICCV), 1999.
10. R.Smith, "Open Dynamic Engine ver. 0.039", User Guide, 2004.
11. G. Muscato, D. Caltabiano, "A New Modular Architecture for the Mobile Robot MORDUC: from the Hardware to the SLAM Algorithm", Proceedings of the 1st International conference on Dextrous Autonomous Robots and Humanoids (DARH2005), 2005.

ASYNCHRONOUS LOCAL POSITIONING SYSTEM BASED ON ULTRASONIC ACTIVE BEACONS AND FEED FORWARD NEURAL NETWORKS*

PABLO ESTÉVEZ, JUAN G. HERNÁNDEZ, JOSÉ CAPPELLETTO AND JUAN C. GRIECO

Mechatronics Group, Electronics and Circuits Department, Simón Bolívar University, Sartenejas, Caracas 1020, Venezuela

This paper describes the design and test of an asynchronous 3D local positioning system architecture for small indoor environments, using ultrasonic bursts emitters as active beacons. Their times of arrival at the device are measured, defining its position in an hyperbolic reference system, and these coordinates pass trough a feed forward neural network which calculates the equivalent Cartesian coordinates. Various simulated setups are used to evaluate the configurability of the system and one 2D setup is physically implemented and tested, showing positioning error measurements for each case. The proposed architecture shows itself easily configurable to a variety of setups, capable of solving the positioning problem with small errors; its low computational cost together with the asynchronous approach allows for a distributed implementation on small devices.

1. Introduction

Spatially distributed applications, such as mobile robots and sensor networks, make good use of localization systems, which allow them to know their current position for tagging their sensor data and planning their navigation through the working space. There are different options for it, depending on the size and characteristics of the working space, the number of devices, and even the cost, many of them referenced in [1]. The work presented on this paper is a part of an investigation on underwater applications with multiple mobile robots, and the special characteristics of that environment ask for a system not depending on electromagnetic waves, since the attenuation factor is high in water; it should also be distributed and asynchronous to be easily scalable to big number of devices, and should be capable of 3 dimensional positioning. This paper describes the design, implementation and tests of an LPS which meets most of the requirements previously mentioned, even if not yet meant for underwater environments. The proposed architecture relays on fixed ultrasonic emitters and

* This work was supported in part by Simon Bolivar University Research & Development Deanship.

one mobile receptor. By using the difference on the ultrasonic bursts time of arrival (DTOA), the position of the device is determined, with a feed forward neural network performing the needed calculation.

The second chapter of this paper briefly describes the theory behind the localization algorithm here used. Also, neural networks (NN) are presented as an option for solving the equations system related to this kind of localization. In the third chapter, its actual implementation is described. The fourth chapter describes the set-up for the system tests, and shows the results gathered from them, which are analyzed in the fifth chapter. The sixth one presents the conclusions about the system performance and gives recommendations for future works.

2. Theory of Operation

2.1. *Hyperbolic localization*

This technique is based on the difference in the travel time of signals (usually radio or ultrasonic waves) between beacons and the device to be located. Differences in the TOA of each signal depend on the difference in distance from the device to each beacon. The time difference (and hence the distance difference) between signals associated with to beacons defines a set of possible locations described by an hyperbolic surface, creating an hyperbolic coordinate for the device. In a system with m beacons B_k, $k=1...m$ located at known positions (x_k,y_k), each pair (B_i, B_{i+1}), $i=1...m-1$ generates an hyperbolic coordinate h_i. Notice that even if there are many other possible pairs, their time differences are a combination of these ones, thus providing no new information to the system. At Cartesian coordinates (x,y), the ith hyperbolic coordinate of the device is:

$$h_i = \sqrt{(x_k - x)^2 + (y_k - y)^2} - \sqrt{(x_{k+1} - x)^2 + (y_{k+1} - y)^2} \quad (1)\ [2]$$

In a three dimensional working space and as the number of beacons increases, the solution space shrinks from surfaces to curves, then a set of points and, finally, a single location. Also, constrains in the working space help on finding the solution. For instance, a two dimensional space requires 4 beacons to locate the device, and if only a section of it is observed, three beacons are enough, as shown in Figure 1.

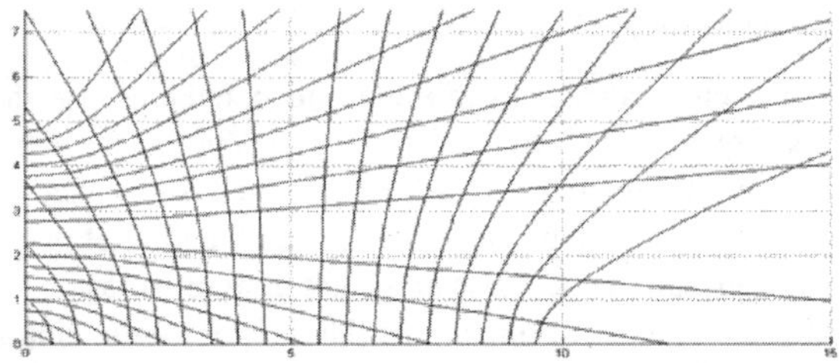

Figure 1: Hyperbolic two-dimensions map for a three beacons system.

2.2. *Neural Networks Solving Method*

The problem faced here is finding a position given a set of measured time intervals (or hyperbolic coordinates), which means finding a common solution for a set of equations of the kind of (1). That equation has no explicit solution. In hyperbolic navigation systems like LORAN and CHAYKA [2], maps are used for graphical identification of the intersection. But for automated systems numerical methods are needed, like the spherical interpolation presented in [3], least-squares methods or the stochastic gradient method used in [4] for a setup similar to this one. In this work, a Neural Networks approximation to the problem is suggested. As can be seen in Figure 1, the problem is that of a nonlinear continuous space transformation. The space described by the "hyperbolic coordinates" of the device, given by the DTOA, has to be transformed into a space described by Cartesian coordinates, and that is a problem that feed forward neural networks can easily solve.

To train the neural network a set of input-output pairs should be provided. These can be obtained from real measurements for a better result, but a simulated approximation can also be used. Using (1) is easy to find the hyperbolic coordinates for each Cartesian one, and these pairs (reversed) can be used to train the NN.

3. Implementation

3.1. *Ultrasonic Channel Sharing Protocol*

For this system, distributivity is a requisite. Therefore, the mobile device should do all of the calculations and needs to act as receiver of the signals emitted by the beacons. Since all the emitters beacons send their ultrasonic bursts at the same frequency, a protocol have to be established to share the channel. In this case, the time division multiple access scheme proposed in [4] is used.

3.2. *Neural Networks algorithm*

To make the system scalable, distributed and untethered, the feed forward neural networks are required to run on small microcontrollers with low computational power. The neural network library used is designed using integer numbers and optimizes the size of the variables and the kind of operations made with them, to ensure small memory use and computing time [5].

The estimate of the number of operations (NOO) per localization is computed using relations (2) and (3). It is important to notice that all of these instructions are performed over integer numbers, thus minimizing their computational complexity. When using a standard neural network library, all these operations become floating point ones and the number of multiplications is increased due to the sigmoid function implementation.

N_{ads} = Number of add operations

N_{mults} = Number of mult. operations

N_{inputs} = Number of inputs

$N_{outputs}$ = Number of outputs

N_{HidenN} = Number of neurons in hiden layer

$$N_{ads} = N_{HidenN} \cdot (1 + N_{inputs} + N_{outputs}) + N_{outputs} \quad (2)$$

$$N_{mults} = N_{HidenN} \cdot (1 + N_{inputs} + N_{outputs}) + N_{outputs} \quad (3)\ [5]$$

For the training process, a set of points is got from a grid in the working space, and given the positions of the emitters, the receiving time intervals are calculated for each one. A NN is created using the functions provided with the library's tool, and the data set is used to train it to solve the localization problem. The resulting network is tested with Matlab® and exported, to include it on the microcontroller code. Also, real inputs-output pairs can be measured at locations on the working space and used as part of the training data.

3.3. *Ultrasonic Transceivers*

For the physical implementation, SFR04 commercial ultrasonic range finders were used as transceivers. These devices are designed to measure distances, but in this case they are used only as burst emitters and device receivers. With the receiver pointing upwards and the emitters fixed some distance above the ground, a good enough signal could be received in most of the work area. Nevertheless, these transceivers have a narrow lobe shape and even if they could cover most of the area when pointing to its center, pointing to the proximity of the device was needed in some areas, to avoid detecting reflections instead of line of sight waves. This doesn't affect the validity of the tests and can be solved using omnidirectional emitters for final implementations.

4. System tests and results

4.1. *Simulation*

The simulated tests ran in Matlab®. A data generator program was written to create the vectors containing the Cartesian coordinates and the corresponding DTOA for various arrangements of beacons. This program generated a training data set from a regular grid and a validation data set of randomly chosen points. Using Matlab®'s nntool and the [5] library transfer functions, different networks architectures were tested, and their mean absolute error (MAE) to the real position over the verification data set were measured after 300 epochs. All the NN have one hidden layer of either sigmoidal neurons (SN) or an integer and computationally cheaper approximation from [5] (ISN), and a linear output layer. The Number of Operations (NOO) is calculated using (2) and (3) for the ISN networks. The SN network uses Matlab®'s functions whose NOO are not reported. The results of these tests are shown in Table 1.

Table 1: MAE for various configurations of the system.

N°	Work Space	Beacons Arrangement	FFNN hidden layer architecture	MAE (cm)	NOO (adds,mults)
1	2D, from [0 0 0] to [3 4.5 0] meters.	[0 0 1], [2 0 1] and [0 3 1]	20 SN	0.06	-----
2			40 ISN	0.55	202, 202
3			20 ISN	1.29	102, 102
4		[0 0 1], [0 2.5 1] and [2.25 1.3 1]	20 ISN	1.49	102, 102
5	3D, from [0 0 0] to [3 3 3] meters.	[0 0 0], [2 0 0], [0 2 0] and [0 0 2]	40 ISN	1.85	283, 283
6			20 ISN	2.81	143, 143
7			10 ISN	6.54	73, 73
8		[0 0 0], [2 0 0], [0 2 0], [0 0 2] and [2 2 2]	20 ISN	2.53	163, 163
9	Same as test 6 with added Gaussian noise on validation inputs.			7.18	143, 143
10	Same as test 8 with added Gaussian noise on validation inputs.			6.45	163, 163

4.2. *Physical setup*

In order to test the system in a real setup, the 3rd and 4th configurations from Table 1 were implemented. Both the emitter's protocol and the receiver code were programmed in separate microcontrollers. A 50 cm by 50 cm grid was painted on the 2m by 3m work space, and the emitters were placed 1 meter above the ground level to have better area coverage with the used transceivers.

This area was free of obstacles, but surrounded by a normal laboratory environment, with walls and other ultrasound reflecting objects. The spatial distribution of MAE in these tests can be seen on Figure 2.

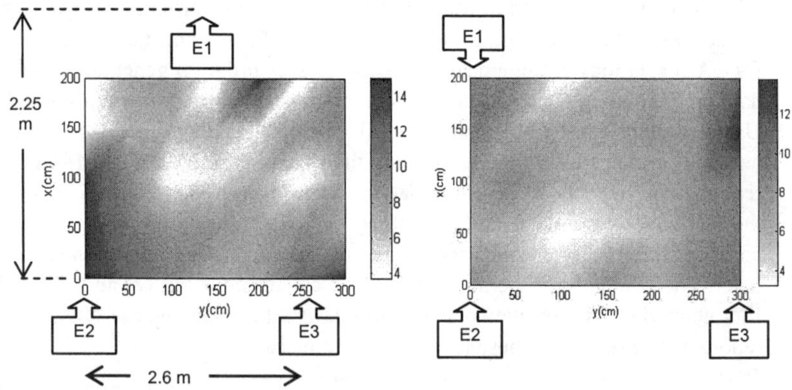

Figure 2: Positioning errors (in cm) for two experimental setups.

Also, 55 measurements were taken on the work space and the FFNN of the 4th test was trained for 10 additional epochs with this data to adjust it to the conditions of the real environment. Then the test was repeated on points not previously seen by the system, and the results are presented in Figure 3.

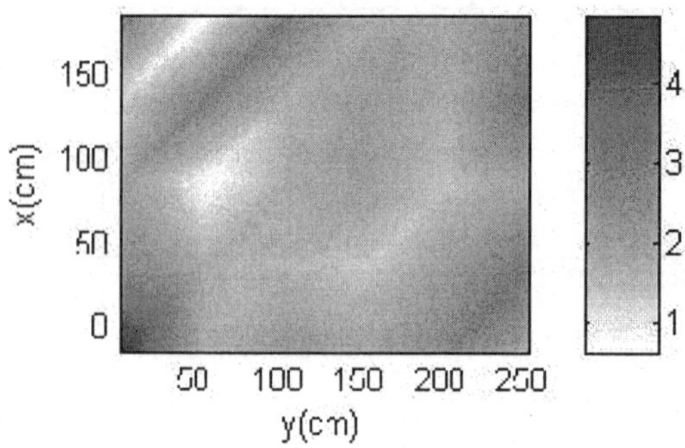

Figure 3: Positioning errors (in cm) for the equilateral setup trained with real measurements.

5. Results Analysis

The first test got very low values for the MAE, which could be expected for such a big and complex FFNN. Nevertheless, given the computational economy requirements of the system it becomes important to place our attention on the following experiences. Their use of the integer approximation of the sigmoidal function yields to accuracy loss on the results, but requiring much less time to compute them.

Tests 2-3 and tests 4-5-6 show an interesting property of the system which is its graceful degradation with the decrease of the computational power. Even if the accuracy of the systems gets worse when the FFNN is shrunk, it is still able to get to a useful result.

Tests 3 and 4, and their corresponding physical implementations, show the reconfigurability of the proposed system. With just a small change in the data generator function and a fast training, the FFNN was able to work with a different beacons-arrangement. Also, the physical tests demonstrate that the system is able to generate appropriate position measurements in the real world, even if trained only with simulated data. The MAE in these cases (7.1cm for test 3 and 8.3cm for test 4) were bigger than those obtained in the simulations, possibly as a consequence of a too simplified training model. Also, small errors on the positioning of the beacons affect the results, especially in their proximity as can be seen on Figure 2. Nevertheless, Figure 3 (with a MAE of 2.2cm) shows how these results can be greatly improved in the whole working space whit just some measurements from the real setup and a few training epochs to avoid over fitting, thanks to the inference capability of the NN. During the test it was observed that the presence of objects or persons in the working space did not affect the results, as long as the line of sight between the receiver and the beacons was not broken, validating the timing scheme of [4] in those conditions.

The 5th to 7th tests show the capability of the system to work on a 3D space, given enough beacons. The growth of the solution space yields to increased positing errors for NN of the same size than those of tests 2 to 4.

Finally, the 8th to 10th tests show the possibility of working on an overdetermined scenario. The small improvement on the MAE from test 6 to 8 is not likely to be due to the extra information from the 4th input, since in the simulation it is just a combination of the other three, but due to the extra network parameters. Nevertheless, it shows that it is feasible and easy to include beacons that over determinate the system, which might be useful in noisy environments. This is proved in test 9 and 10, where the presence of the 5th beacon improves the accuracy of the system when fed with noisy input data.

6. Conclusions and future work

The simulated tests proved the capability of the proposed architecture to solve the positioning problem in a variety of conditions, resulting in an adequate translation of the DTOA to the Cartesian coordinates. The algorithm showed itself useful and effective to do 2D and 3D local positioning, with errors whose magnitude can be handled with the size and complexity of the FFNN used. Nevertheless, when doing so the computing requirement of the system should be taken into account.

Its capacity to work in presence of reflecting objects makes of it a real possibility for indoor environments as long as line of sight between the beacons and the device can be assured across the work space. The system is easily configurable to a specific setup with just a computer simulation, and a small number of measurements improves remarkably its performance.

Finally, the asynchronous approach and the low computational requirements of the solution allows for its application to environments with many robots sharing a working space, since each one of them can do its own position estimation at no big computational, power or even economic cost.

In order to improve the accuracy of the system trained with simulated data, it is suggested to develop a better model of it, taking into account all of the variables that affect the velocity of sound like temperature and air pressure, which can be measured by the device while working and included in its calculations. The optimal placement of the beacons should also be studied since their positions affected the error distribution and values. For a real and useful implementation of the system, omni-directional transceivers have to be implemented. The use of a CDMA channel sharing would greatly improve the position computing time and robustness of the system, specially for overdetermined scenarios.

Given the framework of this investigation, it's planned to test the system in the future in underwater environments using the appropriate transducers.

References

1. J. Hightower and G. Borriello, “Location systems for ubiquitous computing,” IEEE Computer, vol. 34, no. 8, pp. 57–66, August 2001.
2. A. R. Jimenez, F. Seco, R. Ceres, and L. Calderon, “Absolute Localization using Active Beacons: A survey and IAI-CSIC contributions,” Institute for Industrial Automation, CSIC Madrid.
3. J. Smith and J. Abel, “The spherical interpolation method of source localization,” IEEE Journal of Oceanic Engineering, vol. 12, January 1987, pp. 246-252.
4. Fermín, Leonardo; Medina, Wilfredis; Chinea, Ana; Veloz, Nicolás; Grieco, J.. "A Low Cost Local Positioning System using Ultrasonic Sensors". 9th International Conference on Climbing and Walking Robots (CLAWAR 2006). Brussels, Belgium. September 2006. CD.
5. P. Estévez, “Implementación y prueba de algoritmos para reconocimiento de contexto sobre plataformas smart-its” BSc. thesis, Dept. Electron. Eng., Simón Bolívar University., Caracas, Venezuela, 2005.

CONTACT PROCESSING IN THE SIMULATION OF CLAWAR

JUHASZ[1], KONYEV[2], RUSIN[1], SCHMUCKER[1]

[1]*Department Virtual Engineering, Fraunhofer Institute for Factory Operation and Automation, Sandtorstrasse 22, 39106 Magdeburg, Germany*

[2]*Institute for Electrical Energy Systems, University of Magdeburg, P.B. 4120, 39106 Magdeburg, Germany*

Contact processing, including collision detection and collision response, is one of the most difficult, but most important areas in simulation of the multi-body systems. However, the most widespread multi-body simulators, like Matlab/SimMechanics or Modelica/Dymola, don't support the contact processing. Other multi-body simulators, like Vortex or ODE, support the contact processing, but are more limited in the rest of the functionality. This paper presents the implementation of the contact processing in both Matlab/SimMechanics and Modelica/Dymola and the comparison of the implemented functionality with Vortex. It was performed through an example of contact tasks for a six-legged robot.

1. Introduction

In mechanical systems certain machine elements usually interact with each other. When a mathematical model of such system is designed, the interactions between the parts can be divided into two following categories: -*Mechanical joints* are used for defining permanent constraints of motion. -*Mechanical contacts* are almost instantaneous, typically short-time interactions caused by non-penetration contact forces arising between the bodies in the model. The forces occur when the surfaces of bodies touch each other. Two major phenomena occur in the mechanical contacts: collision contacts (causing collision response forces) and friction contacts (causing static or dynamic friction forces).

Contact processing is a difficult task [1,5,6]. The bodies can move in a complicated way, and they can have complex geometries. In the case of contacting bodies, the penetration of them must be prevented. There is a tradeoff between efficiency and accuracy. Accurate methods for computing contact forces are based on finite element methods. Such methods are based on subdivision of bodies into very small fragments. The surfaces of two colliding bodies are to be covered by a mesh and the relevant forces in the contact are to be computed for each point on the mesh. The resulting forces can be defined by integration of all forces acting on the contact surface. These methods are implemented in software packages for FEM-analysis (ANSYS, Nastran etc) or in multi-body simulation (MSC Adams etc). Experiments [2] show that these methods are accurate, but require tremendous computing resources and therefore are very slow. However, many simulation applications do not require extreme accuracy and additional assumptions are taken into account providing high simulation speed, but decreasing the accuracy. As a matter of fact, different assumptions lead to different computation methods but with the same (or nearly the same) computation results. In such cases, it's not important for the application what assumptions and methods were used.

The goal of this paper is to present the ways of adding contact processing (chapter 2) to existing mechanical multi-body simulator Matlab/SimMechanics and to compare them with contact processing in Vortex, which is characterized by internal optimization loop with considering of energy and impulse conservation law. The results are presented though the example of contact tasks for a six-legged robot (chapter 3).

2. Contact Processing

Implementation of the contact processing is based on the following steps. (1) Mechanical models, which describe physical bodies, should be extended to describe contacting physical bodies. (2) There should be a routine that can detect collisions and can return detailed information regarding contact parameters, such as contact points and their velocities. (3) A special routine should calculate the contact response from contact parameters. (4) Each of these components should have an interface that allows replacing its implementation without doing major redesign of the other components.

Figure 1 presents four basic components of contact processing, which are derived from the discussion above and will be described more detailed later.

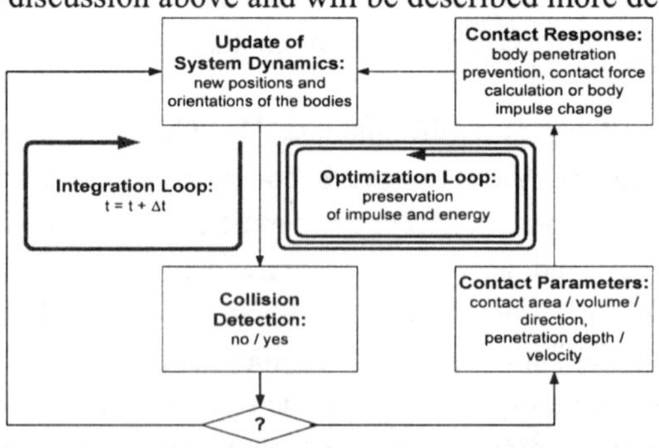

Figure 1. General scheme of contact processing in the mechanical multi-body simulation.

2.1. *Collision Detection*

The collision detection between bodies, which are indirectly described, for example using sets of points in the space, means that the contact points or trajectories between two bodies are to be defined.

In the first part (*collision culling*), the pairs of non-colliding objects should be excluded. These can be performed using the methods of space division like Quadtree/Octree, BSP–Tree, Sweep-and-Prune. The whole space should be divided and the potentially non-colliding objects should be excluded.

In the second part (*broad collision detection*), the possibility of two objects collision by means of so-called Bounding Volumes like AABB, sphere, OBB, k-DOP should be defined. These covers simplify the complex geometry objects and make the collision detection simpler and faster.

The third part (*exact collision detection*) defines the collision between the contacting objects. Spatial, hierarchical data structure, as AABB-tree, OBB-tree, sphere-tree or k-DOP-tree should be used for faster collision definition.

2.2. *Contact Parameters*

The intersection test follows the third part of collision detection and finds the geometrical contact parameters, as contact plane, contact volume, contact normal and penetration, with the aim of the following methods: *intersection point of ray and sphere*, *intersection point of ray and plane*, *intersection point of ray and triangle*, *intersection line of two planes*, *intersection line (point) of two triangles*. The presented on the market algorithms (Lin-Canny Closest Features Algorithm, I-/Q-COLLIDE, V-Clip, OBB-Tree, QuickCD, KDS, GJK, GJK-based EPA) combine "collision detection" and "contact parameters detection" tasks [7-11] and are implemented in the leading softwares, e.g. SWIFT, SOLID, ODE and others.

2.3. *Contact Response and Update of System Dynamics*

The contact response seems to be the most problematic and controversial part of the contact processing, since many computation approaches exist, which require different input information and may produce quite different numerical results. The following two methods are commonly used in the contact processing: the impulse-based method and the force-based method. Both assume that the bodies are rigid. The update of system dynamics is closely connected with calculation of contact response and therefore both parts should be explained together.

2.3.1. *Impulse-based Approach*

The impulse-based approach uses collision impulses between the bodies and changes the velocity vector of the bodies during the contact [3,4]. This method based on an impact law such as Poison's hypothesis. It considers the impulse conservation law and operates with the impulses of the colliding bodies before and after the collision as well as with the restitution coefficient of materials.

The main advantages of this method are that only a few constants are needed for description of the impact law and that the integrator step size is not influenced by the response calculation because it is performed during an infinitely small time instant. However, since the velocity is not continuous in the impulse-based model, the traditional ODE solvers can't be used. The continuous integration process in the solver should be stopped at the instant of collision and should be resumed with a new velocity. The impulse-based approach can be easily used in MBS-based models if the collision impact on the other bodies in the system is negligible (i.e. in the system of free-flying bodies). In the other words, this approach can't be used in the cases of a static objects and structures consisting of several bodies connected by joints. Furthermore such idealized impact laws are only useful for stiff collisions. These properties restrict the applicability of the impulse-based method of the dynamical analysis.

2.3.2. *Force-based Approach*

An alternative approach of contact processing in multi-body mechanical systems is based on the force and torque model of collision. It is assumed that the contacting bodies penetrate each other and the separation forces are caused by this penetration. These forces try to prevent further penetration and to separate the contacting bodies.

The calculation of contact force magnitude is difficult task and is sometimes not motivated by physics, but rather by numeric analysis. The overall result of collision should match physical laws (i.e. preservation of impulse, and preservation of energy). In addition it should be chosen so smooth that numerical methods used in simulation could handle these functions. The many existing methods for the calculation of the force in the mechanical joints and contacts are divided into two following groups:

- *Force based methods with Lagrange multipliers formulation* models the mechanical constraints (contacts and joints) with the reactive forces, which are presented as Lagrange multipliers λ. The constraint forces perform no work on the environment and the physical meaning of the mechanical contact is lost. The mechanical interaction of the bodies caused by the contact is represented by these reactive forces **λ**, which should be optimized between the simulation steps in the additional optimization loop (s. Fig.1) under consideration of the energy or/and impulse preservation laws.

- *Force based methods with penalty formulation* models the mechanical contact with the strong poss. nonlinear spring. The active contact/friction forces (s. Fig.2) perform work on the environment and the physical background of the mechanical contact is not lost. The mechanical interaction of the bodies caused by the contact is represented by the active forces $\boldsymbol{F_{CONTACT}}$ and $\boldsymbol{F_{FRICTION}}$, without any additional optimization between the simulation steps.

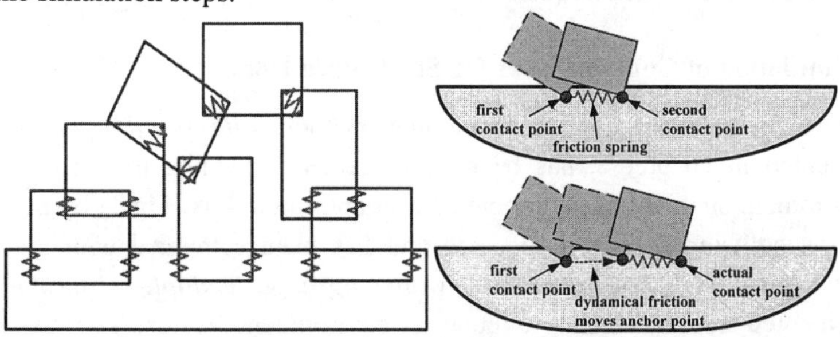

Figure 2. Force based methods with penalty formulation for contact and friction forces calculation.

There are many different equations for determining the contact/friction force magnitudes, which depend on the penetration depth $\boldsymbol{p}$, on the penetration velocity $\boldsymbol{dp/dt}$, on the frictional penetration $\boldsymbol{l}$, and on the Coulomb friction coefficient $\boldsymbol{\mu}$. For the purpose of this work the following equation system is chosen for the calculation of the force magnitudes, which is very stable in the wide range of the simulation sample time and corresponds to all above mentioned requirements. Eq.1 describes the contact force magnitude depending on the stiffness $\boldsymbol{S_{CONTACT}}$ in the contact area, the restitution factor $\boldsymbol{\varepsilon}$ of the materials and the collision velocity $\boldsymbol{v_{COLLISION}}$ at the first time instant of intersection:

$$F_{CONTACT} = \begin{cases} 0, & p < 0 \\ \left[1 + \dfrac{1-\varepsilon}{\varepsilon \cdot \upsilon_{COLLISION}} \cdot \dot{p}\right] \cdot S_{CONTACT} \cdot p, & p \geq 0 \end{cases} \tag{1}$$

$$F_{FRICTION} = \begin{cases} S_{CONTACT} \cdot l, & S_{CONTACT} \cdot l < \mu \cdot F_{CONTACT} \\ \mu \cdot F_{CONTACT}, & S_{CONTACT} \cdot l \geq \mu \cdot F_{CONTACT} \end{cases}$$

The main advantages of the force-based approach are the simplicity and the possibility of using it for stiff and soft contacts. This approach works reasonably well if several contact points are also present at the same instant time. The disadvantage of this approach is that the integrator step size should be reduced in the contact phase in order to catch the rapidly changing contact forces and torques. And similar to the impulse-based method there is necessary to choose the contact parameters (spring, damping, restitution) because the contact force is proportional not only to the penetration depth/velocity but also to the contact area and the contact volume.

Update of system dynamics take place each integration step depending of the acting contact forces/torques.

3. Simulation of Contact Tasks for Six-Legged Robot

The force based contact processing method *with penalty formulation*, represented in chapter 2, has been implemented according to Fig.1 into the multi-domain simulation environments Matlab/Simulink (s. Fig.3) by means of Solid as collision detection software and has been compared with the force based contact processing method *with Lagrange multipliers formulation*, implemented into multi-body simulation environment Vortex. The six-legged robot "SLAIR2", developed at the University of Magdeburg and the Fraunhofer Institute for Factory Operation and Automation, has been used as a test subject. Two modes, *staying on the ground* and *moving on the surface*, have been investigated and compared in the mentioned simulation environments (s. Fig.4).

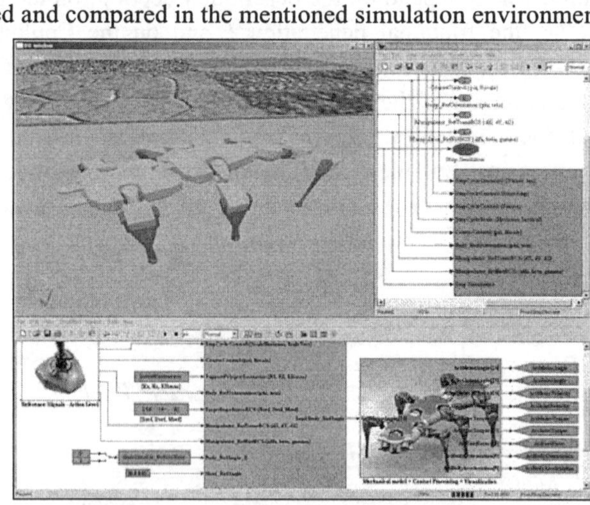

Figure 3. Modular six-legged robot "Slair2" with articulated body in the multi-domain simulation environment "Matlab/Simulink".

Investigations show that the contact processing in Vortex is very fast, robust and reality-near. It is caused by the first-order integration method with adaptive step and by the optimization loop. However in dynamic mode the calculation of the contact forces is not precise, because the number of the optimization steps is limited to 35 by developers.

The developed contact processing environments in Matlab/Simulink are free from aforementioned disadvantages, can use the integration methods of higher order, and are sufficiently precise by calculation of the contact forces, because Eq.1 has been laid out under consideration of the energy/impulse preservation law.

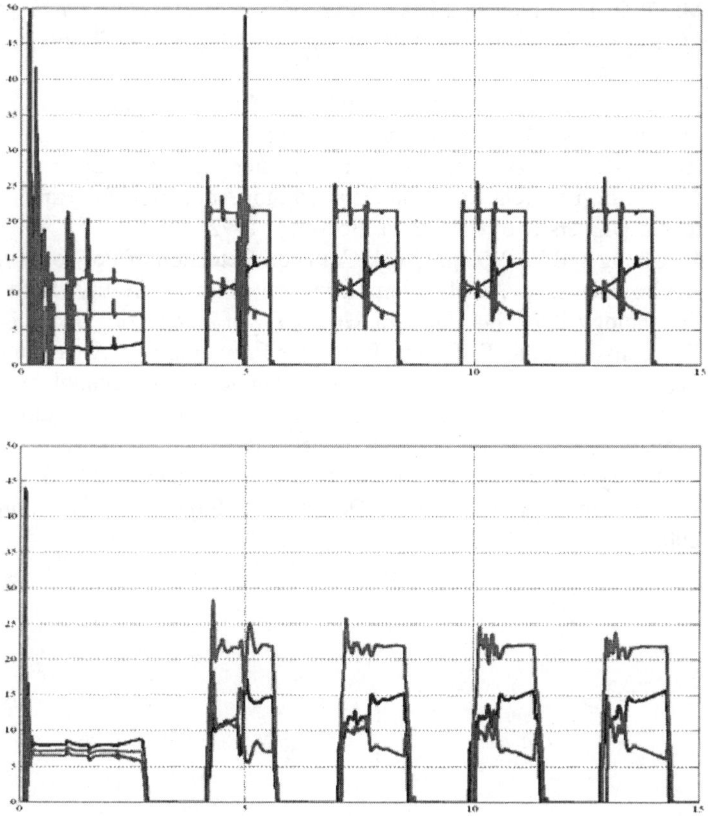

Figure 4. Comparison of the normal forces under the feet of the six-legged robot "Slair2" staying on the ground and then doing four steps on the surface in (top) Vortex and (down) SimMechanics/Solid.

References

1. V. Engelson. *Integration of Collision Detection with the Multibody System Library in Modelica*. Thesis, PELAB, IDA, Linkoping University, 2000.
2. D. Fritzson, P. Fritzson, P. Nordling, T. Persson. *Rolling Bearing Simulation on MIMD Computers*. Int. Journal of Supercomp. Appl. and High Performance Computing, 11(4), 1997.
3. B. Mirtich: *Impulse-based Dynamic Simulation of Rigid Body Systems*. Ph.D. thesis, University of California, Berkeley, 1996.
4. P. Zhang. *Physically Realistic Simulation of Rigid Bodies*. Thesis, Department of Computer Science, Tulane University, 1996. Available via http://www.eecs.tulane.edu/www/Zhang/.
5. M. Otter, H. Elmqvist, J.Diaz Lopez. *Collision Handling for the Modelica MultiBody Library*. 4th International Modelica Conference, pp.45-53, March 2004.
6. N.Galoppo, M.Otaduy, P.Mecklenburg, M.Gross, M.Lin. *Fast Simulation of Deformable Models in Contact Using Dynamic Deformation Testures*. ACM SIGGRAPH Symposium on Computer Animation, 2006.
7. M. C. Lin: Efficient Collision Detection for Animation and Robotics. Ph.D. thesis, University of California, Berkeley, 1992.
8. K. Chung and W. Wang: Quick Collision Detection of Polytopes in Virtual Environments, , ACM Symposium on Virtual Reality Software and Technology 1996, 1-4, July, 96, University of Hong Kong, Hong Kong.
9. D. Schmalstieg, R. F. Tobler: Real-time Bounding Box Area Computation. Institute of Computer Graphics, Vienna University of Technology, 1999.
10. J. Erickson, L.J. Guibas, J. Stolfi and Li Zhang: Separation-sensitive collision detection for convex objects; Proceedings of the tenth annual ACM-SIAM symposium on Discrete algorithms, Pages 327 – 336, 1999.
11. G. van den Bergen: Collision Detection in Interaction 3D Environments. 2003.

CREATING A GESTURE RECOGNITION SYSTEM BASED ON SHIRT SHAPES

PAVEL STAROVEROV, SILVIA MARCOS, DMITRY KAYNOV, MARIO ARBULU, LUIS CABAS, CARLOS BALAGUER

Robotics Lab, Department of System Engineering and Automation, University Carlos III of Madrid, Av. Universidad 30, Leganés 28911(MADRID), Spain

The present article describes the gesture recognition system that has been developed at the University Carlos III of Madrid. The system's functionality is based on the assumption that the person that wants to interact with the robot should wear a shirt which characteristics must have been previously passed on to the program. The algorithm used for the recognition is presented in detail. The results of a number of tests are presented and discussed along with the suggestions for future improvements.

1. Introduction

Gesture recognition can be a very helpful tool that can make a welcome addition to the interaction abilities of any service robot. Humanoid robots would benefit from gesture recognition even more, because this natural way of communication that is often used between humans, can open the door to a new level of interaction, that can be logically continued by imitation and learning.

The objective of this work is to propose a gesture recognition system for the Rh-1 humanoid robot, a system that is easy to implement yet presents an acceptable performance. Rh-1 is a humanoid robot developed at the University Carlos III of Madrid (Fig.1).

It has 21 degrees of freedom (and 2 more degrees of freedom of the built-in camera). It measures approximately 130 cm and weighs in about 45 kg. It already can speak, recognize voice commands, faces and "T-shirts". Those abilities were discussed in one of our previous works [10]. Somewhere in the present proceedings you may find another article concerning the recently added sound source detection system.

The present article is organized as follows: in Sec.2 some information concerning related works can be found, Sec.3 describes the way the system works, experimental results are presented and discussed in Sec.4. Section 5 concludes the paper.

Fig.1. Rh-1 robot and several members of the Robotics Lab.

2. Related works

Recently there has been a lot of interest towards the field of gesture recognition. One the works that we would like to mention is by Kojo *et al.* [1]. It deals with an advanced gesture recognition system for the HRP-2W robot, which is the wheel version of the famous HRP-2P humanoid. The algorithm used is based on the suggested Proto-symbol Space and also uses Hidden Markov Models (HMM), which is a widely used method for gesture recognition [2]-[5]. Calinon and Billard in their research go further and use HMMs for *programming by demonstration* [6]. In their work [7] they also use an algorithm combining PCA, ICA and HMM. Later Calinon and Billard and some colleagues inspired by them made another step forward and used a similar method to reproduce the movements on the robot [8, 9].

The algorithms used nowadays for gesture recognition are quite advanced, but the downside is that they are not very easy to implement. We wanted to use an algorithm that would be easier to implement and also make use of the software created earlier.

3. Development of the system

The design of a robust gesture recognition system can present significant challenges to the researcher. Mostly this happens because of the fact that many gestures are shown in 3D space, thus creating the need to extract spatial information from 2D images. Our robot doesn't have a stereo vision, it has only one camera, so working with 3D gestures would be very hard. So it was decided to limit the gestures that can be recognized by the robot. First of all, only the upper body gestures will be considered. Those gestures are often used in the signals of the traffic police, referees at sporting events and flight directors on the airport runways [11]. The lexicon of those signals is always finite, so the number of upper body gestures to recognize can be limited by several poses. It was decided to select the following gestures (Fig.2): 1) no gesture (both hands down), 2) left hand stretched out to the left, 3) right hand stretched out to the right, 4) left hand up, 5) right hand up, 6) both hands stretched out to the sides, 7) both hands up. At the moment the system works only with static gestures.

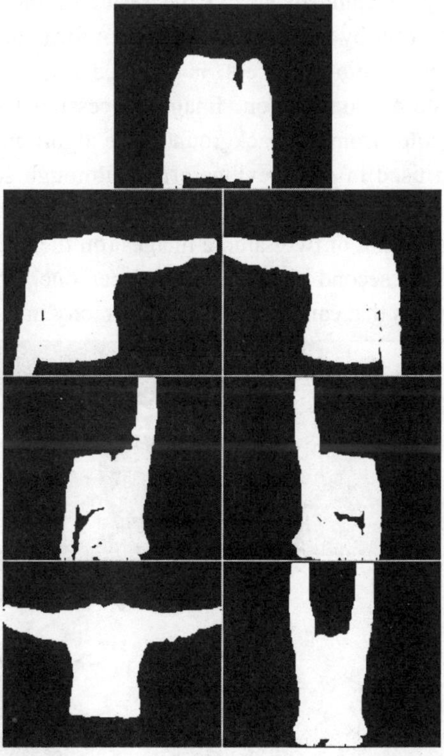

Fig.2. Processed sample pictures with the gestures that can be recognized by the program.

Another assumption that was to be made is that the person that wants to interact with the robot should wear a shirt of a color that must be previously passed on to the robot. It seems quite normal because road police, referees and flight directors usually wear some kind of uniform to be distinguishable from other people. Some initial restrictions do apply to secure the system's acceptable performance, namely, the shirt must be of a color that is different from the objects in the background and the illumination during the calibration and the recognition phases must be as close as possible.

It was decided to modify the "T-shirt based" image recognition system described in one of our previous works [10] to detect gestures using a similar straightforward algorithm. Our idea was to recognize gestures using geometrical properties of shirts (various types of clothes can be used but all of them will be referred to as shirts for simplicity). Shortly, that system is capable of finding T-shirts or similar pieces of clothes in the images using color information and it can also calculate the distance to the person. The execution of the program can be controlled by voice commands, and the result of the processing can be communicated to the user by the robot in the spoken form, too.

The gesture recognition process is divided into two stages: image processing and gesture classification. Image processing tries to separate the object of a known color from the background. The algorithm used is nearly the same that has been used in our previous effort, although some improvements have been made in order to provide a better object separation. In Fig.3 you may see the result of processing of two sample images (on the left), the first image is with a red shirt and the second image is with a green one. On the right you may see the resulting images that can be recognized correctly in the second stage.

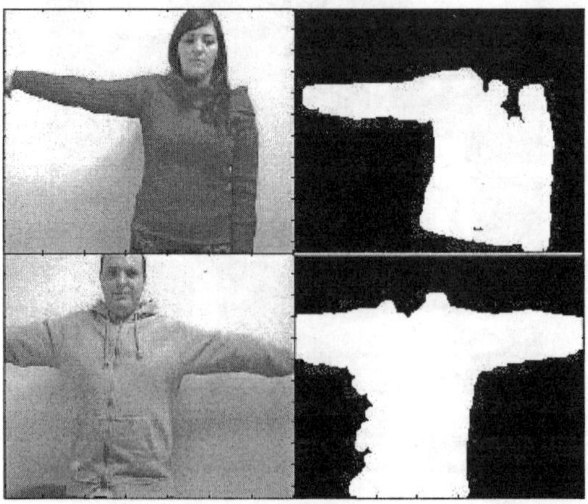

Fig.3. Two sample images and the result of their processing.

In the second example it can be seen that the right hand was clipped because the person was too close and was not at the center of the image. As we will see later, in this case there is a high probability of a recognition error.

After the image has been extracted, the geometrical parameters of the extracted object are analyzed. First of all, we look at the height/width ratio. If it is between 0.95 and 1.2 then it is supposed that there is no gesture (case 1). If it is less than 0.65 then it is supposed that both hands are stretched out to the sides (case 6). If it is between 1.2 and 3 then we verify whether both the top right pixel and the top left pixel are at the same side of the image. If it is true then an extra check is made to make sure that the mirrored hand area is black, after that the corresponding gesture is chosen (cases 4 or 5, left or right hand up). In the contrary case an extra check is made to make sure that the area between the hands is black and after that it is supposed that both hands are up (case 7). If the height/width ratio is between 0.65 and 0.95 then we verify which pixel, the right top one or the left top one, is farther from the vertical centroid and the corresponding gesture is chosen (cases 2 or 3, one of the hands stretched to a side). If it's more than 3 then it's considered a recognition error. The values of the height/width ratio may be adjusted for various persons.

Having detected a gesture, the robot may react to it in a way that we would want to, for example, perform some action like stopping, making one step, turning left, right, around, etc. It is the kind of response action one would expect, for example, when following the signals of police directing traffic. Also, the robot's number of degrees of freedom lets it imitate all of the given gestures if needed, although the robot's response movement has not been implemented yet in any form. Currently the robot just tells the person that a certain type of gesture has been detected.

The suggested algorithm has a lot of assumptions and seems to be too simple to be robust, but let us make the final decision after seeing the experimental results.

4. Results and discussion

In order to evaluate the performance of the system, two series of tests have been made. Two persons were repeating all the gestures three times, one of the persons was wearing a red shirt, and the other was wearing a green one. All the tests were made under normal lighting conditions. The persons intended to remain parallel to the robot and to be reasonably far from it to enter in the image completely. However, due to the dimensions of the room where the test was conducted, it was not always possible. The results are presented in Table 1.

Table 1. Test results for two persons repeating all gestures three times (+ means a positive result, - means a negative one).

GESTURE TYPE	RED SHIRT RESULTS	GREEN SHIRT RESULTS
No gesture	+++	+++
Left hand stretched	+++	++-
Right hand stretched	+++	++-
Left hand up	+++	+++
Right hand up	+++	++-
Both hands stretched	++-	+++
Both hands up	+++	++-

It can be seen that the results of the red shirt are a bit better, resulting only in one miss when both hands were stretched out but the program recognized only one hand stretched. It was due to the fact that the room where the tests were performed was rather small, and the person's hands did not enter in the image completely. In the case of the green shirt, there was a similar case when the right hand was stretched out. There have also been two misses because of the person's inclined pose during the test (it is important to stand up straight) and one miss due to the changed lighting conditions. In general, there have been 37 hits and 5 misses out of 42, which can be considered a satisfactory result. It can be said that the system has an acceptable performance for our type of application. It also has several advantages, namely, it is easy to implement and it doesn't need any training.

Nevertheless, the described system makes a lot of assumptions, depends greatly on the calibration quality, on the lightning conditions and other factors. Also, currently it works with static images only. All of this can be a ground for improvements. First of all, it is recommended to use an algorithm based on HMM. Although it is more difficult to implement, it may also provide better results. Another important improvement would be the possibility to process online video instead of static images. And finally, it is important to program robot's movements in response to those gestures, be they imitation movements or some other movements. Later it would be interesting to make some kind of learning by demonstration as suggested in [6] and [7].

5. Conclusion

In the present paper a gesture recognition system developed at the University Carlos III of Madrid was described. A simple geometrical method is used to recognize shirt shapes. The algorithm was described in detail, the experimental results were presented and discussed. The performance of the system was considered acceptable. Finally, several improvements have been suggested.

Acknowledgments

We would like to thank the members of the Robotics Lab of University Carlos III of Madrid for their cooperation and suggestions. The present work was supported by CICYT (Comisión Interministerial de Ciencia y Tecnología).

References

1. N. Kojo, T. Inamura, K. Okada, M. Inaba “Gesture Recognition for Humanoids using Proto-symbol Space”, 2006 6th IEEE-RAS International Conference on Humanoid Robots (Humanoids 2006), pp.76--81, 2006.
2. S. Eickeler, A. Kosmala, G. Rigoll, “Hidden Markov Model Based Continuous Online Gesture Recognition", In Proc. Int. Conference on Pattern Recognition (ICPR), Brisbane, 1998, pp.1755-1757.
3. Rigoll, G., Kosmala, A., and Eickeler, S. “High Performance Real-Time Gesture Recognition Using Hidden Markov Models.” In Gesture and Sign-Language in Human-Computer Interaction, pages 6980, 1998
4. R. Yang, S. Sarkar, "Gesture Recognition using Hidden Markov Models from Fragmented Observations," CVPR, pp. 766-773, 2006 IEEE Computer Society Conference on Computer Vision and Pattern Recognition - Volume 1 (CVPR'06), 2006
5. J. Yang and Y. Xu, “Hidden Markov Model for Gesture Recognition”, tech. report CMU-RI-TR-94-10, Robotics Institute, Carnegie Mellon University, May, 1994
6. Calinon, S. and Billard, A. “Gesture Recognition and Reproduction for a Humanoid Robot using Hidden Markov Models.” AMI/PASCAL/IM2/M4 Workshop on Multimodal Interaction and Related Machine Learning Algorithms. 2004
7. Calinon, S. and Billard, A. “Recognition and Reproduction of Gestures using a Probabilistic Framework combining PCA, ICA and HMM. “In Proceedings of the International Conference on Machine Learning (ICML). 2005
8. Calinon, S. and Billard, A. “Incremental Learning of Gestures by Imitation in a Humanoid Robot.” in Proceedings of the ACM/IEEE International Conference on Human-Robot Interaction (HRI) 2007.
9. T. Asfour, F. Gyarfas, P. Azad and R. Dillmann “Imitation Learning of Dual-Arm Manipulation Tasks in Humanoid Robots”, IEEE-RAS International Conference on Humanoid Robots (Humanoids 2006), Genoa, Italy, December 2006
10. P.Staroverov; M.Arbulú; L.M.Cabas; D.Kaynov; C.Pérez; C.Balaguer. “A Voice Controlled Image Recognition System.” Proc. of the 9th International Conference on Climbing and Walking Robots. Brussels. Belgium. Sep, 2006.

11. Sclaroff, S., Betke, M., Kollios, G., Alon, J., Athitsos, V., Li, R., Magee, J., and Tian, T. 2005. Tracking, Analysis, and Recognition of Human Gestures in Video. In Proceedings of the Eighth international Conference on Document Analysis and Recognition (August 31 - September 01, 2005). ICDAR. IEEE Computer Society, Washington, DC, 806-810

DESIGN AND DEVELOPMENT OF MICRO-GRIPPING DEVICES FOR MANIPULATION OF MICRO-PARTS

Z.W. ZHONG, S.K. NAH, S.H. TAN

School of Mechanical and Aerospace Engineering, Nanyang Technological University, 50 Nanyang Avenue, Singapore 639798, Republic of Singapore

In this study, design, fabrication and tests of micro-grippers were carried out. The geometry design and the material stresses were considered by means of the finite element analysis. The simulation model was used to study the profiles of von Mises stresses and deformation. Micro-gripper prototypes were made using various machining methods. Micromanipulation tests were conducted to evaluate the performance of the micro-grippers and confirm potential applications of the micro-grippers in handling micro-objects. The simulation and experimental results have proven the good performance of the micro-grippers.

1. Introduction

An effective mechanical micromanipulator should have the ability to grasp objects of different shapes with high positioning accuracy. The manipulators should be able to accurately control grasping forces in order to avoid any damage to the small-size delicate objects, which are less than 1 mm in size [1].

Micro-scale technologies have been developed for many applications in electronics, information technology, optics, medicine and biology covering areas such as diagnostics, drug delivery, tissue engineering and minimally invasive surgery [2]. A microgripper is one key element in micro-robotics and micro-assembly technologies for manipulating micro-objects. Various prototypes of microgrippers have been developed using electrothermal actuators [6–9], electrostatic actuators [10–15], piezoelectric actuators [16,17], electromagnetic actuators [18] and shape memory alloy actuators [19–22].

Micro-parts with major dimensions less than 0.1 mm are often fragile and can be easily damaged during gripping, and thus special grasping techniques are required. The specifications to realize a microgripper are quasi-static motion to have high accuracy in micro-positioning, a large-stroke to grasp the maximum types of object, and the use of special actuation method [23].

In this study, design, fabrication and tests of micro-grippers were carried out. The geometry design and the material stresses were considered by the finite element analysis. The simulation model was used to study the profiles of von Mises stresses and deformation. Micro-gripper prototypes were made using various machining methods. Micromanipulation tests were conducted to evaluate the behavior of the micro-grippers.

2. Basic Studies

2.1. *Actuation of Mechanism*

A simple-structured microgripper with piezoelectric actuation was designed. The actuator utilizes the shear deformation of piezoelectric elements to drive the compliant mechanism of the microgripper. The entire microgripper consists of a compliant gripper mechanism, a shear-mode piezoelectric actuator and a support base. An exploded assembly view of the microgripper system is shown in Figure 1.

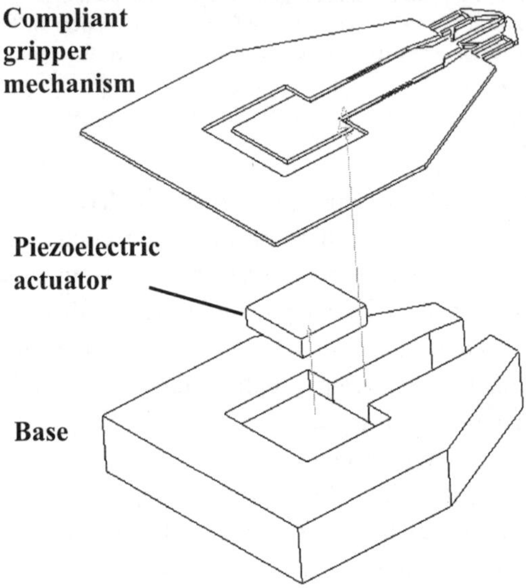

Figure 1. Assembly view of the microgripper system.

The piezoelectric actuator drives the movable arm of the compliant gripper mechanism back and fro, thus providing the desired actuation to grasp and release the target micro-parts or biological cells. For the assembly process, the piezoelectric actuator is secured to a solid base surface. Epoxy glue provides a strong, flexible and non-conducting bond. The flexible nature of a glue bond also eliminates the possibility of vibration-induced fatigue during operation.

2.2. *Finite Element Analysis of Microgripper*

FEA software ANSYS was used as a modeling and calculation tool. The simulation work involved the use of structural elements to model the compliant mechanism and also the use of contact elements to model the interacting bodies. Visual animations and numerical calculation results were obtained from the program. The basic objective of the proposed compliant mechanism design is to obtain desirable output displacements of the tip of the microgripper with known input actuating displacements. The compliant mechanism is an elastic mechanism that performs its kinetic function through its intended flexibility, which is enabled by the notch hinges on the parallel arm links. Thus the mechanism performance largely depends on its material properties and the geometrical design. The design is a planar, monolithic mechanism consisting of parallel links, flexural notch hinges and actuating links.

The maximum stress and the bulk of the stresses are concentrated at the flexural hinges of the compliant mechanism. This is expected as the flexural hinges experience the most bending deformations within the whole mechanism. When the FEA model has a parallel arm length of 1.5 mm and the input displacement is 10 μm, the value of the maximum stress obtained at the flexural hinges is 828 MPa, which is within the elastic region of the mechanism material.

For the simulation study, the variables are the length of the parallel arms, l, and the input displacement value. The ratio of the output displacement and the input displacement is the geometrical advantage. The graph in Figure 2 illustrates that the longer the parallel arms, the greater the geometrical advantage is. The geometrical advantage provided by the microgripper design is in the range of 3.5 to 5.5 depending on the length of the parallel arms.

2.3. *Contact Simulation of Microgripper and Micro-Part*

The circular object is flexible while the microgripper tips are rigid. Thus upon contact, the circular object is deformed while the microgripper tips still retain their shape. For one simulation, material 1, which is the material of the microgripper, is defined as Foturan glass. Material 2, which is the material of the target object, is defined as polystyrene. To avoid slipping, a coefficient of friction value of 0.2 is included in the simulation. Two additional contact element types are also defined to study the contact relationship.

When a deformation of 1 μm is applied on both sides of the circular polystyrene object, the contact pressure is obtained as shown in Figure 3. The maximum pressure occurs at the center of the contact region and decreases exponentially to zero towards the outer end of the contact region. The corresponding stress experienced by the object can be expresses by a parabolic curve with its peak value of 436 MPa at the centre of the contact area, which is the location where the target object experiences the maximum stress. The stress on the target object drops rapidly to zero with increasing distance away from the centre of the contact area.

3. A Prototyped Gripper Actuated Using a PZT Actuator

3.1. *Design and Fabrication*

The basic study results of Section 2 have led to several microgrippers fabricated and tested. The dimensions of one of the microgrippers fabricated were designed in such a way that micro-manufacturing technologies could be used, leading to low cost and an easy adaptation to different micromanipulation needs.

The microgripper was fabricated from a single piece of spring steel using a micro-wire electrical discharge machine (EDM). Another material used to fabricate was aluminium. These materials were selected because they are common engineering materials and are easy to be machined by EDM.

The micro-wire EDM technology can generate the microgripper with good precision. With achievable dimensional tolerances of ± 5 μm and the capability to use ϕ20-μm wires, the wire cutting method is suitable for micro machining. The microgripper prototype is 36 mm long, 30 mm wide and 3 mm thick.

The prototype of the gripping device is shown in Figure 4. A piezoelectric actuator was combined with the microgripper to achieve precise grasping motion due to its very fine positioning feature.

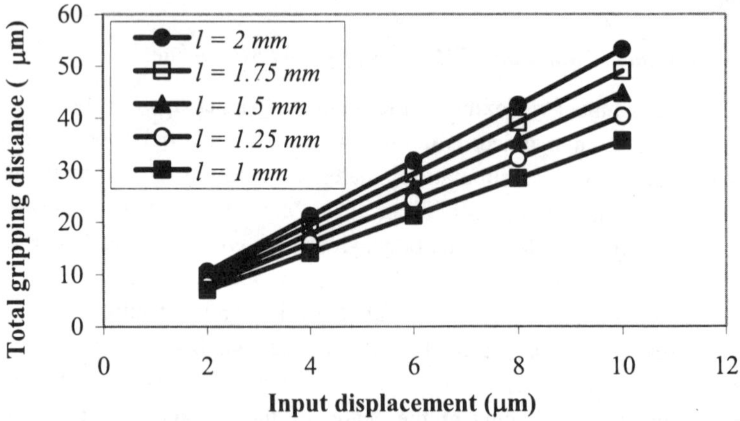

Figure 2. Total gripping distance against input displacement.

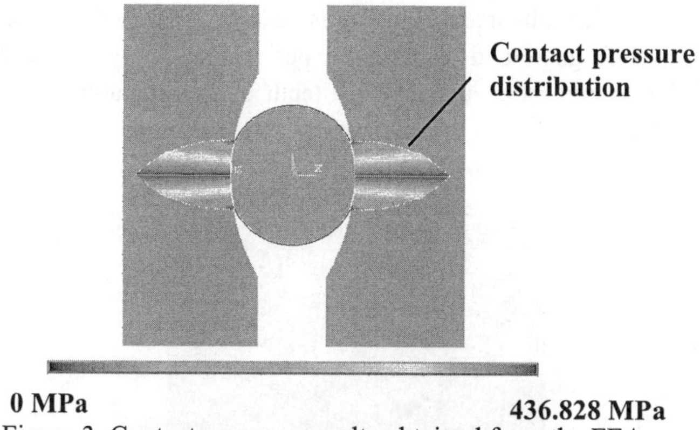

Figure 3. Contact pressure results obtained from the FEA.

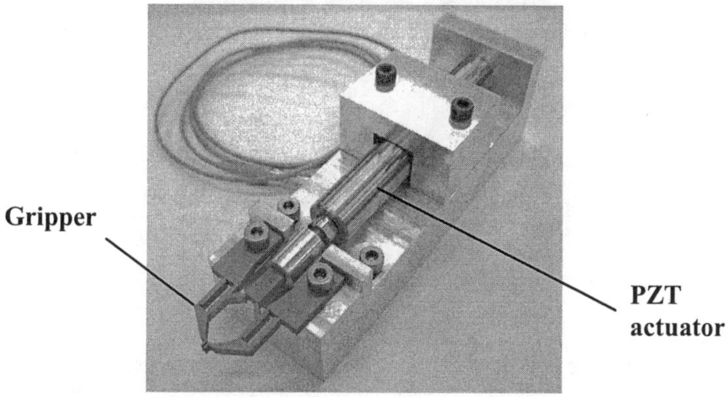

Figure 4. The gripping device with a PZT actuator.

3.2. *Tests*

A movement towards the microgripper tips is transformed by the compliant mechanism kinematics into the tip opening. Retraction of the PZT actuator closes the microgripper tips. With the microgripper as the tool, it is possible to grasp objects of different materials and release them. The mechanism has a mean amplification of 3.0 calculated by the average value of three sets of test readings.

Some grasping actions were made on a 500 μm Teflon wire. Figure 5 shows gripping of the Teflon wire in the horizontal position. The Teflon wire placed vertically can also be gripped by the microgripper tips.

The assembly of miniaturized gear systems, typically with diameters below 2 mm, requires the use of special micro-gripping tools. The miniature gears are typically of some hundreds of microns in size. Tasks of pick and place of miniature watch gears and shafts were performed. The grasping motion was controlled to avoid damaging the tiny teeth of the miniature gears. Figure 6 shows an isometric view of the microgripper tips gripping a shaft used for micro gears.

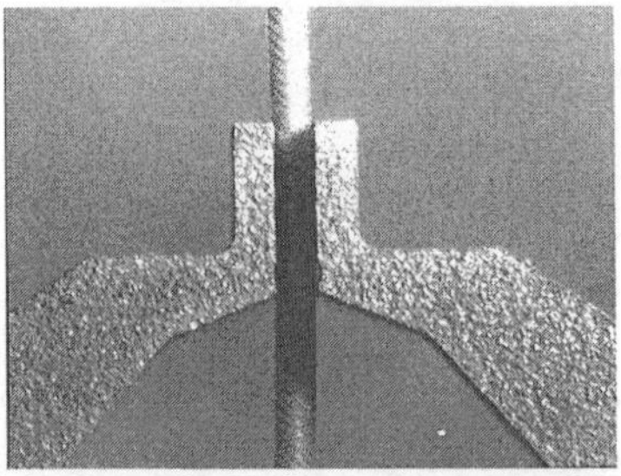

Figure 5. Gripping of a Teflon wire in the horizontal position.

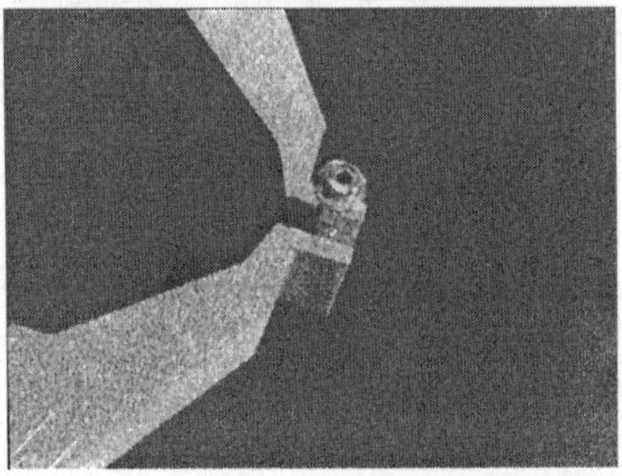

Figure 6. Microgripper tips grasping a micro gear shaft.

4. A Gripper Actuated Using SMA

4.1. *Prototyping*

A gripper (shown in Figure 7) actuated using shape memory alloy (SMA) wire was also fabricated and tested. The gripping force of this gripper can be adjusted by both mechanical and electrical methods.

4.2. *Gripping Force versus Current*

The relationship between gripping force and electrical current was studied. One example of experimental results is shown in Figure 8. The figure and the value of R^2 = 0.9997 indicate that the gripping force increases linearly with the increasing electrical current.

5. Conclusion

Design, FEA, fabrication and tests of micro-grippers were conducted. The results have proven the good performance of the micro-grippers.

Acknowledgments

The first two authors thank the Robotics Research Center of Nanyang Technological University for the support with the piezoelectric actuator.

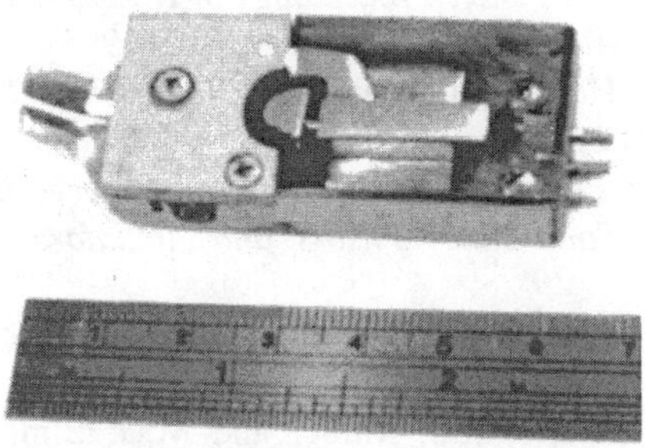

Figure 7. A gripping device actuated using SMA wire.

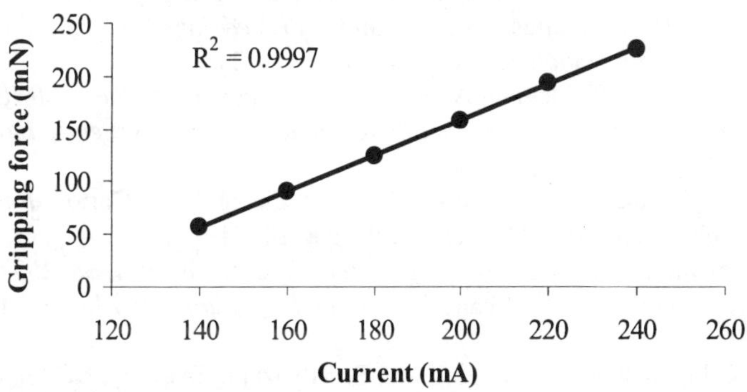

Figure 8. Relationship between gripping force and electrical current.

References

1. S.K. Nah and Z.W. Zhong, *Sensors and Actuators A: Physical* **133**, 218 (2007).
2. P. Dario, M.C. Carrozza, A. Benvenuto and A. Menciassi, *Journal of Micromechanics and Microengineering* **10**, 235 (2000).

6. E.V. Bordatchev and S.K. Nikumb, *Proceedings of the International Conference on MEMS, NANO and Smart Systems*, Canada, 308 (2003).
7. S.H. Lee, K.C. Lee, S.S. Lee and H.S. Oh, *Proceedings of The 12th International Conference on Solid-State Sensors, Actuators and Microsystems*, Boston, 552 (2003).
8. H.Y. Chan and W.J. Li, *Proceedings of the IEEE International Conference on Robotics and Automation*, Taipei, 288 (2003).
9. C.S. Pan and W.Y. Hsu, *Journal of Micromechanics and Microengineering* **7**, 7 (1996).
10. P.B. Chu and K.S.J. Pister, *Proceedings of IEEE International Conference on Robotics and Automation*, San Diego, 820 (1994).
11. B.E. Volland, H. Heerlein and I.W. Rangelow, *Microelectronic Engineering* **61-62**, 1015 (2002).
12. C.J Kim, A.P. Pisano and R.S. Muller, *Proceedings of International Conference on Solid-State Sensors and Actuators*, San Francisco, 610 (1991).
13. C.J Kim, A.P. Pisano and R.S. Muller, *Journal of Microelectromechanical Systems* **1**, 31 (1992).
14. C.J Kim, A.P. Pisano, R.S. Muller and M.G. Lim, *Proceedings of 4th Technical Digest of Solid-State Sensor and Actuator Workshop*, Hilton Head Island, 48 (1990).
15. P. Boggild, T.M. Hansen, K. Molhave, A. Hyldgrad, M.O. Jensen, J. Richter, L. Montelius and F. Grey, *Proceedings of the 2001 IEEE Conference on Nanotechnology*, Maui, 87 (2001).
16. Y. Haddab, N. Challet and A. Bourjault, *Proceedings of the 2000 IEEE/RSJ International Conference on Intelligent Robots and Systems*, Japan, 659 (2000).
17. M.C. Carrozza, A. Menciassi, G. Tiezzi and P. Dario, *Journal of Micromechanics and Microengineering* **8**, 141 (1997).
18. H. Ren and E. Gerhard, *Sensors and Actuators A: Physical*, **58**, 259 (1997).
19. Z.W. Zhong and S.Y. Chan, *Sensors and Actuators A: Physical* **136**, 335 (2007).
20. M. Kohl, E. Just, W. Pfleging and S. Miyazaki, *Sensors and Actuators A: Physical* **83**, 208 (2000).
21. A.P. Lee, D.R. Ciarlo, P.A. Krulevitch, S. Lehew, J. Trevino and M.A. Northrup, *Proceedings of The 8th International Conference on Solid-State Sensors and Actuators, and Eurosensors IX*, Stockholm, Sweden, 368 (1995).
22. Z.W. Zhong and C.K. Yeong, *Sensors and Actuators A: Physical* **126**, 375 (2006).
23. B. J. Pokines and E. Garcia, *Smart Material Structures* **7**, 105 (1998).

DESIGNING OF A COMMAND SHAPER USING MULTI-OBJECTIVE PARTICLE SWARM ALGORITHM FOR VIBRATION CONTROL OF A SINGLE-LINK FLEXIBLE MANIPULATOR SYSTEM

M. S. ALAM [†], M. O. TOKHI

Department of Automatic Control and Systems Engineering, University of Sheffield, UK

M. A. HOSSAIN

Department of Computing, University of Bradford, UK

A new command shaping method is proposed using gain and delay units to shape the reference input in order to reduce vibration of a single-link flexible manipulator system. The values of gain and delay elements can be derived analytically with a priori knowledge of natural frequencies and associated damping ratios of the system, which may not be available for complex flexible systems. Moreover, command shaping in principle causes delay in system's response while it reduces system vibration and in this manner the amount of vibration reduction and the rise time conflict one another. Assuming that, no prior information is available about the system, a new multi-objective particle swarm optimisation (MOPSO) algorithm is applied to optimise the gain values and the amount of delay in order to provide a wide range of solutions that trade-off these conflicting objectives so as to satisfy design goals.

1. Introduction

Flexible manipulators are lighter, faster and less expensive than rigid ones but they pose various challenges as compared to rigid manipulators. In order to achieve high-speed and accurate positioning, it is necessary to control the manipulator's vibratory response in a cost effective manner. A good literature review of different control strategies for flexible manipulators can be found in (Benosman and Vey, 2004).

A feedforward control scheme based on input command shaping, introduced by Singer and Seering (1990), has been applied to the control of different types of flexible systems for vibration reduction or trajectory tracking or occasionally both (Alam, et al., 2006; Md Zain, et al., 2006; Singh and Singhose, 2002). The command shaping technique in practice causes delay in system's response while it reduces vibration and the amount of reduction in vibration and the rise time are found to be in conflict with one another.

This paper presents a new command shaping method using gain and delay units. A new multi-objective particle swarm algorithm (MOPSO) is proposed and applied to optimise the gain values and the amount of delay in order to obtain a wide range of solutions, which trade-off conflicting objectives so as to satisfy design goals.

2. Experimental set-up

A schematic representation of the single-link flexible manipulator system considered in this work is shown in Figure 1, where X_oOY_o and XOY represent the stationary and moving co-ordinates respectively, τ represents the applied torque at the hub. E, I, ρ, V, I_h and M_P represent the Young modulus, area moment of inertia, mass density per unit volume, cross sectional area, hub inertia and payload of the manipulator respectively. In this study, an aluminium type flexible manipulator of dimensions $900\times19.008\times3.2004\text{mm}^3$, E=$71\times10^9\text{N/m}^2$, I=$5.253\times10^{-11}\text{m}^4$, ρ=2710kg/m^3, and I_h=$5.8598\times10^{-4}\text{kgm}^2$ is considered (Tokhi and Azad, 1997).

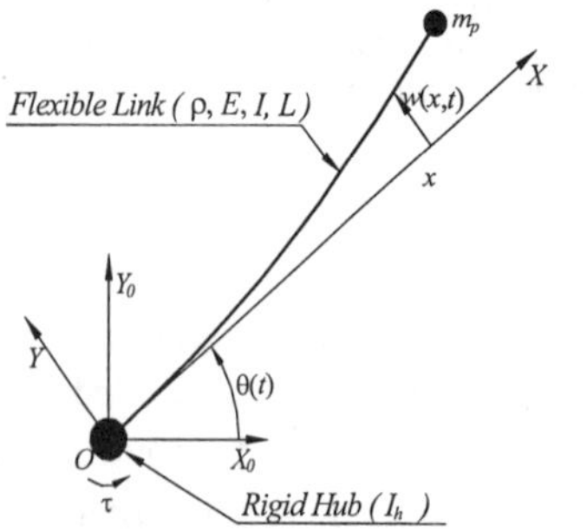

Figure 1: Schematic representation of the single-link flexible manipulator

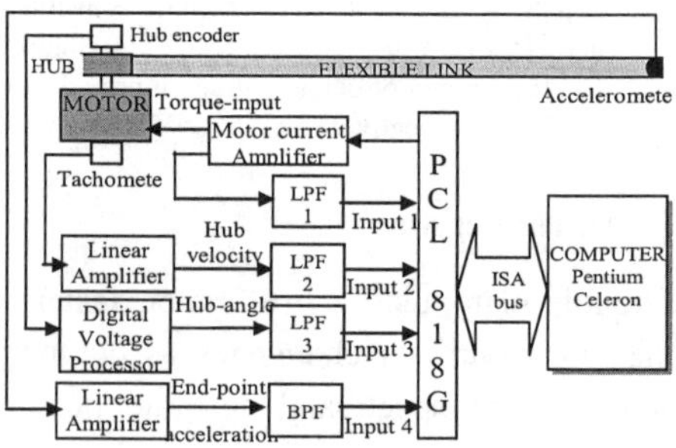

Figure 2: Block schematic diagram of the experimental rig

The experimental rig, as shown in Figure 2, is equipped with a U9M4AT type printed circuit motor driving the flexible manipulator. The measuring devices used to record the various responses of the manipulator are shaft encoder, tachometer and accelerometer along the arm. The shaft encoder is used for measuring the hub-angle of the manipulator. A precision interface circuit

PCL 818G is used to interface the flexible manipulator system with a computer (Pentium celeron- 500MHz).

3. The proposed command shaping for vibration control

A new command shaping method is proposed, as shown in Figure 3, using gain and delay elements to shape the reference input. The unshaped reference signal is passed through multiple delay units, Δ, and then multiplied with gain factors, K. The shaped command is formed by summing up the delayed components. For simplicity, the numbers of delay units and gain elements are kept the same. The gain values are selected in such a way to give 1 when they are added together (Singer and Seering,1990). This can be shown as:

$$\sum_{i=1}^{n} K_i = 1 \tag{1}$$

In order to minimise delay in system's response, the first delay unit is set to zero, i.e., $\Delta_1 = 0$. Assuming that, no prior information is available about the natural frequencies and associated damping ratios, a MOPSO algorithm is used to find a set of solutions (delay and gain values) that trade-off between conflicting design objectives. The proposed command shaping method is described in Figure 4.

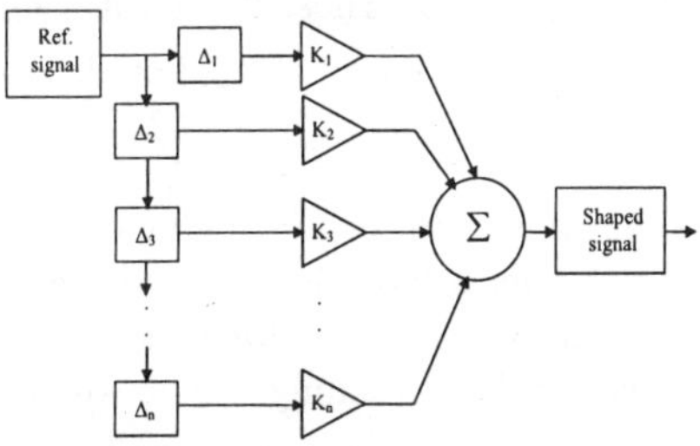

Figure 3: Schematic diagram of command shaping using gain and delay elements

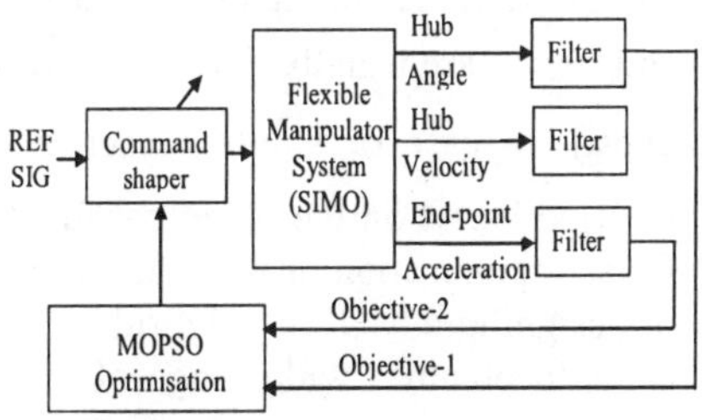

Figure 4: Schematic diagram of MOPSO based command shaper

3.1. *Objective functions and goal values*

Root mean squared value of residual error and rise time of hub angle response are chosen as two objectives of the MOPSO process. The two objectives and the corresponding goal values are shown in Table 1. Objective-1 thus chosen corresponds to reduction in end-point acceleration response as compared to that 70% lower than the open loop response with unshaped bang-bang signal. The maximum goal value of rise time (objective-2) thus chosen was 7.5% higher than that with bang-bang unshaped response.

Table 1: Two objectives and goal values

Objective	Parameter	Goal value
Obj-1	RMS value of end point acceleration	≤ 0.2
Obj-2	Rise time of hub angle response	≤ 1.2 sec

4. The proposed MOPSO algorithm

The dynamic equation of a modified PSO algorithm (Kennedy and Eberhart, 1995; 2001), for d -dimensional searching space is given as

$$v_{id} = \omega \times v_{id} + c_1 \times rand(\bullet) \times (p_{id} - x_{id}) + c_2 \times Rand(\bullet) \times (p_{gd} - x_{id}) \tag{2}$$

$$x_{id} = x_{id} + v_{id} \tag{3}$$

where c_1 and c_2 are positive constants, and $rand(\bullet)$ and $Rand(\bullet)$ are two random functions in the range [0,1]; ω is inertia weight; x_{id} represents the *i*-th particle; v_{id} represents the rate of the position change (velocity) for particle i. p_{id} represents the best previous position of the *i*-th particle and p_{gd} represents the best particle among all the particles in the population. These two terms, p_{id} and p_{gd}, are usually known as 'local guide' (or pbest)' and 'global guide' (or gbest). In case of single objective optimisation problem, p_{id} and p_{gd} are selected based on the objective function either minimum or maximum value as far as minimisation or maximisation problem is concerned. The main challenge, in designing a MOPSO algorithm, is to select pbest and gbest for each particle so as to obtain a wide range of solutions that trade-off among the conflicting objectives.

4.1. *Selection method of global guide and local guide*

In the proposed algorithm, a new technique is introduced that combines external archive and non-dominated fronts of the current population in order to select gbest for each particle. An external archive and associated control mechanism, as used in (Coello et al. 2004), is also employed here.

For a two-objective optimisation problem, Figure 5 shows the state of the external archive and solutions of the current particles in the objective domain. The dark circles inside a 2-D grid structure indicate the non-dominated solutions found so far while circles on the right represent solutions of current particles in a 2-D objective domain and the number associated with them indicate index of the particles in the initial population. The current solutions are sorted based on

Pareto dominance and several non-dominated fronts (ND fronts) are formed as shown in Figure 5. For each particle on ND front-1, the corresponding gbest is selected from the external archive based on fitness sharing and roulette wheel selection method (see Figure 5). Details of this process can be found in Coello et al. (2004). For particles on the remaining fronts, i.e., ND front-2, 3, 4 and 5: gbest of each particle is selected in the following way:

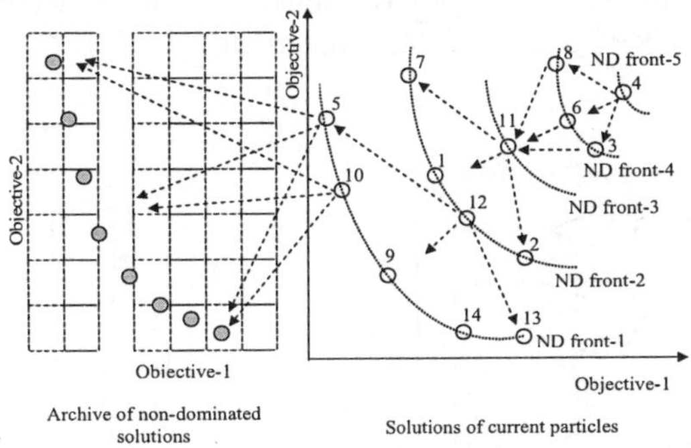

Figure 5: Schematic diagram for finding GBEST guide for particle in MOPSO

At first, shared fitness of each particle in the current population is calculated based on the exact non-dominated sorting GA (NSGA) fitness assignment scheme which was adopted by Srinivas and Deb (1994). Then, for each particle on ND fron-2, the corresponding gbest is selected from particles lying on the immediate lower front (better solutions), i.e., ND front-1, based on shared fitness and roulette wheel selection method (see Figure 5). This process continues for particles residing on the remaining ND fronts. Local guide or pbest for each particle is selected based on Pareto-dominance (Deb, 2001).

5. Implementation

Considering the complexity of MOPSO optimisation process and required amount of computation, the design procedure is implemented off-line. A dynamic model of the single-link flexible manipulator based on finite element method is utilised throughout this work. Although the flexible manipulator has infinite number of modes with associated damping ratios, only the first few modes appear to be dominant contributing to the flexible motion of the system. In order to reduce vibration, mainly, at these resonance modes and to design a command shaper having all attributes of a conventional zero vibration derivative type (Singer and Seering,1990), both the number of gain elements and delay units are chosen as 9. In order to minimise delay in system's response, the first delay unit is set to zero, i.e., $\Delta_1 = 0$. So there are effectively 8 delay units denoted as $\Delta_2, \Delta_3, ... \Delta_9$ and the 9 gain values are denoted as $K_1, K_2, ..., K_9$. The Matlab/Simulink (The MathWorks, Inc., 2006) model (see Figure 6) is chosen as it allows for simple construction of command shaping using gain and delay units. Filters are used with all the three outputs with the aim to filter out flexible motion keeping the rigid body motion intact. The cut-off frequencies of the filters were chosen to be consistent with main vibration modes and rigid body motion of the system. Low pass Butterworth filters were used with hub angle and hub velocity outputs with cut-off frequency of 90Hz.

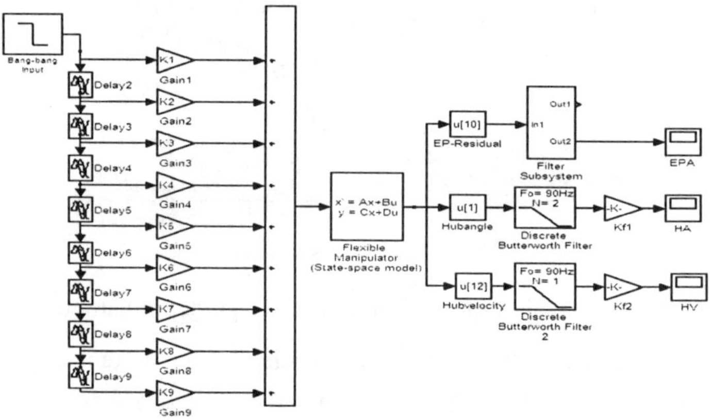

Figure 6: Simulink model of the gain-delay based command shaping control strategy

5.1. *Parameter encoding*

A swarm of 20 particles having 17 elements each, i.e., 10 × 17 is created randomly within the range of [0, +1]. The first 9 elements of each individual are normalised and assigned to gain elements, $K_1, K_2, ..., K_9$ as indicated in the Simulink model (see Figure 6). In Matlab/Similink, the delay units are usually represented in terms of number of samples which are integer. So the remaining 8 elements of each individual are converted into integer numbers by a conversion factor of 0.01 following a 'round' operation and then assigned to $\Delta_2, \Delta_3, ..., \Delta_9$.

6. Solutions and results

The acceleration coefficients of MOPSO algorithm are set as $c_1 = c_2 = 1.5$, and inertia coefficient ω is gradually decreased from 1.4 to 0.1 with generation. The optimisation process is run for a maximum generation of 500. The maximum number of solutions that the external archive can keep is limited to 50. In order to maintain diversity among the solutions in external archive, an adaptive grid mechanism is employed (Coello et al., 2004). The MOPSO algorithm with a population size of 20 individuals was run on this for 500 generations. Out of a total of 10000 points evaluated, only 26 solutions were non-dominated. The Pareto optimal set at generation 500 is shown in Figure 7 where, objective-1 is represented on the x-axis and objective-2 on the y-axis. To validate the design approach as well as solution set, one example solution (solutin-1) is selected on the Pareto front as shown in Figure 7. Unshaped bang-bang input and shaped signal based on solution-1 are shown in Figure 8. The hub angle responses for unshaped bang-bang and shaped signal based on solution-1 are presented in Figure 9. It is clearly evident (in enlarged section) that oscillation in the hub angle response is almost eliminated with shaped command and it seems to settle quickly to the steady state. End-point acceleration of the flexible manipulator due to unshaped bang-bang and shaped signals based on solution-1 are presented in the time domain in Figure 10.

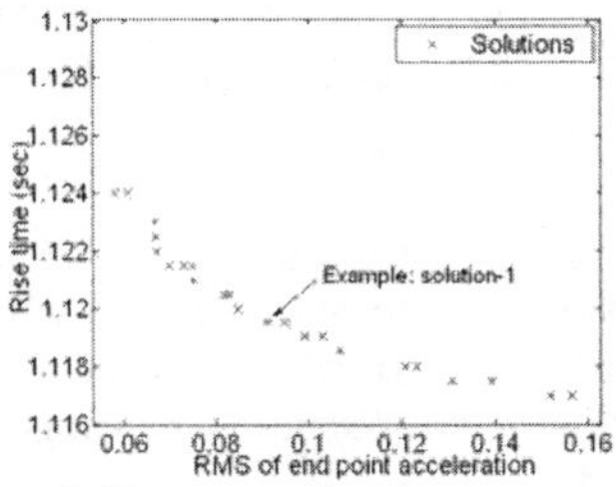

Figure 7: Pareto optimal set at generation 500

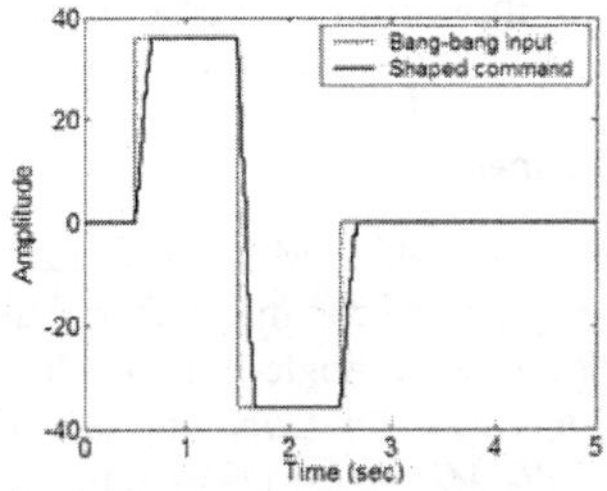

Figure 8: Bang-bang input and shaped command (for solution-1)

Performance measures, such as, peak-to-peak amplitude, rise time, settling time and steady-state error of the hub angle response and RMS value, norm-∞ and settling time of the end-point acceleration for system response with shaped input based on solution-1 are calculated and presented in the Table 2.

Table 2: Performance measures of different solutions

Input of open- loop Control strategy	Hub angle response			
	Max. oscillation (peak-peak)	Rise time (sec)	Settling time (sec)	SS error
Bang-bang	3.1656	1.116	3.9	0
Solution-1	0.108	1.12	2.39	0

7. Conclusion

A new command shaping technique has been presented using gain and delay units to reduce vibration of flexible systems. Considering the conflicting design objectives, such as, reduction of vibration and speed of response, a new MOPSO algorithm has been successfully applied to derive solutions based on two objectives, satisfying the design goals, known as Pareto-optimal set, which describes the trade-off among conflicting objectives. The Pareto front yields a set of candidate solutions, from which the desired one is picked under different trade-off conditions. All the performance measures give a good indication of reduction of vibration in system's response compared to unshaped response.

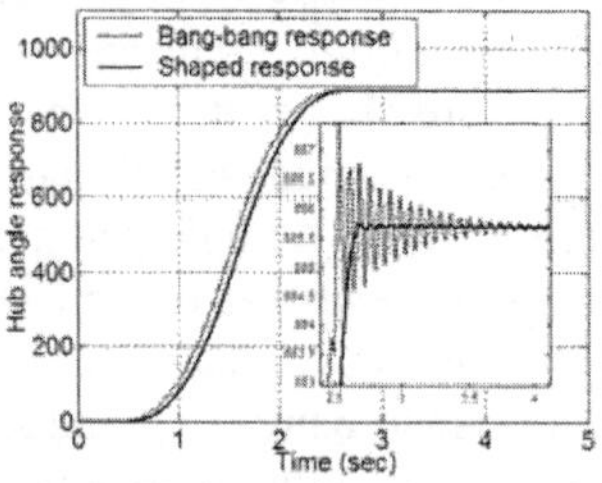

Figure 9: Hub angle response for bang-bang input and shaped command (for solution-1)

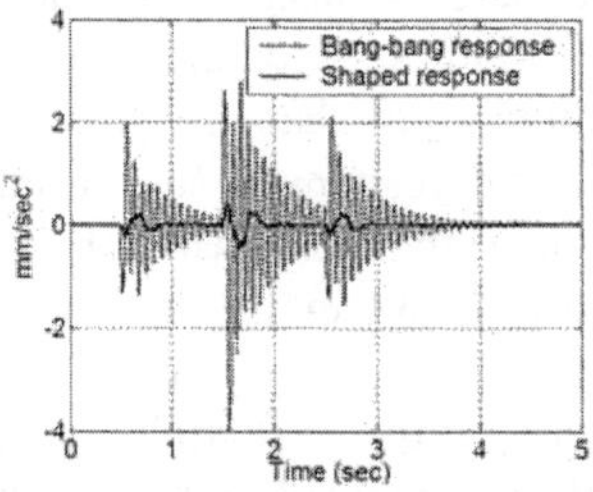

Figure 10: End-point acceleration due to bang-bang input and shaped command (for solution-1)

References

Alam, MS, Tokhi, MO., Siddique, MNH. and Hossain, MA. (2006). Selection and Designing of command shaper using multi-objective genetic algorithms for vibration control of a single-link flexible manipulator, *Proceedings of the 9th International Conference on Climbing and Walking Robots and the Support technologies for Mobile Machines*, 11-14 Sept., Brussels, Belgium.

Benosman, M. and Vey, L. (2004). Control of flexible manipulators: A survey, *Robotica*, 22, 535-545.

Coello, CAC., Pulido, GT. and Lechuga, MS. (2004). Handling Multiple Objectives with Particle Swarm Optimization, *IEEE Transactions on Evolutionary Computation*, 8(3), 256–279.

Deb, K. (2001). *Multi-objective optimization using evolutionary algorithms*. New York; Chichester: Wiley.

Kennedy, J. and Eberhart, R. (1995). Particle swarm optimization, *Proc. of IEEE International Conference on Neural Networks, IV*, 1942-1948, Perth, Australia,

Kennedy, J. and Eberhart, R. (2001). *Swarm Intelligence*, Morgan Kaufmann Publishers.

MATLAB Reference Guide, The Math Works, Inc., 2006.

Md Zain, MZ., Tokhi, MO. and Mohamed, Z. (2006). Hybrid learning control schemes with input shaping of a flexible manipulator system, *Mechatronics,* 16(3-4), 209-219.

Singer, NC. and Seering, WP. (1990). Preshaping command inputs to reduce system vibration, *J. Dynamic Systems, Measurement and Control*, 112(1), 76–82.

Singh, T. and Singhose, W. (2002). Tutorial on input shaping/time delay control of maneuvering flexible structures, *Proceedings of 2002 American control conference*, Omnipress, Madison, 1717–1731.

Srinivas, N. and Deb, K. (1994). Multiobjective Optimization Using Nondominated Sorting in Genetic Algorithms, *Evolutionary Computation*, 2(3), 221–248.

Tokhi, MO. and Azad, AKM. (1997). Design and development of an experimental flexible manipulator system, *Robotica*, 15(Part 3), 283-292.

DETECTING SOUND SOURCES WITH THE HUMANOID ROBOT RH-1

PAVEL STAROVEROV, RICARDO MARTINEZ, DMITRY KAYNOV, MARIO ARBULU, LUIS CABAS, CARLOS BALAGUER

Robotics Lab, Department of System Engineering and Automation, University Carlos III of Madrid, Av. Universidad 30, Leganés 28911(MADRID), Spain

Rh-1 is a humanoid robot under development at the University Carlos III of Madrid. Along with other interaction types, the robot is capable of recognizing voice commands. This article describes the sound localization system that has been added recently. It is independent of the voice recognition system and uses the information from several microphones attached to the robot's head that is further processed in order to estimate the sound source direction. Results are presented and discussed and the ideas how to improve the performance are suggested.

1. Introduction

The ability to detect the direction of sound sources can be a useful addition to the interaction system of a humanoid robot or another type of service robots. In case if the speaking person is not facing the robot, the robot could turn to the speaker or just turn its head, thus making the communication more natural and pleasant for the person.

Rh-1 is a humanoid robot developed at the University Carlos III of Madrid. Rh-1 has 21 degrees of freedom (+2 more degrees of freedom of the built-in camera), is 130 cm tall and weighs in about 45 kg (Fig.1).

From the point of view of interaction, the robot is capable of speech recognition, speech synthesis, face detection / recognition and the so-called "T-shirt based" recognition. Those abilities were discussed in [16]. There must be another paper somewhere in the present proceedings describing the newly added gesture (posture) recognition system based on shirt shapes.

The objective of the work was to develop an easy-to-implement sound source detection system with a good correspondence between the invested time and effort on one hand and with an acceptable performance on the other hand. The present article is organized as follows: in Sec. 2 some related works are discussed, Sec. 3 describes the development of the system, in Sec. 4 the experimental results are presented and discussed. Section 5 concludes the paper.

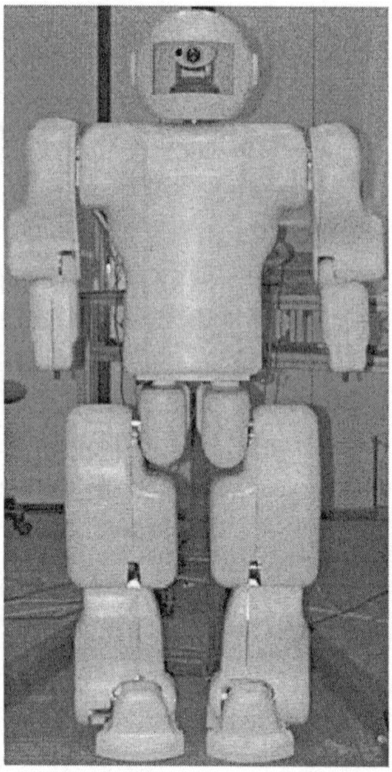

Figure 1. Rh-1 humanoid robot.

2. Related works

One of the first works in the field of acoustic perception for robots made in 1995 at MIT involved the robots Cog and Kismet [1]. SIG and his successor SIG2 (developed at the University of Kyoto since 2000) have auditory systems based on binaural hearing that use 2 pairs of microphones (one pair is used to cancel the noises produced by the motors) [2,3]. ROBITA robot developed at Waseda University can follow a conversation between two persons using his 2 microphones [4]. Lately there has been a trend to use more than 2 microphones. For example, one of the most advanced humanoid robots HRP-2 developed by Kawada Industries and AIST has 8 microphones [5], Sony QRIO small sized humanoid robot has 7 of them [6]. Finally, there is the famous Honda ASIMO humanoid robot that is able to localize sound sources, making difference between voices and other sounds or noises [7]. Concerning more recent works, there is a robot called Wakamaru developed by Mitsubishi that uses microphone array for sound source localization [8].

Speaking of the algorithms used, there are several signal processing techniques that are used in robotics and in other applications as well, among these are: energy based methods [9], methods based on the time difference of the signals received by a pair of microphones like "Cross-power Spectrum

Phase" [10]-[14] and finally there is the "Sum and Delay Beamformer" technique that is used to process the signals from a 128 microphone array [15].

3. Development of the system

The objective was to develop a system that would let the robot detect the position of the speaking person relative to the robot. This implies using several microphones, converting their signals into the digital form and processing the data. First of all, let us take a look at the hardware components that the system is made of. The schematic view of the system is presented in Fig.2.

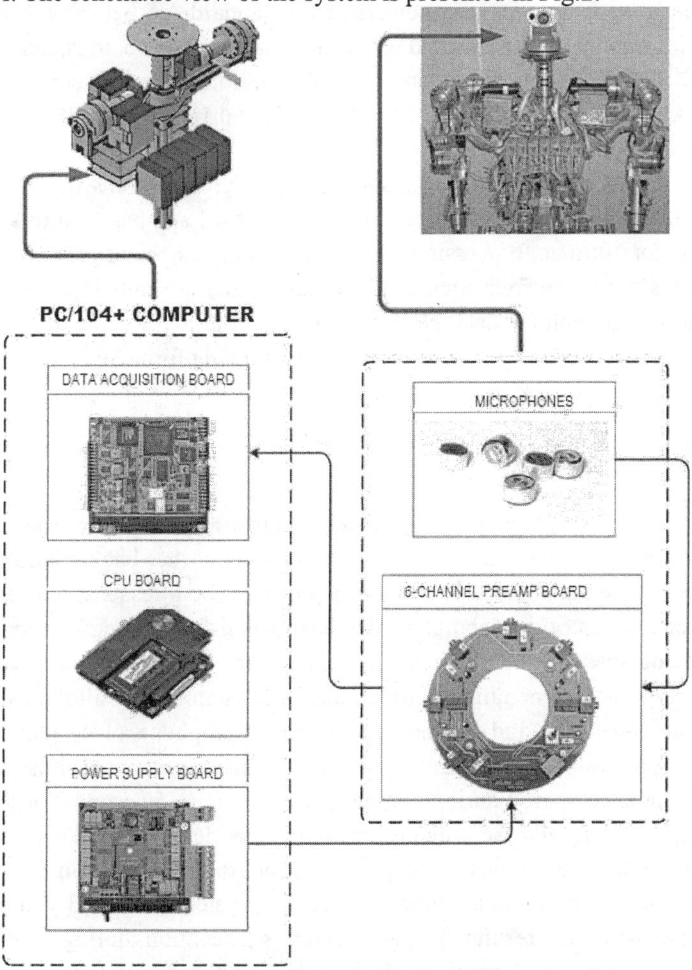

Fig.2. Hardware of the sound localization system.

There are 4 Knowles Acoustics MD9755USZ-1 unidirectional microphones that are located in the lower part of the robot's head in the horizontal plane, one in the front, one in the back and two at both sides. Those microphones feed the signals into the custom made preamp board with. It has the shape of a flat donut (it was necessary to design the board that way to fit it into the robot's head) and can amplify up to 6 microphones, though currently only 4 of them are used. The gain of each channel can be adjusted manually. Some of the voltages needed by the board are provided by the PC/104 power supply (that also powers all the computer components of the robot), but the most noise-sensitive components are powered by 3 standard AA batteries. It helps to reduce the amount of noise picked up by the preamps. The board is connected to the RTD SDM7540 data acquisition board that performs the A/D conversion (12 bits, 22 kHz sampling rate). The CPU board Digital Logic MSM855 is running Windows XP operating system.

Now let us take a look at the software and the algorithms used. A driver written in C reads the data from the acquisition board and passes it to a Matlab program for further processing. First, we apply a band-pass filter that accentuates the voice frequencies and eliminates the unwanted noises and DC offset in the microphone data. Next, we calculate the RMS value of N samples of each microphone's readings $\{x_n\}$ using the following formula:

$$\varepsilon_{RMS} = \sqrt{\frac{1}{N}\sum_{n=0}^{N-1}|x_n|^2} \quad . \tag{1}$$

During the experiments a value of N equal to 1000 samples was used. After that a threshold function is applied to eliminate the background noises. Threshold value equal to 0.015 mV RMS proved to be a reasonable one for the normal conditions of our laboratory that has several functioning PCs and an air conditioning system. Then the readings from all the microphones are compared and the result is assigned in each timeframe of N samples, 0 if all the values are below the threshold, and the number of the microphone (1-4) that has the loudest signal in the contrary case. Finally, the number of hits of each microphone is compared during a larger period of time (5 seconds for the most of the test results), and the winning microphone is determined. This result can be passed to other programs running on the robot that can use it in conjunction with the voice recognition and image processing systems. In Fig.3 you may see a window with the results of the program's execution during one of the experiments, there are 4 graphs with the signal amplitude for each microphone and the last graph showing the winning microphone in each timeframe. The final result is also shown.

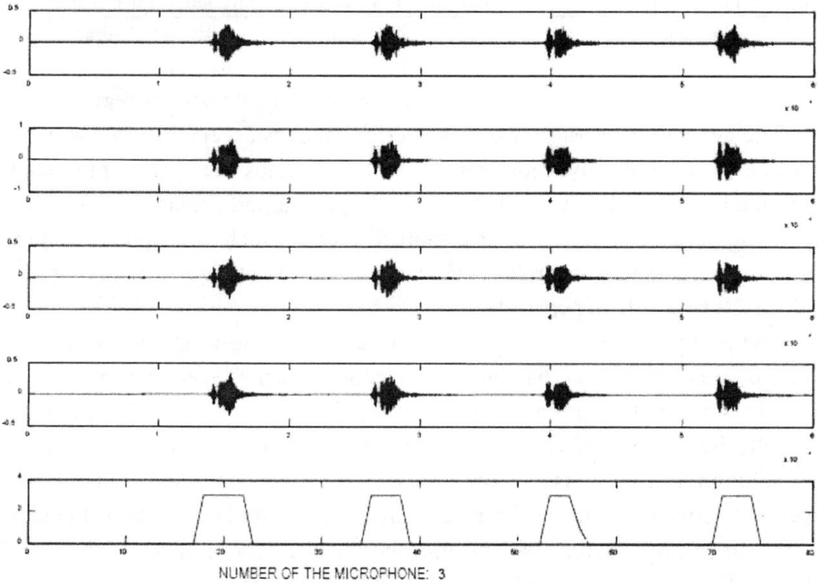

Fig.3. An example of the results.

The method used is very simple and only provides 90 degrees of accuracy. However, that might be enough in case of our application. The main reason why this method was selected is because it is easy and fast to implement. Let us see the experimental results to see whether the performance is acceptable for our application or not.

4. Results and discussion

There have been three series of experiments, in each one of them the speaker was repeating the same sound (or phrase) three times in front of every microphone. The first test was with the person producing 2 handclaps, the second one was with a 1-second-long phrase "Robot, I'm here", and the third one was with a longer, 3-second-long phase "Robot, I'm here, do you know where I am?" Every test was repeated at 1 m distance from the robot and then at 2 m distance. The result was considered positive, if the number of the microphone returned by the program coincided with the speaker's position and negative in the contrary case. The total number of tests was 72. The robot was placed at 2.5 m distance from the walls, nearly in the center of the room. The results are presented in Table 1.

Table 1. Results of the tests.

Type of the sound	Results at 1 m distance	Results at 2 m distance
Two handclaps	7 positive / 5 negative	2 positive / 10 negative
Short phrase	9 positive / 3 negative	7 positive / 5 negative
Long phrase	11 positive / 1 negative	8 positive / 4 negative

The general impression is that the performance is acceptable only at a short distance (1 m) and only with phrases, with handclaps the system presented a high number of misses which is unsatisfactory. The performance of the system with long phrases pronounced at a short distance, which is supposed to be the most frequent case for our robot, may be considered quite good, with more than 90% of the results being correct.

The performance of the system with handclaps and at greater distances is worse because of the fact that under these conditions the reverberations from the walls become quite prominent, masking the direct sound. The algorithm used currently is quite straightforward and makes no effort to cancel those echoes. The solution may be to use a more robust, more advanced time-based algorithm like Cross-power Spectrum Phase described in [11] and [12]. Also, if the sound source direction detection will be used during the robot's motion, the sounds generated by the motors should be canceled like suggested in [2] and [3].

To summarize, it can be said that the performance of the developed system is acceptable for out type of application, however, there are ways to improve it.

5. Conclusion

The present article suggests a sound source direction detection system for a humanoid robot. Sound source direction detection may be a valuable addition to the interaction system of any service robot. The proposed system is based on 4 microphones and a simple energy-based algorithm. The development of the system was explained in detail, along with the algorithms used. Some experimental results were presented and discussed. The performance of the system is considered acceptable for our needs. Several ways to improve the performance of the system are suggested. Future works may include implementation of a more robust time-based algorithm that would also cancel reverberations and noises from the motors.

Acknowledgments

We would like to thank the members of the Robotics Lab of University Carlos III of Madrid for their cooperation and suggestions. The present work was supported by CICYT (Comisión Interministerial de Ciencia y Tecnología).

References

1. R. Brooks, C. Breazeal, M. Marjanovie, B. Scassellati, and M. Williamson. "The Cog project: Building a humanoid robot," in Computation for Metaphors, Analogy, and Agents, C. Nehaniv, Ed. Springer-Verlag, 1999, pp. 52-87.
2. Kazuhiro Nakadai *et al.*, "Active Audition for Humanoid", Kitano Symbiotic Systems Project, ERATO, Japan Science and Technology Corp.
3. K. Nakadai, H. G. Okuno, and H. Kitano, "Real-time sound source localization and separation for robot audition", Proc. IEEE ICSLP, 2002, pp. 193-196.
4. Yosuke M. *et al.*, Perceptual Computing Group, Waseda university, "Modelling of Conversational Strategy for the Robot Participating in the Group Conversation", Eurospeech 2001.
5. Isao Hara, Futoshi Asano et. al., "Robust speech interface based on audio and video information fusion for humanoid HRP-2" Proc. ICIRS, Sendai, September 2004, pp. 2404-2410.
6. Fujita M. *et al.*, Intelligent Dynamics Laboratory, Sony Corporation, "Autonomous behaviour control arquitecture of entertainment humanoid robot SDR-4X", International Conference on Intelligent Robots and Systems, 2003.
7. Sakagami Y. *et al.* "The intelligent ASIMO: System overview and integration" Proceedings of the 2002 IEEE/RSJ Intl. Conference on Intelligent Robots and Systems, EPFL, Lausanne, Switzerland, pp.2478-83
8. Tomonaka T. *et al.* "Computer Vision Technologies for Home-use Robot "Wakamaru", Mitsubishi Heavy Industries, Ltd. Technical Review Vol. 42 No. 1 (Feb. 2005)
9. T. Ikeda *et al.*, "Framework of Distributed Audition", Proc. 13th IEEE International Workshop on Robot and Human Interactive Communication, Sept. 2004, Japan, pp. 77 – 82.
10. M. Brandstein, A Framework for Speech Source Localization Using Sensor Arrays, Ph.D. thesis, Brown University, 1995.
11. M. Omologo, P. Svaizer, "Use of the Cross-Power Spectrum Phase in acoustic event localization", IRST Trento, Technical Report #9303-13.
12. M. Omologo, P. Svaizer, "Acoustic Event Localization using a Crosspower Spectrum Phase based Technique", Proc. ICASSP, Adelaide 1994, pp. II273-II276.

13. M. Omologo, P. Svaizer, "Talker Localization and Speech Enhancement in a Noisy Environment using a Microphone Array based Acquisition System", Proceedings Eurospeech, Berlin, September 1993, pp. 605-609.
14. Amir A. Handzel, P. S. Krishnaprasad, "Biomimetic Sound-Source Localization", IEEE Sensors Journal, Vol.2, No.6, December 2002, pp. 607-616.
15. Y. Tamai *et al.*, "Real-Time 2 dimensional Sound Source Localization by 128-Channel Huge Microphone Array", Proc. IWRHIC, Okayama, Sep 2004, pp. 65-70.
16. P.Staroverov; M.Arbulú; L.M.Cabas; D.Kaynov; C.Pérez; C.Balaguer. "A Voice Controlled Image Recognition System." Proc. of the 9th International Conference on Climbing and Walking Robots. Brussels. Belgium. Sep, 2006.

IN SEARCH OF PRINCIPLES OF ODOUR SOURCE LOCALISATION

E. E. KADAR

University of Portsmouth, Department of Psychology, King Henry Building, King Henry I Street, Portsmouth, PO1 2DY, UK.

G. S. VIRK

Massey University, School of Engineering and Technology, Wellington, New Zealand.

C. LYTRIDIS

Department of Electronic and Computer Engineering, Anglesea Building, Anglesea Road, Portsmouth, PO1 3DJ, UK.

Arguably, chemical sensing is perhaps the most fundamental sensory modality in animals. This modality is far simpler than most other modalities (vision, hearing, heat sensing, electric and magnetic sensors) and yet navigation in a chemical diffusion field is still not well understood. Biological studies have already demonstrated the use of various search methods (e.g., chemotaxis and biased random walk) by various animals, but robotics research could also provide new ways to investigate principles of olfactory-based search skills [1],[2],[3]. The primary aim of the present paper is to demonstrate that similar navigational principles can be observed in various animals ranging from invertebrates to mammals but the use of these principles can lead to complex search strategies. Importantly, search strategies derived from these biological principles tend to result in similar patterns in both simulation studies and robot experiments.

1. INTRODUCTION

After more than a century of research on animal behaviour and numerous recent studies with robots, the principles of navigational strategies in chemical fields are still not well understood. For instance, early behaviourist experiments tried to explain behaviour in terms of reactions to stimuli and introduced various reaction including klinokinesis, klinotaxis, tropotaxis, etc. In general, too many reactions were needed to explain search behaviour for all observable conditions. To simplify theoretical account of search patterns, researchers attempted to use or rely on some intuitive principles. For instance, Fraenkel and Gunn [4] emphasized that chemoreceptors are not direction receptors and they cannot be used to orientate the animal to a gradient in a chemical field. Later, these different animal reactions to chemicals seem to have been forgotten and all search strategies in chemical fields were, in an overly simplistic way, called chemotaxis [5], [6]. During the past few decades, researchers were strongly influenced by the cognitive revolution and tended to created artificial search algorithms by often ignoring the fact that chemical search behaviour is highly influenced by environmental constraints such as the nature of the diffusion field (highly instable and especially noisy far from the chemical source) as well as the biological constraints such as the limitations of sensors (number of sensors, slow sensory response, problem of sensory adaptation). To further complicate matters, robotic researchers tend to confound their search models by using other sensory modalities without being clear on the use of principles and limitations of chemical searching. These problems highlight the importance of having a clear and consistent terminology and use them in relation to fundamental principles in experiment tests.

Motivated by these ideas, we set out to achieve a better understanding of the principles of search strategies in chemical diffusion fields. First we investigated biological strategies and tried to implement them in simulation studies. Biased random walk (BRW) strategies are shown to be robust and yet more efficient than chemotaxis in unstable and noisy chemical fields [7]. In addition to locating the static point odour source in unstable chemical fields, BRW has also been assessed for moving targets [8], odour trails [9] and turbulent plumes [10], [11].

Most of our previous work has focused on testing navigational strategies in simulation studies. In agreement with Webb [1], we believe that robotics research provides new ways to investigate biological strategies. In particular, the principles of olfactory navigational strategies can be distilled from various search patterns (based on biological and simulation studies) and systematically investigated on robotic platforms to overcome some of the limitations of simulation studies (e.g., difficulties in modelling odour diffusion and dispersion processes when a moving agent introduces additional noise in the chemical field).

Although the poor quality of artificial sensors does constrain research and simulation as well as robot-based experiments are insufficient to find accurate models for biological search performance, we argue in the present paper that investigation of animal search patterns, simulation studies and robot experiments provide sufficient evidence that only a few principles are needed to explain complex olfactory search patterns observed in natural settings.

2. BASIC PRINCIPLES OF NAVIGATIONAL STRATEGIES

Traditionally, in the literature chemical search is often called chemotaxis. Recognizing that taxis behaviour require two sensors, it becomes clear, however, that chemotaxis should only be used for search strategies that rely on directional information to orientate the agent towards the direction of a chemical gradient. Given the fact that chemosensors are not direction sensitive [4] and can only use local information (in contrast to visual and auditory sensors) a single sensory reading cannot provide sufficient information on the direction of a gradient [10]. Although many species have bisensory chemical receptor organs (antennae, nostrils), quite often they have to rely on only one of these sensors (e.g., when one of the sensors is impaired). These facts led us to turn our attention to unisensory search mechanisms.

In our early simulation studies, we have investigated unisensory BRW as an alternative search strategy to chemotaxis [7], [8]. Also, we argued that most bi-sensory species are probably rely on unisensory strategies even though they use both unisensory and bisensory (chemotaxis) strategies. Thus, one of the fundamental principles is that a chemical sensor is not direction sensitive but if it is combined with partly random locomotion it could be robust and effective to follow gradient information under various conditions. Another important principle we learned from our investigations is that bisensory capability is important for many species to provide some redundancy for unisensory search but at the same time two sensors could enhance efficiency of unisensory search under a variety of conditions (e.g., high concentration in the neighbourhood of target etc.). The present study provides some examples from animal research as well as simulation and robot studies to support the validity of these principles and demonstrate how they give rise to complex search patterns.

2.1. *Animal behaviour in chemical search*

Search patterns of small worms as well as higher order species exhibit some fundamental similarities. Typically, at far range from target, search patterns are erratic and wildly meandering. Nevertheless, animals gradually drift close to the target. As the target is approached, the movement patterns are getting less meandering and often either getting straight or mildly sinusoid around an axis pointing toward the target. In [12], Koehler has reported movement patterns of Planaria lugubris clearly showing these generic characteristics (see Figure 1a). In a pilot study with a blind cat, we have also observed these patterns (see Figure 1b). Interestingly, the overall characteristics of these search patterns can be observed in bisensory chemotaxis search as well as unisensory search performance.

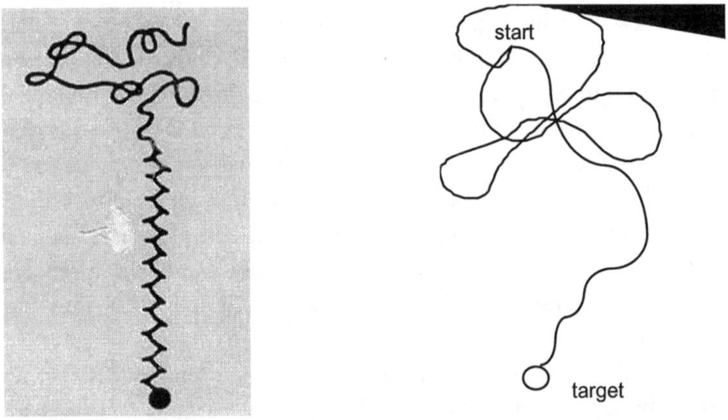

(a) (b)

Figure 1: (a) Schematic drawing of typical search patterns in Koehler's (1932) study with planarians. (b) Search pattern of a blind cat toward a dish with cat-food in it. In both animals, the early stage of search at a distance from target is characterized by meandering patterns and gradual drifts closer to target. At a close to target range, the movement becomes nearly straight with some pendulation of the head.

2.2. *Chemotaxis in simulation and robot research*

Chemotaxis is based on the detection of a concentration difference between two chemical sensors and a steering mechanism toward the direction of higher intensity while moving forward at a constant speed. In principle, chemotaxis-based navigation could result in smooth movement trajectories in smooth chemical fields without much environmental noise. Practically, however, this strategy is not sufficient and, as we have shown [7], often leads to failure due to the intrinsically noisy and highly unstable chemical field conditions. Robotic implementation create additional problem by the fact that robot movement further increases the noise in the diffusion field. Thus, the bisensory orientation becomes ineffective under noisy condition and at low concentration level (i.e. far from target range). However, moving in a high concentration and/or stable smooth diffusion field the bisensory system can facilitate search patterns that are driving the agent toward the target with a nearly smooth trajectory with some addition meandering factors.

Figure 2 shows successful search patterns by chemotaxis from simulation and robot search. Two main characteristics.of these patters can be observed: 1)

At a far range, the search performance is not promising because it is meandering but the agent still drifts closer to the target. 2) Close to the target, the search pattern becomes less erratic and often results in an oscillatory movement while approaching toward the target.

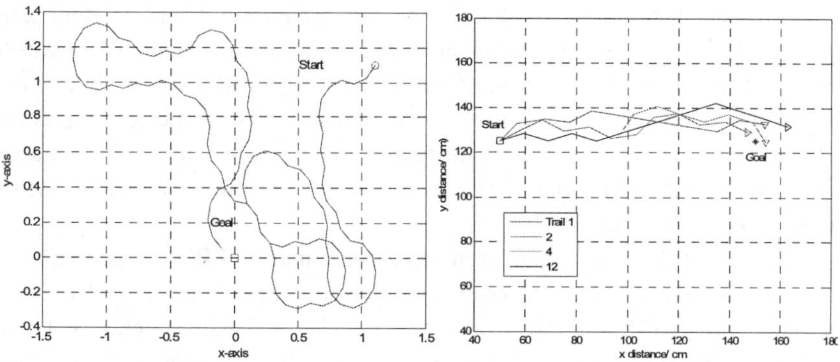

Figure 2: Chemotaxis-based search patterns of robot trials and simulations demonstrating that search performance is meandering in a noisy field at far range but it becomes smooth and slightly meandering driving the agent toward the target if the field is stable [13]

2.3. *Biased random walk*

Lytridis et al. argued in [13] that, in principle, unisensory search could be based on chemotactic principles because unisensory temporal sampling of the chemical concentration field could be interpreted as a method of compensation for the absence of spatially separated sensors in a unisensory organism. But this 'chemotaxis mimicking' strategy would require a memory for registering spatial locations with associated intensity values of the chemical field to calculate a new movement direction. The use of this complicated computational mechanism is unlikely in primitive organisms and it may not be necessary in higher-order species either.

And indeed, studies suggest that bacteria as well as other higher order species use a much simpler unisensory strategy consisting of straight runs with occasional directional changes (tumbling) [5]. Alternating these two modes of motion, the animal generates polygon shaped trajectories. It has been shown that both the run lengths and directional changes can be described by Poisson distributions. The characteristics of movement patterns depend on the concentration distribution in the search environment. In a homogeneous field, there is no gradient information and the resulting motion is a purely random walk. In an inhomogeneous environment, the distribution function of the run lengths is continuously changing since the mean of the distribution depends on the detected concentration gradient (e.g., intensity difference between two different locations). When the concentration level increases, the frequency of directional changes decreases, and therefore the forward runs tend to be longer. The result of this strategy is an overall drift toward the source of the diffusion field. To put it differently, the random walk becomes a biased random walk process. This is a simple but efficient strategy using a fast adaptation process to the local field conditions instead of relying on memory of previously measured intensities at various locations in the search process. During the search process, the moving organism can adjust its sensory organs to the local concentration level and can detect relative increases or decreases in field intensities when it drifts into another region.

A similar approach has been developed into a BRW algorithm at a macroscopic scale. Its use in artificial agents was first demonstrated in noisy Gaussian fields [7]; and this work was extended to dynamic fields [10]. The BRW strategy in odour source localisation has been shown to be a robust method under a variety of conditions, but its efficiency can be improved [7], [8].

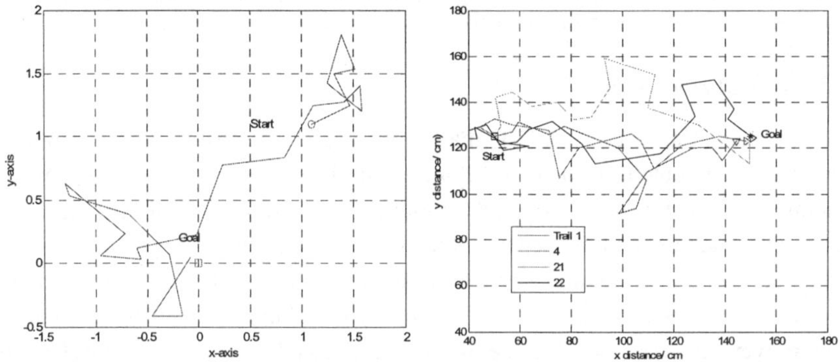

Figure 3: Search patterns of robot trials and simulations using biased random walking

Although BRW is successful in driving the searching agent close to the target from regions of low concentration level, BRW is less efficient in the neighbourhood of the target because it tends to generate longer runs which may lead to 'mistakes' such as overshooting or passing by the target. We have tried to remedy this shortcoming based on the observation that search patterns in animals get less meandering as they get closer to the target. This can be achieved by halting the agent at a certain high intensity threshold near the target. Similar result could also be achieved by reducing the range of random directional changes as well as reducing the run length bias above an intensity threshold typical close to the target. In both cases, search patterns remain unisensory.

Figure 3 shows a few search trials from simulation as well as robotic search. Typically, search patters are irregular especially far from the target because the field intensity values are smaller and the noise makes the field less stable and more difficult to use to detect gradient. Despite this difficulty, the agent gradually drifts toward the target and as it gets closer to the target where field intensity tends to be higher even though the noise is also present. Thus, it is more likely that high concentration field provides better opportunities to detect gradient information on the direction of target during a straight run.

2.4. *Combined strategy*

With intact bisensory systems, animals can benefit from both the unisensory as well as the bisensory search strategies and animal search patterns seem to confirm this hypothesis. This is why we have developed combined search stategies in our simulation studies and robot experiments [13]. Far from target with low field intensity and noisy environment, unisensory BRW search patterns are robust and efficient to drift the agent closer to the target. When the agent reaches a higher intensity threshold characteristic of the region close to the target, the unisensory BRW strategy can still be applied with a small modification. The bisensory system can be used to introduce a directional bias in the otherwise random directional changes. The extent of this bias can be dependent on the concentration level of the chemical field. Figure 4 shows a trial with a combined strategy exhibiting similar movement patterns shown in previous examples (See Figures 1-3).

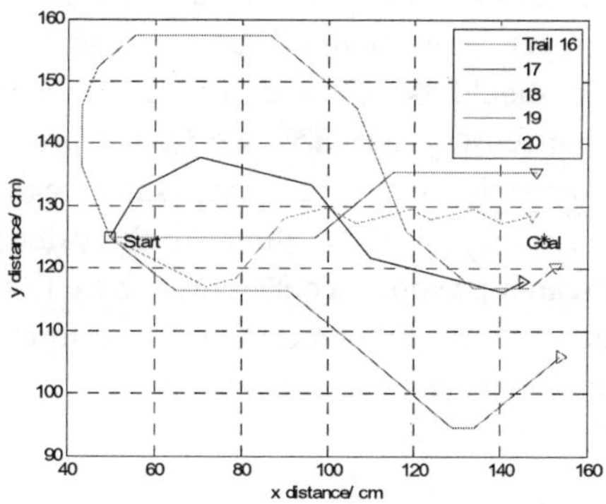

Figure 4: Search patterns of robot trials using the combined strategy

3. CONCLUSION

Recent experiments in robotics are often based on complex algorithms and the performance is assessed by success in artificial environment. In the past decade, we followed a different strategy. We tried to identify fundamental principles of biological search strategies in relation to the intrinsic properties (e.g., weak especially at a far range, lack of stability, and fast changing characteristics, etc.) of chemical diffusion fields [13], [14]. The present study provided a brief but hopefully convincing highlight of a decade long systematic research programme. We have started our research by observing biological search strategies in various species ranging from bacteria to cats. Then we formulated some hypothesis on fundamental principles of search strategies (unisensory versus bisensory in relation to field conditions) and tested them in simulation studies. Finally, we implemented these strategies in robotic platforms and our experiments resulted in similar search patterns observed in animals. In sum, our starting point was observation of animal search patterns to derive fundamental principles of biological chemical search strategies and we concluded our research by implementing these principles in 'artificial search algorithms' on robotic platforms to create search patterns similar to the ones created by animals.

We have to finish this paper, however, with a cautionary note. In biologically inspired robotics research as well as other biological modelling experiments, scientists have to be careful and remain clear about the distinction between simulation and mimicking. Although in this paper we used the word simulation and robotic implementation of simulation, these studies actually cannot be considered proper simulations. They are simply just mimicking biological strategies. We should be allowed to use the word "simulation" properly if the actual robot sensory detection and further processing would be similar to the processes animals use in their remarkable search performance. Nevertheless, this cautionary note does not diminish the value of our research because we believe the search performance should be based on the interaction field of the organism/agent and its environment when a performance is based on principles of search strategies in a natural environment.

References

1. Webb, B. (2000). What does robotics offer animal behaviour? *Animal Behaviour*, 60 545-558.
2. Grasso, F. W., Consi, T. R., Mountain, D. C., & Atema, J. (2000). Biomimetic robot lobster performs chemo-orientation in turbulence using a

pair of spatially separated sensors: Progress and challenges. *Robotics and Autonomous Systems*, 30(1-2), 115-131.

3. Lytridis C, Virk GS and Kadar EE (2005). Search performance of a multi-robot team in odour source localisation, Proceedings of 8th International Conference on Climbing and Walking Robotics (CLAWAR 2005), London, 3-15 September.
4. Fraenkel G. S. and Gunn D. L. (1961). The orientation of animals, Dover, New York.
5. Berg, H. C., & Brown, D. A. (1972). Chemotaxis in Escherichia coli analysed by three-dimensional tracking. Nature, 239(5374), 500–504.
6. Beer, R. D. (1990). Intelligence as adaptive behaviour: An experiment in computational neuroethology. Perspectives in artificial intelligence no. 6. Boston: Academic Press.
7. Kadar, E. E., & Virk, G. S. (1998). Field theory based navigation for autonomous mobile machines, Proceedings of the IFAC Workshop on Intelligent Components for Vehicles (ICV '98), Seville, Spain, 23-24 March.
8. Kadar, E. E., & Virk, G. S. (1998). Field theory based navigation towards a moving target. Advanced Robotics: Beyond 2000: 29th International Sympo-sium on Robotics, Birmingham, England, 27 April - 1 May.
9. Virk, G. S., & Kadar, E. E. (2000). Trail following navigational strategies. Proceedings of the 3rd International Conference on Climbing and Walking Robots (CLAWAR 2000), Madrid, Spain, 2-4 October.
10. Lytridis, C., Virk, G. S., Rebour, Y., & Kadar, E. E. (2002). Odour-based navigational strategies for mobile agents. *Adaptive Behaviour*, 9(3-4), 171-187.
11. Mielle, P., Marquis, F., & Latrasse, C. (2000). Electronic noses: specify or disappear. *Sensors and Actuators B: Chemical*, 69(3), 287-294.
12. Koehler, O. (1932). Sinnesphisiologie der Süsswasserplanarian. *Zeitung Vergl. Physiologie*, 16, 606-756.
13. Lytridis, C., Kadar, E. E., and Virk, G. S. (2006) A systematic approach to the problem of odour source localisation. *Autonomous Robots*, vol. 20, 3, pp. 261-276.
14. Kadar E. E. (1996). A field theoretic approach to the perceptual control of action. PhD Dissertation. University of Connecticut, USA.

GA TUNED CLOSED-LOOP CONTROL OF SPRING BRAKE ORTHOSIS

M SAIFUL HUQ, RASHA MASSOUD, M SHAFIUL ALAM AND M O TOKHI
University of Sheffield, Sheffield S1 3JD, United Kingdom

Abstract: Spring brake orthosis (SBO) is a kind of hybrid orthosis system (HOS) which generates the swing phase of paraplegic gait by employing a spring at the knee joint to store energy during the knee extension through quadriceps stimulation, which is then released to produce knee flexion. The acceptance of any HOS and hence its degree of success depends largely on its performance in generating an acceptable gait trajectories. In this paper the performances of PID and Fuzzy logic controller (FLC) is studied in generating the swing phase in an SBO equipped human leg. A leg model of an average sized (183 cm height) is developed for simulation purposes. The leg model includes segmental dynamics, passive joint viscoelastic properties, an electrically stimulated knee extensor quadriceps muscle and an SBO. An optimal knee joint trajectory to be used as the reference knee joint trajectory is also derived. All the controller parameters for both PID and FLC are tuned using genetic algorithms (GAs) for minimum tracking error.

1. Introduction

Complete or partial loss of ability to walk or stand due to lower limb paralysis is a very common as well as drastic result of thoracic level spinal cord injury (SCI). The loss of lower limb function and inability to walk or stand significantly reduce the quality of life of disabled individuals and carry with them psychological and physiological effects. The spectrum of problems that may interfere with locomotory performance after and SCI includes hyperactivity of spinal reflex (muscle spasticity) [1], alternation in the muscle activation patterns, including weaknesses and difficulty in coping with weight bearing, balance and gait speed [2].

The electrical stimulation of nervous system below the level of the spinal lesion can produce powerful muscle contractions and thus can be used to generate primitive movements. This is termed as 'functional electrical stimulation' (FES) and is used to obtain a functional, useful movement by evoking artificial contraction of the muscles deprived of nervous control. The aims of the restoration varies with the ambition of the researchers and range from assistance with wheelchair transfers to the ability to stand up and sit down, to take few steps, to walking for some distance. Most work has concentrated on the correction of foot drop in hemiplegia and on the restoration of standing and walking in paraplegia.

The main challenge in FES assisted movement arises from the fact that the artificially stimulated muscle fatigues very quickly because of the 'reversed recruitment order' of the artificially stimulated motoneurons. The consequences are twofold: (*a*) limited duration of the FES assisted movement, especially standing and walking, (*b*) drastic changes in the actuator (muscle) properties lead to poor movement control. One of the major approaches to overcome these limitations is to reduce the use of active muscle, where possible, through the use of passive braces and is called hybrid orthosis systems (HOS) [3]. Several research groups have tested many HOS in the laboratory [4]. The spring brake orthosis is a kind of HOS, which generates the swing phase of paraplegic gait.

Some open- and closed-loop FES control strategies that were designed and tested are described in [5]. The open-loop FES control performance was however found unsatisfactory for accurate movement control due to existing parameter variations (e.g., muscle fatigue), inherent time-variance, time delay, and strong nonlinearities present in the neuromuscular-skeletal system (plant). Closed-loop control is thus necessary for accurate control of movement, and adaptive closed-loop strategies are obviously the most suitable candidates to tackle the FES control problem.

1.1. *The Spring Brake Orthosis Concept*

The quadriceps muscles, when artificially stimulated, could produce much more torque than is required just to extend the knee. During knee extension of swing phase, SBO exploits this feature of the quadriceps through partially storing FES generated quadriceps force as potential energy in a torsion spring attached to the knee joint. A brake is then employed to maintain the knee extension without any muscle contraction. Then knee flexion is achieved by releasing the brake and letting the spring to return to its resting position (approx 70^0 – 80^0). The hip flexion is simultaneously produced as a result of consequent shift in the centre of mass (CoM) of the overall leg segment during this knee flexion and is maintained throughout the required duration by applying a brake/ratchet at the hip joint. This results in a hybrid orthosis combining electrical stimulation of quadriceps muscle, spring and brake at the knee joint and brake/ratchet at the hip joint, with the activation of each of them at appropriate instant and for appropriate period [6]. Bearing in mind that an acceptable gait can be achieved with a locked ankle [7], the ankle joint is kept fixed with an ankle-foot-orthosis (AFO).

2. Methods

2.1. *Model*

A forward dynamic planner (sagittal) human leg consisting of segmental dynamics [8], passive properties at the joints [9], electrically stimulated muscle model of quadriceps [10] and SBO was modelled through a combination of visualNastran® (VN) software and Simulink®. The SBO design parameters are taken from a previous study which employed Genetic algorithms (GAs) to find the best parameters.

2.2. *Control of SBO*

Since the SBO aided swing phase involves no control at the hip joint, its control in generating swing phase refers to the control of knee joint trajectory (mainly during knee extension) through quadriceps stimulation. Both PID and PD type Fuzzy logic controller (FLC) are tested in the closed-loop form. All the parameters of the controller are optimised using GAs for carefully chosen objective functions. Cycle-to-cycle control [11] is also investigated and optimised due to its increasing popularity and evident superiority over closed-loop control in FES control [12]. Once again, a GA is employed to search for the best parameters for the cycle-to-cycle control strategy.

2.2.1. *Reference Trajectory*

Attention is paid to derive the best trajectory for the closed-loop controllers. Two factors affected the choice of trajectory, (a) maximizing the energy efficiency following suggestion from [13], so as to increase energy efficiency as well as reduce muscle fatigue, (b) its degree of ability to drive the knee joint all the way to full knee extension. Since cycle-to-cycle control takes care of only the end-of-phase orientation of the concerned limb joint rather than taking it through a predefined trajectory, the joint trajectory thus traversed is of significant importance in terms of achieving full knee extension. Bearing this in mind, the knee joint trajectory traversed under an optimised cycle-to-cycle control strategy is used as reference trajectory in closed loop control strategy.

2.2.2. *Controller Tuning with GAs*

The controller parameters are initialised using Ziegler-Nichols Method (for PID) and heuristic method (for FLC). Several search operations are performed to find the best values for the controller parameters using GAs.

GAs, first proposed by Holland in 1975 [14], are search and optimisation techniques inspired by two biological principles of natural evolution, viz. the process of "natural selection" and the mechanics of "natural genetics". Contrary to regular search algorithms, GAs manipulate not just one potential solution to a problem, but a collection of potential solutions, called a population.

Real world problems, in most cases, involve multiple objectives to be achieved simultaneously. Due to the conflicting nature of the objectives, it is often difficult to find a single optimal solution for a problem, and makes more sense to seek for a set of nondominated perato-optimal solutions for which an improvement in one of the objectives will lead to degradation in one or more of the remaining objectives. GAs are a suitable technique for multiobjective optimisation due to there population based nature which enables them to support multiple solutions concurrently. One of the first approaches to utilise the concept of pareto optimality in GAs was Fonseca and Fleming's multiobjective GAs (MOGA) [15].

3. Results

3.1. *Closed-loop control (Reference Trajectory Derived from Passive Oscillation)*

The reference trajectory for the knee, being inspired by [13], is derived from simple passive oscillation of the shank in absence of any stimulation whatsoever. Both PID controller and FLC are used in an attempt to drive the knee joint through the given trajectory and implement the swing phase. All the controller parameters for both PID and FLC are optimised using a GA. For the PID controller, only 3 parameters, viz. P, I and D are optimised whereas the with FLC, it is 73 parameters, including the weightings of 25 rules (5X5 rule base) and parameters associated with triangular member functions (MF). For each of the controllers the GA was run for 100 generations with 50 individuals in each generation, thereby evaluating 5000 points for each of the parameters. The integral of the 'time-weighted squared error' (ITSE) is defined as the objective function to be minimized. This time weighting in the objective function is to emphasize the tracking error nearer to the knee extension so as to prioritise the necessity of full knee extension.

Figure 1 shows the decay of the objective functions with generation while Figure 2 shows the resultant controlled trajectories achieved with the optimised controller parameters. Although it's obvious from both the figures that the FLC performs much better than the PID controller in terms of input tracking, both of them are unacceptable in practice due to their inability to produce full knee extension. This suggests that, given this particular trajectory both these controllers are incapable of producing full knee extension. To put it in a different way, with the given plant (human leg) with all its limiting characteristics, this particular trajectory is probably unsuitable for this particular purpose.

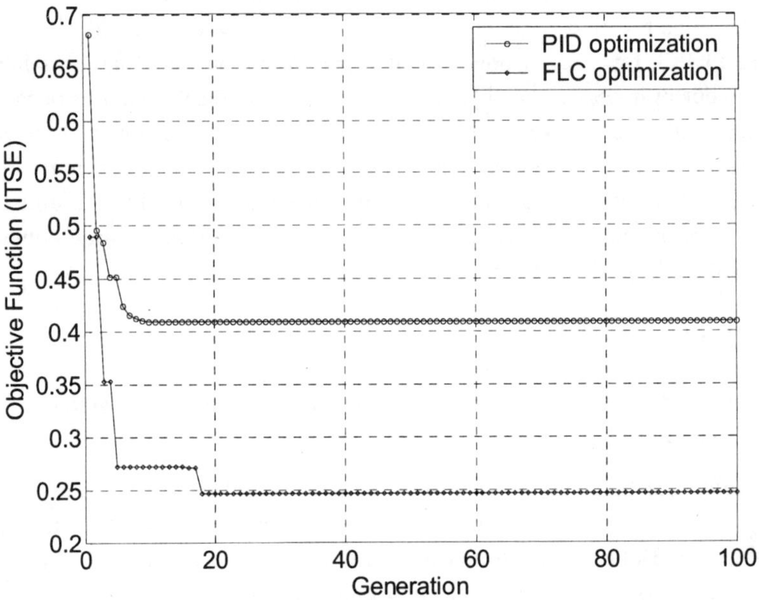

Figure 1. GA optimisation of PID controller and FLC – decay of objective function with generation

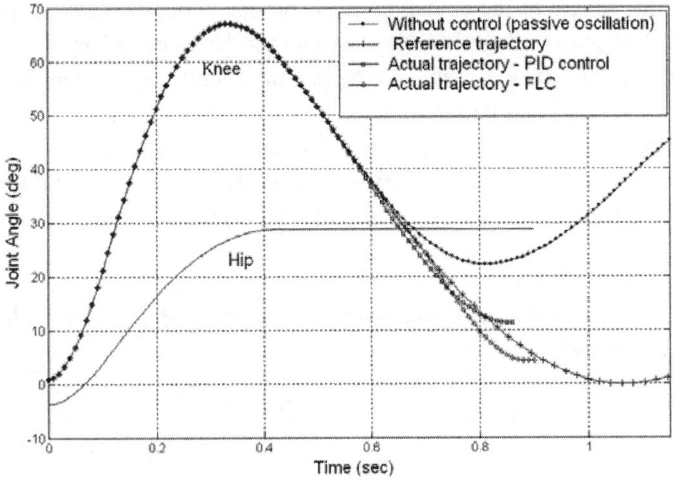

Figure 2. Knee and hip joint trajectories – reference trajectory derived from plain passive oscillation, actual controlled trajectories under PID control and FLC.

3.2. *Cycle-to-cycle Control*

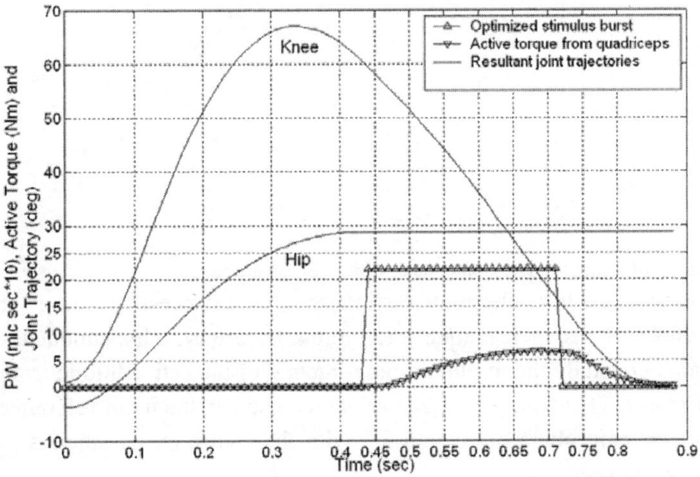

Figure 3. Optimised cycle-to-cycle control

The model, at this stage, is subjected to cycle-to-cycle control with a fixed pulse with of 220 ms, while the burst duration as well was its position in the time scale is optimised formulating a minimization problem using a MOGA for two objectives, (a) square of the deviation of the produced knee extension from 0° (objective 1), (b) time-integral of the active torque (from quadriceps) (objective 2). Minimizing 'objective 1' helps producing an exact 0° knee extension, while it contributes towards less muscle fatigue in case of 'objective 2' [16].

From the obtained pareto-optimal solution set, the one with a minimum value of 'objective 1' (near 0° knee extension) is chosen and the result is shown in Figure 3. Also presented in the same plot the optimised stimulus burst and resultant active torque.

3.3. *Closed-loop Control with Reference Trajectory Derived from Cycle-to-cycle Control*

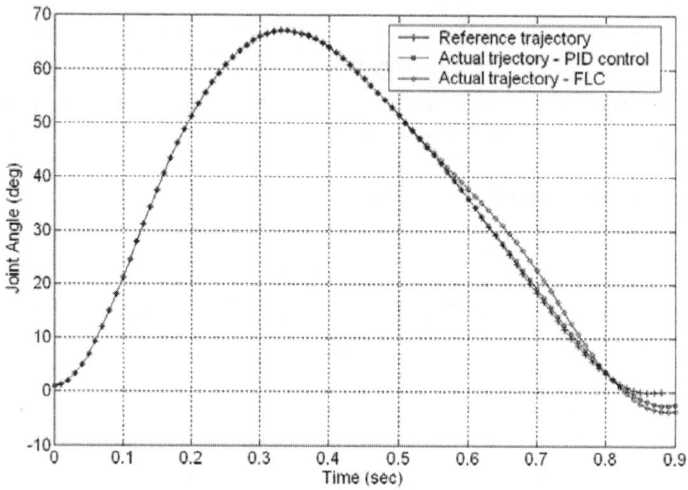

Figure 4. PID and FL control of knee joint trajectory for SBO swing phase

The optimised cycle-to-cycle control strategy with a successful end-of-phase knee extension results in a unique knee joints trajectory. This uniqueness is mainly characterized by the minimum time-integral of the active torque from the stimulated muscle. This unique trajectory is then used as the input reference for the closed-loop controllers of section 3.1 with the controller parameters tuned again in the same way.

The success in introducing the new reference trajectory is very much obvious in Figure 4, which presents the resultant knee joint trajectory under FLC and PID control along with the reference trajectory. The result, besides asserting a high degree of suitability of the given trajectory, also reveals the variation in the degree of suitability amongst arbitrary trajectories for FES applications.

Hip joint trajectory is found to be satisfactory (25°±4°) in all the cases, and that's why it is not presented in all the results.

3.4. *Comparison between the Controllers*

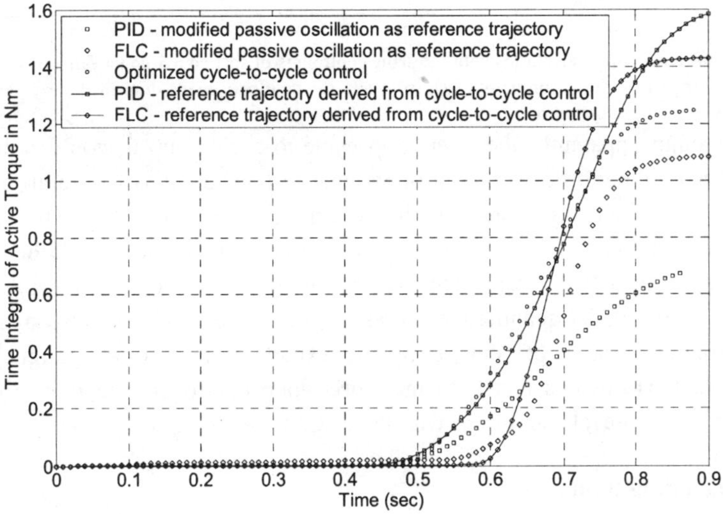

Figure 5. Accumulation (integral) of active muscle torque during swing phase with different controllers

In FES applications, muscle fatigue is always an inevitable pitfall and hence always desirable to keep it as minimum as possible. Figure 5 is a brief representation of performances of the controllers investigated so far, in terms of fatigue immunity. The figure shows time-integral of the active quadriceps torque generated under each control strategy which is related to the energy consumption of the muscle and thus to the fatigue.

The two curves indicating least amount of time-integral of active torque correspond to the initial closed-loop control approach with modified passive oscillation as reference trajectory. Although they seem to produce potentially less fatigue, they are of no significance due to their inability to produce full knee extension. Among the other 3 curves cycle-to-cycle control seems to accumulate least active muscle torque. FLC appears to accumulate less active torque between the rest 2 curves of which both are closed-loop.

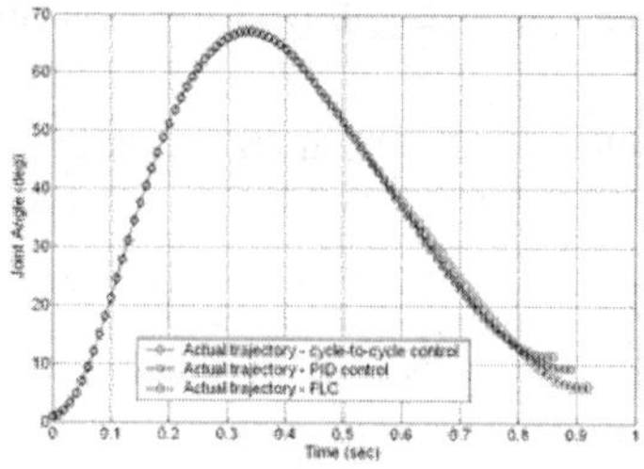

(a)

Control	% Normal Excursion
Cycle-to-cycle	≅ 83%
PID	86.57%
FLC	91.1%

(b)

Figure 6. (a) Knee joint trajectories with the same controllers of section 3.3 under fatigue condition (50% of normal quadriceps torque output), (b) Table showing percentage drop in the excursion of the knee under fatigue condition.

Although apparently the open-loop cycle-to-cycle control would produce least fatigue, the closed-loop controllers, due to their obvious nature are more robust against plant variations, e.g. due to fatigue. This is revealed in the result in Figure 6 (a) and (b), where Figure 6 (a) shows the trajectories with the same controllers, but active torque halved, simulating fatigue condition, while the table in Figure 6 (b) quantitatively presents the same result in terms of knee joint excursion as a percentage of the excursion in normal (fatigueless) condition. It is obvious that both the closed-loop controllers are more robust than the cycle-to-cycle controller with the FLC at the top.

4. Concluding Remarks

This work primarily concentrates on closed-loop tracing control of SBO assisted swing phase with a view to benefit for the advantages of feedback tracking control. One of the main attractions of cycle-to-cycle control lies in achieving the necessary end-of-phase orientation required for successful functional movement without the need of any trajectory. Being discrete-time in nature, it also suffers from the lack of real time online supervision of states [16]. In principle, the current work combines the advantages of both the control strategies. A successful practical implementation of any FES control must include adaptation mechanism for the controller parameter(s), especially to cope with fatigue related changes. Such an optimised controller, as the one in this work, can serve as an initialised part of an adaptive controller through the addition of an online adaptation mechanism.

Reference

1. Eltorai, I. and R. Montroy, *Muscle release in the management of spasticity in spinal cord injury*. Paraplegia, 1990. **28**(7): p. 433-40.

2. Waters, R.L., et al., *Recovery following complete paraplegia.* Arch Phys Med Rehabil, 1992. **73**(9): p. 784-9.
3. Goldfarb, M., et al., *Preliminary evaluation of a controlled-brake orthosis for FES-aided gait.* IEEE Trans Neural Syst Rehabil Eng, 2003. **11**(3): p. 241-8.
4. Durfee, W.K. and A. Rivard, *Design and simulation of a pneumatic, stored-energy, hybrid orthosis for gait restoration.* J Biomech Eng, 2005. **127**(6): p. 1014-9.
5. Crago, P.E., et al., *New control strategies for neuroprosthetic systems.* J Rehabil Res Dev, 1996. **33**(2): p. 158-72.
6. Gharooni, S., B. Heller, and M.O. Tokhi, *A new hybrid spring brake orthosis for controlling hip and knee flexion in the swing phase.* IEEE Trans Neural Syst Rehabil Eng, 2001. **9**(1): p. 106-7.
7. Goldfarb, M. and W.K. Durfee, *Design of a controlled-brake orthosis for FES-aided gait.* IEEE Trans Rehabil Eng, 1996. **4**(1): p. 13-24.
8. Winter, D.A., *Biomechanics and motor control of human movement.* 2nd ed. 1990, New York ; Chichester: Wiley. xvi, 277p ill 25cm.
9. Amankwah, K., R.J. Triolo, and R. Kirsch, *Effects of spinal cord injury on lower-limb passive joint moments revealed through a nonlinear viscoelastic model.* J Rehabil Res Dev, 2004. **41**(1): p. 15-32.
10. Riener, R. and T. Fuhr, *Patient-driven control of FES-supported standing up: a simulation study.* IEEE Trans Rehabil Eng, 1998. **6**(2): p. 113-24.
11. Franken, H.M., et al., *Cycle-to-cycle control of swing phase of paraplegic gait induced by surface electrical stimulation.* Med Biol Eng Comput, 1995. **33**(3 Spec No): p. 440-51.
12. Arifin, A., T. Watanabe, and N. Hoshimiya, *Design of Fuzzy Controller of the Cycle-to-Cycle Control for Swing Phase of Hemiplegic Gait Induced by FES.* Ieice Transactions on Information and Systems E Series D, 2006. **89**: p. 1525-1533.
13. Williamson, M.M. *Exploiting natural dynamics in robot control.* in *Cybernetics and systems research.* 1998. Vienna: Austrian Society for Cybernetic Studies.
14. Holland, J.H., *Adaptation in Natural and Artificial Systems.* 1975, Ann Arbor: The University of Michigan Press.
15. Fonseca, C.M. and P.J. Fleming. *Genetic algorithms for multiobjective optimisation: Formulation, discussion and generalization.* in *Fifth International Conference on Genetic Algorithms.* 1993. Morgan Kaufmann, San Mateo, CA.
16. Veltink, P.H., *Control of FES-induced cyclical movements of the lower leg.* Med Biol Eng Comput, 1991. **29**(6): p. NS8-12.

Hidden Markov Model based Fuzzy Controller for Flexible-link Manipulator

M.N.H. SIDDIQUE[1]
[1]*School of Computing and Intelligent Systems, University of Ulster*

M.A. HOSSAIN[2]
[2]*Department of Computing, University of Bradford*

M.S. ALAM[3] AND M.O. TOKHI[4]
[3,4]*Department of Automatic Control and Systems Engineering, University of Sheffield*

A major problem with fuzzy rule-based systems is that with an increasing number of inputs and linguistic variables, the possible number of rules for the system increases exponentially. Unfortunately, there is no systematic approach to learning of rule-base of fuzzy logic (FLC) controller if there is no control expert available, then it must be constructed from the controlled environment or a suitable data set should be available. The adaptive neuro-fuzzy inference system (ANFIS) proposed by Roger Jang, which reduces the development time involved in constructing the rule-base requires a set of input-output data. The problem is now how to cope with developing an FLC where *a priori* information such as a set of input-output behaviour or expert knowledge is not directly available. The hidden Markov model (HMM) is a probabilistic finite-state machine used in finding structures in sequential data. A rule-base of an FLC can be compared to a finite state machine which can produce a sequence of output MFs. Therefore, the main interest of this research lies in finding a functional mapping from a rule-base of FLC to a hidden Markov model (HMM) and train the HMM using the available data source. The developed controller is then applied to a flexible-link manipulator to verify the performance of the methodology.

1. Introduction

In most existing applications, the fuzzy rules are generated by an expert in the area, especially for the control problems with only a few inputs. As is well recognized, rule acquisition has been and continues to be regarded as a bottleneck for implementation of any kind of rule-based systems. A major problem with fuzzy rule-based systems is that with an increasing number of inputs and linguistic variables, the possible number of rules for the system increases exponentially, e.g. for a three-inputs and single-output system with

7 primary sets for each input will require $n \times n \times n = 7 \times 7 \times 7 = 343$ rules. This huge number of rules makes it difficult for experts to define a complete set of rules and associated membership functions for a sufficient level of system performance. The difficulty arises if there is no control expert available for constructing the rule-base, then it must be constructed from the controlled environment or a suitable data set should be available. Another well-known problem with fuzzy rule-based systems is that the processing of such a rule-base is time consuming. Consequently, most of the time is spent in calculating the control output using center of gravity method of defuzzification, which in some applications can cause slow system response. The problem of such defuzzification methods has been eliminated by the use of Sugeno-type fuzzy systems, where each consequent fuzzy set is replaced by a linear function [1]. This imposes a further set of consequence parameters to be estimated. Current neural-fuzzy systems are mainly Sugeno-type fuzzy systems with the rule-base replaced by a neural network. The most widely used neuro-fuzzy system is the adaptive neuro-fuzzy inference system (ANFIS) proposed by Roger Jang, which reduces the development time involved in constructing the rule-base [2]. The learning of the parameters of the neuro-fuzzy system requires a set of input-output data and uses several passes of backpropagation and/or LMS algorithm to estimate the antecedent parameters of MFs and consequent parameters of the system [1-3]. The problem is now how to cope with developing a fuzzy logic controller (FLC) for a flexible-link manipulator, which is a highly nonlinear system [4,5] and *a priori* information such as a set of input-output behaviour or expert knowledge for constructing rule-base is not directly available. The hidden Markov model (HMM) is a probabilistic finite-state machine used in finding structures in sequential data [10]. A rule-base of an FLC can be compared to a finite state machine which can produce a sequence of output MFs. The control input is thus generated by defuzzifying output MFs using usual methods.

The objective of this research is to minimize the processing time required for the control input by avoiding Mamdani-type inferencing using rule-base and centre of gravity defuzzification procedure. Therefore, the main interest of this research lies in finding a conceptual mapping from a rule-base of FLC to a hidden Markov model (HMM) and train the HMM using the available data source. The developed controller is then applied to a flexible-link manipulator to verify the performance of the methodology.

2. HMM-Fuzzy controller

The basic principles of the HMM-fuzzy controller is that for a given rule-base as defined by a global linguistic association constrained by fuzzy sets, a functional mapping from the fuzzy logic-based controller to the HMM-based approach is established. In other words, the inference mechanism (or approximate reasoning mechanism) of a rule-based FLC is implemented by a HMM. A block diagram of the proposed HMM-fuzzy control system is shown in Figure 1.

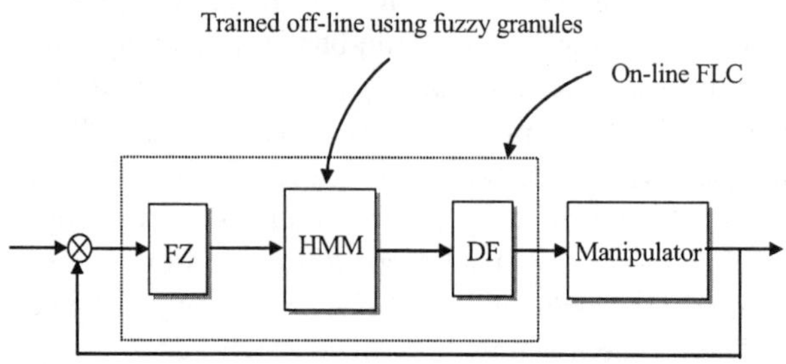

Figure 1: Block diagram of the HMM-Fuzzy Controller.

2.1 Fuzzification

A membership function (MF), such as large, medium or small, is typically a fuzzy set. The linguistic variable is defined by the following Gaussian membership function (MF).

$$\mu_i(A_i) = \exp\left[-\frac{1}{2}\left(\frac{m_i - x_1}{\sigma_i}\right)^2\right] \tag{1}$$

Where m denotes the central value and σ denotes the spread of a Gaussian MF.

2.2 HMM as Rule-base Mapping

Unfortunately, there is no systematic approach to learning of rule-base of fuzzy controller. Efforts have been made to automate the construction of rule-bases in various ways using neural networks and genetic algorithms [6-8]. In most of the cases, either the rule-base is fixed and the parameters of the membership functions are adjusted or membership functions are fixed and the rule-base is optimised by genetic algorithms [5]. The rule-base of a Mamdani-type fuzzy controller with two inputs (mostly error and change of error) and a single output (torque input) consists of the following rules.

$$R\text{: IF } X_1 \text{ is } A_i \text{ AND } X_2 \text{ is } B_j \text{ THEN } U \text{ is } C_k \tag{2}$$

where $X_1 \triangleq$ error, $X_2 \triangleq$change of error, $U \triangleq$ control input, A_i, $i = 1,2,\cdots,N$, B_j, $j = 1,2,\cdots,M$ are the input linguistic variables and C_k, $k = 1,2,\cdots,L$ are the output linguistic variables, N, M and L are the maximum number of linguistic variables (primary fuzzy sets defined by membership functions). The rule-base of the Mamdani-type fuzzy controller is shown in Table1, where C_r is defined as $C_r \in C_k = \{C_1, C_2, \cdots, C_L\}$ and each C_r is to be found from available information granules.

Table 1: Rule-base *R*

		Input X_2			
	U	B_1	B_2	$\cdots$	B_M
Input X_1	A_1	C_r	C_r	$\cdots$	C_r
	A_2	C_r	C_r	$\cdots$	C_r
	$\vdots$	$\vdots$	$\vdots$	$\ddots$	$\vdots$
	A_N	C_r	C_r	$\cdots$	C_r

The rule-base R in Table 1 can be shown as a mapping Φ from input fuzzy sets to output fuzzy sets defined by the equation (2). The mapping Φ can be visualized as mapping of information granules from inputs to output pictorially as shown in Figure 2. More descriptively, the particular interest in this research is the development of possibility functions μ_{A_i}, $\mu_{B_{ji}}$ from information granules $\{A_i\}$, $\{B_j\}$[11], which will compute the probability function $P(O \mid \lambda^i)$ from the MHH model λ^i, i.e. HMM will produce $\{C_k\}$ such that it satisfies the requirements of the mapping Φ in (3).

$$\Phi : \{A_i, B_j\} \rightarrow C_k \tag{3}$$

The functional mapping $\Psi : R \rightarrow \Phi$ can be represented by HMM, defined by

$$\lambda^i = \{TR, E, \pi\} \tag{4}$$

where TR is the transition probability matrix, E is the emission matrix and π is the initial probability. $TR = [a_{ij}]$ is an $N \times N$ matrix and the elements should satisfy $a_{ij} \geq 0$ and $\sum_j a_{ij} = 1$. Such an HMM is shown in Figure 3, where each a_{ij} is comparable to $\Gamma\{\mu_{A_i}(x_1), \mu_{B_j}(x_2)\}$. The function $\Gamma\{\mu_{A_i}(x_1), \mu_{B_j}(x_2)\}$ is so defined such that it satisfies the condition $\sum_N \Gamma\{\mu_{A_i}(x_1), \mu_{B_j}(x_2)\} = 1$. In other words, the inference mechanism of a rule-based Mamdani-type FLC is implemented by the HMM in Figure 3. Once the HMM is trained, it is then inserted in the control loop for on-line operation.

2.3 Defuzzification

The most time-consuming operation in Mamdani-type FLC is the defuzzification procedure using center of gravity method. To minimize the time required for defuzzification, Gaussian MFs are used in the consequent part. Since the Gaussian MFs are symmetric and invertible, defuzzified value is calculated using equation (5) while ignoring the left half of the MFs, shaded area, shown in Figure 4.

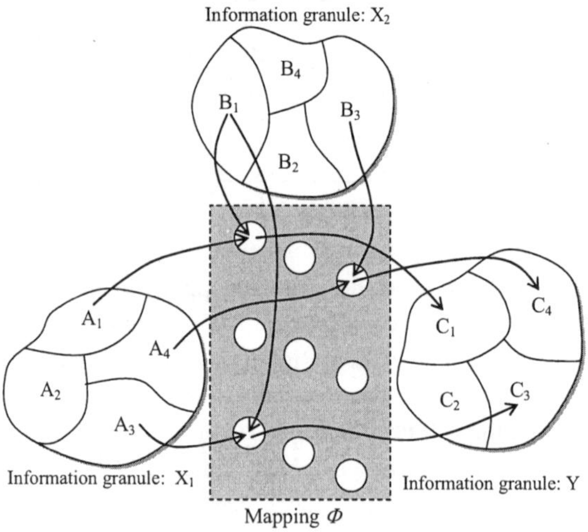

Figure 2: Rule-base as a Mapping

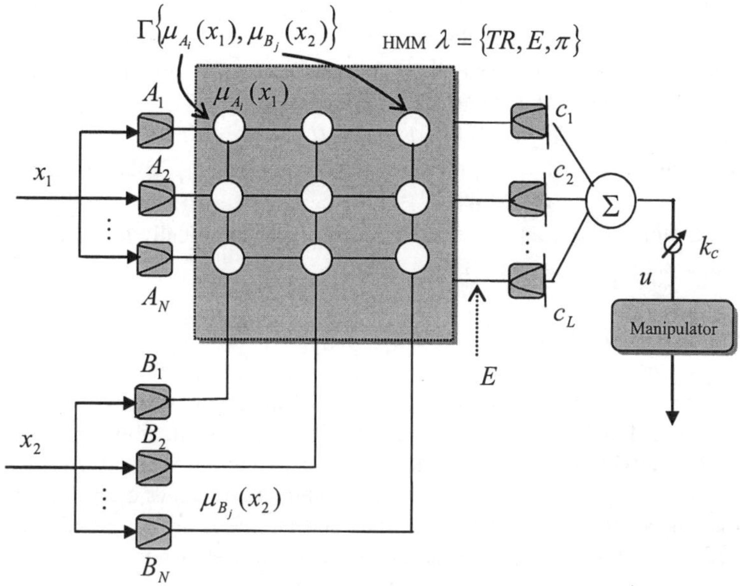

Figure 3: HMM as a rule-base mapping.

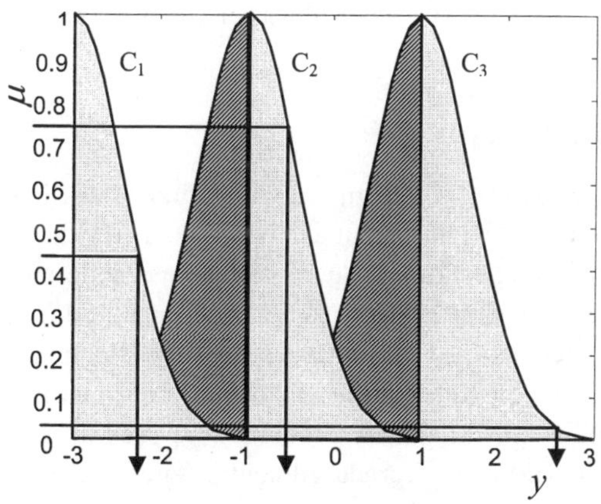

Figure 4: Defuzzification of output fuzzy sets.

$$y_i = m_i + \sigma_i \sqrt{-2\ln[\mu(C_i)]} \tag{5}$$

Where m denotes the central value, and σ denotes the spread of a Gaussian linguistic variables. A suitable support set for output fuzzy sets C_i is to be defined so that the aggregated crisp value does not fall outside the universe of discourse. That is, the support comprises those elements of y of the universe such that $\mu_C(y) > 0$. In this case, support set is defined as

$$Support(C_i) = \{y \mid \mu_{C_i}(y) \geq \mu_\delta\} \tag{6}$$

Where μ_δ is chosen arbitrarily for each output fuzzy set C_i and defined as

$$\mu_\delta = \{\mu_{C_i}(y) \mid m_i \leq x < m_{i+1}\} \tag{7}$$

It is straightforward and easy to find the value of μ_δ empirically for all output fuzzy sets, which satisfies equation (7).

3. Experimental rig

The experimental rig constituting the flexible manipulator system consists of two main parts: a flexible arm and measuring devices. The flexible arm contains a flexible link driven by a printed armature motor at the hub. The measuring devices are shaft encoder, tachometer, accelerometer and strain gauges along the length of the arm. The shaft encoder, tachometer and accelerometer are essentially utilised in this work. The experimental flexible-link manipulator is shown in Figure 5.

The flexible arm consists of an aluminium-type beam. The outputs of the sensors as well as a voltage proportional to the current applied to the motor are fed to a computer through a signal conditioning circuit and an anti-aliasing filter for analysis and calculation of the control signal. Physical parameters of the flexible arm are given in Table 2.

4. Experimental results

The HMM-fuzzy controller is constructed with 6 fuzzy sets (linguistic variables) for two inputs and 3 fuzzy sets for output and hence the HMM has 3 states as shown in Figure 3. HMM is trained using the control torque collected from a bang-bang controller applied to the flexible-link manipulator, quantised in fuzzy interval NS, Z and PS, labelled them as C_1, C_2, and C_3 respectively. The transition matrix TR and emission matrix E are initialised arbitrarily using simple heuristic which is the main advantage of training HMM. 800 data points were used to train HMM to an error goal of 0.001. The HMM-FLC produced control torque is shown in Figure 6. The endpoint acceleration is shown in Figure 7.

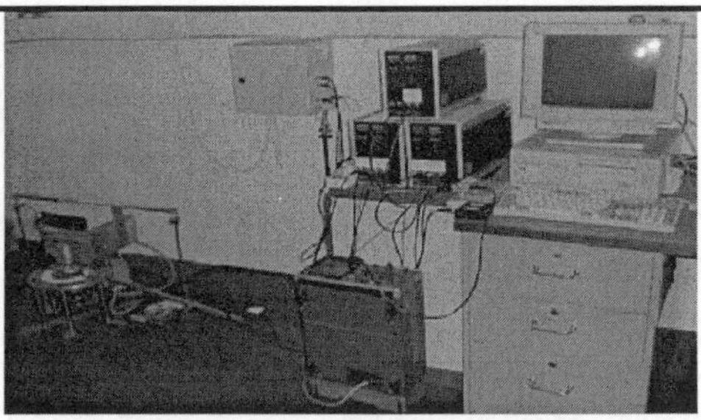

Figure 5: Flexible-link manipulator

Table 2: Physical parameters.

Parameter	Value
Length	960.0 mm
Width	19.008 mm
Thickness	3.2004 mm
Mass density/ unit volume	2710 kgm^{-3}

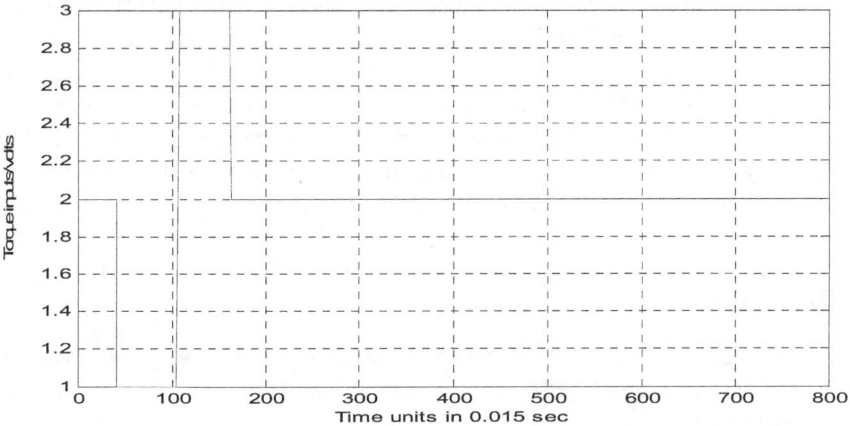

Figure 6: Control torque produced by HMM-FLC

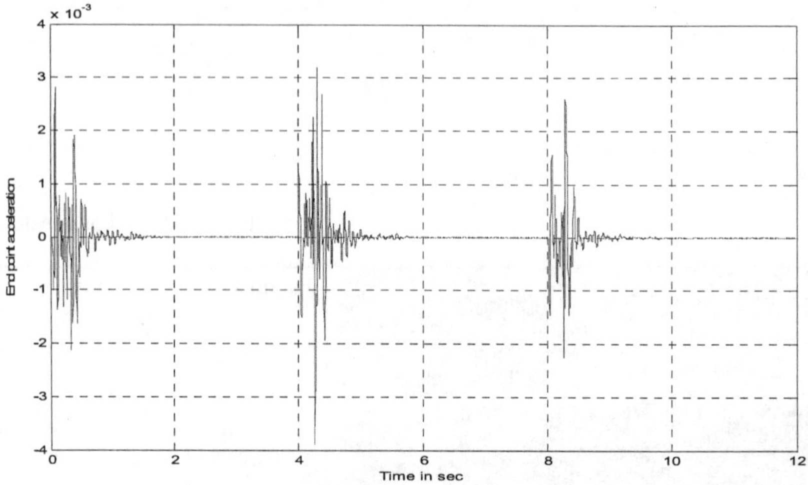

Figure 7: Endpoint acceleration.

5. Conclusion

This paper has been presented an investigation to minimize the processing time required for the control input by avoiding Mamdani-type inferencing using rule-base and centre of gravity defuzzification procedure. A conceptual mapping from a rule-base of FLC to train the HMM using the available data source has been developed to demonstrate the merits of the proposed system. The developed controller was applied to a flexible-link manipulator to verify the performance of the methodology through a set of experiments. It is noted that the proposed controller performed reasonable well and very much similar to the traditional FLC.

References

1. T. Takagi and M. Sugeno (1985). "Fuzzy identification of systems and its applications to modeling and control", *IEEE Transaction on System, Man and Cybernetics*, Vol. 15, 1985, pp. 116 -132.
2. J.-S. Roger Jang, (1993). "ANFIS: Adaptive-network-based fuzzy inference system", *IEEE Transaction on Systems, Man and Cybernetics*, Vol. 23, No. 3, 1993, pp. 665-685.

3. D.A. Linkens and J Nie (1994). "Backpropagation Neural Network based Fuzzy Controller with a Self-learning Teacher", *International Journal of Control*, vol. 60, No. 1, 1994, pp. 17-39.
4. S. Yurkovich, F.E. Pacheo and A.P. Tzes, (1989). "On-line Frequency Domain Information for Control of a Flexible-link Robot with varying Payload", *IEEE Transaction on Automatic Control,* Vol. 34, No. 12, 1989, pp. 1300-1304.
5. M.N.H. Siddique, (2002). "Intelligent Control of Flexible-link Manipulator Systems", PhD Thesis, Department of Automatic Control and Systems Engineering, The University of Sheffield, England, UK.
6. Lin, C.-T. and Lee, C.S.G. (1995). "A neural fuzzy control system with structure and parameter learning", *Fuzzy Sets and Systems*, Vol. 70, pp. 183-212.
7. Nauck, D and Kruse, R (1993). "A fuzzy neural network learning fuzzy control rules and membership functions by fuzzy error backpropagation", *Proceeding of IEEE International Conference on Neural Networks*, pp. 1022-1027.
8. Nauck, D and Kruse, R. (1996). "Designing neuro-fuzzy systems through backpropagation." *In W. Pedrycz, ed. Fuzzy Modelling: Paradigms and Practice*, pp. 203-228, Kluwer, Boston.
9. H Bourlard and C. Wellekens, (1990). "Links between Hidden Markov Models and Multilayered Perceptrons", IEEE Transaction on Pattern Analysis and Machine Intelligence, vol. 12, pp. 1167-1178, 1990.
10. L. R. Rabiner (1989). "A Tutorial on Hidden Markov Models and Selected Applications in Speech Recognition", Proceedings of the IEEE, Vol. 77, No. 2, pp. 257-286.
11. L.A. Zadeh (1978). "Fuzzy Sets as a Basis for a Theory of Possibility", Fuzzy Sets and Systems, Vol. 1, pp. 3-28.

HIL/SIL BY DEVELOPMENT OF SIX-LEGGED ROBOT SLAIR2

DZHANTIMIROV[1], PALIS[2], SCHMUCKER[1], TELESH[2], ZAVGORODNIY[2]

[1]*Department Virtual Engineering, Fraunhofer Institute for Factory Operation and Automation, Sandtorstrasse 22, 39106 Magdeburg, Germany*

[2]*Institute for Electrical Energy Systems, University of Magdeburg, Universitätsplatz 2, P.B. 4120, 39106 Magdeburg, Germany*

The development process of CLAWAR is a complex task. The Hardware- and Software-in-the-Loop frameworks, which are used by development of new 22 DoF six-legged robot SLAIR2 are presented. A novel universal real-time communication bridge system is introduced and test results are presented. The ability to real-time communication and service tasks related control of the robot is discussed.

Keywords: Embedded and Real Time Systems, Modelling and Simulation Languages, Automatic Code Generation.

1. Introduction

The control system of the full or part of autonomous legged robot is almost controlled by embedded system. Commonly the embedded systems are designed to control complex plants such as engines, satellites, vehicles, spacecrafts, and of course CLAWAR. They generally require a high level of complexity within the embedded system to manage the complexity of the plant under control.

Development and test of complex real-time embedded systems consists of many steps from modelling and simulation of the plant till the implementation of the source code in the real hardware. Hybrid techniques, that are used increasingly, are the Hardware-in-the-Loop (HiL) and Software-in-the-Loop (SiL) simulations (see Fig.1). In our work we use these methods within the development and testing process of six-legged robot.

The embedded system (i.e. real electronic control device or real mechatronical component) within a HiL-framework is connected to corresponding HiL-simulator (which emulates real system response) via its IOs building a closed loop. The system to be controlled (i.e. legged robot) is then simulated to test the correct performance of the control system (i.e. drive control). The inputs of control system under test are stimulated with the model-based sensor outputs. To close the control loop is the reaction of the control system outputs (i.e. motor voltage) guided back into robot model.

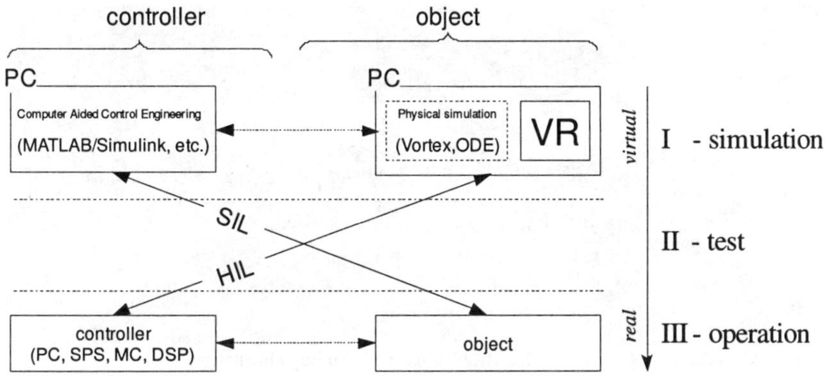

Figure 1. Common scheme of the HiL/SiL structure of the mechatronical system.

In the pure simulation phase a virtual control system is connected to a virtual object. Both software systems communicate via pure software interface. In the hybrid simulation phase for such communication a hardware interface needed. The development process is through communication inconsistence more complex and needs more effort to transfer of results [5]. And moreover with the change of communication interface is even more effort to expect.

The paper presents the developed six-legged robot "SLAIR2" [3] as well as the environment, which was used in the pure simulation phase and the hybrid HiL/SiL phase by development and test separate robot components. Moreover the flexible communication bridge for real-time communication between the virtual/real control system and the virtual/real robot is presented as well.

2. Modular Six-Legged Robot "SLAIR2"

Figure 2 represents the multi-legged robot with articulated body "SLAIR2" that has been developed at the Fraunhofer Institute for Factory Operation Otto-von-Guericke University both from Magdeburg in Germany [4]. The robot mechanics, sensor system and control system make it possible to maintain the additional flexibility in the body, to measure and control the support reactions as well as to control and forecast the robot motion stability.

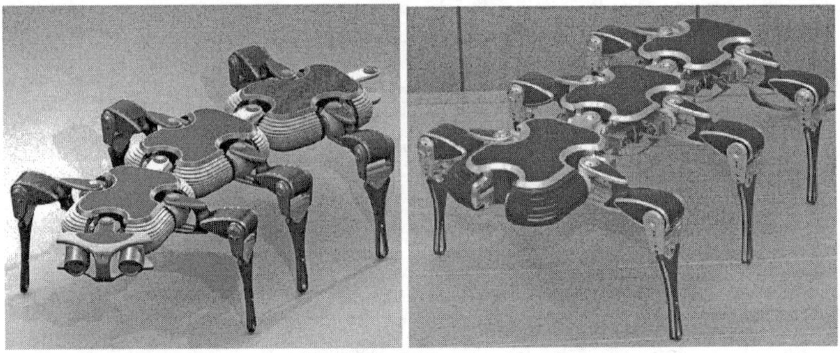

Figure 2. Modular six-legged robot "SLAIR2": CAD and real implementation.

2.1. *Mechanics and Sensor Systems*

The robot construction consists of $n=3$ modular segments (shoulders) linked to each other through two DOF joints and 6 legs. Each shoulder includes one articulated body segment linked with two 3-DOF-insectomorphic legs. It is possible to extend the construction to the case of $n>3$ shoulders. The main mechanical parameters are shown in the Table 1. The robot drives are servomotors, with maximal power $P_{DC}=4.5\,\mathrm{W}$, with potentiometer with angular range $\varphi_{POTI}=\pm 90^\circ$ and gears with ratio $i_{GEAR}=251$ and efficiency $\eta_{GEAR}=85\%$.

Table 1. The main mechanical parameters of the robot

leg lengths	total: l_{LEG} = 300 mm thigh: l_1 = 120 mm shank: $l2$ = 180 mm
body dimensions	total length: l_B = 660 mm segment length: l_{BS} = 220 mm
total robot mass	m_{ROBOT} = 3,20 kg
maximal speed	V_{ROBOT} = *1 km/h*
max power consumption	P_{MAX} = 90 W

The sensor system of the robot consists of components that are standard for mobile robots and that make it possible to achieve autonomous robot functions in an environment. It includes:

- 22 potentiometers and 22 current sensors (installed in each robot joint),
- 6 three-component force sensors (mounted in each leg's shank),
- 2-axis gyroscopic sensor (located in body), and

- further it will be equipped with near navigation sensor system (stereoscopic camera).

In accordance with the requirements on measurement and control of the support reactions, the developed force sensor consists of two parts: the core measuring lateral components of support reaction, and the elastic parallelogram module for measurement of the longitudinal component. The sensor is designed for loadings up to $F_{CONTACT} = 50\,\text{N}$; interference between channels does not exceed 1%.

2.2. *Control System und Control Structure*

The hierarchically organized modular control structure (Fig.3a,3b,3c) is completely located on PC-side that implement additionally the interaction with user and produces the control signal for robot drives as well as monitors all actuators and sensors of the robot. The robot-side is implemented by fast and flexible FPGA, includes the hardware abstraction layer (HAL) for drive and sensors. The real-time connection between two parts is made via proposed netX® [1] based communication bridge described below. All this provides flexibility and simplifies the development of control algorithms. The control system can also be extended for the additional shoulders in exactly the same manner as the mechanical structure.

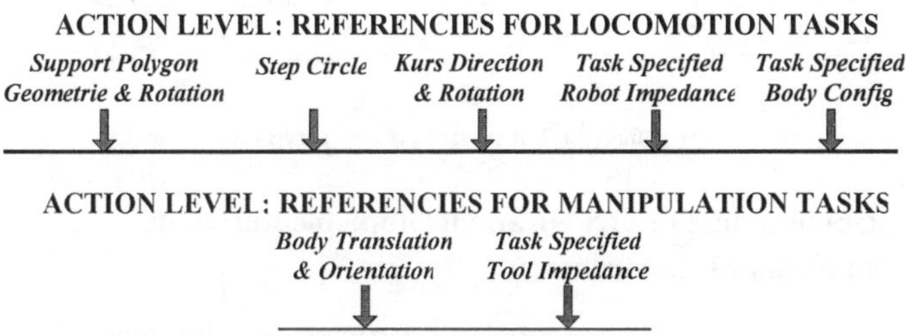

Figure 3a. The hierarchically organized robot control system: action level

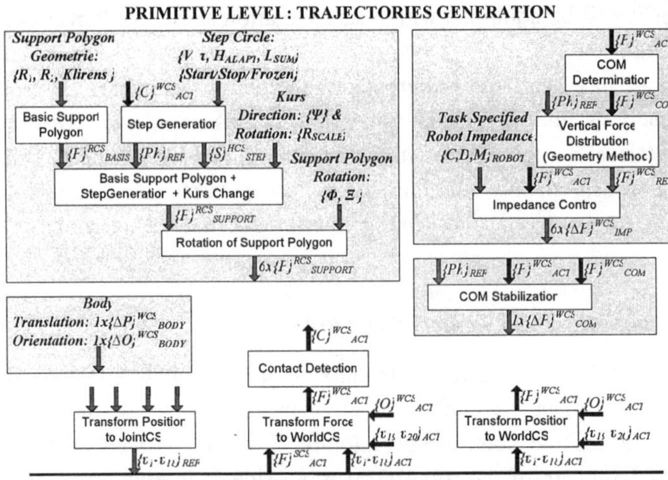

Figure 3b. The hierarchically organized robot control system: primitive level

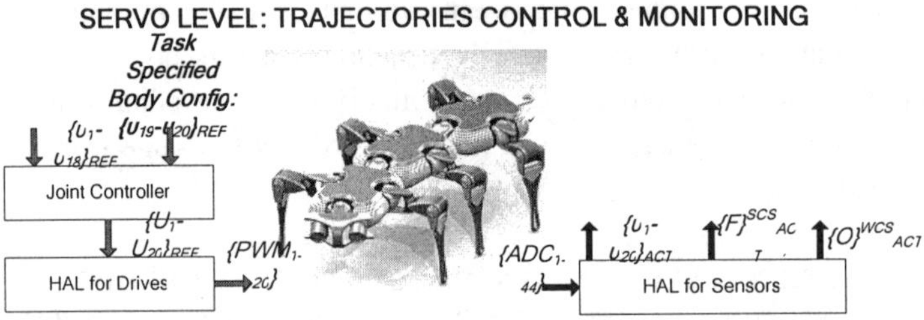

Figure 3c. The hierarchically organized robot control system: servo level

3. NetX Communication Bridge and its implementation in Matlab/Simulink Environment

The communication bridge emulates a router, which connects the control system with all the components of the control object (i.e. legged robot). These components can be connected to the bridge via diverse bus systems (Ethernet, CAN, UART, SPI etc). The robot components in such system can also be represented in virtual environment. Through a common physical interface is the virtual or real control system isolated from the object, and therefore the development effort of communication interface to the legged robot is reduced. The output signals from the control system (reference signals) are de-multiplexed and sent to corresponding actuators with the help of the communication bridge. On the other side the outputs from the robot (actual sensor signals) are multiplexed by bridge and sent to the control system. These routing tasks are entirely performed by the communication bridges. The physical implementation has been done using ARM9-based netX - SoC [1], which is able to communicate with all well known fieldbus and real-time Ethernet systems.

Figure 4 represents the proposed flexible communication system based on netX communication processors.

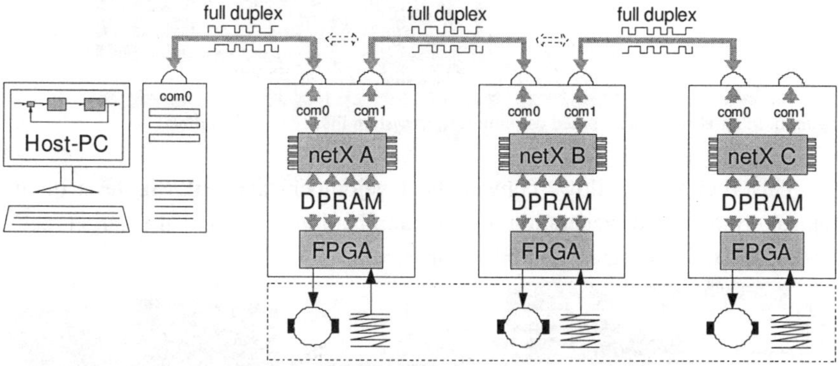

Figure 4. Flexible communication system based on netX processors.

The parameterisation of the communication bridge takes place in control program of embedded ARM processors. This program can be compiled with conventional tools, but in order to simplify the reconfiguration of the bridge within simulation framework the programming via RTW and GUI in Matlab/Simulink was selected.

For these purposes a special software stack (s.Fig.5) was developed and the communication bridge was integrated into Matlab/Simulink framework (s. Fig.6) under Real-Time Workshop [2].

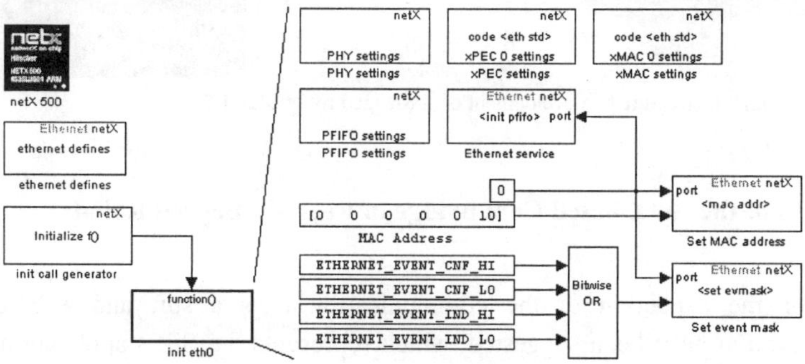

Figure 5. Software Ethernet communication stack implemented in Matlab/Simulink.

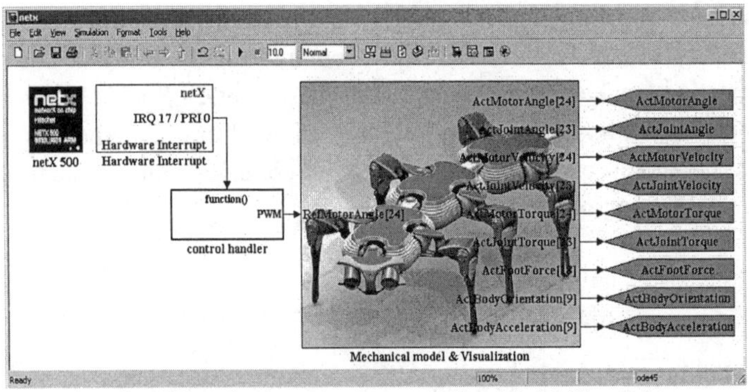

Figure 6. Integration of netX based communication system into Matlab/Simulink.

This framework allows convenient development and HiL/SiL tests of the control system and generation of the application code for the target netX processors under Matlab/Simulink environment only (s.Fig.7).

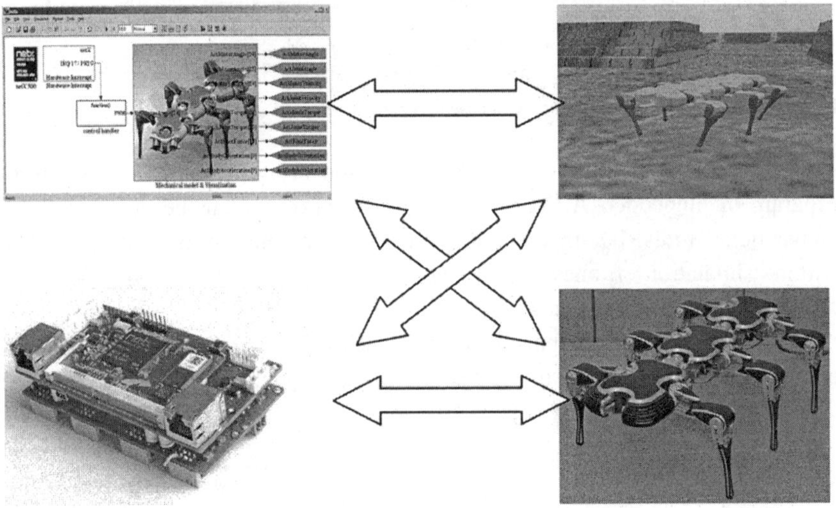

Figure 7. HiL/SiL structure by development of six-legged robot SLAIR2

4. Test of the NetX based Com Bridge in Real Six-Legged Robot

For the execution of the communication tests a soft and a hardware configuration have been suggested, which represents the SiL - application. The hardware part of the test platform consists of a host PC and two netX units. The units are interconnected via serial communication chain.

The communication is implemented in full duplex mode, whereby simultaneous sending and receiving of the packets are possible. The task of the host PC is in the periodic sending of request and the collecting of all response packages in the same communication cycle with all netX units (s.Fig.8)

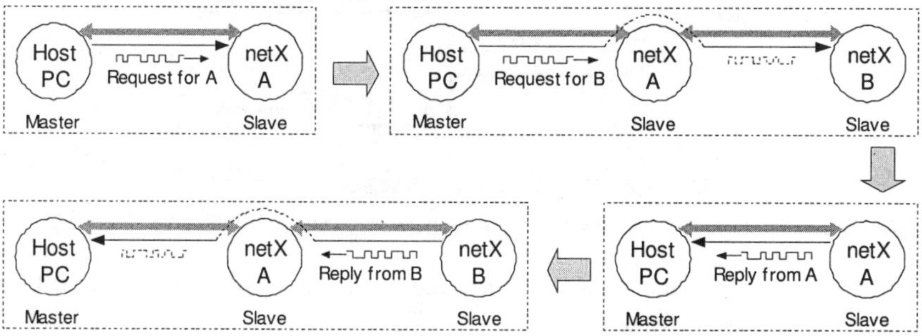

Figure 8. Data flow within communication cycle between Host-PC and netX chain.

The packets are led to and from the second netX unit "B" over the first netX unit "A". In order to reduce the turn-around time of the those Ethernet frames, whose goal MAC address does not agree with the own MAC address of the current unit, the first netX unit "A" functions in the so-called switch mode.

To test purposes the response time of netX units is interesting, which corresponds to the performance of the program generated by the RTW, as well as the entire cycle time for the operation of the netX chain with the Simulink model, running at the host PC. To check the performance of the whole system the forward and backward communication flow with distributed control units have been used (s.Fig.9).

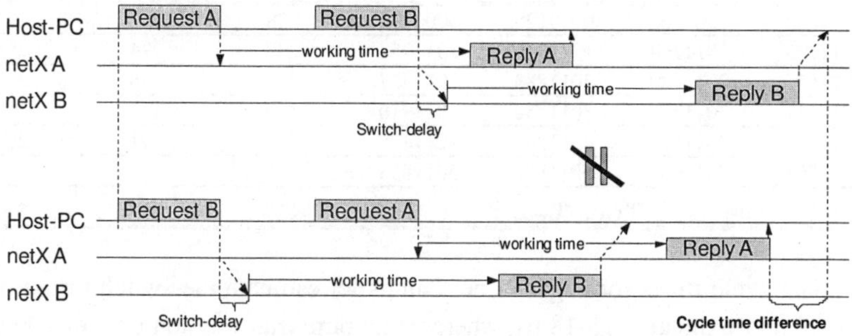

Figure 9. Forward and backward communication sequence with netX 'A' and netX 'B'

The timing measurement is accomplished in certain places over GPIO exits and observed thus at the oscilloscope (s.Fig.10). The measurement shows the incoming and processing times of the request packages for both netX units.

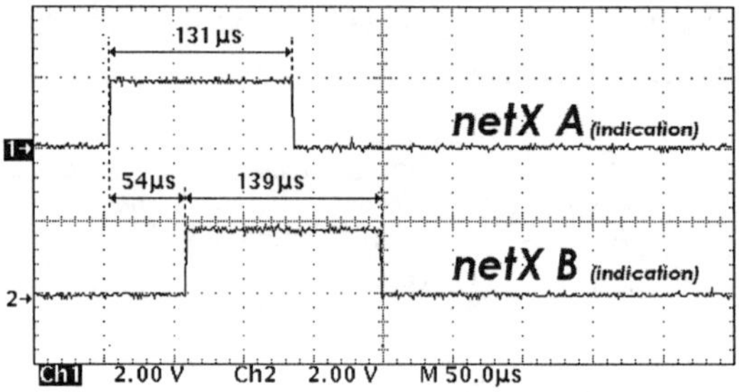

Figure 10. Packet incoming and processing time for netX 'A' and netX 'B'

The test results (table 2 and 3) show the maximum cycle rate of about 3 kHz for the presented test system.

Table 2: Network monitor records by 'direct' communication (PC - netX A - netX B)

A_{out} [µs]	B_{out} [µs]	A_{in} [µs]	B_{in} [µs]	$A_{out}...A_{in}$ [µs]	$B_{out}...B_{in}$ [µs]	$A_{out}...B_{in}$ [µs]
0000000	0000066	0000200	0000293	200	227	293
1016545	1016638	1016747	1016863	202	225	318
2016530	2016593	2016727	2016819	197	226	289
3016509	3016570	3016706	3016794	197	224	285
4016503	4016589	4016701	4016826	198	237	323
5016486	5016570	5016684	5016802	198	232	316
Average runtime: T_{Bround} /T_{Around} /$\mathbf{T_{round}}$ ➔				**199**	**229**	**304**

Table 3: Network monitor records by 'inverse' communication (PC - netX A - netX B)

B_{out} [µs]	A_{out} [µs]	B_{in} [µs]	A_{in} [µs]	$B_{out}...B_{in}$ [µs]	$A_{out}...A_{in}$ [µs]	$B_{out}...A_{in}$ [µs]
0000000	0000100	0000228	0000297	228	197	297
1011693	1011764	1011926	1011958	233	194	265
2011657	2011729	2011883	2011927	226	198	270
3011631	3011720	3011850	3011910	219	190	279
4011627	4011697	4011860	4011888	233	191	261
5011598	5011666	5011833	5011853	235	187	255
Average runtime: T_{Bround} /T_{Around} /$\mathbf{T_{round}}$ ➔				**229**	**193**	**271**

The cycle times for separate netX units let estimate the switch turn-around time of approximately 15-18 µs, whereby the pure transmission time of 100 byte package via 100 Mbit/s Ethernet is about 10 μs. This indicates the fact that the passing package is completely buffered by the netX switch.

The source code generated by the RTW for the netX units is not optimal concerning to the response time. Therefore the additional manual optimization of this source code allows the reduction of the cycle time from about 270 μs down to 90 μs.

References

1. http://www.hilscher.com Hilscher GmbH, Hattersheim, Germany
2. http://www.mathworks.com/products/rtwembedded/
3. http://www.uni-magdeburg.de/ieat/robotslab
4. F.Palis, V.Rusin, U.Schmucker, A.Schneider, Y. Zavgorodniy. Legged Robot with Articulated Body in Locomotion over Complex Terrain. 7th Int. Conference on CLAWAR, 22-24 September 2004, Madrid.
5. F. Kanehiro et.al. Distributed Control System of Humanoid Robots based on Real-time Ethernet. IEEE/RSJ Int. Conference on Intelligent Robots and Systems, 9-15 October 2006, Beijing

IMPROVING POWER TO WEIGHT RATIO OF PNEUMATICALLY POWERED LEGGED ROBOTS

G. MCLATCHEY, J. BILLINGSLEY

Faculty of Engineering and Surveying, University of Southern Queensland, West St, Toowoomba, QLD 4350, Australia

Legged Robots require actuators with a high power to weight ratio. Although pneumatic actuators do not perform well in this regard, they have other attractive characteristics which are useful in Legged Robots. This paper describes a mechanical solution for significantly improving the payload capacity of a robot powered with pneumatic cylinders, Robug IV, and reports on the theoretical design and experimental outcomes.

Keywords: Pneumatic cylinders, legged robots.

1. State of Pneumatically Powered Legged Robot Research.

The supporting legs of legged robots form part of multiple closed kinematic chains in which antagonistic forces pose a problem. Such a system is much better suited for force control, which can accommodate the over-constrained system [1], [2], [3], than position control where small errors can increase antagonistic forces [4], [5], [6].

Fluid based actuators such as hydraulics and pneumatics are ideally suited for force control, however they are less popular than electric motors because their natural compliance makes position control difficult and they are low in power density [2], [7].

Control of pneumatic cylinders has been advanced as much as possible by the authors' work in Nested Loop Control [4], however, the inherent problem of low power to weight ratio remains. The most promising solution is to re-direct some of the torque available from the actuator to lift a leg into pushing the leg down to hold the body up. This has been implemented in Robot V but not investigated [8].

Here, such a solution is designed, implemented and reported for the robot Robug IV. Also, the maximum increase of performance and the extent of the benefits of such a solution are discussed.

2. Design

Figure 1a) shows the original Hip joint for Robug IV. Note the Hip actuator between points F and E and the Knee actuator (which is not considered further) between points G and I. The Hip on a Robug IV leg is the major load-bearing joint and must balance the largest force acting on the robot, which is its weight due to gravity. On uneven terrain the Abductor and Knee joints must provide higher torque to balance Robug's weight, but a strategy of re-directing more torque in a particular direction is unsuitable since similar torque is required in both directions of joint movement.

The simplest way of re-directing torque is to use a spring element. The basic design of a spring inclusion is shown in Figure 1b). The original link between *B* and *D* is replaced with a plate that provides a spring mounting point *C*. A tension spring stretches between points *C* and *I*, and provides a positive torque around the Hip axis at *I*. Please note that an identical spring and plate is provided on the other side of the Hip.

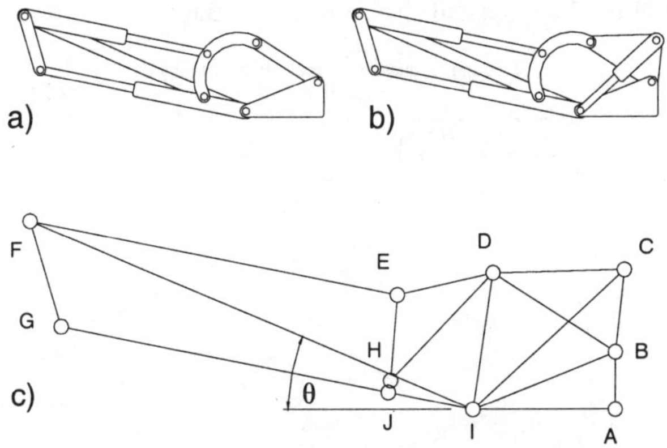

Figure 1. a) Robug IV Hip Joint, b) With Spring, c) Geometric Lengths and Angles Identified.

2.1. *Modelling*

Due to a simplification in the kinematic model of the Hip joint described in [9] and multiple errors in reproducing the equations, a more accurate kinematic model was derived by the first author for the Hip joint. Kinematic equations were also derived that describe the torque around *I* due to the spring extension. The force of the spring was modelled as:

$$Force = (L_f - L_c)k + F_1$$

L_f = final length of spring

L_c = initial length of spring

k = spring stiffness

F_1 = minimum force to extend spring

2.2. *Method of Determining Optimal Parameters*

The new kinematic models for the Hip joint and spring were coded in Matlab. Design characteristics of the springs were matched to available springs that were thought to have adequate stiffness and be of suitable length for the dimensions of the leg. Preliminary findings resulted in a spring with twice the stiffness of the original being chosen. Final specs for the springs were L_c = 90 mm, k = 6.54 N/m and F_1 = 58.86 N.

Several experiments were conducted to determine the available lifting and downward torque of an unaltered Hip joint. Figure 2 shows the original lifting torque of the Hip joint at system air-pressure of 6 Bar.

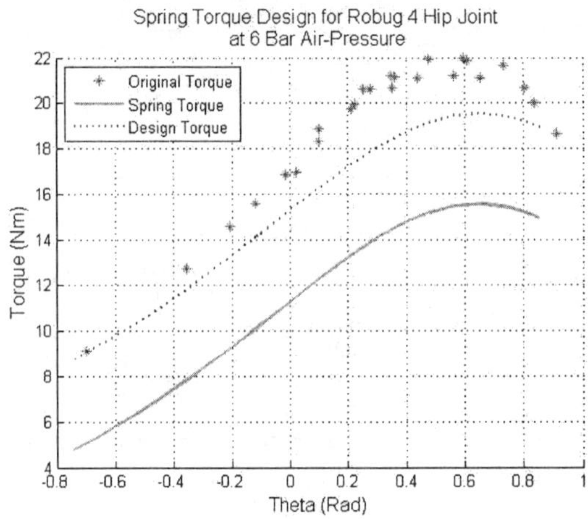

Figure 2. Spring and Plate Design

The Design Torque was then calculated as the Spring Torque plus a constant Buffer to ensure adequate lifting torque remained. The Spring Torque was modified by changing the spring support position, C, until the Design Torque was as close as possible to the original lifting torque without exceeding it anywhere along the range of θ. This ensured the maximum possible Spring

Torque without reducing the resultant lifting torque below the Buffer (in this case the Buffer was set equal to 4 Nm).

Once the spring support position, *C*, was determined, a mild steel plate was made that replaced the original link *BD* and provided support for a mild steel shaft at *C* that the springs would be attached to (see Figure 1).

3. Results

The original and increased supporting torque can be observed in Figure 3. The increase in supporting torque due to the springs is substantial. At $\theta = 0$ there is at least 36% more torque available to the Hip. From $\theta = 0.2$ there is at least 40% more torque. Since the leg design is of the insect type, where the Knee is usually located above the Hip joint, the increase of supporting Hip torque for the useable range of $\theta \geq 0$, is at least 36%.

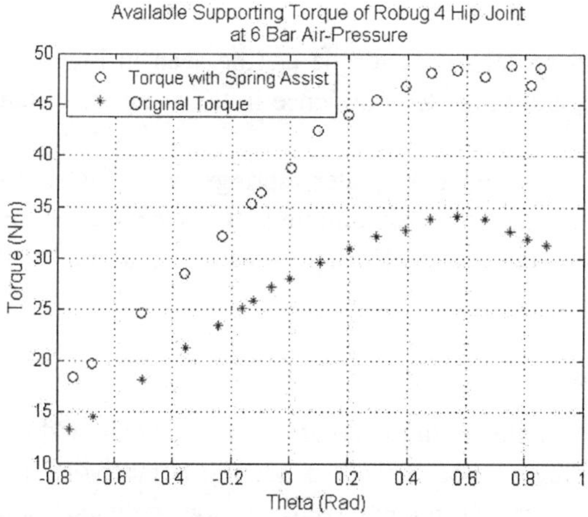

Figure 3. Increase in Available Supporting Torque of Robug IV Hip Joint at 6 Bar Air-Pressure

The increase in supporting torque is irrelevant unless adequate lifting torque remains to relocate a leg. Figure 4 shows the original and decreased lifting torque after adding springs. Lifting torque is drastically reduced, but remains above zero. This ensures the leg can still be lifted, although with an extended response time. In the preferred operating range of $\theta \geq 0$ the minimum lifting torque is 4 Nm. Please note that the torque plotted in Figure 4 remains after the weight of the leg is accounted for.

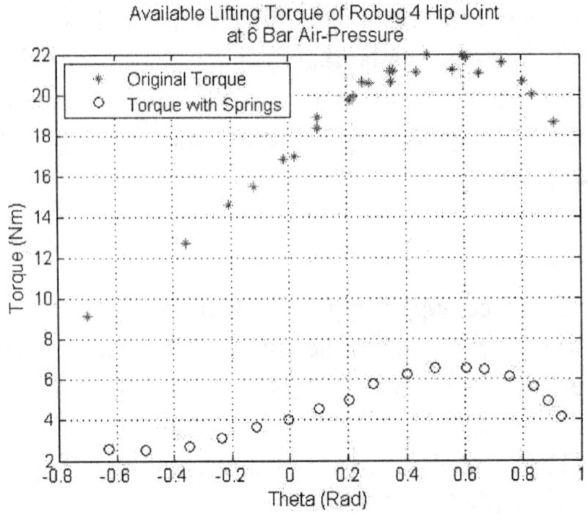

Figure 4. Available Lifting Torque of Robug IV Hip Joint at 6 Bar Air-Pressure

The designed and actual spring torque can be seen in Figure 5. The two are remarkably close, and promote confidence in the accuracy of the kinematic and computer models.

The weight of the new plates springs and shafts total 1.155 kg. Manufacturing the plates from aluminium will reduce the weight to 0.627 kg or less. Using hollow stainless steel shafts instead of solid mild steel would reduce weight further.

4. Observations

The final spring and aluminium plate arrangement would add an extra 3.8 kg at the most to the 22.2 kg robot. This does not impact much on leg performance, as the extra weight is near the body; however it does mitigate the 36% increase in supporting torque (minimum) by a 17% increase in weight. This trade-off is expected to become more favourable as higher air-pressure is used. This is because only a small increase in the weight of the springs should be necessary over the current springs of lower stiffness.

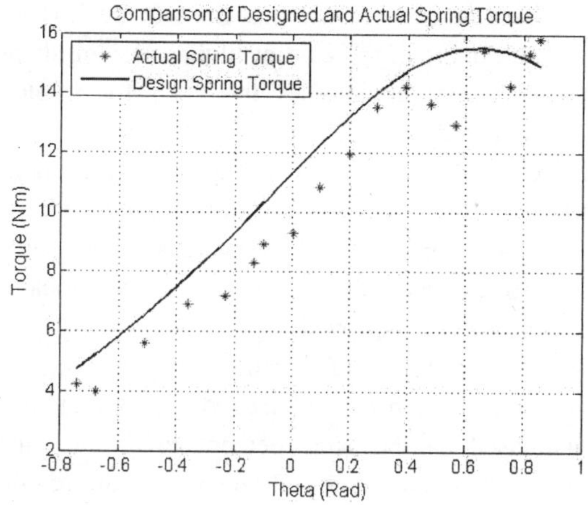

Figure 5. Comparison of Designed and Actual Spring Torque

Increased friction due to the springs can be observed in some plots. Actual Spring Torque was close to but less than the Designed Spring Torque (see Figure 5). However, Figure 4 shows that the Lifting Torque with Springs is less than the Buffer of 4 Nm for $\theta < 0$. This disparity can be explained by increased friction due to the large tangential force of the spring at pivots *D* and *B*. At most, this friction accounts for a loss of 2 Nm of torque. At higher system air-pressures this friction will increase, and may require a change of strategy from nylon bushes to bearings.

While the data for actual spring torque in Figure 5 does follow the shape of the designed spring torque, it does seem to be affected by a small random factor. This is most pronounced for $\theta > 0.4$ and reaches a maximum value of 2 Nm at $\theta = 0.5697$. This phenomenon can be explained by stiction that is introduced by the friction between the springs and the shafts they are attached to. This can cause the springs to bend rather than stretch thus producing a torque opposing the motion of the system. This could be reduced by separating the spring and shaft with a nylon bush.

5. Conclusions

Redirecting some of the excessive lifting torque available to a pneumatically powered leg to assist in supporting the robot is feasible and very beneficial. An increase in supporting torque of 36% to well above 40% was found for the Robug IV Hip configuration.

Friction and stiction mitigation methods need to be investigated. These contributors will become more significant at higher system air-pressures. The importance of this discovery emphasises the need for implementation and experiment, not just theoretical analysis.

The performance of this system will improve with higher system air-pressures, provided the friction and stiction factors are reduced. Not only should higher air-pressure result in a more significant increase of supporting torque than 36%, but they should also experience decreased time for relocating a leg (since the lifting response time will be improved).

Increasing the power to weight ratio of pneumatic cylinders will significantly improve the performance of legged robots that incorporate them. Payload can be increased by the same percentage as the supporting torque increase, or extra equipment such as a pressure vessel and regulator could free pneumatically powered robots from their tether, making them much more mobile. Different gaits could also be possible along with corresponding increases in speed and stability.

6. Acknowledgment

The authors would like to extend their gratitude to the University of Southern Queensland (USQ) for funding this research.

7. References:

1. Zielinska, T., *Synthesis of Control System - Gait Implementation Problems.* 4th International Conference on Climbing and Walking Robots, 2001: p. 489.
2. Pratt, J., B. Krupp, and C. Morse, *Series Elastic Actuators for High Fidelity Force Control.* Industrial Robot: An International Journal, 2002. **29**(3): p. 234-241.
3. Colbrunn, R., G. Nelson, and R. Quinn. *Modeling of Braided Pneumatic Actuators for Robotic Control.* in *2001 IEEE/ RSJ International Conference on Intelligent Robots and Systems.* 2001.
4. McLatchey, G.J. and J. Billingsley. *Force and Position Control Using Pneumatic Cylinders.* in *9th International Conference on Climbing and Walking Robots.* 2006. Belgium.
5. Jiang, W.Y., A.M. Liu, and D. Howard. *Foot-force Distribution in Legged Robots.* in *4th International conference on Climbing and Walking Robots.* 2001.
6. Kar, D.C., *Design of Statically Stable Walking Robot: A Review.* Journal of Robotic Systems, 2003. **20**(11): p. 671-686.

7. Cubero, S., *Force, compliance and position control for a pneumatic quadruped robot*, in *Faculty of Engineering*. 1997, University of Southern Queensland: Toowoomba.
8. Kingsley, D.A., *A cockroach inspired robot with artificial muscles*, in *Department of Mechanical and Aerospace Engineering*. 2005, Case Western Reserve University. p. 231.
9. Galt, S., *Soft Computing Based Motion Control for an Eight Legged Robot", Thesis, Portsmouth: , Dept. of, 1998.*, in *Electrical and Electronic Engineering*. 1998, University of Portsmouth: Portsmouth.

INTEGRATED INTELLIGENT MECHROBOT SYSTEM

LIQIONG TANG AND GURVINDER SINGH VIRK

Institute of Technology and Engineering
Massey University, New Zealand

This paper presents the design and implementation of an Integrated Intelligent MechRobot System (IIMRS). The framework and the functionality of IIMRS are developed based on intelligent control, robotics, shop floor integration, and factory automation. Modern control software and hardware can be easily implemented with IIMRS. The system consists of autonomous robot, embedded system, PC-based control, image processing, and functionalities commonly used for robotic control and system automation. IIMRS aims to develop a real-time physical training and learning environment for the design, simulation, and testing of mechatronic systems, robotic control, and industrial automation.

1. Introduction

With the increased computing power and the advances in artificial intelligence, smart robots are now marching into every possible corner of the industrial field and our daily lives. Brainless machines and hard automation systems are being replaced by intelligent systems with sophisticated sensors and flexible functionalities. The design and development of such modern systems sits on the foundation of mechatronics, robotics, computer systems, artificial intelligence, machine vision, automation control, etc. An intelligent system, a robot, or a control system follows the same product development cycle just like any other product on the market. Design analysis, modifications, testing and retesting are unavoidable for any new product development. The aim of the Integrated Intelligent MechRobot System (IIMRS) is to provide a physical platform to support the design and development of intelligent machines and automation systems, which is becoming an increasingly complex and difficult task.

IIMRS is an integrated system. It mimics an automated shop floor [1]. The system consists of two robots, one microcontroller-based autonomous robot, the other is a 4-axis pick-and-place robot controlled through a PC. A pneumatic transport frame with two pallets and two robotic grippers is for transferring components between different locations. The pneumatic system is controlled using two industrial PLCs. Quality data acquisition and control relies on a computer-based vision system. All the subsystems in IIMRS can also be linked through a network. The schematic diagram of the whole IIMRS system is illustrated in Figure 1. The first version of IIMRS has been successfully implemented in the Institute of Technology and Engineering at Massey

University. It is currently used for undergraduate and postgraduate teaching and research projects. IIMRS provides a physical environment with features and functionalities commonly used in robot control, mechatronic system design, and industrial automation. IIMRS provides a platform a robot, a control system, or an assembly line to development new control strategies and test new functionalities. It also simulates and motivates designers to further improve existing systems and develop new ideas. The system has been used in studying mechatronics applications and the design and testing of autonomous robots.

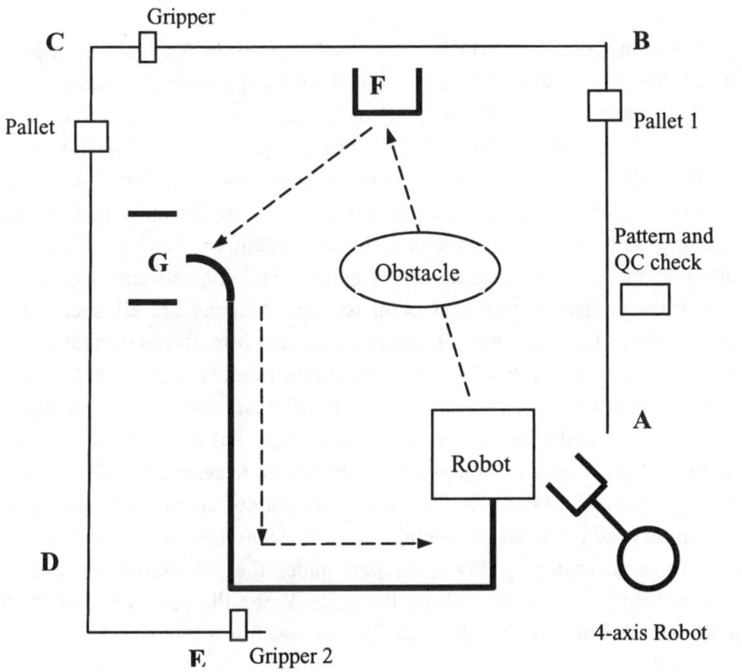

Figure 1. The schematic diagram of IIMRS

2. Autonomous Robot Design

Automated Guided Vehicles (AGV) have become common facilities on the factory floor for transportation and material handling. The autonomous robot running under IIMRS mimics the common functionalities of a real AGV [2], [3]. The "brain" of the vehicle is a microcontroller with embedded programs. It is designed to be able to perform the functions commonly implemented in an AGV such as [4], [5]:

1) Auto-recognize part loading and unloading
2) Give warning signals
3) Avoid obstacles on its path
4) Transfer parts to different destinations
5) Accurately position the parts for the robot to pick up

The starting position of IIMRS is located at position A as shown in Fig 1. At the beginning the autonomous robot will stay at position A waiting for the part to be loaded. The 4-axis pick-and-place robot is able to sense the appearance of the autonomous robot. Once the part is placed on top of the autonomous robot, it will trigger the robot to move from location A towards G via F. The route from A to G simulates an AGV on an open factory floor delivering parts to different locations with the possibility of having to overcome obstacles on the way. Different sensors and methodologies such as proximity sensors, laser guidance, electrical compass, and beacons are all successfully used for guiding the autonomous robot to avoid the obstacle and correctly guide the robot to arrive at position G. The autonomous robot is required to be parked at position G between the two dark lines for a few seconds. This imitates the situation of a part under some service. Once the time has elapsed, the robot will move towards position E by following a defined line to send a part for the robot gripper at location E to pick up. Such a functionality of an autonomous robot is often seen on production and assembly lines. The autonomous robot is required to be able to accurately position the part under the pick-and-place robot at location E for the gripper to pick up the part. When the part is unloaded, the autonomous robot will follow the path back to position A and position itself for the next cycle.

The brain of the autonomous robot is a C8051F020 microcontroller from Silicon Laboratories (SiLab). The SiLab C8051F020 is a fully integrated microcontroller with mixed-signal system on a chip. The main features of the microcontroller are listed in Table 1.

Table 1. SiLab C8051F020 Microcontroller

Peak throughput	25 MIPS
FLASH Program Memory	64K
On-chip Data RAM	4352 bytes
Full-duplex UARTS	x2
16-bit Timers	x5
Digital I/O Ports	64 pins
12-bit 100ksps ADC	8 channels
8-bit 500ksps ADC	8 channels
DAC Resolution	12 bits
DAC Outputs	x2
Analog Comparators	x2
Interrupts	2 Levels
PCA (Programmable Counter Arrays)	

The following example is just one of the many autonomous robots running under IIMRS. This robot uses proximity detection for obstacle avoidance. Modulated IR sensors are used as the proximity detection units. The autonomous robot has two such units, one on each front corner. It is programmed to turn away from the object when an object comes within 6 cm. If both the proximity units are activated at the same time, the robot will back away from the obstacle and turn. The guidance for the robot to move from location A to G via F is provided by beacons using IR sensors. Each beacon unit has two outputs, one for directing the robot and the other for stopping the robot when it reaches the destination. The robot identifies the beacon units through different frequencies. The receivers on the robot are linked to a microcontroller and the microcontroller is able to identify which beacon it is communicating with. Using beacon units make the guidance for the autonomous robot very flexible. As the beacon units can be placed at different locations, it is easy to expand the system to have more destinations at different locations for the robot to move around. In Figure 2, location F and G each has a beacon unit to communicate with the autonomous robot. The ability of the autonomous robot communicating to the beacons and then using the beacon's positions correctly to direct the robot to location G via F can be easily demonstrated.

Figure 2. Communication using beacon units

3. Pick and place robot for loading and unloading

A PC-based pick-and-place robot shown in Figure 3 is integrated in IIMRS for part loading and unloading [6],[7]. A motion control card from National Instruments is implemented in the robot control system. The motion control card is a combination of a servo motor and stepper control card. It can provide full motion control for independent as well as coordinated four-axis motions. The servo axes always operate in closed loop control. All the stepper axes support half and full step as well as microstepping possibilities. The driver software called NI Motion is also from National Instruments, which allows creating motion control applications using the graphical programming Motion Assistant. With Motion Assistant, it is possible to transfer the application program to either a LabView program or a program in other computer languages such as Visual Basic, C, or C#.

The robot has 4 motors and the motion control card can cope up to 4-axis control. The 4 motion control axes for the robot are defined as:

Motion control axis 1 - X axis (robot base rotation)
Motion control axis 2 - Y axis (lower arm rotation)
Motion control axis 3 - Z axis (upper arm rotation)
Motion control axis 4 – Gripper movement

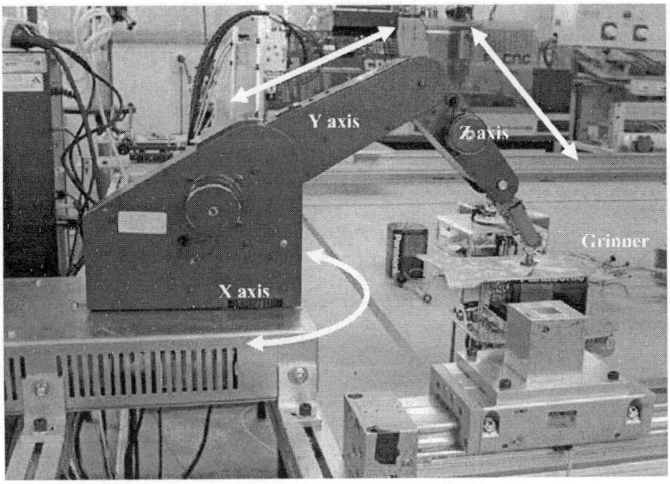

Figure 3. A 4-axis PC-based pick-and-place robot

A software program for the control of the robot is developed using LabView. The control panel for the robot arm is presented in Figure 4. The control system has two modes, automatic and manual, which can be toggled through the push button on the panel. The control panel also provides a user friendly interface to teach the robot and define the pick and place positions. There are two indicators on the panel, one for indicating that a part has arrived at location A and another for signalling the autonomous robot is ready for loading. Only until both indicators are switched on can the robot arm be activated. In other words, the Boolean "AND" of the two indicators trigger the pick and place movement of the robot arm.

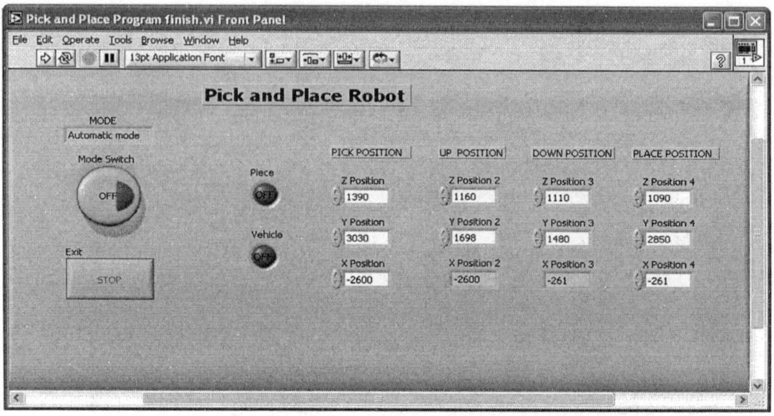

Figure 4. Pick and Place Robot Control Panel

The robot works closely with a part data collection system. Based on the data of the part such as geometry and dimensions, the robot is able to identify differences and perform part sorting functions. Such processes are commonly used for product quality control. In IIMRS, each time a part is delivered to location A by pallet 1, the robot will load the good part on top of the autonomous robot and put the bad ones into a bin.

4. Part Handling System

The part handling system is a pneumatic system controlled by two Omron Programmable Logical Controllers (PLC). The pneumatic frame consists of two pallets and two pneumatic grippers refer to Figure 1. Apart from picking and placing functions, one of the grippers can also rotate to place components at different orientations. Each PLC controls one pallet and one pneumatic gripper. Sequential control is implemented in the control system. The pneumatic system is activated by the autonomous robot. Each time the autonomous robot arrives at location E and correctly positions the component under Griper 2, the pneumatic system will be trigged by a sensor. Griper 2 will come down and pick up the part and travel to Location D to place it into the holder on top of Pallet 2. After Gripper 2 is returned, Pallet 2 will move the part from Location D to C. The controls are realized through an Omron PLC. The corresponding state diagram is illustrated in Figure 5. The other PLC is for the actions from Location C to A via B. The process functions are very similar except the griper is able to rotate up to 90 degrees. Once the part is transfer to Location C, Gripper 1 will pick up the part and rotate 90 degrees first, and then transfer the part to Location B. When Gripper 1 puts the part in Holder 1, the part is in a new orientation. Pallet 1 will then send the part to Location A.

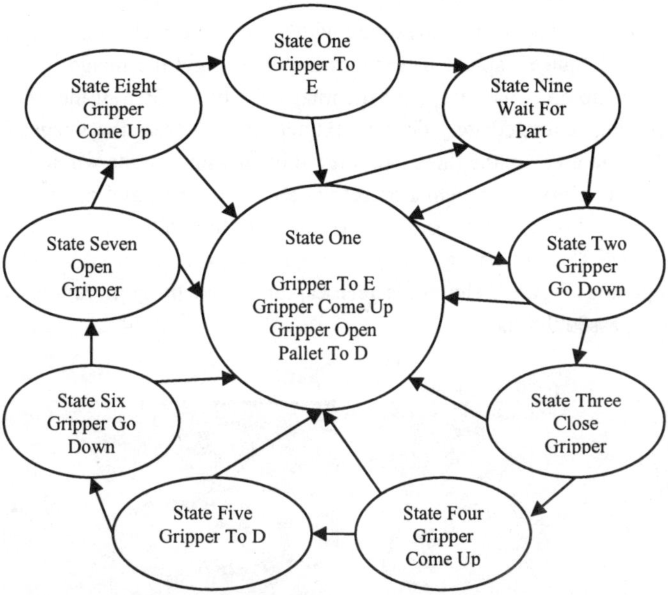

Figure 5. The state diagram for the control from position E to C.

5. Real-time data collection

A real-time data collection system is implemented in IIMRS. The system is running on a PC. Each time when a part is transferred from Location B to A, it will pass the data collection point where the information of the part is captured through different sensors or by a camera. The part data is processed in real time and then output the control information the pick and place robot at Location A. If the part does not have the correct geometry or wrong size, the robot at Location A will pick it out. Otherwise it will be put on top of the autonomous robot to activate the next cycle.

The following example uses a camera in the data collection system. While pallet 1 travels from position B to A, the system controlled by the PC will recognize the badly made components. When a problem part is recognized, the system will give an alarm signal. To implement such a system a computer-based vision system is developed. This system visually checks whether parts on pallet 1 moving from position B to A are faulty or not. This vision system uses a Logitech QuickCam webcam with a USB plug. The system is developed under Visual Basic 6.0 (VB). When the VB program is run the system creates a capture window to display the streaming video from the webcam. Through Microsoft WDM Image Capture driver it is possible to get the image from the webcam to the capture window. Once the image has been taken of the part the image processing is immediately done to extract the data from the image. Part length, width, and area are the three variables to be measured. The whole image is scanned and any pixel less than a preset variable is the part and coloured red and any other pixel is said to be background and set to light blue.

A Graphical User Interface (GUI) is designed to give a visual display of what is happening. Figure 7 shows a screen capture of the program running and a correct part passed through.

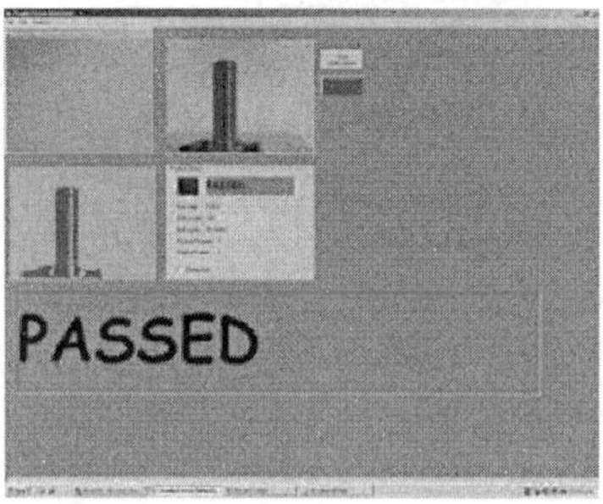

Figure 7: Screen capture of a good part.

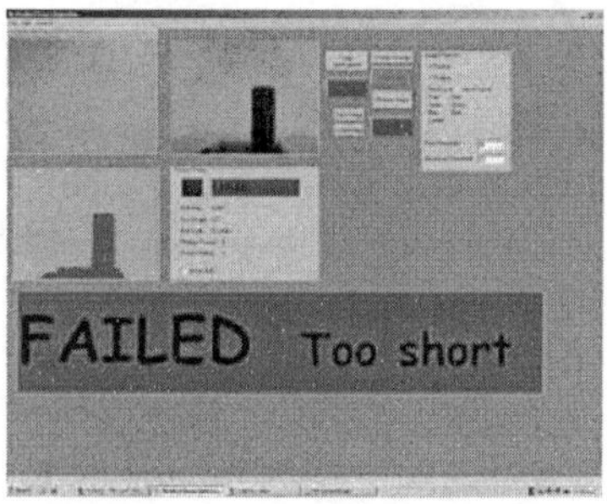

Figure 8: Screen of a part that was short.

The top left window shows the live streaming image from the webcam, the top right window shows the captured image that is to be processed. The bottom left window shows the processed image and note the shape of the bolt has been brought out in red. The bottom right frame show data about parts being passed through. It displays the number of parts passed through, showing parts failed and parts passed. Also for each image that is processed it displays the value calculated for bolt length, bolt width and bolt area. Then there is a large clearly visible text area which displays the result of each part processed as shown in Figure 7 and 8.

6. Conclusions

The framework and the functionality of IIMRS has been developed and implemented in the Mechactronics Labortory at the Institute of Technology and Engineering, Massey University. The system covers the use of microcontroller, PLC, pneumatic equipment, motor control, sensing system, vision, and intelligent control. It provides a useful platform for engineering students to study both hardware and software. It is also an ideal test rig for robot design, automated pneumatic system, simulation of shop floor control, and intelligent factory automation. The system is now used for Mechatronics and engineering undergraduate teaching and as a test rig for autonomous robot design projects.

Acknowledgments

The authors are grateful to all the 4th year mechatronics students who contributed to IIMRS system. In particular, thanks are due to: John West, James Barkwith, Lyndsy Homes, Lara Christian, Glenn Gregory, Alex Wishart, Kieran Orchard, Rob Paddison.

References

1. J. A. Rehg, "Introduction to robotics in CIM systems", *Prentice Hall* (2000).
2. Robert J. Miller, "Robotics : future factories, future workers", *Sage Publications* (1983).
3. Joseph A. Angelo, "Robotics: a reference guide to the new technology", *Greenwood Press* (2006).
4. T. Bräunl, "Embedded robotics : mobile robot design and applications with embedded systems", *Springer* (2003).
5. U. Nehmzow, "Mobile robotics : a practical introduction", Springer (2003).
6. John J. Craig, "Introduction to robotics: mechanics & control", *Addison-Wesley Pub. Co.* (1986).
7. 7340 User Manual, National Instruments (2003).

MCA2 - AN EXTENSIBLE MODULAR FRAMEWORK FOR ROBOT CONTROL APPLICATIONS

Klaus Uhl* and Marco Ziegenmeyer

FZI Forschungszentrum Informatik, Interactive Diagnosis and Service Systems (IDS), Haid-und-Neu-Str. 10-14, D-76131 Karlsruhe, Germany
**E-mail: uhl@fzi.de*

For the rapid development of different robot prototypes the adoption of reusable software components with standardized interfaces is a very important aspect. In this context reusability can mean reuse on a different robot or in a different environment (e.g. user space vs. realtime). MCA2 provides an extensible modular framework for robot control applications in order to fulfill these needs.

Keywords: Control architecture; Robots

1. Introduction

Efforts in software development should be restricted to create new components and - in the context of robotics - new control methods which have not been implemented so far. A framework which encourages the developer to organize the whole controller system as a set of small modules with standardized interfaces which can be individually reused in other projects or other execution environments is a huge step into this direction. If the communication between the modules and the synchronization between the different parts of the software are implemented by the framework then the developers can focus their work solely on the methods and algorithms which are necessary for controlling the robot.

The *Modular Controller Architecture Version 2* (*MCA2*)[1] was developed to meet these requirements. It is based on *MCA*[2] but provides additional enhancements. MCA2 is written in C++, runs on Linux and Windows and uses RTAI/LXRT[3] for hard realtime operation. The graphical tools are written using Qt.[4]

This paper presents the architecture and the core components of MCA2, introduces its tools and shows some projects which were implemented using

MCA2. In addition it gives an overview of other frameworks for robotic applications.

2. Architecture

MCA2 targets at the design and development of low and medium level robot control software. Therefore it uses a strictly hierarchical control structure and a two-way data flow. The lowest level comprises the hardware interface while the units get more and more abstract when you ascend the system hierarchy. The data flow is split into two parts:

- *Sensor data* is flowing upwards from the hardware interface units through all units which need to process it. Intermediate units may alter the sensor data or create new "virtual" sensor data from it.
- *Control data* is flowing downwards from the highest level control units or the user interface back to the hardware interface units. Control data may also be altered or derived by intermediate modules.

Each unit can participate in both the sensor and the control data flow.

The basic units of an MCA2 application are *Modules*, *Groups* and *Edges*. Modules are the smallest units and contain basic functional blocks. As complex control systems usually contain many functional blocks which are combined to even bigger entities modules can be put together into groups. They can be connected with edges and are then able to communicate with each other.

2.1. *Modules*

Figure 1 shows the schematic structure of a module. Sensor data flows into a module through the *Sensor Input* interface. It is processed in the `Sense()` function which writes its results into the *Sensor Output* interface. From there the sensor data flows upwards to the higher levels of the system hierarchy. On the other side the control data flow enters a module through the *Controller Input* interface. Control data is processed in the `Control()` function. It writes the altered or newly created control data to the *Controller Output* interface from where the control data flows further down the system hierarchy.

The `Sense()` and `Control()` functions can exchange information via local variables which comprise the *internal state* of the module. They are called periodically by the control loop of the enclosing thread container (see section 2.4) in configurable time intervals.

The four sensor and controller interfaces are modeled as vectors of type `double`. Each interface entry has a description which defines its meaning. Furthermore each interface owns a *changed flag* so that `Control()` and `Sense()` can determine in each cycle of the control loop whether the controller or sensor input needs to be processed.

As many algorithms depend on certain parameters which are mostly static and are not influenced by other modules each module has a *Parameter* interface. In contrast to the sensor and controller interfaces parameters are not restricted to `double` values but can be boolean values, integer values, floating point values or strings. Parameters can be set from within the MCA2 application itself, e.g. with values read from a configuration file. Another more powerful option to alter parameters is the browsing and debugging tool described in section 3.2. With this tool it is possible to access and change parameters at runtime and thus fine-tune the application without the need to restart it.

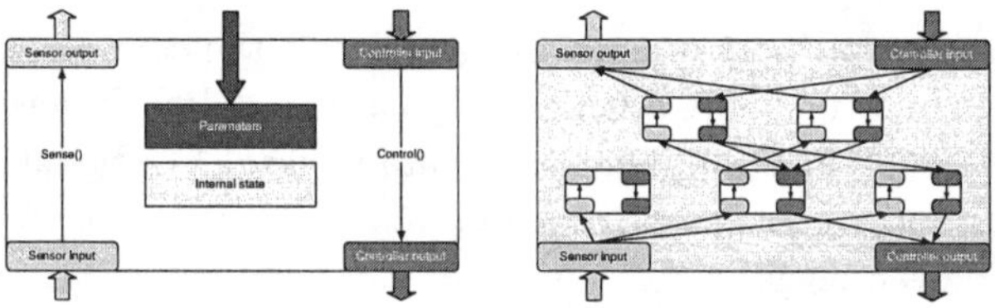

Fig. 1. A *Module* with its five interfaces (left) and a *Group* containing several communicating modules in a two-layer hierarchy (right).

2.2. *Groups*

To build a bigger functional block from basic functional blocks which are implemented in individual modules an instance of each module is put into a group. Then the modules are connected by *Edges*. An edge always connects an output interface of one module with a corresponding input interface of another module (controller output with controller input and sensor output with sensor input). Either complete interfaces or parts of them can be connected. Arbitrary permutations of interface elements are allowed. Edges carrying sensor data are always directed upwards while edges carrying control data are directed downwards as can be seen in figure 1.

The graph of modules and edges within a group can be organized in a multilayered hierarchy in the following way:

- All modules which either have no connected sensor input and controller output interfaces or which are only connected to the group's sensor input and controller output are placed on the lowest level $l := 0$.
- Each module is placed one level above the highest level module to which its sensor input or controller output is connected.
- Modules on the same layer are arranged in the order in which they were added to the group.

Groups behave like modules implementing bigger functional blocks. Therefore they also have controller input, controller output, sensor input and sensor output interfaces (the parameter interface is not used for groups) and can themselves be nested into other groups to build even bigger functional blocks. Moreover, groups also have a `Sense()` and a `Control()` function. They manage the processing and data flow inside the group. The `Sense()` function performs the following steps:

(1) Copy the group's sensor input data to all modules connected to it by edges.
(2) Set the current level $l := 0$.
(3) Call the `Sense()` function in all modules on the l'th level.
(4) Copy the sensor output data of all modules on the l'th level which have set their sensor output's changed flag to the modules which are connected to them by edges. If there is an edge to the group's sensor output then copy the data there.
(5) Set $l := l + 1$.
(6) If there are modules on the l'th level go to step 3. Otherwise processing is finished.

The group's `Control()` function performs virtually the same steps but starts with the group's controller input data and the highest level modules. It then descends the module hierarchy and passes data from controller output to connected controller input interfaces.

2.3. *Parts*

A *Part* is the basic execution environment of an MCA2 application. It either manages a single module or – usually – a group. It initializes the execution environment, communicates with other parts over a TCP/IP connection and runs the control loop of the main thread.

Parts expose the interfaces of the managed group to other parts which can use this information to establish *inter-part edges*. These are similar to regular edges as described in section 2.2 but connect sensor (or controller) outputs of one part with sensor (or controller) inputs of another part.

A part's control loop is executed periodically in configurable time intervals and consists of the following steps:

(1) Pass all sensor data received from lower-level parts to the sensor input interface of the managed group.
(2) Call the `Sense()` function of the managed group.
(3) Pass the sensor output data of the managed group to all connected higher-level parts.
(4) Pass the control data received from higher-level parts to the controller input interface of the managed group.
(5) Call the `Control()` function of the managed group.
(6) Pass the controller output data of the managed group to all connected lower-level parts.

2.4. *Multithreading*

Thread containers allow for splitting an MCA2 application into multiple concurrent threads of execution. Like parts thread containers manage a module or a group and expose their sensor and controller interfaces to other modules. Additionally they run their own control loop in a separate thread. The control loop is similar to the part's control loop. But instead of sending and receiving data from other parts the interfaces of the managed module are only copied to the exposed interfaces of the thread container so that a thread-safe access is guaranteed.

2.5. *Realtime and network transparency*

MCA2 offers complete realtime and network transparency. From the viewpoint of the implementation of groups and modules it makes no difference where a module is executed, be it in the same thread, in the same process, in a different process or even on a different host. It can also be run either in a realtime or a non-realtime thread without modifications.

This transparency makes it possible to develop a controller system in a standard Linux or Windows environment using the usual set of development tools like compilers, editors and debuggers. Most bugs can be fixed with those tools before the system is run in a realtime environment. Due to the use of RTAI with its LXRT extension as the realtime operating system of choice it is only a matter of setting special thread priorities to execute specific thread containers in hard realtime.

2.6. *Blackboards*

The controller and sensor interfaces are modeled as vectors of `double` values and are therefore only suited for small to mid-size data. But sometimes there is the need to exchange big and complex chunks of data between modules which may even dynamically change its size. MCA2 contains *blackboards* for this purpose.

A blackboard is merely a memory buffer which can be accessed as a vector of elements with a user specified size. The blackboard elements are not typed but can be converted into anything the application demands. The only restriction is that the types which should be stored in blackboards must be placable in a continuous memory block.

Blackboards are also network transparent and can be used in any part of the system whether they were originally created in the same process or on a different host. The access to blackboards is protected with read-write locks which work even across host boundaries.

3. Tools

A framework is only as good as the tools which it provides to assist developers. Because of this MCA2 comes with two powerful tools for browsing, debugging and controlling MCA2 systems. Both tools connect to an MCA2 system via TCP/IP and can therefore also be used from a distant location for remote operation or diagnostics.

3.1. *User interface*

The *mcagui* allows to quickly click together a control user interface and to connect it to an MCA2 system. The mcagui runs atop the highest-level part in the system, reads its sensor outputs, writes its controller inputs and accesses its blackboards. It comes with a set of general-purpose widgets like buttons and value displays but can also be extended through plugins which provide special functionality for a certain application.

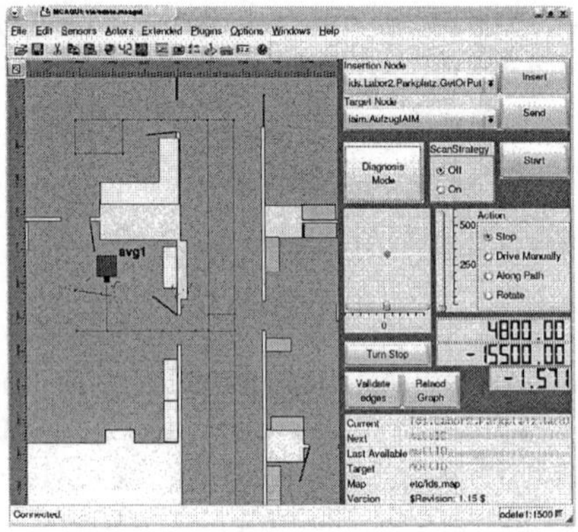

Fig. 2. The *mcagui* showing the control interface for the mobile research platform *Odete*.

3.2. *Browsing and debugging*

The *mcabrowser* is a browsing and debugging tool. It can be used to browse the structure and components of a running MCA2 system. This includes all modules, groups, edges, parts, component interfaces and blackboards. One can inspect the current values of interface entries and alter them even while the system is running. Additionally individual modules can be stopped and restarted, debugging messages which are written to the console can be switched on and of on a per-module basis and the current debug level, i.e. the amount of debug messages that are generated, can be changed.

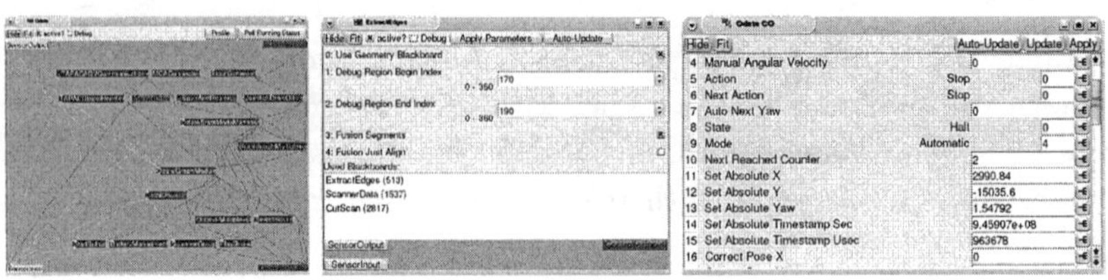

Fig. 3. The *mcabrowser* showing the module graph of a group, a module with its interfaces, parameters and blackboards and an interface window to view and alter the controller output of a module.

4. Projects

During the years MCA2 has been successfully used for many robotic projects within our group and in other research groups. Some of those projects have made their way into industrial applications which are deployed worldwide. Figure 4 shows a collection of current robotic projects which have been developed using MCA2.

(a) The autonomous sewer inspection robot MAKROplus.[5]

(b) The six-legged walking machine Lauron IV.[6]

(c) The mobile research platform "Odete" with a "Puma 200" manipulator.

(d) The healthcare industry automatic guided vehicle "LTC2" from Swisslog Telelift.

(e) The autonomous "Smart Roadster".[7]

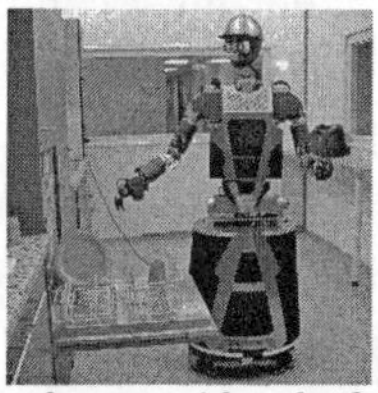

(f) The humanoid platform for perception-action integration *Armar III*.[8]

Fig. 4. Robotic projects which were implemented using MCA2.

5. Related work

Besides MCA2 there exist several other frameworks for robotic applications. Each of them takes a slightly different approach and sets a different focus. In the following a collection of popular frameworks is presented. In contrast to MCA2 they all don't support hard realtime operating systems and are thus not suited for robot control systems in which at least certain parts need to run in a predictable realtime environment.

ORCA[9] follows a Component Based Software Engineering (CBSE) approach to develop robotic systems. ORCA provides the developer with a framework for developing components which communicate via interfaces. ORCA does not define a system architecture but only provides a means to build components and implements the component communication. ORCA comes with a library of standard components but has no powerful browsing and control tools.

In *Miro*[10] components are implemented as CORBA services. On the lowest level Miro is less more than an easy to use wrapper around CORBA. But Miro adds several layers atop the basic system which provide generic services for communication, sensor and actuator interfacing and so-called "frameworks" which can be specialized or extended for given robotic tasks. Although Miro comes with several graphical development tools none of them is able to inspect or modify a running system.

The *Player/Stage/Gazebo*[11] project consists of the network transparent robot device server *Player*, the lightweight 2D robot simulator *Stage* and the high-fidelity 3D multiple robot simulator *Gazebo*. It is designed to support virtually any architecture and language for developing robot control applications. But Player's support ends at the device layer. It provides no support for higher-level components.

MARIE[12] also follows a CBSE approach but unlike most other robotic frameworks it mainly acts as a mediator between components from different frameworks. The advantage of this approach is that the best available implementations of the needed functions can be tied together to build a robot control application. The drawback is the large communication and computation overhead and the big memory footprint.

6. Summary and outlook

In this paper *MCA2*, a scalable modular controller architecture for robotic applications, has been presented. It allows to easily reuse components and thus makes the process of getting new robot prototypes working faster. Development and testing can be performed in Linux or Windows user space before the system is deployed to a realtime environment. The network transparent communication between individual parts of a complex controller system makes it possible to distribute such a system over several hosts without any additional effort. The user interface *mcagui* provides a fast and flexible way to control a system and visualize data from it. The use of the *mcabrowser* enables the developer to browse the complete system while it is running and fine tune system parameters.

Although MCA2 has been successfully used for many robotic projects some aspects of the design of MCA2 can still be improved. The next major release will include typed interfaces and blackboards. Moreover an event mechanism for asynchronous communication will be introduced.

MCA2 is released under the terms of the GNU General Public License and is available at **http://www.mca2.org**.

References

1. Modular controller architecture version 2 http://www.mca2.org.
2. K.-U. Scholl, J. Albiez and B. Gassmann, MCA – An Expandable Modular Controller Architecture, in *3rd Real-Time Linux Workshop*, 2001.
3. RTAI - the RealTime Application Interface for Linux http://www.rtai.org.
4. Qt – The Cross-Platform C++ Development Framework http://www.trolltech.com/products/qt.
5. C. Birkenhofer, S. Studer, J.-M. Zöllner and R. Dillmann, Hybrid impedance control for multi-segmented inspection robot kairo-ii, in *International Conference on Informatics in Control, Automation and Robotics (ICINCO 2006)*, 2006.
6. B. Gaßmann, Modellbasierte, sensorgestützte navigation von laufmaschinen im gelände, PhD thesis, Fakultät für Informatik, Universität Karlsruhe (TH)2007.
7. J. Schröder, U. Müller and R. Dillmann, Smart Roadster Project: Setting up Drive-by-Wire or How to Remote-Control your Car, in *Proceedings of the 9th International Conference on Intelligent Autonomous Systems*, 2006.
8. T. Asfour, K. Regenstein, P. Azad, J. Schröder and R. Dillmann, ARMAR-III: A HUMANOID PLATFORM FOR PERCEPTION-ACTION INTEGRATION, in *HCRS, Second international workshop on Human-Centred Robotic Systems 2006*, October 2006.
9. A. Makarenko, A. Brooks and T. Kaupp, Orca: Components for robotics, in *2006 IEEE/RSJ International Conference on Intelligent Robots and Systems (IROS'06)*, Dec 2006.
10. H. Utz, S. Sablatnög, S. Enderle and G. Kraetzschmar, *IEEE Transactions on Robotics and Automation, Special Issue on Object-Oriented Distributed Control Architectures* **18**, 493(August 2002).
11. B. Gerkey, R. T. Vaughan and A. Howard, The player/stage project: Tools for multi-robot and distributed sensor systems, in *Proceedings of the 11th International Conference on Advanced Robotics (ICAR 2003)*, June 2003.
12. C. Côté, Y. Brosseau, D. Létourneau, C. Raïevsky and F. Michaud, *International Journal of Advanced Robotic Systems* **3**, 55(March 2006).

MOTION ESTIMATION AND SELF-LOCALIZATION BASED ON COMPUTER VISION AND ARTIFICIAL MARKER DEPOSITION

SAVAN CHHANIYARA, KASPAR ALTHOEFER, LAKMAL D SENEVIRATNE

Department of Mechanical Engineering, King's College London
Strand, London WC2R 2LS, UK

This paper presents a new approach of self-localization utilizing artificial markers deposited from the mobile robot during motion. The main idea of this system is to temporarily place artificial markers in the environment such as the surface the robot is maneuvering over. Once the markers are placed in the environment, they have zero speed, i.e. they represent the speed of the environment over which the robot is traversing. The speed of these markers with respect to the moving robot is then measured using a two dimensional (possibly three dimensional) sensor affixed to the robot. The proposed sensor system is composed of two main subsystems. The first subsystem generates the markers and places them on the surface. The second subsystem is a receiving element, which continuously acquires relative position signals from the markers placed on the surface. This new sensor concept is envisaged to be applied in the following areas, automotive sector, planetary exploration and underwater seabed exploration. This approach also can be extended easily for walking machines. Initial experiments employing a camera system as sensor have been conducted and results are presented.

1. Introduction

To accurately estimate the motion employing on-board sensors is still a very important research topic for both mobile & walking robot applications. At research centers world-wide, efforts aim at developing alternatives to conventional sensors i.e. wheel odometer, GPS/DGPS and inertial sensors. Over the past decade, several researchers have made significant advances in utilising various vision and image processing techniques for navigation, obstacle avoidance and control of ground vehicles [1]. Most of the recent approaches employ forward facing cameras and use image processing techniques for generating 2D and also 3D spaces from 2D images, detecting obstacles or extracting motion from the image sequence [2]–[7]. Different methods can be used to calculate the relative motion based on the translation and rotation of points in these 2D or 3D spaces [8]. Although significant improvements have been made in motion estimation from 3D points (including optical flow, Stereo Odometry), it has been noted that a considerable amount of processing time is spent on point matching and outlier removal [9]. In the field of walking machines, it is very important to reach target location accurately and autonomously. Problem of localization get worse in walking machines as slippage and small errors in position & orientation can lead to failures. Several sensors and techniques been used for navigation and localization, like laser sensors, inertial sensors and vision [16] & [18].

2. Recent Relevant Research

In [10-14], researchers introduced dropping of artificial markers in to the environment. Mostly, these approaches addressed the issue of robotic exploration in known/unknown environments, where no distance or metric information taken in to consideration. The approach taken in this paper and by few research groups is to use droppable markers in to the environment and localize the robot pose [13-14]. In [13], main focus was to develop evidence grid map and topological maps. The main assumptions in [13] for perceived position of artificial markers were based on robot's global position using wheel encoders. These can generate large error in accurate position estimations. In [14], author presented a system capable of navigation without prior beacon locations for autonomous underwater vehicle. Global translation/rotation errors were around 2 meters and a few degrees of heading error, which is too large for terrain based robot localization. In the field of walking machines, popular humanoid robot ASIMO [17] uses artificial landmarks arranged forehand on ground for navigation using camera looking down. These arrangements can constrain robot motion.

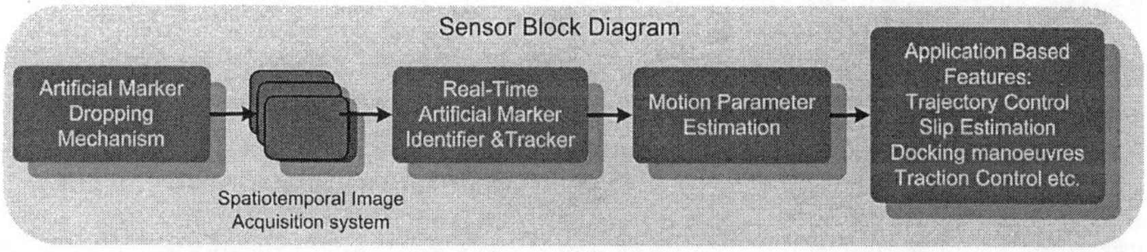

Figure 1 Sensor Block Diagram

3. SENSOR CONCEPT

3.1. *Sensor architecture*

This paper proposes a novel sensor concept to measure a vehicle/robot's trajectory and motion. It is envisaged that the sensor system is mounted on the robot. Sensing comprises a 2-stage process. In a first step, the system ejects small, but distinct markers which attach themselves to the environment the robot is operating in. In a second step, the on-board sensor which is designed to be particularly receptive to the strewn markers senses the distance and orientation of the robot with respect to these markers see Figure 1. This paper focuses on the research towards such an active sensor system for the installation on a ground vehicle. Real-time image processing algorithms are developed to locate the prominent marker features in the successively acquired camera images and allow the on-line calculation of relative vehicle speed, position and orientation with mm accuracy. Initial experiments have been conducted and results are presented.

3.2. *Artificial Marker selection and discharge*

Here in this paper, we propose to place specially designed objects 'Artificial Markers' in the working environment and subsequently and used as landmarks. In an attempt to simplify and speed up the process of locating the markers in an image, circular (rotationally invariant) markers have been chosen in this implementation. This proves to be more computationally efficient. One issue that needs consideration is that the environment is actively altered through the sensing process and depending on the application the appropriate type of marker and on board marker dropping mechanism needs to be selected. In most cases, a temporary (decaying) marker is preferable. The marker needs only to be active during the acquisition process by the on-board sensor. After this it may decay by itself or be actively removed by the vehicle.

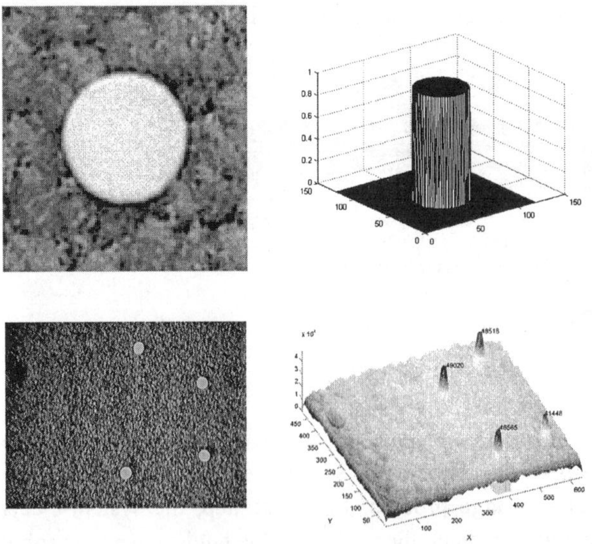

Figure 2 Image of marker used in this study (top left), simplified representation of marker image used as kernel (top right), input image for convolution and convoluted image plot indicating 4 clear picks (bottom right).

3.3. *Feature Identification and Tracking*

The core of the developed program to detect the markers in the image is a 2-D convolution algorithm [15]. Since it is known which markers are used for this process and at what distance from the camera they will appear, the dimension and size of the marker image can be computed and this knowledge can be exploited to create the appropriate convolution kernel (filter mask). The markers used in this study are circular paper shavings as shown in Figure 2.

Circular markers are rotational invariant. Thus, a single convolution kernel (which is effectively a simplified, graphical representation of a marker image) is sufficient to give a strong response at a marker's location, even if an image rotation has occurred. The convolution process outputs an image whose peaks represent their locations of the markers in a camera frame. Also, an intensity value is stored for each marker, which can be used for distinguishing between markers if different colour markers used in process.

The following standard convolution algorithm has been employed:

$$C(i,j) = \sum_{m=0}^{(Ma-1)} \sum_{n=0}^{(Na-1)} A(m,n) * B(i-m, j-n) \tag{1}$$

Where $0 \leq i < Ma + Mb - 1$ and $0 \leq j < Na + Nb - 1$

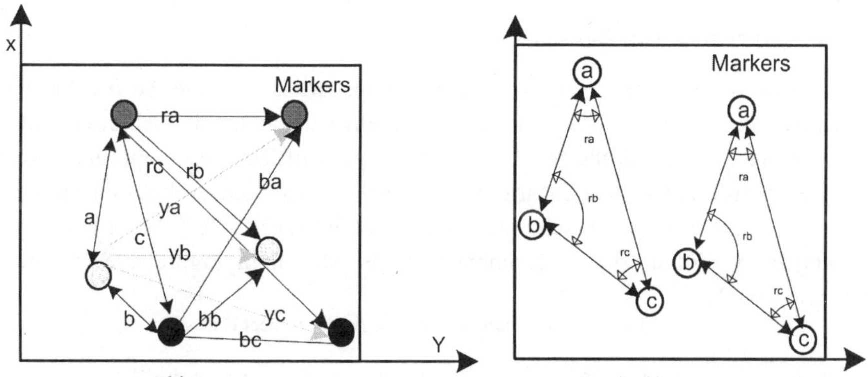

(a) (b)

Figure 3 (a) Distances between marker points in two subsequent frames. Distances between all marker point pairs are shown; matching distances between frames are highlighted. (b) Angle between marker points in one frame and angle between marker points in second frame are compared and this function is minimized to obtain the vehicle's change of orientation from one frame to another.

3.4. *Marker Grouping*

In a second stage of the proposed process, the markers which are identified in one frame need to be paired with the corresponding markers in the subsequent frame.

An algorithm has been devised that attempts to pair all marker points in the first frame with all marker points in the second frame and picks those pairs which would lead to a "reasonable" movement. The first feature is the distance between marker points in the first frame. Only those pairs of marker points in first frame can be considered which have a corresponding pair of markers in the second frame which are the same distance apart. The program calculates the distance between each pair of marker points in the first frame. This set of distances is then compared to the set of distances computed for the 2nd frame. Marker points pairs whose distance varies strongly between frames are excluded from the further process see Figure 3.

To increase the robustness of the approach, the rotations between the straight lines spanned by the selected pairs of marker points in the first frame and corresponding straight lines in the second frame are compared to the computed rotation of the previous frame. The marker pair with the least change in orientation is selected. Hence the algorithm eliminates marker pairs which would lead to an unlikely abrupt rotation from previous frames to the current frame transition.

4. Experiments & Results

The main objective of the experimental procedure is to prove the feasibility of the proposed approach. A number of tests were conducted in a lab environment in order to establish the feasibility and accuracy of the chosen approach. Tests were performed using a linear test rig as well as on a mobile robot Pioneer P3Dx for longer distances. Results are summarized in Table 1 & Table 2 below and trajectory plot of first experimental set are shown in Figure 4 & Figure 5 respectively.

Table1: Result Summery for Test rig Experiment Data set

d (mm) Encoder	d'(mm) Marker	% Error	Avg. Encoder velocity v (mm)	Avg.Marker Velocity v'(mm)	% Error
516.21	530.97	2.85	13.489	13.56	0.526
444.65	442.40	0.506	18.8	19.2	2.13
454.75	449.63	1.125	19.50	20.106	3.01
426.23	429.12	0.678	24.56	24.12	1.79
439.47	435.54	0.894	32.65	32.24	1.255
440.94	434.89	1.37	42.22	41.88	0.805

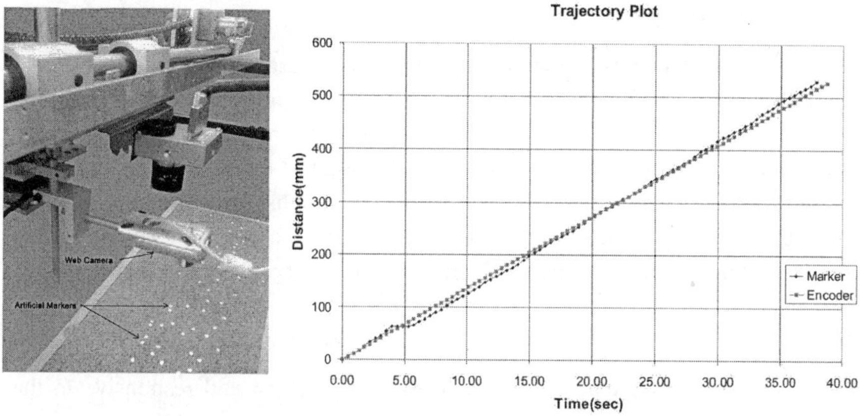

Figure 4 (a) Test Rig (b) Estimated trajectory and encoder readings on test rig experiment set1

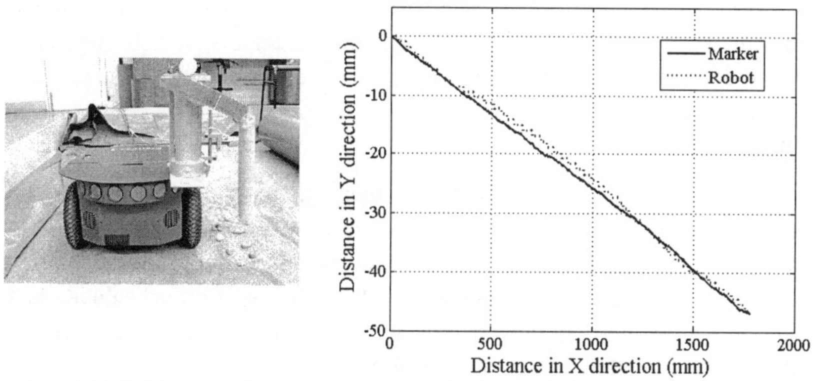

Figure 5 (a) Mobile robot with marker dropping mechanism (b) Estimated trajectory and encoder readings on mobile robot experiment set1

Table2: Result Summery for Robot Experiment Data set

d_m (mm) Marker Distance	d_r (mm) Robot Distance	% $Error_{me}$
1775	1776.1	-0.0619
1775.37	1786.1	-0.6008
1862.002	1897	-1.8449
1867.903	1825.367	2.33027
1880.8	1822.9	3.17626

4.1. *Discussion of results*

The position estimates computed by our system are shown in table I and compared with on board encoder of Linear Test rig. The results from the experiments show that high accuracy can be achieved at each step. The average error over the number of experiments is 1.237%. The error in average velocity of the carriage motion is 1.6% (Table 1). The results from the mobile robot experiments are also promising. The proposed system correctly computed the distance with only a relatively small error. The overall error is less then 3.2%.

5. Conclusion

The proposed sensor concept proved to be fairly robust and responsive to the markers even when experiencing changes in environment, lighting conditions and image quality. It is observed that even using a standard web camera with a low frame rate of 5-20 fps, good results were achieved. These initial experiments suggest that the proposed artificial marker disposition concept is feasible and robot position and speed can be accurately and robustly estimated.

6. Future work

Research is underway to improve affine motion estimation between frames and to develop a robust self-localization system. Emphasis is also given to minimizing computational time. Further efforts will aim at improving the tracking accuracy and analyzing the effect of height variation on velocity and position estimation. Further, the approach will be extended to walking machines for both indoor as well as outdoor applications.

7. References

1. DeSouza GN, Kak AC, " Vision for mobile robot navigation: A survey ", IEEE Transactions On Pattern Analysis And Machine Intelligence 24 (2): 237-267 Feb 2002
2. A. Davison, "Real-Time Simultaneous Localization and Mapping with a Single Camera," IEEE International Conference on Computer Vision, pp. 1403–1410, 2003.
3. Y. Takaoka, Y. Kida, S. Kagami, H. Mizoguchi and T. Kanade, "3D Map Building for a Humanoid Robot by using Visual Odometry," IEEE Int. Conf. on Systems, Man and Cybernetics, pp. 4444–4449, 2004.
4. E. Marchand, P. Bouthemy, F. Chaumette and V. Moreau, "Robust real-time visual tracking using a 2D-3D model-based approach," IEEE

International Conference on Computer Vision, pp. 262–268, September 1999.

5. B. Jung and G. S. Sukhatme, "Detecting Moving Objects using a Single Camera on a Mobile Robot in an Outdoor Environment," In the 8th Conference on Intelligent Autonomous Systems, pp. 980–987, March 2004.
6. M. Betke, E. Haritaoglu and L. Davis, "Multiple Vehicle Detection and Tracking in Hard Real Time," IEEE Symposium on Intelligent Vehicles, pp. 351–356, September 1996.
7. J. M. Ferryman, S. J. Maybank and A. D. Worrall, "Visual Surveillance for Moving Vehicles," In Journal of Computer Vision, pp. 187–197, 2000.
8. Niko Sünderhauf, Thomas Krause, Peter Protzel, "Bringing Robotics Closer to Students -A Threefold Approach," Proceedings of the 2006 IEEE Int. Conf. on Robotics and Automation, Orlando, Florida - May 2006.
9. David Fernandez and Andrew Price, " Visual Odometry for an Outdoor Mobile Robot", Proceedings of the 2004 IEEE Conference on Robotics, Automation and Mechatronics, Singapore, 1-3 December, 2004
10. Dudek G. ; Jenkin M. ; Millios E. ; Wilkes D., "Robotic exploration as graph construction", IEEE transactions on robotics and automation, Vol. 7, No. 6, pp. 895-865, 1991
11. S. Caselli, K. L. Doty, R. R. Harrison, F. Zanichelli, "Mobile Robot Navigation in Enclosed Large-Scale Space", Proceedings IECon94, Bologna, Italy, Sept. 5-9, 1994.
12. Ioannis M. Rekleitis, Vida Dujmovic, Gregory Dudek, "Efficient Topological Exploration.", IEEE International Conference on Robotics and Automation, Detroit, Michigan, USA, pp. 678-681, May 1999,
13. Maxim A. Batalin, Gaurav S. Sukhatme, "Efficient exploration without localization", IEEE International Conference on Robotics and Automation, Taipei, Taiwan, September, 2003.
14. B. Tovar, S.M. LaValle, R. Murrieta, "Locally-optimal navigation in multiply-connected environments without geometric maps", IEEE/RSJ International Conference on Intelligent Robots and Systems, (2003).
15. John C. Russ, "The Image Processing Handbook ",Published 2002, CRC Press, ISBN 084931142X
16. S. Thompson, S. Kagami, and K. Nishiwaki, "Localisation for autonomous humanoid navigation," in In Proc. IEEE-RAS International Conf. Humanoid Robots, pp. 13–19, 2006.
17. K. Hirai, M. Hirose, Y. Haikawa, and T. Takenaka, "The development of honda humanoid robot," in In Proc. IEEE International Conf.Robotics and Automation, pp. 1321–1326, 1998.
18. K. Sabe and M. Fukuchi and J. S. Gutmann and T. Ohashi and K. Kawamoto and T. Yoshigahara, "Obstacle avoidance and path planning for humanoid robots using stereo vision," in In Proc. IEEE International Conf. Robotics and Automation, 2004.

NEW STANDARDS FOR NEW ROBOTS

GURVINDER SINGH VIRK[†]
CLAWAR Ltd, UK and Massey University
Institute of Technology and Engineering, Wellington, New Zealand
Email: gsvirk@clawar.org, g.s.virk@massey.ac.nz

Robotics is a relatively new discipline, which came into existence when the first production line robots were commercialised by Unimates in 1962. Since then the industrial robot market has matured but has been restricted to manufacturing applications where humans and robots have been kept largely separated due to safety concerns. As part of this, industrial robot standardization has been developed to support the sector. In recent years new service robots and new service robot applications have been emerging; these do not fit into the "so-called" industrial environments and scenarios and there are no standards for these new situations. This is causing a bottleneck in the proper development of robotics and the International Standardization Organisation (ISO) has recently set up new initiatives to address these concerns. This paper focuses on these latest activities in robot standardization for the emerging areas of service robotics where close human-robot interactions are essential. In particular, the work of the new ISO TC184/SC2/PT2 Project team on Robots in personal care and the Advisory Group AG1 on Service robotics are presented.

1. Introduction

It is widely acknowledged that robots will soon begin to play a significant role in our everyday lives to assist us in performing a huge range of service and personal care tasks. As such, robotics is seen as likely to become a booming business area and major companies, such as Microsoft, have already started to invest in it (see Gates [1] in January 2007 issue of Scientific American, www.sciam.com). UNECE and IFR have forecast that the worldwide market for robots will be in excess of $66B by 2025, [2]. The main barrier to rapid growth is that there are no standards (especially no safety standards) that cover the new service robotic sectors and organisations fear litigation in the event of accidents occurring (see Virk [3]). In addition there are some ethical issues arising in the use of robots in the new sectors (see Virk et al [4]). Current robot standards only apply to "industrial robots" but these do not extend to the new types of robots

[†] Work partially supported by the UK's DTI.

being developed for the new applications and non-industrial environments because the design and operational aspects are quite different.

The current ISO 10218-1 safety standard [5] has been recently revised to update the safety requirement for industrial robots so that some of the concerns can be addressed. The changes includes the provision of a "collaborative" robot operational mode where robots and humans can work together. It states that the robot must provide a visual indication when it is in the "collaborative operation" mode and comply with one or more of the following:

1. The robot must STOP when a human is in the collaborative workspace
2. In hand-guiding mode, the robot shall operate at a reduced speed determined by a risk assessment, but <250mm/sec
3. An appropriate separation distance (ISO 13855 [6]) from the operator shall be maintained
4. The robot shall be designed to have a maximum dynamic or static powers of 80W or 150W respectively
5. The robot's maximum power and force shall be limited to the above constraints by a control system.

2. Missing standards for new robots

This main problem is that scenarios where robots are designed to have close human interaction are not included in the standardization process; in particular cases where a robot has a "human as its workpiece" are not covered. In order to identify the issues for these new scenarios and environments, it is useful to think of the different applications using working distances between the robot and humans as the basis for formulating the requirements. In this respect we can consider the various cases as "FAR", "CLOSE", "TOUCHING" and "INVASIVE". These classifications lead to the following assessment in terms of their state as regards robot standardization:

- FAR: This is the traditional class of industrial robots for which the current robot standards have been formulated. These standards rely on separating the robots and humans for safety
- CLOSE: The ISO 10218-1 (2006) [5] industrial robot safety standard covers these applications as stated above by monitoring and controlling the output of the robot when the robot is in collaborative mode. It is felt that some new applications are not adequately addressed; for example, when a mobile manipulator has to perform a manipulation task while its mobile base is in motion has not been included in the new safety standard.
- TOUCHING: Robots designed to perform tasks that involve touching a human (continuously or intermittently) are not currently covered in robot standardization.

- INVASIVE: Robots designed to perform tasks that involve performing invasive tasks in a human body are not currently covered in robot standardization even though a few robotic devices exist and are in full commercial operation with FDA approval (e.g, da Vinci robot, see www.intuitivesurgical.com).

To address this shortcoming, ISO (see www.iso.org) has been investigating the area to identify what is needed to develop the new standards for the new robots that are emerging. The Technical Committee that is responsible for robot standardization is TC184 under sub-committee SC2. An Advisory Group entitled Standards for mobile service robots was set up in June 2005 to investigate the overall standardization requirements and identify the priority area to focus the initial efforts. The results of the work have been reported in Virk [3] and the main recommendation was that the safety standard for the new type of robots should be developed as a matter of urgency. Following this work, and other recommendations, the need for change within ISO has been accepted and it has been decided to broaden the application base for robot standardization and several decisions have been made:

- The "industrial" focus of the ISO robot committees should be widened to cover the new non-industrial environments. Details of this widening include:
    - TC184 is proposing to change it current title, namely "**Industrial automation systems and integration**" to "**Automation systems and integration**" and its new scope is likely to be: *Standardization in the field of automation systems and the integration of those systems for design, sourcing, manufacturing and delivery, support, maintenance and disposal of products and their associated services. This standardization encompasses the application of multiple technologies, such as information systems, machines, equipment, robotics for fixed and mobile robots in industrial and non-industrial environments, automation & control software, multi-media capabilities, and multi-modal communications networks. Excluded are base standards in the following areas, 1) Electrical and electronic equipment as dealt with by IEC TC 44; 2) PLCs for general application as dealt with by IEC TC 65 and 3) multi-media capabilities as dealt with by IEC TC100.*
    - TC184/SC2 has recently changed its name to "**Robots and robotic devices**" from its old name of "**Robots for industrial environments**" and its new scope is: *Standardization in the field of automatically controlled, reprogrammable, manipulating robots and robotic devices, programmable in more than one axis and either fixed in place or mobile. Exclusions not covered are toys and military applications.*

- A new Project Team PT2 (Robots in personal care) has been set up to develop a new safety standard in the field of robots in personal care applications, including healthcare applications and excluding entertainment applications under the chairmanship of Prof GS Virk of CLAWAR Ltd, UK.
- A new Advisory Group AG1 (Service robots) has been set up to further explore needs for standardization in the field of service robots under the chairmanship of Prof S Moon of Sejong University, Korea.

3. PT2 on Robots in personal care

The new TC184/SC2 Project team PT2 on Robots in personal care (including healthcare) was set up in June 2006 to develop a safety standard for robots that are designed to work closely with humans. The committee comprises robot experts from France, Germany, Hungary, Japan, Korea, Sweden, UK and USA as well as liaisons with IEEE and IFR, OMG and other ISO Technical Committees (TCs). PT2's work is in its early stages but as it covers a new area of robot standardization it is important that there are wide ranging discussions, and views are sought from all relevant sectors on what would be acceptable to include in the new standard. From the early discussions, within the committee, it appears that personal care robots can be usefully categorized by some form of "degree of closeness" as discussed above to determine appropriate safety measures. It is clear that the "Far" category has been adequately addressed in the latest ISO 10218 revisions. Hence these standards could also apply to personal care robots designed to operate far from humans. Therefore, it is only necessary to focus on the "Close", "Contact" and "Invasive" categories. Furthermore, ISO 10218-1 covers some aspects of "Close" operating robots and the cases which are not adequately handled need to be identified in order that they can be included in the new standard. This is discussed further next.

3.1. *"Close" personal care robots not covered in ISO 10218*

Although work is continuing within PT2, some safety issues which are not covered within the current ISO 102180-1 standard or nor the planned 10218-2 been identified. Issues that are relevant here include the following:

- The operational modes where hands on (master/slave)/hands off operations, as well as operations involving unplanned contact need to be addressed.
- Proximity sensing measures need to be formulated. These need to cover human presence, relative positions, velocity and acceleration. In addition, human body parts sensing will also be needed.

- Human contacting measures will need to be formulated covering appropriate forces and pressures that can be imparted on the human.
- Measurement procedures for monitoring various human parameters by robots will also need to be formulated, eg EKG, EMG, EEG
- Additional hazards to those considered within ISO 10218 will also be needed. These include vibrations and jerking (for repetitive strain), levels of noise, loss of stability or overturning of robot (falling out or falling on human), unexpected or unallowed usage (over speeding or overloading).

3.2. *Classification of personal care robots for PT2*

Personal care (PC) robots cover an extremely wide are and to facilitate the development of a safety standards it has been decided to classify PC robots into different categories or robot types. Originally six categories were formulated, namely, 1) Minimally invasive surgical robots, 2) People carrier robots, 3) Treatment and examining robots, 4) Rehabilitation robots, 5) Physical assistance robots and 6) Mobile manipulators. However, after detailed discussions it was felt that there was considerable overlap in the groups and hence the categories have been reduced to the following four:

1. Surgery and medical robots: this group includes all PC robots (non-invasive and invasive) for monitoring and treatment of persons for achieving and maintaining good health and providing good healthcare functions.
2. Mobile manipulator robots including the provision of a "zero level service": This group covers PC robots that need to move in their environment while performing specific tasks of manipulation and gripping. The functionality that is included here is that the robot could actually perform the required tasks for persons rather than assisting the person to do it. This is referred to as a "zero level PC service".
3. Physical assistance robots (including rehabilitation): These robots assist a person to perform some required tasks. The assistance is aimed at providing supplementation capabilities only that bring the functionalities of a disabled/ injured/ elderly person to what can be performed by a "normal person". It is felt this could be referred to as providing [0-1] level PC service and the "0" is included as for mobile manipulator robots.
4. People carrier robots (allowing for augmentation capabilities): these robots are aimed at assisting the mobility of humans. The power requirements can be large as it may be necessary to provide functionalities that augment human abilities (eg., moving very fast, lifting heavy loads, etc) are included. It is felt that this is should focus on providing >1 level PC services but [0-1] level PC services could also be included; in this case clearly there would be some overlap with physical assistance roots. In view of the large power

outputs that could arise, it is likely that more stricter safety standards may be needed for these type of robots.

3.3. *Issues for PT2 robots*

Discussions within the project team have highlighted the issues that needed to be included in developing the safety standards for PT2 robots. These issues are important in defining the situations as well as provide the basis on which appropriate safety requirements could be formulated. The main issues identified thus far are the following:

1. **Definitions & terminology:** As the types and uses of robots is growing very fast it is important that the new situations and scenarios have to be defined so that the terminology is understood throughout the international community. In this respect, it is necessary to formulate a variety of definitions for the terms and terminology (robot types, tasks, usages, etc). AG1 is also likely to recommend that updating the vocabulary is the most important issue for service robots after safety. PT2 will focus on the terms and definitions directly to robots in personal care.
2. **Classification of PT2 robots on the needs of human:** It is clear that the PT2 robots presented above could be used by different people, namely, (a), physically impaired/disabled/ injured people, (b) elderly people and (c) (normal) able-bodied people to perform supplementary or augmentary tasks. In addition, the persons could be using the robots in a professional capacity (this could require some level of competence via formal training of the professional user as part of his employment) or for personal use (no formal technical competence can be assumed unless governments decide to introduce legislation into this area and start asking citizens to train and pass a "robot operating" test to obtain a robot licence along the lines of a car driving licence).
3. **Classification on closeness level:** The "closeness" categories appropriate for PT2 could be (a) far, (b) close (non-contacting), (c) contacting (intermittent), (d) contacting (continuous) and (e) invasive. Having such a fine resolution is useful but the key difference in all these groups is felt to whether the application is non-invasive or invasive if safety considerations are the primary concern. Hence all the categories could be usefully grouped into non-invasive or invasive cases.
4. **Classification on the operational environment:** The different environments identified include (a) domestic, (b) industrial/manufacturing, (c) health and care centres, (d) workplace (office), and (e) public places (shops, leisure centres, parks, etc.).

It is clear that in formulating the safety standard it will be necessary to consider all the above issues and adopt a risk analysis approach. The types of hazard and associated protective measures that need to be adopted exist and are covered in the international standard ISO 14121-1999 [8]. These include mechanical hazards (shape, relative location, mass and stability, mass and velocity, etc), crushing hazard, shearing hazard, cutting or severing hazard, entanglement hazard, drawing-in or trapping hazard, impact hazard, stabbing hazard, friction or abrasion hazard, electrical hazards, thermal hazards, noise hazards, vibration hazards, radiation hazards, hazards due to materials and substances, hazards due to non-ergonomic designs, combinations of hazards, etc. Other general guidelines for safety and risk assessment issues relevant to PT2 include IEC GUIDE 51 [9], IEC GUIDE 63 [10], and IEC GUIDE 73 [11].

Work to develop the safety standard for personal care robots is progressing well as reported above and the structure of the new safety standard has also been discussed. It has been agreed that the new safety standard could be in two parts along the following lines:

- Part 1: Non-invasive personal care robots (including healthcare)
- Part 2: Invasive personal care robots (including healthcare).

The initial work will focus on Part 1 and a roadmap for developing it has been formulated. The details are as follows:

By June 2007 a report on the structure of the Part 1 standard will be presented to the SC2 plenary meeting. This should contain the main section headings, sub-sections and possibly some draft contents based on the work carried out by PT2. A new work item is expected to be proposed at the 2008 plenary meeting of ISOTC184/SC2 where a Committee Draft of the Part 1 safety standard should be available. A Draft standard should be ready for distribution in 2009, followed by a Final Draft by the 2010 SC2 plenary meeting. The new standard is then expected to be formally offered for voting and acceptance by the ISO Members in 2011. The main sections of the Part 1 is likely to be along the following lines:

1. Scope
2. Normative References
3. Terms and Definitions
4. Hazard Identification and Risk Assessment
5. Design Requirement and protective measures
6. Information for use
7. Annex A (Normative: list of significant hazards).

In addition to the above work needed for the Part 1 standard, activities will also be initiated for the Part 2 standard (Invasive personal care robots). For this, it is vital to ensure good and extensive liaisons are put in place with the relevant groups involved (especially medical staff and medical equipment manufacturers). The intention is to organise an event between specialised medical staff and PT2 members in 2008/09 to highlight the main issues that need to be included. By the 2009 SC2 plenary committee meeting, it is expected that a report containing the structure of the Part 2 standard's sections, sub-sections and draft content will be prepared. A new work item is then expected to be made to the 2010 SC2 plenary meeting where a committee draft document should be available. A similar time-scale as for the Part 1 is then expected to follow.

Each part will have its own risk assessment based approach which will utilise IEC Guide 51 [9] and IEC Guide 63 [10] and it will be essential that correct links and liaisons are put in place to ensure that each part is developed in as comprehensive a manner as possible.

4. AG1 on Service robotics

The task of the Advisory Group (AG1) is to explore the needs for standardisation in the field of service robots so that important areas not already addressed may be identified and put forward as new work items for ISO. AG1 is working in parallel with PT2 and meetings have been arranged in co-operation so that experts can be involved in both committees. The work of AG1 has been organised into the following four Work Groups:

1. **WG1 (Vocabulary):** This WG is likely to form the new work item as it is felt to be the most important after safety. The service robotics area is requiring new definitions to be made in light of these developments. For example the current ISO 8373 [7] definition of a robot applies only to industrial robots and this is not wholly appropriate. This states that an industrial robot *is an automatically controlled, reprogrammable multipurpose manipulator, programmable in three or more axes, which may be either fixed in place or mobile for use in industrial automation applications*.

   In fact the definition is not adequate when looking at the new scope of SD2, where robots are seen as "*automatically controlled, reprogrammable, manipulating robots and robotic devices, programmable in more than one axis and either fixed in place or mobile*". The current trend is to define a robot as a system that has "motion" and "intelligence" rather than being "reprogrammable", "multi-purpose", etc as defined in ISO 8373. In view of this AG1 has started to formulate some new definitions that are required. For example, several definitions of a *robot* have been produced. These are as follows:

   1. Reprogrammable machine with a degree of autonomy, programmable in more than one axis, either fixed in place or mobile, and able to perform task(s). (Note: Degree of autonomy can be from manual (including tele-operation) to fully autonomous).

   2. A machine that perceives and manipulates its own state or the state of its environment under automatic control.
   3. Reprogrammable machine with a degree of autonomy which can purposefully manipulate its own state and the state of its environment and which is programmable in more than one axis, either fixed in place or mobile.
   4. A reprogrammable machine (or system) with a degree of autonomy (or intelligence) that perceives and manipulates its own state or the state of its environment under automatic control.

   In addition, two definitions of a *service robot* have also been formulated; these are as follows:

   1. A robot which provides "usefulness" for the well-being of humans, society and equipment, excluding manufacturing operations.
   2. A robot which provides "usefulness" for the well-being of humans, society and equipment, excluding direct manufacturing operations.

2. **WG2 (Performance):** The WG is aimed at formulating standards to assess the performance of service robots. It will consider how to objectively assess the quality of the services that are provided. For example the quality of the cleaning that a vacuum robot provides is clearly needed. The WG also includes maintenance issues.
3. **WG3 (Software):** The WG includes all software issues including architecture, communication, middleware, HRI and robot-environment communications.
4. **WG4 (Other topics):** This WG includes areas such as technical tasks and modularity.

5. Conclusions

The paper considers the area of standardization for new types of service robots that are emerging and the current standardization activities that are on-going under ISO. The paper focuses on PT2's work on formulating the safety standard for robots in personal care (including healthcare) and the latest results of the project team have been presented. This standardization work is vital to remove the barriers that exist for widening the application base of robots and robotic systems to new non-industrial environments.

Acknowledgments

The author wishes to acknowledge the valuable contributions to the robot standardization work reported here by the international experts on the ISO

Project team PT2 on Robots in personal care as well as the Advisory Group AG1 on Service robotics.

References

1. B. Gates, Dawn of the age of robots, Scientific American (2007).
2. World Robotics 2006, United Nations Economic Commission for Europe, UNECE and IFR (2006).
3. G.S. Virk, Standards for mobile service robots, Proceedings 9th International Conference on Climbing and Walking Robotics (CLAWAR 2006), Brussels, Belgium, 12-14 September (2006).
4. G.S. Virk, C. Sjöström, M. Engström and W. Trinius, Standards and/or ethics for service robots?, Proceedings Ethics of human interaction with robotic, bionic, and AI systems: Concepts and policies, Naples, 17-18 October (2006).
5. ISO 10218-1:2006, Robots for industrial environments – safety requirements – Part 1: Robot, 34 pages (2006).
6. ISO 13855:2002, Safety of machinery – Positioning of protective equipment with respect to the approach speeds of parts of the human body (2002).
7. ISO 8373:1996, Manipulating industrial robots — Vocabulary (1996).
8. ISO 14121-1999, Safety of machinery — Principles of risk assessment (1999).
9. IEC GUIDE 51, Safety aspects — Guidelines for their inclusion in standards (1999).
10. IEC GUIDE 63, Guide to the development and inclusion of safety aspects in International Standards for medical devices (1999).
11. IEC GUIDE 73, Risk management — Vocabulary — Guidelines for use in standards (2002).

PARALLEL PARTICLE SWARM OPTIMIZATION FOR NETWORKED CLAWAR SYSTEM COOPERATION

FABIO P. BONSIGNORIO

RTD, Heron Robots s.r.l., V.R.Ceccardi, 1/18
Genova, I-16121, Italy

This paper shows an example of cooperation between CLAWAR robots based on particle swarm optimization. The approach is based on a parallel version of multiobjective pso and is fit to be implemented on a parallel computing backbone.

The benefits and trade-offs of this approach for managing swarms of CLAWAR systems are compared and discussed against alternative solutions.

1. Introduction

This paper exposes a model for the cooperation within networks of CLAWAR autonomous agents, based on particle swarm optimization.
At the same time Reynolds' algorithms are used for flocking.
Particle swarm optimization (PSO) is a population based stochastic optimization technique first proposed by Eberhart and Kennedy in 1995 [1,2], inspired by the modeling of bird flocking by Reynolds [19].
This method shows some similarities with evolutionary computation techniques like genetic algorithms . The system is initialized with a population of random solutions and searches for optima by updating generations. PSO has no, yet evolution operators like crossover and mutation.
In PSO, the potential solutions, called particles, fly through the problem space by following the current optimum particles.
Each particle keeps track of its coordinates in the problem space which are associated with the best solution (fitness) it has achieved so far ('pbest'), the coordinates in problem space are called 'lbest'. The best value of all the solution space is called 'gbest'.

The particle swarm optimization algorithm at each time step, changes the velocity of (accelerating) each particle toward its pbest and lbest locations by means of a random procedure.
Particle swarm optimization is a simple algorithm that has been proposed for optimizing a wide range of functions.

Nethertheless it was originally proposed to model social behaviour of birds and other animals.
In this paper we apply a particle swarm optimization method to manage the cooperation between a group of networked CLAWAR agents with very reduced computing capabilities, which in turn 'flock' according to a similar algorithm.
From a certain standpoint we apply to the managements of a flock of autonomous agents the behavior rules supposedly working in natural flocks of birds following the original example from Reynolds and at a different level we use PSO to optimize the behavior parameters of the single CLAWAR agent.
PSO are comparatively simple algorithms, although they tend to be computing expensive.
To deal with this fact we have choosen a parallel implementation which can be used on a cluster of computers or a computing grid.

The usage of a quite similar algorithmic approach for the management of the agent collective behavior and for the online optimization of the behavior of the single agent allows a simplification of the computing backbone architecture, increasing the robustness of the overall architecture.

2. Flocking algorithm

The flocking algorithm is basicly a parallel version of the one proposed for the flocking of birds.
Reynolds proposed a simple but effective model to model the behavior of swarms, such as a flock of birds or school of fish, which are able to move in a synchronized manner without any central planning or control.
The simulations of bird-like objects flying around, called 'boids', from bird-oids, based on his model looks very realistic and recently has been applied to swarm of robots [20].
His model, which can considered a multi agent example of 'behavior based' control strategy, consisted of four behaviors applied to all the 'boids':

1. separation: to make the boids avoid colliding with one another
2. alignment: to make a boid steer in the common direction of his neighbors
3. cohesion: to make a boid steer towards the common position of his neighbors
4. avoidance: to make a boid steer away from other obstacles [Rabin 2002].

In our case the Reynolds' algorithm is applied to a flock of surveillance robot with a fitness function related to threshold emission in the infrared and sound simulating an 'intruder', the generalization of the approach is quite easy.

The standard PSO algorithm we use for flocking, implements the following procedure [16]:

1. Set d equal to the dimension of the fitness function.
2. Create n particles, 0 to n-1, each with a position vector, x^i, of dimension d.
3. Assign random position values to each particle
4. For each particle, i :

    1. Evaluate fitness (pass position vector values into fitness function and assign return value as fitness)

    2. If fitness better than personal best fitness, then assign current position as personal best position (p^i)

    3. If fitness better than global best fitness, then assign current position as global best position ($p^g{}_k$)

5. For each particle, i:

    1. Update velocity, v^i:

$$v_{k+1}^{i} = wv_{k}^{i} + c_1 r_1 \frac{\left(p^{i} - x_{k}^{i}\right)}{\Delta t} + c_2 r_2 \frac{\left(p_{k}^{g} - x_{k}^{i}\right)}{\Delta t} \tag{1}$$

    2. Update position, x^i:

$$x_{k+1}^{i} = x_{k}^{i} + v_{k+1}^{i} \Delta t \tag{2}$$

6. If criteria met, end simulation, else repeat from step 4.

The criteria for ending the simulation is usually based on whether the global best fitness is sufficient, or whether the simulation has run for the maximum amount of epochs (iterations). It is also important to note that in step 3, the personal best fitness of each particle, and the global best fitness need to be initialized to very poor values. At this stage, each particle's velocity is also usually set to zero, although they may be initialized to random values.

The number of particles, n, inertia weight, w, and the local and global component variables, l and g, are all system parameters of the PSO algorithm.

The local and global component variables, l and g, control the influence of the personal best and global best positions respectively. They are defined as $l = c_1 r_1$ and $g = c_2 r_2$, where r_1 and r_2 are random values between 0 and 1, c_1 and c_2 are acceleration constants that are usually set to values close to .
It has been demonstrated that we convergence if:

$$w > \frac{1}{2}(c_1 + c_2) - 1 \tag{3}$$

The standard algorithm as described above is also called the global best (gbest) algorithm, because the globally best particle is followed by all the particles. The local best (lbest) algorithm, however, follows more closely the original model created by Reynolds. It differs from the gbest algorithm in that the whole swarm is divided into a number of neighborhoods, where each particle is only aware of the particles in its neighborhood. There is no global best particle, but rather a local best particle for each neighborhoods. Each particle is therefore only influenced by its neighbors, and not the whole swarm. The gbest algorithm can be seen as the lbest algorithm with one neighborhood consisting of all the particles.
The size of neighborhoods plays an important role in the lbest algorithm. It has been shown that smaller neighborhoods result in slower convergence, but generally lead to better results, since a larger part of the search space is explored and particles are less likely to be trapped in local optima [16].

3. PSO behavior optimization

When the flock algorithms is used as an optimization tool there are several possible improving including dealing with constrained optimization problems, introducing a craziness operator to increase the likelihood of escaping from a local minima, and dynamically changing the inertia value w.
The PSO (Particle Swarm Optimization) is a stochastic population based process depending on the memory of each agent as well as the knowledge gained by the population as a whole.
The population is called the "swarm' while the single agent is called a 'particle'
The craziness operator adds additional randomness to the swarm improving algorithm's exploitation and reduce the likelihood of premature convergence and is similar to the mutation operator of a genetic algorithm. Typically a small subset of particles are selected at each design iteration, and the position and/or velocity vectors are randomly modified.
The inertia parameter w controls the exploration properties of the algorithm. It is a common practice to start with a larger inertia value (a more global search) that is dynamically reduced towards the end of the optimization (a more local search).
PSO is a comparatively recent example of non-gradient based, probabilistic search algorithms. Other examples are evolutionary algorithms and simulated annealing algorithms.
This class of optimization algorithms have several appealing characteristics.
They are generally easy to implement, can be implemented on a large numbers of parallel processors, are efficient for finding global or near global solutions, they

don't need the computation of derivatives.
The main disadvantage is the weak local search capabilities and the computational cost.
Thanks to their similarity to evolutionary algorithms, like for example GA (Genetic Algorihms) they have been applied to design optimization [13,14,15]

As told, we base our research on a multi objective particle swarm optimization algorithm, which is comparatively easy to implement in the form of a parallel algorithm, in order to make possible to harness the computing capability of a parallel backbone computing facility.
The multi-objective online optimization of an exapode gait control parameters is used is perfron. The motivation for using PSO in the present design problem is the presence of discrete choice variables and numerical noise coming from agent sensors, for this kind of application a gradient based optimizer would be comparatively not very effective.
In order to apply the PSO strategy for solving multi-objective optimization problems, we aim to find a set of different solutions (the so-called Pareto optimal set).
A new particle replaces its pbest particle usually when this particle is dominated or if both are incomparable (i.e., they are both nondominated with respect to each other).
The PSO algorithm is well fit for a coarse-grained parallel implementation on a parallel or distributed computing network and, in perspective e grid implementation [21] or on dedicated microprocessors. For each time step (agent control set iteration), all particles (control set points) are independent of each other and can be computed in parallel.
The sirect PSO parallel implementation is to simply evaluate the design points within control set point iteration in parallel, without changing the overall logic of the algorithm itself. In this implementation, all particles within a design iteration are sent to the parallel computing environment, and the algorithm waits for all the analyses to complete before moving to the next design iteration. This implementation is referred to as a synchronous implementation. This implementation is not very efficient as it is difficult to obtain that all processors finish their iteration before the beginning of the next. Moreover we think that an asyncronous implementation can be more suitable to be didteiubet to an on demand computing backbone (a 'grid').
The needs of different agents on the environment, by they way, can require optimization of control set point leading to vvery different computation loads.
In order to implement an asynchronous parallel PSO algorithm is necessary to separate the update actions associated with each point and those associated with the particle swarm as a whole.
Two operators have to be applied when the iteration is finished for all the swarm particles: the mutation (crazyness) and the inertia operator.

4. Application scenario

The method have been tested in simulation in Matlab, for algorithm tuning, and Webots.

Each agent representing a CLAWAR was programmed to evaluate its present pbest, lbest values, which in our case have a clear interpretation in terms of physical position of a particle representing the CLAWAR (modelled by a simple exapode) estimated center of mass position in the 'real' 3d space.

In Webots simulation multibody kinematical additional variable evolution are simulated .

The considered , semi-structured, scenario is the identification and surrounding of an intruder in a secured area, the neighborings of a building in an exhibition area. The robots flock around the intruder, modeled as an infrared and sound source (see fig.1)

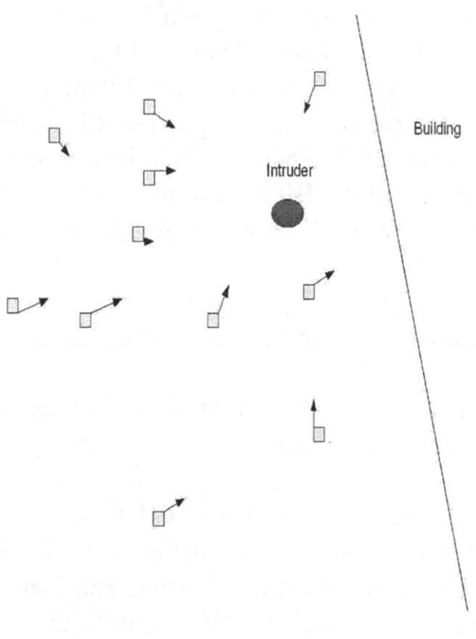

Figure 1. Example of agents flocking around an intruder.

5. Further Research

We are considering to apply this system in a real world scenario using wheeled low cost robot appliances and connecting them to a parallel computing backbone in a first stage and furtherly of newly designed low cost robotic appliances with legged locomotion.

It could also be interesting to blend the PSO procedure appropiately with some kind of evolutionary algorithms.

6. Conclusions

We have designed a control and coordination system for networked CLAWAR system for security applications based on a parallel multi objective particle swarm optimization algorithm and a widely used , in research and even in video game application, flocking model.
This approach seems very suitable to applications integrating networked robotic appliances with a parallel computing backbone.

References

1.J. Kennedy, R. C. Eberhart, and Y. Shi,Swarm Intelligence, (2001).
2.J. Kennedy, R. C. Eberhart, "New optimizer using particle swarm theory," in Proceedings of the 1995 6th International Symposium on Micro Machine and Human Science, pp. 39– 43, (1995).
3. J. Kennedy and R. C. Eberhart, "Particle swarm optimization," in Proceedings of the 1995 IEEE International Conference on Neural Networks, vol. 4, (Perth, Australia, IEEE Service Center, Piscataway, NJ), pp. 1942–1948, (1995).
4. P. J. Angeline, "Evolutionary optimization versus particle swarm optimization: Philosophy and performance differences," in Evolutionary Programming VII, 1998.
5. R. C. Eberhart and Y. Shi, "Comparison between genetic algorithms and particle swarm optimization," in Evolutionary Programming VII, (1998).
7. M. Løvbjerg, T. K. Rasmussen, and T. Krink, "Hybrid particle swarm optimizer with breeding and subpopulations," in GECCO 1998: Genetic and Evolutionary Computation Conference, San Francisco, CA , (1998).
8. B. Al-kazemi, C. K. Mohan, "Multi-phase discrete particle swarm optimization," in FEA 2000: Fourth International Workshop on Frontiers in Evolutionary Algorithms, (2000).
9. R. G. Reynolds and C. J. Chung, "The importance of function complexity in regulating the amount of information required to guide self-adaptation in cultural algorithm," in Proceedings of the Seventh International Conference on Genetic Algorithms (ICGA97) (T. B¨ack, ed.), San Francisco, CA, Morgan Kaufmann, (1997).
10. R. G. Reynolds, "Cultural algorithms: Theory and applications," in New Ideas in Optimization (D. Corne, M. Dorigo, and F. Glover, eds.),

pp. 367–377, London: McGraw-Hill, (1999).
11.Z. Michalewicz, D. Dasgupta, Evolutionary Algorithms in Engineering Applications, Springer Verlag, (1997).
12. G.L. Nemhauser, G. L., L.A. Wolsey, Integer and Combinatorial Optimization, Chapter 3, John Wiley &Sons, (1988).
13. Fourie, P. C. and Groenwold, A. A., The Particle Swarm Optimization Algorithm in Size and Shape Optimization, Structural and Multidisciplinary Optimization, 23, pp. 259–267, (2002)
14. Fourie, P. C. and Groenwold, A. A., Particle Swarms in Topology Optimization, In Proceedings of the Fourth World Congress of Structural and Multidisciplinary Optimization, Dalian, China, 2001.
15. G. Venter, G., B. Watson, Efficient Optimization Algorithms for Parallel Applications, Proceedings of the 8th AIAA/USAF/NASA/ISSMO Symposium on Multidisciplinary Analysis and Optimization, AIAA-2000-4819, Long Beach, CA, (2000)
16. A.P. Engelbrecht, Computational Intelligence: An Introduction, Wiley, (2002).
17. S. Rabin, AI Game Programming Wisdom, Charles River Media, 2002.
18. F. van den Bergh, An Analysis of Particle Swarm Optimizers, PhD Thesis, Department of Computer Science, University of Pretoria, (2002).
19. C.W.Reynolds, Flocks, Herds, and Schools: A Distributed Behavioral Model,Computer Graphics, 21(4), pages 25-34 ,(1987).
20. J.M. Hereford, "A Distributed Particle Swarm Optimization Algorithm for Swarm Robotic Applications", IEEE Congress on Evolutionary Computation, CEC 2006. Volume , Issue , 16-21, Page(s): 1678 – 1685, (2006).
21. F.P. Bonsignorio, "A grid based distributed multiagent enabling system for intelligent autonomous robot swarms", ISR 2005, http://www.isr2005.com/,Tokyo, (2005).

Performance metrices for improving human–robot interaction

YIANNIS GATSOULIS*†

School of Mechanical Engineering, University of Leeds,
Leeds, West Yorkshire, LS2 9JT, UK
** E-mail: menig@leeds.ac.uk*

GURVINDER SINGH VIRK

School of Engineering and Technology, Massey University
63 Wallace Street, Wellington, New Zealand

The increasing use of robots working together with humans has stressed the importance of research in human-robot interaction where the robot is seen as a team member. This paper investigates the human-centric characteristics of situational awareness, telepresence and workload and how these affect the overall performance in a task which a human is teleoperating a simulated robot system in the critical domain of urban search and rescue. We present an overview of some new methods being developed by the authors together with performance metrics that can be used to assess the effectiveness of the human-robot interactions proposed. The experimental results, from a set of users consisting of professional paramedic rescuers, has shown that situation awareness and telepresence affect to some extent performance. Bivariate regression models were also shown to better predict performance than simply using a mean and error model. Workload, which was also hypothesised to affect performance, was shown to have no effect in the end.

Keywords: Human–robot interaction, teleoperation, performance, situation awareness, telepresence, workload.

1. Introduction

The increasing use of robots working together with humans in critical application domains, such as search and rescue, scientific exploration, etc., has led to major initiatives to research human-robot interactions (HRI) where the robots are considered more as members of a team. Moreover, it has been widely proposed that a human-centric perspective should be adopted in the

†The author has been supported by an EPSRC scholarship.

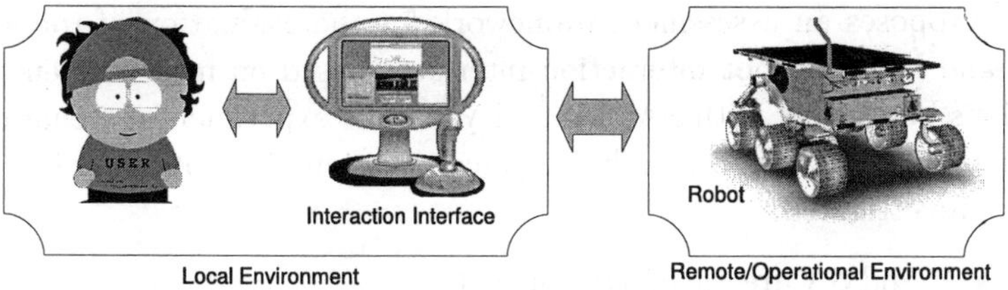

Fig. 1: Robot teleoperation

design of robotic systems particularly in the research and development of HRI interfaces.[1]

Although levels of robot autonomy are increasing in research machines, tele-operation (Figure 1) is still by far the most common form of robots used in practice mainly due to reliability and safety reasons. In such situations, the human operator needs to be provided with high quality information about the robot and its environment for effective control to be carried out. In order to utilise a human-centric approach certain human characteristics that affect task performance (P) should be taken into account, these being situation awareness (SA), telepresence (TP) and workload (WL). Good SA can lead to good decision making and hence better performance. However, given the wide range of definitions and theories,[2] more research is needed to develop a clear consensus on what are the main technical issues and how these can be solved. TP and WL are two other characteristics that have been speculated to affect task performance, and despite their long research their relations to performance are still vague.[3,4] As shown in Figure 1 the main elements are the user, the human-robot interface, the robot and the local and remote environment, and they are determining SA, TP and WL. From all these, the system designer has mainly control over the robot and the human-robot interfaces; to some less extent over the local environment; and very little over the remote environment and the users.

The very few studies on SA measurement in co-operative intelligent robots that exist in the literature have focused on specific features of the robot assisted system, such as the user interaction interfaces,[5] the use of a camera,[6] autonomous features, e.g. mapping and localisation,[7] robot health awareness,[8] etc. A more comprehensive study is that of J. Riley et al[9,10] which looks at the overall relations between P, SA, TP and WL, but with emphasis to TP.

On the same lines of this former study but with emphasis to P, this paper proposes an assessment framework for the evaluation of robot systems and human-robot interaction interfaces based on measurements and relations between P with SA, TP and WL. The experimental scenario is a simulated robotic urban search and rescue mission. The following hypotheses are investigated:

- P and SA are *positively* correlated.
- P and TP are *positively* correlated.
- P and WL are *negatively* correlated.

It could be argued that these hypotheses are not new. However, there have been very few empirical and formal studies including them, and they suggest further investigation.[9] Understanding the magnitude of their effect to performance, as well as developing a prediction model is vital for better human-centric design and development of robotic systems.

2. Measurement Methods

The response or dependent variable is task performance (P), while the explanatory variables or predictors are situation awareness (SA), telepresence (TP) and workload (WL).

The experimental task is a computer simulation of a robotic urban search and rescue scenario. Sixteen subjects were required to teleoperate a robot and search a floor of a building for any possible casualties, while maintaining high levels of SA and protecting the robot from any collisions. They were free to choose any course of action. At the end of the mission or before the batteries were depleted they were requested to bring the robot into the exit point. At the end of the trial they were requested to answer to some questions regarding their SA, TP and WL. All subjects are professional paramedic rescuers in the Greek health system. Although the majority had low computer experience, prior training was provided with the system to ensure that all subjects had a clear understanding of the mission and the system and were feeling comfortable with using it.

Bivariate correlation analysis (one-sided Pearson's ρ) and bivariate linear regression model analysis of the forementioned hypotheses were calculated with the aid of R, a statistical analysis and computing software.[11] The linear model has the form:

$$y = mx + b \tag{1}$$

Altough statistical significance is measured at the 95% level, the 90% level is also reported, and hence a statistical insignificant result is reported on this latter one.

2.1. *Performance measurement*

Performance is measured according to Equation 2[a]:

$$P \propto \frac{AC + RS + EP}{t} \tag{2}$$

where,
P: performance score, $P \in \mathbb{R}\{1.00, \ldots, 100.00\}$
AC: percentage of the area covered
RS: a reward for bringing the robot safely back
EP: exceptional performance, a reward for extensive use of robot, awarded if the percentage of the area covered is more than a threshold value
t: mission time

In contrast to the RobocupRescue competition where task performance measurement is mainly based upon the number of victims and their difficult in being located,[12] we preferred the area covered over time as the main variable affecting performance as a more objective measure. The reason is that based on the assumption that the more area searched, the more likely it is to locate any existing casualties, or in the case than none exists time can be saved as the area can be marked as "clear". Also number of casualties located and the level of difficulty in finding them is an inappropriate measurement in this case as the casualties are placed in clearly visible open spaces without being hidden behind any obstacles. As a result the casualties are relatively easy to identify once seen in the camera. Furthermore, they do not provide any additional visual (e.g. thermal signature) or audio cues. This actually means that locating a victim is merely a function of visiting the area close to the casualty. For this reason, the area covered is a more objective measurement of performance.

2.2. *Situation Awareness Measurement*

Situation awareness has mainly been investigated in the area of avionics and air traffic management. However, the majority of these methods are either too specific or too simple to be used in the domain of robotics. For this reason we developed a post-trial self-rating questionnaire to address the needs of SA in a robot tele-operation scenario. The questions have been

[a]The actual equation is a bit more complex involving some normalising constants, and is beyond the scope of the paper.

designed to cover the different dimensions of the user's SA, such as localisation awareness, time awareness, coverage awareness, spatial awareness, etc.[13,14] along the three levels of SA as proposed by M. Endsley,[15] namely having a clear perception of the available data, comprehension of the data to meaningful information and being able to make good predictions of future states. Example questions of the measurement methods are shown in Table 1. The user is asked to rate herself on a 6-point scale.

Table 1: Example questions of the situation awareness measurement method

–How easy was it to avoid the hazards (e.g. narrow doors)?
–How easy was it to keep track of the status of the various modules of the robot
–How easy was it to keep track of the time?
–How easy was it to predict what would happen next?

2.3. *Telepresence Measurement*

Telepresence is measured after the trial through a self-rating questionnaire developed by Witmer and Singer,[16] which was modified to meet the requirements of our search and rescue experimental task. Example questions are shown in Table 2. The user is asked to rate herself on each question on a 7-point scale.

Table 2: Example questions of the Witmer–Singer presence questionnaire

–How much were you able to control the various events?
–How responsive was the environment to actions that you initiated?
–How aware were you of the events occurring in the real world around you?
–How often do you player video/computer games?

2.4. *Workload Measurement*

Workload is measured after the trial through NASA-TLX,[17] which is a multi-dimensional weighted rating method. The different dimensions are mental demand, physical demand, time demand, performance, effort, stress and frustration. From those the physical demand dimension was omitted as it is inappropriate for this computer simulation task.

3. Results

The Lilliefors normality test scores were for P, $D(16) = 0.15, p > .05$; for SA, $D(16) = 0.12, p > .05$; for TP, $D(16) = 0.12, p > .05$; and for workload WL, $D(16) = 0.09, p > .05$; indicating that all samples are similar to a normal distribution, and hence allowing parametric statistic tests to be used. Descriptive statistics of the distribution for each dataset of each variable are shown in Table 3.

Table 3: Descriptive statistics of the distribution of the dataset of each variable

variable	mean	σ	min	max	valid-range
Performance	51.25	17.82	19.44	89.01	$[1-100]$
Situation Awareness	4.24	0.83	2.82	6.00	$[1-6]$
Telepresence	4.44	0.37	3.78	5.13	$[1-7]$
Workload	65.95	13.53	44.00	87.00	$[1-100]$

Scatterplots of P/SA, P/TP, and P/WL are shown in Figure 2. The term "outlier" is used in the context in which a case seems to largely deviate from the cluster of the rest of the values. Correlation and linear regression model analysis were carried out for both the complete set of values ($N = 16$) as well as when omitting these outliers. In each case the sample size is reported to signify whether the results are for the complete set or the filtered one.

The results of the correlation analysis for the complete set show that there is a significant positive correlation between P and SA ($N = 16$, $\rho = .57$, $p < .05$). This positive correlation is even stronger when using the filtered dataset ($N = 15$, $\rho = .71$, $p < .01$). A significant positive correlation also appears between P and TP ($N = 16$, $\rho = .47$, $p < .05$). For the complete set it seems that there is no relationship between P and WL ($N = 16$, $\rho = .09$, $p > .1$), while for the filtered dataset a small positive correlation appears to exist ($N = 14$, $\rho = .34$, $p > .1$); this is also verified by the wide spread of points in the scatterplot shown in Figure 2.(c). Both results are not statistically significant, not even at the 90% level. Table 4 summarises all these results.

In the case of a bivariate linear regression model of P/SA, SA can account for 31.0% of the variation of P in the complete dataset ($N = 16$, $R^2 = .31$), while in the filtered dataset this increases to 50% ($N =$

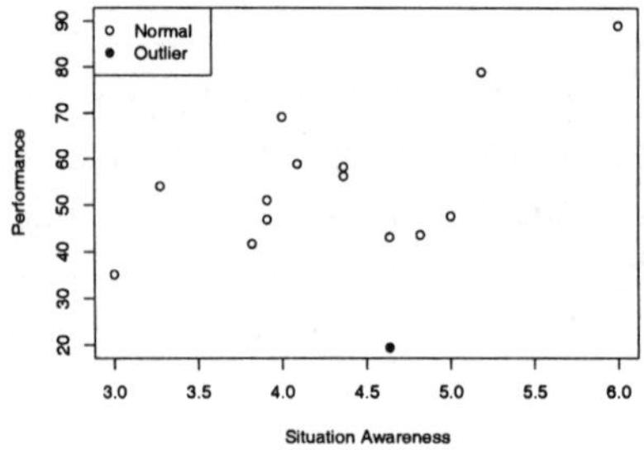

(a) Situation Awareness

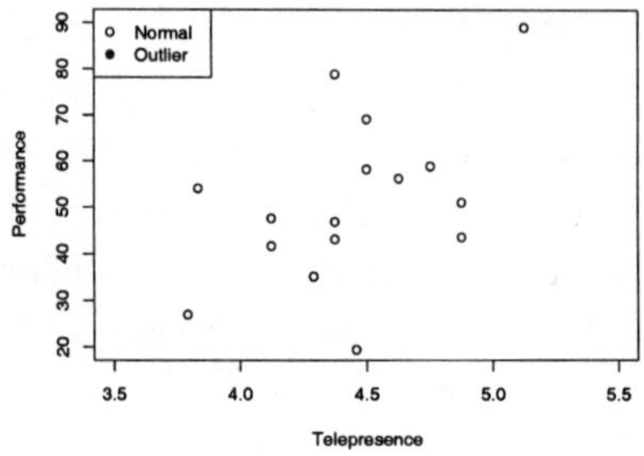

(b) Telepresence

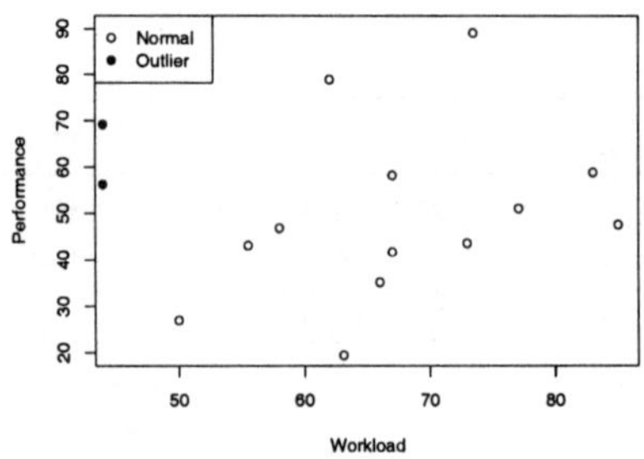

(c) Workload

Fig. 2: Scatterplots of Performance with

Table 4: Bivariate correlation, Pearson's ρ, one-sided (P: Performance, SA: Situation Awareness, TP: Telepresence, WL: Workload)

Y/X	N	ρ	R^2
P/SA	16	.57**	.31
	15	.71***	.50
P/TP	16	.47**	.22
P/WL	16	.09	.01
	14	.34	.11

**significant at the .05 level
***significant at the .01 level

15, $R^2 = .50$). In both cases the linear model predicts P significantly better than the mean value model ($N = 16$, $F = 6.30$, $p < .05$; $N = 15$, $F = 13.14$, $p < .05$). A bivariate linear regression model analysis of P/TP shows that TP can account for 22% of the variation of P ($N = 16$, $R^2 = .22$)

and is (significant at 90% level) better than the mean model ($N = 16,\ F = 4.04,\ p < .1$). For P/WL, a bivariate linear regression model analysis shows that WL has no or very little effect on the variation of P and the linear model is no better to the mean model for the complete dataset ($N = 16,\ R^2 = .01,\ F = 0.12,\ p > .1$) or the filtered one ($N = 14,\ R^2 = .11,\ F = 1.55,\ p > .1$). These results are summarized in Table 5, while Table 6 shows the bivariate linear model coefficients.

Table 5: Summary of bivariate linear regressions

Y/X	N	R	R^2	adj R^2	F
P/SA	16	.57	.31	.26	6.30**
	15	.71	.50	.46	13.14**
P/TP	16	.47	.22	.17	4.04*
P/WL	16	.09	.01	−.06	.12
	14	.34	.11	.04	1.55

*significant at the .1 level
**significant at the .05 level
***significant at the .01 level

Table 6: Coefficients of bivariate linear regressions ($y = mx + b$)

Y/X	N	b	SE_b	t_b	m	SE_m	t_m
P/SA	16	0.25	20.67	.01	12.03	4.79	2.51**
	15	−3.82	16.07	−.24	13.58	3.75	3.62***
P/TP	16	−51.05	51.05	−1.00	23.05	11.47	2.01*
P/WL	16	43.27	23.57	1.84*	.12	.35	.35
	14	11.49	31.02	.37	.55	.44	1.24

*significant at the .1 level
**significant at the .05 level
***significant at the .01 level

4. Discussion

The experimental results have verified the hypotheses that the human-centric characteristics of situation awareness (SA) and telepresence (TP) are two main factors that affect performance (P) in a positive related manner, and should be taken into account when designing robot systems and human-robot interaction interfaces. For the complete datasets they both together account for 53% of the variance of P, while the filtered datasets show an even higher contribution (80%)[b]. The hypothesis that P and workload (WL) are negatively correlated should be rejected as it appears to be no correlation at all in the studies carried out. However, this can be explained by the fact that only very low and very high levels of WL have a negative impact on P. In the former case the user may have not allocated sufficient attentional resources while in the latter it might be beyond the person's processing capabilities. On the other hand, in moderate levels of WL the user is kept occupied without being overloaded. The task difficulty of this particular experimantal scenario can be considered as easy to intermediate, with the values of WL verifying this[c] ($mean = 63.95$, $\sigma = 13.53$), and this can be the reason why workload seems to have no relation to P.

Bivariate analysis has the limitation that it investigates performance and a variable without limiting the effect of the rest. A more accurate analysis and improvement is semi-partial correlation and multiple regression, as well as partial correlation to see if there are any relations between the explanatory variables themselves. It would also be interesting to further explore which particular dimensions of SA, which are already measured by the assessment technique, have a bigger impact to the variance of P, as SA in this analysis was treated as an end product.

5. Conclusions

This study has shown that the human-centric characteristics of situation awareness and telepresence affect task performance to some extent. These are within the sphere of influence of a system designer unlike remaining variables that might influence performance such as for example task difficulty, training, experience and cognitive capabilities of the user. By taking into account these and by using an assessment framework as the one proposed, system designers can compare their developed systems and human-robot interaction interfaces, as well as being able to predict to some extent their task performance. By further analysing the individual dimensions of each

[b] $R^2 = .30$, $F = 5.53$, $p < .05$ of P/TP calculated for the same dataset as the filtered one used for P/SA

[c] Although it has to be mentioned that the subjects had not experienced very easy or very difficult scenarios, something that may have influenced these results.

variable, system designers can focus on particular aspects of the overall system that support the ones that have a bigger impact, resulting in a more efficient research and development.

References

1. J. L. Burke, R. R. Murphy, E. Rogers, V. J. Lumelsky and J. Scholtz, *IEEE Transactions on Systems, Man and Cybernetics* **34** (2004).
2. C. Dominguez, Can SA be defined?, in *Situation Awareness: Papers and Annotated Bibliography Interim Report No. AL/CF-TR-1994-0085*, eds. M. Vidulich, C. Dominguez, E. Vogel and G. Mcmillan 1994 pp. 5–15.
3. T. B. Sheridan, *Presence* **1**, 120 (1992).
4. P. Tsang and G. F. Wilson, Mental workload, in *Handbook of Human Factors and Ergonomics*, ed. G. Salvendy 1997 pp. 417–449, 2nd edn.
5. J. C. Scholtz, B. Antonishek and Young, *IEEE Transactions on Systems, Man and Cybernetics-Part A: Systems and Humans* **35** (2005).
6. S. B. Hughes and M. Lewis, *IEEE Transactions on Systems, Man and Cybernetics - Part A: Systems and Humans* **35**, 513 (2005).
7. H. A. Yanco and J. Drury, "Where am I?" Acquiring situation awareness using a remote robot platform, in *Proc. of the IEEE Conference on Systems, Man and Cybernetics*, Oct. 2004.
8. K. M. Reichard, *Robotics and Autonomous Systems* , 105 (2004).
9. J. M. Riley, The utility of measures of attention and situation awareness for quantifying telepresence, PhD thesis, Department of Industrial Engineering, Missisipi State University2001.
10. J. M. Riley, D. B. Kaber and J. V. Draper, *Human Factors and Ergonomics in Manufacturing* **14**, 51 (2004).
11. R Development Core Team, *R: A Language and Environment for Statistical Computing.* R Foundation for Statistical Computing, Vienna, Austria, (2007).
12. A. Jacoff, B. Weiss, E. Messina, S. Tadokoro and Y. Nakagawa, Test arenas and performance metrics for USAR robots, in *Proc. of the 2003 IEEE/RSJ International Conference on Intelligent Robots and Systems*, (Las Vegas, 2003).
13. Y. Gatsoulis and G. S. Virk, Modular situational awareness for CLAWAR robots, in *Proc. of CLAWAR 2005, 8th International Conference on Climbing and Walking Robots*, 2005.
14. Y. Gatsoulis, G. S. Virk, M. Parack and A. Kherada, "What's going on?" an alternative approach into investigating human-robot interactions, in *Proc. of CLAWAR 2006, 9th International Conference on Climbing and Walking Robots*, 2006.
15. M. R. Endsley, *Human Factors* **37**, 32 (1995).
16. B. G. Witmer and M. J. Singer, *Presence: Teleoperators and Virtual Environments* **7**, 225(June 1998).
17. S. G. Hart and L. E. Stavenland, Development of NASA-TLX: Results of empirical and theoretical research, in *Human Mental Workload*, eds. P. A. Hancock and N. Meshkati 1988 pp. 139–183.

REAL-TIME COMPUTATIONAL COMPLEXITY OF THE ALGORITHMS FOR A SINGLE LINK MANIPULATOR SYSTEM

M.A. HOSSAIN[1]

[1]*Department of Computing, University of Bradford*

M.N.H. SIDDIQUE[2]

[2]*School of Computing and Intelligent Systems, University of Ulster*

M.O. TOKHI[3] AND M.S. ALAM[4]

[3,4]*Department of Automatic Control and Systems Engineering, University of Sheffield*

This paper presents an investigation into the real-time computational complexity of the algorithms for a single link manipulator system. A dynamic simulation algorithm of a single link manipulator system using finite difference (FD) method is considered to demonstrate the real-time computational complexity. The simulation algorithm of the manipulator system is analyzed, designed in various forms and implemented to explore the impact in implementing real-time. Finally, a comparative real-time computing performance of various forms of the new and previously reported algorithms is presented and discussed to demonstrate the merits of different design mechanisms through a set of experiments.

1. Introduction

Although computer architectures incorporates fast processing hardware resources, high performance real-time implementation of a complex algorithm requires an efficient design and software coding of the algorithm so as to exploit special features of the hardware and avoid associated shortcomings of the architecture. In practice, more than one algorithm exists for solving a specific problem. Depending on the formulation, each can be evaluated numerically in different ways. As computer arithmetic is of finite accuracy, different results can evolve, depending on the algorithm used and the way it is evaluated. On the other hand, the same computing domain could offer different performances due to variation in the algorithm design and in turn, source code implementation. The choice of the best algorithm for a given problem and for a specific computer is a difficult task and depends on many factors, for instance, data and control dependencies of the algorithm, regularity and granularity of the algorithm and architectural features of the computer domain [1], [2].

The ideal performance of a computer system demands a perfect match between machine capability and program behaviour. Program performance is the turnaround time, which includes disk and memory accesses, input and output activities, compilation time, operating system overhead, and central processing unit (CPU) time. In order to shorten the turnaround time, one can reduce all these time factors. Minimising the run-time memory management, efficient partitioning and mapping of the program, and selecting an efficient compiler for specific computational demands, could enhance the performance. Compilers have a significant impact on the performance of the system [3]. This means that some high-level languages have advantages in certain computational domains, and some have advantages in other domains. The compiler itself is critical to the performance of the system as the mechanism and efficiency of taking a high-level description of the application and transforming it into a hardware dependent implementation differs from compiler to compiler [3].

The performance demand in modern real-time signal processing and control applications has motivated the development of advanced special-purpose and general-purpose hardware architectures. However, the developments within the software domain have not been at the same pace and/or level as within the hardware domain. Thus, although advanced computing hardware with significant levels of capability is available in the market, these capabilities are not fully utilised and exploited at the software level. Efficient software coding is essential in order to exploit the special hardware features and avoid associated shortcomings of the architecture. There has been a substantial amount of effort devoted to this area of research over the last decade [4], [5], [6], [7], [8].

This paper presents an investigation into the algorithm analysis, design and software coding for real-time computational complexity. It is worth mentioning that this is an extension of the research reported earlier in [9]. Three new analyses and design models are explored, implemented, tested and verified to demonstrate the computational complexity of the algorithms of a manipulator system implemented in real-time. These new design approaches are then compared with the previous three approaches to demonstrate the merits and capabilities of the algorithms.

2. The Single Link Manipulator System

Figure 1 shows a schematic representation of a single-link manipulator system. A control torque τ is applied at the pinned end (hub) of the arm by an actuator motor. θ represents the hub angle, POQ is the original co-ordinate system (stationary coordinate) while $P^{'}OQ^{'}$ is the co-ordinate system after an angular rotation θ (moving coordinate). I_h is the inertia at the hub, I_p is the inertia associated with a payload M_p at the end-point and u is the flexible displacement (deflection) of a point at a distance x from the hub. The dynamic equation of the flexible manipulator, considered as an Euler-Bernoulli beam equation, can be expressed as [10]:

$$\rho \frac{\partial^2 y(x,t)}{\partial t^2} + EI \frac{\partial^4 y(x,t)}{\partial x^4} = \tau(x,t) \tag{1}$$

where, $y(x,t)$ is the displacement (deflection) of the manipulator at a distance x from the hub at time t, ρ is the density per unit length of the manipulator material, E is Young modulus, I is the second moment of inertia, $\tau(x,t)$ is the applied torque. The product EI represents the flexural rigidity of the manipulator.

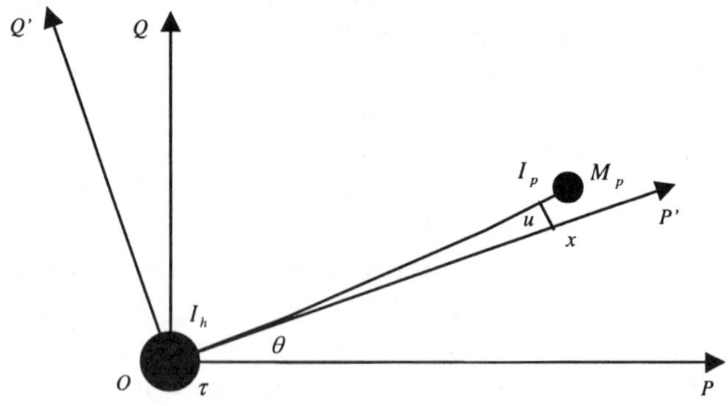

Figure 1. Schematic representation of the flexible manipulator system

Discretising the manipulator in time and length using central finite difference methods, a discrete approximation to equation (1) can be obtained that can be stated in matrix notation as [10]:

$$\mathbf{Y}_{i,j+1} = \mathbf{A}\mathbf{Y}_{i,j} + \mathbf{B}\mathbf{Y}_{i,j-1} + \mathbf{C}\mathbf{F} \tag{2}$$

where $\mathbf{Y}_{i,j+1}$ is the displacement of grid points $i = 1,2,\cdots,n$ of the manipulator at time step *j+1* , $\mathbf{Y}_{i,j}$ and $\mathbf{Y}_{i,j\text{-}1}$ are the corresponding displacements at time steps *j* and *j-1* respectively. **A** and **B** are constant $n \times n$ matrices whose entries depend on the flexible manipulator specification and the number of sections the manipulator is divided into, **C** is a constant matrix related to the given input torque and **F** is an $n \times 1$ matrix related to the time step Δt and mass per unit length of the flexible manipulator.

3. Algorithm Design

The FD simulation algorithm of the manipulator system as shown in equation 2 is considered to demonstrate the different design impact in implementing the algorithm in real-time. As mentioned earlier, three different approaches have been reported earlier [9]. This section presents three new design approaches of the algorithm with their advantages and disadvantages.

Algorithm–4: The 'Algorithm–4' is listed in Figure 2, in which the new methods of Algorithm–4 were applied with the concepts of Algorithm–3 reported in [9]. Three distinct calculation runs are performed during each iteration, but instead of listing the instructions for each segment separately, nested loops are used to limit the number of instructions (source code lines) in the main program loop. The benefits of employing this technique are identical with those listed in the description of Algorithm–3. However, it possesses the same disadvantage of overhead produced by the complex substitutions required.

Figure 2. Algorithm-4

```
Initialise the parameters of the Manipulator.
Step 1: Calculate Current position of different segments y[i][2]; where i=0-19;
Loop {
y0[2]= (A[1][1]*y0[1]+ A[1][2]*y1[1] + A[1][3]*y2[1])- (B[1][1]*y0[0]+
B[1][2]*y1[0]) + tau[j]*C;
y[1][2]= (A[2][1]*y[0][1]+ A[2][2]*y[1][1]+A[2][3]*y[2][1]+A[2][4]*y[3][1])-
(B[2][1]*y[0][0]+B[2][2]*y[1][0] + B[2][3]*y[2][0]);
Loop{
y[i][2]=(A[i+1][i-1]*y[i-2][1]+A[i+1][3]*y[i][1]+A[i+1][i+2]*y[i+1][i]
+A[i+1][i+3]*y[i+2][i])-(B[i+1][i]*y[i-1][0]+B[i+1][i+1]*y[i][0]
+B[i+1][i+2]*y[i+1][0]);
}
```

```
y[18][2]=(A[19][17]*y[16][1]+A[19][18]*y[17][1]+A[19][19]*y[18][1]+A[19][
20] *y[19][1])- (B[19][18]*y[17][0] + B[19][19]*y[18][0]+
B[19][20]*y[19][0]);
y19[2]= (A[20][18]*y17[1]+ A[20][19]*y18[1] + A[20][20]*y19[1])-
B[20][20]*y19[0];

//Step 2 : Calculate y[i][0] ; where i = 0-19 ;
y0[0]= (A[1][1]*y0[2]+ A[1][2]*y1[2] + A[1][3]*y2[2])- (B[1][1]*y0[1]+
B[1][2]*y1[1]) + tau[j]*C;
y[1][0]= (A[2][1]*y[0][1]+ A[2][2]*y[1][1]+A[2][3]*y[2][1]+A[2][4]*y[3][1])-
(B[2][1]*y[0][0]+B[2][2]*y[1][0] + B[2][3]*y[2][0]);
 Loop{
y[i][0]=(A[i+1][i-1]*y[i-2][1]+A[i+1][3]*y[i][1]+A[i+1][i+2]*y[i+1][i]
+A[i+1][i+3]*y[i+2][i])-(B[i+1][i]*y[i-1][0]+B[i+1][i+1]*y[i][0]
+B[i+1][i+2]*y[i+1][0]);
 }
y[18][0]=(A[19][17]*y[16][1]+A[19][18]*y[17][1]+A[19][19]*y[18][1]+A[19][
20] *y[19][1])- (B[19][18]*y[17][0] + B[19][19]*y[18][0]+
B[19][20]*y[19][0]);
y19[0]= (A[20][18]*y17[2]+ A[20][19]*y18[2] + A[20][20]*y19[2])-
B[20][20]*y19[1];

// Step 3 :  Calculate y[i][1] ; where i = 0 -19.
y0[1]=(A[1][1]*y0[0]+ A[1][2]*y1[0] + A[1][3]*y2[0])- (B[1][1]*y0[2]+
B[1][2]*y1[2]) + tau[j]*C;
y[1][1]= (A[2][1]*y[0][1]+ A[2][2]*y[1][1]+A[2][3]*y[2][1]+A[2][4]*y[3][1])-
(B[2][1]*y[0][0]+B[2][2]*y[1][0] + B[2][3]*y[2][0]);
 Loop{
y[i][1]=(A[i+1][i-1]*y[i-2][1]+A[i+1][3]*y[i][1]+A[i+1][i+2]*y[i+1][i]
+A[i+1][i+3]*y[i+2][i])-(B[i+1][i]*y[i-1][0]+B[i+1][i+1]*y[i][0]
+B[i+1][i+2]*y[i+1][0]);
 }
y[18][1]=(A[19][17]*y[16][1]+A[19][18]*y[17][1]+A[19][19]*y[18][1]+A[19][
20] *y[19][1])- (B[19][18]*y[17][0] + B[19][19]*y[18][0]+
B[19][20]*y[19][0]);
 y19[1]= (A[20][18]*y17[0]+ A[20][19]*y18[0] + A[20][20]*y19[0])-
B[20][20]*y19[2]; }
---------------------------------------------------------------------------------------
```

Algorithm–5: The 'Algorithm–5' is listed in Figure 3. This makes use of the fact that access to the oldest time segment is only necessary during re-calculation of the same segment. Hence, it can directly be overwritten with the new value as shown in Fig. 4. The conventional re-calculation algorithm in Fig. 2 requires three memory segments in the time domain. In contrast, Algorithm–5 is optimized for the particular discrete mathematical approximation of the governing physical formula, exploiting the previously observed features.

It is noted that this particular algorithm is not suitable for applications for which the previous assumption does not hold. This technique may offer a major performance advantage over the conventional rotation method, in particular when the number of segments is increased.

Figure 3. Algorithm-5

```
Initialise the parameters of the Manipulator.
Loop{
Step 1: Calculate Current position of different segments y[i][1]; where i=0-19;
y0[1]= (A[1][1]*y0[1]+ A[1][2]*y1[1] + A[1][3]*y2[1])- (B[1][1]*y0[0]+
B[1][2]*y1[0]) + tau[j]*C;
y[1][1]= (A[2][1]*y[0][1]+ A[2][2]*y[1][1]+A[2][3]*y[2][1]+A[2][4]*y[3][1])-
(B[2][1]*y[0][0]+B[2][2]*y[1][0] + B[2][3]*y[2][0]);
/ Continue ------
y[18][1]=(A[19][17]*y[16][1]+A[19][18]*y[17][1]+A[19][19]*y[18][1]+A[19][
20] *y[19][1])- (B[19][18]*y[17][0] + B[19][19]*y[18][0]+
B[19][20]*y[19][0]);
y19[1]= (A[20][18]*y17[1]+ A[20][19]*y18[1] + A[20][20]*y19[1])-
B[20][20]*y19[0];

//Step 2 : Calculate y[i][0] ; where i = 0-19 ;
y0[0]= (A[1][1]*y0[1]+ A[1][2]*y1[1] + A[1][3]*y2[1])- (B[1][1]*y0[0]+
B[1][2]*y1[0]) + tau[j]*C;
y[1][0]= (A[2][1]*y[0][1]+ A[2][2]*y[1][1]+A[2][3]*y[2][1]+A[2][4]*y[3][1])-
(B[2][1]*y[0][0]+B[2][2]*y[1][0] + B[2][3]*y[2][0]);
/Continue -------
y[18][0]=(A[19][17]*y[16][1]+A[19][18]*y[17][1]+A[19][19]*y[18][1]+A[19][
20] *y[19][1])- (B[19][18]*y[17][0] + B[19][19]*y[18][0]+
B[19][20]*y[19][0]);
y19[0]= (A[20][18]*y17[1]+ A[20][19]*y18[1] + A[20][20]*y19[1)-
B[20][20]*y19[0]; }
```

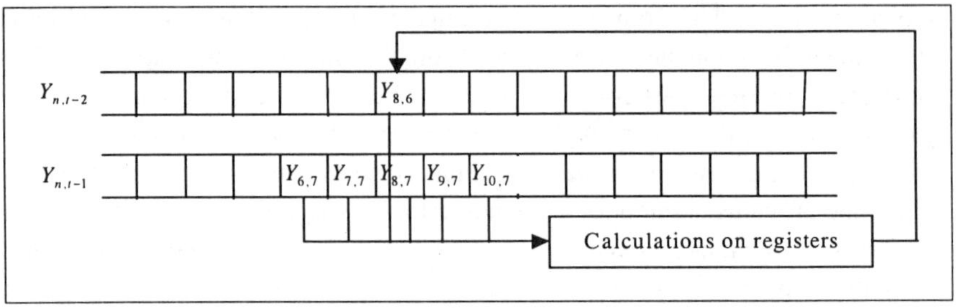

Figure 4. Recalculation with reduced memory allocation

Algorithm–6: Algorithm–6, as shown in Fig. 5, is based on improvements achieved with Algorithm–6. Additionally, the notion of nested loops was incorporated. The advantages and disadvantages of this approach were identified earlier and remain true for this particular algorithm.

Figure 5. Algorithm-6

```
Initialise the parameters of the Manipulator.
Loop{
Step 1: Calculate Current position of different segments y[i][1]; where i=0-19;
y0[1]= (A[1][1]*y0[1]+ A[1][2]*y1[1] + A[1][3]*y2[1])- (B[1][1]*y0[0]+
B[1][2]*y1[0]) + tau[j]*C;
y[1][1]= (A[2][1]*y[0][1]+ A[2][2]*y[1][1]+A[2][3]*y[2][1]+A[2][4]*y[3][1])-
(B[2][1]*y[0][0]+B[2][2]*y[1][0] + B[2][3]*y[2][0]);
 Loop{
y[i][1]=(A[i+1][i-1]*y[i-2][1]+A[i+1][3]*y[i][1]+A[i+1][i+2]*y[i+1][1]
+A[i+1][i+3]*y[i+2][1])-(B[i+1][i]*y[i-1][0]+B[i+1][i+1]*y[i][0]
+B[i+1][i+2]*y[i+1][0]);
 }
y[18][1]=(A[19][17]*y[16][1]+A[19][18]*y[17][1]+A[19][19]*y[18][1]+A[19][
20] *y[19][1])- (B[19][18]*y[17][0] + B[19][19]*y[18][0]+
B[19][20]*y[19][0]);
y19[1]= (A[20][18]*y17[1]+ A[20][19]*y18[1] + A[20][20]*y19[1])-
B[20][20]*y19[0];

//Step 2 : Calculate y[i][0] ; where i = 0-19 ;
y[0][0]= (A[1][1]*y0[1]+ A[1][2]*y[1][1] + A[1][3]*y2[1])- (B[1][1]*y[0][0]+
B[1][2]*y[1][0]) + tau[j]*C;
y[1][0]= (A[2][1]*y[0][1]+ A[2][2]*y[1][1]+A[2][3]*y[2][1]+A[2][4]*y[3][1])-
(B[2][1]*y[0][0]+B[2][2]*y[1][0] + B[2][3]*y[2][0]);
```

```
Loop{
y[i][0]=(A[i+1][i-1]*y[i-2][1]+A[i+1][3]*y[i][1]+A[i+1][i+2]*y[i+1][1]
+A[i+1][i+3]*y[i+2][1])-(B[i+1][i]*y[i-1][0]+B[i+1][i+1]*y[i][0]
+B[i+1][i+2]*y[i+1][0]);
}
y[18][0]=(A[19][17]*y[16][1]+A[19][18]*y[17][1]+A[19][19]*y[18][1]+A[19][
20] *y[19][1])- (B[19][18]*y[17][0] + B[19][19]*y[18][0]+
B[19][20]*y[19][0]);
y[19][0]= (A[20][18]*y[17][1]+ A[20][19]*y[18][1] + A[20][20]*y[19][1])-
B[20][20]*y[19][0]; }
```

4. Experiments and Results

Three new design methods were implemented for similar specification to demonstrate the merits and capabilities of the approaches. Same compiler (C++) and computing domain is used to implement the simulation environment for testing and verification. It is worth noting that a fixed number of segments for various iterations were considered in implementing all the algorithms for the sake of consistence. Moreover, the sampling time *0.000217904 sec* is considered for each iteration, therefore, required real-time performance of *5000* iterations, as an example, should be *5000X0.000217904 = 1.0895 sec.*

Figure 6 shows the comparative performance of the three new algorithms for 20 segments. It is observed that the execution time for the algorithms increases almost linearly with the increment of iterations. It is also noted that the Algorithm-5 performs best among the algorithms. In contrast, Algorithm-6 performs waste among the three algorithms. It is also observed that Algorithm-4 and Algorithm-6 have not achieved required performance to implement in real-time.

Table 1 demonstrates the relative performance of the six algorithms, including previously reported [9] three. It is worth mentioning that the performance in Table 1 is presented as ratio of the Algorithm-1, Algorithm-3, Algorithm-4, Algorithm-5 and Algorithm-6 relative to the Algorithm-2 to keep consistence of the previous report [9]. It is observed that the Algorithm-4, Algorithm-5 and Algorithm-6 offer relatively better performance as compared to the Algorith-1 and Algorim-3. On the other hand, the Algorithm-5 is about 1.21 times (average) slower as compared to the Algorithm-2. Execution time of the Algorithm-6 is more then double as compared to the Algorithm-5 and very similar level of Algorithm-4. Thus, the design mechanism employed in Algorithm–2 and Algorithm-5 can offer potential advantages for real-time implementation.

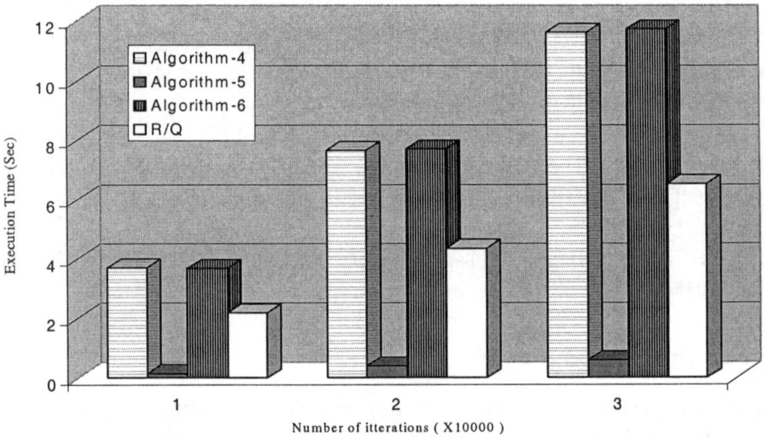

Figure 6. Performance comparison of the various forms of the Algorithms and real-time requirement (R/Q)

Table 1: Performance of the Algorithms relative to the Algorithm-2 (Alg-2)

Iterations	30000	5000	10000	15000	20000	25000	30000	Avg
Alg-1 /Alg-2	11.0	20	22.5	13.8	15.5	16.71	15.67	16.5
Alg-3 /Alg-2	26	52	55.99	33.59	38.99	42.42	39.56	41.23
Alg-4 /Alg-2	2.87	2.66	2.45	2.55	2.39	2.89	2.48	2.61
Alg-5 /Alg-2	1.22	1.25	1.33	1.09	1.13	1.19	1.25	1.21
Alg-6 /Alg-2	2.75	2.16	2.35	2.78	2.23	2.54	2.21	2.43

5. Concluding Remarks

This paper has presented an extension of the investigation reported in [9]. The investigation has focused into the algorithms analysis, design, software coding and implementation so as to reduce the execution time and, in turn, enhance the real-time performance. Three new design approaches for real-time implementation have been proposed and demonstrated experimentally. It has been observed that the execution time and in turn, performance of the algorithms varies with different approaches in a real-time implementation context. It is also noted that only Algorithm-2 (previously reported) and Algorithm-5 have achieved real-time performance for various number of iterations. Although, the other four algorithms can also be implemented in real-time by using high performance computing domain, however, identification of the suitability of Algorithm design and implementation mechanism for best performance is a challenge, in particular for a specific architecture.

References

1. A. U. Thoeni, *Programming real-time multicomputers for signal processing*, Prentice-Hall, Hertfordshire, UK (1994).
2. M. O. Tokhi and M. A. Hossain, CISC, *RISC and DSP processors in real-time signal processing and control*, Journal of Microprocessors and Microsystems, **19**(5), UK. pp. 291-300 (1995).
3. G. Bader and E. Gehrke, *On the performance of transputer networks for solving linear systems of equation*, Parallel Computing, 1991, **17**, pp. 1397-1407 (1991).
4. M. O. Tokhi, M. A. Hossain, M. J. Baxter and P. J. Fleming, *Heterogeneous and homogeneous parallel architectures for real-time active vibration control*, IEE Proceedings-D: Control Theory and Applications, **142**, (6), pp. 1-8 (1995).
5. B. N. Bershad and D. Lee, T. H. Romer and B. J. Chen, *Avoiding conflict misses dynamically in large direct-mapped caches*, Proceedings of Sixth International Conference on Architectural Support for Programming Languages and Operating Systems (ASPLOS'94), San Jose, Canada, pp. 158-170 (1994).
6. B. Clader, C. Krintz, S. John and T. Austin, *Cache -concious data placement*, Proceedings of Eighth International Conference on Architectural Support for Programming Languages and Operating Systems (ASPLOS'98), San Jose, Canada, pp. 139-149 (1998).
7. K. Hwang, *Advanced computer architecture – Parallelism, scalability, programmability*, McGraw-Hill, California, USA (1993).
8. M. A. Hossain, U. Kabir and M. O. Tokhi, *Impact of data dependencies for real-time high performance computing*, Journal of Microprocessors and Microsystems, **26**(6), pp. 253 – 261 (2002).
9. M. A. Hossain, M. N. Siddique, M. O. Tokhi and M. S. Alam, *Design Constraints in Implementing Real-time Algorithms for a Flexible Manipulator System*, CLAWAR 2005: Climbing and Walking Robots, M O Tokhi, G S Virk and M A Hossain (eds.), Springer, Germany, pp. 583-590 (2005).
10. A. K. M. Azad, *Analysis and design of control mechanisms for flexible manipulator systems*, PhD Thesis. Department of Automatic Control and Systems Engineering, The University of Sheffield, UK (1994).

SOFTWARE AND COMMUNICATION INFRASTRUCTURE DESIGN OF THE HUMANOID ROBOT RH-1

D. KAYNOV, M. ARBULU, P. STAROVEROV, L. CABAS, C. BALAGUER

Robotics Lab,
Department of Systems Engineering and Automation,
University Carlos III of Madrid, Av. Universidad 30,
Leganés, Madrid, Spain

This paper presents an advanced control system for the humanoid robot. The main advantage of the proposed control architecture is wide use of the standardized and frequently used in the automation industry solutions. It provides scalability, modularity and the application of standardized interfaces and brings the design of the complex control system of the humanoid robot from a closed laboratory to the industry. The main parts of the proposed software control system architecture and communication infrastructure are presented. As well as some aspects of the humanoid robot walking control from the automation side are discussed. The designed software control system was implemented with the Rh-1 humanoid platform, the second phase of the Rh project, which have been launched by the Robotics Lab at the University Carlos III of Madrid at 2002.

1. Introduction

Recently the development of sophisticated humanoid robots has increased and it has become very active research area. There is growing interest not only in academic area but in some industrial areas too. Several humanoid robots have been developed in these years. One of the best humanoid robots is HRP-2 designed by Kawada Industries [1]. The most impressive humanoid robot should be ASIMO constructed by HONDA [2]. It is necessary to mention that the great success of ASIMO makes the current research of humanoid robots to become very promising working field for scientists and engineers. One of the recently presented successful projects is KHR-2 humanoid robot constructed by KAIST (Korea Advanced Institute of Science and Technology) [3].

In the major part of works on the humanoid robot problem, the design is impartially concentrated on the walking problem. The software architectures description is superficial and does not give the clear idea about criterions used to design the system and its validity. The goal of this paper is to determine the humanoid robot software systems as the individual type of the architecture inside the general control problem.

This paper presents the Rh-1 robot, introduces its communication infrastructure and software control architecture. Some aspects of the humanoid robot control from the automation side are also discussed.

2. Humanoid robot Rh-1

Rh-1 is a new humanoid robot in the second phase of the Rh project. The robot is developed for several principal tasks such as human care, maintenance of dangerous for human health plants, entertainment etc Figure 1 shows mechanical configuration of the Rh-1.

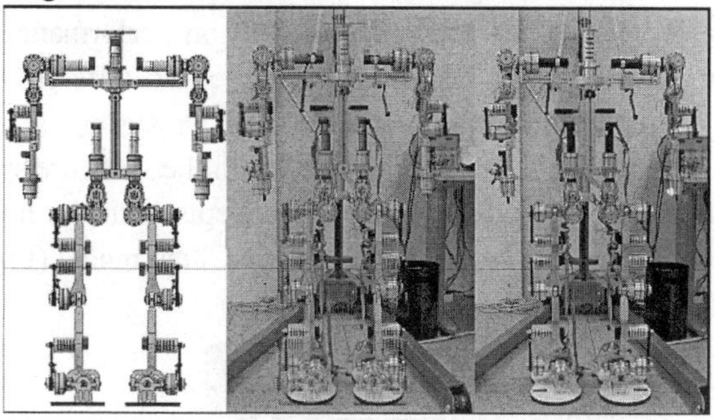

Figure 1 Mechanical design and developed Rh-1 robot.

Electrical design of Rh-1 robot is based on distributed motion control philosophy there each control node is an independent agent on the network. Figure 2 shows the physical distribution of the hardware inside the humanoid robot.

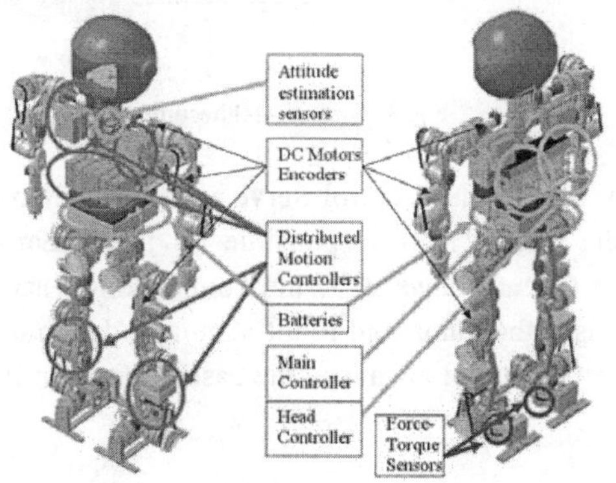

Figure 2 Hardware distribution inside the humanoid robot.

The mechanical and dynamical design of Rh-1 is completed. The detail software and communication infrastructure design will be presented in followings sections.

3. Software Architecture

A humanoid robot can be considered as a plant were the shop floor consists of a series of cells (intelligent motion controllers and a sensors) managed by controllers (Main Controller, Communication Supervisory Controller, etc.) In general, there are two basic control tasks for the control system of a humanoid robot. The first goal is to control all automation and supervise the data transmission. And the second goal is to control and monitor the entire floor to detect failures as early as possible and to report on performance indicators. In the automation industry, in order to perform these two different control tasks a PLC with its own program inside and a SCADA system are used. In this context, the humanoid robot Rh-1 is provided with the software system allowing the implementation of the industry control concepts with the humanoid robot. The Software architecture is based on the Server-Client model (Figure 3).

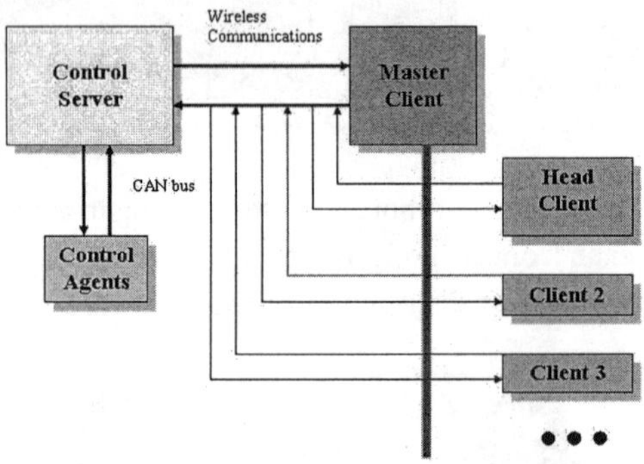

Fig. 3. System Architecture.

For the security reason, the Control Server accepts the connections of other clients, such as the Head Client, responsible for the human-robot interaction, only if the Master Client allows it. If the connection is accepted, the Master Client only supervises the humanoid robot state and data transmission between the robot and other Client, but always in the case of any conflict it has the major priority.

According to Server-Client model, the humanoid robot is controlled by the passive Server, which waits for requests and upon its receipt, processes them and then serves replies for the Client. But on the other hand, the Server controls all Control Agents which reside on the CAN bus network. In that case, the Control Server is no longer a slave. It is a network master for Control Agents which performs their operations (motion control or sensing) and reply for the Server.

When the robot is working in the operational interaction mode (walking, objects manipulation, etc.), there are several computing and communication tasks that need to be performed in a cyclic mode and fast enough to avoid any possible loss of the control. The periodic (with period Ts) chain (Figure 4) begins with the sensing task, taking the time interval t_{att} for attitude estimation gyros and accelerometers readings and t_{zmp} for ZMP force-torque sensors readings.

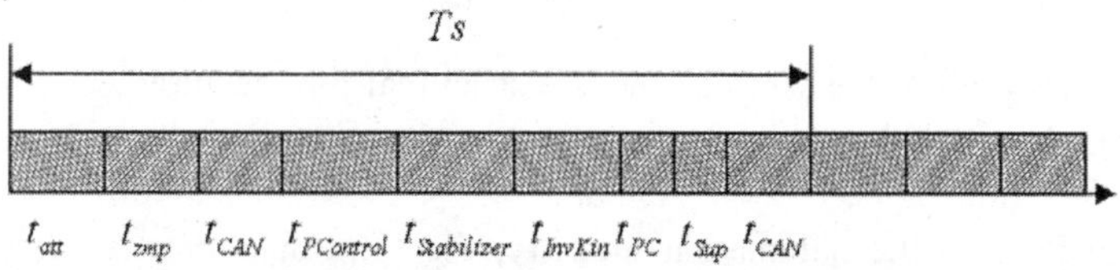

Figure 4 Main computing and communication tasks for the Rh-1.

These tasks are followed by the tasks performing CAN bus communications, Posture Control and then Stabilizer and Inverse Kinematics Calculator computing, internal PC bus communications, Supervisory controller, and CAN bus transmission of new reference for each joint of the humanoid robot.

The period Ts should be small and compatible with the dynamics of the humanoid robot movement. On the one hand, Ts cannot be made arbitrarily small because various computing and communication tasks with execution times t_{att}, t_{zmp}, t_{CAN}, t_{PC}, $t_{Stabilizer}$, etc., cannot be made arbitrarily small. Also, a small value for Ts would generate too many messages in communication lines (RS-232, PC bus, CAN bus) that would overload it. On other hand, Ts cannot be made arbitrarily large because of the dynamics of the robot (Nyquist criterion). Thus,

$$Ts > t_{att} + t_{zmp} + t_{CAN} + t_{PControl} + t_{Stabilizer} + \\ + t_{InvKin} + t_{PC} + t_{Sup} + t_{CAN} \tag{1}$$

And

$$Ts < \frac{1}{2 \cdot Fr} \tag{2}$$

Where Fr is the highest movement frequency on any robot link.
Assuming that the robot walks about the same speed of a normally walking human, $Fr = 2$ Hz.
Thus,

$$Ts < 250 \text{ ms} \tag{3}$$

A reasonable value for the sum of the time intervals taken by of all computations and calculations is $0.75 \cdot Ts$, thus

$$\begin{aligned} & t_{att} + t_{zmp} + t_{CAN} + t_{PControl} + t_{Stabilizer} + \\ & + t_{InvKin} + t_{PC} + t_{Sup} + t_{CAN} = 187.5 \ ms \end{aligned} \tag{4}$$

This upper limit of the sample time is a strict real-time requirement considering the complexity of the computing and communication tasks to be performed with this time limit.

As a PLC in the automation industry, the Control Server is designed and programmed as finite state automata with a sample time Ts. Figure 5 shows the state diagram of the humanoid robot Control Server functioning.

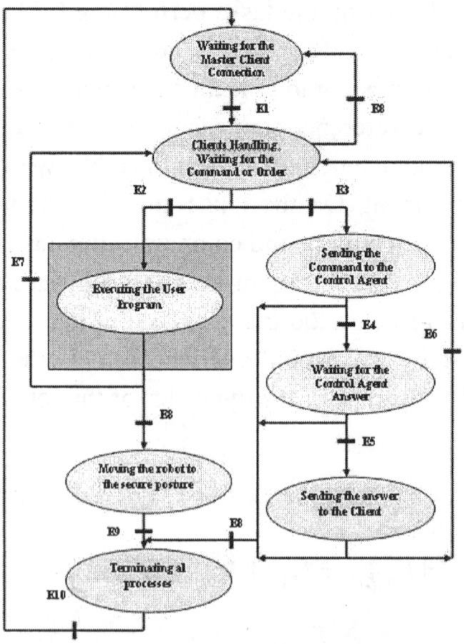

Figure 5 Server functioning state diagram.

Two basic types of the incoming data are processed. A command is the simple data, which can be executed by one Control Agent. The order is a complex command which needs the simultaneous action of many Control Agents and sensors which possesses the humanoid robot. After the connection of the Master Client, the humanoid robot stays at the Client Handling state waiting for an order or a simple command. The arrival of an order launches the User Program. The User Program is the core of the humanoid robot Control Server software. It performs the trajectory execution at the synchronized multi-axis walking applications, controls the posture and ZMP errors at the dynamic walking mode, reads the sensors state etc. Developed software provides the set of the C-based function to work with the robot and to generate user's motions and control procedures not only for walking, but for implementing different human-robot cooperation tasks.

To provide the robot Rh-1 with bottom level control, a SCADA system for the humanoid robot, titled HRoSCoS (Humanoid Robot Supervisory Control System) Client was developed.

The HRoSCoS Client provides the trending of different parameters of the robot, such as for example the joints velocities, accelerations, currents, body inclinations, forces and torques appears during humanoid robot walking. Real-time and historical trending is possible. Logged data is time-stamped and can be filtered when viewed by a user. Also it is possible to generate different reports on the humanoid robot state at any time.

The HRoSCoS Client system presents the information to the operating personnel graphically. This means that the operator can see a representation of the humanoid robot being controlled (Figure 6).

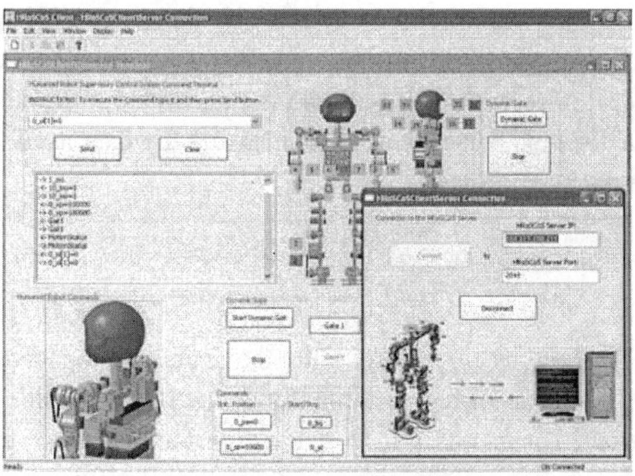

Fig. 6. HRoSCoS Client views.

The HMI supports multiple screens, which can contain combinations of synoptic diagrams and text. The whole humanoid robot is decomposed in "atomic" parameters (e.g. a battery current, its maximum value, its on/off status, etc.) to which a Tag-name is associated. The Tag-names used to link graphical objects to devices. Standard windows editing facilities are provided: zooming, re-sizing, scrolling, etc. On-line configuration and customisation of the HMI is possible for users with the appropriate privileges. Links are created between display pages to navigate from one view to another.

4. Communication Infrastructure

When building a control application, communication with the host is often a crucial part of the project. Nodes of the network always function as data servers because their primary role is to report information (status, acquired data, analyzed data, etc.) to the host at constant rates.

Hardware Architecture consists of three basic levels of automation which uses its own communication systems. The upper (Control) level uses a TCP/IP based communication protocol. Ethernet communication is one of the most common methods for sending data between computers. The TCP/IP protocol provides the technology for data sharing, but only the specific application implements the logic that optimizes performance and makes sense of the data exchange process. When data transmission begins, the sender should packetize each piece of data with an ID code that the receiver can use to look up the decoding information. In that way, developed communication protocol hides the TCP implementation details and minimizes network traffic by sending data packages only when it is needed. When a data variable is transmitted by the sender, it is packetized with additional information so it can be received and decoded correctly on the receiving side. Before each data variable is transmitted, a packet is created that includes fields for the Data Size, the Data ID and the data itself. Figure 7 shows the packet format.

Data Size (8 bits)	Data ID (16 bits)	Data (32 bytes)

Fig. 7. The package format.

The Data ID field is populated with the index of the data array element corresponding to the specified variable. Since the receiving side also has a copy of the data array, it can index it to get the properties (name and type) of the incoming data package. This very effective mechanism is implemented to provide data exchange between the Control Server and different Clients on the Control level of automation of the humanoid robot.

Bottom level (Sensorial and Field) communications are realized using CAN and CanOpen protocols. These communication protocols provide data transmission in broadcast type of communication. This guarantees data integrity as all devices in the system use the same information. The sensorial system of the humanoid robot makes data exchange under the lower CAN protocol and the intelligent motion controllers use the upper CANOpen protocol.

The communication implemented on the bottom level involves the integrating of CANopen (Drives and Motion Control Device Profile) and the introduction of new functionality which is not contained within the relevant device profiles for the sensorial data processing.

5. Conclusions

This paper presented the development of the control system of the humanoid robot Rh-1. The control system was designed using software development standards from the industrial automation field.

It was proved experimentally, that Rh-1 robot provided with the current mechanical, hardware and software control architectures can walk stably.

This work makes an effort to show that it is possible to bring some basic aspects of the industrial automation programming to other, more sophisticated fields of robotics in order to extend the further standardization and unification of the design processes. Moreover, proposed approach allows consider a humanoid robot locomotion inside the global control problem.

In the future a human - humanoid robot interaction from the software control system design side will be considered. Also, the analysis and improvements of the presented software control architecture will be continued.

References

1. K. Kaneko, F. Kanehiro, S. Kajita, K. Yokoyama, K. Akachi, T. Kawasaki, S. Ota and T. Isozumi, "Design of prototype humanoid robotics platform for HRP", *Proc. of IEEE/RSJ Int. Conference on Intelligent Robots and Systems,* pp. 2431-2436, (2002).
2. Y. Sakagami, R. Watanabe, C. Aoyama, S. Matsunaga, N. Hikagi and K. Fujimura, "The intelligent ASIMO: system overview and integration", *Proc. of IEEE/RSJ Int. Conference on Intelligent Robots and Systems,* pp. 2478-2483, (2002).
3. I. Park, J. Kim, S. Park, J. Oh, "Development of humanoid robot platform KHR-2", *Proc. of IEEE Int. Conference on Humanoid Robots,* pp. 292-310, (2004).
4. D.Kaynov; M.A.Rodríguez; M.Arbulú; P.Staroverov; L.M.Cabas; C.Balaguer. "Advanced motion control system for the humanoid robot Rh-0", *8th International Conference on Climbing and Walking Robots*, (2005).

SPARBOT – A ROBOTIC FOCUS MITT TRAINING PLATFORM

RICHARD STOKES, LIQIONG TANG and IBRAHIM AL-BAHADLY
Institute of Technology and Engineering
Massey University
New Zealand

This paper describes the design and implementation of a prototype robotic training platform for focus mitt training called 'SparBot'. Focus mitt training is an important aspect of the Martial Arts. The goal of SparBot is to emulate the focus mitt pad-holding trainer and introduce a level of automation to focus mitt training. The system has two movable arms with rotating wrists, which makes SparBot able to move its arms into different positions to receive the strikes. Sensors throughout SparBot monitor arm position and robot location on the linear translation drive. The robot is capable of giving audio cues for strike patterns and can track training progress in the control software. SparBot seamlessly integrates modern robotic technologies with Martial Art training.

1. Introduction

Intelligent robotic and mechatronic systems form the core of industrial and consumer products in the 21st century. The use of robotic apparatus in the domestic consumer sector is on the rise. The core disciplines of Robotics and Mechatronics allows engineers to quickly turn an idea into a physical prototype. The integration of robotics, precision mechanical systems, intelligent control, and reliable electrical and electronic systems has made many previously unattainable ideas a commercial reality. The work presented in this paper describes the development of a prototype robotic Martial Arts training system – SparBot. It is the first prototype of a Robot Martial Arts Trainer for focus mitt training.

Martial Arts Training is part of the training course for army soldiers and police officers in many countries. It brings benefit to both physical and emotional health. Martial Arts Training improves self-defense capability and foster self-confidence that allows people to diffuse volatile situations without resorting to force. The strict mental and physical training schemes also promotes cardiovascular and mental health, which enable people cope better with the high stress level in today's competitive society. Nowadays, Martial Arts Training courses become popular especially for young people. The objective of the project is to develop a prototype robotic trainer for focus mitt training and domestic sports equipment for daily exercise.

Different training schemes, tools, and equipment are used in Martial Arts Training. Focus mitts are a type of training pad. Martial Arts trainers fit the0 pad over their hands during focus mitt training. By calling out strikes and placing the pads in certain designated positions, the trainer can control how their partner moves and strikes. The presented prototype robotic trainer is designed to emulate the pad-holding partner for focus mitt training. The robot is able to position its arms to receive strikes, has some degree of lateral movement and can provide audio cues for the user. The entire robot is controlled via a graphical user interface running on a local PC.

2. Robot Mechatronic System Design

The main strikes in focus mitt training are the jab, cross and hook. To mimic the pad-holding partner in these strikes, the robot trainer must have a mechanism which can realise wrist and arm movement as well as the ability to change its location around. In addition the system is able to give commands to conduct the training. Thus, the final machine must have the following fundamental components and features to be a basic robot focus mitt trainer.

- Two movable arms, each with rotating wrists
- Translational movement of the robot body
- The ability to give audio cues
- Auto and manual mode

During focus mitt training, the pad-holding trainer must hold his arm in space at a right position to withstand the strong impact forces [1]. In order to maximise the area moment of inertia to provide a strong mechanism for impact force opposition, the robot arms are made from aluminium tubing. This also facilitates placement of the wrist stepper motor, cable routing and coupling inserts.

The structure of the robot arm is designed in the form of a human being's arm. The robot arm assembly as shown in Figure 1 has a wrist and a shoulder insert.

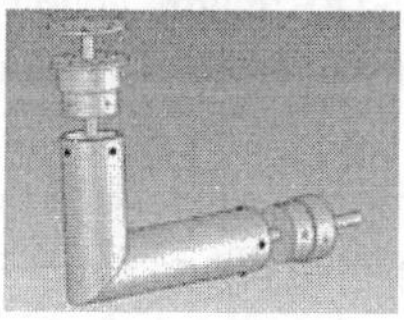

Fig 1: Robot Arm Assembly

Fig 2: Bearing assembly

A rotating pad plate is fixed in the wrist via a bearing housing shown in Figure 2. Such a design provides the rotating movement for the pad through a stepper motor inside the wrist and transfers strike forces away from the rotor shaft into the arm housing. The shoulder insert connects the arm assembly to the shoulder drive shaft inside the upper torso and also transfers the impact forces to the robot base [2].

The pad drive stepper motor is coupled to the pad base plate and mounted inside the forearm. On top of the pad base plate is the pad assembly as illustrated in Figure 3. Providing a comfortable strike surface, the pad backing plate is fitted with composite foam, behind which is embedded a micro switch that provides a means for registering that a strike has occurred. If an off-centre strike is apparent at the pad surface, the stepper holding torque is not sufficient to withstand such an impact. Thus, a solenoid operated locking pin is implemented. The system engages when the pad is in place locking the rotating plate to the base plate on top of the bearing housing. The stepper motor facilitates accurate pad positioning for the jab, cross and hook punches. Positional accuracy is of the utmost importance to ensure correct locking pin alignment.

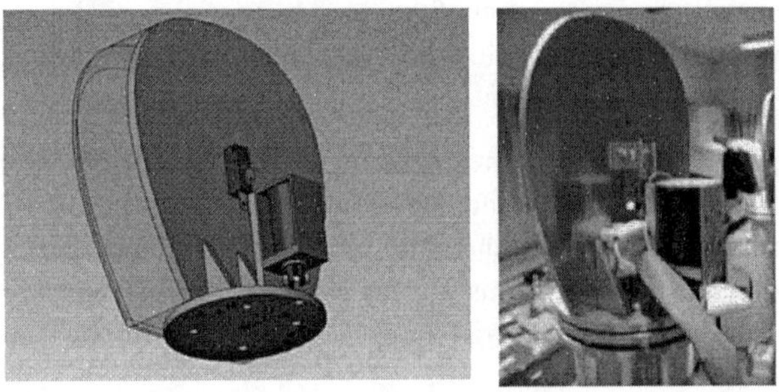

Figure 3. Robot Pad Assembly

Arm rotation is facilitated by two 24V DC motors coupled to the shoulder drive shaft via a 2.5:1 chain-driven gear arrangement. To remove strike forces from the DC drive gearbox and drive train, a braking system is employed. The braking system is operated by solenoid through a pin to lock the brake drum and hold the arm at the correct positions as shown in Figure 4.

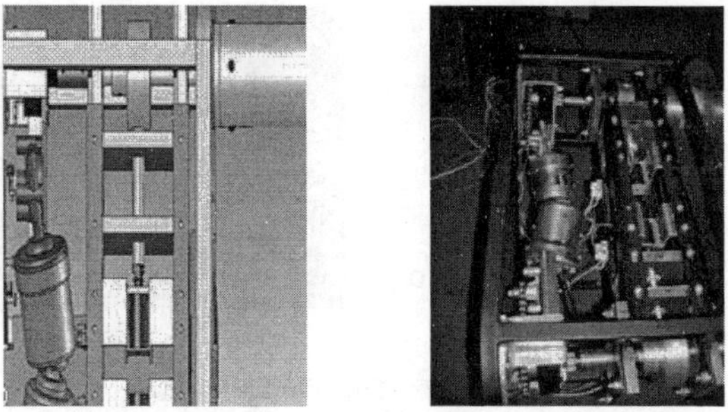

Figure 4. Robot Arm Brake System

The ideal robot trainer should be able to change its location and move around, which is a two-dimensional movement. As this is the first stage of the project, an existing linear translation drive system is used to provide another dimension to SparBot's movement. The intention is to emulate the linear movement of a pad-holding partner. The system has a three phase 0.18 kW drive motor and a power screw. After installed some sensors and limit switches, it was nicely integrated with the robot body and realise the linear movement.

3. Electrical System

The robot trainer electrical system consists of two main parts, the Low Voltage (LV) subsystem and the Functional Extra Low Voltage (FELV) subsystem. The LV subsystem contains the three phase and associated single phase power systems. The LV scheme is shown in Figure 5. The incoming three-phase supply is run through a local isolator then passed to the live side of both the forward and reverse contactors. The motor supply is run through a three-phase overload relay then to the motor. A red phase tap supplies, via a local MCB, two solid state relays (SSR) that are switched via the DIO of the Labview card. This provides the electrical isolation necessary for FELV classification. The switching current to each contactor coil is interlocked via an auxiliary contact mounted on the opposite contactor to prevent accidental engagement of both contactors simultaneously, which would result in a phase-to-phase fault.

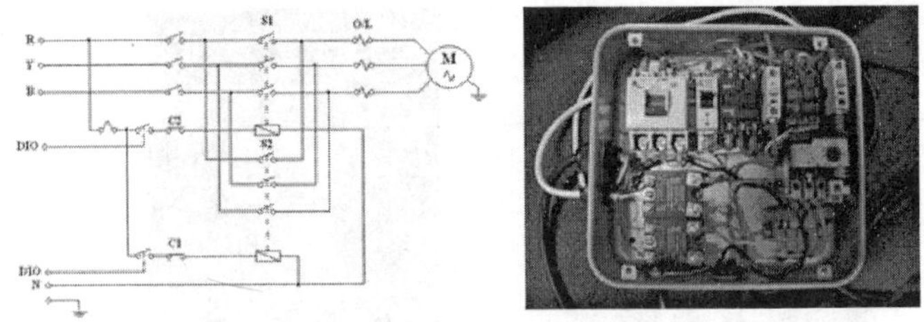

Figure 5. The Low Voltage Subsystem

The FELV subsystem contains the three DC supply rails of 5, 6 and 24 VDC and the DIO signals from the Labview control card. The DC rails facilitate the following functions:

- **5V rail**: logic supply
- **6V rail**: supply for DC solenoids and stepper motors
- **24V rail**: supply for the DC servos

The control card for SparBot contains drivers for the DC servos stepper motors as well as MOSFET's for solenoid control. The physical connections from the control card to the board and out to the field are facilitated by standard DIN rail terminals with blade disconnects. This allows one to connect each subsystem in turn during commissioning, allowing for easier debugging, as shown in Figure 6.

Figure 6. Electrical Layout

4. Robot Control System

The robot trainer control system is developed using LabView via the PCI-6229 multi-functional card and associated breakout board. The card provides 48 digital I/O lines and several analog I/O. The software developed allows SparBot to run under two modes: manual and automatic.

Under manual mode the SparBot Test Centre is invoked. Once the Test Centre is activated the software controls on the user interface are linked to each field device or control chip via the SCB-68 breakout box. Each field device may be manually driven, and the status of each sensor is continually polled. Figure 7 is the user interface of the Test Centre.

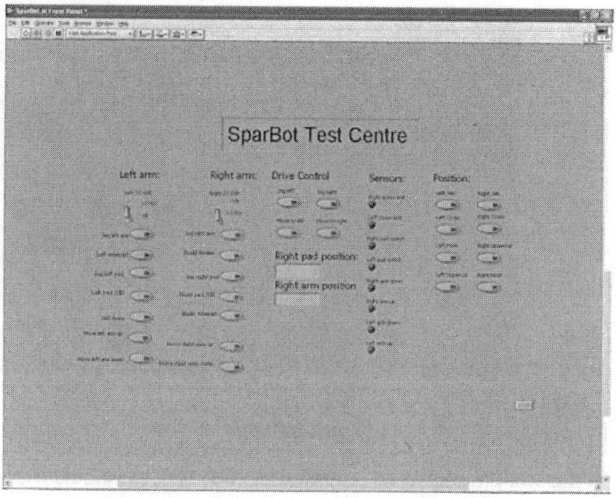

Figure 7. SparBot Test Centre User Interface

From the Test Centre, the user can perform linked functions, such as moving the arm to a desired position. From the click of a button, all necessary control actions are performed in sequence. This allows the function and interaction of each hardware and software subsystem to be fully tested and tuned. For instance, to move from the "left jab" position (arm up, pad forward) to the "left hook" position, the following actions are required:

- Check current position, play audio cue.
- Disengage pad lock, set stepper direction, enable stepper for turn time, *t*.
- Disengage arm lock, set DV drive direction, enable DC drive.
- When *t* has elapsed, disable stepper and engage pad lock
- When "arm down" sensor reports true, de-energise DC drive, enable brake.

When the system is used under the automatic mode, after a short audio welcome note, the user is presented with a keypad as illustrated in Figure 8 into which the user enters the PIN number. This allows easy identification and tracking of each individual user. The PIN is compared with a lookup table and if the user has a data file it is loaded with the last session details, and the software automatically configures itself for the correct level. If no record exists, one is created, and the user is supplied with a PIN.

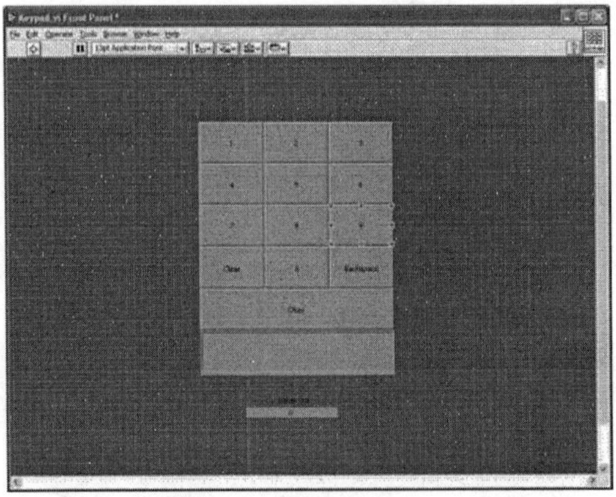

Figure 8. Keypad Screen

Once the set-up phase is completed, the user may select a round time and level of difficulty (if they do not want the defaults), then the training will begin. This consists of concatenations of the basic movement building blocks discussed in the 'manual mode' section. These subroutines are called as needed. For instance, when performing a 'double-punch, hook-uppercut' manoeuvre, the arms are placed for the jab-cross combination then once the last cross punch has been registered, the arms automatically move to the hook-uppercut positions. Only until the uppercut punch is registered, can the next sequence begin. During training, certain parameters such as punch timing, accuracy, number of punches, etc are recorded and appended to a database file. This information associated with the user via the PIN is stored in the database and can be retrieved at a later time.

By using a text-to-speech (TTS) program [3], any typed text can be created from a selection of voice types and stored on the local disk as a PCM wave file. A wave file player was developed in LabView that takes a single filename parameter pointing to the file location on disk. Once the control movements are complete, the wave player is invoked, informing the user of the strike type and order required. The termination of the wave player sub-VI indicates the end of the movement, and control is returned to the main loop.

5. Conclusion

A prototype robotic training platform for focus mitt training has been developed. The system can demonstrate the major functions used in focus mitt training. Different training programs can be stored in the system through the software. The system is also able to store personal training data and track the training history. However, the prototype does reveal some problems. Further improvements will focus on mobility, safety, and wireless control.

6. References

1. John D. Pierce Jr., Kirk A. Reinbold, Barry C. Lyngard, Robert J. Goldman, and Christopher M. Pastore, "Direct Measurement of Punch Force During Six Professional Boxing Matches," *Journal of Quantitative Analysis in Sports*: Vol. 2: No. 2, (2006). Article available at: http://www.bepress.com/jqas/vol2/iss2/3.

2. The Engineerng Toolbox, "Steel modulus of rigidity," *The Engineering Toolbox* (2006), http://www.engineeringtoolbox.com/modulus-rigidity-_d_946.html.

3. Ultra HAL Text-to-Speech Reader version 1.0, 10th September, 2006, www.zabaware.com, Zabaware, Inc.

ACTUATORS AND ORTHOSES TO ASSIST FES-CYCLING

R. MASSOUD, M. O. TOKHI, M. S. ALAM, M. S. HUQ

Department of Automatic Control and Systems Engineering, University of Sheffield, UK

In this paper, mechanical springs are used to assist quadriceps-FES-cycling, where only the knee extensors are electrically stimulated to drive a stationary bicycle. Three types of mechanical springs are tested: torsion spring, linear extension spring, and constant-force spring. Utilizing torsion spring does not improve FES-cycling performance, while FES-cycling performance shows significant improvement when other spring types are used. This study shows that the improvement of FES-cycling performance depends on the type, position, and parameters of the spring. Different spring situations were tested and the performance was evaluated in terms of minimizing the error in crank cadence and maximizing the efficiency. It is noted that selection spring's situation is not easy, as this is a multi objective optimization problem. Non-dominated solutions were obtained, and the decision was taken for the solutions for which the error and efficiency were within acceptable ranges.

1. Introduction

FES-cycling is a rehabilitation cycling exercise for people with spinal cord injury (SCI), where functional electrical stimulation (FES) is applied to their still functioning leg muscles to generate functional movements and enable them drive the rehabilitation bicycle. FES-cycling has significant impact on an SCI person's physical and psychological health [1]. It has proven that FES-cycling performance can be enhanced when auxiliary motors are utilized. Therefore, by using these motors, the overall power can be greatly increased, loss of power due to muscle weakness or fatigue can be compensated for, and leg cycling motion can be maintained even at very low leg power levels [2]. People with very weak muscles at the beginning of their training period will be able to exercise FES-cycling assisted by an auxiliary motor [3].

Pons et al. [4] indicated that it is possible to have the motor assist the pedalling during part of the cycle and retard it during another part of the same cycle. The motor resistance can be replaced with an energy storage device such as mechanical spring, which releases its energy when assistance is needed. In his ergonomic study of cycling for normal people, Rasmussen et al. [5] tried to eliminate the dead centres of the pedal cycle by providing the bicycle frame with a spring arrangement. They used two springs attached to the bicycle frame, which allow to store elastic energy that help to overcome the dead centre of the pedal cycle. The springs' stiffness, slack length, fixation points on the frame, and fixation points on the crank were optimized. This lead to uniform effort over the cycle period, allowing the rider to produce an even crank torque. Furthermore, the maximum muscle activity was less than half of the case with no springs mounted [5].

To the authors' knowledge, the use of mechanical spring to assist FES-cycling has not been reported in the literature. This paper discusses the possibility to replace conventional auxiliary electric motor with an energy storage actuator, so as to increase the FES-cycling efficiency and obtain even FES-cycling. Mechanical springs may come in a wide variety of different types, such as helical compression or tension springs, helical torsion springs, leaf springs and flat springs. Generally, mechanical springs have many shapes, different specifications, and various sizes. Moreover, the can be made from a wide range of materials. This allows to produce springs with mechanical properties very similar to that in human muscles, to produce similar forces produced by human muscles when released after being distorted.

Gharooni et al. [6] used a spring to mimic the knee flexors (hamstrings muscle group) during paraplegic gait restoration. They extended the knee joint by stimulating the quadriceps muscle group, and used the spring for knee joint flexion.

2. Simulation set up

A humanoid and bicycle model is designed via Visual Nastran software. The humanoid dimensions are described in [7]. The bicycle consists of a fixed seat and a crank with constant load resistance of 0.18 Nm. The distance between the centres of the humanoid trunk and the crank is 0.7m, and the seat and the crank are at the same height (see Figure 1).

In order to simulate FES, Riener's muscle model is used [8]. The model is for quadriceps muscle group, and it depends on the activation and contraction dynamics of the paralyzed muscle. The bicycle model is driven by the knee and hip joints torques, produced by the muscle model. The muscle model produces the torques according to the level of the FES, which is controlled by a proportional-derivative (PD) type fuzzy logic (FL) controller. The controller's design and the FL control strategy are described in [7].

Since only the quadriceps (knee extensors) is stimulated in this work, the springs are used to mimic the knee flexors (hamstrings). Each spring has two work shifts; energy storage and energy release. During energy storage shift the spring stores the exceeded energy produced by quadriceps stimulation and the lower limbs dynamics, while in the energy release shift the FES is off for both legs allowing the spring to flex the knee joint and drive the bicycle. Accordingly, FES pedalling cycle is divided into three essential phases: the left quadriceps stimulation phase, the right quadriceps stimulation phase, and the relax phase where no quadriceps are stimulated allowing the springs to release their potential energy, and to take part in the driving process.

3. FES cycling with torsion springs

Two linear torsion springs have been mounted very close to the knee joints of the humanoid as shown in Figure 1. In pratice, this can be done by following the same arrangement as implemented by Gharooni et al. [6].

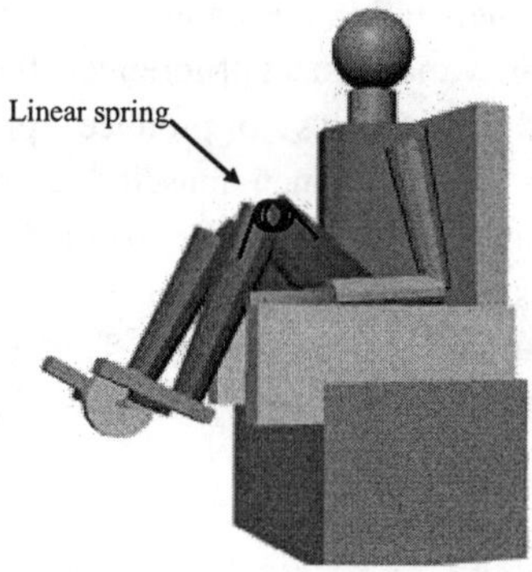

Figure 1: Humanoid with bicycle designed in visual Nastran software.

Figure 2 shows that the best FES-cycling performance is obtained when no torsion spring is used, error in crank cadence increases as the spring constant increases. Conversely the gross efficiency (the ratio between the output power and the muscle power) decreases with the increment of the spring constant. This can be seen clearly in Figure 3. Due to these results, it is clear that using torsion springs positioned at the knee joint does not enhance the performance of FES-cycling, in terms of efficiency and percent error.

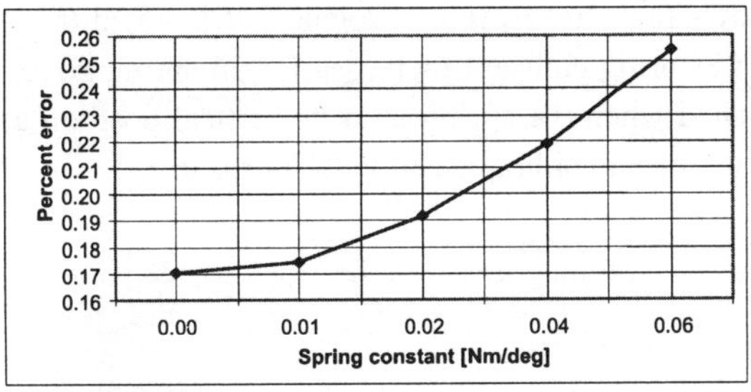

Figure 2: Percent error for torsion spring implementation

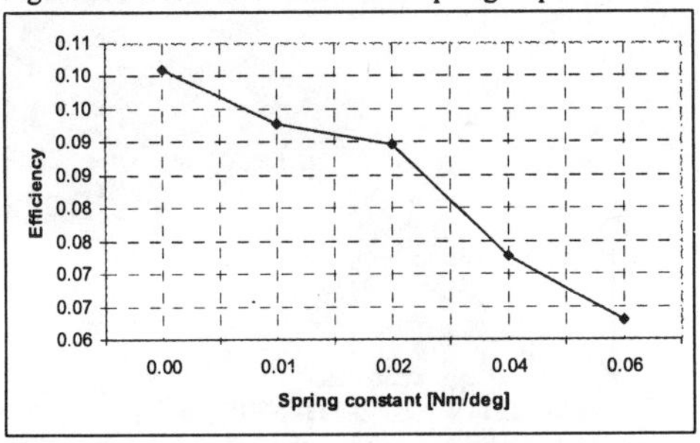

Figure 3: Relation between efficiency and torsion spring's constant

4. FES cycling with linear extension springs

In this case, the torsion springs are replaced by two linear extension springs mounted on the frame of the bicycle. The spring's one end (say, first end) is fixed to the pedal and the other (say, second) end is fixed to the bicycle frame in a semi vertical position. Figure 4 shows the spring position according to the bicycle. The vertical distance between the centre of the crank and the second end of the spring (VDCS) is 0.29m. This vertical distance is still constant while the horizontal distance between the centre of the crank and the second end of the spring (HDCS) is changed from 0m (vertical position) to 0.25m to the right, and from 0m to 0.1m to the left by a step of 0.05m. Eight spring positions were considered. At each spring position, the spring constant was changed from 0 N/m (no spring used) to 50 N/m, at steps of 5 N/m. At the spring positions (–0.2, 0.29) and (–0.25, 0.29) the spring constant was increased to 60 N/m. Accordingly 92 different situations for the spring were tested.

FES-cycling performance was evaluated in terms of efficiency and error in crank cadence. Figure 5 shows that the percent error in crank cadence decreases with increase in the spring constant until it reaches a point which has the lowest error value. After this point the error starts to increase with increase in the spring constant. Moreover, the error decreases as the spring leans to the right (towards the humanoid).

Figure 6 shows the FES-cycling efficiency. It is noted that the efficiency increases when the spring constant increases for all spring positions; maximum efficiency is reached when the spring is in its vertical position and it decreases when the spring inclines to either direction (left or right).

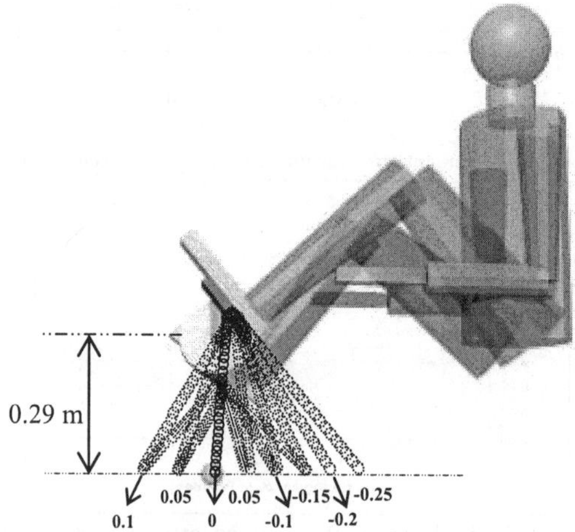

Figure 4: Linear spring implementation

5. FES cycling with constant force extension springs

The constant-force extension spring is basically a high stress, long deflection device and offers great advantages such as very low or zero gradient, large extension, small package size. Two force–constant springs were fixed to the bicycle; the first end of the spring was fixed to the pedal while the second was attached to the bicycle frame. Figure 7 shows that the horizontal distance between the centre of the crank and the second end of the spring (HDCS) is constant (0.5m), while the vertical distance is changed up and down from –0.1m to 0.2m in steps of 0.05m.

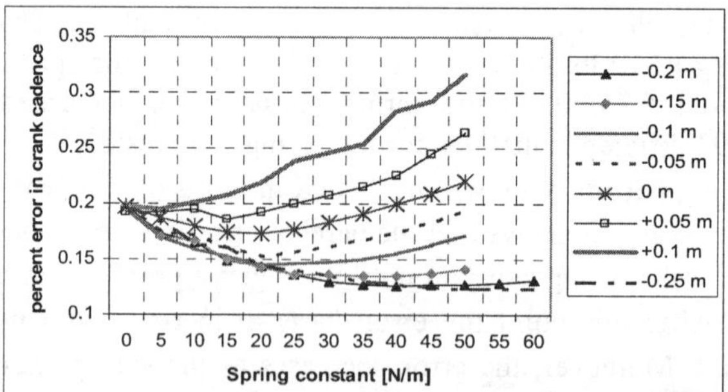

Figure 5: Relation between percent error and linear spring constant at different HDCS

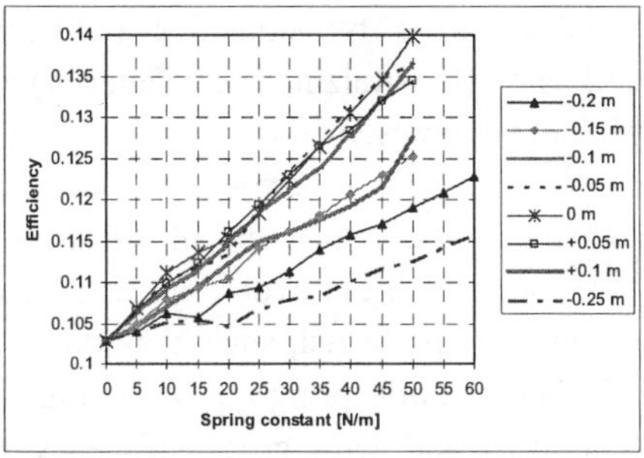

Figure 6: Relation between FES-cycling efficiency and linear spring constant at different HDCS

At each spring position, the spring constant was changed from 0N (where no spring was used) to 25N in steps of 5N. Therefore, 42 different possibilities for the spring position and parameter were tested. The FES-cycling performance was also evaluated in terms of efficiency and percent error in crank cadence.

Similar to the case of using linear spring in FES-cycling, the percent error decreases when the spring constant increases until it reaches the turning point, which has the lowest percent error. Beyond this point the error increases with an increase in the spring constant, as shown in Figure 8. Gross efficiency obtained at different situations of the spring is shown in Figure 9. It is noted that the relation between the efficiency and the spring constant is nearly linear.

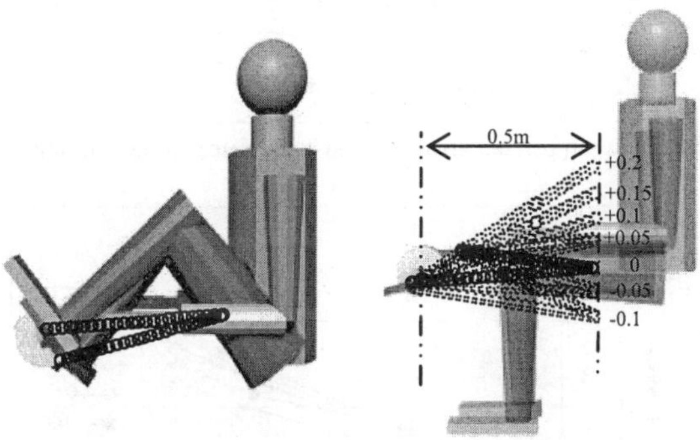

Figure 7: Constant-force spring implementation

6. Optimization of the spring parameters and fix points

The tested spring parameters and positions were chosen heuristically. And the results obtained from applying theses parameters show the difficulty in selecting the optimum spring situation. In order to find the best solution, the efficiency should be as high as possible while the percent error should be as low as possible. Since both objectives are in conflict with each other, the optimization problem can be converted to a two-objective minimization problem by minimizing both, 1/efficency (maximizing the efficiency) and the percent error in crank cadence. This can be represented by a two –objectives space, where the horizontal axis represents the percent error and the vertical axis represents 1/efficiency. Figure 10 shows the results for all the spring situations for both linear and constant-force springs in the two –objectives space. Solutions which are very close to the origin are preferred, since they dominate the others. From the non-dominated solutions of both spring types (see Figure 10), solutions with percent error less than 0.13 and with gross efficiency more than 12% are chosen. Three points are satisfied; two of them represent two linear spring cases where the spring's HDCS= 0.2m and VDCS= 0.29m and the spring rate for the first case is 55N/m, and for the second case is 60N/m. The third point represents the constant force spring where the HDCS= 0.4m and the VDCS= 0.2m and the spring constant is 15N.

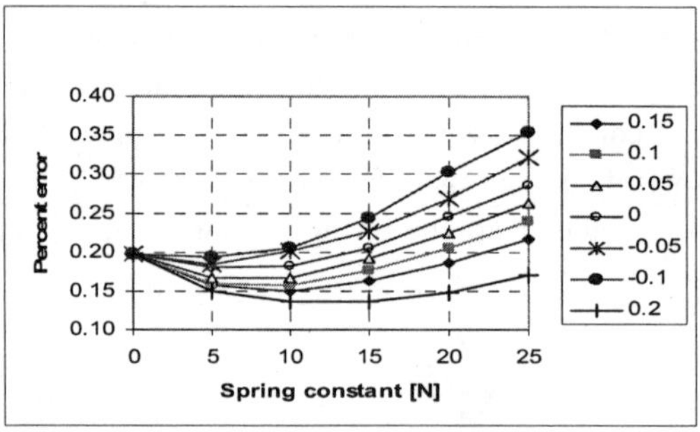

Figure 8: The relation between percent error and constant –force spring constant at HDCS=0.5m and at different VDCS

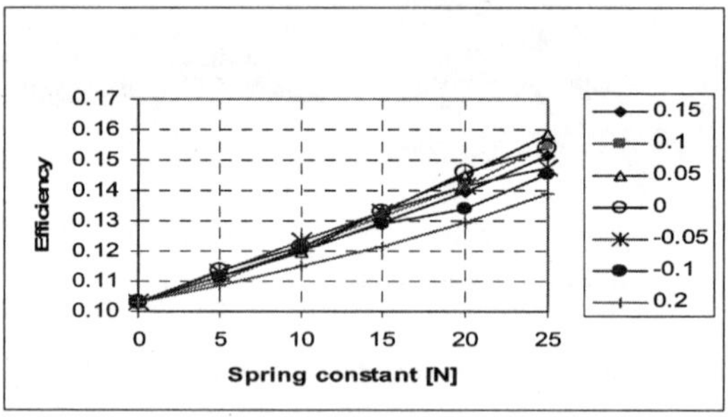

Figure 9: The relation between the efficiency and the constant-force spring constant at HDCS=0.5m and at different VDCS

7. Conclusion

This study concludes that using torsion springs mounted near the knee joints does not enhance FES-cycling, while using linear or constant force extension springs mounted on the bicycle frame can increase the efficiency and reduce the percent error in crank cadence. Therefore, it has significant impact on the overall performance. It is clear from this study that optimizing spring position and parameters leads to better FES-cycling performance.

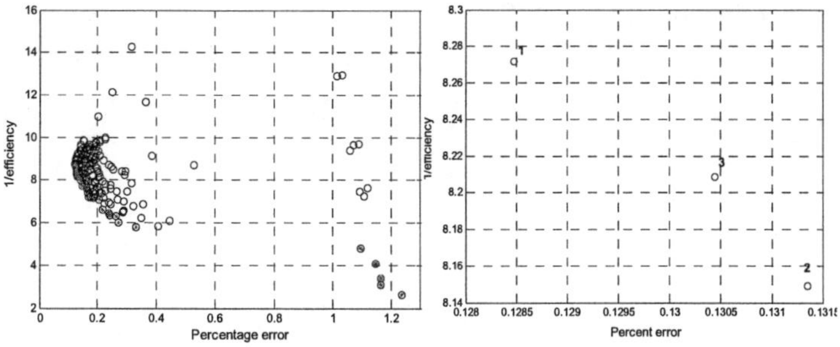

Figure 10: Spring situations for linear and constant –force springs; non-dominated solutions are marked with filled points.

References

1. Petrofsky, J. S., Laymon, M. (2004). The effect of previous weight training and concurrent weight training on endurance for functional electrical stimulation cycle ergometry. *European Journal Applied Physiology*, **91** (4), 392-8.
2. Hunt, K. J., Stone, B., Negard, N. O., Schauer, T., Fraser, M. H., Cathcart, A. J., Ferrario, C., Ward, S. A. and Grant, S. (2004). Control strategies for integration of electric motor assist and functional electrical stimulation in paraplegic cycling: utility for exercise testing and mobile cycling. *IEEE Transactions on Neural Systems and Rehabilitation Engineering,* **12** (1), 89-101.
3. Gföhler, M., Loicht, M., Lugner, P. (1998). Exercise tricycle for paraplegics. *Medical & Biological Engineering & Computing*, **36,** 118-121.
4. Pons, D. J., Vaughan, C. L. and Jaros, G. G. (1989). Cycling device powered by the electrically stimulated muscles of paraplegics. *Med Biol Eng Comput,* **27,** (1), pp. 1-7.
5. Rasmussen, J., Christensen, S. T., Damsgaard, M., Zee, M. (2005). Ergonomic optimization of a spring-loaded bicycle crank. *6th World Congresses of Structural and Multidisciplinary Optimization*, Rio de Janeiro, 30May-03June, Brazil.

6. Gharooni, S., Tokhi, M.O. (2002) Synthesising the swing phase in orthotic gait using a hybrid FES spring brake orthoses. *PhD thesis*, University of Sheffield, Sheffield.
7. Massoud, R., Alam, M. S., Tokhi, M. O., Gharooni, S. C. (2006). GA optimized spring orthosis to enhance FES-cycling performance. *CD Proceedings of the 9th International Conference on Climbing and Walking Robots (CLAWAR2006)*, Brussels, Belgium, pp 334-337.
8. Riener R., and Fuhr T. (1998), Patient-driven control of FES-supported standing up: A simulation study, IEEE Transactions on Rehabilitation Engineering, 6(2), 113-124.

A NOVEL MINIATURE ATTITUDE MEASUREMENT SYSTEM FOR CLIMBING AND WALKING ROBOTS

GUANGLONG WANG, CHUNXI ZHANG

School of Instrument Science and Opto-electronics Engineering, Beijing University of Aeronautics and Astronautics, Beijing 100083, China

ZHAOYING ZHOU, RONG ZHU

MEMS Lab, Department of Precision Instrument, Tsinghua University, Beijing 100084, China

A novel design of miniature attitude measurement system for climbing and walking robots is presented. By integrating three-dimensional magnetometers and three microelectromechanical accelerometers, the system provides high-frequency position and attitude data. And it takes advantages of the solid-state configuration, small size, light weight, high reliability, low power consumption, rapid startup and relatively low cost. It can be used to test the transient attitude of the moving body such as robots. The three-axis accelerometers detect the three orthogonal components of the gravity, by which tilt of the robots including pitch and roll can be derived. Otherwise a three-axis magnetometer which are also assembled as an orthogonal placement to measure the three components of the Earth's magnetic field. The calculation and simulations indicate that the formalism proposed in this paper is able to reach an excellent long-term performance. Even if there is various error sources existing in the measurement, the non-orthogonal assembling error is the most dominating. We analyze the representative features of this error, establish a mathematic model for it, and successfully compensate this error via arithmetic. Finally, real experiments performed with a mobile robot are presented and analyzed. The results of the test indicate that the system can measure the freewill attitude in three-dimensional space. It is expected that this system will be adopted in many fields.

1. Introduction

In the past decade, precise attitude measurement performed with mechanical accelerometers and gyroscopes based on solid-state sensor technology have become readily available and have supplanted mechanical sensors in some climbing and walking robots applications. Unlike mechanical devices, these microelectromechanical solid-state sensors do not rely on numerous and large moving parts. Instead, they use miniature sensing elements made from quartz crystals or micro-machined silicon. Knowing the initial orientation and position of the robot, the mathematical integration of the acceleration and rotation rate measurements yields the attitude and the trajectory of the robot. Such inertial techniques are completely autonomous and need no external reference [1]. But they will induce the inevitable long-term drift by dead reckoning, and have relatively slow start-up for its initial alignment. This paper provides a novel miniature attitude measurement system, which is composed of micro-electromechanical accelerometers and magnetometers. We use a kind of absolute measurement to abandon the problems of the conventional inertial measurement mentioned above. Micromachining is one of the most important emerging technologies for sensors because it provides unique advantages due to offering small size and lowing cost [1] [2]. The main benefits that silicon based microelectromechanical system derive from its common base with the IC industry are "low-cost", "massproduction", and "monolithic integration". The important requirements for micromachining technology is that it should be simple, easily accessible to a wide range of users, be low cost and provide sufficient flexibility so that different types of sensors can be fabricated on the same chip without significant redesign[3]. During the past few years, two generic technologies have become mainstream in the field of micro electromechanical systems. These technologies are commonly known as bulk micromachining and surface micromachining. Among them, the surface micromachining technology attracts more attention by its simplicity, high reliability, relatively low cost and high finished product probability. As result of this, we adopt a dual-axis sensor and a single-axis sensor, which take benefit of silicon surface micromachining technology[4] [5].

2. The Composing of the System

The attitude measurement system is made up of a suit three-dimensional accelerometers, a suit three-dimensional magnetometers, A/D converter and microprocessor etc. The analog signals readings from the robot attitude sensors are converted to the digital signals through the A/D converter. Then they are processed in real-time by microprocessor to get the three attitude angles. Finally the attitude angles are conveyed through serial port into control unit to display or drive servos actuators. Figure 1. shows the system composing of the system.
The three-dimensional orthogonal accelerometers measure the components of acceleration of gravity, which deduce the tilt. We select ADXL202 which has dual-axis and surface micromachined polysilicon structure as the acceleration sensor because it has the advantages of fully integrated monolithic, low cost, low power and faster response. And the three-dimensional orthogonal magnetometers measure the components of the earth's magnetic field, by which the robot attitude is computed. For magnetic sensors, a common type of it for

the robot systems is the fluxgate sensor. It provides a low cost means of magnetic field detection, but it also tends to be bulky, somewhat fragile, and has a slow response time. Another type of magnetic sensor is the magnetoresistive sensor [4]. This sensor is made up of thin strips of permalloy whose electrical resistance varies with a change in applied magnetic field. The magnetic product we prefer is HMC1001 and HMC1002. They are single and dual-axis magnetoresistive microcircuits， provide extremely sensitive magnetic sense in small package outlines. These sensors offer low cost and small size, and high sensitivity.

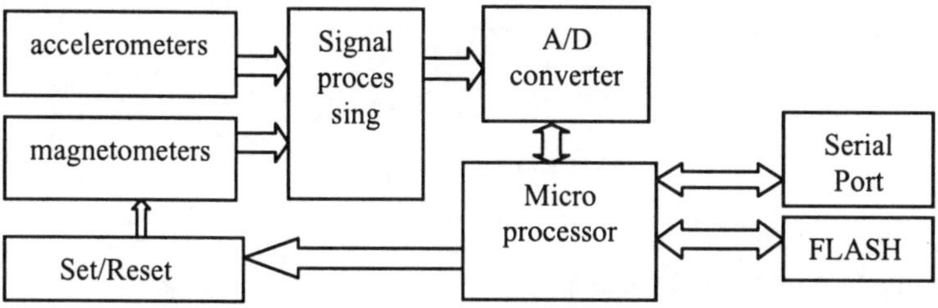

Figure 1. Shows the system composing of the system.

3. *Principle of the Attitude Measurement*

The earth's gravity field and magnetic field in geographic coordinate system (GCS) is shown in Figure 2. Therein to β is the dip angle of the magnetic field. The three axes of the magnetometers are mounted along the X, Y and Z. And the magnetic readings x_m, y_m and z_m which figure the X, Y and Z components of the earth's magnetic field, can be transformed back to the geographic coordinates by applying the Cosine Matrix C_n^b shown below:

$$\begin{bmatrix} x_m \\ y_m \\ z_m \end{bmatrix} = C_n^b \begin{bmatrix} \cos\beta \\ 0 \\ \sin\beta \end{bmatrix} + \begin{bmatrix} e_{mx} \\ e_{my} \\ e_{mz} \end{bmatrix} \tag{1}$$

where

$$C_n^b = \begin{bmatrix} \cos\psi\cos\theta & \sin\psi\cos\theta & -\sin\theta \\ \cos\psi\sin\theta\sin\gamma - \sin\psi\cos\gamma & \sin\psi\sin\theta\sin\gamma + \cos\psi\cos\gamma & \cos\theta\text{sion}\gamma \\ \cos\psi\sin\theta\cos\gamma + \sin\psi\sin\lambda & \sin\psi\sin\theta\cos\gamma - \cos\psi\sin\gamma & \cos\theta\cos\gamma \end{bmatrix}$$

e_m represents the disturbance and error in magneto measuring.

The three axes of accelerometers are also along the X, Y and Z to measure the components x_g, y_g and z_g of the earth's gravity field. The earth's gravity field f_g in

GCS has very simple pattern, and it is always pointing to downword direction and the horizontal components are zero. Then

$$\begin{bmatrix} x_g \\ y_g \\ z_g \end{bmatrix} = C_n^b \begin{bmatrix} 0 \\ 0 \\ f_g \end{bmatrix} + \begin{bmatrix} e_{gx} \\ e_{gy} \\ e_{gz} \end{bmatrix} \tag{2}$$

where e_g represents the disturbance and error in gravity measuring.

Putting into the C_n^b and eliminating the disturbance and error by signal processing, the attitude angles can be derived as:

$$\gamma = arctg(y_g / z_g) \tag{3}$$

$$\theta = -arctg(\cos\gamma \cdot x_g / z_g) \tag{4}$$

$$\psi = arctg(y_h / x_h) \tag{5}$$

where $x_h = x_m \cos\theta + (y_m \sin\gamma + z_m \cos\gamma)\sin\theta$,

$y_h = z_m \sin\gamma - y_m \cos\gamma$

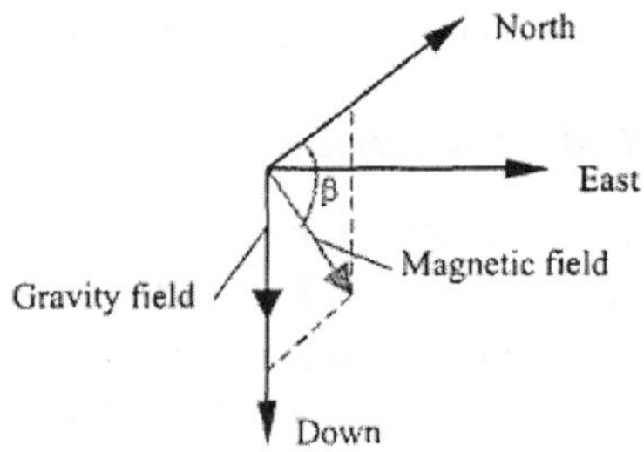

Figure 2. The earth's gravity field and magnetic field in GCS

4. Error Analysis

There are several of errors existing in system, including random noise and measured deviation. Filter technology has been applied to restrain the random noise. And for measured deviation, it mainly consists of the non-orthogonal assembling errors came from the separate single-axis senor. This error exhibits a kind of regular characteristics and is distributed in the attitude measuring. The most straightforward scheme for eliminating this error is avoiding non-orthogonal, as adopting a three-axis monolithic sensor. But a three-axis sensor has more complex by surface micromaching technology. The cost is unavoidably highly raised, but the performance is restricted. Take account of this reason, we select a dual-axis sensor and a single axis sensor to compose an orthogonal three-axis placement. For eliminating the non-orthogonal assembling error, a software compensation scheme based on error modeling is designed[6] [7] [8].

5. *Discussion and Conclusion*

Theory analysis and experimental results indicate that this kind of attitude measurement system can be used in climbing and walking robots attitude test and control. Also the system can be packaged in a immature size and fixed in the body of small robots. The performance of the accelerometer and magnetometer is shown to be acceptable as a short-duration distance-measuring device for robot platform. It can be combined with a gyroscope and odometer to form a dead-reckoning positioning system for a robot or platform. Further research would focus on the proper modeling of the accelerometer in order to reduce the effect of random bias drift and error.

Acknowledgments

The authors would like to thank Dr. Rong Zhu for her help in developing some of the theory analysis and calculation. The authors would also like to thank the MEMS Lab, Department of Precision Instrument, Tsinghua University to facilitate the success of the experimental work. The research work was financially supported by a China National Science Foundation.

References

1. Bing L. Luk, David S. Cooke, Stuart Galt, Arthur A. Collie and Sheng Chen, *Robotics and Autonomous Systems,* Vol. 53, Issue 2, 30 November, 142-152(2005).
2. Demoz gebre-egziabher, Roger C. Hayward. Hayward, J. David Powell, *IEEE Transactions on Aerospace and Electronic Systems*, Vol. 40, No. 2 627-649(2004).
3. Gebre-Egziabher, D.; Hayward, R.C.; Powell, J.D., *IEEE Transactions on Aerospace and Electronic Systems,* Vol. 40, Issue 2, 627-649 (2004).
4. Rong Zhu, Zhaoying Zhou, Sha Li and Xuefeng Sun, *Microelectromechanical Systems Conference*, 50-53 (2001).
5. Ph. Bonnifait, G. Garcia, *Control Engineering Practice 7th* 383-390 (1999).
6. C. Lemaire, *in Proc. IEEE ITSC'97*, 1068–1072. (1997).
7. A. Nishi, *Computers & Electrical Engineering*, Vol. 22, Issue 2, 123-149(1996).
8. B. Barshan and H. F. Durrant-Whyte, *IEEE Trans. Robot. Automat.*, Vol. 11, 328–342(1995).